TELE**MEDICINE** AND ELECTRONIC **MEDICINE**

THE E-MEDICINE, E-HEALTH, M-HEALTH, TELEMEDICINE, AND TELEHEALTH HANDBOOK
VOLUME I

TELEMEDICINE AND ELECTRONIC MEDICINE

Edited by
Halit Eren
John G. Webster

CRC Press
Taylor & Francis Group
Boca Raton London New York

CRC Press is an imprint of the
Taylor & Francis Group, an **informa** business

CRC Press
Taylor & Francis Group
6000 Broken Sound Parkway NW, Suite 300
Boca Raton, FL 33487-2742

First issued in paperback 2017

© 2016 by Taylor & Francis Group, LLC
CRC Press is an imprint of Taylor & Francis Group, an Informa business

No claim to original U.S. Government works

ISBN-13: 978-1-4822-3658-3 (hbk)
ISBN-13: 978-1-138-89359-7 (pbk)

Contents

Section I Integration of eMedicine, Telemedicine, eHealth, and mHealth

Section II Wireless Technologies and Networks

Section III   Sensors, Devices, Implantables, and Signal Processing

Section IV   Implementation of eMedicine and Telemedicine

Preface

Introduction

The purpose of the *Telehealth and Mobile Health* handbook is to provide a reference that is both concise and useful for biomedical engineers in universities and medical device industries, scientists, designers, managers, research personnel, and students, as well as healthcare personnel, such as physicians, nurses, and technicians, who use technology over a distance. The handbook covers an extensive range of topics that comprise the subject of distance communication, from sensors on and within the body to electronic medical records. It serves the reference needs of a broad group of users—from advanced high school science students to healthcare and university professionals.

Recent development in digital technologies is paving the way for ever-increasing use of information technology and data-driven systems in medical and healthcare practices. Hence, this handbook describes how information and communication technologies, the Internet, wireless technologies and wireless networks, databases, and telemetry permit the transmission of information and control of information both within a medical center and between medical centers. Recent developments in sensors, wearable computing, and ubiquitous communications have provided medical experts and users with frameworks for gathering physiological data on a real-time basis over extended periods of time. Wearable sensor-based systems can transform the future of healthcare by enabling proactive personal health management and unobtrusive monitoring of a patient's health condition. Wireless body area networks permit a comfortable tank top with sensors to use wireless local area networks (e.g., Wi-Fi and Bluetooth) to continuously transmit to other systems such as smartphones, and then from any location, such as home, away from home, on the streets, or a nursing home, to a medical center for analysis of cardiac arrhythmias and ventilation. For example, a simple remotely located base unit can continually collect and locally integrate many incoming signals such as electrocardiography, oxygen saturation, heart rate, noninvasive blood pressure, temperature, and respiration, and provide the information required for detecting any possible emergency cases for the patients. The medical center can then accommodate all complementary and bulky systems, including telemedicine-enabled equipment such as intensive care units, intelligent analyzers, and automatic recorders plus a professionally managed database system supported by a professional service provider.

Today's technology allows clinical processes to be conducted at a distance; hence, it is an enabler, but in itself, the technology is not telemedicine. Telemedicine can be thought of as the tasks that the clinician carries out (such as observing, consulting, interpreting, and providing opinions), assisted by information and communication technologies, in circumstances where there is distance between the patient and the provider. Put succinctly, modern telemedicine is simply medicine at a distance.

This handbook also intends to bridge the gap between scientists, engineers, and medical professionals by creating synergy in the related fields of biomedical engineering, information and communication technologies, network operations, business opportunities, and dynamically evolving modern medical and healthcare practices. It includes how medical personnel use information and communication technologies, as well as sensors, techniques, hardware, and software. It gives information on wireless data transmission, networks, databases, processing systems, and automatic data acquisition, reduction, and analysis and their incorporation for diagnosis.

The chapters include descriptive information for professionals, students, and workers involved in eMedicine, telemedicine, telehealth, and mHealth. Equations in some chapters also assist biomedical engineers and healthcare personnel who seek to discover applications and solve diagnostic problems that arise in medical fields not in their specialty. All the chapters are written by experts in their fields and include specialized information needed by informed specialists who seek to find out advanced applications of the subject, evaluative opinions, and possible areas for future study.

Organization

The handbook is organized into two volumes and each volume in sections.
   The sections in Volume 1 are

   Section I: Integration of eMedicine, Telemedicine, eHealth, and mHealth

   Section II: Wireless Technologies and Networks

   Section III: Sensors, Devices, Implantables, and Signal Processing

   Section IV: Implementation of eMedicine and Telemedicine

Section I contains information on the integration of modern eMedicine, telemedicine, eHealth, and telehealth. The interactions between these practices are explained and examples are given. Wireless technology, an essential part of telemedicine, is explained in Section II, with a particular emphasis on the fast deploying wireless body area networks. The state of the art on sensors, devices, and implantables is explained in Section III, while Section IV is dedicated to practical applications of all the information given in Sections I through III, ranging from telecardiology and teleradiology to teleoncology and acute care telemedicine.
   The sections in Volume 2 are

   Section I: Medical Robotics, Telesurgery, and Image-Guided Surgery

   Section II: Telenursing, Personalized Care, Patient Care, and eEmergency Systems

   Section III: Networks and Databases, Informatics, Record Management, Education, and Training

   Section IV: Business Opportunities, Management and Services, and Web Applications

   Section V: Examples of Integrating Technologies: Virtual Systems, Image Processing, Biokinematics, Measurements, and VLSI

We are all aware of telesurgery implementation as a routine process while the patient and the surgeon may be continents apart. This is explained in Section I, with emphasis on medical robotics and image guidance. Remote patient care, personalized care, and telenursing can be found in Section II. For an effective remote care, the use of networks, data management, record management, and the education and training aspects of personnel are given in Section III. Implementation of new technologies in eMedicine and eHealth bring many business, management, and service opportunities as explained in Section IV. Examples of emerging technologies, developing engineering, and scientific contributions are given in Section V. For example, the sound understanding of biokinematics has led to successful implementation of brain-controlled bionic human parts such as bionic arms and hands.

Locating Your Topic in the *Telehealth and Mobile Health* Handbook

Select your topic, skim the Contents, and peruse the chapter that describes your topic. Consider the alternative methods of distance communication with each of their advantages and disadvantages prior to selecting the most suitable method. For more detailed information, consult the Index, since certain principles of eMedicine, eHealth, mHealth, telemedicine, and telehealth may appear in more than one chapter.

MATLAB® is a registered trademark of The MathWorks, Inc. For product information, please contact:

The MathWorks, Inc.
3 Apple Hill Drive
Natick, MA 01760-2098 USA
Tel: 508 647 7000
Fax: 508-647-7001
E-mail: info@mathworks.com
Web: www.mathworks.com

Acknowledgments

We thank all 69 authors in this volume for sharing their expertise and sparing their valuable time to contribute to this volume. We also gratefully acknowledge the CRC Press team for their patience and tireless effort in putting everything together. We also thank all our readers in selecting this book to advance their knowledge and technical skills.

Editors

Halit Eren received BEng, MEng, and PhD degrees in 1973, 1975, and 1978, respectively, from the University of Sheffield, Sheffield, United Kingdom. He obtained an MBA degree from Curtin University, Perth, Australia, in 1999.

After his graduation, Dr. Eren worked in industry as an instrumentation engineer for two years. He held a position as assistant professor at Hacettepe University, Ankara, Turkey, in 1980–1981 and Middle East Technical University, Ankara, Turkey, in 1982. He has been at Curtin University since 1983, conducting research and teaching primarily in the areas of control systems, instrumentation, and engineering management. Currently, Dr. Eren holds an Adjunct Senior Research Fellow position at Curtin University.

Dr. Eren held an associate professor position at the Polytechnic University in Hong Kong in 2004, visiting professor position at the University of Wisconsin, Madison, Wisconsin, in 2013, and a visiting scholar position at the University of Sheffield, UK, in 2015. He is a senior member of the Institute of Electrical and Electronics Engineers, taking roles in various committees for organizing conferences and as a member of editorship in transactions. Dr. Eren has over 190 publications in conference proceedings, books, and transactions. He is the author of *Electronic Portable Instruments: Design and Applications* (2004) and *Wireless Sensors and Instruments: Networks, Design, and Applications* (Boca Raton: CRC Press, 2006). He coedited the fourth edition of *Instrument Engineers' Handbook, Volume 3: Process Software and Digital Networks,* in 2011 with Bela Liptak, and the second edition of the two-volume set *Measurement, Instrumentation, and Sensors Handbook* (Boca Raton: CRC Press, 2014) with John G. Webster. Dr. Eren is active in researching and publishing on wireless instrumentation, wireless sensor networks, intelligent sensors, automation and control systems, and large control systems.

John G. Webster received a BEE degree from Cornell University, Ithaca, New York, in 1953, and an MSEE and a PhD degree from the University of Rochester, Rochester, New York, in 1965 and 1967, respectively.

Dr. Webster is professor emeritus of biomedical engineering at the University of Wisconsin, Madison, Wisconsin. He was a highly cited researcher at King Abdulaziz University, Jeddah, Saudi Arabia. In the field of medical instrumentation, he teaches undergraduate and graduate courses and does research on an intracranial pressure monitor, electrocardiogram dry electrodes, and tactile vibrators.

Dr. Webster is author of *Transducers and Sensors,* an Institute of Electrical and Electronics Engineers/Educational Activities Board Individual Learning Program (Piscataway: IEEE, 1989). He is coauthor, with B. Jacobson, of *Medicine and Clinical Engineering* (Englewood Cliffs: Prentice-Hall, 1977); with R. Pallas-Areny, of *Sensors and Signal Conditioning,* second edition (New York: Wiley, 2001); and with R. Pallas-Areny, of *Analog Signal Conditioning* (New York: Wiley, 1999). He is editor of *Encyclopedia of Medical Devices and Instrumentation,* second edition (New York: Wiley, 2006); *Tactile Sensors for Robotics and Medicine* (New York: Wiley, 1988); *Electrical Impedance Tomography* (Bristol: Adam Hilger, 1990), *Teaching Design in Electrical Engineering* (Piscataway: Educational Activities Board, IEEE, 1990), *Prevention of Pressure Sores: Engineering and Clinical Aspects* (Bristol: Adam Hilger, 1991); *Design of Cardiac Pacemakers* (Piscataway: IEEE Press, 1995); *Design of Pulse Oximeters* (Bristol: IOP Publishing, 1997); *Encyclopedia of Electrical and Electronics Engineering* (New

York, Wiley, 1999); *Minimally Invasive Medical Technology* (Bristol: IOP Publishing, 2001); *Bioinstrumentation* (Hoboken: Wiley, 2004); *Medical Instrumentation: Application and Design*, fourth edition (Hoboken: Wiley, 2010); and *The Physiological Measurement Handbook* (Boca Raton: CRC Press, 2015). He is coeditor, with A. M. Cook, of *Clinical Engineering: Principles and Practices* (Englewood Cliffs: Prentice-Hall, 1979) and *Therapeutic Medical Devices: Application and Design* (Englewood Cliffs: Prentice-Hall, 1982); with W. J. Tompkins, of *Design of Microcomputer-Based Medical Instrumentation* (Englewood Cliffs: Prentice-Hall, 1981) and *Interfacing Sensors to the IBM PC* (Englewood Cliffs: Prentice Hall, 1988); with A. M. Cook, W. J. Tompkins, and G. C. Vanderheiden, of *Electronic Devices for Rehabilitation* (London: Chapman & Hall, 1985); and with H. Eren, of *Measurement, Instrumentation, and Sensors Handbook* (Boca Raton: CRC Press, 2014).

Dr. Webster has been a member of the Institute of Electrical and Electronics Engineers–Engineering in Medicine and Biology Society Administrative Committee and the National Institutes of Health Surgery and Bioengineering Study Section. He is a fellow of the Institute of Electrical and Electronics Engineers, the Instrument Society of America, the American Institute of Medical and Biological Engineering, the Biomedical Engineering Society, and the Institute of Physics. He is the recipient of the Institute of Electrical and Electronics Engineers Engineering in Medicine and Biology Career Achievement Award.

Contributors

Shaftab Ahmed is an associate professor and chair of Computer and Software Engineering at Bahria University, Islamabad, Pakistan. His primary field of expertise, electrical engineering and computer engineering, encompasses computer system design and development and e-applications. His teaching experience spans almost four decades. He has contributed in planning and setting up training facilities, curriculum design, and faculty development activities. He has vast experience in design of data acquisition and control systems, very large-scale integration, and microprocessor design for real-time applications. His current activities include teaching in bachelor's and master's degree courses in computer and software engineering along with direction of academic programs of the university. His areas of interest include eHealthcare, telemedical sensor networks, and informatics for healthcare data management services in a virtual hospital environment over cloud architecture. His interests also cover the role of social networking in patient monitoring and the ubiquitous services with knowledge-based decision-support solutions.

Arni Ariani received a BE degree in electrical engineering from Universitas Kristen Maranatha, Bandung, Indonesia, in 2004, and an ME degree in electrical engineering from the Institut Teknologi, Bandung, Indonesia, in 2006. Her PhD degree in biomedical engineering was obtained in 2013 from the University of New South Wales, Sydney, Australia.

Dr. Ariani is currently a researcher at the Indonesian Biomedical Engineering Society, under the supervision of Professor Soegijardjo Soegijoko. Her current research interests include development works covering areas of expertise from optical wavelength division multiplexing technology, fraud management systems, fall-detection systems, ambient assistive living technologies, and telehealthcare. So far, she has published one book chapter, five conference papers, one journal, and one provisional Australian patent. Dr. Ariani is a member of the Indonesian Biomedical Engineering Society and the Institute of Electrical and Electronics Engineers.

Nigel R. Armfield's primary research interest is in the development and formal evaluation of sustainable pediatric clinical telemedicine applications and services. He has particular interests in designing and conducting studies to assess the feasibility, efficacy, and clinical effectiveness of telemedicine for delivering neonatal and pediatric critical care at a distance.

As a PhD candidate, Dr. Armfield developed and evaluated a novel real-time telemedicine application to support remote clinical consultation between a tertiary neonatal intensive care unit and four peripheral hospitals in Queensland, Australia. In addition, Dr. Armfield has a research interest in health geography, particularly in formally assessing and describing the effect of geography and regionalization on the availability, accessibility, and utilization of specialist health services for children.

Dr. Armfield is involved in telemedicine service delivery at the Queensland Children's Hospital in Brisbane and is a coordinator for the Indigenous Health Screening Programme.

Tanya Baldacchino is the telehealth nurse manager for the Nepean Telehealth Technology Centre within the Nepean Blue Mountains Local Health District, New South Wales, Australia. Baldacchino's role has been vital to the implementation of telehealth across the health district, engaging stakeholders and ensuring that telehealth is embedded into work practice. Baldacchino is a registered nurse with over 12 years of nursing experience in various roles including extensive experience within the emergency department, on bed management, and as a project officer. Baldacchino has also completed a postgraduate certificate in nursing in clinical studies at the University of Sydney, New South Wales, Australia and a diploma in project management at the Agency for Clinical Innovation.

Twan Basten is a professor in the electrical engineering department at Eindhoven University of Technology, Eindhoven, the Netherlands, where he chairs the Electronic Systems group. He is also a senior research fellow of TNO Natuurwetenschappelijk Onderzoek– Embedded Systems Innovation in the Netherlands. He holds an MS (1993) and a PhD (1998) degree in computing science from Eindhoven University of Technology. His research interests include embedded, networked, and cyber-physical systems; dependable computing; and computational models. Basten served in over 55 technical program committees. He (co)authored 1 book and over 160 scientific publications, of which 4 received the best paper awards. He (co)supervised 13 PhD degree students. Basten is a senior member of the Institute of Electrical and Electronics Engineers and a life member of the Association for Computing Machinery. Contact him at http://www.es.ele.tue.nl/~tbasten.

Natalie K. Bradford has a clinical background as a nurse specializing in pediatric oncology and palliative care. In Queensland, Australia, families with children who require treatment for childhood cancer or life-limiting conditions are required to travel from throughout the large state to Brisbane to access healthcare and treatment. It was witnessing the burden of travel placed on families that sparked Natalie's interest in telehealth. Her PhD thesis investigated the use of home telehealth to support families and healthcare clinicians in local areas. Natalie has a keen interest in research, particularly health services research, investigating models of care for optimal healthcare delivery. Natalie also has an interest in promoting the integration of telehealth into models of care with the purpose of improving outcomes and increasing equity of access.

Liam Caffery is a senior research fellow and director of telehealth technology for the University of Queensland's Centre for Online Health, Queensland, Australia. He is an executive member of the Australasian Telehealth Society. Liam is an associate investigator of the National Health and Medical Research Council's Centre for Research Excellence in Telehealth. He is actively involved in telehealth service delivery via his work programs, including RES-e-CARE, Health-e-regions, Princess Alexandra Hospital Telehealth Centre, and Queensland Telepaediatric Service. He has an active research agenda in health services research and health informatics with a special interest in imaging informatics, indigenous health, and rural health.

João Paulo Carmo was born in Maia, Portugal, in 1970. He received a BS and an MS degree in electrical engineering from the University of Porto, Porto, Portugal, in 1993 and 2002, respectively, and a PhD degree in industrial electronics from the University of Minho, Guimarães, Portugal, in 2007. His PhD thesis was on transceivers for integration in microsystems to be used in wireless sensor network applications.

Since 2008, Dr. Carmo has been an assistant researcher at the Algoritmi Center, University of Minho, where he is also currently with the Department of Industrial Electronics. He is involved in research on micro-/nanofabrication technologies for mixed-mode/radio-frequency systems, solid-state integrated sensors, microactuators, and micro-/nanodevices for use in wireless and biomedical applications. He is also involved with the supervision of PhD students from the Massachusetts Institute of Technology Portugal program.

Dr. Carmo is a member of the Institute of Electrical and Electronics Engineers Industrial Electronics Society.

Andrea Cadei was born in Brescia (Italy) where he graduated in electronics engineering. During his research activity, he has worked on several projects related to implantable medical devices, printed sensors, and energy-harvesting systems that use human body motion and has published a number of papers on these subjects. Presently, he works on data acquisition and monitoring software for heterogeneous sensors, real-time controllers, and in-vehicle systems.

Alex Chun Kit Chan has more than 15 years of hardware circuit design and software development experience and is the chief technology officer of AffordSENS Corporation, which was founded in November 2013 for designing wearable healthcare-monitoring devices. Earlier, he was a research engineer of the Maenaka Human-Sensing Fusion Project of the Japan Science and Technology Agency, focusing on the low-power embedded software design and radio-frequency module development. Prior to that, Chan worked as a developer and hardware design engineer at a liquid crystal display company, where he was involved in the development of low-power consumption drivers, liquid crystal display modules, evaluation circuits, hardware description language design, and testing software. He is also interested in mobile application development and digital photography. He received his BS degree in electronic engineering from the Chinese University of Hong Kong, Hong Kong.

David A. Clifton is a tenure-track member of faculty in the Department of Engineering Science of the University of Oxford, United Kingdom and a Governing Body fellow of Balliol College, Oxford, United Kingdom. He is a research fellow of the Royal Academy of Engineering. A graduate of the University of Oxford, Dr. Clifton trained in information engineering and founded the Computational Health Informatics lab when he was appointed to the faculty in 2013. His research includes big data and mobile health projects, where systems derived from his research are used to monitor over 20,000 patients every month in the United Kingdom National Health Service. His previous work includes the world's first (and, so far, only) Food and Drug Administration–approved multivariate patient-monitoring system based on machine learning, and where his related health-monitoring systems are used on the engines of the Airbus A380, Boeing 787 Dreamliner, and Eurofighter Typhoon.

Lei Clifton received a BS and an MS degree in electrical engineering from Beijing Institute of Technology, China, and a PhD in electrical engineering from Manchester University, United Kingdom. After six years of postdoctoral research at the University of Oxford, United Kingdom, she was appointed as a medical statistician at the Centre for Statistics in Medicine, University of Oxford. Her research interests include statistical signal processing and machine learning for intelligent health-monitoring systems. Her current work in the Centre for Statistics in Medicine involves performing statistical analysis in medical trials and advising on research design for grant applications to the National Institute for Health Research.

José Higino Correia received a BS degree in physical engineering from the University of Coimbra, Coimbra, Portugal, in 1990, and a PhD degree from the Electronic Instrumentation Laboratory, Delft University of Technology, Delft, The Netherlands, in 1999, working in the field of microsystems for optical spectral analysis.

Correia is a full professor in the Department of Industrial Electronics, University of Minho, Guimarães, Portugal. His professional interests are micromachining and micro-fabrication technology for mixed-mode systems, solid-state integrated sensors, micro-actuators, and microsystems.

Professor Correia is a member of the Institute of Electrical and Electronics Engineers Industrial Electronics Society. He was the general chairman of Eurosensors 2003 and Micromechanics and Microengineering Europe 2007 held in Guimarães, Portugal.

Derrick W. Crook is a professor of microbiology in the Nuffield Department of Medicine Oxford University, United Kingdom. He is a practicing clinical microbiologist and infec-tious diseases physician in the Oxford University Hospitals National Health Service Trust where he also has responsibility for Infection Control. He is codirector of the Infection Theme of the Oxford Biomedical Research Centre, and leads a large research consortium, Modernising Medical Microbiology, focused on translating whole-pathogen sequencing and data linkage as well as undertaking intervention studies investigating pathogens or infectious diseases of major public health importance.

Crook studied medicine at the University of Witwatersrand, Johannesburg, South Africa; obtained a diploma in tropical medicine (London); specialized in internal medicine at the University of Virginia, Charlottesville, Virginia, United States; and completed a fellowship in infectious diseases at the Tufts New England Medical Center, Boston, Massachusetts. He obtained his board certification in internal medicine and infectious diseases. He trained in clinical microbiology at the John Radcliffe Hospital Oxford, obtaining Fellowships of the Royal College of Physicians and the Royal College of Pathologists.

Mohammad N. Deylami received a BS degree in computer engineering from Tehran University, Tehran, Iran, in 2002. He received his MS degree in computer engineering science from Amirkabir University of Technology, Tehran, Iran, in 2008. He received his PhD degree from the University of Alabama in Huntsville, Alabama, United States, in 2013. He is currently working at A10 Networks, San Jose, California. The focus of his research was on the coexistence between health-monitoring wireless body area net-works. His research interests include wireless networking, multimedia, embedded sys-tems, development of wearable medical devices, and wireless body area networks for telemedicine.

Alessandro Dionisi was born in Brescia, Italy, in 1985. He obtained a laurea degree with honors in electronic engineering for automation at the University of Brescia in 2011. He is cur-rently a PhD student in technology for health at the Department of Information, University of Brescia. His PhD study focuses on the study and design of wearable sensors and in par-ticular on the planning of the systems for monitoring biomedical parameters. This study also aims at supporting systems based on emerging techniques for persons with disabilities. A wearable object detection system for the blind and an autonomous smart shirt for monitor-ing vital signs are the main contributions that he has developed.

Tim Donovan is a senior neonatal consultant in the Grantley Stable Neonatal Nurseries at the Royal Brisbane and Womens Hospital and holds an appointment as an associate professor of neonatal medicine at the University of Queensland, Brisbane, Australia.

Donovan completed his undergraduate degree at the University of Queensland and postgraduate training at the Children's Hospital Brisbane, Royal Children's Hospital Melbourne, and the Royal Women's Hospital Victoria, and a fellowship at the University of California, San Francisco, California, before returning to the Mater Hospital and Royal Brisbane and Women's Hospital in 1989. Currently his research interests include telehealth in newborn care, the epidemiology and outcome of gastroschisis, congenital anomaly surveillance, and body composition during neonatal growth.

Robert H. Eikelboom is the head of the eHealth Group, Ear Science Institute Australia; an adjunct professor at the Ear Sciences Centre, The University of Western Australia, Crawley, Australia; and an extraordinary professor at the Department of Speech-Language Pathology and Audiology, University of Pretoria, South Africa. He serves on the Tele-audiology Committee of the International Society of Audiology and the Committee of the Australasian Telehealth Association. He has published more than 115 peer-reviewed articles and book chapters.

Dagan (David) Feng received his ME degree in electrical engineering and computing science from Shanghai Jiao Tong University, Shanghai, China, in 1982 and MS degree in biocybernetics and PhD degree in computer science from the University of California, Los Angeles, California, in 1985 and 1988, respectively. After briefly working as an assistant professor at the University of California, Riverside, California, he joined the University of Sydney, Australia, at the end of 1988 as a lecturer, senior lecturer, reader, professor, and head of the Department of Computer Science and also the head of the School of Information Technologies. He is also an honorary research consultant in the Royal Prince Alfred Hospital in Sydney; a chair-professor of Information Technology, Hong Kong Polytechnic University, Hong Kong; an advisory professor in Shanghai Jiao Tong University, China; and a guest professor at Northwestern Polytechnic University, Fremont, California, Northeastern University, Boston, Massachusetts, and Tsinghua University, Beijing, China. Professor Feng is a fellow of the Australian Computer Society, the Australian Academy of Technological Sciences and Engineering, the Hong Kong Institution of Engineers, the Institution of Engineering and Technology, and the Institute of Electrical and Electronics Engineers.

Jacob Fraden is an electronic engineer, inventor, entrepreneur, and technical educator. He is an expert in sensors and instruments for monitoring and measuring a broad range of signals. Among his over 50 patents are an instant ear thermometer, a smartphone with infrared sensors, a motion light switch, and a home blood-pressure monitor. Dr. Fraden founded and currently serves as president of SensorJacket, Inc., a U.S. company that develops mobile phones with various sensors.

Dr. Fraden is the author of the best-selling *Handbook of Modern Sensors* (five updated editions since 1994) and a book of short stories, *Adventures of an Inventor*. Dr. Fraden authored over 100 technical and scientific papers and book chapters. He is teaching the course Sensors for Science and Industry at the University of California, San Diego, California.

Marc Geilen is an assistant professor in the Department of Electrical Engineering at Eindhoven University of Technology. He holds an MS degree in information technology and a PhD degree from the Eindhoven University of Technology, Eindhoven, Netherlands, in formal run-time verification techniques and modeling languages and semantics. His research interests include modeling, analysis and synthesis of dynamic stream processing systems using dynamic data flow models, multiprocessor systems-on-chip, cyberphysical and networked embedded systems, automated design-space exploration techniques, and multiobjective optimization. He has been involved in multiple national and European programs and projects.

Manfred Glesner, born in 1943 in Saarlouis, Germany, is an internationally recognized professor in the area of integrated electronic circuits and systems, which are used in communication and automation technologies, as well as consumer and medical electronics.

Glesner's current research interests include advanced design and computer-aided design for micro- and nanoelectronic circuits, reconfigurable computing systems, and architectures and organic circuit design. He is a member of several technical societies and is active in organizing international conferences. Professor Glesner is the recipient of four honoris causa doctoral degrees, as well as a honoris causa professorship. He was a recipient of the honor/decoration of Palmes Academiques in the order of Chevalier by the French minister of National Education (Paris) for distinguished work in the field of education in 2007/2008. He is a fellow of the Institute of Electrical and Electronics Engineers.

Rajarshi Gupta is currently an assistant professor at the Instrumentation Engineering Section of the Department of Applied Physics, University of Calcutta, India. He obtained his graduate, postgraduate, and PhD degrees in instrumentation engineering from the University of Calcutta. He is the principal author of the book *ECG Acquisition and Automated Remote Processing* published in 2014 by Springer. His research areas are biomedical signal compression, health-monitoring systems, and biomedical signal analysis. He has published 30 papers in international journals and conferences. Apart from this, his academic interests include process automation, embedded systems, and smart sensors. Dr. Gupta is a senior member of the Institute of Electrical and Electronics Engineers, a life fellow of the Institution of Electronics and Telecommunication Engineers (India), a member of the Institution of Engineering and Technology (United Kingdom), and a life member of the Biomedical Engineering Society of India. He is also a reviewer of a few Institute of Electrical and Electronics Engineers Transactions and Elsevier Science journals.

Jyri Hämäläinen is currently an associate professor and vice-head of the Department of Communications and Networking of the Aalto University School of Electrical Engineering, Espoo, Finland. Dr. Hämäläinen earned his PhD degree in applied mathematics at the University of Oulu, Finland, in 1998 and DSc (technology) degree in signal processing for communications at the Helsinki University of Technology, Finland, in 2007. Dr. Hämäläinen is author or a coauthor of 150 scientific publications and 35 U.S. patents or patent applications. His research interests include multiantenna transmission and reception methods, radio resource scheduling, relays, small cells, and the design and analysis of wireless networks in general. Furthermore, Dr. Hämäläinen also serves as chair of the Degree Programme Committee for Information and Communications Technology at Aalto University School of Electrical Engineering. Additionally, Dr. Hämäläinen works part time for Ericsson on a research and development collaboration.

Thomas M. Helms, MSc, MD, PhD, is a medical specialist in internal medicine and cardiology with many years of clinical experience.

Since 1998, Dr. Helms has been the head of the Peri Cor Cardiology Working Group, Hamburg, Germany. This group is concerned with clinical electrophysiology and invasive cardiology, carrying out clinical and experimental research and providing training courses as well as consultation in the health sector.

Since 2004, Dr. Helms has been chairman of the board of the German Foundation for the Chronically Ill in Fürth, Germany. He also is a member of numerous national and international societies, including the European Society of Cardiology, the International Medical Science Academy, and the German Cardiac Society and Cardiovascular Research, where he also participates in different working groups, for example, as a member of the nucleus AG 33—Telemonitoring.

Dr. Helms published numerous scientific papers and scientific posters. He is the author of a number of medical books, e.g., *Herzschrittmacher- und ICD-Kontrolle* (*Pacemaker and ICD Control*, published in December 2006 and *Kursbuch Kardiologische Elektrophysiologie* (a course book on cardiac electrophysiology) published in May 2009, both by Thieme Verlag.

Kohei Higuchi is the chief executive officer of AffordSENS Corporation, which was founded in November 2013 for designing wearable healthcare-monitoring devices. He is also a visiting researcher of University of Hyogo, Kobe, Japan. He received a PhD degree in physics from Tokyo University, Japan, in 1978. Then he joined the Central Research Laboratories of NEC Corporation and was involved in the development of large-scale integrated circuits. Since 1984, he has been working on the research and development of charge-coupled device image sensors, infrared image sensors, and complementary metal–oxide–semiconductor integrated microelectromechanical system sensors for pressure and acceleration. He also engaged in the development of an ultrasonic imager for an Advanced Robot for Hazardous Environment, which was one of the Japanese National Big Projects supported by the Ministry of International Trade and Industry. From 2004 to 2005, he was Japan's delegate to the International Electrotechnical Commission TC105/Micro Fuel Cells Standardization WG8 (safety). From 2005 to 2006, he worked for Direct Methanol Fuel Cell Corporation, Pasadena, California. From 2008 to 2013, he joined the Maenaka Human-Sensing Fusion Project of Japan Science and Technology Agency as a research manager.

Hossein Hosseini-Nejad received a BS degree from the Mazandaran University, Babol, Iran, an MS degree from the K.N. Toosi University of Technology, Tehran, Iran, and a PhD degree from the Tarbiat Modares University, Tehran, Iran, in 1999, 2001, and 2013, respectively, all in electrical engineering. He joined the Faculty of Electrical Engineering at the K.N. Toosi University of Technology as a lecturer in 2001, where he is currently an assistant professor of electrical and computer engineering and the director of the Field-Programmable Gate Array Laboratory. He was with the Application-Specific Integrated Circuit Design group at the University of Lund, Sweden, as a visiting PhD researcher from April to September 2013. His research interests include the design and development of digital systems and application-specific integrated circuit/field-programmable gate array implementation of signal-processing algorithms.

Kuei-Fang Hsiao is an associate professor of the Department of Information Management and head of the Augmented Reality Team in E-learning Lab at Ming Chuan University, Taipei, Taiwan. Dr. Hsiao received a BS degree in educational audiovisual technology from the National Taiwan University, Taipei, Taiwan, in 1991. She pursued her postgraduate studies in information education at the University of Manchester, United Kingdom, where she was awarded MEd and PhD degrees in 1993 and 1998, respectively. Dr. Hsiao joined

Ming Chuan University in August 1998 to build her own research area on applications of computerized systems of using the augmented reality to elevate the deteriorating health of students in the very tense and competitive educational environment dominating the country. Her lengthy challenges include finding new practical solutions for applications of computerized systems to including hands-on experimental test runs and pilot studies during 2006–2014 for various groups of local high school and university undergraduate students. Further extension of these applications and sensor technologies includes medical and psychological health of elderly patients and use of smartphones. Some of her work is published in various technical and educational journals and conference proceedings, including 50 research and review papers.

Helen Irving currently holds the positions of senior medical officer in oncology in Children's Health Queensland and senior lecturer in the School of Medicine of the University of Queensland, Australia, specifically in the discipline of medical ethics, law, and professional practice. In 2008, Dr. Irving was awarded preeminent specialist status in recognition of her achievements and performance across multiple fields including advanced clinical/professional knowledge, innovation in service development, professional leadership, health quality improvement activities, and clinical governance and outstanding contributions to clinical research, teaching, clinical training, and education. Dr. Irving developed the state-wide pediatric oncology palliative care on call service and led the development of and successfully obtained funding for the Queensland Pediatric Palliative Care Service. Telehealth has become an integral component of the state-wide service in both oncology and palliative care and has successfully erased some of the barriers related to the problems of distance across Queensland, a state approaching two million square kilometers.

Sundaresan Jayaraman is Kolon professor at the Georgia Institute of Technology in Atlanta, Georgia. He and his research students have made significant contributions in the following areas: healthcare information systems and technologies including wearable biomedical systems; enterprise architecture and modeling methodologies for information systems; and engineering design and analysis of intelligent textile structures and processes. His group's research has led to the realization of the world's first Wearable Motherboard™, also known as Smart Shirt (http://www.smartshirt.gatech.edu). This invention was featured in a special issue of *LIFE* magazine entitled "Medical Miracles for the New Millennium" (fall 1998) as one of the "21 Breakthroughs That Could Change Your Life in the 21st Century." The first Smart Shirt is currently housed in the Smithsonian Museum in Washington, DC. Professor Jayaraman is a recipient of the 1989 Presidential Young Investigator Award from the National Science Foundation.

Emil Jovanov (M '98–SM '04) received Dipl.-Ing, MSc, and PhD degrees in electrical and computer engineering from the University of Belgrade, Belgrade, Serbia. From 1984 to 1998, he was with the Mihajlo Pupin Research Institute, Belgrade, Serbia. Between 1994 and 1998, he was an adjunct assistant professor at the University of Belgrade. In 1998, he joined the University of Alabama, Huntsville, Alabama, where he is currently an associate professor at the electrical and computer engineering department. He is the originator of the concept of wireless body area networks for health monitoring. He has been engaged in developing intelligent sensors for personal health monitoring and mobile computing for more than 15 years. His current research interests include ubiquitous and mobile computing, biomedical signal processing, and health monitoring. He has authored or coauthored 14 book chapters and more than 180-peer reviewed publications. Dr. Jovanov is a member of the editorial boards of the *International Journal of Telemedicine* and *Application and Applied Psychophysiology and Biofeedback*. He is an associate editor of *Transactions on Information Technology in Biomedicine* and *Transactions on Biomedical Circuits and Systems*. He is a member of the Institute of Electrical and Electronics Engineers–Engineering in Medicine and Biology Society Technical Committee on Wearable Biomedical Sensors and Systems and a member of the Institute of Electrical and Electronics Engineers Medical Technology Policy Committee.

Mohamed Khadra is a professor of surgery at the University of Sydney, Australia. He has had a successful and varied career as a leader in education and medicine, internationally and in Australia. He has a degree in medicine, a PhD degree, and a fellowship of the Royal Australasian College of Surgeons. He also has a postgraduate degree in computing and a master of education degree. He has won several research prizes, including the Noel Newton Prize for Surgical Research and the Alban Gee Prize in Urology.

Professor Khadra is both a member of the Board of the Health Information Bureau, New South Wales Ministry of Health, and a member of the Board of the Faculty of Engineering and Information Technology, which exemplifies his strength in connecting medicine and surgery to engineering and information technology. He has recently received substantial funds to develop a state-of-the-art telehealth research lab and its integration to the Nepean Hospital.

Jinman Kim, PhD (2006), is a senior lecturer at the School of Information Technologies, Faculty of Engineering and Information Technologies, University of Sydney, Australia. In 2013, Dr. Kim was appointed as the director of the state government-funded Nepean Telehealth Technology Centre to collaborate with the hospital in clinical research and innovation. He is also a theme leader (imaging, visualization, and information technology) of the faculty-wide institute of Biomedical Engineering and Technology. Dr. Kim's contributions in telehealth technology have been instrumental in becoming a guest editor to the special section of the Institute of Electrical and Electronics Engineers' *Journal of Biomedical and Health Informatics* (published in early 2015). Prior to his telehealth research, Dr. Kim was an Australian Research Council postdoctoral fellow, working on medical imaging and visualization research closely with the Royal Prince Alfred hospital. In 2009–2010 (for one year), he joined the MIRALab research group, Geneva, Switzerland, as a Marie Curie Senior Research Fellow.

Mohit Kumar received a BTech degree in electrical engineering from the National Institute of Technology, Hamirpur, India; an MTech degree in control engineering from the Indian Institute of Technology, Delhi, India; a PhD degree (summa cum laude) in electrical engineering from the University of Rostock, Rostock, Germany; and the Dr.-Ing. habil. degree with venia legendi in automation engineering from the same university in 1999, 2001, 2004, and 2009, respectively.

Dr. Kumar is currently a head of the Computational Intelligence in Automation Research Group at the Center for Life Science Automation, Rostock, Germany. He is serving as adjunct faculty at the College of Nursing of the University of Alabama in Huntsville, Alabama, and adjunct research scientist at Institute of Preventive Medicine of the University of Rostock.

Dr. Kumar's research topics include system theoretic approach to machine learning; signals modeling and analysis; intelligent data interpretation; stochastic models for process identification; modeling applications in eHealth and mHealth; modeling applications in preventive medicine and public health; and modeling applications in chemistry and drug design.

Christopher Lemon graduated from the University of Sydney, Australia, in 2012 with a bachelor of arts degree, majoring in psychology and philosophy. Since then he started to study master of philosophy (medicine) at the Nepean Telehealth Technology Centre through the Sydney Medical School at the Nepean Hospital. The topic of his research is evaluating and optimizing the human compatibility of mobile telehealth systems by integrating and applying ideas from usability and psychology studies. He is also currently a medical student at the University of Notre Dame, Sydney, Australia, with a broad range of future career interests that enable a synthesis of efficient and compassionate clinical care, effective multidisciplinary public health leadership, and innovation through mobile technology infrastructure.

Nigel H. Lovell received BE (Honors) and PhD degrees from the University of New South Wales, Sydney, Australia. He is currently at the Graduate School of Biomedical Engineering of the University of New South Wales, Australia, where he holds a position of Scientia Professor. He has authored more than 203 journal papers. His research work has covered areas of expertise ranging from cardiac modeling, telehealth technologies, biological signal processing, and visual prosthesis design. Through a spin-out company from University of New South Wales, TeleMedCare Pty. Ltd., he has commercialized a range of telehealth technologies for managing chronic diseases and falls in the older population. He is also one of the key researchers leading a research and development program in Australia to develop a retinal neuroprosthesis or bionic eye.

Vincenzo Luciano in 2007 received his laurea degree in ingegneria meccanica at the Università degli Studi di Napoli Federico II, Naples, Italy, with the thesis "Analysis and design of a Eddy current dynamometer bench test." After an employment as design engineer in a manufacturing company of paper machinery, he received a master universitario di II livello in fluid power at the Università degli Studi di Modena e Reggio Emila Modena, Italy, in 2010 and a PhD in mechatronics at the Università degli Studi di Bergamo, Bergamo, Italy, in 2013 with the thesis "Design and construction of an energy harvester to supply an electronic measure circuit implanted in a human total knee prosthesis." Currently, he teaches mechanical engineering at a high school.

Kazusuke Maenaka received his BE, ME, and PhD degrees from Toyohashi University of Technology, Japan, in 1982, 1984, and 1990, respectively. Since 1993, he has been with the Department of Electronics of the Himeji Institute of Technology, Hyogo, Japan. By unification of universities in the Hyogo prefecture in April 2004, he joined University of Hyogo, Japan, where he is presently a professor. From 2008 to 2013, he was the project leader of the Maenaka Human-Sensing Fusion Project supported by the Japan Science and Technology Agency. His research interests include microelectromechanical system devices and technology, especially silicon mechanical sensors and their integration.

Axel Müller, MD, PhD, studied medicine at the University of Leipzig, Saxony, Germany, from 1983 to 1989. In 1990 he received a doctor of medicine degree.

After having worked for a year at the Institute of Pathology, Dr. Müller began his residency in internal medicine. In 1996 he was assigned the specialist title for internal medicine. In 1999 he was able to complete advanced training on special internistic intensive care medicine. Dr. Müller acquired a subspecialization in cardiology in 2003. He works as a deputy senior physician at the Clinic of Internal Medicine, Department I (main focus on cardiology, angiology, and internistic intensive care medicine) at the clinical centre Chemnitz gGmbH.

Dr. Müller is member and vice speaker of the nucleus of the working group Telemonitoring of the German Cardiac Society and Cardiovascular research e.V. In the field of telecardiology, he conducted various studies and published numerous scientific papers.

Edward Mutafungwa is a staff scientist at the Department of Communications and Networks of Aalto University School of Electrical Engineering, Espoo, Finland. He received a DSc (technology) degree in communications engineering from the Helsinki University of Technology (now Aalto University) in 2004 and an MS degree in telecommunications and information systems from the University of Essex, United Kingdom, in 1997. Dr. Mutafungwa has alternatively been a lecturer, a researcher, and a project manager in various national projects and international projects (European Commission, Celtic) implemented at the Department of Communications and Networks, Aalto University. Dr. Mutafungwa is an author or coauthor of over 50 scientific publications and his research interests lie in the areas of telehealth, broadband access networks, public safety communications, information and communication technology for emerging markets, green information and communication technology, user-centric design, and optical networks.

Majid Nabi is an assistant professor in the electrical engineering department at Eindhoven University of Technology, Eindhoven, Netherlands. He received his MS and PhD degrees in electrical and computer engineering from the University of Tehran, Iran, and Eindhoven University of Technology, respectively, in 2007 and 2013. Dr. Nabi has several years of experience in design and implementation of real-time data acquisition and processing systems. His research interests include design, modeling, and performance analysis of efficient and dependable networked embedded systems, wireless sensor and ad hoc networks, and wireless body area networks. He is a member of the Institute of Electrical and Electronics Engineers.

Katherine E. Niehaus is currently a PhD student at the University of Oxford, United Kingdom, within the Institute of Biomedical Engineering. She is supervised by Dr. David Clifton and her current work focuses on using machine learning to analyze heterogeneous genomic and electronic health record data sets. She is looking at clinical applications for both infectious diseases and inflammatory bowel disease. She holds an undergraduate degree in biomechanical engineering and a master's degree in bioengineering, both from Stanford University, Stanford, California.

Sungmee Park is vice president and head of Future Strategy at Kolon Corporation in South Korea. Park is engaged in identifying new growth engines and strategic opportunities for the Kolon Group. During 2010–2012, she served as the president of research and development at Kolon Glotech, Inc., and transformed it into Korea's leader in the field of smart textiles. Park served as the chair of the Textile Information Technology Innovation Center (TIIC) that was established in May 2011 by the government of South Korea to promote and advance research in the field of information technology-enabled textiles. Prior to joining Kolon in 2008 as vice president of research and development, Park was a senior research scientist at the Georgia Institute of Technology in Atlanta, Georgia, United States. She is a coinventor of The Smart Shirt. Among Park's numerous awards is the prestigious Jang Yeong-Sil Award from the South Korean government for the development of Hea Tex in 2008.

Timothy E. A. Peto is a professor of medicine at the University of Oxford, United Kingdom. He is the coleader for the Infection Theme of the Oxford Biomedical Research Centre and is a National Institute for Health Research senior investigator. Peto's research has included combination therapy for acquired immune deficiency syndrome, the search for an effective acquired immune deficiency syndrome vaccine, the transmission of methicillin-resistant *Staphylococcus aureus* in hospitals, and transmission mechanisms for *Clostridium difficile* infections.

François Philipp received a double degree in 2009 from the Ecole Nationale Supérieure de l'Electronique et de ses Applications, Cergy, France, and from the Technische Universität Darmstadt, Germany, in the field of computer engineering. In 2014, he completed his PhD thesis with highest honor on the topic of run-time hardware reconfiguration in intelligent wireless sensor networks within the microelectronic systems research group at Technische Universität Darmstadt. He was involved in a research center on adaptronics and in the European Union Seventh Framework Programme project Maintenance on Demand. His research interests include the development of novel smart systems based on distributed data processing and energy-efficient processor architectures. Applications of his work can be found in condition-monitoring systems and personal health-monitoring devices. He is now part of Intel as an expert in power management for mobile devices.

Marco A. F. Pimentel moved to Lisbon from the Azorean islands after high school to pursue his studies and training in medical technology and engineering. He received BS and MS degrees in biomedical engineering from Universidade Nova de Lisboa, Lisbon, Portugal, in 2009. His master's degree focused on medical imaging and neuroscience in biomedical engineering. After working in different institutions in Portugal and China for a few months as a research assistant, he was awarded a scholarship from the Research Councils United Kingdom in 2010 and joined the Institute of Biomedical Engineering at the University of Oxford, United Kingdom, where he is currently pursuing his PhD in computational health informatics.

Pimentel's main research interests include the development of signal-processing algorithms and machine learning methods that cope with the complexity and heterogeneity of electronic health record data.

M. Yasin Akhtar Raja is professor of physics and optical science and electrical and computer engineering (adjunct) at the University of North Carolina at Charlotte, North Carolina, United States. Since 1990, he played a leading role in several planning and program committees for establishing new PhD programs, research centers, and academic units. His research expertise on optical science and engineering spans photonics devices and components for optical communication networks and information and communication technology e-applications. He has several patents and has published over 180 articles in journals, book chapters, and refereed proceedings. He has instituted the international conference series High-Capacity Optical Networks and Enabling/Emerging Technologies (http://honet-ict.org) with cosponsorship of the National Science Foundation and the Institute of the Electrical and Electronics Engineers since 2004. Professor Raja received his PhD degree in 1988 from the University of New Mexico, Albuquerque, New Mexico, where he coinvented resonant periodic gain semiconductor lasers (today's vertical-cavity surface-emitting lasers) and conducted pioneering research. He obtained MPhil and MS degrees from Quaid-i-Azam University, Islamabad, Pakistan. He is a senior member of the Institute of Electrical and Electronics Engineers and the Optical Society of America and a member of the Society of Photo-Optical Instrumentation Engineers and the Fiber to the Home Council.

Habib F. Rashvand, CEng, LIEEE, following his distinguished engineering qualifications from the University of Tehran, Iran, in early 1970s, in association with the Prime Time Telecom, Inc., Nippon Telegraph and Telephone Corp., and other Japanese industries, headed an international project for building the Iran Telecommunication Research Center as distinct national resources of the country. His doctorate of philosophy from the University of Kent, United Kingdom, in 1980 showed his interest for a new world of data communications in the 1980s for modems and in the 1990s for wireless and mobile technologies until his professorship on networks, systems, and protocols was granted in 2001 by the German Ministry of Education. After 25 years of industrial research and development with Racal, Vodafone, Nokia, and Cable & Wireless plc. with academic positions at Tehran University, University of Zambia, Coventry University, Open University, and Magdeburg German Universities, with over 100 projects, three books, and a large number of research papers, chapters and lecture notes as guest speaker, course manager, and industrial collaborator, he has been heading a special operation as the director of Advanced Communication Systems in association with the University of Warwick, United Kingdom, since 2004, involving academics, industries, and professional institutions for innovative information and a communication technology solutions paradigm to help to build a sustainable future global village upon a new info-rich global infrastructure.

Hermawan Nagar Rasyid, MD, SpOT(K), MT, PhD, obtained his medical doctor (MD) degree from the Faculty of Medicine of the Universitas Padjadjaran, Bandung, Indonesia. He finished as an orthopedic surgeon in mid-January 1993 from the same university. In 2002, he obtained his master's degree in biomedical engineering from the School of Electrical Engineering and Informatics, Institut Teknologi Bandung, Indonesia. After completing a double degree doctorate program, he received a doctorate degree (Cum Laude) from Institut Teknologi Bandung in 2008 and a doctorate degree from University Medical Center Groningen, Rijksuniversiteit Groningen (Netherlands), in February 2009. Currently, Dr. Rasyid, is a member of the teaching staff at the Department of Orthopaedics and Traumatology, Faculty of Medicine Universitas Padjadjaran, Hasan Sadikin Hospital, Bandung, Indonesia.

Sedigheh Razmpour received BS and MS degrees from the Khajeh Nasir Toosi University of Technology, Tehran, Iran, in 2009 and 2012, respectively in computer hardware engineering and electrical engineering. She has been with the Research Laboratory for Integrated Circuits and Systems at the Khajeh Nasir Toosi University of Technology as a graduate student member from 2010 to 2012 and as an associate member since 2012. She has also been with TPI Inc., Tehran, Iran, as a research and development engineer. Razmpour's research interests include implantable biomedical circuits, biological signal processing, and digital/mixed-signal systems.

Stephen J. Redmond is an Australian Research Council Future Fellow at the Graduate School of Biomedical Engineering at the University of New South Wales, Australia. He completed his bachelor's degree in electronic engineering at the University College Dublin, Ireland, in 2002. He also completed his PhD degree in biosignal processing at the same institute in 2006. His primary research interests revolve around the application of signal processing and pattern recognition techniques to solve or understand biomedical engineering problems. The principal application areas for these signal processing and pattern recognition techniques include telehealth, fall detection and prediction for older people, monitoring physical activity using wearable technologies, and tactile sensing and physiology.

Emilio Sardini graduated in 1983 in electronic engineering from the Polytechnic University of Milan, Italy. Since 1984, he has conducted research and teaching activities at the Department of Electronics for Automation, University of Brescia, Italy. Since November 1, 2006, he is a full professor of electrical and electronic measurement. He has been a member of the Integrated Academic Senate and of the Board of Directors of the University of Brescia and a deputy dean of the faculty. Now he is the coordinator of the Technology for Health PhD program, a member of the College of Mechatronics PhD program at the University of Bergamo, Italy and the director of the Department of Information Engineering. He has done intensive research in the field of electronic instrumentation, sensors, and signal-conditioning electronics. Recently, his research work has addressed the development of autonomous sensors for biomedical applications with some specific interest toward devices that are implantable inside the human body. He is an author or coauthor of more than 100 papers published in international journals or proceedings of international conferences.

Jörg Otto Schwab, MD, PhD, studied medicine at the University of Giessen, Germany, from 1989 to 1995 and started to work as a doctor of medicine in 1996.

After having worked at the cardiology department of the University of Giessen until 2001, he was appointed to the Department of Medicine—Cardiology at the University of Bonn, Germany, until to the end of 2014. He became a consultant on internal medicine in 2002 and specialized in cardiology in 2003. He was assigned as a head doctor in 2003 and was responsible for the intensive care unit as well as the defibrillator and heart failure unit. While focusing his research on sudden cardiac death, device, and remote monitoring therapy, he received his PhD degree in 2004.

Currently, Dr. Schwab is a member and speaker of the nucleus of the Telemonitoring working group of the German Society of Cardiology—Heart and Circulatory Research e.V. In the field of telecardiology, he participated in randomized trials, conducted various studies, and published numerous scientific papers.

Johannes Schweizer, MD, PhD, studied medicine at the University of Leipzig, Germany and University of Erfurt, Germany, from 1977 to 1982. In 1986 he received a doctor of medicine degree. In 1987 and he was assigned as a specialist in internal medicine.

Dr. Schweizer acquired a subspecialization in cardiology in 1991 and a subspecialization in angiology in 1995. Since 1990, he has participated in different clinical studies for prevention of reocclusion after angioplasty.

Dr. Schweizer is working in different public health projects on medical care in cardiology. He is an extraordinary professor at the Dresden University of Technology, Germany and the head of the Clinic of Internal Medicine, Department I, Hospital Chemnitz gGmbH.

Mauro Serpelloni received a laurea degree (summa cum laude) in industrial management engineering and a research doctorate degree in electronic instrumentation from the University of Brescia, Italy, in 2003 and 2007, respectively. He is currently an assistant professor of electrical and electronic measurements in the Department of Information Engineering, University of Brescia. He has worked on several projects relating to the design, modeling, and fabrication of measurement systems for industrial applications. His research interests include biomechatronic systems, contactless transmissions between sensors and electronics, contactless activation for resonant sensors, and signal processing for microelectromechanical systems.

Mohammad Ali Shaeri received BS and MS degrees in electrical engineering from Khajeh Nasir Toosi University of Technology, Tehran, Iran, in 2008 and 2011, respectively. From 2009 to 2011, he was a graduate student member of the Research Laboratory for Integrated Circuits and Systems at the Khajeh Nasir Toosi University of Technology. He is the recipient of the 2011 Research Laboratory for Integrated Circuits and Systems Outstanding Researcher Award. From 2010 to 2012, he was with Allameh-Majlesi University, Qazvin, Iran, as an Instructor. Since 2012, he is working toward a PhD degree in cognitive neuroscience at the School of Cognitive Sciences, Institute for Research in Fundamental Sciences, Tehran, Iran. Mohammad Ali Shaeri's research interests include processing and compression of neural signals and design and development of digital systems.

Amir M. Sodagar received a PhD degree in electrical engineering from the Iran University of Science and Technology, Tehran, Iran, in 2000. From 2001 to 2009, he was with the National Science Foundation's Engineering Research Center for Wireless Integrated Microsystems, University of Michigan, Ann Arbor, Michigan, United States, as a postdoctoral research fellow, assistant research scientist, and subsequently as the technical director for biomedical microsystems. Currently, he is with Khajeh Nasir Toosi University of Technology, Tehran, Iran, as an associate professor of electronics and biomedical engineering and founding director of the Research Laboratory for Integrated Circuits and Systems. He is also with the School of Cognitive Sciences of the Institute for Research in Fundamental Sciences, Tehran, Iran, as a member of the Scientific Board, and with Polytechnique Montréal, Montreal, Quebec, Canada, as a professeur associé. He is the author of three books, including the textbook *Analysis of Bipolar and CMOS Amplifiers* (Boca Raton: CRC Press, 2007). His research interests include implantable microsystems, biomedical circuits and systems, biological signal processing, and analog and mixed-signal integrated circuits.

Soegijardjo Soegijoko received his engineering degree in telecommunication engineering from the Department of Electrical Engineering, Institut Teknologi Bandung, Indonesia, in 1964, and his doctorate degree from the Universite des Sciences et Techniques du Languedoc, Montpellier, France, in 1980.

Since 1966, Soegijardjo Soegijoko is a member of the teaching staff in the Department of Electrical Engineering Institut Teknologi Bandung. From 1994, he has been actively involved in the development of the Biomedical Engineering program (for the undergraduate, master, and doctorate levels), the Biomedical Engineering Laboratory, and Biomedical Engineering research activities at the Institut Teknologi Bandung until his retirement in 2007. He is currently an adjunct professor of biomedical engineering at the School of Electrical Engineering and Informatics, Institut Teknologi Bandung. His current research interests include telemedicine and eHealth systems, biomedical instrumentation, and biomedical engineering education.

Professor Docteur Ingenieur Soegijardjo Soegijoko is a member of the Indonesian Biomedical Engineering Society, the Indonesian eHealth and Telemedicine Society, the Institute of Electrical and Electronics Engineers Engineering in Medicine and Biology Society, and the Institute of Electrical and Electronics Engineers Circuits and Systems Society.

M. B. Srinivas is currently a professor and Dean of Administration at Birla Institute of Technology and Science, Hyderabad Campus, India. His research interests include high-performance logic circuit design and synthesis, medical body area networks, and mHealth. His passion for enabling billions to have access to affordable healthcare made him design and develop a variety of portable diagnostic devices that operate in conjunction with Smartphones and mobile networks. Dr. Srinivas was a recipient of Microsoft Research Digital Inclusion Award in 2006 and his work has been funded by Microsoft Research, Redmond, Swedish International Development Agency, and Indian Council of Medical Research, among many.

Norbert Stoll studied technical cybernetics and automation technology at the University of Rostock, Germany, graduating in 1979 and receiving his PhD in the field of measurement in 1985.

Starting as a scientific coworker and later on as group leader at the Central Institute for Organic Chemistry of the Academy of Sciences of the German Democratic Republic, Dr. Stoll worked in a number of automation technology– and analytical chemistry–related subjects. He was appointed as an associate director of the Institute of Organic Catalysis Research in 1991. Dr. Stoll contributed greatly to the development of the Institute of Organic Catalysis Research e.V. In 1994, he was appointed as a professor of process measurement at the University of Rostock and worked as the managing director of the Institute of Automation from 1998 to 2000.

From 2000 to 2002, Dr. Stoll served as the dean of the College of Engineering of the University of Rostock. As an initiator and founding member of the Institute for Measurement and Sensor Systems e.V., he managed it from its foundation in 1996 until 1998 as president. Dr. Stoll is one of the directors of the Center for Life Science Automation. He is one of the founding members of the Rostock–Raleigh e.V., which is mainly based on his remarkable personal involvement.

Dr. Stoll's research topics include process and environmental measurement technology; computer-based Measurements and automation systems, and robotic systems.

Regina Stoll studied human medicine at the University of Rostock, Germany, from 1974 to 1980. In 1980, she graduated as approbated and diploma doctor. Through 1985 she worked as assistant doctor at the Sport Medicinal Department. She received her PhD in occupational medicine and only one year later passed her medical specialist exam on sports medicine and became a specialist at the Institute for Occupational Medicine at the University of Rostock. In 1988, she became a group leader in work physiology and sports medicine and passed her second specialist exam on occupational medicine. In 1992, she was appointed as an associate director of the Institute for Occupational Medicine.

Since 1996, Professor Stoll has also been group leader for medical automation at the Institute for Measurement and Sensorsystems e.V. and was appointed head doctor at the Institute for Occupational and Social Medicine. In 2002, she habilitated and was awarded venia legendi for occupational and sports medicine from the University of Rostock (PD, MD, habil.). Also in 2002, she was appointed as director of the Institute for Occupational and Social Medicine and serves as its managing director until today. Since 2004, Professor Stoll is an adjunct associate professor and faculty member at the North Carolina State University at Raleigh, North Carolina, United States.

Professor Stoll is member and chairperson of numerous boards, committees, and associations as well as an associated faculty member of the Rostock Colleges of Philosophy as well as of the Faculty of Computer Science and Electrical Engineering. Professor Stoll is vice president of Rostock–Raleigh's Sister City Association.

Professor Stoll's research topics include occupational physiology and preventive medicine.

Sugiyantini received her medical doctor (MD) degree from the Faculty of Medicine, University of Gadjah Mada, Yogyakarta, Indonesia, in 1979, and her dermatovenereology specialization in 1990 from the same university. She has a number of professional experiences in the field conducted in various hospitals. From 1991, she has worked as a medical specialist at the Panembahan Senopati General Hospital in Bantul, Yogyakarta, until her retirement in 2010. Currently, she is with the PKU Muhammadiyah General Hospital in Bantul and Ludira Husada Tama Hospital, Yogyakarta. Dr. Sugiyantini, SpKK, is a member of Ikatan Dokter Indonesia (Indonesian Medical Association) and Perdoski (Indonesian Society of Dermatology and Veneorology). Her practical professional experiences have supported the writing of book chapters.

De Wet Swanepoel is a professor of audiology at the Department of Speech-Language Pathology and Audiology, University of Pretoria, South Africa, and a senior research fellow at the Ear Science Institute Australia and has adjunct positions at the University of Texas at Dallas and the University of Western Australia. He has published more than 90 peer-reviewed articles, books, and book chapters and has received numerous awards in recognition of his work. His research capitalizes on the growth in information and communication technologies to explore, develop, and evaluate innovative solutions to improve access to ear and hearing healthcare. Professor Swanepoel is an associate editor of the *International Journal of Audiology* and serves on a number of international boards. He is currently an executive board member of the International Society of Audiology, where he serves as president-elect, and he also co-chairs the telehealth task force for the American Academy of Audiology.

Kerstin Thurow studied chemistry at the University of Rostock, Germany, and graduated far quicker than the standard period time of study. Afterward she received her PhD from the Ludwig Maximilian University, Munich, Germany (Professor Dr. Lorenz), working on metal–organic sulfur compounds. In 1999 she finished her habilitation at the Department of Electrical Engineering at the University of Rostock and obtained a faculty membership with veni legendi on measurement and control. In October 1999, Professor Dr. Thurow was appointed as Germany's youngest university professor and obtained the worldwide unique professorship on laboratory automation. In December 2004 she was appointed for a novel professorship on life science automation. This chair is connected with the Center for Life Science Automation management directorate. Besides many awards, such as for the foundation of a start-up company amplius—Screening Technologies and Analytical Measurement, Dr. Thurow was awarded the highly renowned Joachim Jungius Award for Science in 2004.

Currently Dr. Thurow is the managing director of the Institute of Automation at the University of Rostock, working as the managing director for the Center for Life Science Automation, and the president of the Institute for Measurement and Sensorsystems e.V. As the founding member and president of the Rostock–Raleigh e.V.—a sister city association—Dr. Thurow is striving for cultural, sportive, but also scientific and economic relations to one of the most important life science regions in the United States.

Dr. Thurow's research topics include life science engineering; laboratory and process automation; high-throughput screening; spectroscopic measurement systems; and instrumental analytics.

I-Hen Tsai is currently a senior software programmer in Delta Research Center, Delta Electronics. He is responsible for developing and building mobile biomedical and healthcare applications. He has also worked extensively with universities and research institutions to promote collaborative research between the industry and academia. I-Hen received his master's degree in medical informatics from National Cheng Kung University, Taiwan, and has a bachelor's degree in computer science and information engineering from National Tainan University, Taiwan, Republic of China. His research interests are currently focused on applications of computational sciences toward improving healthcare, in particular applications which leverage modern mobile devices such as smartphones and tablets. A more detailed profile may be found on LinkedIn: http://lnkd.in/-q9Qwf.

Kejia Wang is a PhD student in the Graduate School of Biomedical Engineering at the University of New South Wales, Sydney, Australia, and is supervised by Scientia Professor Nigel Lovell, Dr. Stephen Redmond, and Dr. Lauren Kark. She graduated with a BE (Honors) degree in biomedical engineering, first class, from the University of Auckland, New Zealand, in 2012. Her current research interests include wearable sensors for activity monitoring and fall prediction, gait biomechanics in aging, and telehealth technologies.

Peter J. Watkinson is a specialist in acute medicine at the University of Oxford Hospitals National Health Service Trust, and a principal investigator within the Kadoorie Centre for Critical Care Research. Dr. Watkinson's research interests include patient monitoring, evidence-based early-warning systems, and the use of technology for improving patient outcomes in critical care environments.

Ronald S. Weinstein, MD, FCAP, FATA, is a Massachusetts General Hospital–trained pathologist. In 1968, as a pathology resident at Harvard University, Cambridge, Massachusetts, he participated in rendering early television-microscopy diagnoses for the groundbreaking Massachusetts General Hospital–Logan International Airport multispecialty telemedicine program. Today, Dr. Weinstein is often referred to as "the father of telepathology." In 1986, Dr. Weinstein invented, patented, and commercialized robotic telepathology and introduced the term *telepathology* into the English language. Telepathology systems of his design were used for the first sustainable telepathology clinical services in the United States. In 2001, Dr. Weinstein co-invented the array microscope, a critical component of the next generation of digital slide scanners. He and his coworkers established a new through-put standard ("one-minute" scans) for digital slide scanners. Dr. Weinstein is a professor of pathology at the University of Arizona's College of Medicine, Tucson, Arizona and the founding director of the national award–winning 160-site state-wide Arizona Telemedicine Program.

Christian Zugck, MD, PhD, is a medical specialist in internal medicine, cardiology, angiology, and intensive care. He was a practitioner in cardiology (Straubing) and an interventional cardiologist (head of a catheterization laboratory in Bogen), and a research associate at the University of Heidelberg, Germany and in Charité, Berlin, Germany. Previously, he held the position of senior physician at the Department of Cardiology, University of Heidelberg (2006–2012). Professor Zugck is a fellow of the European Society of Cardiology. He is also a member of the nucleus of the Telemonitoring working group of the German Cardiac Society and Cardiovascular research e.V.

Section I

Integration of eMedicine, Telemedicine, eHealth, and mHealth

1

Integrating Telemedicine and Telehealth— Advancing Health at a Distance

Habib F. Rashvand and Kuei-Fang Hsiao

CONTENTS

1.1 Introduction

In an earlier work [1] we addressed telemedicine (TLM*) to some extent as "a service-oriented development" process owing its success to the two key economics of the electronic and healthcare industries, showing that the telemedicine service opportunity is basically coming from adoption of the ubiquitous wireless access technology. Since then, in the following 7 years, we have seen considerable progress in the technological developments of telemedicine and its sister technologies telehealth (TLH) and telecare (TLC†) as an indication of its impact on public awareness for a new beginning of business market development across the world.

Now, the number of viable applications of telemedicine and telehealth technologies is growing large and increasing. There are many new and emergency healthcare services for ambulances, rural hospitals, and remote areas of settlements demanding a better telemedicine infrastructure. Remotely based intensive care patients need constant monitoring, and home telecaring for patients suffering from chronic diseases such as heart disease is continuously on the increase, putting scarce medical resources under pressure. It is, however, quite easy to imagine a single but fully automated clinic running under a handful of experts to be able to look after as many as 100 patients equipped with portable and wearable devices. For example, a simple remotely located base unit can collect and locally integrate many incoming signals such as an electrocardiography (ECG or EKG) signal, for continuous monitoring of the oxygen saturation (SpO_2), heart rate (HR), noninvasive blood pressure (NIBP), invasive blood pressure (IBP), temperature, and respiration (Resp) and provide the information required for detecting any possible emergency cases for the patients. The clinic can then accommodate all complementary and bulky systems such as telemedicine-enabled equipment, such as intensive care units (ICUs), intelligent analyzers, and automatic recorders, plus a professionally managed database system supported by a professional service provider.

After a brief refreshing of our understanding of telemedicine as a concept in order to facilitate this chapter's main objective of integrating telemedicine and telehealth for a more prosperous future, we introduce new and enhanced versions of their definitions. That is, for a smoother progress it is essential to clear the ever-thickening mist dominating these terms to bring a better understanding of these terminologies *telemedicine* and *telehealth* and their commonalities. Removing their existing fuzzy differences is important, especially when they address ways of facilitating medical treatment and healthcare at a distance. More importantly, they both should be able to provide sophisticated professional services that run smoothly under various complicated operations restricted by the laws. For example, a remote nursing or monitoring service for the elderly is more likely to be classified as "telehealth" rather than "telemedicine," while a remote, interactive surgical operation is the other way around.

It is interesting to see that the most important aspect of these technologies resides in their traditional feature of challenging and removing the critical factor of distance from the healthcare center. The problem of distance in healthcare is very important. The remoteness, lack of access, and distance and its critical time factor could become a serious threat

---

* We use the terms *telemedicine* and *telehealth* from the industrial point of view, for which this chapter aims to establish their definitions, as well as associated industrial sectors.

† Telecare stands for care services at a distance, which in our view corresponds to a lightweight subclass of the telehealth service industry.

to life when commonly known medical solutions cannot be made available within the required time frame. It could happen in busy towns where transportation of the patient to the specialized clinic becomes the obstacle and an unexpected heart attack may result in serious side effects on patients' future health even if they are fortunate enough to survive.

Although by the end of Section 1.1 we should have a solid set of terminologies to work with throughout the chapter, we take every opportunity to shed further light on telemedicine and telehealth terms so that readers can digest our proposed methods for a deployment process for the integration of these technologies as discussed in Section 1.4, through clear differentiation between telemedicine and telehealth. This would also give us a deeper insight into their technological differences before addressing the management of an integrated solution. We also admit that we do not engage in any technical aspects of distance, which we hope other chapters of this book cover in detail; but as distance is an essential part of operation, we include a few discussions for practical implications of it as why distance is important and how far our technological capabilities can facilitate reducing risks and managing a practically acceptable and controllable way of resolving the problem of distance.

1.1.1 Telemedicine versus Telehealth

We can easily claim that we are already rich with abundantly available intensive mature information and communication technologies (ICTs), including commonly available items such as video, smartphone, data, Internet, and e-mail services over broadband wired and wireless telecommunication technologies. We envisage in the near future that telemedicine capabilities would extend these even further through improved, new developments in smart and intelligent biosensors and related electronics, integrated with the Internet of Things (IoT), enabling new, safe, and viable applications through energy-efficient systems, distributed systems, safe implantable devices, and improved high-speed connectivity under unstructured topologies that are dynamic and ubiquitous and have many other features [2–5].

1.1.1.1 Definitions

Now, in order to separate these two much overlapped sectors of industry, let us informally analyze the terms *telemedicine* and *telehealth* for their required technologies and operational management with respect to their service provision. From a technological point of view, we consider medicine a professional and intensive part of the healthcare system, but when it comes to the health industry and associated services, we have a traditional problem of identifying any hard boundaries between telehealth and telemedicine. For example, some old dictionaries define *telemedicine* as "the use of telecommunications technology to provide, enhance, or expedite healthcare services, as by accessing off-site databases or linking clinics or physicians' offices to central hospitals for transmitting x-rays or other diagnostic images for examination at other sites," which clearly lacks vision of our recent decades of advancement. Then, we have a better definition for *telemedicine* coming from the University of Miami, "telehealth is the remote provision of healthcare services and health education, mediated by technology." This indicates that telehealth, as a matter of fact, has been emerging from our traditional telemedicine technology to deal with the business and service applications of the industry. Then, based on a definition from the American Telemedicine Association (ATA), we see an improved definition of *telemedicine* as "technology-based services tailored to use for handling medical information exchanged from one location to another via electronic communication systems to improve a patient's clinical health status" [6]. For the sake of simplicity, we try not to probe any further into the definitions, but it may be worth mentioning that due to the multidisciplinary nature of services supported by either TLM or TLH (TLM/H), almost every day we have a new addition to the ever-growing TLM/H services. For empowering the real practical progress, we make our own definition to help our readers understand this chapter's objective of integrating telemedicine and telehealth for a systematic implementation harnessing these prime interrelated technologies. Further clarification of the differences between telemedicine and telehealth services is identifiable in Figure 1.1. This figure shows practical boundaries between these two systems.

1.1.2 Technologies versus Services

For a rewarding but very complex new service paradigm like medication and healthcare at a distance, today we have a choice of two approaches of "building the right set of technologies before providing services" or "building a set of agile companies upon demands and requirements to develop the right products and services," which looks like the chicken-and-egg puzzle.

In order to reestablish our earlier basic understanding of telemedicine and telehealth for their differences, similarities, and compatibilities, we can take one of two equally valid approaches: (a) Expand the existing pile of contradictory definitions for their systems and services over the last 50 years and then group them to establish two overlapping but meaningful areas for telemedicine and telehealth. This would require some follow-up in cross-boundary refinements on each side for solid meaningful starting definitions. Or (b) based on our understanding of their conceptual meanings, separate them to play their own dynamic roles so that they can develop independently upon

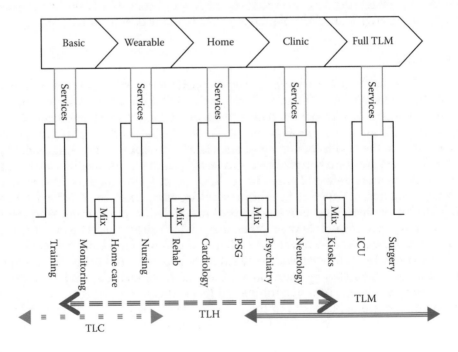

FIGURE 1.1
Hanging-tree-style differentiation between telemedicine and telehealth services.

their own full strengths into a few self-controlled optimized viable industries before reintegration to cooperate for the purpose of some integrated set of strong and trustworthy service industries.

In this case we go for the second option, for which we refine our previous definitions of *telemedicine* and *telehealth* in two sets of capabilities as required to enable each sector to grow freely and independently upon its own market possibility aimed at being ready for ease of integration, as and when required. Therefore, we use the term *telemedicine* for addressing the *infrastructural* aspects of the provision of services at a distance, i.e., similar to the traditional telecoms industry. Telemedicine addresses the *technological requirements of providing such services as a long-term process so that all health and medical services are freed from the barrier of distance*, that is, provision of distance-independent technological solutions for health, medication, and interaction services between the users and providers at the infrastructural level; whereas, telehealth is treated as the *time-dependent risk-free business-style health services offered to people, communities, individual clients, and their patients at each stage of development.*

That is, while telemedicine takes responsibilities over the infrastructure, standards, and core technological requirements, telehealth entails the market viability and demand readiness aspects of medical and health services to be built upon a set of selective required technologies. That is why provision of health services heavily depends on the market where the success involves financial viabilities and sustainability of the service. There are normally many nontechnological aspects, including safety of clients, and trust and security aspects of providers for which the telehealth industry should take full responsibility. On top of these, service providers should demonstrate their technical competence and managerial trustworthiness as well as their commitments to the local and international laws associated with the health service at both ends of the service and all over places wherever the end points of service provision and receivers are involved.

Due to the nature of this book and our chapter's objective, we cut down our nontechnical discussions to a fairly basic and generic level; see Section 1.4. Although we firmly believe our definitions can serve their purpose throughout this chapter and, of course, our final stage of integration, in Sections 1.2 and 1.3 we bring some examples to show the differences and similarities between telemedicine and telehealth. Before closing this section, let us reiterate our view once again in slightly different style: *Without loss of generality, we can define* telemedicine *as "a technologically viable infrastructural sector dedicated to provision of healthcare applications at a distance"—i.e., telemedicine solutions render themselves to "capabilities required for medical health treatments and intensive-care applications"; whereas telehealth is the industry sector ultimately responsible for "providing useful commercial and feasible services built upon telemedicine technologies."*

A technology-enabled healthcare industry built upon upcoming electronic health (eHealth) technological developments influencing all other industries is to form a new structure for health services due to its leading position in the medical business environment. Telemedicine is breaking new barriers and providing new technologically feasible and economically viable distance compatible eHealth systems supporting potentially massive large-scale telehealth and telecare services, which could enable new economies to have a better quality of life throughout the world. As shown in Figure 1.2, TLM in effect is guiding other sectors toward a new health industry paradigm. Figure 1.2 shows the position of each of these four main technology-based health sectors, namely, eHealth, telemedicine, telehealth, and telecare, as displayed upon their cost versus technology enrichment perceived by the authors.

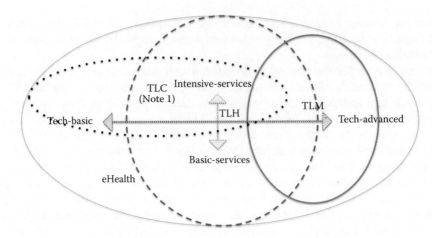

FIGURE 1.2
Distant-health service environments, telemedicine, telehealth, and telecare, in comparison to eHealth. Dimensions represent service capability (axis *y*) versus technological strengths (axis *x*). *Note:* Telecare services compromise the low-tech end of telehealth services.

1.2 Technology-Enabled Distant Health

In order to analyze the developing aspects of providing health at a distance through emerging technologies, we divide the process into three parallel complementary health-enabling parts, namely, eHealth-technology enabled, TLM-technology enabled, and TLH-technology enabled. The following describes these:

a. The eHealth-enabled technological development process deals with digital electronic technologies and can be brought reliably into all fundamental medical and health systems and devices. Due to integrated capabilities coming with digital systems, they can be also used anywhere and at any time, at a distance. That is why eHealth is often regarded as a technological development rather than a service. Due to lesser constraints of digital technologies, adding communication technologies should enable it to embrace a wider scope of the industry.

b. The TLH-enabled technological development process is a service platform of combined medical tools and systems and communication services standards, system capabilities, and service availability that can guarantee a reliable, flexible, and smooth flow of service at the required rate and agreed quality. Therefore, telemedicine services account for medical and professional health applications for distant clinics.

c. The telehealth-enabled technological development process ensures that all services are designed for financially viable application scenarios. Often limited due to the lack of availability of required remotely operated facilities and human expert resources, telecare services are usually considered part of telehealth; but due to their extensive financial scope and low-tech requirements without proper professional quality, they are not usually counted as part of TLM/H technologies. Obviously, there is no objection to telecare services making use of the professional and advanced health capabilities of telemedicine.

Combining all above three processes together, we then have the required infrastructure in place interworking in all dimensions and environments such as clients, experts, equipment, and devices, all physically synchronized and harmonized between the users to share eHealth, telemedicine, and telehealth systems providing any required health service through a single or multiple links as shown in Figure 1.3. Figure 1.3a shows the basic link configuration as required for connecting the central service's facilities and experts, called the complex, a TLM/H clinic, a well-equipped hospital, or a rural clinic center located in somewhere suitable with appropriate facilities for local clients so that patients may easily visit for further treatment when required. This service normally makes a good use of the following system components:

- E as the experts' center tech-rich facilities.
- N as the center's networking system.
- S as secure service exchange and a control provision point for serving the clients.
- Com as ubiquitous communication systems.
- G as gateway for patient/client's high-tech environment.
- N as patient/client's network enabling local sensors and medical devices.
- P as patient/client's wearable and fixed medical and general devices.

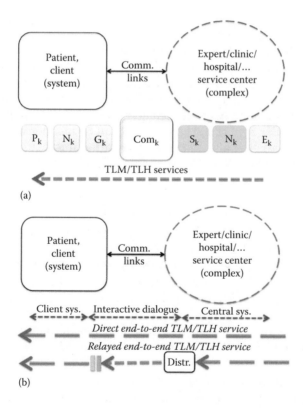

FIGURE 1.3
Simplified model for delivering services: (a) main interconnecting components and (b) direct end-to-end versus relayed linked service delivery.

Figure 1.3b shows two possible direct and relayed alternative connections normally as required in ubiquitous access in a telemedicine infrastructure.

1.2.1 Telemedicine Technological Requirements

Telemedicine in many cases can be regarded as creation of a safe medical haven in an extreme or hostile environment. That is, in practice we enjoy a large number of medical care technologies, which can be made available in remote and inaccessible places on earth and beyond. One of the greatest challenges of telemedicine is handling health issues within extreme environments. Wired or wireless, we can break the greatest ever barriers of distance. Typical examples are cases happening in high mountains, in deserts, in the bush, in space, underwater, underground, and in the extremely cold climate at the North Pole. For example, a high-tech capable surgeon can do a life-saving simple operation remotely using a dependable communication link using a highly reliable automated robotic tool.

Regarding the infrastructural requirements for provision of health and associated medical services, we can categorically consider a two-way exchange of information to deliver any scenario-based application service with two fundamental parts: (a) telecommunication technologies for a dependable, ubiquitous access to remove any possible concerns associated with the distance and (b) a collection of interactive medical devices and systems (sensor, actuator, and accelerometer) enabling the provision of useful capabilities to enable a response to good health.

For the first part of this technology, we have come a long way from a century ago; i.e., we have removed the basic voice and data exchange networking systems. Now, with media-independent capability services, we can have a large number of competitive alternative solutions for signal-guided conducting media, such as cables, and new and progressive types of broadband wireless solutions for a wide range of distances from very short to very long. However, it may be worth mentioning that in practice a telemedicine system needs to integrate many short- and long-distance connections in a relay or multilayer style as they fit one or many particular application scenarios. It can be mentioned that most telemedicine services are still growing and successfully merging into market. The following are sample scenarios suitable for telemedicine, as shown in Figure 1.1:

- **(Tele)Surgery**: Also known as remote surgery, telesurgery enables an expert surgeon to perform an operation on a patient without being there. It makes use of telepresence and some sophisticated robot surgical systems to physically apply the task at a distance. Laparoscopic and thoracoscopic surgeries are specialized techniques offering great potentials for telesurgery applications.

- **(Tele)ICU**: Tele-intensive-care-unit or TeleICU is basically a remote service where the sickbeds are located at a remote place or at patient's home, emergency, or local basic clinics running with minimum staff. TeleICU service could be especially useful for severe and life-threatening illnesses and injuries, which require constant monitoring and medication with remotely accessible specialist equipment if required.

- **(Tele)Kiosk**: Telekiosks, as explained in the example case in Subsection 1.3.2, are a wide range of remote, rural and distant units where a range of medical health services are provided for the local and surrounding communities by using already established infrastructure for working information and communications technologies.

- **(Tele)Neurology**: In dealing with diagnosis and treatment of diseases associated with the nervous system, teleneurology has potential scope for both telemedicine and telehealth services with a clear indication of professionalism, where some cases such as lumbar puncture should be adopted only in emergencies in adequately equipped and resourced locations under direct responsibility of at least one specialized medical surgeon.

1.2.2 Telehealth Technological Requirements

Based on our definitions, telehealth, on one hand, should make maximum use of capabilities and infrastructural scenarios made available by telemedicine. That is, in an economy-driven environment no idealistic research and development (R&D) can continue without proving its usefulness and returns for the time of investment. On the other hand, usefulness and adoption of a technology can prove its place in society only when it is adopted, thus saving lives and improving health in large dimensions. That is, national, international, and health organizations and charities cannot survive without showing factual results as they must show their usefulness in solving real problems for their supporters. The need for telehealth services has never been as high with services merging in the market. The following are examples of telehealth-suitable services (see Figure 1.1):

- **(Tele)Psychiatry**: This is a potential but complex set of tele-behavioral-health treatments that professionally trained medical experts should be able to expand easily into a greater number of patients.
- **(Tele)Polysomnography (PSG)**: This is a clever monitoring method for studying patient's biological behavior during sleep to look for medication signs such as heart rhythm, brain, eyes, and muscle activities during sleep that due to time-consuming activities and comfort lends itself to telehealth services.
- **(Tele)Cardiology**: The heart as the center of human-life behavior can be analyzed and medicated for possible failure through electrical and biological dysfunctions. Telecardiology comes with large potential for distant-health treatments.
- **(Tele)Rehabilitation**: This is a well-established method with two main functions of clinical assessment and therapy. It requires a long-term process for treatment and all interactions between a patient and medical professionals can be done at a distance.
- **(Tele)Nursing**: Telenursing is a well-known healthcare professional service aimed to help or enhance people's and children's quality of life so that delivery at a distance can save patients, professional staff, and the industry by cutting down non-essential visits; it is normally categorized as telecare.
- **(Tele)Homecare**: This is a telecare sector of telehealth. With a wide range of: old-age patients, pensioners, cancer patients, and disabled patients, the demand is subject to rapid increase in the near future.
- **(Tele)Monitoring**: Telemonitoring covers a very wide range of special cases and use of smart shirts, wearable medical-care devices, the demand for which is on the rise.
- **(Tele)Training**: As with the telehealth industry, distant training covers a wide variety of applications from on-site training to training new nurses, new helpers, medical equipment training, and much more with an extension to education and medical schools.

1.3 Typical Distant-Health Examples

In order to classify and position telemedicine systems and telehealth services on the technology–business plane, here we examine some popular application cases.

1.3.1 Smart Medical Shirts

Smart clothing can facilitate the use of pervasive healthcare technology on some mass-production scales. It can replace or interwork with the growing mobile health (mHealth) telecare services of smartphone devices in a much easier way with the intention of removing awkward, bulky, and inconvenient improperly designed wearable devices. If properly designed for the purpose, wearable devices and new implants would have the advantage of always being on and have networked computational artifacts that can assist moving patients and those under any kind of health conditions or risks to be connected and monitored virtually all the time everywhere and under any situation. Smart textiles can collect various vital parameters to be delivered wirelessly to patients' smartphones, the clinic, or an emergency center, e.g., to activate a call for an ambulance or on-watch doctor [7]. Smart shirts are now heavily supported by advanced medical and fashion industries. They are expected to revolutionize the new and old industries in a few decades. The integration of light sensors into textiles and wearable clothing, using new materials with new functions, has shown a way to bring about new significant growths in the market. Their recent developments in telemedicine systems are fully supported by telecare, telehealth, and, of course, the eHealth services. All are in favor of and can make use of these shirts to enable patients to stay in close contact with their doctors and consultants all the time. Those with chronic diseases now can stay medically fully connected with some centralized supported equipment that continuously collects a selective number of medical health signals, analyzes them, and alerts the experts when any inappropriate signals of any risks to patients arrive. This is something that has a positive impact on the quality of life for individuals in large-scale organizations such as the World Health Organization (WHO) and other health research and educational institutions, using these medical health signals for their studies and case investigations. The uses of these informative signals can be expressed in two main groups:

A. Short-term testing investigations
1. Medical disease monitoring.
2. Clinical monitoring.
3. Obstetrics and pregnancy period monitoring.
4. Infant and biofeedback monitoring.
5. Athletics and exercise uses.

B. Long-term wearing
1. Maintaining a healthy lifestyle.
2. Athletes' health and sports performance optimization.
3. Continuous home monitoring.
4. Remote patient examination.
5. Infant vital signs monitoring.

6. Sleep problems monitoring.
7. Monitoring vital signs of mentally ill patients.

OMsignal, for example, in its early production made their tiny sensors woven into the fabric for capturing ECG activities, breathing patterns, and emotional states on a continuous basis where data are transmitted either via an application on a mobile device or directly relayed into centralized analyzing equipment. We expect to see some surge followed by saturation in the market with outer jackets, shirts, pants, footwear, shoe soles, head wear, and undergarments and all the foregoing come with embedded health-monitoring sensors. In general smart shirts are well accepted by patients with serious health problems, especially by those whose health could be in danger due to heart disease or other critical cases that require continuous observations. Due to lack of trust in the technology and today's industries, women seem less inclined to use them than men [8–9].

1.3.2 Haptic Platform

One of the important application areas of using sensors and accelerators that normally count as professional (high tech) is haptic technology. Haptic technology is somehow related to the smart shirt at the technology end, but more closely we can connect it to smart space technologies, such as smart homes, where an eHealth platform with up to three complementary components of (a) activator; (b) smart environment, the platform; and (c) receptor creates a high-tech manipulative function that needs to be established, interworking for the service. Having the smart environment fully controlled and suitable for reliable operation, then all essential TLM-enabled systems for an operation or serious examination could be designed easily for remote interactions, each with two complementary components with the freedom of becoming mobile, located anywhere, and split or integrated without users' involvement. As far as TLM/H is concerned with the possibility of two-way interactions and the supply of a return channel, the uses of this kind of application for health and medication are significantly advanced with many potential futuristic application scopes ranging from physically functional remote-controlled augmented- and virtual-reality scenarios, including telesurgery operations and so on, down to very light scenarios which can be configured for smart-home applications. Haptic technology, in particular, is a tactile feedback technology that takes advantage of a user's sense of touch by applying forces and motions. Other medical and health applications, to name a few, are medical simulation, minimally invasive training, laparoscopy, interventional radiology, robotic rehabilitation, and remote examinations and light surgical operations. Haptic technology allows expert surgeons to operate, assisted with local experts, across the world, increasing availability of experts for their medical services.

With over 100 research publications every year, in the last decade the use of haptic and other wearable technologies for health is leading to their higher visibility in the near future. Let us scan a few recent publications from the Institute of Electrical and Electronics Engineers (IEEE):

1. A new validated method for improving the master–slave robot interacting with a human eyeball model was used in "Transparency optimized interaction in telesurgery devices via time-delayed communications" [10].
2. Improved upper limb rehabilitation requiring careful and reconstructed information around stroke patients' muscle activation characteristics and kinematic

features in functional movement was carried out in "Using body sensor network (BSN) for telehealth application in upper limb rehabilitation" [11].

3. An ontology-based flexible and scalable architecture for addressing main challenges presented in home-based telemonitoring scenarios and, thus, providing a means to integrate, unify, and transfer data supporting both clinical and technical management tasks was utilized in "Designing an architecture for monitoring patients at home" [12].

4. A method for acceleration-based bilateral control to achieve haptic feedback, visual serving is used to compensate for organ motion in "Heartbeat synchronization with haptic feedback for telesurgical robots" [13].

5. A forced feedback robot-assisted surgery substituting haptic stimuli composed of a kinesthetic component and a skin deformation with cutaneous stimuli only, which generates a subtraction between the complete haptic interaction and cutaneous and kinesthetic stimuli, was performed in "Sensory subtraction in robot-assisted surgery" [14].

6. In surgical teleoperation, haptic devices need to have high degrees in rigidity and accuracy; here, new spherical parallel manipulator architecture enables the use of robot as master in a teleoperation system [15].

7. Modeling the physical-level antenna for the systems used in wireless body area networks (WBANs) enables sensors to provide error-free and domain-tolerant performance [16].

8. With wireless personal area networks (WPANs) adopted in most wearable systems playing a key role in the future of eHealth industry and wireless access, WPANs have the best ability to provide the required ubiquitous connectivity for health monitoring under various conditions. However, interactions between the two fields of WPAN and Wi-Fi (wireless local area network [WLAN], worldwide interoperability for microwave access [WiMAX], etc.) interrupt patient vital signals seriously; this issue is now being resolved in this contribution [17].

9. In an open surgery the clinicians press their fingers on the patient's soft-tissue organs to assess interaction forces in locating subsurface anatomical structures and assessing tissue properties. A laboratory prototype tactile probe head for detecting abnormal tissues during palpation in open and minimally invasive surgery (MIS) has been developed to help with the probe's tactile sensing accuracy [18].

10. Robot-assisted minimally invasive surgery (RMIS) can offer potential medical manipulations at much higher accuracy and lesser patient trauma. This review paper scans recent advancements and challenges in the development of tactile sensing devices designed for surgical applications [19].

That is to say, haptic technology is very fresh and advancing continuously, bringing many new devices and systems into a new era of telemedicine. We expect to see many random and sporadic uses of it in telehealth services, but its true, integrated telehealth services we expect to go well beyond the next decade.

1.3.3 Overgrown Cities

As the world is shrinking, following our dreams of hope for a better lifestyle, we are now facing the side effects of our old ideas. The first technological revolution some 300 years ago initialized the idea of centralization as a way to improve life. This encouraged the growth of the cities at the cost of depriving and abandoning small villages. This tendency supported by other sociopolitical and economic factors caused this process to spread all over the world, hurting humanity at large. Many youngsters left their families and their homes with fresh air, lovely landscapes, and healthy natural living environments, migrating to large and overpopulated places to face new realities of life. On the top of stressful social life and economic problems, we have two main continuously deteriorating factors in many large cities; these are the uncontrollable problems of pollution and healthcare emergency resourcing. "Out of control smog and dirty weather in large cities with pollution 50 or 60 times higher than safe levels may lead to many unhealthy populations in a few decades" [20]. The old-age healthcare resourcing problem became global a decade ago with serious life dangers in overpopulated cities. Reports and estimates come with many facts and figures. Some samples are "EU old age population reaches 57 million in 2050," "USA old age population goes beyond 60 million, 20% in 2050," and "two seniors for each junior worker in China in 2020" [21]. The old-age problem is just the tip of the iceberg. The pressing issues are too many, including the following:

1. Epidemic outbreaks spread much faster in heavily populated areas.
2. Disasters and emergency cases put medical and care resources temporarily but heavily in very short supply.
3. The emotional reactions of professionals, experts, and staff in these cases are such that when there is a short supply, they panic and often take risks, much more when they lack sleep, are mentally tired, and are physically exhausted.
4. In some cases professionals need to take on nonprofessional and administrative tasks, causing extra distractions and shortages.
5. Noise, traffic, and pollution create special types of sickness.
6. Although sickness in a polluted environment is mainly mild and curable, in practice it can consume considerable healthcare resources.
7. The psychological and dimensional problems have been on the increase extensively, mostly in the large cities with sporadic and distressful effects on the system.
8. With the fast rise in average age, the number of visits to hospitals, clinics, and distress calls has increased and become the main time consumer of medical and health resources in the cities.

That is, although in most of the above cases the distance between patients' homes to clinics and hospitals is not very far, under limited local government budgets new problems such as reducing sickbeds and shortage of experts and professional resources would lead to large-scale deterioration in the quality of life in the near future. To solve the problem of shortages in resources, sickbeds, and professional availability, we need to adopt the superior services supported by TLM, TLH, or TLC (TLM/H/C) technologies. Smart homes, smart spaces, and sometimes patients living in their own normal places but equipped with specially designed smartphones and, lightweight wearable, haptic, or smart shirts could both save their lives and keep them secure and in the comfort of their homes while scarce resources are saved and shared for maximum usefulness. One most significant saving on the overall cost comes from the impact on professionals' traveling time and expenses for visiting the patients who are not critically ill enough to be hospitalized but seriously ill enough to justify often and regular visits or, in case of epidemics, insufficiently available sickbeds.

1.3.4 Rural Health

The author's personal long interest in science and technology for humanity, as it is now the IEEE mission, enabled him to get involved in helping develop community telecenters (CTCs), multipurpose community telecenters (MTCs), wireless telecenters (WTCs), remote communities services telecenters (RCSTs), and rural communications systems (RCSs) for breaking down the ever-growing digital divide to improve communities' local economies through sustainability and self-developments. Through many public speeches, lectures, conferences, and publications and the proposed set of strategic development technology, deprived communities were required to make use of ICTs through adoption of community telecenters, MTCs, RCSs, and RCSTs at their first stage of development under "introduction to technology." This stage normally lasts about a decade or sometimes longer until it reaches a viable market at its maturity before a local telecommunication operator can take over. Then, at a second stage that begins as soon as a local telecommunication business is established, a strategic restructuring should take place transforming the locally established center (CTC, MTC, WTC, or RCS) into a remote health center by upgrading it into a telehealth center (THC) enhancing the quality of life. As there is usually a considerable time between closure of the community telecenters and opening of the telehealth center phases, the place can be used by small businesses for Internet users and professional training complex units. Although a telehealth center may start with relatively basic technologies but could grow extensively from a very basic to a well-advanced and complex, tech-rich center with new ICT and medical devices, in general the telehealth center should be treated as an advanced hi-tech clinic for advancing humanity, earning trust, and building reliability—three prime objectives of the operators. There are normally two strategic alternatives for this transformation: (a) small-business-style natural conversion where a professional-training phase fills the gap between community telecenters and telehealth center or (b) a well-planned multipurpose community telecenter takes the process through engaging the community with a wide range of activities for a seamless transformation [22–24]. Figure 1.4 shows two practical sequences of strategic transformation stages for converting remote community telecenters into a telehealth center.

1.3.5 Satellite Telehealth

Broadband connectivity for one-way broadcasting is an old technology but two-way broadcasting is mature, assisted by hundreds of active satellites already established in orbit. Now

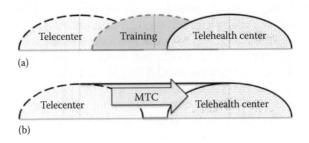

FIGURE 1.4
Strategic transformation stages converting an ICT telecenter to a telehealth center: (a) with a training and associated small business to fill the gap or (b) with a professional multipurpose community telecenter for seamless transformation.

they can cover most economically viable spots around the globe. In its framework, the European Space Agency (ESA) has developed digital video broadcasting with return channel through satellite (DVB-RCS) technology, enabling potential locations to gain access to broadband services using low-cost satellite interactive terminals (SITs). Their potential uses for telemedicine and telehealth applications are extensive and can be easily established, supporting emerging healthcare programs and connecting patients remotely for various uses of monitoring, consultancy, and interactive services such as telepathology, teleradiology, telecardiology, teledermatology, teleophthalmology, telepsychiatry, and telesurgery, which are some of many more possibilities in the near future. Satellite telehealth technology can enable the availability of expert medical care promptly in understaffed areas like rural health centers, ambulance vehicles, ships, trains, and airplanes as well as homes and uses rich satellite network coverage under various telemedicine platforms but unfortunately is still at relatively high operational cost. A proposed topology for the satellite-based telemedicine platform is shown in Figure 1.5. This system consists of some remote sites (RSs), each equipped with some required special communication systems to connect medical data acquisition units to some regional access points (RAPs) [25].

However, despite a continuously increasing number of satellites in orbit, due to their high cost and limited bandwidths in many isolated and rural spots on earth, connections remain not as cost effective as in urban areas. Satellite communications supporting worldwide TLM/H services are globally approved by the law under "legal aspects of satellite telemedicine." However, in 2003, the United Nations (UN) Commission on Human Rights,

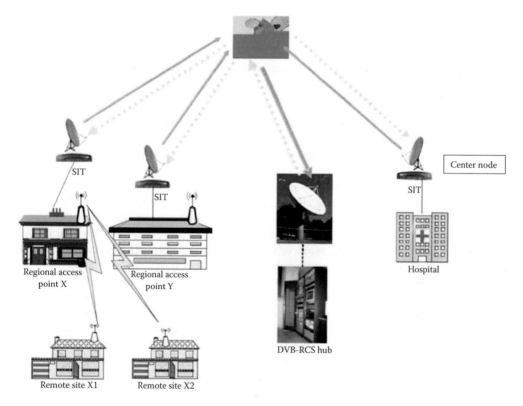

FIGURE 1.5
A typical satellite technology platform for telemedicine applications.

on the right to health, presented a report on the global availability of healthcare. Article 25 of the Universal Declaration of Human Rights (UDHR) together with the International Covenant on Economic, Social, and Cultural Rights (ICESCR), the Convention on the Rights of the Child (CRC), and the Convention on the Elimination of All Forms of Discrimination against Women (CEDAW) encapsulates individual rights to healthcare and corresponding state obligations to ensure the availability of health facilities and required technologies. But international telecommunications law protects state sovereignty and explicitly authorizes states to erect barriers to satellite-based telemedicine, a technology that could provide access to medical professionals on a nondiscriminatory and cost-effective basis to patients otherwise lacking access to healthcare.

The satellite-based telemedicine technology is now becoming a double-edged sword. While broadband Internet and other abundant communication networks have opened new frontiers for treating some patients, this new disparity in availability of telecommunications threatens to further widen the existing gaps in the quality of life through provision of healthcare. Satellites, however, are different as it is to where a user is located with the capacity to fill the gap without any significant costs for the infrastructure with large footprints so that they can serve urban and rural communities for the same cost. In general, it is a shame that satellites now can play only a limited role in upcoming global telemedicine paradigm. National restrictions on the use of terrestrial satellite technology perpetuate barriers to entry for satellite technologies that could facilitate satellite telemedicine (SATMED). National licensing also inhibits the aggregation of a global market of remotely located medical providers to justify any commercially viable SATMED industry [26].

1.4 Integrated Service Management

In the previous sections of this chapter, we have discussed the practical meanings of the terms *telemedicine* and *telehealth* and also analyzed the differences between them with clear conclusive decisions of associating telemedicine with technology-rich activities and telehealth with the service ends. Now, if we take the term *telemedicine* further to be regarded as a technology enabler, it should play a significant role in a technology-based infrastructure, which can help with the provision of intensive health services to the operational organizations, while telehealth can conveniently build and deliver its refined service-based technologies, taking on business and end-user responsibilities, i.e., delivery of the health products and associated medical services at a distance. In this section we look at the real world of implementation where users, technocrats, and business managers come together for their final challenges to set up a smooth deployment of the great idea of delivery of sustainable medical health services through integrating sensor-rich health systems without any worries or risks for the obstacles of distance and lack of service availability. That is, in order to reach such an operational capability, we need to establish a set of infrastructural platforms for these two main complementary organizations to work together, one optimized for building the technology (telemedicine company takes the full responsibility) and the other one for interfacing the customers, including the health clients, consultants, service providers, and the third party if not the same (with the full responsibility of telehealth company), who should all be able to work closely together.

Due to the sensitivity of this operation, we, however, need to recommend two important management structural procedures ensure that:

a. These two organizations are independent professional service providers.

b. Both are highly competent, fit, and optimized in their roles to adopt advanced management of technology (MOT) in their own specialized fields [27–28].

As telehealth companies are those actually facing the customers and clients, they need to earn the customers' and clients' full trust by demonstrating competence and care. The level of technology in a telehealth company must have clear boundaries. Innovation and research add extra complexity to the operation and, therefore, unpredictable risks, so if not planned properly, they could create customer distrust and associated financial black holes. Therefore, it is advisable to build the main services for such operations upon less risky and well-established technologies. For this we group telehealth companies into three categories of *mature* and *reliably developed* and *innovative*, where each category should be addressed properly. In other words, those companies developing with potential to have significant impacts in the provision of viable services in the future may bring good publicity and trust, but due to the possible risk of failure and slow development possibilities, innovative services should ensure the supply of the required funds to ensure that the process can be handled until a new breakthrough idea appears as a product or service. Figure 1.6 shows the telemedicine and three telehealth complementary services corresponding to the above categories. In this figure, a telehealth basic services (TLH-0) company takes no more risks at the cost of being slow and a follower with innovation, while a telehealth mainstream services (TLH-1) company is prepared for small risks by providing some good new service ideas, and a telehealth innovative services (TLH-2) company is taking any risks for developing highly innovative services and is also known as a front-runner.

On the legal side of the business without getting into any details, we need to look at national and international restraints. Here, we need to know if the local laws are supporting or restricting any of the services. For a fair telecommunications competition market environment, we divide the digital communication services such as telemedicine systems and telehealth services into the four layers of content, application, logical, and physical.

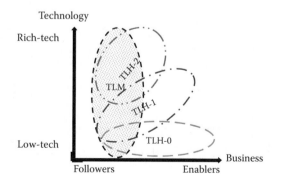

FIGURE 1.6
Relative technology intensity business professionalism capabilities for three typical telehealth innovative service options provided upon a high-tech telemedicine infrastructure.

Here, the content layer represents the human usable information. As shown in Figure 1.7, under the content layer we have the application layer for dealing with the software and systems' interactive dialogues for a service, a logical layer for digital protocols to enable devices to communicate with others, and finally the physical layer for the actual physical media where the physical signals are to be transmitted [26,29].

More importantly Figure 1.7 effectively shows the business-level interactive coopera-tion between telemedicine and telehealth organizations as effective business strength in a competitive global market. That is, two distinct and nonoverlapping telemedicine and telehealth business organizations complement each other by sharing the market at differ-ent ends to avoid any unnecessary service complexity so that they both have their own maximum competitive edge against other global market players. Here one with the tech-nological competence builds the infrastructural platform while the other one designs and delivers tailored services to the end users such as patients, clinics, hospitals, and other healthcare organizations built upon the first cooperative company.

These layers could also become important when some legal impediments apply to the provision of telemedicine or telehealth services. That is, the telemedicine service bundles prepared for delivering the service are subject to local laws and specific legal rules and reg-ulations including national censorship, international human rights law, patient–physician confidentiality, professional ethics, and any other content sensitive information affecting data security, antitrust competition, commercial licensing, and domestic and international copyright and patent protection. At the pivotal point of the process integrating telemedi-cine and telehealth, the success can be demonstrated only with actual provision of the services to the customers. That is, when deploying a large-scale tech-rich service to the user of medical and healthcare, many business factors must be investigated; these include capability (company), competence (personal), quality, and trust. We then can divide the deployment program into three stages:

  a. Analysis of telemedicine critical technologies as few as possible—selectivity based on the requirements and not luxury of complexity.
  b. Analysis of telehealth marketing process for decisive discovery of user needs.
  c. Creation of innovative and trustworthy companies for provision of viable and competitive services.

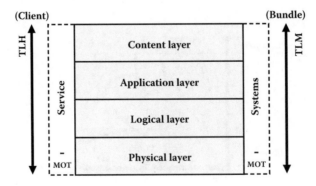

FIGURE 1.7
Interfacing telemedicine and telehealth through innovative technological developments—the four layers of digital divide competition.

1.4.1 Telemedicine Critical Technologies

It is essential to know that the design and interface of communication devices to medical equipment and medical systems is neither easy nor readily available as the other common ICT counterparts. Tools and gadgets used by practitioners are very specialized. They are mostly small and handheld and may not allow any attachments that could inhibit their perfected convenience, or practitioners and staff see them as interfering or odd to use with comfort. This aspect mostly lies on the telemedicine side of development. However, close communications with the users, staff, and patients would be essential and any approach should be professional, being either direct or via telehealth sector companies, who should have their communications already established. Some of these technologies have been discussed in earlier examples. Here we list a few more. Those with many alternative options or lesser medical intensive uses would, however, better be left for telehealth sector companies to develop their service optimization purposes. We are sure that the book would be able to cover all prime and the majority of viable existing diagnostic tools that need to be electronically enabled and interconnected. Typical telehealth-enabled physician's examination instruments are video otoscope (ear), dermascope (skin), and electronic stethoscope (heart, lung, and bowel sounds). Similarly, medical diagnostic equipment with standard video outputs is required in both telemedicine and telehealth. Examples include ultrasound for pediatrics, echocardiogram, abdominal, B-scan (eye), fundus camera, tomography imaging, rhinolaryngoscope, colposcope, endoscopes, sigmoidoscope, ECG, electroencephalography (EEG), electromyography (EMG), glucometers, weighing scales, and spirometers [30].

In order to succeed in building a reliable telemedicine infrastructure upon a complementary set of basic but fully functional technologies, spanning from microelectromechanical systems (MEMSs) and nanotechnology-based medical devices to well-established ICT systems and smart sensors, we need to consider the correct selection of technologies. The technology components are plentiful and mostly available at viable production cost, for which selectivity in and choice of the right set of service products is becoming a new profession to be decided as needed for implementing popular services and essential for provision of telehealth services. Therefore, a practical system requires only a selective set of complementary technologies to enable any smooth and successful operational service. That is, any additional or alternative service should be avoided as it adds some complexity to the system if excessive or nonessential, and then management of the operation loses its required agility. For the detailed technologies, in order to avoid possible duplication in the book, here we can mention only a few samples but we feel it is useful to provide a compact list by dividing them into four groups. The following list shows these four basic technologies as the main nuts and bolts for building the required system platforms for deployment of TLM/H services [2–4]:

1. Enhanced superior telecoms access—"Ubiquity of access" to enable a set with seamless connectivity from one device to another should be provided. Such direct link connectivity is not always feasible; therefore, normally we need to establish a set of interconnecting links through selective technologies, often called ad hoc networking.

2. Enhanced medical and clinical technologies such as capsules, endoscopy devices, and robotically enhanced equipment and systems required for telesurgery and other interactive health and medical application scenarios.

3. Creation of smart enabling and working environments at both ends of TLM/H by using rich sensing and actuating systems including a set of suitable wearable devices to ensure a highly visible working environment for experts and patients.

4. Enhanced signal, data, and information processing systems to aid the medical diagnostics and treatment of patients, providing a safe working environment.

1.4.2 MOT: Marketing

Many start-up businesses may find it easier to begin with lightweight services. A good source of details is the work of Oudshoorn [31], which describes possible administrative procedures under "good practice." Although the approaches are useful as long-term business strategies and could thrive in a noncompetitive market environment, however, due to the global dimension of TLM/H business, they could be short lived unless this practice is extended by taking some further steps using one of the following two activities: (a) continuously moving into new marketplaces, or (b) under nontechnical built-in superior work in cooperation with larger high-tech global organizations, either as research and developer units or as capital-style venture projects at the delivery end of the service.

1.4.2.1 Marketing Orientation

Historically, if high-tech industries decided to deploy and manage a TLM/H-type service as a sustainable business, they would need a couple of decades for adopting their technology push-style global innovative firms, but with new rapid changes soon they would realize that they needed to change their systems once again before they could take any further actions. To this effect then, in the early part of the second half of the 20th century, these companies had to change their strategies toward adopting the market pull style, empowering their sales and customers, where the concept of marketing orientation (MO) as a new strategy starts to grow. This strategy is based on its definition, *"a marketing-oriented business begins with communicating with potential and typical customers to find out what they really want, and then prepares to produce the exact products for them."* Using the adaptive customer intelligence system, companies can produce more sustainable products that support their overall business strategy, which enables the company to compete effectively in an increasingly global competitive market, and deliver exact solutions for current and also their future customers. Although this technique is helpful with the TLM/H mainly due to its service nature; however, MO shows its usefulness even more in identifying "what users want," which in the case of TLM/H services the needs are better defined in the community and commonly known to all including governments and medical industries. MO strategies here can also help with the basic need for identification of more sophisticated customer supportive factors such as trust, political, and financial implications of the service within the nation in general and the community in particular. More strategic details, variants, and models are widely available in the literature and business-interested readers are encouraged to realize this before taking any action in this area. The simplest and most effective marketing tool is strengths, weaknesses, opportunities, and threats (SWOT) analysis. Kotler groups these four into two subgroups of external and internal environments. The external analysis looks into the *opportunities* and *threats* as these two are normally out of management control about which a business should have reliable information. Then the internal capabilities and restraints (*strengths* and *weaknesses*) are the company's responsibility to ensure they have sufficient resources and confidence to turn the opportunity into a long-term business success [32].

1.4.2.2 Holistic Marketing

Market globalization in the 21st century comes with new competition forces and abandons new technologies, expanding in new dimensions unknown to many experts. This market environment dictates adoption of a so-called holistic approach to marketing that enables a fuller control of the process. This approach in practice goes far beyond classic marketing. By definition it consists of the *four* basic components "relationship marketing," "integrated marketing," "internal marketing," and "social responsibility marketing." That is, for managing services we can mention Kaiser performance indicator empowerment for using new sources such as *"health-maintenance organizations (HMO) for its provision of low running cost for the award-winning website to give free-registered customers the possibility of any required office visit, and e-mail enquiries, with response in 24 h with professionally monitored discussions supported with a rich supply of information from databases and encyclopedia"* [32].

In general, professional marketing, which has nothing to do with sales forces who can be used only for contacting major accounts, enables taking correct strategic decisions for many critical business factors, including (a) minimizing the cost of investment and reducing the operational cost of the services, (b) providing sustainable and long-lasting service, and (c) preparing the market environment for a natural diffusion of the products; i.e., properly designed products do not require an expensive and impressive sales force department. Table 1.1 shows market-diffusion behavior and user influence for deployment of telemedicine and telehealth services.

1.4.3 MOT: Innovative Deployment

All technological revolutions in history have been enabled by a timely set of correct technological innovations. Managing a successful TLM/H at global scale is not an easy task. To succeed it requires the four distinct phases of (1) inventive breakthroughs, (2) innovative developments, (3) acceptable rate of diffusion, and (4) provision of services or products unique and advantageous for a fair viable share of the market in the competitive globalization market. For the first phase of patentable inventive ideas, the company requires to succeed with some invention processes. Traditionally it may need to follow some key stages of the process with either four stages of Usher's model or five stages of Lawson's model. Converting the potential ideas into innovative products and services in the second phase requires knowledge of some "marketing processes." The viability of the product can take various strategies for using telemedicine technologies mentioned in previous sections of this chapter. For the third phase, diffusion, one may adopt some proactive but natural

TABLE 1.1

Market-Diffusion Process for TLM/H Services

	TLM	TLH	Dependents	Notes
Core process	Technology enabling	Provision of service	Scenario based	Maturity rate
Diffusion	Slow adoption	Fast adoption	Cultural and financial	Used by (human)
Users' influence	Slow process	Marketing process	Marketing and agility	Patients, experts

marketing strategies, relying on its superior features tested, for example, by Rogers' crite-ria with five basic components: relative advantage, compatibility, complexity, observabil-ity, and trialability. As explained later in this subsection, for a successful fourth phase deployment, TLM/H requires a new paradigm-based tailored structure enhanced with the traditional-style pilot service. Pilot services can provide the best chance to test these factors, where a viable set of critical technologies is an introduction to the service as well as a guide to true deployment of integrated eye-catching TLM/H services as a way ahead to a successful future upon building trust and a communication interface with the customers [33]. However, at every phase of the progress we need to examine any significant changes in the development and emerging ideas to satisfy the following classic seven checkpoints originally known for designing innovative products:

Stage 1: Originality of the ideas

Stage 2: Demand for the ideas

Stage 3: Workability of the ideas

Stage 4: Selling worth

Stage 5: Intellectual property

Stage 6: Financial expectations

Stage 7: Commercial strategy

In the last quarter of the 20th century, political upheaval throughout the world coincided with the advent of the Internet when we witnessed an unexpected approach to the dawn of the global village. The changes are not limited to the market behavior but most of them are in the volume of unpredictable changes that are mainly due to poor management in most sec-tors of the industry. The regular modules such as "industrial movers and shakers" processes that shaped the global industries in the 19th and 20th centuries did not respond and the gross domestic product (GDP) of industrial nations is less effective as an accurate indicator of the quality of life, leaving many developing nations in the darkness of ambiguities to worry for their future. There is no doubt that these changes will have an impact on the ever-growing demand for medical care and business-associated industries for the telemedicine and most definitely the telehealth services at the core of the process, where MOT and innovative services are expected to play prime roles in the provision of successful and sustainable services.

For clearing the air, we look at two main bottleneck factors:

a. Complex and restricted market of medical care leading the thriving business.

b. Nature of the medical industry

The first factor, negative from a business point of view, inhibits any real free-market activi-ties. Two extremes may happen. In developed nations, most of the medical businesses are con-trolled by states and tightly ruled by restrictive laws to ensure that minimum risks are taken on citizens' lives and no personal patient information is exposed to third parties. However, bureaucratically controlled states are known to suffer from poor and inefficient governance, hence preventing private businesses to engage in an effective and meaningful role for thriv-ing, smooth, efficient, and swift business operations. On the other extreme side, in less devel-oped countries, human rights are often suppressed, wealth became a symbol of power, and businesses rule the economy, resulting in a statistically widespread distribution in the quality of life, thus restricting the development of effective health-related private business.

The second factor is related to the financial aspects of the operation, i.e., running a business. One interesting case study looks into 12 different industrial sectors for business competencies by using return on assets managed (ROAM) and asset turnover (ATO) for comparison. This shows that the medical and healthcare industry comes with the lowest figures of –24.18% and 0.00 for these two indicators, while the telecoms industry holds figures of +18.85% and 1.4 and sits much higher in the competency position [27].

The above factors and associated figures indicate deploying medical services as running a business is extremely difficult, one reason that we do not see many "purely medical" companies in operation. However, introduction of the ICT and telecoms in particular could change all this. That is, a new form of making a medical industry sector enhanced with added features from ICT, telecoms, and other featured service industries can carry much higher indicating figures of ROAM and ATO. This could move the traditional medical operations toward service industries, which should be able to change the rigid nature of medical care forever and definitely for future generations. This task understandably is neither an easy process nor a smooth one to handle. As we have already mentioned in various parts of this chapter, TLM/H opportunity is unique but it comes with a rigid medical hard core that needs to be enhanced as soon as possible. One approach is to take it through a well-planned reform process for many critical but sustainable stages. In order to shed a glimmer of light in the process, first we need to free the medical economy by taking the center of control away from the hard-core traditional mentality and through four important modernization stages, namely:

1. Introduction of electronics, automation, and information technology (IT) for improved performance and reliability, which are then equipped with universal standard interactive interfaces—to help with production and deployment toward fully eHealth-enabled medical systems and devices.

2. Networkable medical systems and devices can become more effective and very helpful in many ways, including (a) saving space, (b) reduced labor efforts, (c) sharing, (d) remote diagnostics, and (e) quality of operation. Fortunately, most recent generations of eHealth devices come already equipped with medically approved communications capability (wireless fidelity or wireless networking [WiFi], Bluetooth, etc.). If missing then, those that need to be connected should be modified or enhanced with the required modules.

3. Professional pilot service introduction sites: The true acceptance for many users can be achieved only through engaging with some small-scale services with a view to expanding in size and improving the quality to earn the full trust of the users. Best starting points could be located somewhere meaningful such as rural and remote places and not right in the middle of a busy town—where facts and positive views could be easily observed; i.e., the service as a product should be able to speak for itself.

4. It is important to take the services under the motto of "quality," definitely not "cost"—the greatest misleading concept that most politicians try to justify their move for introducing new projects, which fail every time. Therefore, a large number of telehealth services could be starting for those who need the care and not those who can pay. The need for quality is unlimited and so the market demand strongly exists as long as patients trust: (a) the systems and most of the existing standards, which appear to have been designed to support the businesses and, therefore, too weak to support the patients needs; and, more importantly, (b) the service management and

business operations with experts and professionals in charge. It is advisable to have all services, from very large to very small, monitored and recorded to generate full trust of the users. Aiming to expand the limits for such pilot schemes, one should approach the change using their historical development, i.e., integrated TLM/H process through a professionally managed business.

The first stage enables us to perform professionally and to save medical experts' valuable time from wasteful minor issues by avoiding a number of simple and trivial human errors through systematic safeguarding of all known possible watchdog cycles that can be embedded in the system. This stage would also help to convince medical experts to trust electronic systems further. This trust has not been very high due to an earlier, opportunistic rush in the past decades. This very first step is the most critical stage of acceptance of electronic systems by experts and is essential to diffuse the trust as an icebreaker effect into the medical communities: medical doctors, nursing professionals, and, of course, the patients should all be considered important for eHealth and beyond. The rate of adoption and diffusion of acceptability is an extremely important factor at this stage mainly due to the fact that there are always random casualties in every system due to human neglect that often systems have been blamed for.

Following a successful deployment of this, the follow-up stages would become much easier as they merge automatically at their own pace. The rate of acceptance, however, is expected to vary extensively depending on many practical factors including the deployment scenario, level of investment, location, society, culture, and, of course, the resistance of the hard-core medical profession in practice and associated industries. We should not forget that the sluggish medical industrial era of the late 20th century has passed, and now with the advent of the global village, only a number of true globally competent industries have survived the tough competition dominating the medical markets. To be an honest survivor in this business, the organization needs to be agile and open with clear and visible strategic objectives. One way to construct such competent industries is to build it upon, e.g., a known "structural formation paradigm" working under a smooth and integrable modular system with a cellular structure. Being rich with innovative experts and technologies while fully connected to the market is advantageous but only if managed properly and professionally. Figure 1.8 shows a typical approach of forming a suitable structural model for creating a MOT-style company conversion paradigm [34].

In Kotler's design and management of services for shaping the market, it is explained that the Mayo Clinic manages a clear set of visible experimental objectives by using their sustainable services. With their motto of patient comes first, they have gathered a trustworthy business environment, so that patients and their accompanying people feel better after the visit. Every branch offers something unique; e.g., all laboratory buildings come with unique user-friendly design features, so that they say "patients feel a bit better before seeing their doctors." The Gonda Building has wide-open spaces, Scottsdale's lobby has indoor waterfall and mountain views, and many more features provide added features for the visitors. It is often said that "Mayo has set a new standard in the healthcare industry" [32].

1.4.3.1 Integrated TLM-TLH Service Paradigm

Following our discussions throughout the chapter, we are now in position to propose an innovative approach to adopting the integrated TLM-TLH service provision. As shown in Figure 1.9a, building a telemedicine infrastructural platform by using the foundation for

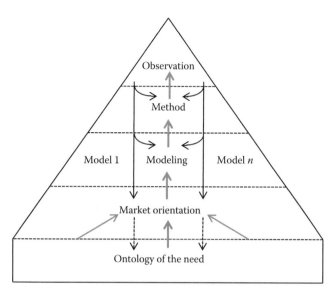

FIGURE 1.8
Paradigm-based approach for constructing of an agile technology-rich service company.

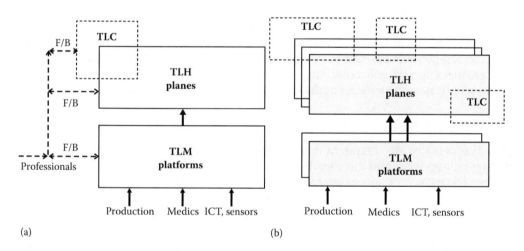

FIGURE 1.9
TLM infrastructural platform as foundation for providing TLM-TLH services: (a) the structure and (b) the service planes and associated business service bundles.

the service via a pilot or real service would enable well-structured communications with the clients and professional bodies for developing exact products and the telemedicine level to feed telehealth service planes to tailor most suitable service bundles for different target markets, as shown in Figure 1.9b. Here telecare could be subcontractors or value-added business partners who share the service for their own particular business ends. Figure 1.10 shows the areas of telemedicine and telehealth service planes where the service providers can make best use of available service elasticity for configuring a variety of services upon their aimed cost and quality.

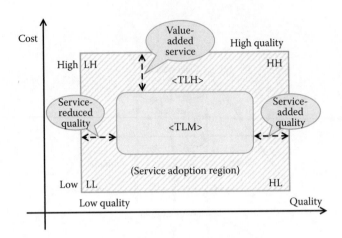

FIGURE 1.10
TLM/H service plane for cost versus quality of service.

1.4.3.2 Holistic Approach

Management of technology can also adopt a holistic approach to its strategic management, so then the company could grow in size for handling a selective set of technology-rich TLM-TLH services by taking some small and manageable business risks for the speed of change, uncertainty, trust, and other management parameters to shift the company slightly beyond its competitive edge. In order to maintain the required operational agility with the competitiveness and sustainability, then additional management tools such as local partnership and subcontracting and other professional business cooperation techniques will be necessary for an optimized smooth operation if required.

1.4.3.3 Building Trust

TLM-TLH services are unique in many ways, which are often overlooked by most slow-moving management and business institutions mainly due to the fact that these services are scenario based and emerging from designed application cases that could vary extensively from one place to another in size and in the degree of expertise and professionalism. Provision of such services commonly requires long-term commitment to the maintenance to guarantee delivery of the minimum quality of service along with other associated basic values. Building trust is a long-term process of demonstrating the company's capability through the profession and through engaging with the customers and patients. One good source of details in engaging the patients is a book from Oldenburg et al. [35], where in some sections of the book authors expect patients to build their trust within the operation such that they can consider the doctors and, therefore, themselves as partners, which goes far beyond being merely sources of authorities or professional advisors.

1.5 Conclusions

The need for a "better health at a distance" has never been as high as it is today. Following an extensive discussion for redefining and, therefore, positioning the telemedicine and telehealth industrial sectors through examples by analyzing different points of view, we have identified their most suitable relative positions for them to grow independently in both dimensions of technological development and operational service provision. We have then shown how an optimum deployment can be achieved through a smooth professional interworking of these two health industrial sectors. In this chapter, we, therefore, have analyzed these two key technologies to point out the most important business and technology management factors for taking advantage of many existing and rapidly maturing technologies to be used for developing new superior, sustainable health services at a distance. These include (a) the need for an integrated approach for interworking professionals, innovators, and product champions, and (b) establishment of new standards and regulations to support telemedicine and telehealth via effective use of available resources including new generations of communication technologies being wireless, mobile, or satellite for an ultimate goal of "health at a distance." We, therefore, propose a set of well-structured professional companies to begin an integrated development of telemedicine and telehealth services for a new implementation breakthrough for harvesting this unique opportunity.

For the integration stage we then recommend establishment of two main agile professional companies with well-established interworking business protocols between the two industrial sectors to operate under two different but complementary management development approaches of "infrastructural led" and "service led" to help with a timely delivery of these two prime business opportunities. In here, the telemedicine, in building firm infrastructural technologies, adopts a top-down management approach while the complementary telehealth business service sector takes on the need-based bottom-up approach built upon the telemedicine infrastructure tuned for satisfying the need of the customers and associated businesses to serve humanity for a healthier and more prosperous future.

Acknowledgments

The authors would like to thank the medical and engineering communities for their useful comments. In particular our thanks go to helpful individuals of the Advanced Communication Systems Limited and the University of Warwick. Special thanks go to J. K. Elmes for her dedicated help with the writing and to the Advanced Communication Systems Limited for the financial support during the preparation and compilation of the materials.

Abbreviations and Nomenclature

ATA	American Telemedicine Association
ATO	asset turnover
BSN	body sensor network
CEDAW	Convention on the Elimination of All Forms of discrimination against Women
CRC	Convention on the Rights of the Child
CTC	community telecenter

DVB-RCS digital video broadcasting with return channel through satellite
ECG   electrocardiography
EEG   electroencephalography
eHealth   electronic health
EKG   electrocardiography
EMG   electromyography
ESA   European Space Agency
HMO   health-maintenance organization
HR   heart rate
IBP   invasive blood pressure
ICESCR   International Covenant on Economic, Social, and Cultural Rights
ICT   information and communication technology
ICU   intensive care unit
IEEE   Institute of Electrical and Electronics Engineers
IoT   Internet of Things
MIS   minimally invasive surgery
MO   marketing orientation
MOT   management of technology
MTC   multipurpose community telecenter
NIBP   noninvasive blood pressure
R&D   research and development
RAP   regional access point
RCS   rural communications system
RCST   remote communities services telecenter
Resp   respiration
RMIS   robot-assisted minimally invasive surgery
ROAM   return on assets managed
RS   remote site
SATMED satellite telemedicine
SIT   satellite interactive terminal
SpO$_2$ oxygen saturation
SWOT   strengths, weaknesses, opportunities, and threats
THC   telehealth center
TLC   telecare
TLH   telehealth
TLH-0   telehealth basic services
TLH-1   telehealth mainstream services
TLH-2   telehealth innovative services
TLM   telemedicine
TLM/H   services supported by either TLM or TLH
TLM/H/C services supported by TLM, TLH, or TLC
UDHR   Universal Declaration of Human Rights
WBAN   wireless body area network
WHO   World Health Organization
WiFi   wireless fidelity (wireless networking)
WiMAX   worldwide interoperability for microwave access
WLAN   wireless local network
WPAN   wireless personal area network
WTC   wireless telecenter

References

1. Rashvand, H. F. et al., Ubiquitous wireless telemedicine, *IET Communications*, 2, 2, pp. 237–254, 2008.
2. Rashvand, H. F. and Kavian, Y. S. Eds., *Using Cross-Layer Techniques for Communication Systems*, IGI Global Press, 2012.
3. Rashvand, H. F. and Alcaraz-Carelo, J. M., *Distributed Sensor Systems—Practice and Applications*, John Wiley & Sons, 2012.
4. Rashvand, H. F. and Chao, H.-C. Eds., *Dynamic Ad Hoc Networks*, Institution of Engineering and Technology Press, 2013.
5. Rashvand, H. F. and Hsiao, K.-F., Smartphone intelligent applications: A brief review, *Multimedia Systems*, pp. 1–17, Aug. 2013.
6. What Is Telemedicine?, American Telemedicine Association, http://www.americantelemed .org/about-telemedicine/what-is-telemedicine#.U4rl, accessed on Feb. 8, 2014.
7. Schaar, A. K. and Ziefle, M., Smart clothing: Perceived benefits vs. perceived fears, *IEEE 5th International Conference on Pervasive Computing Technologies for Healthcare*, Dublin, pp. 601–608, 2011.
8. Smart Garments, Future of Garments, Intelligent Textiles in Medical, http://smartgarments .blogspot.co.uk, accessed on Feb. 8, 2014.
9. Yishuang, G. et al., An empirical channel model for the effect of human body on ray tracing, *IEEE 24th International Symposium on Personal, Indoor and Mobile Radio Communications*, pp. 47–52, 2013.
10. Mahdizadeh A. et al., Transparency optimized interaction in telesurgery devices via time-delayed communications, *IEEE Haptics 2014*, pp. 603–608, 2014.
11. Tan, B. and Tian, O., Short paper: Using BSN for telehealth application in upper limb rehabilitation, *IEEE World Forum on Internet of Things (WF-IoT)*, pp. 169–170, 2014.
12. Lasierra, N. et al., Designing an architecture for monitoring patients at home: Ontologies and web services for clinical and technical management integration, *IEEE Journal of Biomedical and Health Informatics*, 18, 3, pp. 896–906, 2014.
13. Nakajima, Y. et al., Heartbeat synchronization with haptic feedback for telesurgical robot, *IEEE Transactions on Industrial Electronics*, 61, 7, pp. 375–386, 2014.
14. Meli, L. et al., Sensory subtraction in robot-assisted surgery: Fingertip skin deformation feedback to ensure safety and improve transparency in bimanual haptic interaction, *IEEE Transactions on Biomedical Engineering*, 61, 4, 1318–1327, 2014.
15. Saafi, H. et al., Development of a spherical parallel manipulator as a haptic device for a teleoperation system: Application to robotic surgery, *39th Annual Conference of IEEE Industrial Electronics Society, IECON 2013*, pp. 4097–4102, 2013.
16. Crepaldi, M. et al., A physical-aware abstraction flow for efficient design-space exploration of a wireless body area network application, *Euromicro Conference on Digital System Design (DSD)*, pp. 1005–1012, 2013.
17. Wang, Y. et al., WiCop: Engineering WiFi temporal white-spaces for safe operations of wireless personal area networks in medical applications, *IEEE Transactions on Mobile Computing*, 13, 5, pp. 1145–1158, 2014.
18. Xie, H., Liu, H., Seneviratne, L., and Althoefer, K., An optical tactile array probe head for tissue palpation during minimally invasive Surgery, *IEEE Sensors Journal*, 2014, Early access on June 22, 2014.
19. Konstantinova, J., Jiang, A., Althoefer, K., and Dasgupta, P., Implementation of tactile sensing for palpation in robot-assisted minimally invasive surgery: A review, *IEEE Sensors Journal*, pp. 2490–2501, Aug. 2014.
20. Busch, S., And the world's most polluted city is . . ., *CNN*, January 31, 2014.
21. Rashvand, H. F., Keynote speech: Role of computer in "third technological evolution," *International Conference on Computer Technology*, CV Raman College of Engineering, Bhubaneswar, December 3–5, 2010.

22. Rashvand, H. F., Wireless community telecentres, *ITU Telecom '01*, Johannesburg, 2001.
23. Rashvand, H. F. et al., Integrated telehealth-telecentre development—Innovative remote community care & development, *4th IET Seminar on Appropriate Healthcare Technologies for Developing Countries*, pp. 121–128, 2006.
24. Rashvand, H. F. et al., Integrated telehealth—Requirement and implementation, *IET 3rd International Conference on Advances in Medical Signal and Information Processing, MEDSIP 2006*, Glasgow, UK, pp. 1–4, July 17–19, 2006.
25. Pasias, V. et al., E-health performance assessment of an interactive satellite network infrastructure, *The Journal on Information Technology in Healthcare*, 4, 6, pp. 356–365, 2006.
26. Rooke, S., SATMED: Legal aspects of the physical layer of satellite-telemedicine, *SSRN, Michigan Journal of International Law*, 34, 1, 2012, Electronic: http://ssrn.com/abstract=2061392, accessed on Feb. 8, 2014.
27. Lochner, F. C., *A MOT-Based Cost Management Competency Index: Formulation and Testing of Association with Financial Performance*, Business administration dissertation, Cape Town, UNISA, 72 p., 2005.
28. Sahlman, K., *Elements of Strategic Technology Management*, Academic dissertation, University of Oulu, 2010.
29. Nuechterlein, J. E. and Weiser, P. E., *Digital Crossroads: American Telecommunications Policy in the Internet Age*, The MIT Press, Cambridge, Massachusetts, and London, England, 2005.
30. Nafchi, A. R. et al., High performance DOA/TOA-based endoscopy capsule localization and tracking via 2D circular arrays and inertial measurement unit, *IEEE International Conference on Wireless for Space and Extreme Environments (WiSEE)*, pp. 1–6, 2013.
31. Oudshoorn, N., *Telecare Technologies and the Transformation of Healthcare*, Palgrave Macmillan Press, 2011.
32. Kotler, P. and Keller, P., *Marketing Management*, 14e, Prentice Hall, 2011.
33. Rashvand, H. F., ICT & the 3rd technological revolution, Invited speech, *Open University Summer School, T302-IDEA*, University of Bath, July 2004.
34. Lochner, F., *The Functionality Model: An Experiment to Expose Paradigm Formation*, University of Stellenbosch Business School, 16th EDAMBA Summer Academy, Soreze, France, July 2009.
35. Oldenburg, J. et al. (Eds.), *Engage! Transforming Healthcare through Digital Patient Engagement*, HIMSS, 2013.

2

Readying Medical Applications for Telehealth

I-Hen Tsai

CONTENTS

2.1 Introduction

Telehealth is defined by several groups as the use of ICTs to provide healthcare and related services [1]. Sometimes the term *telehealth* is also used interchangeably with *telemedicine* to describe similar services. mHealth is one aspect of telehealth that specifically focuses on using mobile devices such as mobile phones to provide healthcare services [1]. Services commonly provided to the patients by telehealth include medical knowledge compendiums, and teleconsultations, as well as digital logging of factors such as diet, glucose levels, and menstrual cycles. For medical professionals, services such as providing access to patient medical imagery and health records and prescription verification are seeing increased adoption.

2.2 History and Recent Developments in Telehealth and mHealth

2.2.1 History of Telehealth and mHealth

The opening of the Internet to the general populace and the development of the short message service (SMS) in 1990s produced the earliest tools for telehealth. Medical and engineering experts strove to integrate available technology with healthcare to provide better services. Early examples of telehealth pilot research include the usage of the Internet to remind patients to take a prostate examination and to manage a diabetic patient who has moved to a new location of residence [2]. As the availability and speed of the Internet improved, various different web-based applications accessible from a personal computer (PC), had been created to provide support for patients at home to manage their own conditions [3–4]. By the mid-2000s, mobile phones had incorporated the ability to access the Internet, and at the same time laptop computers as well as personal digital assistants (PDAs) became widely available. Mobile phones [5] and PDAs [6] were the choice of many telehealth studies for their mobility and ability to access the Internet. Some consumer product makers have also attempted merges of healthcare devices with mobile phones to leverage telecommunication capabilities and consumer needs [7]. At the same time in development was the wireless protocol Bluetooth [8], which was available in many laptops and mobile phones. Healthcare device makers have also manufactured products with Bluetooth onboard, to be used with management software for patients' self-care. This led to some studies as to whether a framework involving telemonitoring of patients would lead to better outcomes in patient health [9]. Other makers of professional medical devices have also gradually followed suit, providing medical professionals with new devices featuring wireless data capabilities [10–11].

The emergence of smartphones and tablets from the late 2000s led by Apple's iOS products and followed by Google's Android system in recent years has created a new frontier for providing healthcare services. These new mobile platforms come with decent computing power and an intuitive interface and have already highly penetrated the consumer market.

2.2.2 Recent Developments

The coming of iOS and Android mobile systems has added a lot of fuel to the telehealth fire. Although as mentioned before, telehealth has been a subject of many research domains, the medical community and patients have not really been adopting telehealth at scales larger than single-institution deployments for research purposes. The meteoric rise of smartphones provided a highly penetrated mobile platform for telehealth to explore as a deployment platform.

Large numbers of startups in the past 3 years have looked at building medical devices to utilize smartphones as a platform for data aggregation and data processing. Some of the most notable products include the Scanadu Scout [12], a hockey puck–sized device that could measure the user's vital signs when placed against the side of the head. The measurements could tell users their health condition and can be transmitted to designated clinicians to prepare for incoming patients. Medical device companies such as pacemaker provider Boston Scientific [13] have been marketing telehealth systems for clinicians to manage cardio patients wirelessly over the web.

Wearable tech has seen many new products moving towards health and fitness purposes. Recent trade shows have showcased several products including consumer-grade electrocardiograms for fitness monitoring [14] and smart watches [15], as well as wireless earbuds [16] with pulse oximetry and heart-rate sensors. Samsung recently joined the group with their Galaxy Gear series, which includes activity measurements and heart-rate monitoring. Apple is rumored to provide similar fitness sensors in their next series of product lineup, with the iPhone playing the role of a digital logbook and health counselor. Google Glass is also at the forefront of telehealth developments; its Explorer edition has already seen use in daily-operation scenarios [17] as well as teaching of medical students [18].

Another trend of interest occurring in parallel now is big data, particularly on using the data to improve the human quality of life. Medical care has always produced vast amounts of records regarding its patients, and such records have not yet been fully utilized due to a lack of real-time communication options and resources to follow-up on each patient. Telehealth enables existing services to have the potential of gathering better data in greater amounts in a timely fashion. Companies providing analytics solutions, such as IBM, have demonstrated several data-centric telehealth projects to improve healthcare in many scenarios, including, for example, newborn care [19], patient prognosis [20], and personalized care [21].

Although a lot of research and solutions point to ICT as a generally beneficial factor in the healthcare and medical process, there is a glaring example that reminds everyone of what can go wrong with large ICT projects: HealthCare.gov. The website crashed mere hours after being launched and after weeks of emergency repair, only a portion of users were able to access the website. Postcrash analysis indicated multiple factors of failure: bad design and outdated infrastructure [22], factors which will weigh in even more heavily in a healthcare scenario.

2.3 Present Challenges and Benefits

The benefits of telehealth lie in using ICTs to increase the reach of health services involving patients. ICTs have successfully allowed health services that were once limited to being performed on the premises of a health institution, as shown in the concept in Figure 2.1, to be conducted over large distances as long as there is a sufficiently stable network connection and requisite hardware. Whereas most interactions between the clinicians and patients occur only within a clinic or hospital, ICT allows clinicians to monitor patients outside of the hospital and also provides patients a convenient channel to relay feedback regarding their condition to their physician. The latter is particularly useful, as there are many cases where patients feel unwell but do not exhibit described signs of discomfort at the time when they are being examined by a clinician. However, ICT does require networking infrastructure at the location of deployment and a reliable networking environment to operate.

FIGURE 2.1
Typical procedural flow when a patient visits a hospital or clinic. Depending on the disease being identified, the patient may be asked to return home to wait for test results or return at a later date for further treatment. Communication with the patient between visits to the health institution is rare due to a lack of real-time communication options and resources to follow-up on each patient.

Many of the developments in telehealth and mobile health are heavily based on using a digital medium to replace the traditional physical media of printouts and written records. Some of the well-known examples include the handling of X-ray imagery, which used to be displayed by photographic film and is now shown digitally. The digitalization of such medical imagery coupled with telehealth developments allowed the imagery to easily be accessible to physicians that needed to refer to the imagery. Moreover, digital readouts reduce the need for paper and at the same help ensure record legibility and reduce associated errors in medical operations.

These developments allowed health data to be easily stored away for future reference and be quickly retrieved when the data need to be referred to again. Data-exchange models based on the ideals of cloud computing have been mentioned often in research and by various IT solution providers to the medical sector (Figure 2.2). However, at the same time, this new interaction model also creates a lot of new work and data in the process. Furthermore, existing fragmentation in recording procedures between different health institutions remains the largest obstacle to be resolved before patients can truly benefit from its convenience.

2.3.1 Deployment Problems

In the event new procedures or equipment are introduced, medical personnel would have to go through training to familiarize themselves with their new tools. Although these new tools may present a good solution to the problem they are designed to solve functionally, they are often not user-friendly to the people who actually use them. This is almost always a domain-knowledge issue, where the engineers are not familiar with the usage scenario of the medical device. Integration of new systems into the present framework always poses a large challenge, as in a hospital taking systems offline is usually not an option. As a result, a common complaint from users is that the device may be equipped to communicate wirelessly and transmit information, but it is not connected to the present

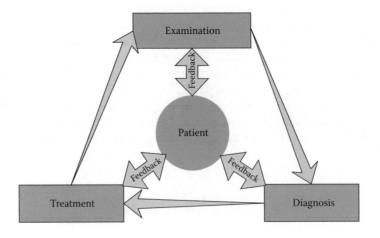

FIGURE 2.2
With ICT, it is possible to conduct some of the medical procedures outside the confines of the health institution and to also involve patients in the healthcare process. This can help reduce missed opportunities from lack of communication between visits, in addition to chances of ensuring patient compliance with the clinician's orders.

health IT systems of the institution. In order to relay outputs from the device to relevant parties, the user ends has to resort to traditional mediums, often pen and paper. The ideals presented in the Internet of Things concept provide a good foundation for planning ahead in deployment scenarios. Many older institutions would need to update their network infrastructure to provide the connectivity needed for telehealth applications.

2.3.2 Technical Challenges

Nobody likes to use a very bulky device or platform unless it is absolutely necessary. Much of the bulk and the weight came from the power supply. Recent developments in the wearable tech field show another power-related problem: uptime per charge as many devices had only less than half a week of uptime when used continuously. The battery is often integrated into the device to minimize the size of the device, thus rendering the option to keep spare batteries impossible. Given the trade-off between the uptime of the power supply and the size and weight of the platform, displays are one of items that are first sacrificed. Some devices choose to use minimal light-emitting diodes (LEDs) of different colors to attempt to indicate information, while some require the user to use a smartphone in order to facilitate any form of interaction between the user and the device.

Furthermore, the host platform, commonly the smartphone or computer, often lacks the preferred wireless communications hardware to interact with the device. This limits the options for wireless communications: either only the users who have a device with the corresponding wireless receiver can use the service, or the service must supply an adequate adaptor. A current dilemma example is the Bluetooth low energy (BLE) protocol, which is more energy efficient than past versions. Many fitness peripherals developed for use with smartphones opt for BLE to extend the uptime of the device, but due to Android smartphones mostly lacking BLE functionality, such devices are usable only with Apple's line of products or paired with a receiver on a computer. In addition, the present network capabilities of many hospitals and clinics are not ready to provide patients access to their data, especially not the potentially large number of users attempting to access the data, and to provide access to data from an external network as well as to protect the data.

2.3.3 Handling Data and Privacy

In essence, telehealth aims to provide better health services than existing models by collecting relevant evidence and more data. Therefore, by enabling telehealth for an existing health service, methods to handle the data will need to be conceived. A good start is to understand the properties of data you will need to process; medical domain knowledge is needed to build a correct model of understanding. The architecture of the algorithm or model should reflect the constraints and implications put forth by medical knowledge. To properly process incoming data, the infrastructure of the telehealth system should consider the four Vs of big data: volume, variety, velocity, and veracity. Medical imagery has some of the largest sets of data that are transferred daily within hospitals and clinics; the volume of data in a single X-ray slide is about 30 MB, and three-dimensional (3D) computerized tomography (CT) scans can go up to 1 GB or more. Imagery is but one kind of evidence that clinicians work with to understand the situation of the patient. Many different kinds of chemical tests for blood, urine, and various bodily fluids; assessment questionnaires; and vital signs all provide data from which clinicians must draw conclusions. Each clinician has to handle multiple patients, meaning that telehealth will instantaneously update clinicians with all manner of data from the patient. But data itself can be unreliable, tests can go wrong if controlled incorrectly, and patients may report altered numbers to avoid being scolded by doctors for bad self-management.

While telehealth can help medical care by providing more evidence to work with, it is at the same time overloading medical personnel with raw material. There are many available methods in the computer sciences domain that can be leveraged to quickly and efficiently refine the raw data into something that is understandable by humans in a short time. Machine learning and data mining are two of the most popular methods used in research presently; there are also the statistical methods favored by those with a medical background. Each approach of data analysis has its pros and cons; it is usually best to build a method in an ensemble to leverage the strengths of chosen methods. Domain knowledge must be present as the foundation of the analysis; otherwise, the empirical nature of computational approaches is prone to discovering results that are already known in medical knowledge.

By collecting large amounts of patient data, we are at the same time storing very sensitive information about the target individual. Private knowledge such as those of disabilities, genetic diseases, and sexually transmitted diseases (STDs) may still put people at a social disadvantage in many situations. Utmost caution must be taken in storing the information; encryption, access privileges, and de-identification are routine procedures of protecting the data. Given that the telehealth scenario is also likely to allow patients access to their own data, sufficient guidance must be given to patients to help them protect their own privacy and inform patients regarding your data policies when applicable.

2.4 Groundwork for a Good Telehealth Application

2.4.1 Communicating and Understanding Needs

Even though in today's developments we are seeing a lot more of cross domain interactions, it is still quite difficult for many to grasp the requirements of members in different roles. This becomes a very dangerous obstacle, particularly dealing with the bridging of technology and medical applications in provisioning telehealth. There are still limits to what technology can do to provide a desirable solution, and the same goes that there are certain criteria for solutions aimed at medical applications.

It is easy for engineers to break down a medical application need and come up with technical solutions given enough time and resources. However, without understanding the real need that the medical problem requires us to solve, it is quite likely that the solution would end up only migrating the paperwork to routines to be done with a computer workstation.

2.4.2 Build on Familiar User Experiences

Personal computers are what everyone is most familiar with and have been a platform for telehealth experimentation since the Internet was made available to the general public. Telehealth applications on the personal computer often use a Web-based interaction scenario with a browser or software program. Given that the hardware of the personal computer could easily be expanded, telehealth applications have also used different connectors to link peripheral devices to transfer or collect needed data. Smartphones and tablets are popular choices for host platforms since 2009 for their mobility, computing capability, and connectivity. Their wide acceptance and adoption by the masses created a ready platform for deployment of telehealth applications, especially if the application wishes to monitor the person continuously, for which smartphones and tablets already have some sensors. A number of companies have demonstrated that smartphones can host professional health devices by creating peripherals for medical diagnostics. Finally, the host platform can also be a single-purpose device, specifically designed for the selected telehealth application.

2.5 Enabling Telehealth for Your Existing Medical Application

Whether you are enabling telehealth support for an existing function or building something new from the ground up, there are several key items that must be considered. These items may differ in each scenario, such as giving devices telecommunication capabilities via external add-on modules or providing clinical consulting service outside of institutional premises. Here we shall discuss some key common items that should be considered in the process.

Security is an issue that should be first thought about and planned for accordingly since in the telehealth scenario, the users' data will be transferred over the Internet and be exposed to all sorts of potential danger if not properly handled. In order to identify the user, it is often required for the user to provide their name or reachable e-mail address. Depending on the service provided, it is likely even more sensitive information may be divulged by the user. For example, tracking services for seniors with dementia often require knowledge of the senior's home/work addresses, contact details of close family members, and financial information to bill for the provided service. The billing information alone could already expose the senior as a target for unwanted attention. Good security measures such as encryption can help protect sensitive data and deter those with malicious intent. This could be further enhanced by proper partitioning of access privileges to the accumulated patient data. Restricting access to patient records only to the assigned physician and allowing only fields of data relevant to helping the physician care for the patient to be viewable is one possible approach.

Enabling telehealth for existing medical needs allows in many cases the collection of more data which hopefully will provide physicians a better understanding of their patient. Patients can use the data to understand their own progress and be given suitable advice on how to improve or maintain their health. However, this can create a lot of data feedback, for which certain mechanisms must be designed to make it easy for the person who is to act upon the data. Statistics, data mining, and machine learning techniques all provide good tools to filter and refine collected raw data into evidence. Even better, such techniques could provide clinicians with supportive evidence to screen disease development possibilities and predict patient prognosis. Also helpful is the presentation of refined evidence in visual format, such as graphs of change over time or annotated images. Without proper data tools that preprocess collected telehealth data before access, the end users will simply be overwhelmed and have great difficulty making meaningful use of telehealth-enabled services and devices. A good companion data analytics tool for the telehealth application should be able to help the user identify critical points of interest at a glance, for example, a cluster of irregular ECG data points, abnormal vitals, or conflicting medical prescriptions.

With these two items in mind, the engineering team and the medical team should discuss these together to understand and set the usage scenario of the telehealth service. The engineering team must understand the usage requirements of the medical team, such that their technical solution takes these into consideration. The medical team should be well informed by the engineering team to understand the current capabilities and limitations of technology; this is to ensure that the usage scenario is feasible and to design sensible specifications.

Several specifications should be discussed between the medical and engineering teams to ensure the conception of a good telehealth service and/or device: The first is the issue of power; if the service involves electronics in any form, it is unavoidable. Power supplies have a limited charge, as well as weight and space overheads. To medical personnel out in the field, a bulky device is an undesirable addition to the already numerous tools they have to manage. Alternatively, the power supply may have a shorter uptime in return to reduce weight and space, although this possibly means spare cells must be prepared. The key here is to balance these factors, based on the procedure medical personnel will perform using the telehealth service or device. Power-supply charges should be able to last through several sequences of the procedure before being recharged. This should be a reasonable number dictated by the medical team.

Without data communications capability of some form, the provided service/device would just be a traditional medical application. The selection of the communications approach is crucial as it ties in directly to the power-supply challenge that was just discussed. Wireless communications is a very power-consuming option; with current technological developments there are more low-power options available. BLE is one option that has seen much discussion recently; ANT+ is another choice that has been used in many fitness trackers. On the other hand, the device may also use a physical connection such as the universal serial bus (USB) to pass collected data to a computer to be transferred over the web. No matter which method is chosen, keep in mind that there is always a power cost involved if a wireless option is preferred. Moreover, there may not always be a matching receiver on the user's computer or host platform; therefore, either provide multiple options for wireless communications or provide an adapter to a popular interface such as the USB.

Together with choosing the protocol for data communications, the host platform and the target audience must be considered properly. The host platform is crucial at it affects the choices you have available for data communication and changes the way the user experience should be designed. The current host platforms can largely be placed in to three groups: personal computer, smartphones/tablets, and stand-alone devices. Personal computers in the form of a desktop or a laptop have the benefit of versatility; their hardware can be easily expanded to fit requirements and at the same time also provide operational power. However, desktops are not mobile and their counterpart, the laptop, while possible to carry around, is still bulky compared to the other categories of host platforms. Smartphones and tablets are the current hype of all manner of telecommunications-based developments of existing services in many sectors. Within their compact form, these mobile platforms pack decent computing power along with intuitive controls, allowing average users to become quickly familiar with their manipulation. The shallow learning curve has helped smartphones and tablets become an everyday appliance, thus becoming a ready platform for deploying telehealth applications. Lastly, stand-alone devices are usually single-purpose-dedicated medical devices that have telecommunications properties. It is common that this host platform is chosen for existing medical devices, as it requires only minor changes to inner components and an interface to become a telehealth-ready device.

We have covered some of the major factors to consider when making an existing medical application telehealth ready; the general concept is shown in Figure 2.3. Designing a telehealth version of an existing service or device is not a simple task; it is even more

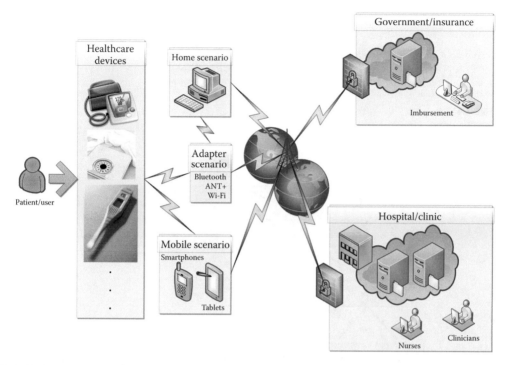

FIGURE 2.3
Interaction between users and devices over an ICT telehealth framework.

difficult if starting from scratch. Always remember that in the end, it is a tool for the users: medical professionals, caregivers, and patients; it should not be a tool that creates more difficulties than it provides benefits. We will discuss some past examples in the following section to look at different approaches of making existing medical applications telehealth ready.

2.6 Case Study—Panic Disorder Self-Therapy System

First, we shall discuss a telehealth device for panic disorder treatment. Patients suffering from a panic attack can often be seen in the emergency room [23]. These patients may exhibit symptoms such as headaches, hyperventilation, or heart palpitations. This may lead to the patients being misdiagnosed and subjected to unnecessary treatments. If not properly treated, it is likely that the patient would deteriorate, leading to depression and possible substance abuse in attempt to control the fear of losing control over oneself. The telehealth scenario is designed to provide the patient a means to perform biofeedback therapy at home, without having to visit a health institution for treatment. A ring-shaped device is given to the patient to measure their skin temperature; this device can wirelessly upload collected data to a personal computer over an USB receiver for the PC. We call this ring device the "emotion ring," shown in Figure 2.4. The patients can be managed using a web-based system, where access privileges are tiered and segregated so only relevant therapists could access their patients' information as shown in Figure 2.5.

FIGURE 2.4
The emotion ring.

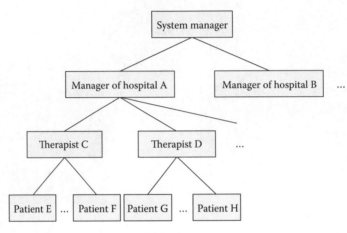

FIGURE 2.5
System and data management structure for the therapy telehealth system.

Patients are asked to take temperature measurements with the emotion ring every day. A companion software program was built as a Java applet for PCs, which serves to sync the emotion and provide patients therapy material as well as self-evaluating questionnaires. The emotion ring is worn to measure temperature changes during the daily relaxation course, which uses soothing music pieces and guided muscle-relaxation techniques. At the end of the daily course, patients are asked to complete a self-evaluation questionnaire, recording their moods before and after the course. By coupling the temperature data and the patient's self-evaluation records, the therapists were able to better identify which part of the relaxation course required more adjustment for the patient. Ultimately, the data are to provide an evidence basis for the therapists to design a suitable relaxation course for the patient and to teach the patient how to perform the relaxation course on their own; the methodology is shown in concept in Figure 2.6. Course progress was reevaluated weekly or monthly, depending on the severity of the patients, and all records were stored digitally and available to be browsed over a web interface. Both the therapists and the patients themselves could look at the course history to understand how they are performing along the treatment process.

This telehealth scenario was successful in providing patients with a method to quantify their self-help progress outside of the hospital and to help therapists understand their patients. The emotion ring was designed to be small and portable, so patients were not encumbered by a large device. Subsequent experiments and field trial results both showed that the accumulated data had been helpful in tuning the therapy because therapists were presented with more evidence regarding the patient. Based on these test results, a further

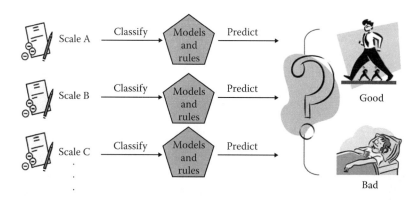

FIGURE 2.6
Clinical outcome prediction based on patient input from questionnaires and evaluation scales.

evidence model based on the biofeedback data of the emotion ring was built for the therapists to quickly discover suitable therapy procedures for panic attack patients.

However, the device had some minor problems with signal propagation. The signal sometimes failed to transmit data properly, partly because of bad signal strength and partly because the USB receiver was not plugged in properly. Due to designing the device to be compatible with different operating systems on the PC, Java was chosen as the target programming environment. Java, though having strengths in cross operating systems compatibility, is weaker in interaction with the hardware layer of computers. In order to remedy this situation, specialized software libraries had to be built to facilitate the data transfer from the USB to the PC. This sometimes led to resource allocation problems on the PC when attempting to set up a data connection between the emotion ring and host PC. The emotion ring itself did not have a display, requiring the patient to have access to a PC loaded with corresponding software and network access to display measurements. This is a very inconvenient requirement, especially for people who are not adept with computers.

For this scenario, the device was helpful in allowing patients to take care of themselves without medical help. To medical professionals, it provided an easier way to keep track of patient progress. It was overall a good example of building a new device as a provision for enabling telehealth of an existing service that was previously restricted to the patients visiting the therapist in person.

2.7 Case Study—Diabetes Telehealth Framework

In this case study, we shall discuss a telehealth framework built to care for diabetic patients [24]. The purpose of this framework is twofold: allowing medical care professionals to monitor their patient progress and allowing patients themselves to manage their own condition. The framework, shown in Figure 2.7, is built upon an architecture consisting of a service server for providing web-based access to patients and clinicians; a smartphone companion app for patients to record glucose measurements and view health education advice regarding diabetes self-care; a PC-based software for uploading and browsing records;

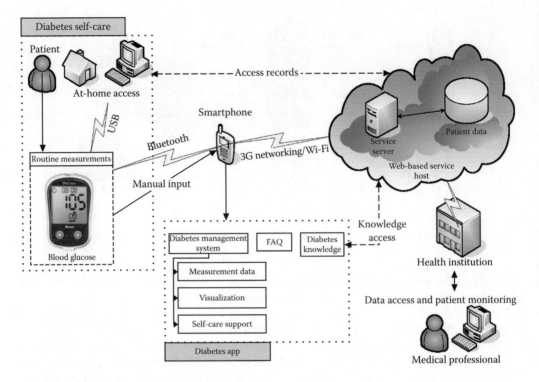

FIGURE 2.7
Framework for diabetes telehealth scenario.

and an adapter module, shown in Figure 2.8, with Bluetooth transmission that allows patients to directly sync measurements from the glucometer with the app.

The web-based access is a series of web pages containing records of the patients, which can be selected and visualized as a series of graphs and tables to observe the trends of the patients' glucose levels. Clinicians and nurses are able to access records of the patients they are managing; patients may view only their own records. The records stored on the server are uploaded from the smartphone app at the preset time. Patients may also upload records manually in the app and are reminded by the app to upload their new measurements if they have not done so over a period of time.

Traditionally, patients used written records such as little booklets provided with the glucometer or paper if the booklets were out of the room. These records were shown to their clinicians during routine checkups to allow clinicians an understanding of how the patient was doing in managing their glucose levels. However, patients often forget to record the measurement value, forget to bring their records with them, or even lose their record booklet. The companion smartphone app provides an option to digitally store glucose measurements for the patients. Patients may sync their glucometer via Bluetooth using the adapter module or enter their measurements manually. Measurements stored within the app can be visualized as a line graph to show the patient their glucose values over time. Self-care information for diabetics is also provided within the app, so that patients may have health education material to refer to outside of the hospital. If the patient's glucose levels are outside the suggested range, then the app will also prompt the patient to provide additional information from a selected list of common causes. Based on their selection, the app will provide some feedback as to how to improve the management of their glucose

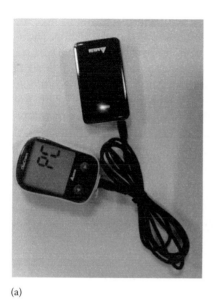
(a)

(b)

FIGURE 2.8
(a) Glucometer with connectivity dongle (USB and Bluetooth) and (b) dongle connected to PC for charging.

levels. Alternatively, patients may also use the PC program to upload data to the server via the USB port on the adapter module.

The adapter module provides the glucometer with wireless communication capabilities over Bluetooth. It also has an USB port for charging and data transfer to the PC. The original glucometer did not have any data transmission capabilities, so this module was built to support data functions. A lithium cell powers the module, allowing several weeks of use on one charge.

Patients with type 2 diabetes are normally asked to take at least two measurements a day after meals, with type 1 diabetics under a more strict suggestion of three to four measurements. It was inconvenient for most patients to carry the testing kit and recording materials around each day. They found that by using the smartphone app, they could avoid having to carry and keep track of their records. The sync function was what they found to be the most helpful, because they could take several measurements and sync the records after a period of time. Since the glucometer already contains time stamps for the measurements, the app is able to organize the records according to time when syncing new measurements. Patients also found the visualization of records a good addition, since it is much easier to see changes and understand trends than viewing lists of numbers. On the contrary, younger patients found that the feedback during abnormal glucose levels was quite menial, and would prefer to not have advice that they are already familiar with from suggested guidelines. Other patients hoped that the adaptor module could be part of the glucometer so that they could just carry one device to be even more convenient. In some cases the nurse was surprised when viewing the patients' uploaded records; collected measurement records from patients did not match what the patients were saying when returning for routine checkups. Some patients' records showed dangerously high levels of blood glucose but they had told the nurse they were in good control. The largest hurdle was deployment to elderly patients, many of whom were unfamiliar with electronics and needed another member of the family to help do glucose tests.

In this framework, we can see that using readily available platforms to host telehealth services is very approachable for the end user, particularly if the platform is widely available and familiar. One key issue raised here about the hardware is crucial to user acceptance of the device: if possible allow the users to carry just one device for their needs. If the telehealth service is difficult to use, has a steep learning curve, or is not convenient enough, users will eventually be turned away from the system. One may argue that usage can be enforced via rule and regulations, but as already seen here, users can and will lie to avoid being scolded by nurses and clinicians. For data, users may opt to record only numbers that would look good. If the telehealth system is full of this kind of record or the end users are issuing this kind of feedback, then the system needs to be seriously reevaluated.

2.8 Case Study—Telehealth Support for Unit Care

Here we shall discuss building telehealth support for nurses in unit care [25]. Nurses routinely check up on inpatients at hospital units, monitoring their health situation and dispensing medication. Several nurses cover multiple patients across different shifts daily, and usually the nurse-to-patient ratio is higher than recommended, increasing the difficulties in management. As nurses change shifts, information must be exchanged and patients handed over to the new shift, which presents a data-exchange problem. Sometimes patient assignment may change between shifts and the incoming shift nurse is unfamiliar with the patient's conditions, making the situation even more complicated and stressful.

The solution approach here is to use a tablet as the host platform with a developed software application, and use Bluetooth-enabled devices to provide vital signs measurements. A web-based access portal is used at the nurse station to allow viewing of patient data. From the web-based access, nurses are able to browse patient records and generate graphs and, tables needed for reports. In addition to collecting vital signs measurement data, the tablet also has an app, pictured in Figure 2.9, to provide other functions that nurses

FIGURE 2.9
Patient data management app on an Android tablet, connecting with a glucometer/blood pressure monitor combi-device (Fora D40b) over Bluetooth for data synchronization.

use on a daily basis. Primarily, the app allows the nurse to measure vital signs by using Bluetooth-enabled hardware or to input measurements manually if using other measurement devices. Patients are identified by a quick response (QR) code at their bedside; nurses can scan the code to call up the corresponding file to start the measurement routine. Each patient's data are recorded accordingly and uploaded to a server so that the records are shared with other shifts, coworkers at the nurse station, and the clinician in charge of the patient. A large number of assessment questionnaires are included for the nurses to cover all kinds of situations for which they have to assess the health condition of the patient. The assessment questionnaire results are also recorded accordingly where applicable. Because each nurse has to manage a number of patients and each patient has different conditions, a checklist system was also designed to help nurses verify that they have completed the required checkups for the current patient, and to remind them when they have missed an item. Also included in the app is a searchable database of guidelines and procedures for unit care. For the Bluetooth-enabled measurement devices, two off-the-shelf items were chosen: the first device is a pulse oximeter; the second is a blood pressure and glucometer combination device. The whole system is part of the set of tools on the medical trolley that the nurse uses when visiting each of the units.

Nurses that have participated in the trial deployment of this system found that it was a big improvement from their current operations standard. Normally they would have to jot down notes on logbooks or other available paper when making their rounds, then fill out corresponding forms at the end of their shift as well as typing all the notes and measurements again in their present online system. Having access to fill in the corresponding paperwork immediately at the end of each patient's routine checkup was much more efficient and reduced the chances of forgetting patient's conditions because notes were in abstract form. In addition, the reminders and checklists helped reduce mistakes and oversights in routines, as nurses were prone to being interrupted by family members with questions. For younger nurses, the procedures and guidelines database was something they really needed, partially because they have to take more night shifts where fewer senior personnel are around to support in the event of irregularities.

To protect the integrity of the hospital's existing information systems, the data collected from the tablet app were uploaded to another server set aside for this study. These requirements from the hospital led us to also discuss with their information systems management department the current network infrastructure. As recent trends in medical care call for institutions to digitalize care procedures and to use online systems, it is becoming increasingly difficult for hospitals to create and manage good telehealth systems. Part of the problem comes from the fact that many institutions are quite old, and the networking requirements to deploy large numbers of wireless systems for telehealth would require a large upgrade to the internal network. To further complicate the problem, government departments in charge of healthcare also issue their own requests as to how they think telehealth or digitalized healthcare systems should be implemented. The first complaint that most medical professionals often voice is that the user interface is unfamiliar and difficult to use. In this study, similar feedback has also been received, hoping that the tablet app interface can be designed to be as similar to existing internal hospital information system (HIS) interfaces as possible. Other suggestions from the nurses are mainly oriented to unit's user experience; for instance, the interactive interface is small and sometimes causes nurses to wrongly activate other functions; the font size is too small and difficult to read. One feedback suggestion serves nicely to illustrate the importance of understanding medical needs: a senior nurse suggested that the assessment questionnaires should primarily be simple ones, since in emergencies they often do not have time to go through the full assessment. One last important issue raised by the nurses was that whether the devices or tablets are sufficiently waterproof. In the hospital setting, it is a mandatory procedure to disinfect all tools regularly to prevent contamination and for patients' safety. Most professional medical devices are able to be wiped down with 70% alcohol solutions, but consumer electronics such as tablets are usually not conforming to these requirements.

Reflecting on these case studies, we can find that in the process of providing telehealth for existing medical applications, user experience is a key factor. For medical professionals, they are already busy enough in their normal routines; what they want is a system that can streamline their work and is easy to pick up. For patients, they want an indication of progress and help in understanding their condition with minimal effort. These are key factors in building a successful telehealth system. The engineering team should work with end users whenever possible, especially medical professionals, to design and implement telehealth systems around these key factors within good technological constraints.

2.9 Conclusions

Although telehealth is becoming a popular tool for medical services, it still needs some more developments on many fronts. Often we see a service or device for a telehealth purpose that is very well intentioned, but then the key infrastructure needed to host the service is not present. The intention of extending the reach of medical care to more needy regions by telehealth often succumbs to the fact that Internet infrastructure is nonexistent in the target region or that the signals are unstable, making data connections impossible to run the service. This often restricts telehealth deployment to more urban areas, where medical coverage is already comparatively more accessible. At the same time, the target audience such as the elderly is not too keen on adopting such a service because they would need to spend extra effort to learn how to operate the device or service. On top of that, the cost of purchasing the device or subscribing to the service is quite high, whereas in some cases users need to spend on both. How the telehealth service can be designed and implemented to provide maximum benefit to patients and medical professionals alike is an ideal that must be always kept in mind.

Like all new technology developments, initial deployments are almost always met with skepticism. With increasing evidence of effectiveness, more people are embracing the use of telehealth as an option to deploy health services. Particularly members of the medical community have been adopting more and more telehealth versions of present healthcare scenarios. Using smartphones and tablets as a host platform, many hospitals have deployed their consultation services outside the confines of the institution. Medical-grade devices such as electrocardiographs are also being certified by the Food and Drug Administration (FDA) in recent years, giving medical personal more telehealth tools to help patients. The holy grail of such telehealth devices would be something like the *Star Trek* medical tricorder, one device which can diagnose and provide information on-site. Already there are some prototype telehealth devices that show this potential purpose. The adoption of telehealth has progressively increased in pace over the past few years, and we believe that it will soon become the new standard for healthcare.

Acknowledgments

Case studies on diabetic telehealth and unit care are funded and jointly developed with Delta Electronics, Inc. Many thanks go to the medical professionals at Chi-Mei Medical Center, National Cheng Kung University Hospital, for sharing their expertise with us and taking time to support the field trials of these frameworks.

References

1. World Health Organization, "Global observatory for eHealth," 2011. [Online]. Available: http://www.who.int/goe/publications/en/. Accessed January 18, 2015.
2. P. C. Jones, B. G. Silverman, M. Athanasoulis, D. Drucker, H. Goldberg, J. Marsh, C. Nguyen, D. Ravichandar, L. Reis, D. Rind, and C. Safran, "Nationwide telecare for diabetics: A pilot implementation of the HOLON architecture," in *Proceedings of the AMIA Symposium*, 1998.
3. C. Ruggiero, R. Scile, and M. Giacomini, "Home telecare," in *Journal of Telemedicine and Telecare*, vol. 5, pp. 11–17, 1999.
4. P. Daras, D. K. Bechtsis, and M. G. Strintzis, "A Web/WAP-based system for remote monitoring patients with data mining support," in *NEUREL*, 2002.
5. F. Sufi, Q. Fang, S. S. Mahmoud, and I. Cosic, "A mobile phone based intelligent telemonitoring platform," in *IEEE EMBS*, Boston, 2006.
6. M. Rigla, E. M. Hernando, E. J. Gomez, E. Brugues, G. Garcia-Saez, V. Torralba, A. Prados, L. Erdozain, J. Vilaverde, and A. de Leiva, "A telemedicine system that includes a personal assistant improves glycemic control in pump-treated patients with type 1 diabetes," in *Journal of Diabetes Science and Technology*, vol. 1, no. 4, pp. 505–510, 2007.
7. J. Fruhlinger, "LG KP8400 cellphone with blood tester for diabetics," Engadget, October 8, 2004. [Online]. Available: http://www.engadget.com/2004/10/08/lg-kp8400-cellphone-with-blood-tester-for-diabetics/. Accessed January 23, 2015.
8. Bluetooth SIG, Inc., "Bluetooth Technology Website," Bluetooth SIG, Inc., 2013. [Online]. Available: http://www.bluetooth.com/Pages/Bluetooth-Home.aspx. Accessed January 23, 2015.
9. R. S. Istepanian, K. Zitouni, D. Harry, N. Moutosammy, A. Sungoor, B. Tang, and K. A. Earle, "Evaluation of a mobile phone telemonitoring system for glycaemic control in patients with diabetes," in *Journal of Telemedicine and Telecare*, vol. 15, no. 3, pp. 125–128, 2009.
10. Abbott Laboratories, "i-STAT System," Abbott Point of Care Inc., 2014. [Online]. Available: https://www.abbottpointofcare.com/. Accessed January 22, 2015.
11. Koninklijke Philips N.V., "IntelliVue Clinical Network (ICN)," Philips Global, 2014. [Online]. Available: http://www.healthcare.philips.com/main/products/patient_monitoring/products/intellivue_clinical_network/. Accessed January 23, 2015.
12. Scanadu, "Scanadu," Scanadu, 2014. [Online]. Available: http://www.scanadu.com/scout/. Accessed January 21, 2015.
13. Boston Scientific Corporation, "LATITUDE Remote Monitoring," Boston Scientific, 2014. [Online]. Available: http://www.bostonscientific.com/lifebeat-online/live/latitude-remote-monitoring.html. Accessed January 23, 2015.
14. Under Armour, INC, "Armour39™ fitness and heart rate monitor, workout tracking system," Under Armour, INC, 2014. [Online]. Available: http://www.underarmour.com/shop/us/en/armour39. Accessed January 22, 2015.
15. BASIS Science, Inc., "Basis—Health and heart rate monitor for wellness and fitness," BASIS Science, Inc., 2014. [Online]. Available: http://www.mybasis.com/. Accessed January 23, 2015.

16. BRAGI LLC., "The Dash—Wireless smart in ear headphones by BRAGI LLC.—Kickstarter," Kickstarter, Inc., 2014. [Online]. Available: https://www.kickstarter.com/projects/hellobragi /the-dash-wireless-smart-in-ear-headphones?ref=48hr. Accessed January 23, 2015.

17. C. Boulton, "Hospital Okays Google Glass in the Emergency Department," Dow Jones & Company, Inc., March 13, 2004. [Online]. Available: http://blogs.wjs.com/cio/2014/03/13/hospital -okays-google-glass-in-the-emergency-department/. Accessed March 13, 2014.

18. J. Nosta, "How Google Glass Is Changing Medical Education," Forbes.com LLC™, June 27, 2013. [Online]. Available: http://www.forbes.com/sites/johnnosta/2013/06/27/google-glass -teach-me-medicine-how-glass-is-helping-change-medical-education/. Accesed February 2014.

19. IBM, "The fusion of big data and little babies," IBM, 2012. [Online]. Available: http://www .ibmbigdatahub.com/blog/fusion-big-data-and-little-babies. Accessed January 20, 2015.

20. D. Patnaik, P. Butler, N. Ramakrishnan, L. Parida, B. J. Keller, and D. A. Hanauer, "Experiences with mining temporal event sequences from electronic medical records: Initial successes and some challenges," in *ACM SIGKDD Conference on Knowledge Discovery and Data Mining*, 2011.

21. H. Neuvirth, M. Ozery-Flato, J. Laserson, M. Rosen-Zvi, J. Hu, M. S. Kohn, and S. Ebadollahi, "Toward personalized care management of patients at risk—The diabetes case study," in *ACM SIGKDD Conference on Knowledge Discovery and Data Mining*, 2011.

22. M. Kelly, "Why the Obamacare Website Failed in One Slide," Business Insider, Inc., November 19, 2013. [Online]. Available: http://www.businessinsider.com/why-the-healthcaregov-website -failed-at-launch-in-one-slide-2013-11. Accessed December 2013.

23. B.-E. Shie, F.-L. Jang, and V. S. Tseng, "An intelligent and effective mechanism for mental disorder treatment by using biofeedback analysis and web technologies," in *International Journal of Software Engineering and Knowledge Engineering*, vol. 21, no. 1, pp. 55–72, 2011.

24. I.-H. Tsai, Y.-F. Lin, and V. S. Tseng, "A mobile framework for personalized diabetes telecare," in *Conference on Technologies and Applications of Artificial Intelligence*, 2012.

25. Y.-F. Lin, H.-H. Shie, Y.-C. Yang, and V. S. Tseng, "A real-time and continua-based care guideline recommendation system," in *Asia-Pacific HL7 Conference on Health Care Information Standards*, 2013.

3

Virtual Hospitals: Integration of Telemedicine, Healthcare Services, and Cloud Computing

Shaftab Ahmed and M. Yasin Akhtar Raja

CONTENTS

3.1 Introduction

The enabling technologies developed in the past decade have made it possible to remain connected to social networks in the cyberdomain while being mobile in the physical domain. Ever-increasing human mobility and changing lifestyles demand ubiquitous services on an anywhere-and-anytime basis [1–2]. New paradigms of eHealthcare provide services through virtual hospitals where patients can communicate with physicians and the medical staff effortlessly using Internet-enabled cell phones, handheld PDAs, or fixed network devices. The doctors, physician assistants, patients, nursing staff, and emergency-handling units are integrated in communities over metropolitan and wide area networks for remote monitoring, consultation, and rapid response. The decision-support and work-flow systems enable the physicians to handle triage as well as provide prompt advice to patients and second opinions to their peers [3–4]. Various new paradigms are emerging; e.g., medical tourism [5] is among the interesting developments where the patients can get inexpensive services in both surgical procedures and postoperative care.

The other part of the spectrum is healthcare for physically frail and weak/geriatric patients unable to withstand the hassle of visiting medical clinics or hospitals; but they require continuous health monitoring. Admission in hospital wards may not be practicable, cost effective, or affordable. Hence, patients would like to be in their homes while being looked after by the paramedic staff remotely. A little bit of training and adaptability can help in such situations. The patients can provide their biological and physiological measurements as periodic or aperiodic uploading of clinical data as prescribed by the physicians in the respective medical therapy/rehabilitation program. The proactive study of clinical measurements and symptoms/indications from embedded software agents in the clinical data acquisition modules can predict medical problems or imminent health failure and issues [6]. Doctors and nursing staff can get advance alerts in order to take care of such situations while patients may not have noticed any abnormality. In many old and ailing patients, medical symptoms may go unnoticed; hence, eHealthcare solutions can be of great help.

Medical sensor networks are being used to provide continuous nonobtrusive patient monitoring from nursing stations and emergency mobile units. Sensors are of various types and can be set up in varying topologies. In case of *in-body* or *on-body* sensors, suitable ways of clinical measurements can be set up which might be automatic or semiautomatic [6–7]. Wireless sensor networks are often integrated in healthcare solutions, effortlessly enabling remote and mobile access. There can be a mix of wireless sensor networks at the patient end, using Bluetooth or an ad hoc network protocol followed by wireless personal area network (WPAN), wireless application protocol (WAP), and WiMAX connectivity leading to fixed networks [8]. In case of large-scale disasters, physical networks may be knocked out, leaving thousands of victims helpless. Wireless sensor networks can be rapidly setup as mobile ad hoc networks (MANETs) to reach out to such patients and victims.

eHealthcare data management is a core issue in contemporary hospitals and related systems. The data generated are enormous due to newer and multiple diagnostic techniques; indeed, healthcare data are the source of recent interest in large-scale data storage over the cloud architecture outsourcing services to cloud service providers (CSPs) [9]. Physicians, healthcare providers, and patients demand a greater lifetime for medical data records; they require fast data storage/retrieval and presentation services for clinical decision-support systems (CDSSs) [3–4]. The cost of maintaining a data service center in a typical hospital may be substantial for both infrastructure and ICT staff. Cloud computing is a relatively new discipline for exploiting the virtual environment over the Internet for such data-intensive applications. CSPs offer large infrastructure, computing power, and software services configurable by the client at low up-front costs. But earlier reports of vulnerability of the cyberdomain have led to serious security concerns of clients regarding services and data management [9–10] and vulnerability issues still remain to be fully addressed. Electronic patient record (EPR) management requires foolproof security. It is extremely important to develop methods to relieve clients from concerns and problems of security, integrity, availability, and cross CSP mobility.

Cybersecurity mechanisms have evolved over the past two decades to handle these problems at data-transport and operation levels. In order to maintain and enhance user confidence in cloud computing, end-to-end security, data integrity, proof of retrievability (POR), and third-party audit (TPA) along with forensic methods have been proposed. Security solutions offered by a trusted platform group (TPG) can be used for cloud computing through integration with trusted platform modules (TPMs) embedded in the client and server machines [11–12]. The scheme addresses the physical security concerns posed by portable data access units like mobile phones, notebooks, iPads, and laptops. Research is being carried out to address bandwidth and security issues with minimal battery power and low computational requirements while meeting acceptable security standards.

3.2 Related Work

mHealth is becoming a widely accepted means of supporting patient mobility and medical assistance beyond the conventional boundaries. eHealthcare applications are being developed using embedded platforms. Smartphones, wireless motes, and telecommunication solutions have enabled researchers to initiate physical, physiological, and cognitive/ behavioral studies. Examples of some of the projects being pursued at leading universities and research institutes are summarized below:

- CodeBlue is a Harvard University project focusing on wireless sensor network development for medical applications. An integrated system for sensing of medical data, e.g., ECG, oxygen saturation (SpO_2) measurement, pulse rate, and electromyography (EMG), are being developed with Wi-Fi communication to a server through wireless motes [13].

- MIThril is a project of Massachusetts Institute of Technology (MIT) Media Lab for studying human behavior by using wearable medical sensors to measure ECG, skin temperature, galvanic skin response (GSR) along with data for step-and-gait analysis [13].

- The European Microsoft Innovation Center (EMIC) in Aachen, Germany, has started the innovative project Emergency Monitoring and Prevention (EMERGE) [14] to improve emergency assistance through early detection, proactive prevention, and unobtrusive sensing. The sensor network technology through wireless connectivity lets the patients lead normal lives with automated data collection round the clock [15].

- Prototype intelligent homes have been built at many research institutes and universities to test concepts and make suitable products to have continuous monitoring of aging and seriously ill or disabled persons [16].

- AlarmNet is used for pervasive, adaptive eHealthcare at homes and addresses the requirement of aging population requiring proactive monitoring for early detection of abnormalities and maintaining a medical therapy/exercise schedule through alerts [15].

- Intel Research Seattle and the University of Washington have built a system to monitor daily living activities. The University of Rochester is building a smart home; Georgia Tech has built the "Aware Home" and MIT is working on the "PlaceLab" initiative [14].

- Some of the web-based software used to enable and support eHealthcare communities includes e-learning spaces, e.g., HealthStream, and web meetings using WebEx. Similarly, discussion groups on Webcrossing and Project Spaces utilizing Outlook and knowledge workers depend on tools such as Plumtree. They are integrated with inference engines providing artificial intelligence to add knowledge-based assistance for the user.

- The Healthcare Compunetics Special Interest Group is working on developing expertise in "20 identified areas" and is planning to integrate these in third-generation (3G) and fourth-generation (4G) networking. Hospital information systems (HIS) today integrate individually optimized sections along with home care, ambulatory care, and long-term nursing [17–18]. To handle the diverse nature of clinical protocols, three tiers of implementation have been used by the European Development on Indexing Techniques for Databases with Multidimensional Hierarchies (EDITH) [19]. The tiers are organizational, operating, and technological. The EDITH architecture provides an open and modular environment, using network-independent communication environment (NICE) and distributed hospital environment (DHE).

3.3 Service Integration in Virtual Hospitals

In this section, various service components are listed and briefly discussed that enable mHealth delivery. The mHealth data include personal medical history, diagnostic reports, therapy schedule, and guidance tutorials through audio/video and text presentation on handheld devices. The mobile apps vary in their scope for patients, nursing staff, and the doctors. Service integration for virtual hospitals would involve social networking, data accessibility via cloud infrastructure; telemedicine via ubiquitous cloud services [15]; and data acquisition, archiving, and retrieval along with the work-flow and decision-support systems (DCSs) in a cloud architecture. That would expand to include agent-based proactive study using knowledge-based expert systems.

3.3.1 Social Networking

The migration of software applications to user-centric handheld devices has been a major shift in the ICT industry in the new millennium. Rapid developments in communication technologies and the availability of services at the user edge, like general packet radio service (GPRS), Wi-Fi, and global positioning system (GPS), have enabled usage of mobile applications over smart phones. The mobile communication has become an essential requirement of the modern society.

Developers are concentrating on making personal mobile device clients for large data storage and computation service providers in the virtual domains well known today as the cloud [20–21]. *Mobile application development* and *social networking applications* have become the buzz words; mobile apps are software packages to be run on mobile devices and smartphones, PDAs, and tablets. Some mobile apps are preinstalled on the mobile devices, while various others are downloadable. A few of the common applications that offer such services are iCloud, Dropbox, Google Drive, and Microsoft (MS) SkyDrive.

3.3.2 Data Accessibility in the Cloud

As noted earlier, mHealth data include personal medical history, diagnostic reports, therapy schedule, and guidance tutorials through audio video and text presentation on the handheld devices. Apps vary in their extent and scope for patients, nursing, and paramedic staff as well as for physicians. The security issues and some potential solutions will be discussed later in Section 3.6.

The ubiquitous availability of information on portable/mobile devices is a key feature of promoting use of a variety of human-centric applications. Healthcare applications are popular because they integrate well in daily activities providing not only alerts and tips but also access to information in a virtual hospital environment hosted over the ubiquitous cloud. Patients are becoming part of treatment planning; hence, they display better behavior and responsibility. Self-care can be supported by mHealth services that are proactive; e.g., they prompt the patients to take medicine or maintain regular report and therapy schedules. The availability everywhere and all the time is very useful. Comments/remarks of doctors and staff and access to lab records, etc., are easily available through mobile and cloud services. Location of nearby hospitals and pharmacies can be found or alerts can be embedded in GPS mobile apps. Virtual hospitals [18–22] are a concept rapidly evolving to provide services to indoor, outdoor, regular, or geriatric patients. They can use cloud services to access secure information.

3.3.2.1 Telemedicine for Ubiquitous Healthcare

Telemedicine provides medical services remotely [15,23]. It is used to extend medical services, e.g., consultancy, real-time engagement of patients with doctors, continuous patient monitoring, and ubiquitous availability. In developing countries, live services are provided to rural and remote areas through telemedicine. A few applications of telemedicine are the following:

- Teleconsultation
- Telediagnosis
- Teletherapy
- Telemonitoring
- Telerehabilitation

The scope of telemedicine has been extended with the advancement of ICT to include medical education services and eHealthcare. A broader term used today is *telehealth*, which includes telemedicine, eHealthcare, and health education. Telehealth systems are composed of biomedical sensor nodes attached to patients or installed in the rooms/environment, which acquire data aggregated through body area networks (BANs) or body sensor node (BSN) and forwarded to a personal area network (PAN) or wearable PAN [22,24]. The data may then be presented on nursing consoles, archived or transmitted live to physicians through Internet-enabled connectivity.

3.3.3 Data Acquisition, Transmission, and Archiving

Sensor networks exploit personal wearable devices used to measure various clinical parameters of patients and subjects. Such an exercise is useful for diagnosis, postoperative

monitoring, and rehabilitation in cases of serious injuries, heart attacks, or brain trau-
matic conditions. A number of physiological and biomedical sensors are available for temperature, humidity, blood pressure, and heartbeat measurements and recording. Wireless
sensor networks collect biomedical data by using BAN or wearable wireless body area
network (WWBAN) [25].

The interconnectivity between sensor nodes can easily be provided by ad hoc networking without the wires considered to be obtrusive and uncomfortable by the patient. Data
records are maintained on monitoring stations through PAN. Sensor networks may use
continuous, event-driven, or query-based methods for data collection and transport to
nursing stations for data aggregation. Figures 3.1 and 3.2 show some components of data
acquisition and communication in wireless-enabled hospital management [26–28].

The challenges faced by wireless sensor networks (WSNs) are development of robust,
tamper-resistant, secure protocols, and battery power management. Mobile telemedicine
includes the possibility of collection of sensor data on PDAs or mobile phones and sending the data over cellular networks. Such solutions might be suitable for chronic patient

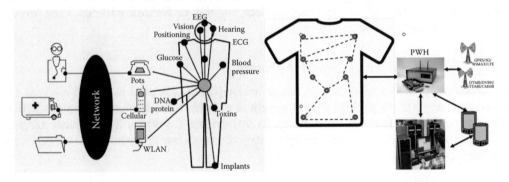

FIGURE 3.1
Wearable wireless body area network (WWBAN).

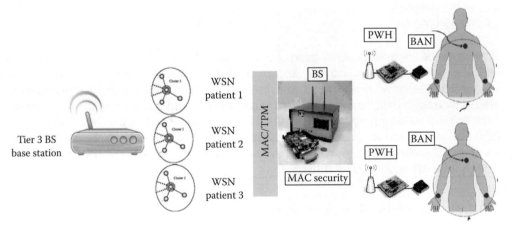

FIGURE 3.2
Components of data communication in wireless-enabled hospital management.

monitoring, those suffering from cardiac problems, diabetes, and asthmatic conditions, etc. GPS features can be integrated to add patients' traceability as well.

Telemedic diagnostics and treatment are extremely important in the situations involving disasters, wars, and other emergency conditions. Wireless-enabled sensors can easily be integrated into ad hoc networks for patient monitoring. Data connectivity to the upper hierarchy of services architecture can then be provided through mobile communication, which eventually leads to the nearest medical facility or hospitals with accessibility to physicians and specialists. Developing countries are also trying to take advantage of experience of advanced medical systems but a lot more has yet to be done in terms of infrastructures, education, and training.

3.3.4 Work-Flow and Decision-Support Systems in Cloud Architecture

As described earlier, the cloud is a virtual computing environment which provides applications, platforms, and software support as services [2,18]. The applications are extended over the Internet domain to the CSP, which maintains computer systems in clusters; they usually have large storage capacity utilizing storage area networks (SANs) [29] connecting the data centers (DCs). The CSPs support fault-tolerant systems meeting the user requirements of hardware, software, and platforms in scalable and elastic service architectures through contracts. A typical data center architecture is illustrated in Figure 3.3.

Enterprises using data centers' support outsource their computational needs and loads on the "pay-as-you-go" basis. Hence, beginning an application in a smaller scope, then extending with proven viability is a great advantage of such a new paradigm in ICT. The CSPs take care of all the day-to-day maintenance of systems and software usage, i.e., logging and accounting procedures. The clients do not maintain computing facilities; instead, they have access to the data centers over the Internet connectivity on anywhere, anytime basis [29–32]. Cloud computing architecture, opportunities, and concerns are briefly reviewed in this section. Software and services offered by leading software developers are used as references.

Enterprises like data banks, medical facilities and hospitals, and scientific, industrial, and research organizations usually need large storage facilities to maintain archives,

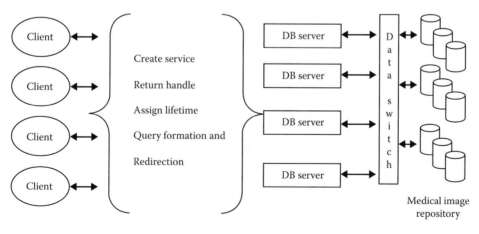

FIGURE 3.3
Typical data center for medical image repository.

which run into petabytes in size. It is, therefore, feasible to outsource this activity to data centers. In many cases, hierarchical ordering based on usage coupled with intelligent search utilities with caching support improves data availability. In the cloud environment, a developer leases data center facilities, saving time for research and innovative development from handling routine jobs.

Foster et al. [33] defined the *cloud* as "a computing paradigm which is a pool of abstracted, virtualized, dynamically scalable, managed, computing, power storage platforms, and services for on demand delivery over the Internet." The cloud services can be implemented at private, enterprise, and external levels. The clouds may also be federated into virtual private clouds. The community clouds may also be formed by joining trustful domains of similar activities together to share computing and data storage resources. Figure 3.4 displays hierarchical data access and control to field cloud services.

A variety of applications infrastructure and platforms is offered by the leaders of the ICT industry. Some of these are summarized below [34–39]:

Google:

Applications:	Google Docs, Google Calendar, MapReduce
Infrastructure:	BigTable, a multiserver proprietary database system using Google File System with a storage capacity in petabytes
Platform:	Google App Engine

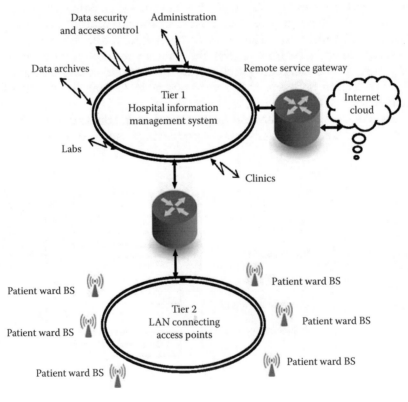

FIGURE 3.4
Hierarchical data access and control to field cloud services.

Salesforce:

Applications:	Customer Relationship Management (CRM)
Infrastructure:	Force, which offers multitenant database and application services
Platform:	Integrates application development in public and private domains; it uses a sandbox to replicate Salesforce over cloud architecture

Microsoft:

Applications:	Windows Live, SharePoint, Exchange
Infrastructure:	Azure
Platform:	.NET Framework

Amazon:

Applications:	E-commerce, Flexible File System, business entities to expand product services
Infrastructure:	Amazon Web Services (AWS), Elastic Compute Cloud (EC2), and Simple Storage Service (S3)
Platform:	Database model for heterogeneous platforms

EMC/VMware

Applications:	Uses Web 2.0 services application programming interface (API) for application integration; it supports Open Virtualization Format (OVF)
Infrastructure:	vCloud, vApp, and multitenant support
Platform:	VMware

The Apache Hadoop project [40] has developed open-source software which may be used for clinical data processing involving large data archives. It provides reliable, scalable software, platform, and infrastructure support over the cloud. Some of the products available include the following:

HBase:	A scalable, distributed database for supporting storage of large tables
HDFS:	Hadoop Distributed File System; the primary storage system containing replicas of data blocks in clusters to provide fault tolerance and high availability
Hive:	A data warehouse infrastructure that provides data summarization and ad hoc querying
MapReduce:	A framework for distributed processing of large data sets in clusters to enable parallel processing
Pig:	A high-level data flow language and execution framework for parallel computation

3.3.5 Agent-Based Proactive Study and Knowledge-Based Management Systems (KBMSs)

Both of these topics are briefly discussed below.

3.3.5.1 Agent-Based Proactive Study

eHealthcare is rapidly becoming an important component in medical systems, especially for the aging population [22,41], which is on the rise due to increased life expectancy. Sometimes, patients require round-the-clock monitoring and support. In other instances, vital signs of some patients are so weak that the medical disorder symptoms may go unnoticed. In these scenarios the sensor networks can play an important role. Biomedical sensors are attached to the patient's body and sometimes in the patient's premises/environment or the medical ward to monitor the patient's activity, condition, and movements. Sensors keep transmitting useful information to monitoring stations conveniently located in the range of wireless transmitters. Data collected are routed to the servers, which maintain logbooks that, in turn, could trigger alerts on various checkpoints. Alerts generated by the server can be routed to the nursing console or to physicians on shift duty or the medical specialist on call according to servicing policy and severity of the alerts. In extremely important cases, these alerts can be delivered to the mobile phones/devices of the concerned doctors or specialists.

Patient monitoring and eHealthcare systems are highly promising for effective medical treatment and lifesaving. The patient history log can be used for preventive medical prescriptions and early diagnosis of imminent emergency, e.g., a heart attack. Intelligent agents can be embedded in such systems, analyzing patients' records in the background and raising alerts or messages to the doctors and specialists. Electronic data received through an event-monitoring system are analyzed online to raise alerts for requesting services. In many cases, the data received may be used with archived data to proactively generate warnings and alarms. Arabshian and Schulzrinne [6] proposed a medical event-monitoring system using session initiation protocol (SIP). This helps in establishment, modifications, and terminations of sessions between the participants. For this purpose, "invite-and-subscribe" procedures are used. An event may be ascribed to particular signals; in turn, the monitoring system may execute procedures to generate alerts and appropriate actions.

3.3.5.2 Knowledge-Based Management Systems

Information availability and access has increased greatly in recent years due to penetration of computers/smart devices, enabling technologies (e.g., cyberphysical systems based on optical networks) in every domain. Social networks have become a preferred way of interaction between people over the abstract domains run by Internet applications. These applications are ubiquitous and readily available through Wi-Fi connectivity. Medical diagnostic procedures generate large chunks of data as text and pictures to archive for ready-to-access EMR [42]. Decision-support systems for medical doctors, researchers, and nursing staff may use the repositories of medical data collected over time.

KBMSs for eHealthcare allow users to interact with a huge information base through intelligent tools. They have the capability to transform user requests into appropriate queries for knowledge extraction and presentation in suitable form. Expert systems and data repositories are integrated in the sublayers over the ubiquitous cloud infrastructure and services architecture [43–45]. The cloud has the advantage of global availability of anywhere, anytime accessibility, which has revolutionized society today beyond healthcare.

The architecture of eHealthcare systems has to capture, archive, and present information through foolproof security procedures in order to meet the requirements of the Health Insurance Portability and Accountability Act (HIPAA) [46]. KBMSs go beyond mere database or information management by providing opportunities of collaboration, consultation, and learning from experience to improve procedures of treatment and healthcare. Those sometimes have integrated support of expert systems for assisting in medical prescriptions to avoid errors and omissions. In order to make the virtual hospital concept a real success, researchers are actively working on development of suitable procedures and security architectures meeting the challenges of cybersecurity [45,47–49]. Figures 3.5 and 3.6 show a typical KBMS.

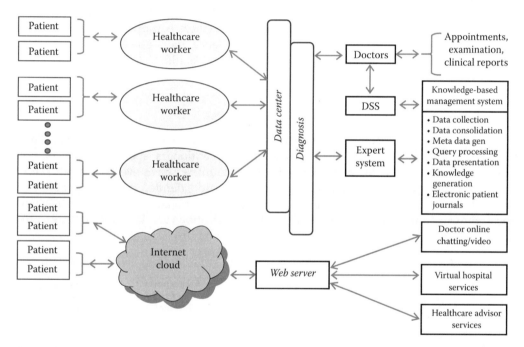

FIGURE 3.5
Knowledge-based management system for eHealthcare.

External wireless	GPRS/3G/ WiMAX	Digital TV	GPS
Vehicular wireless gateway			
In-vehicle wireless	Bluetooth	WPAN/ ZigBee	Wi-Fi	Car, bus
Portable devices like PDA, earphone, etc.		LCD display, speaker, radar, sensors, temp	

FIGURE 3.6
Layers of cyberdomain data-exchange architecture.

3.3.6 Disaster Management and Emergency Response Services

For disaster-stricken areas, social networking and telemedicine offer a quick response which is not dependent on any physical infrastructure. The mHealth cloud services are offered beyond the bounds of location at all times. Disaster management requires quick establishment of healthcare centers to organize and coordinate teams of medical doctors and staff [50]. In emergency conditions, ad hoc networks play an important role in providing quick healthcare services to victims. The wireless sensor networks for these purposes would use Bluetooth or other emerging technologies to communicate with local hosts which support services offered by the hospitals and clinics [51–52]. Ubiquitous availability is used to provide first-response/first-aid and medical services to patients and disaster-stricken victims. In case of triage and ambulatory services, the patients' clinical data and physical/health conditions augmented with audio and video streaming may be used by physicians and doctors to issue instructions to the ambulatory paramedic staff in advance. Preparation for the procedures and strategy to handle patients on arrival could be made based on the live information received.

The inclusion of mobile vehicles for ambulatory services equipped with mHealth systems is required to handle emergency conditions. Hence, the scope of mHealth activities is broader and it can be generally implemented in tiers as illustrated in Figure 3.7 and as explained in the subsequent sections.

3.3.7 Recovery Rehabilitation and Medical Tourism

Medical treatment is becoming extremely expensive in developed countries, e.g., United Kingdom and United States. Physicians, medical practitioners, specialists, and pathologists who have worked for many years in these countries have eventually returned to their homeland mostly because of cultural bondage and extended family responsibilities. Internet connectivity enables them to remain in touch with advancements in their fields and at the same time provides access to research publications and journals. mHealth offers

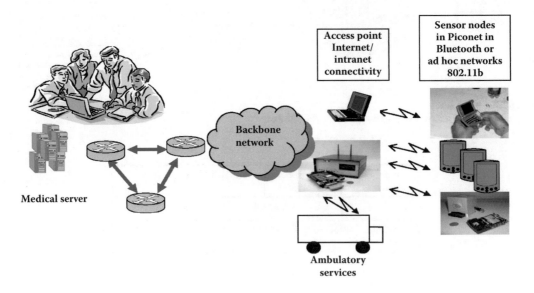

FIGURE 3.7
Typical mHealth network tiers and network services.

an opportunity for patients to move temporarily to Asian hospitals for surgical procedures followed by physiotherapy and healthcare services. Hence a new term, *medical tourism*, has emerged and become popular [5]. Patients remain connected to their families and loved ones in their respective homelands through ubiquitous Internet connectivity during medical treatment and rehabilitation. Patients also use such trips to visit historical and new geographical regions with diverse populations and cultures.

3.4 Medical Data Storage and Presentation Standards

Medical image repositories provide clinical data for various case studies. Images are archived using compression techniques and available tools. Images are accessible over the web and have proved very useful for diagnosis and therapy prescription. Two main challenges for repository services are browsing for selection of images interactively and downloading under a secure policy. A major task in developing the archive is to provide fast access to images through decompression tools. For this purpose, data size is reduced by judicious image processing, cropping, and compression.

Various standards have emerged in recent past for picture archiving and communication system (PACS), including digital imaging and communication in medicine (DICOM), developed by the National Electrical Manufacturers Association (NEMA) [53–54]. It is a widely used standard for interoperability of medical imaging and instrumentation. The layered service stack has three tiers, i.e., application, collective, and fabric layers. PACS uses these layers to provide services for images stored under a classical archiving and sharing model. Enhancements are required to provide acceptable security for using medical data and provision of real-time services like voice over Internet protocol (VoIP), video and audio streaming, and bidirectional request servicing.

3.5 Nursing Stations for Remote Patient Monitoring

Nursing stations for remote patient monitoring as displayed in Figure 3.8 may be used for medical data acquisition through wearable sensors and a video camera for each patient in a room or ward. The software generates alerts at the nursing station on conditions set up by the doctor/nurse interactively. Online data, history, medical prescriptions, etc., are readily available at the nursing station through interactive console software. Real-time monitoring by video cameras is often required for postsurgical recovery and very ill patients. Authorized doctors may also monitor patients from remote locations and doctors may be consulted online. Project SENSITIVE was developed at Bahria University to provide remote services for a nursing station [55]. Some of the features included are the following:

- Patient registration and electronic medical record (EMR) management [42]
- Interactive control panel functionality
  - Real-time physiological monitoring: display pulse rate, body temperature, blood pressure, and ECG for selected patient

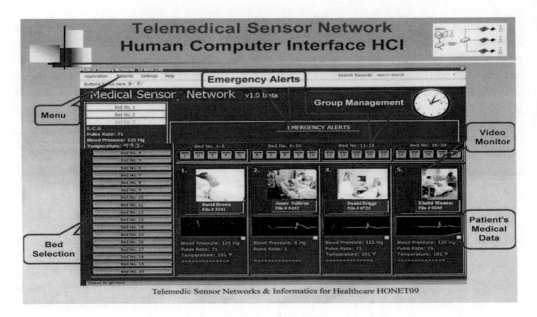

FIGURE 3.8
Nursing station for remote patient monitoring presented in the International Symposium on High Capacity Optical Networks and Enabling Technologies (HONET) 2009. (From J. Saeed and H. Asim, "SENSITIVE— Medical sensor network for monitoring and recording physiological data on a nursing console," FYP 2010, Computer and Software Engineering Department, Bahria University, Islamabad, Pakistan, 2010.)

- Alarm generation and proactive analysis: audio warnings in case of critical situations in the form of beeps, along with visual warning indicating patient bed number
- Live video streaming feature
- Medical data acquisition and communication
- Medical history management

3.6 Cybersecurity Issues

3.6.1 Hierarchical Security Management

EMR management is a highly sensitive issue which requires foolproof security. It is extremely important to develop methods of relieving the clients from problems of security, integrity, availability, and cross CSP mobility [56–64]. eHealthcare models for ubiquitous services for data acquisition archiving and presentation in the cloud are briefly described. The management issues and security concerns in cloud domains are addressed by the proposed services architecture. The model includes wireless sensor networks besides communication and storage systems for typical hospitals taking advantage of the cloud services architecture (CSA) [65].

The proposed hierarchical security management scheme uses biometric data, eHealth cards, or radio-frequency identification (RFID) tags for patient or doctor authentication to access healthcare services. The healthcare server issues a session key for one-time usage only. The upper layers use Internet protocol (IP) security [66] along with the symmetric session key for data transport. In cases of disaster, temporary arrangement can be made based on bypassing the requirements necessary to access the healthcare servers to be confirmed at a suitable later time. Security at the lowest level can be further strengthened by including the concept of root-level trust embedded in the sensors which are registered with a trusted coordinator (TC).

3.6.2 Root-Level Security for Physical Identity-Based Accessibility

In this section, we discuss a model medical community cloud for hosting services on a trusted computing environment of a cloud service provider such as Amazon, MS Azure, and Ubuntu [34–39,67–69]. The security architecture in a medical community (SAMC) protocol over the cloud are depicted in Figure 3.9. SAMC would have embedded hardware-level certification, periodic software attestation, key management, and a virtual secure machine (VSM) to monitor the activities of host infrastructure and services provided by the CSP, for intrusion detection or contract violations [70].

A qualified trusted node (TN) has to boot under a protocol exercised by embedded TPM. It produces platform certification, conformance certification keys, and endorsement certificates. TC authenticates data and program activity in cyberdomain through attestation of trusted nodes attested by a trusted third party (TTP). It maintains a list of active TNs along with their encryption keys. TCs can inspect and verify the availability and integrity of these nodes through exchange of nonce periodically.

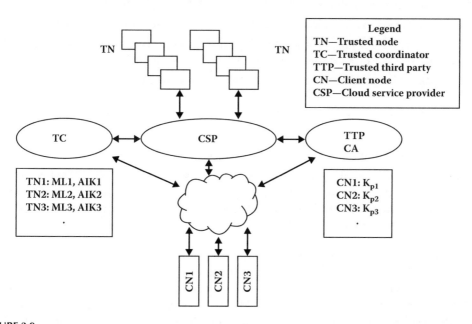

FIGURE 3.9
SAMC—trusted computing in the cloud. (From S. Ahmed et al., "Cyber security: Vulnerabilities and challenges in telemedicine over cloud," Fourth Networked Digital Technologies 2012, Dubai, United Arab Emirates, 2012.)

To ensure dynamic security and integrity management, the CSP is bounded with trusted platform control (TPC) procedures and service-level agreement (SLA).

3.6.3 Proposed Trusted Computing Protocol

The proposed protocol for trusted computing in SAMC in the cloud domain [70] is highlighted as follows:

1. Client node (CN) has the public key of cloud service provider ($K_{Pub\ CSP}$).

2. TN available to the CSP has an integrated TPM module to compute a measurement list (ML) at boot time, which is used to bind the TPM with the processor under a protocol. It is basically a sequence of "hashes generated," which is stored for subsequent attestation.

3. A TC maintains a pool of TNs by issuing signed attestation identity keys (AIKs) to TNs by a protocol [70]. TC maintains the MLs for trusted nodes available along with AIKs.

4. CN acquires a certificate of secure boot/remote attestation of TN from the TC by a protocol as summarized below:

    a. CN challenges TC by a nonce encrypted by public key of TC ($K_{Pub\ TC}[n_U]$).

    b. TC replies by its bootstrap measurements along with nonce encrypted by its private key ($K_{Priv\ TC}[ML, n_U]$).

    c. CN attests TC's authenticity using public key $K_{Pub\ TC}(K_{Pub\ CN})$.

    d. TC attests CN through a similar protocol.

    e. On successful exchange, the public key of CN is saved in TC's database.

5. Client requests TC to assign required number of trusted nodes in the cloud.

6. TC services client request by assigning TNs from a pool. TC provides their ML keys to the client for monitoring activities of the assigned TNs.

7. Periodic testing/validation may follow to monitor the TN activities.

NOTE: The Amazon EC2 machines offer root-level access for this purpose.

3.6.4 Executing Client Application in Secure Environment

The client-level security monitoring can be enhanced by deploying a VSM on the TN under a mutually agreed protocol. The VSM proposed will be deployed on the CSP end to monitor the activities and send useful information regarding SLA compliance. The VSM will operate in a glass-box environment or secure zone, having its own intrusion detection system.

The cloud architecture supported by a CSP, such as Amazon, provides virtual machine instances (VMIs) for hosting the client applications. The VMI can be modified by integrating the proposed protocol of TPM; it may be called trusted virtual machine instance (TVMI). To enhance the security level, a VSM can be transported to the TVMI which is self-revealing and installing. After successful installation VSM ensures that even the system administrator at the service provider end will not be able to inspect or tamper with TN. The research and development of embedded design of TPM and dynamics of VSM has been initiated and is ongoing.

3.7 Conclusion and Discussion

The mHealth industry has established its roots and is growing rapidly due to developments in the ICT industry and the paradigm shift in the healthcare system. The tendency to take advantage of cloud computing environment in all walks of life has provided an opportunity for software developers to extend services to the medical community for mHealth. The availability of high-capacity and high-speed Internet connectivity in the wireless and mobile communications (enabled by fiber-optic-based backhaul) has been a source of interest for developing applications for mobile devices/phones and PDAs. These applications are user friendly and require common sense to operate for a host of useful functions [71–73].

In the mHealth domain, connectivity to the nearest service providers or the home hospitals at remote places is now possible through touch screens and a few clicks. Besides this, a large number of services for navigation, weather updates, news alerts, and social networking and microblogging through Facebook, Twitter, MySpace, LinkedIn, Bebo, Friendster, Ning, etc., are available. It is no longer necessary to carry a laptop. Instead, an intelligent mobile, iPad, or PDA is usually enough as the information is easily accessible on request from the host servers. The mobility of human beings and all-time service through wireless-enabled sensors, PDAs, and mobile devices are the realms of the fast-growing ICT industry. Network protocols like WiMAX, WAP, Bluetooth, and MANET have made it possible to dynamically configure the communication paths. Small and smart operating systems like Android with tools and software available in freeware have made mobile apps development an interesting area for the ICT industry [8,17,74–76].

While the scenarios discussed provide flexibility, mobility, and a large scope of functions in the cloud domain, data security has assumed greater importance. It is no longer limited to secure content transmission and delivery; the new dimensions of provenance and verifiability have to be included. A number of security concerns regarding service providers themselves have arisen. Secure web services governed by SLAs have to be carefully articulated. Various aspects of the newer technologies and services for providing HIPAA-compliant secure healthcare have been briefly discussed. A protocol supported by the virtual machine concept for operating in cloud domain has also been presented.

Pervasive healthcare models are rapidly replacing conventional systems. mHealth has made a transition in healthcare from being doctor centric to being patient centric, sampling to monitoring, and reactive response to continuous proactive and preventive care [77]. These systems provide solutions easily adjustable to the lifestyle of patients. In contrast to cities and urban regions, rural areas have fewer doctors, hospitals, and other medical services. The population in rural areas is generally poorer, with a higher percentage of elderly citizens. Pervasive mHealth systems promise to be very effective in such areas.

eHealthcare is available to more people along with hospital staff; advanced diagnostic techniques, e.g., CT, magnetic resonance imaging (MRI), and positron emission tomography (PET), generate huge amount of data [78]. Physicians/researchers demand longer medical history of patients; hence, larger information systems or data centers are required. To make the data storage and servicing cost effective, outsourcing data storage to CSPs is often recommended. CSPs such as Amazon S3 [67] are offering large storage spaces along with virtual computing facility on pay-as-you-go basis, for example, the EC2 services of Amazon [79–80].

Researchers and medical doctors are often interested to see case histories of similar patients to infer from clinical symptoms while making prescriptions/clinical decisions and therapy schedules. The big data [81] is an initiative to develop efficient software for archiving, fast searching, retrieval, and presentation of information to the clients. It is not the volume of data alone; but rather its complexity, diversity, and timeliness are important for the DSSs that physicians or researchers often require. SLAs with CSPs have to be catered for such quality of service requirements for hosting medical repositories. The researchers and medical doctors can easily perform data mining for selected clinical conditions, symptoms, and other parameters interactively for prescribing medical treatment for patients. The EMRs for such repositories are made through procedures conforming to the HIPAA recommendations to maintain confidentiality of patient history. Software tools are capable of removing the personal information from EMRs before transferring the big data archives.

The advances in cloud services have given rise to new directions to the ICT industry through collaborative services spanning over a number of clouds called "sky computing" [82]. Data archiving in data centers and access in an integrated environment of clouds is termed *the sky*. Cooperating CSPs make trusted domains and SLAs covering all the stakeholders. Microsoft Windows has offered SkyDrive, and other giants of the ICT industry are expected to join the new technology with innovative services. At present sky computing has to address many issues and challenges to earn acceptability and entrench especially in the healthcare domain.

List of Acronyms and Abbreviations

CDSS   clinical decision-support system
CSA   cloud services architecture
CSPs   cloud service providers
DHE   distributed hospital environment
DICOM   digital imaging and communication in medicine
ECG   electrocardiography
EDITH   European Development on Indexing Techniques for Databases with
              Multidimensional Hierarchies [19]
EMG   electromyography
EPR   electronic patient records
GSR   galvanic skin response
HIPAA   Health Insurance Portability and Accountability Act [46]
HIS   hospital information systems
KBMS   knowledge-based management system
MANET   mobile ad hoc network
NEMA   National Electrical Manufacturers Association
NICE   network-independent communication environment
PACS   picture archiving and communication system
PDAs   personal digital assistants
POR   proof of retrievability
SAMC   security architecture in a medical community
SLA   service-level agreement
SpO$_2$ oxygen saturation
TPA   third-party audit

TPG   trusted platform group
TPM   trusted platform module
TVMI   trusted virtual machine instance
VSM   virtual secure machine
WAP   wireless application protocol
Wi-Fi   wireless fidelity
WiMAX   worldwide interoperability for microwave access
WPAN   wireless personal area network

References

1. J. Varia, "Cloud Architectures," Technology Evangelist Amazon Web Services, http://jinesh varia.s3.amazonaws.com/public/cloudarchitectures-varia.pdf, Accessed on Aug. 11, 2014.
2. M. Armburst et al., "Above the Clouds: Berkeley View of Cloud Computing," Adaptive Distributed Systems Laboratory, University of California, Berkeley, Feb. 10, 2009, http://www .eecs.berkeley.edu/Pubs/TechRpts/2009/EECS-2009-28.pdf, Accessed on Aug. 11, 2014.
3. S. Ponedal and M. Tucker, "Understanding decision support sytems," *J. Managed Care Pharmacy*, vol. 8, no. 2, Mar. 2002.
4. N. Shadbolt and P. Lewid, "MIAKT: Combining Grid and Web Services for Collaborative Medical Decision Making," University of Southampton, http://www.aktors.org, doi=10.1.1.121.6798, Accessed on Aug. 11, 2014.
5. L. Turner, "Medical tourism," *Can. Fam. Physician,* vol. 53, Oct. 2007.
6. K. Arabshian and H. Schulzrinne, "A SIP-Based Medical Event Monitoring System," Department of Computer Science, Columbia University, http://www.cs.columbia.edu/~knarig/sipMed .pdf, Accessed on Aug. 11, 2014.
7. R. Kohno, B. Zhen, and H.-B. Li, "Networking issues in medical implant communications," *Int. J. Multimed. Ubiq. Eng.*, vol. 4, no. 1, pp. 23–37, Jan. 2009.
8. A. Milenković, C. Otto, and E. Jonanov, "Wireless sensor networks for personal health monitoring issues and implementation," *Comp. Comm.*, vol. 29, no. 13, pp. 2521–2533, 2006.
9. R. Chow, P. Golle, M. Jakobsson, E. Shi, J. Standon, R. Masuoka, and J. Molina, "Controlling data in the cloud: Outsourcing computation without outsourcing control," CCSW '09, Chicago, Illinois, Nov. 13, 2009.
10. "Security Guidance for Critical Areas of Focus in Cloud Computing V2.1," Cloud Security Alliance, Dec. 2009, http://www.cloudsecurityalliance.org/guidance/csaguide.v2.1.pdf, Accessed on Aug. 11, 2014.
11. Z. Shen, F. Yan, and X. Wu, "Cloud computing system based on trusted computing platform," *Int. Conf. Intelligent Computation Technol. Automation*, 2010.
12. A. Koehler, "Trusted Computing: From Theory to Practice in the Real World," Utimaco Safeware AG, Oberursel, Germany.
13. A. Wood, G. Virone, T. Doan, Q. Cao, L. Selavo, Y. Wu, L. Fang, Z. He, S. Lin, and J. Stankovic, "ALARM-NET: Wireless Sensor Networks for Assisted-Living and Residential Monitoring," University of Virginia Technical Report CS-2006-13, http://www.cs.virginia.edu/~gv6f/tr06 -alarmnet.pdf and http://fiji.eecs.harvard.edu/CodeBlue, Accessed on Aug. 11, 2014.
14. S. Prueckner, C. Madler, D. Beyer, M. Kleinberger, and T. Becker, "Emergercy Monitoring and Prevention EU Project EMERGE," European Microsoft Innovation Center, Fraunhofer Institute for Experimental Software Engineering (*IESE*) Report No. 089.07E, Sept. 2007.
15. C. Otto, A. Milenković, and C. E. Jjovanov, "System architecture of a wireless body area sensor network for ubiquitous health monitoring," *J. Mob. Multimed.*, vol. 1, no. 4, pp. 307–326, 2006.

16. M. Sukor, S. Ariffin, N. Fisal, S. Yusof, and A. Abdallah, "Performance study of wireless body area network in medical environment," *Modeling & Simulation AICMS '08*, 2008.

17. P. Bobbie, C. Deosthale, and W. Thain, "Telemedicine: A mote-based data acquisition system for real time health monitoring," The Second IASTED International Conference on TELEHEALTH 2006, July 3–5, 2006, Banff, Alberta, Canada, pp. 22–28, 2006.

18. A. Marsh, S. Laxminarayan, and L. Bos, "Healthcare compunetics," *Stud. Health Technol. Inform.*, vol. 103, pp. 3–11, 2004.

19. http://www.edith.in.tum.de, EDITH project 2000.

20. G. Graschew, T. Roelofs, S. Rakowsky, and P. Schlag, "The Concept of the Virtual Hospital as ICT tool for E-Health and U-Health," University Medicine Berlin, Germany.

21. "Virtual Hospital," http://www.vh.org, Accessed on Aug. 11, 2014.

22. M. Horstmann, M. Renninger, J. Hennenlotter, C. C. Horstmann, and A. Stenzel, "Blended e-Learning in a web-based virtual hospital: A useful tool for undergraduate education in urology," *Educ. Health*, 22/2(269), 2009.

23. S. Kim and Y. Woo, "Ubiquitous community system for medical information," *Int. J. Multimed. Ubiq. Eng.*, vol. 1, no. 1, pp. 1–5, Mar. 2006.

24. P. Pandian, K. Safeer, P. Gupta, D. Shakunthala, B. Sundrsheshu, and V. C. Padaki, "Wireless sensor network for wearable physiolgical monitoring," *J. Networks*, vol. 3, no. 5, pp. 21–29, 2008.

25. E. Jonanov, D. Raskovic, J. Price, J. Chapman, A. Moore, and A. Krishnamurthy, "Patient monitoring using personal area networks of wireless intelligent sensors," *Biomed. Sci. Instrum.*, vol. 37, pp. 373–378, 2001.

26. R. Mangharam, "High-confidence platforms for personalized continuous care," *Third IEEE Int. Conf. Sensors, Mesh Ad Hoc Communications Networks*, IEEE SECON 2006, Reston, Virginia, Sept. 2006.

27. R. Mangharam, "ZipCare End-to-End Platforms for Personalized Continuous Health Care," Department of Electrical and Systems Engineering, University of Pennsylvania, Philadelphia, Presentation, Feb. 2008.

28. G. Virone, A. Wood, L. Selavo, Q. Cao, L. Fang, T. Doan, Z. He, R. Stolero, S. Lin, and J. A. Stankovic, "An assisted living oriented information system based on a residential wireless sensor network," *Proc. 1st Distributed Diagnosis Home Healthcare (D2H2) Conf.*, Arlington, Virginia, April 2–4, 2006.

29. C. Wang, Q. Wang, K. Ren, and W. Lou, "Ensuring data storage security in cloud computing," *17th Int. Workshop Quality Service IWQoS*, 2009.

30. K. Bowers, A. Juels, and A. Oprea, "HAIL: A High-Availability and Integrity Layer for Cloud Storage," RSA Laboratories, Cambridge, Massachusetts.

31. P. Feresten, "Storage multi-tenancy for cloud computing," *SNIA*, Mar. 2010.

32. F. Fedele and G. Srl, "Healthcare and distributed system technology," ANSAworks '95, Cambridge, United Kingdom, 1995.

33. I. Foster, Y. Zhao, I. Raicu, and S. Lu, "Cloud computing and grid computing 360-degree compared," *Grid Computing Environments Workshop, GCE '08*, 2008.

34. Amazon Elastic Compute Cloud (EC2), http://www.amazon.com/ec2, Accessed on Aug. 11, 2014.

35. Google App Engine, http://appengine.google.com, Accessed on Aug. 11, 2014.

36. Windows Azure Platform, http://windowsazure.com, Accessed on Aug. 11, 2014.

37. K. Pijanowski, "IaaS, PaaS, and the Windows Azure Platform," Microsoft, Oct. 2009.

38. S. Wardley, E. Goyer, and N. Barcet, "Ubuntu Enterprize Cloud Architecture," Technical white paper, Aug. 2009.

39. "Aneka: Enabling .NET-Based Enterprise Grid and Cloud Computing," http://www.manjrasoft .com/products.html, Accessed on Aug. 11, 2014.

40. Apache Software Foundation information at http://hadoop.apache.org, Accessed on Aug. 11, 2014.

41. C. R. Baker, K. Armijo, S. Belka, M. Benhabib, V. Bhargava, and N. Burkhart, "Wireless sensor networks for home health care," *21st Int.Conf. Advanced Information Networking Applications Workshops*, AINAW '07, 2007.

42. D. Garets and M. Davis, "Electronic Patient Records, EMRs and EHRs," Oct. 2005, http://www.providersedge.com/ehdocs/ehr_articles/electronic_patient_records-emrs_and_ehrs.pdf, Accessed on Feb. 2, 2015.
43. S. Ahmed and M. Y. A. Raja, "Telemedical sensor networks and informatics for healthcare services," *HONET 2009*, Alexandria, Egypt, Dec. 2009.
44. H. Abangar, P. Barnaghi, K. Moessner, A. Nnaemego, K. Balaskandan, and R. Tafazolli, "A service oriented middleware architecture for wireless sensor networks," *Future Network & MobileSummit 2010 Conf. Proc.*, 2010.
45. T. Hsia and L. Lin, "A framework for designing nursing knowledge management systems," *Interdisciplinary J. Information, Knowledge, Management*, vol. 1, p. 13, Jan. 2006.
46. "Protecting Personal Health Information in Research: Understanding the HIPPA Privacy Rule", Department of Health and Human Services, USA, NIH Publication Number 03-5388 http://www.hhs.gov/ocr/hippa.
47. J. Guptill, "Knowledge management in health care," *J. Health Care Finance*, vol. 31, no. 3, pp. 10–14, 2005.
48. H. Owaied, M. Abu-A'ra, and H. Farhan, "An application of knowledge-based system," *Int. J. Comput. Sci. Network Security*, vol. 10, p. 208, Mar. 2010.
49. C. Mubaraka, M. Salisu, and K. Priscilla, "Towards knowledge based medical prescription system in health sector in Uganda: A case of MBUYA military hospital," *Global J. Commerce Management Perspective*, vol. 2, no. 3, pp. 136–143, 2013.
50. K. Lorincz, D. Malan, T. Fulford-Jones, A. Nawoj, A. Clavel, V. Shnayder, G. Mainland, M. Welsh, and S. Moulton, "Sensor networks for emergency response: Challenges and opportunities," *IEEE Pervasive Comput.*, vol. 3, no. 4, pp. 16–23, Oct.–Dec. 2004.
51. S. Garg, "Efficient data sharing and its application in mobile ad hoc networks (MANET)," *J. Info. Sys. Commun.*, vol. 3, no. 1, pp. 96–101, 2012.
52. J. R. Swati and K. Priyanka, "Wireless sensor network (WSN): Architectural design issues and challenges," *Int. J. Comp. Sci. Eng.*, vol. 2, no. 9, pp. 3089–3094, 2010.
53. DICOM (Digital Imaging and Communications in Medicine), http://medical.nema.org, Accessed on Aug. 11, 2014.
54. R. Choplin and J. Boehme, "Picture archiving and communication systems: An overview," *Radiographics*, vol. 12, no. 1, pp. 127–129, Jan. 1992.
55. J. Saeed and H. Asim, "SENSITIVE—Medical sensor network for monitoring and recording physiological data on a nursing console," FYP 2010, Computer and Software Engineering Department, Bahria University, Islamabad, Pakistan, 2010.
56. A. J. Duerinckx and E. J. Pisa, "Filmless picture archiving and communication system (PACS) in diagnostic radiology," *Proc. SPIE*, vol. 318, pp. 9–18, 2003.
57. T. Ristenpart, E. Tromer, H. Shacham, and S. Savage, "Hey, you, get off of my cloud: Exploring information leakage in third-party compute clouds," *Computer and Communications Security—CCS 2009*.
58. R. Hasan, R. Sion, and M. Winslett, "Introducing secure provenance: Problems and challenges," *StorageSS '07 Proc. 2007 ACM Workshop on Storage Security Survivability*, pp. 13–18, 2007.
59. A. Barth et al., "Robust defenses for cross-site request forgery," Proc. 15th ACM Conf. Comput. Communications Security CCS '08, pp. 75–88, 2008.
60. "Amazon Web Services: Overview of Security Processes," June 2014, http://aws.amazon.com/security.
61. G. Fox, O. Balsoy, S. Pallickara1, A. Uyar, D. Gannon, and A. Slominski, "Community grids," *Int. Conf. Computational Sci.—ICCS 2002*. Vol. 2329, pp. 22–38, 2002.
62. S. Kinney, *Trusted Platform Module Basics: Using TPM in Embedded Systems*, Newnes Publications, 2006.
63. R. Lu, X. Lin, X. Liang, and X. Shen, "Secure provenance: The essential of bread and butter of data forensics in cloud computing," *Proc. 5th ACM Symp. Information, Comput. Communications Security ASIACCS '10*, pp. 282–292, 2010.

64. Q. Wang, C. Wang, J. Li, K. Ren, and W . Lou, "Enabling public verifiability and data dynam-ics for storage security in cloud computing," *Proc. 14th European Conf. on Research in Computer Security*, ESORICS '09, IIT, Chicago, Illinois, pp. 355–370, 2009.

65. F. Naseem and Z. Manzoor, "Patient Monitoring via Wireless Nodes and LAN," FYP 2012, Computer and Software Engineering Department, Bahria University, Islamabad, Pakistan.

66. W. Stallings, "IP security," *The Internet Protocol Journal*, vol. 3, no. 1, Mar. 2000, http://www.cisco.com/ipj, Accessed on Aug. 11, 2014.

67. "Amazon Elastic MapReduce: Developer Guide," http://aws.amazon.com/ElasticMapReduce/latest/DeveloperGuide/emr-what-is-emr.html, 2010, Accessed on Feb. 2, 2015.

68. "Creating HIPAA-Compliant Medical Data Applications with Amazon Web Services," April 2009, http://aws.amazon.com, Accessed on Feb. 2, 2015.

69. W. Belmans, S. Puopolo, and S. Yellumahanti, "Network Service Providers as Cloud Providers," Cisco Internet Business Solutions Group (IBSG); Nov. 2010; www.cisco.com/web/about/ac79/docs/wp/sp/Third_Generation_WP_Find.pdf, Accessed on Feb. 2, 2015.

70. S. Ahmed, A. Abdullah, and M. Y. A. Raja, "Cyber security: Vulnerabilities and challenges in telemedicine over cloud," Fourth Networked Digital Technologies 2012, Dubai, United Arab Emirates, 2012.

71. N. Ain and M. Gul, "Web portal TELEMED for a doctor," FYP 2011, Computer and Software Engineering Department, Bahria University, Islamabad, Pakistan.

72. M. Raza, O. Sarwar, and A. Alvi, "Design and implementation of EEG data acquisition module," FYP 2011, Computer and Software Engineering Department, Bahria University, Islamabad, Pakistan.

73. P. Kumar, Y. D. Lee, and H. J. Lee, "Secure Health Monitoring Using Medical Wireless Sensor Networks," Department of Ubiquitous IT, Graduate School of Design & IT, Dongseo University, Busan, South Korea.

74. J. Adams, "Building Low Power into Wireless Sensor Networks Using ZigBee Technology," Industrial Embedded Systems Resource Guide, Freescale Semiconductor, 2005.

75. A. Jangra, Swati, Richa, Priyanka, and Kurukshetra, "Wireless sensor network (WSN): Architectural design issues and challenges," *Int. J. Comp. Sci. Eng.*, vol. 2, no. 9, pp. 3089–3094, 2010.

76. V. Shnayder, B. Chen, K. Lorincz, T. F. Jones, and M. Welsh, "Sensor Networks for Medical Care," Technical Report TR-08-05, Division of Engineering and Applied Sciences, Harvard University, 2005.

77. D. Vouyioukas and I. Maglogianni, "Communication Issues in Pervasive Healthcare Systems and Applications," book chapter, http://www.icsd.aegean.gr/dvouyiou/dmdocuments/pubs/BOOK%20CHAPTERS/IGI/978-1-61520-765-7.ch010.pdf, Accessed on Aug. 11, 2014.

78. H. Zaidi, "Medical Imaging: Current Status and Future Perspectives," Division of Nuclear Medicine, Geneva University Hospital, Geneva, Switzerland, 2001.

79. J. Varia, "Amazon Web Services: Architecting for the Cloud: Best Practices," Jan. 2010.

80. J. Varia, *Cloud Computing: Paradigms and Patterns*, John Wiley & Sons, Inc., 2010.

81. P. Groves, B. Kayyali, D. Knott, and S. Kuiken, "The Big Data Revolution in Healthcare: Accelerating Value and Innovation," Business Technology Office, Center for US Health System Reform, McKinsey & Company, http://www.mckinsey.com/insights/health_systems_and_services/the_big-data_revolution_in_us_health_care, Accessed on Feb. 2, 2015.

82. K. Keahey, M. Tsugawa, A. Matsunaga, and J. Fortes, "Sky computing," *IEEE Internet Comp.*, vol. 13, no. 5, pp. 43–51, Sept. 2009.

4

Intelligent Electronic Health Systems

David A. Clifton, Marco A. F. Pimentel, Katherine E. Niehaus, Lei Clifton,
Timothy E. A. Peto, Derrick W. Crook, and Peter J. Watkinson

CONTENTS

4.1 Introduction

Healthcare systems worldwide are entering a new phase: ever-increasing quantities of complex, massively multivariate data concerning all aspects of patient care are starting to be routinely acquired and stored [1], throughout the life of a patient. This exponential growth in data quantities far outpaces the capability of clinical experts to cope, resulting in a so-called data deluge, in which the data are largely unexploited. There is huge potential for using advances in large-scale machine learning methodologies* to exploit the contents of these complex data sets by performing robust, scalable, automated inference to improve healthcare outcomes significantly by using patient-specific probabilistic models, a field in which there is little existing research [2] and which promises to develop into a new

---

* Sometimes termed *"big-data"* methods.

industry supporting the next generation of healthcare technology. Data integration across spatial scales, from molecular to population level, and across temporal scales, from fixed genomic data to a beat-by-beat electrocardiogram (ECG), will be one of the key challenges for exploiting these massive, disparate data sets.

Electronic health records (EHRs) are being rapidly adopted for use by healthcare providers, particularly in hospital settings. These EHRs typically contain heterogeneous physiological data that have been deposited throughout routine care of patients and which may be of varying provenance and fidelity. These disparate data sets often represent a wide range of data types, acquired across a wide range of measurement scales; for example, we might encounter the following:

- Background demographic data, including categorical data (sex and ethnicity), discrete data (number of admissions to hospital and severity scores for long-term conditions), and continuous data (weight and age)

- Diagnostic data, describing long-term or acute conditions, and other similar clinical indicators

- Surgical and treatment data, describing interventions and their details

- Pharmacological data, describing treatments prescribed

- Imaging and radiological data of varying resolutions

- Low-rate physiological data, such as those measurements of the vital signs made manually by clinicians during routine clinical care of a patient (perhaps several times per day)

- Higher-rate physiological data, such as those parameters recorded by bed-side monitors and ICU monitoring systems (perhaps several times per minute)

- Very high-rate physiological data, such as waveforms acquired by sensors (perhaps many times per second)

- Genomic data and proteomic data, which can be massively multivariate and which are often represented in binary or discrete format

Current interest [3–4] focuses on linking such disparate databases within the EHR by using conventional statistical methods. However, conventional methods do not scale to these incomplete, noisy, terabyte-scale, massively multivariate data sets. Additionally, the contents of these data sets, when linked together, may offer contradictory information concerning the patient, and these contradictory data typically confound conventional statistical methods. This chapter describes the construction of complex multiscale models for use with EHRs, for so-called *intelligent EHRs*, and their use in the extraction of clinically useful information from these very large healthcare data sets for the improvement of patient outcomes.

Models can be built using either a bottom-up approach (for example, using stochastic processes for modeling multivariate time-series data) or a top-down approach based on massively multivariate probabilistic models. In both cases, we describe how learning can take place within a machine learning (and, ideally, Bayesian) framework, which provides the optimal approach for quantifying the uncertainty associated with the noisy and missing data typical of large healthcare data sets.

4.1.1 Objectives

The primary objective of this overlapping area of medical and engineering research is to develop novel, automated methods for improving patient outcomes by exploiting the contents of the EHR (including genomic data), augmented with multivariate high-rate data acquired from wearable sensors and other hospital devices. Researchers in this field aim to, for example, (i) demonstrate a reduction in adverse outcomes for high-risk patients, including outcomes such as cardiac arrests, unscheduled admission to ICU, and death; (ii) provide automatic stratification of patients according to perceived risk or efficacy of treatment, allowing more effective use of healthcare resources; and (iii) construct novel patient-specific models that allow treatments to be tailored to the physiological condition of the individual patient, rather than relying on population-based, generic models.

These exemplar primary objectives will be, in the work described by this chapter, addressed in two complementary ways, which represent the two extremes of EHR analysis: (i) performing analyses across low-rate, but massively multivariate, data sets in the EHR, incorporating genomic and demographic background data, and (ii) performing online dynamical modeling of multivariate, high-rate time series of physiological data from patients who are at risk of deterioration, such as those in acute hospital wards.

4.1.2 Themes Considered in This Chapter

We will introduce exemplar research themes within the field of intelligent electronic health systems, considering a number of case studies to demonstrate the potential of such research:

- Theme I: using the broad range of data sets within the EHR, for improving understanding of infectious disease
- Theme II: augmenting the EHR with sensor data, for continuous monitoring of high-risk ambulatory patients
- Theme III: EHRs in the developing world, for improving access to affordable healthcare

4.2 Theme I: Using the Broad Range of Data Sets within the EHR

EHRs have great potential for improving our understanding of, and ability to tackle, large-scale problems in a manner previously impossible. One such example is the emerging global threat posed by infectious disease. The chief medical officer (CMO) of the UK noted that "infectious disease is as great a threat to national security as climate change" (with more than one new life-threatening disease identified each year) and that "the challenge in identifying future threats is not the acquisition of data, but in using these huge databases with new computer science methods" [5]. Tuberculosis infects one in three of the world's population and claims over one million lives per year; deaths from infectious disease, including methicillin-resistant *Staphylococcus aureus* (MRSA) and *Escherichia coli*, account for as many deaths as road traffic accidents in Europe each year [6]. As the CMO also observed recently, no new general classes of antibiotics have been discovered since 1990, and resistance to existing antibiotics is far outpacing our ability to produce new drugs. This high death rate is exacerbated by the fact that conventional methods for identifying pathogens take up to 6 *weeks* to perform. There is, therefore, an urgent need to improve our ability to fight infectious disease, and it is in this respect that EHRs can play a key role.

EHRs provide the data required to develop predictive models that, once validated, can be used in clinical support systems (i.e., an intelligent EHR) to provide real-time information for clinicians. Such models will also inform our scientific understanding of patient response to bacterial infection and appropriate treatment strategies. The benefit of EHRs will only be further compounded when gene-sequencing platforms become more widely available for routine hospital sequencing of bacterial specimens [6]. Sequencing the bacterial genome provides immediate benefit to the patient by enabling quick identification of the bacterial species and known mechanisms of drug resistance, as well as enabling surveillance of hospital-acquired infections. Combining both genomic and patient-based data sources from the EHR will allow for prediction of changes to bacterial virulence, patient risk, drug resistance, etc. This interplay is illustrated in Figure 4.1.

Whole-genome sequencing (WGS), together with information typically contained in the EHR, such as patient admission, length of stay, and movement within the hospital, has already been used to investigate the spread of infectious outbreaks within the hospital [7]. Tools such as phylogenetic tree-building, which maps the relationship between bacterial isolates based upon their genetic differences, can illustrate whether cases of infection are most likely being spread within the hospital or if they came from outside the community. Such analysis is typically retrospective in nature. Several recent proof-of-concept studies have also illustrated the potential of using WGS to identify known resistance-conferring mechanisms within the bacterial genome [8–9]. This approach has been shown to have very competitive performance for detecting resistance with a minimal number of false alarms, as assessed against gold-standard phenotypic techniques.

Early prospective prediction of changes to bacterial virulence also becomes possible through the use of intelligent EHR systems. As described in Section 4.1, EHR data sets also typically contain information such as patient laboratory test results and medications

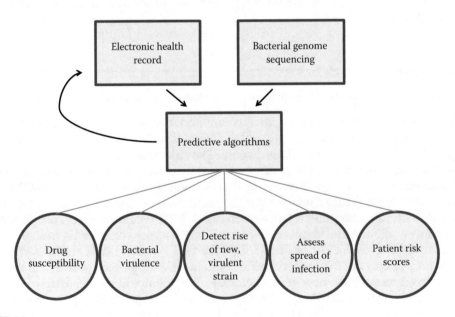

FIGURE 4.1
Illustration of the feedback process implicit in using an intelligent EHR system. The EHR provides the data required to develop predictive algorithms (with outputs shown in circles), which can then be used as clinical support tools while continuing to be modified as new data are obtained.

prescribed. Using machine learning approaches, these disparate sources of information can be combined to form an underlying model of "normal" patient characteristics during a bacterial infection. Subtle departures from a normal trajectory, which may be difficult for clinicians to identify in the course of routine care, can be detected through such modeling techniques. This can allow the early detection of unusually virulent strains of bacteria and necessary escalations in treatment and infection containment. An applicable branch of machine learning for constructing such models of normality is that of novelty detection, particularly, when examples of abnormality are few or when abnormality is poorly understood [10–11].

Patient risk scores can also be determined through similar modeling techniques. The association between a bacterial genotype and its virulence can be established using supervised machine learning [12]; this type of analysis will be described in the case study to follow. Once a patient has become infected by a bacterial infection and the bacterial genome has been obtained, the predicted virulence level could be used to produce a baseline patient risk score. As a patient remains in the hospital and more clinical information, such as additional lab test results, becomes available, the risk score can be updated to provide a real-time indicator of his or her infection severity.

4.2.1 Case Study: Prediction of Bacterial Drug Susceptibility

We illustrate how information from an intelligent EHR system, including bacterial genome sequencing, can be used to develop a predictive algorithm for the bacterial phenotype. We will examine *Mycobacterium tuberculosis* (MTB), a bacterial species with a very stable genome. As a report from the World Health Organization warns, bacterial drug resistance is becoming an imminent threat to global healthcare systems [13], making appropriate antibiotic prescription essential. While genetic mechanisms underlying resistance are well-known for some antibiotics, others are less well understood. For example, isoniazid is a drug with very well-characterized resistance mechanisms, and yet it is estimated that up to 20% of isoniazid drug resistance remains unexplained by known genotypic factors [14]. Furthermore, drug resistance using conventional phenotypic techniques (i.e., growing samples in a culture impregnated with antibiotics) can take up to 2 months for slow-growing bacteria such as MTB. Genome sequencing, with results available in the range of a few hours to a few days, offers a much faster alternative. Algorithms that can quickly and automatically process the entire sequenced bacterial genome to produce a resistance prediction would be very useful in clinical practice and are the focus of much current research. We therefore, consider drug resistance as our outcome phenotype of interest in the following example.

There may be interactions between different genomic mutations, and machine learning provides an appropriate and principled method for examining the relationship between the genetic pattern of variation and the associated drug-resistance profile. Here, we explain how features may be generated from the genome and provide examples of machine learning classification algorithms that underpin intelligent EHRs in this context. We will describe an example set of data containing 1800 MTB isolates from the Midlands, UK. Figure 4.2 illustrates the steps involved in EHR-based analysis, which are described in detail below.

4.2.2 Features

With genomic data being increasingly present in EHRs, it is important for data scientists to understand the provenance and process by which such data are obtained, and to gain an

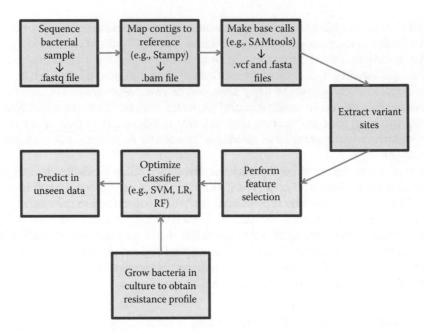

FIGURE 4.2
Illustration of the analysis steps that are involved in the process of predicting a bacterial phenotype from geno-type data. The main text describes genomic terms used (e.g., *contig*).

appreciation of the vocabulary used in this discipline. A "read" of a genomic sequence will typically contain thousands of unaligned contiguous regions of DNA (also termed *contigs*). As MTB has a very stable genome, these contigs can be aligned to a reference MTB sequence (such as the Hv37Rv reference), using software such as Stampy [15]. Following this align-ment, bases can be determined ("called") using SAMtools [16] or similar software. There will be some regions in the genome that are sequenced with poor quality; this may be due to a small number of reads available, or repeated regions of DNA in the genome. Such bases are represented in the sequencing output as "null calls."

As our goal in this example is to link genomic variation to the *in vitro* phenotypic response of the bacteria to antibacterial drugs, as described in the EHR, it makes sense to extract all of the sites where a single bacterial isolate differs from the reference. These differences between the isolate and the reference genome are possible mechanisms for resistance, and these variant sites are termed single-nucleotide polymorphisms (SNPs). As a first step in this example, we focus on 23 genes that are suspected of being involved in antibiotic resistance. For any given MTB strain, there are an average of 5 SNPs across these 23 genes. Many of these are "private" SNPs, meaning that they are not shared by other isolates in the sample. In this EHR example, with a data set of 1800 isolates, there were 1621 variant sites in total, including null calls. Removing null calls that most likely corre-sponded to the reference sequence (based upon the majority proportion of reads when the call is uncertain), and removing private mutations, resulted in 301 variant sites remaining for analysis. These SNPs will compose our feature set for the subsequent example.

We note in passing that, while MTB represents one extreme as a stable, clonal bacteria (in the sense that one generation looks much like the last), other examples such as *E. coli* and *Klebsiella pneumoniae* are much more promiscuous. These bacteria tend to gain and lose plasmids, and therefore genes, very easily between successive generations. With such organisms, we might choose, therefore, a *de novo* assembly of the sequence, rather than comparison against a reference genome. Tools such as Velvet [17] may be used to perform this *de novo* assembly for entities with such widely-varying genomes. For these bacteria, alternative feature options, such as using gene presence/absence, are often more appropriate: gene presence/absence involves the alignment of genes to known bacterial genes by using the Basic Local Alignment Search Tool (BLAST) [18]. Alternatively, a *k*-mer approach may be used, which steps through the sequence, defining each set of *k* bases (*k* is usually between 20 and 40) to be a feature. The *k*-mer approach is beneficial in that it should capture both gene presence/absence and SNP changes in the sequence. However, it leads to a very large number of features, which can be difficult to interpret. The *k*-mer approach has been used within a genome-wide association study (GWAS) context [19]. As presented here, SNPs, *k*-mers, and genes are all binary features.

4.2.3 Supervised Learning Algorithms for the EHR

From the EHR, we can also obtain a set of phenotypic labels defining whether our isolates are resistant or susceptible to first-line tuberculosis antibiotic drugs. MTB phenotypic testing typically involves an initial screen for resistance by growing the bacteria (obtained from patient sputum) in liquid culture mycobacterium growth indicator tubes (MGITs). Positive samples are confirmed by growing the bacteria on Löwenstein–Jensen (LJ) media (LJ slopes) impregnated with antibiotics. This presents us with a supervised classification task: we have a set of features (our SNPs) to describe our bacteria and a set of labels (our phenotypic test results from the EHR) that we wish to predict. Supervised machine learning classification algorithms such as the support vector machine (SVM) and random forests (RFs) have been used for performing this classification.

We will consider a subset of N isolates to represent a training set of examples $x_1, \ldots, x_N$ with labels $\ell_1, \ldots, \ell_N, \ell_i \in \{-1, 1\}$, with 1 indicating drug resistance for a given drug and -1 indicating susceptibility. Each example x_i is composed of a vector of J binary features indicating the presence ($x_{ij} = 1$) or absence ($x_{ij} = 0$) of a given SNP.

The SVM is a classification algorithm that attempts to separate two groups by the widest margin possible. The hyperplane defining this separation is determined by maximizing the distance between it and the closest training points from each class, which are termed the *support vectors*. Through the so-called kernel trick, an SVM can be used to project data into a high-dimensional space, in which the classes may be linearly separable. Common kernels include the linear kernel and the radial basis function (i.e., Gaussian distance) kernels. We defer to Bishop [20] for a detailed description of this well-understood algorithm.

RFs are ensemble learners, which means that the random forest prediction is based upon the votes of a committee of "weak" base learners. The base learner for a RF is a decision tree. Each of the decision trees in the random forest is formed from a (random) subset of features and a subset of examples from the data set. After all of the trees have been built, the classifier's prediction is based upon majority voting of the trees. For problems involving genomic loci as features, building 40–400 trees and using a random selection of half of the features have been found to be useful starting points [21].

There are many additional types of classifiers that have been used for EHR-based analysis in this field. For instance, logistic regression is a linear discriminative classifier that provides relatively easily interpretable weightings of features, and which is a commonly-used tool in conventional medical statistics. The Bayesian product-of-marginals (BPM) model is a generative classifier that assumes that the input features are independent, conditional upon the class. Although this is a strong assumption, a fully Bayesian treatment of BPM also provides a probabilistic distribution over the probability of each SNP feature belonging to each class. It is often beneficial to compare the predictive performance across these different classifiers to understand how well the assumptions of each (e.g., linear combinations of features and independence of the features) are substantiated in the data. More details concerning such algorithms and their assumptions can be found in Goldstein et al. [21].

4.2.4 Feature Selection

Using all variant sites across the genome, or a subset of the genome, results in a large number of features. Many of these SNPs may be irrelevant to the classification problem at hand. It may, therefore, be desirable to perform some form of feature selection to obtain a smaller, more parsimonious feature set, which often results in better-performing classifiers. Preprocessing steps that are often performed in such studies include removing features found within fewer than 5% of the isolates in the sample and removing features that appear to be in linkage disequilibrium (i.e., they co-occur together in samples more often than would be expected by chance) [22–23].

Following any such preprocessing steps, there are three main categories of feature selection methods: filters, wrappers, and embedded methods. Filtering methods use criteria to select features, an independent step that is taken before performing classification. An example of a filtering method is the identification (and removal) of SNPs that are not significantly associated with the outcome of interest (for instance, using a chi-squared test and an adjusted p value for multiple comparisons). Wrapper methods use the machine learning algorithm itself to select relevant variables. A common example is SVM-based recursive feature elimination, in which a linear SVM is trained on a set of features and those that have the lowest weightings are removed, and the process repeats until predictive performance decreases by some margin.

Embedded methods refer to classifiers that have some sort of automatic relevance determination (ARD) for features incorporated within the classification algorithm itself. For instance, the *least absolute shrinkage and selection operator* (LASSO) regularization method for logistic regression shrinks feature weightings down to zero if they are not found to be important in determining a classification. From a Bayesian perspective, this is equivalent to putting a zero-mean Laplacian prior on the feature weightings, meaning that the prior assumption is that the feature is not important until the training data show otherwise [24].

Feature selection becomes even more important when moving from an initial subset of genes to looking at SNPs found across the entire genome, at which point the number of features begins to require increasingly prohibitive computational resources.

4.2.5 Generalization

As in any machine learning setting, it is important to avoid overfitting the learned parameters to the training data set. Performance is commonly assessed by training on a subset of data (e.g., 80%) and testing on the remainder. Parameter tuning (as is required for SVM classifiers) can be accomplished by performing a grid search over the range of expected parameter values within cross validation folds of the training set. Alternatively, Bayesian methods (as will be discussed in Section 4.3) often incorporate penalty terms within the training procedure that aim to avoid overfitting.

4.2.6 Summary of Theme I

Machine learning techniques can be used to create predictive classifiers for drug resistance based upon data contained within the EHR, including genomic data. A machine learning approach provides a predictive outcome, which is in contrast to traditional approaches, which are designed primarily to find genomic variants that have strong linear associations with the desired phenotype. Machine learning approaches also have the capability to identify important, possibly causative, variants through feature selection methods, although the interpretability of such weightings typically depends upon the type of classifier used. Furthermore, machine learning classifiers are well-equipped to handle nonlinear interactions between features. In the context of genomic data within the EHR, in which complex regulatory mechanisms are biologically plausible, this is an advantageous approach.

As this case study has illustrated, bacterial phenotype profiles of resistance from the EHR can be combined with the bacterial genome sequence to develop predictive algorithms for drug resistance. Once established, such an algorithm would ideally be continually updated as new data are collected; this is possible through automated EHR systems. Similar methods can be employed to predict other bacterial phenotypes, such as virulence. Ongoing research in this active field of EHR-based work will combine these phenotypes with clinical data to provide patient risk scores and identify the rise of new, more virulent strains.

4.3 Theme II: Augmenting the EHR with Sensor Data

As we have described in Section 4.1, the increasing use of electronic health records in healthcare systems, especially in hospitals, results in the acquisition and storage of large quantities of patient-confidential data [25]. Typically, the size and heterogeneity of these data mean that only elementary analysis is undertaken in most existing systems. Furthermore, there is a trend toward augmenting the EHR with diagnostic data from point-of-care devices, and physiological measurements (such as blood pressure, heart rate, and temperature) acquired from EHR-compatible medical devices. Automated methods for modeling and analyzing the data in these augmented EHRs are urgently required, such that clinicians may be provided with the results of inference for decision support. This theme introduces a representation of trajectories for the time-series data in the EHR, using Gaussian process (GP) regression, which may be used for the recognition of normal and abnormal patterns by generating a trajectory that provides representative physiological trends, even though the training examples may be unevenly-sampled and noisy. The latter are key factors that must be dealt with in the analysis of real EHR data and which Bayesian systems, in particular, are well-equipped to address.

4.3.1 Case Study: Early-Warning Systems

An estimated 20,000 patients have unplanned admissions to the ICU in the UK each year, which could be avoided if physiological deterioration was identified early [26]. Such patients are at a high risk of morbidity and mortality (and have a 40% higher mortality rate than planned ICU admissions). The delay in detection of physiological deterioration is exacerbated by the fact that acute patients outside the ICU typically have their physiological data observed every 2–4 h [27]. There is a need for reliable, continuous data analysis between nurse observations, to provide early warning of deterioration and improve outcomes. Existing techniques [28] are limited by (i) not taking into account the dynamics of the time series of the physiological data in the EHR and (ii) comparing all patients (regardless of demographics or physiology) to a model constructed from a global population. This theme will illustrate the use of patient-specific probabilistic time-series analysis by using EHR data. Our description will cover Bayesian nonparametric processes, which offer a robust and principled framework for performing inference in the presence of incomplete and noisy data, as is typical for data within the EHR.

We here introduce this theme with a framework for probabilistic analysis of time-series data within the EHR, which may be used for the *functional* characterization of vital-signs trajectories; that is, we will treat the time series within the EHR as being whole *functions*, rather than sequences of individual data points. We will demonstrate the utility of this approach by using Gaussian process models for discovering clusters in the functional data in an unsupervised manner, such that "prototype functions" corresponding to known and unknown modes of physiological behavior of the physiology of postoperative patients may be revealed in the EHR. We demonstrate that our approach may be used to discriminate between "abnormal" trajectories corresponding to patients who deteriorate physiologically and are admitted to a higher level of care and those belonging to patients with a normal recovery. Such systems are immediately useful for providing early warning of deterioration with EHR data.

4.3.2 Estimating Vital Signs with Probabilistic Models

EHRs augmented with high-rate waveform data from patient-worn sensors are becoming possible with the increasing prevalence of high-bandwidth wireless networks in hospitals and other healthcare environments. For example, the ECG is now available from disposable "sticking plaster" patches, which may then be transmitted to a central EHR system, while the photoplethysmogram (PPG) is available from lightweight mobile sensors [29]. Both sensors typically provide point estimates of a number of derived physiological variables, such as respiratory rate (one of the key indicators of impending physiological derangement), because both the ECG and PPG are modulated in several ways by the respiratory process [30]. However, the resulting estimates have no probabilistic interpretation and are typically considered to be frequently artifactual, limiting the efficacy of their use when combined with other data in the EHR, which is our ultimate goal in this theme.

Taking a probabilistic approach to estimating the vital signs, we can assume the modulation of these waveforms to take the form of a Gaussian process [31–32]. We consider the regression model $y = f(x) + \epsilon$, which expresses a dependent variable y in terms of an independent variable x via a latent function $f(x)$ and a noise term $\epsilon \sim N(0, \sigma^2)$. The function f can be interpreted as being a probability distribution over functions, $y = f(x) \sim GP(m(x), k(x, x'))$, which is a GP, and where $m(x)$ is the mean function of the distribution and k is a covariance function which describes the coupling between two values of the independent variable as a function of the (kernel) distance between them [33]. A process is a GP if the joint distribution of the output variable (e.g., respiratory rate) at different time points is multivariate Gaussian.*

---

* Note that this is not assuming that the time series of, for example, heart-rate values are drawn from a Gaussian distribution—only that the joint distribution over all time points is Gaussian. The practical consequence of this simplifying assumption is that the predictive distribution of the heart rates at any time point is univariate Gaussian, but where we note that the mean and variance of that Gaussian may vary with time.

The nature of the GP is such that, conditional on observed data, predictions can be made about the function values y^* at any "test" location of the index set x^*. We hereafter assume that the index set corresponds to time.

The posterior density for a test point x^* is Gaussian,

$$y^* \mid x^*, x, y \sim N\left(\overline{f^*}, \mathrm{var}[f^*]\right), \tag{4.1}$$

where the mean and variance are given by

$$\overline{f^*} = k(x^*, x^*)^T k(x, x)^{-1} y, \tag{4.2}$$

$$\mathrm{var}[f^*] = k(x^*, x^*) - k(x, x^*)^T k(x, x)^{-1} k(x, x^*), \tag{4.3}$$

respectively. Here, we have assumed that the mean function is zero for simplicity, which is commonly assumed to be true.

The covariance function encodes our assumptions concerning the structure of the time series that we wish to model. There exists a large class of well-suited covariance functions, many of which are parameterized by a length-scale parameter (which determines the typical timescale over which the time series varies) and an amplitude (which determines the typical amplitude of deviation from the mean).

A key advantage of the Bayesian framework is our ability to incorporate the significant quantity of prior clinical knowledge that we may have. In the case of estimating respiratory rate, our prior knowledge of the periodicity of the respiratory effect (in this case, the range of values of respiratory rate that are feasible for a patient) and the rate at which respiratory rate can be expected to change may be encoded within an appropriate covariance function,

$$k(r) = \sigma_0^2 \exp\left\{-\frac{\sin^2\left[(2\pi/P_L)r\right]}{2\lambda^2}\right\}, \tag{4.4}$$

in which the hyperparameters σ_0 and λ of the Bayesian model give the amplitude and length scale of the latent respiratory function (where r is the Euclidean distance between two time points of the input waveform). P_L is the length of the period and is the key parameter for estimation of the respiratory rate. Taking a fully Bayesian approach, we assume the log posterior distribution of the hyperparameters to be multivariate Gaussian, allowing us to arrive at a distribution over P_L and, thus, obtain a fully probabilistic estimate of the respiratory rate [34]. An example is shown in Figure 4.3, which shows an ECG time series, obtained from wearable sensors connected to an EHR and where the amplitude modulation caused by respiratory rate is explicitly modeled using a Gaussian process.

The resulting probabilistic estimates of the vital signs may subsequently be used by further EHR-based models that fuse multiple vital signs with other data from the EHR, as described below.

4.3.3 Learning Data Trajectories

With probabilistic estimates of physiological data obtained where possible, as described above, we can subsequently combine these data with other time series available in the

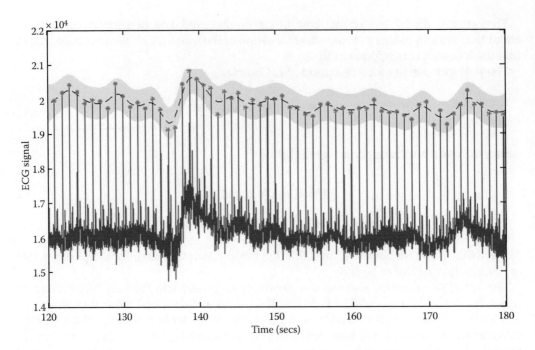

FIGURE 4.3
An example of the probabilistic estimation of the vital signs from waveform data in an augmented EHR in which the respiration process modulates the ECG (dark grey solid lines). The mean function of the Gaussian process (black dashed line) with a 2σ interval around the mean (light grey and shaded area) is superimposed over the peaks of the ECG (asterisks), which describe the respiratory waveform. One of the governing hyperparameters of the GP model used in this figure is P_L, the respiratory rate, over which we have a distribution describing our confidence in the value of respiratory rate provided.

EHR (such as manual clinical observations of the vital signs) by describing a Gaussian process over a collection of the variables. We have previously proposed an extension of extreme value statistics (which is typically used to estimate the distribution of extrema for uni- or bivariate data spaces) to the infinite-dimensional case of functional data. We first construct a Gaussian process with hyperparameters of a squared-exponential covariance function optimized according to maximizing the joint likelihood of the training data, where the training data are examples of "normal" time series. We can then define an "extreme function distribution" over the functional space corresponding to that Gaussian process [35], which effectively allows us to determine if a given function (time series) was generated from the normal model or not according to some given threshold probability. We have shown that time-series EHR data can be classified in this manner, allowing us to construct models of normality based on multivariate physiological time-series data, and then use these to detect potentially abnormal time series.

According to physiological understanding and results from previous analyses of vital-signs data [36], two dominant length scales are apparent in the EHR data: the first corresponds to physiological changes from one day to the next and the second is associated with the periodicity of the data, known as the circadian rhythm, i.e., the variability between daytimes and nighttimes. Analysis shows that these effects are additive [36] and, hence, can be modeled by a sum of two covariance functions. Denoting $r = ||x_p - x_q||$ as the Euclidean distance between two independent variables x_p and x_q, we model day-to-day variations by using a squared-exponential kernel,

$$k_L(r) = \sigma_L^2 \exp\left(-\frac{r^2}{2\delta_L^2}\right),$$ (4.5)

with length scale δ_L and amplitude σ_L. For modeling daily variability we multiply by a covariance function which is periodic:

$$k_S(r) = \sigma_S^2 \exp\left(-\frac{r^2}{2\delta_S^2}\right) \exp\left\{-\frac{\sin^2\left[(2\pi/P_L)r\right]}{2}\right\},$$ (4.6)

with length scale δ_S, amplitude σ_S, and period length P_L. Combining both models by using the sum of covariance functions, we obtain

$$k(x_p, x_q|\theta_k) = k_L(x_p, x_q|\sigma_L, \delta_L) + k_S(x_p, x_q|\sigma_S, \delta_S, P_L),$$ (4.7)

where $\theta_k = \{\sigma_L, \sigma_S, \delta_L, \delta_S, P_L\}$ refers to the hyperparameters of the final covariance function, values for which were selected using a grid search in which we maximized the marginal log likelihood [33] and where the period P_L was allowed to vary between 0.25 and 1 day.

4.3.4 Similarity between Vital-Signs Trajectories

A separate Gaussian process may be trained for each patient's D time-series variables by using the approach described in the preceding section, which results in D univariate trajectories (for one patient). We can now compute a similarity metric to reveal common latent behavior in patients' trajectories. Assuming that we have a number of Gaussian process models X_k, for $k = 1, \ldots, N$ and where each X_k is a set of D time-series EHR data y for one of N patients, we can compare a test time series $X^* = (t, y)$, for times t, to the D Gaussian processes X_k by using a local likelihood evaluated at point i,

$$\left. X_i^*(t_i, y_i)\right|_{X_k} = -\log \prod_{j=1}^{D}\left(y_i^j \middle| t_i, X_k^j\right),$$ (4.8)

and then define a global likelihood over the local likelihoods,

$$L_k(X^*) = n^{-1}\sum_i\left[\left. X_i^*(t_i, y_i)\right|_{X_k}\right].$$ (4.9)

Finally, a global similarity may be obtained by normalizing this global likelihood,

$$S_k(X^*) = L_*/L_k(X^*), \qquad (4.10)$$

where the numerator is the self-global likelihood (i.e., evaluated for the test time series with respect to its own GP) and where $L_k(X^*)$ appears in the denominator because it is the negative log likelihood. If the test data X^* are similar to process X_k, then $S_k \to 1$; else, $S_k \to 0$. Finally, we can then perform hierarchical clustering by using the similarities S_k between processes for all N patients to identify clusters of time-series data that are similar.

We evaluate our example method by using a data set containing manual observations of vital signs acquired from 100 patients who have a normal recovery from cancer surgery in the Oxford University Hospitals National Health Service (NHS) Trust. These observations comprise measurements of heart rate, systolic blood pressure, temperature, blood oxygen saturation, and breathing rate made by clinical staff every hour or every 2 hours on the days immediately following surgery (depending on the patient's condition) and approximately every 4 hours on the last few days of the patient's stay on the postoperative ward. These patients were discharged home after their stay on the ward (median length of stay of 10 days).

The first day of patients' vital-signs trajectories corresponds to the day on which surgery took place. We initially selected two vital signs (systolic blood pressure and temperature), $D = 2$, to determine the mean vital-signs trajectory and then determine the similarity between all trajectories by using the methods described in the preceding sections. Hierarchical clustering revealed four main functional clusters, the mean of the mean functions for each of which is plotted in Figure 4.4. Table 4.1 summarizes the number of patient trajectories included in each cluster. All prototype trajectories reveal the "expected" recovery from surgery, in which blood pressure rises then stabilizes and temperature decreases. However, the range of blood pressures covered in each prototype trajectory varies from one subgroup to the other.

In order to study the influence of the other vital signs in the patient trajectories, we performed the same analysis by using all vital signs ($D = 5$) and determined how many patients were assigned to same cluster found in the previous analysis. The results summarized in Table 4.1 show that most patients (approximately 70%) were assigned to the same clusters, which suggests that the physiological trajectory for normal recovery is primarily determined by the temperature and systolic blood pressure, with other vital signs contributing a secondary level of additional information.

Figure 4.4g and h shows the trajectory of an example patient who deteriorated and was deemed by clinicians to be sufficiently abnormal for admission to an intensive care facility 9 days after surgery. It may be seen that the physiological trajectory of this patient is

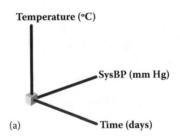

FIGURE 4.4
(a) Functional prototypes revealed by similarity S_k for systolic blood pressure (SysBP) and temperature data.
(Continued)

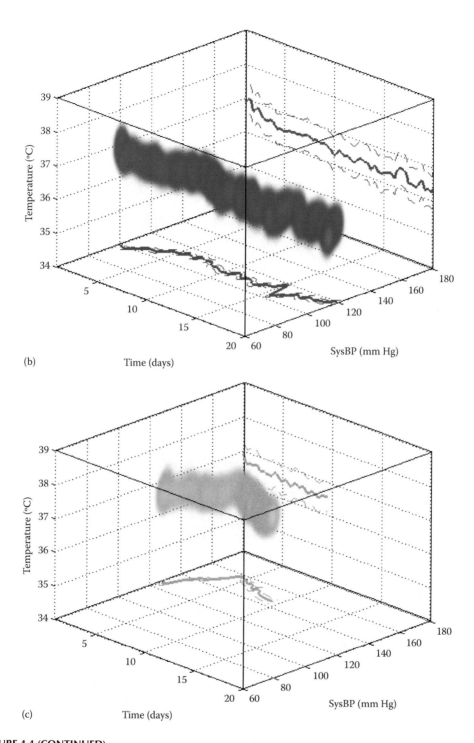

(b)

Time (days)

(c)

Time (days)

FIGURE 4.4 (CONTINUED)
(b, c) Functional prototypes revealed by similarity S_k for systolic blood pressure (SysBP) and temperature data.

(*Continued*)

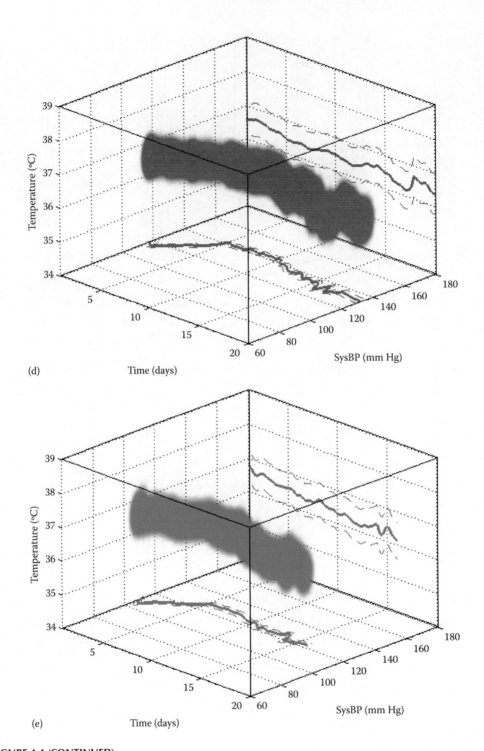

FIGURE 4.4 (CONTINUED)
(d, e) Functional prototypes revealed by similarity S_k for systolic blood pressure (SysBP) and temperature data.
(*Continued*)

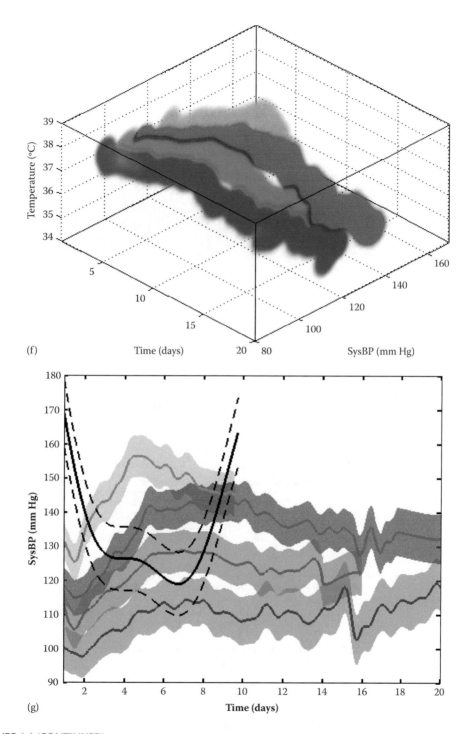

FIGURE 4.4 (CONTINUED)
(f) Functional prototypes revealed by similarity S_k for systolic blood pressure (SysBP) and temperature data; (g) two-dimensional (2-D) representation of the functional prototypes and the mean function, with 95% confidence band (solid and dashed black lines) of an abnormal patient. *(Continued)*

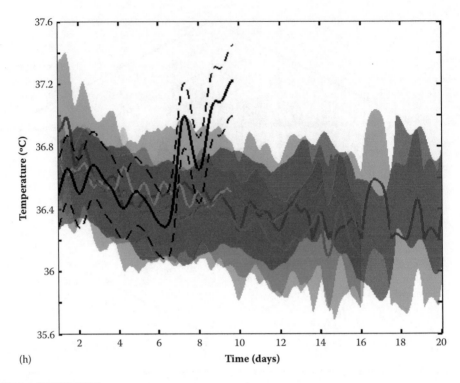

(h)

FIGURE 4.4 (CONTINUED)

(h) Two-dimensional (2-D) representation of the functional prototypes and the mean function, with 95% confidence band (solid and dashed black lines) of an abnormal patient.

TABLE 4.1

Number of Patients Assigned to Each Cluster

Cluster	$D = 2^a$	$D = 5^b$	Overlap
Prototype A	12	14	8
Prototype B	20	19	12
Prototype C	34	35	26
Prototype D	34	35	24

[a] Temperature and systolic blood pressure.
[b] All five vital signs.

different from those of the patients who had a normal surgery recovery and which, therefore, has low similarity S_k to all k normal prototypes.

4.3.5 Summary of Theme II

We have described an example method by which vital-signs data from the EHR may be used to provide improved clinical understanding of patient condition. We have also shown that it is able to discriminate normal from abnormal physiological trajectories, which may be further extended by combining it with other recently proposed methods [35]. Ongoing work involves the incorporation of models in which the distributions over the process hyperparameters are nonstationary and where data from additional sources, particularly higher-rate data from wearable sensors, may be incorporated.

4.4 Theme III: EHRs in the Developing World

While the use of large-scale EHRs is becoming increasingly prevalent in the healthcare systems (and, in some cases, being mandated by national legislation), the power of intelligent EHR systems also has great potential to improve the standard of, and access to, care in the developing world. OpenMRS [37] is one example of an open-source EHR that is frequently used in research for developing regions. Frameworks such as Sana/Moca [38] have been constructed that build upon open-source EHRs to allow direct access to medical records via mobile devices. The latter may be used for connection with physiological sensors, allowing healthcare workers to record data from patients and upload it directly to open-source EHRs [39]. Such studies typically use the data to perform screening for risk of long-term conditions, such as cardiovascular disease, allowing workers with low levels of healthcare training to perform elementary assessment of patients.

4.4.1 Fusing Data from Noisy Time Series

The research challenge for this branch of work is to provide analysis techniques (building on methods described in themes I and II) that are sufficiently robust in the face of the extreme sparsity and noise present in EHRs from resource-constrained countries. Using technology developed from large-scale studies of EHRs in developed-world healthcare systems [40–42], periods of missing or artifactual data may be analyzed in a principled manner, using probabilistic analyses of the kind described in this chapter.

An example is shown in Figure 4.5, which are physiological data stored in an EHR after being acquired by physiological sensors and where the EHR also contains manual observations made by clinicians. A 4-day interval is shown, in which the patient experiences repeated episodes of tachycardia (elevated heart rate) and frequent desaturations (decreases in SpO_2). The physiological EHR data acquired via sensors have been modeled using a multitask Gaussian process or MTGP [43–44]. Figure 4.5 shows extended periods of disconnection from the physiological sensors, particularly in the middle of the interval shown, where the MTGP is able to interpolate effectively to provide a principled estimate of the missing data.

MTGPs are a natural development of the univariate GPs considered in previous sections, in which multiple time series can be considered simultaneously. The covariance function for MTGPs may be written as

$$k_{\mathrm{MTGP}}(x, x', l, l') = k_c(l, l') \times k_t(x, x'),\qquad(4.11)$$

where k_c and k_t represent the correlation *between* two time series l and l' and the temporal covariance functions *within* a time series at times x and x', respectively. That is, the model explicitly takes into account the dynamics of each time series, as well as explicitly modeling how the various time series covary.

The task of learning with MTGPs, therefore, becomes estimating suitable values of the various hyperparameters for the kernels k_c and k_t. As before with the univariate GPs

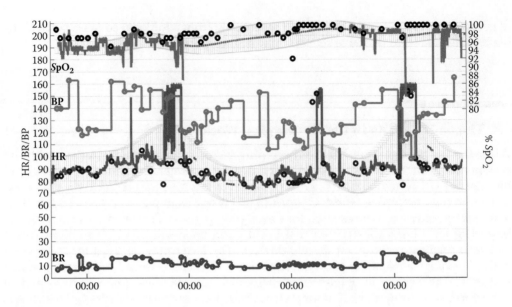

FIGURE 4.5
EHR data, with a multitask Gaussian process model tracking multiple physiological time series in parallel. The correlation between the various time series has been learned by the model, which is able to cope with periods of missing data by using principled interpolation from those time series that are available at any one time. Manual observations, made by a nurse, and continuous observations, made by physiological sensors, are shown by circles and solid lines, respectively. The underlying Gaussian process is shown by dashed lines, and its uncertainty is shown by the gray shaded regions. The horizontal axis shows midnight for four consecutive days; the vertical axes shown the scales for heart rate (HR), breathing rate (BR), systolic blood pressure (BP), and SpO_2.

described in previous sections, this can be performed by optimizing the log marginal likelihood of the hyperparameters given the training data [44]. A toolbox for performing this optimization has been implemented by the authors of this chapter and which accompanies the article by Dürichen et al. [45].

These methods have been formulated for online learning [46], such that models may be sequentially updated as new data arrive in the EHR. While such models are computationally demanding and, therefore, would be best suited to run on the servers that provide the EHR itself, some computation may be performed on local devices by using techniques developed to make processing more computationally efficient [47–48].

4.5 Conclusions and Future Directions

This chapter has highlighted current research directions in the growing field of intelligent EHR systems. We have presented an overview of existing research, with an emphasis on the machine learning techniques that offer a realistic means of tackling the many challenges that exist in the analysis of realistic, disparate, and often contradictory data in typical EHRs.

While advances in machine learning, such as those described, are critical to underpinning the technical success of projects involving intelligent EHR systems, it is important to note that this is just part of the complex task involved in implementing intelligent systems within healthcare practice. A key step in producing appropriate technology for solving clinical problems is an understanding of the data that are being modeled, and close collaboration between data scientists and clinicians is, therefore, *sine qua non* for large-scale biomedical projects. We have described in this chapter those systems that we have found to be suitable for incorporation of the clinical prior knowledge that makes a computational system effective: these techniques are typically Bayesian, whereby clinical information can be explicitly incorporated within the model in two ways.

Firstly, appropriate systems for performing complex analyses (such as those involved in the production of intelligent EHR systems) are often nonparametric; that is, formally, the parameter space of the model is infinite. In practice, this typically means that the number of parameters in the model can grow with the number of data observed—this is a key feature of large-scale data analysis, in which one wishes to update models through time. In healthcare systems, as in the analysis of many other complex systems, it cannot be assumed that the patient is unchanging. Therefore, models need to be able to "grow" and evolve with the changing dynamics of the patient, perhaps as a patient recovers from a condition or perhaps as a patient's condition begins to deteriorate with time. Bayesian nonparametric methods, such as the GP, allow this growth of the model with respect to the data, while allowing us to incorporate key clinical insights in the form of (for the GP) the covariance functions that we use. We showed, in the probabilistic estimation of vital signs, for example, that our knowledge of the respiratory process led to the use of periodic covariance functions, the hyperparameter of which corresponds to the quantity that we aim to estimate: the respiration rate.

Secondly, prior clinical knowledge can be directly incorporated by specifying prior distributions over our hyperparameters. This is a "strongly Bayesian" approach, in which a potentially infinite number of values of the hyperparameter are considered and where inference is typically performed such that the distributions over the hyperparameters are optimized. These distributions over the hyperparameters are often specified using hyper-hyperparameters; for example, our distribution over the respiration rate hyperparameter P_L in the case described above could be chosen to be Gaussian, which will have its own mean and variance hyper-hyperparameters. These analyses are often no longer analytically tractable, and so approximation schemes are typically performed, such as sampling-based methods, or deterministic approximations such as variational Bayesian methods [20,22].

Future research for intelligent EHR systems proceeds on many fronts, worldwide: researchers are currently working on principled means for the large-scale fusion of data sets across the varying data types that exist in the EHR. A key theme is the integration of genomic, proteomic, and metabolomics with other data types; we have described a case study in this chapter, applying such methods to infectious disease. Other researchers have considered fusion of "omic" data with images [49], with applications in cancer treatment. Fusion of gene expression data, including genomic and time-series data, has been performed [50]; this study used GPs for modeling the gene expression time-series data, similar to that which could be obtained within an EHR, along with multinomial models for the categorical data sets, including discretized gene expression levels. Other machine learning techniques have also been considered for the analysis of inpatient EHR data [51], which considered the incorporation of data from other sources peripheral to the EHR, such as data from health insurance providers. The potentially transformative effect of intelligent health systems has been considered as a means of both reducing the cost of healthcare and improving patient outcomes [52], both inside and outside the hospital.

As more data scientists collaborate with clinical teams, this rapidly growing field of research has the promise to deliver the analysis tools required to underpin the next generation of healthcare technologies embedded within clinical practice as "intelligent EHRs." With global challenges to the sustainability of healthcare systems worldwide, technology has a key role to play in bringing about this new generation of tools—and clinicians that are best placed to use and exploit such tools.

Acknowledgments

The authors gratefully acknowledge the support of the Centre of Excellence in Medical Engineering funded by the Wellcome Trust and Engineering and Physical Sciences Research Council (EPSRC) under Grant Number WT 088877/Z/09/Z; the National Institute for Health Research (NIHR) Biomedical Research Centre, Oxford; the Research Councils United Kingdom (RCUK) Digital Economy Programme (Oxford Centre for Doctoral Training in Healthcare Innovation); and the Fundação para a Ciência e Tecnologia (FCT). DAC was funded by a Royal Academy of Engineering Fellowship and by Balliol College, Oxford, and the Balliol Interdisciplinary Institute. KEN was supported by the Rhodes Trust.

References

1. House of Commons Health Committee, United Kingdom, "The electronic patient record, report HC 422-I," White Paper, 2007.
2. I. Buchan, J. Winn, and C. Bishop, "A unified modeling approach to data-intensive healthcare," *The Fourth Paradigm: Data-Intensive Scientific Discovery*, Microsoft Research, Technical Report, 2009.
3. Secretary of State for Health, United Kingdom, "The government response to the health committee report on the electronic patient record, report Cm 7264," White Paper, 2007.
4. Medical Research Council, "e-Health informatics research: Securing the UK as a world leader," http://www.mrc.ac.uk/Ourresearch/ResearchInitiatives/E-HealthInformaticsResearch/, accessed: June 2014.
5. Annual Report of the UK chief medical officer, "Infections and the rise of antimicrobial resistance," White Paper, 2011.
6. X. Didelot, R. Bowden, D. J. Wilson, T. E. Peto, and D. W. Crook, "Transforming clinical microbiology with bacterial genome sequencing," *Nature Reviews Genetics*, vol. 13, no. 9, pp. 601–612, 2012.
7. C. U. Köser, M. T. Holden, M. J. Ellington, E. J. Cartwright, N. M. Brown, A. L. Ogilvy-Stuart, L. Y. Hsu, C. Chewapreecha, N. J. Croucher, S. R. Harris et al., "Rapid whole-genome sequencing for investigation of a neonatal MRSA outbreak," *New England Journal of Medicine*, vol. 366, no. 24, pp. 2267–2275, 2012.
8. N. Stoesser, E. Batty, D. Eyre, M. Morgan, D. Wyllie, C. D. O. Elias, J. Johnson, A. Walker, T. Peto, and D. Crook, "Predicting antimicrobial susceptibilities for *Escherichia coli* and *Klebsiella pneumoniae* isolates using whole genomic sequence data," *Journal of Antimicrobial Chemotherapy*, vol. 68, pp. 2234–2244, 2013.
9. N. Gordon, J. Price, K. Cole, R. Everitt, M. Morgan, J. Finney, A. Kearns, B. Pichon, B. Young, D. Wilson et al., "Prediction of *Staphylococcus aureus* antimicrobial resistance by whole-genome sequencing," *Journal of Clinical Microbiology*, vol. 52, no. 4, pp. 1182–1191, 2014.

10. L. Tarassenko, D. Clifton, P. Bannister, S. King, and D. King, "Novelty detection," *Encyclopaedia of Structural Health Monitoring*, pp. 653–675, 2009.

11. M. Pimentel, D. Clifton, L. Clifton, and L. Tarassenko, "A review of novelty detection," *Signal Processing*, vol. 99, pp. 215–249, 2014.

12. M. Laabei, M. Recker, J. K. Rudkin, M. Aldeljawi, Z. Gulay, T. J. Sloan, P. Williams, J. L. Endres, K. W. Bayles, P. D. Fey et al., "Predicting the virulence of MRSA from its genome sequence," *Genome Research*, vol. 24, no. 5, pp. 839–849, 2014. Also available on http://genome.cshlp.org/content /early/2014/04/02/gr.165415.113.full.pdf+html. Last visited on June 6, 2015.

13. World Health Organization, "Antimicrobial resistance: Global report on surveillance 2014," World Health Organization, 2014, http://www.who.int/drugresistance/documents /surveillancereport/en/.

14. M. H. Hazbón, M. Brimacombe, M. B. del Valle, M. Cavatore, M. I. Guerrero, M. Varma-Basil, H. Billman-Jacobe, C. Lavender, J. Fyfe, L. García-García et al., "Population genetics study of isoniazid resistance mutations and evolution of multidrug-resistant mycobacterium tuberculosis," *Antimicrobial Agents and Chemotherapy*, vol. 50, no. 8, pp. 2640–2649, 2006.

15. G. Lunter and M. Goodson, "Stampy: A statistical algorithm for sensitive and fast mapping of illumina sequence reads," *Genome Research*, vol. 21, no. 6, pp. 936–939, 2011.

16. H. Li, B. Handsaker, A. Wysoker, T. Fennell, J. Ruan, N. Homer, G. Marth, G. Abecasis, and R. Durbin, "The sequence alignment/map (SAM) format and SAMtools," *Bioinformatics*, vol. 25, pp. 2078–2079, 2009.

17. D. R. Zerbino and E. Birney, "Velvet: Algorithms for de novo short read assembly using de Bruijn graphs," *Genome Research*, vol. 18, no. 5, pp. 821–829, 2008.

18. Z. Zhang, S. Schwartz, L. Wagner, and W. Miller, "A greedy algorithm for aligning DNA sequences," *Journal of Computational Biology*, vol. 7, nos. 1–2, pp. 203–214, 2000.

19. S. K. Sheppard, X. Didelot, G. Meric, A. Torralbo, K. A. Jolley, D. J. Kelly, S. D. Bentley, M. C. Maiden, J. Parkhill, and D. Falush, "Genome-wide association study identifies vitamin B5 biosynthesis as a host specificity factor in Campylobacter," *Proceedings of the National Academy of Sciences*, vol. 110, no. 29, pp. 11923–11927, 2013.

20. C. M. Bishop, *Pattern Recognition and Machine Learning*. Berlin: Springer-Verlag, 2006.

21. B. A. Goldstein, A. E. Hubbard, A. Cutler, and L. F. Barcellos, "An application of random forests to a genome-wide association dataset: Methodological considerations and new findings," *BMC Genetics*, vol. 11, no. 1, p. 49, 2010.

22. T. A. Manolio, F. S. Collins, N. J. Cox, D. B. Goldstein, L. A. Hindorff, D. J. Hunter, M. I. McCarthy, E. M. Ramos, L. R. Cardon, A. Chakravarti et al., "Finding the missing heritability of complex diseases," *Nature*, vol. 461, no. 7265, pp. 747–753, 2009.

23. W. S. Bush and J. H. Moore, "Genome-wide association studies," *PLoS Computational Biology*, vol. 8, no. 12, p. e1002822, 2012.

24. K. P. Murphy, *Machine Learning: A Probabilistic Perspective*. Cambridge: MIT Press, 2012.

25. T. Bonnici, D. Clifton, P. Watkinson, and L. Tarassenko, "The digital patient," *Clinical Medicine*, vol. 3, no. 3, pp. 252–257, 2013.

26. National Patient Safety Association, "Safer care for acutely ill patients: Learning from serious accidents," White Paper, 2007.

27. L. Tarassenko, D. Clifton, M. Pinsky, M. Hravnak, J. Woods, and P. Watkinson, "Centile-based early warning scores derived from statistical distributions of vital signs," *Resuscitation*, vol. 82, no. 8, pp. 1013–1018, 2011.

28. M. Hravnak, M. de Vita, A. Clontz, L. Edwards, C. Valenta, and M. Pinsky, "Cardiorespiratory instability before and after implementing an integrated monitoring system," *Critical Care Medicine*, vol. 39, no. 1, pp. 65–72, 2011.

29. G. Clifford and D. Clifton, "Annual review: Wireless technology in disease state management and medicine," *Annual Review of Medicine*, vol. 63, pp. 479–492, 2012.

30. D. Meredith, D. Clifton, P. Charlton, J. Brooks, C. Pugh, and L. Tarassenko, "Photoplethysmographic derivation of respiratory rate: A review of relevant respiratory and circulatory physiology," *Journal of Medical Engineering and Technology*, vol. 36, no. 1, pp. 60–66, 2012.

31. L. Clifton, D. Clifton, M. Pimentel, P. Watkinson, and L. Tarassenko, "Gaussian processes for personalised e-health monitoring with wearable sensors," *IEEE Transactions on Biomedical Engineering*, vol. 60, no. 1, pp. 193–197, 2013.

32. L. Clifton, D. Clifton, M. Pimentel, P. Watkinson, and L. Tarassenko, "Predictive monitoring of mobile patients by combining clinical observations with data from wearable sensors," *IEEE Journal of Biomedical and Health Informatics*, vol. 18, no. 3, pp. 722–730, 2014.

33. C. Rasmussen and C. Williams, *Gaussian Processes for Machine Learning.* Cambridge: MIT Press, 2006.

34. M. Pimentel, D. Clifton, L. Clifton, and L. Tarassenko, "Probabilistic estimation of respiratory rate using Gaussian processes," in *IEEE Engineering in Medicine & Biology Conference, Osaka, Japan*, 2013, pp. 2902–2095.

35. D. Clifton, L. Clifton, S. Hugueny, D. Wong, and L. Tarassenko, "An extreme function theory for novelty detection," *IEEE Journal of Selected Topics on Signal Processing*, vol. 7, no. 1, pp. 28–37, 2013.

36. M. Pimentel, D. Clifton, L. Clifton, P. Watkinson, and L. Tarassenko, "Modelling physiological deterioration in post-operative patient vital-sign data," *Medical & Biological Engineering & Computation*, vol. 51, no. 8, pp. 869–877, 2013.

37. C. Seebregts, B. Mamlin, P. Biondich, H. Fraser, B. Wolfe, D. Jazayeri, C. Allen, J. Miranda, E. Baker, N. Musinguzi, D. Kayiwa, C. Fourie, N. Lesh, A. Kanter, C. Yiannoutsos, and C. Bailey, "The OpenMRS implementers network," *International Journal of Medical Informatics*, vol. 78, no. 11, pp. 711–720, 2009.

38. L. Celi, L. Sarmenta, J. Rotberg, A. Marcelo, and G. Clifford, "Mobile care (Moca) for remote diagnosis and screening," *Journal of Health Informatics in Developing Countries*, vol. 3, no. 1, pp. 17–21, 2009.

39. M. Tian, "The simplified cardiovascular management in India and China study (Simcard)," http://www.georgeinstitute.org/projects/simplified-cardiovascular-management-in-india-and-china-study-simcard, 2014.

40. D. Clifton, D. Wong, L. Clifton, R. Pullinger, and L. Tarassenko, "A large-scale clinical validation of an integrated monitoring system in the emergency department," *IEEE Transactions on Information Technology in Biomedicine*, vol. 17, no. 4, pp. 835–877, 2013.

41. L. Clifton, D. Clifton, Y. Zhang, P. Watkinson, L. Tarassenko, and H. Yin, "Probabilistic novelty detection with support vector machines," *IEEE Transactions on Reliability*, vol. 62, no. 2, pp. 455–467, 2014.

42. D. Clifton, S. Hugueny, L. Clifton, and L. Tarassenko, "Extending the Generalised Pareto distribution for novelty detection in high-dimensional spaces," *Journal of Signal Processing Systems*, vol. 74, pp. 323–339, 2014.

43. K. Yu, V. Tresp, and A. Schwaighofer, "Learning Gaussian processes from multiple tasks," in *Proceedings of the 22nd International Conference on Machine Learning*, 2005, pp. 1012–1019.

44. E. V. Bonilla, K. M. A. Chai, and C. K. I. Williams, "Multi-task Gaussian process prediction," in *Advances in Neural Information Processing Systems (NIPS)*, 2008, pp. 153–160.

45. R. Dürichen, M. Pimentel, L. Clifton, and D. Clifton, "MTGP—A Matlab toolbox for multi-task Gaussian processes," http://www.robots.ox.ac.uk/~davidc/code.php, 2014.

46. G. Pillonetto, F. Dinuzzo, and G. De Nicolao, "Bayesian online multitask learning of Gaussian processes," *IEEE Transactions on Pattern Analysis and Machine Intelligence*, vol. 32, no. 2, pp. 193–205, 2010.

47. M. A. Álvarez and N. D. Lawrence, "Computationally efficient convolved multiple output Gaussian processes," *Journal of Machine Learning Research*, vol. 12, pp. 1459–1500, 2011.

48. M. A. Osborne, S. J. Roberts, A. Rogers, and N. R. Jennings, "Real-time information processing of environmental sensor network data using Bayesian Gaussian processes," *ACM Transactions on Sensor Networks*, vol. 9, no. 1, pp. 1–32, 2012.

49. J. Phan, C. Quo, C. Cheng, and M. Wang, "Multiscale integration of -omic, imaging, and clinical data in biomedical informatics," *IEEE Reviews in Biomedical Engineering*, vol. 5, pp. 74–87, 2012.

50. P. Kirk, J. Griffin, R. Savage, Z. Ghahramani, and D. Wild, "Bayesian correlated clustering to integrate multiple datasets," *Bioinformatics*, vol. 28, no. 24, pp. 3290–3297, 2012.
51. D. Neill, "Using artificial intelligence to improve hospital inpatient care," *IEEE Intelligent Systems*, vol. 28, no. 2, pp. 92–95, 2013.
52. M. Pavel, H. Jimison, H. Wactlar, T. Hayes, W. Barkis, J. Skapik, and J. Kaye, "The role of technology and engineering models in transforming healthcare," *IEEE Reviews in Biomedical Engineering*, vol. 6, pp. 156–177, 2013.

5

Wearable Biomedical Systems and mHealth

Sungmee Park and Sundaresan Jayaraman

CONTENTS

5.1 Introduction

The lack of access to affordable, timely, and high-quality healthcare is negatively impacting the quality of life of individuals in today's society. In a landmark study, the Institute of Medicine concluded, "The U.S. healthcare delivery system does not provide consistent, high-quality medical care to all people" [1]. On a global level too, healthcare is threatened by a confluence of powerful trends increasing demand, aging populations, chronic diseases, rising costs, uneven quality, and misaligned incentives. According to a PricewaterhouseCoopers (PwC) study, "If ignored, these trends will overwhelm health systems, creating massive financial burdens for businesses and governments as well as health problems for current and future generations" [2].

5.1.1 The Healthcare Reality

Consider the following facts:

- Healthcare spending in the United States was $2.8 trillion in 2012. It is projected to reach over $5 trillion in 2022, growing at an average annual rate of 5.8% during the forecast period 2012–2022 [3].

- As a share of GDP, health spending is projected to reach 19.9% by 2022, up from its 2012 level of 17.9% [3]. In contrast, healthcare spending is around 11% of GDP in developed countries in Europe, around 5% in China, and around 4% in India [4].

- The medical care costs of people with chronic illnesses typically represent 75% of the U.S. annual healthcare spending [5].

- With the commissioning of the health insurance marketplace as part of the Patient Protection and Affordable Care Act of 2010 (PPACA), over eight million people have selected a plan [6]. This is likely to increase access to healthcare for individuals while simultaneously triggering the need for more providers (doctors/nurses).

- Experts estimate that as many as 98,000 people die in any given year from medical errors that occur in hospitals [7]. Add the financial cost to the human tragedy, and medical error easily rises to the top ranks of urgent, widespread public problems.

- Private insurers are following Medicare/Medicaid's lead in implementing a system that denies payments for treating so-called never events—serious and costly errors such as pressure ulcers, mismatched blood transfusion, object left in patient after surgery, and patient falls that should have never happened in the first place [8].

- Between 1998 and 2008, the total number of hospital-based emergency departments (EDs) declined by 3.3%, from 4771 to 4613; in this same period, ED visits increased by 30%, from 94.8 million visits to 123 million visits annually [9]. ED use by publicly insured and uninsured patients increased at an even faster pace, largely driven by loss of access to care in other settings.

- A new healthcare delivery channel has emerged in the form of retail or in-store clinics staffed by nurse-practitioners; it is gaining increased acceptance among those pressed for time when needing treatments for common ailments such as colds and ear infections; it is expected to grow by 20%–25% in the years 2013–2015, leading to the establishment of over 2800 clinics by 2015 [10].

- The average bill for treating simple pneumonia was $14,610 in one hospital in Dallas, Texas, while it was over $38,000 in a nearby hospital—a significant variation within a small geographic area [11].

In short, today's healthcare system is afflicted with these symptoms and is diagnosed with the following primary ailments: fragmented with misaligned incentives, lack of transparency in terms of pricing and quality of care, and varying degrees of access to care often limited by an individual's ability to pay for care. The healthcare system is at the proverbial *tipping point*; unless the industry transforms itself to address this *silent*, yet potent, crisis effectively, it will become unsustainable, further eroding the quality of life of individuals around the world.

5.1.2 Technology, Innovation, and the Emergence of the "Patient-Consumer"

The keys to the necessary transformation lie in technology and innovation. According to International Data Corporation (IDC), over one billion smartphones were shipped worldwide in 2013 [12]. With smartphones and broadband becoming ubiquitous, information is being instantaneously accessed anytime, anywhere by anyone. Consequently, people's lives are being significantly transformed—for the better—since they are better informed and are in greater control in decision making, which is based on real data and evidence, not just conjecture. Moreover, the knowledgeable consumer is more demanding and discriminating in making choices. This paradigm is spreading to healthcare, where today's empowered patient is beginning to act like any typical consumer of goods and services and, in fact, a new kind that can be characterized as a "patient-consumer" is emerging.

5.1.3 The Healthcare Bill of Rights

Just as the Bill of Rights guarantees certain freedoms to US citizens and there are similar documents in other nations with guarantees to their citizens, today's patient-consumer has a set of minimum expectations from the healthcare system. The patient's bill of rights can be listed as:

- Choose the provider
- Receive safe care
- Receive quality care leading to best outcomes
- Receive timely care
- Receive affordable care
- Receive preventative care
- Access and control personal health information
- Refuse care
- Be informed of and have access to all options for treating a specific condition or illness
- Transparency of costs and capabilities of providers and other stakeholders influencing care

Similarly, the healthcare provider has a certain set of expectations from the healthcare delivery system including those from the patient and these are reflected in the provider's bill of rights is listed as:

- Access the required resources (e.g., equipment, instruments, medicines, and personnel) necessary to deliver the best care to the patient.
- Practice medicine in the best interest of the patient based on the best available evidence and experience and not be forced to make choices based on cost, third-party interests (e.g., employer and insurance company), or the particular demands of patients.
- Assume that the patient is honest and is sharing all information.
- Define criteria and guidelines for accepting and working with patients.
- Access patient medical information necessary to diagnose, prescribe, and monitor treatment including contacting patients when needed for status updates.
- Fair and timely reimbursement for services.
- Protection against frivolous lawsuits.
- Deny preferred treatment sought if it were against personal beliefs and/or deemed harmful to patient and society (e.g., removing life support and requesting certain types of drugs).
- Terminate relationship with patients who exhibit abusive behavior or fail to follow through on treatment or are not willing to pay.
- Not be held responsible for a patient's noncompliance with recommended treatment.

Taken together, these two sets of *unalienable* rights of patients and providers present a framework for designing a healthcare system that addresses the ailments of the current system and meets the expectations of two of the key stakeholders. The Institute of Medicine's statement of purpose provides one such model: a healthcare system that is patient centric and delivers safe, timely, effective, efficient, and equitable care that does not vary in quality with the geographical location of the patient [1].

5.1.4 mHealth: Key to Enhancing Quality of Life

An individual's health is influenced by factors in five domains: genetics, social circumstances, environmental exposures, behavioral patterns, and healthcare [13]. In terms of proportional contribution to premature death, 40% was due to behavioral patterns, 30% due to genetic predisposition, 15% due to social circumstances, 10% due to healthcare, and 5% due to environmental exposures. Therefore, the critical step in enhancing the quality of life of individuals is to assess the impact of the majority 90% of those factors and let the findings *drive* the type of healthcare that is provided to the individual. So the challenge goes *beyond* what happens in the healthcare setting; it is the entire life cycle and lifestyle of the individual that should be addressed.

And that is the paradigm and promise of mobile health or mHealth.* Information about the individual can be harnessed in real time (and whenever needed) by utilizing the advancements in, and convergence of, microelectronics, materials, optics, and biotechnologies, coupled with miniaturization; these have led to the development of small, cost-effective intelligent sensors that are intimately interwoven into the fabric of our lives;

---

* The term *mHealth* is used throughout this chapter for consistency; it is broader than *eHealth* and *telehealth*.

they are pervasive and also operationally "invisible" to the end user. Using these digital ecosystems, large amounts of data from individuals and surroundings can be collected and harnessed to develop appropriate solutions and strategies to enhance and optimize health outcomes through personalized care and, thereby, fulfill the promise of mHealth. It is important to remember that mHealth should take a holistic perspective and focus on health and not on technology that will enable mHealth.

5.1.4.1 The Desired State

According to a recent PricewaterhouseCoopers study, 69% of consumers surveyed are willing to communicate with doctors and nurses by using e-mail, 49% via online web chat or portal, and 45% by using text messages [14]. In another survey commissioned by PwC, 59% of physicians and insurers believed that the widespread adoption of mobile health applications in the near future was "unavoidable" [15].

Thus, the future lies in mHealth—taking care of the "whole" individual by delivering personalized, affordable healthcare to anyone, anytime and anywhere. The path forward is to act on the *diagnosis* mentioned earlier by viewing technology and innovation as "medication" to treat the *ailments* afflicting the healthcare industry; it must effectively harness technology and innovation to transform itself so that it can successfully meet the growing demands of the patient-consumer by improving access to care, delivering high-quality care in a timely manner at an affordable price, and enhancing transparency, which are the key metrics of healthcare. Park, Chung, and Jayaraman have coined the term *WOW* for "world of wearables" and have discussed their critical role as exemplars of technology and innovation in transforming healthcare and enhancing the quality of life of individuals [16].

5.1.4.2 Organization of the Chapter

The remainder of the chapter is organized as follows: We present the concept of a healthcare continuum and the transformations in healthcare delivery driven by technology and innovation that lead to the concept of *patient-centric value-based* care. We analyze the importance of data in the context of mHealth by using a few scenarios related to the Patient Protection and Affordable Care Act, occupational health, and chronic disease management. We then discuss the WOW paradigm that harnesses patient data and facilitates this transformation from the traditional *point-of-care* model to a *continuum-of-care* model, which is the underlying theme of mHealth delivering value to the patient through affordable healthcare anytime, anywhere for anyone. We present the highlights of the Wearable Motherboard or smart shirt, the wearable biomedical system, which spawned today's wearables revolution and we discuss its role in enabling mHealth. Finally, we look at the future of mHealth powered by WOW that is transforming healthcare by delivering patient-centric value-based care.

5.2 The Healthcare Delivery Model and Its Transformation

The principal stakeholders in the healthcare continuum are shown in Figure 5.1. The patient, the most important stakeholder, is at the center of the system to reflect the ultimate objective of having a successful *patient-centric* healthcare system [17]. The patient is cared for by

FIGURE 5.1
The healthcare continuum: stakeholders and enablers.

the provider either in an ambulatory setting (the doctor's office) or in an acute care setting (the hospital) and this is shown in Figure 5.1. The other major stakeholders affecting the patient include the pharmacies and suppliers that are also in the front line of patient care. In reality, the patient's employer and insurance company have a significant say in the *level* and *cost* of care received since 55.1% of US population had employment-based health insurance coverage in 2011; moreover, among the employed population aged 18 to 64, 68.2% had health insurance through their own employer or another person's employer [18]. So, in reality, the patient today is not a true "consumer." The typical consumer of personal goods and services has complete control over selecting from the options available during the purchase process, which the healthcare consumer, i.e., the patient, does not. However, as mentioned earlier, this is beginning to change and the concept of a patient-consumer is gaining ground, albeit slowly.

The biotechnology companies in the healthcare continuum are responsible for innovative drugs, which are initially funded by venture capitalists and subsequently developed through clinical trials and brought to market for the patient by pharmaceutical companies shown in Figure 5.1. In addition, the patient's *quality* and *cost* of care are influenced by the research and teaching institutions that are developing the required human resources and technological innovations. The last major stakeholder is the government, which strives to ensure the *safety* and *privacy* of the patient and the proper functioning of the health-care continuum. The effects of these interactions on the patient, and, in fact, between the other stakeholders themselves, will significantly influence the healthcare delivered to the patient. For instance, the healthcare provider's reimbursement is often dictated by the insurance company and the hospital's cost structure is influenced by the effectiveness of its supply chain. This integrated view of the healthcare continuum is critical for creating a patient-centric healthcare system that delivers value: the *right* care to the *right* patient in the *right* place, at the *right* time, and at the *right* cost. In turn, this requires access to the right *information* at the right place, at the right time, and at the right cost, thus making healthcare delivery an information business.

5.2.1 The Enablers in the Healthcare Continuum

The key *enablers* that drive this continuum are innovation, capital (finance), technology (medical/information), and legal (laws/regulatory); these are also shown in Figure 5.1. Innovation is the lifeblood of the continuum, whether they are technological innovations in pharmaceuticals and medical devices or management innovations in pay-for-performance systems and patient care. Information technology—in the form of EHRs, computerized physician-order-entry (CPOE) systems, and wireless communications technologies—can streamline the information flow and enhance the care of the patient while minimizing the cost of care. Latest advancements in medical technologies and procedures can speed up recovery and significantly curtail the duration of hospital stay, which, in turn, can reduce the possibility of never events, such as bedsores and other hospital-acquired infections (HAIs). Capital is critical for the smooth flow of operations in the continuum that is made safe through regulations by the other enabler legal.

The ideal expression of the healthcare continuum is one in which information flows seamlessly between the stakeholders (through the pipeline shown in Figure 5.1), thereby ensuring that value is ultimately delivered by and to each of the stakeholders, resulting in a patient-centric value-based healthcare delivery model. We will now examine the transformations that have occurred in healthcare delivery with the advent of technology and innovation and the need for mHealth to meet the changing needs of patients and society over the years.

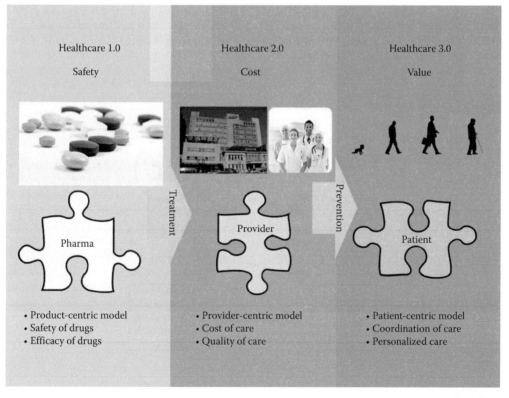

FIGURE 5.2
The transformation of healthcare: emergence of value-based paradigm.

5.2.2 Transformation of Healthcare

Figure 5.2 shows the various stages in the transformation of healthcare from a product-centric model to today's patient-centric model. In healthcare 1.0, the focus was on *safety*, namely, preventing harm to the patients by ensuring the safety and efficacy of drugs, an issue that came into limelight following the infamous thalidomide tragedies [19]. Gradually, the focus shifted from this product-centric model to a provider-centric model (healthcare 2.0) in which the *cost* of care delivered by the provider became the driver, with quality of care also attracting attention. The interventions were focused on operational efficiencies that minimized waste and reduced the cost of care.

As shown in Figure 5.2, the next transformation has been the shift from treatment to prevention and delivering *value* to the patient. This patient-centric model, healthcare 3.0, is driven by emphasizing coordination of care that can, in turn, lead to better patient outcomes through predictive and personalized interventions and treatments. This value-based model is characterized by the optimal use of resources (e.g., eliminating duplicate diagnostic tests through shared results), improved quality, timeliness of care, and greater transparency. Here, technology and innovation are used as enablers to addresses the key symptoms afflicting the healthcare industry discussed earlier.

5.2.3 The Attributes of Patient-Centric Value-Based Care

Figure 5.3 is a conceptual representation of value-based care showing the relationship between the cost of care and patient outcomes. Value is essentially the ratio of the output of the administration of healthcare, namely, the patient outcomes, to the input that goes to administer the care, namely, the cost in terms of resources, such as time and money. Yet another facet of value-based care is represented in the four As and the five Ps in Figure

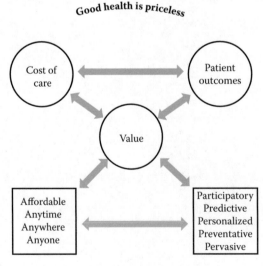

Patient-centric value-based care

FIGURE 5.3
mHealth and the attributes of value-based care.

5.3. The overarching theme is that the value-based paradigm promotes good health, which is priceless. The active participation of the patient is critical [20]; it aids in personalizing treatments that can also be predictive and preventive, all contributing to better patient outcomes, the key goal of value-based care and mHealth. Transparency in cost and capabilities in the healthcare delivery process is another key facet of value-based care.

5.2.4 Data–Value Transformation and mHealth

Typically, a patient receives four types of care: ambulatory care for routine illnesses such as a strep throat; acute care requiring hospitalization, say for appendicitis; chronic care, say for the treatment of diabetes and hypertension; and preventive care involving annual physical examinations. The care may be received in different geographic locations, from different providers and at different times; i.e., the care is pervasive (the fifth P in Figure 5.3). The coordination of care in each setting is critical for delivering value to the patient and is the central tenet of mHealth; this is denoted by the four As in Figure 5.3: affordable healthcare for anyone, anytime and anywhere.

Figure 5.4 shows the data generated during each type of care. Typically, the data from these four *points of care* remain distinct and fragmented; they are not shared and so, the care delivered to the patient may not be holistic. For instance, due to the time sensitivity of the situation, if the surgeon performing the appendicitis does not have detailed information about the patient's chronic condition of diabetes or the medications being taken or the personal living arrangements of the individual, the outcome may be less than optimal. It is, therefore, important to harness the data from each of the interactions and transform the discrete points of care into a *continuum of care* for the individual; the result will be timely care of a higher quality at a lower cost and it will fulfill the promise of mHealth. Thus, data are the *fuel* that powers the coordination of care and its frictionless exchange in this continuum of care "ambulatory–acute–preventive–chronic" is at the heart of mHealth. The desired system depicting the smooth flow and sharing of data is also shown in Figure 5.4.

Central to the success of this transformation is the acquisition of the data from the individual, in real time and continuously, if necessary to provide personalized and predictive care associated with mHealth. The major sources of data in an individual's daily life are

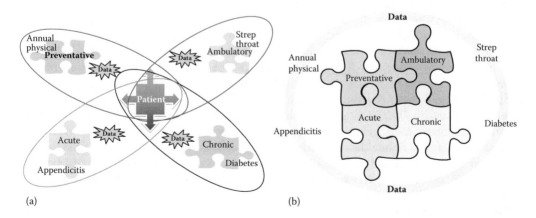

(a) (b)

FIGURE 5.4
Data in the continuum of ambulatory–acute–preventive–chronic care: (a) typical fragmented system and (b) desired ideal system.

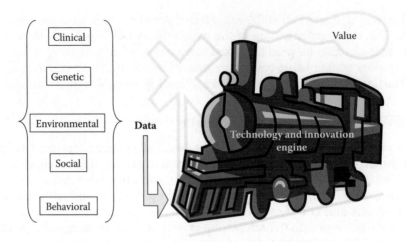

FIGURE 5.5
The technology and innovation engine: data–value transformation in healthcare.

shown in Figure 5.5. These include the medical or clinical data during medical interactions with the provider; the genetic history with links to known predispositions and potential outcomes due to specific health risks; environmental data including occupation-related risks; and social and behavioral data including those related to family, social circles, eating habits, and living conditions. Figure 5.5 also depicts the technology and innovation engine that the provider can harness to transform the raw individual data, convert the fuel to deliver value to the individual through better health outcomes, and enhance the quality of life—the output of power.

5.3 Big Data, mHealth, and Emerging Trends in Healthcare

With the rapid proliferation of smart mobile devices (phones and tablets), sensors, and other smart devices, large amounts of data are being generated, providing greater contextual or situational awareness in real time. *Big data* refers to these huge amounts and varieties of fast-moving data from individuals and groups that can be processed, analyzed, and integrated over periods of time to create significant value by revealing insights into human behavior and activities. At the heart of the concept of big data in healthcare is the individual who is both the source of the data and the recipient of the resulting "value" after the processing/harnessing of the data. These large data sets are beyond the ability of typical database management tools to capture, store, analyze, and manage. According to McKinsey, if the US healthcare system could use "big data creatively and effectively to drive efficiency and quality," the potential value from data in the sector could be more than $300 billion every year, two-thirds of which would be in the form of reducing national healthcare expenditures by about 8% [21].

The key attributes or dimensions of big data are shown in Figure 5.6. The data are coming in at high velocities from real-time transactions; the volume of data is very large and has been likened to drinking from a "fire hose" because it is being generated by a multitude of sources—individuals, transactions, and sensors/devices, among

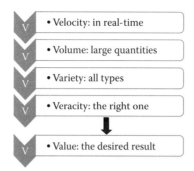

FIGURE 5.6
The attributes or dimensions of big data and delivering value.

others. The variety of data is heterogeneous and diverse; depending on the source, data are structured and unstructured. The former (e.g., vital signs of an individual) can be processed using traditional database management systems, while the latter such as text, image (e.g., food), video, and social media postings (e.g., "likes" on Facebook) are unstructured and highly valuable but also more difficult to process and harness in real time. Finally, it is important to ensure the veracity of the data so that the resulting decisions are made based on noise-free verifiable data. Thus, these four important dimensions of big data must be considered in harnessing the data to deliver value, the ultimate objective, as shown in Figure 5.6.

5.3.1 mHealth and the Patient Protection and Affordable Care Act

The importance of mHealth has been magnified in the context of PPACA since one of its overarching objectives is the delivery of value-based healthcare. As a result, many new innovative paradigms of healthcare delivery are emerging. The three important ones are accountable care organizations (ACOs), the Bundled Payments for Care Improvement (BPCI) initiative, and evidence-based medicine. Each of these is driven by the following tenets: reduce the cost of care, enhance the coordination of care, promote greater risk sharing amongst the stakeholders in the care process, and engage the patient through better case management. These objectives are similar to those of mHealth. The prerequisite for the success of any of these paradigms is the effective harnessing of information from the patient and utilizing it in the continuum of the patient's care.

5.3.1.1 Accountable Care Organization

An ACO is any group of doctors, hospitals, and other healthcare providers who come together voluntarily to give coordinated high-quality care to their Medicare patients [22]. An ACO drives the transformation to a patient-centric model that promotes access and coordination across the continuum of care through better collaboration amongst all the parties in the care process (ambulatory, acute, preventive, and chronic). It aims to avoid unnecessary duplication of services and prevent medical errors to successfully manage the health of the participating population of patients and thrive in a value-based reimbursement environment. When an ACO succeeds both in delivering high-quality care and spending healthcare dollars more wisely, it will share in the savings it achieves for the Medicare program.

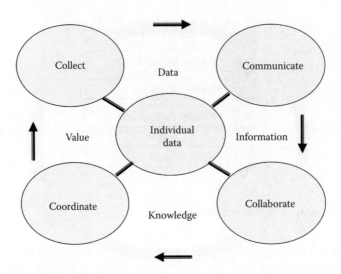

FIGURE 5.7
Individual data and major facets of mHealth.

5.3.1.1.1 Role of mHealth

The coordination of care for better outcomes requires the practice of mHealth to ensure that the participating patient is monitored, diagnosed, treated, and counseled optimally during the entire care process be it in the provider's office during visits, at home, and in other work and social settings. Figure 5.7 shows the central role of data in the context of mHealth. Information about the individual (the specific data suite will be personalized) must be *collected* continuously or periodically in various settings (geographic) and activities (stationary and mobile), *communicated* (securely and in a timely manner and without loss of fidelity) to the *collaborating* entities in the ACO so that the care is *coordinated* to ensure better outcomes for the individual.

The personalized data suite could include typical vital signs (e.g., heart rate, body temperature, and electrocardiogram), food intake (e.g., types, calories, quantity, and frequency), physical activities (e.g., regular movement and exercise), sleep (e.g., timing, duration, and quality), environmental (e.g., outside and inside workplace settings), social interactions (family, friends, coworkers, gatherings, and online) and medication/treatment/therapy regimens. By addressing the entire life cycle, the mHealth paradigm facilitates the success of ACOs. Moreover, all this information represents big data and harnessing it will catalyze the continuous enhancement of healthcare delivery and provide greater value to the patient and the stakeholders, the overarching theme of mHealth. The transformation of data to information to knowledge and value is also shown in Figure 5.7.

5.3.1.2 Bundled Payments for Care Improvement Initiative

Traditionally, Medicare has followed a fee-for-service model and made separate payments to providers for each of the individual services they furnish to beneficiaries for a single illness or course of treatment [23]. This approach has perpetuated the fragmented nature of healthcare delivery with misaligned incentives with very little focus on the quality of care and patient outcomes. The BPCI initiative attempts to align the incentives for all the provider hospitals, postacute care providers, physicians, and other practitioners, allowing them to work closely together across all specialties and settings. Thus, the focus is on delivering the value for better outcomes to the patient. BPCI also promotes greater accountability and transparency in terms of finance and patient care performance during the process and thus has the potential to reduce the overall cost to the Medicare and to the patient.

5.3.1.2.1 Role of mHealth

The success of BPCI depends on the coordination of care and, therefore, requires the practice of mHealth. The same factors related to the importance of information discussed earlier regarding ACOs are applicable to this new paradigm of transforming healthcare delivery by harnessing big data.

5.3.1.3 Evidence-Based Medicine

PPACA provides a robust environment for comparative effectiveness research, systematic reviews, and evidence-based medicine, and implementation of evidence-based medicine should lead to improved quality of care [24]. According to Sackett et al., "Evidence based medicine is the conscientious, explicit, and judicious use of current best evidence in making decisions about the care of individual patients. The practice of evidence-based medicine consists of integration of individual clinical expertise with the best available external clinical evidence from systematic research" [25]. The goal is to enhance the use of data and evidence in delivering value-based care to the individual patients.

5.3.1.3.1 Role of mHealth

Once again, the practice of mHealth is critical to the success of evidence-based medicine as it requires the creation of a warehouse of observational data in the entire care process that will serve as the basis for decision making, leading to value-based care with better patient outcomes.

5.3.2 Occupational Health and mHealth

According to a study from the Institute of Medicine, the type of work and the workplace environment can have a significant impact on an individual's health [26]. For instance, hazards in the physical work environment can lead to injury and illness: nearly 20% of the onsets of new asthma cases are estimated to be caused by occupational exposures [27]. In fact, occupational illness and injury together are the eighth leading cause of death [28]. Organizational attributes of the workplace, such as stress and other psychosocial elements, can also negatively impact health. Other work-related factors such as benefits and salary at work can indirectly influence personal health and lifestyle choices. Therefore, there is significant value for the clinician in having access to an individual's occupational information for diagnosis and clinical care.

5.3.2.1 Role of mHealth

This entire life cycle and lifestyle monitoring of the individual is at the heart of mHealth and it can enhance the quality of life, especially for those significantly impacted by the workplace environment. It can reduce health disparities due to occupation by (1) informing diagnosis for workers exposed to occupational hazards (e.g., chemicals, vapors, and noise); (2) improving treatment and inform plans for return to work by understanding the impact of work practices on health (e.g., irregular shift schedules and the exacerbation of diabetes, and heavy-equipment operators and choice of nondrowsy medication); and (3) providing education opportunities and connections to wellness programs such as company-sponsored health coaching and other community resources [26].

5.3.3 The Value of mHealth

Thus, it is clear from each of these scenarios that mHealth has a critical role to play in the transformation of the healthcare enterprise. The benefits are not only for the patient in terms of better outcomes but also for providers, hospitals, and other stakeholders in the healthcare continuum (Figure 5.1). For instance, a pharmaceutical company developing a new drug to treat diabetes can adopt the mHealth paradigm to continuously assess the impact of the drug on the patient over long periods of time. The results will be more representative and valuable than those obtained in a snapshot assessment of blood sugar level or in laboratory-type controlled conditions.

Likewise, a hospital can implement data analytics to its big-data trove to drive operational improvements by identifying care gaps. By stratifying risks in its patient population, it can redesign treatment protocols that will improve the quality of care and enhance patient outcomes. In the process, it can reduce waste in its operations and the cost of care, all of which will lead to delivering value-based care that is tailored to the individual patients. Recently, the State of Washington demonstrated the benefits of harnessing big data pertaining to ED visits and implementing intervention mechanisms or best practices to prevent abuse of ED by patients. As a consequence, the rate of emergency department visits declined by 9.9%; the rate of visits by frequent clients (who visited five or more times annually) decreased by 10.7%; the rate of visits resulting in a scheduled drug prescription decreased by 24.0%; and the rate of visits with a low-acuity diagnosis decreased by 14.2% [29].

Thus, the value proposition for mHealth in enabling value-based healthcare is significant.

5.3.4 The Grand Challenge: Harnessing Big Data for mHealth

At the core of the success of mHealth is harnessing data—of different types, collected in real time and over long periods of time from large numbers of individuals in different places and engaged in different activities in a range of social, environmental settings. The grand challenge is in collecting or acquiring this data from the individual. The data acquisition process should meet the following major requirements:

- Easy to learn, understand, and use
- Reliable
- Practical
- Natural
- Engaging
- Unobtrusive
- Protects the integrity and security of the data
- Ensures individual's privacy
- Convenient

It should also address the four dimensions of velocity, volume, variety, and veracity of big data discussed earlier (Figure 5.6). So what is needed is an effective and efficient digital ecosystem that is designed based on recognizing the role of the human as being the information node generating the data and the eventual consumer or target for the value being harnessed from the data. In other words, it should be individual centric.

5.4 The WOW

The ideal means for the acquisition of the data is a suite of sensors and devices that would sense, collect, process, store, and transmit the data [30]. The sensor suite could include vital-signs sensors for the individual, a camera for the pictures, a microphone for voice/audio to record foods consumed and social interactions, and environmental sensors for the ambient conditions (for instance, the dust in a metal grinding and polishing facility might affect the person's respiratory health). The sensors and devices must be on or in close proximity to the individual; ideally, they should be wearable by the user to facilitate mobility, one of the key facets of mHealth.

5.4.1 The Value Proposition for Wearables for mHealth

Figure 5.8 shows the value proposition or attributes that wearables must offer for their effective use in mHealth applications.

From a physical standpoint, the wearable must be lightweight and the form factor should be variable to suit the wearer. For instance, if the form factor of the wearable for monitoring the vital signs of an infant prone to sudden infant death syndrome prevents the infant from (physically) lying down properly, it could have significant negative implications. Aesthetics plays a key role in the acceptance and use of any device

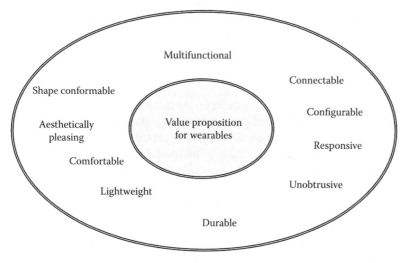

FIGURE 5.8
The value proposition for wearables for mHealth.

or technology. This is especially important when the wearable is used in a workplace or social setting to collect the required data for care coordination and is also visible to others. Therefore, if the wearable on a user is likely to be visible to others, it should be aesthetically pleasing and, optionally, even make a fashion statement while meeting its functionality.

As shown in Figure 5.8, the wearable must have multifunctional capability; this is especially important for mHealth applications where multiple parameters (e.g., heart rate, body temperature, and sleep patterns) must be simultaneously monitored; the wearable must also be configurable to suit the end-use application. The wearable's responsiveness is critical, especially when used for real-time data acquisition and control, e.g., for patients recovering from cardiac surgery. Thus, the design of wearables for desired end-use applications in mHealth must be driven by these attributes.

The current set of wearables in the market such as Fitbit, Nike FuelBand, and Jawbone typically perform a single function (e.g., measuring heart rate) and, so, have a niche or targeted application such as wellness or activity tracking [31–33].

5.4.2 The Metawearable Paradigm for mHealth

The following are the key requirements for a wearable sensor system for mHealth applications that meet both the data acquisition process constraints and the value proposition of wearables [34]:

1. Different *types* of sensors will be needed for the various parameters to be monitored *simultaneously*; for instance, the sensors for monitoring the various vital signs (e.g., heart rate, body temperature, pulse oximetry, and blood glucose level) are of different types. Likewise, for monitoring hazardous gases in the workplace, another class of sensors (e.g., carbon monoxide detection) will be required. Accelerometers will be required to continuously monitor the posture of an elderly person to detect and prevent falls, a never event for which the healthcare provider is not reimbursed by insurance.

2. Different *numbers* of sensors may be needed to obtain the signals to compute a single parameter (e.g., at least three sensors are required to compute ECG).

3. The sensors need to be positioned in *different* locations on the body to acquire the necessary signals (e.g., sensors for ECG go in three different locations on the body, whereas the pulse-ox sensors and accelerometers go in other locations on the body).

4. Different *subsets* of sensors and devices may be used at different times, necessitating their easy attachment and removal, or *plug and play*. For instance, a diabetic patient participating in a drug development study may want to record how the body feels and reacts soon after taking the medication.

5. The signals from the various sensors and in different physical locations (e.g., home, workplace, and restaurants) have to be *sensed, collected, processed, stored,* and *transmitted* to the remote control and coordination location (Figure 5.7).

6. Signals from different types of sensors (e.g., body temperature, ECG, and accelerometers) have to be processed in parallel to evaluate the various parameters in real time.

7. The large numbers of sensors to be deployed would require that the sensors be of low cost and, hence, will have minimal built-in (onboard) processing capabilities.

8. The sensors should be power aware (i.e., have low power requirements).

9. Power must be supplied (distributed) to the various sensors and processors.

Thus, there is a need for a platform that has both a physical form factor and an integrated *information infrastructure*. In addition to serving as a wearable in its own right, the platform must be able to *host* or hold other wearables or sensors in place and provide *data buses* or pathways to carry the signals (and power) between sensors and the information processing components in the wearable networks [35–36]. Simply attaching different types of sensors and processors to different parts of the body is not the ideal solution as it will not be "natural" and will impair the routine activities of the individual. What is required is a *metawearable* [30].

5.4.3 Textiles as a Metawearable

That *metawearables* are textiles because textile meets all the desired attributes of wearables in Figure 5.8. The textile yarns, which are an integral part of the fabric, can serve as data buses or communication pathways for the sensors and processors and can provide the necessary bandwidth required for interactivity. The topology of these data buses can be engineered to meet the desired sensor surface distribution profile, making it a versatile technology platform. Moreover, they can readily accommodate redundancies in the system by providing multiple communication pathways in the network and enable easy power distribution from one or more sources through the conducting textile yarns integrated into the fabric, thereby minimizing the need for onboard power for the sensors. Thus, from a technical performance perspective, a textile fabric (or clothing) is a true metawearable, making it an excellent platform for the incorporation of sensors and processors to harness data from the individual [37].

Humans are used to wearing clothes from the day they are born; in general, no special training is required to wear clothes, i.e., to use the interface. In fact, it is probably the most *universal* of interfaces and is one that humans need, use, and have familiarity with and which can be easily customized [38]. This universal interface of clothing is in contrast to typical computer interfaces/systems (e.g., Windows, Linux, Android, and iOS), each of which has unique characteristics and requires time and effort to learn to use. Moreover, humans enjoy clothing and this universal interface of clothing can be tailored to fit the individual's preferences, needs, and tastes, including body dimensions, budgets, occasions, and moods. Textiles can also be designed to accommodate the constraints imposed by the ambient environment in which the user interacts, i.e., different climates or operating requirements, and still keep the user comfortable.

Textiles are pervasive: they span the continuum of life from infants to senior citizens and from astronauts' space suits (denoting *functionality*) to elegant evening dresses (denoting *fashion*). In today's harried world, an individual is likely to be forgetful and leave the personal mobile communication device (e.g., smartphone) behind, but is unlikely to walk out of the home without clothes! This is yet another compelling reason to view textiles as a metawearable and as an infrastructure for a wearable sensor network. In addition to the two dimensions of functionality (or protection) and fashion, if intelligence can be embedded or integrated into textiles as a *third dimension*, it would lead to the realization of clothing as a personalized and flexible *wearable information infrastructure* [39].

In the process of integrating sensors and processors into textiles, the traditionally passive, yet pervasive, and aesthetically pleasing textiles can be transformed into an interactive infrastructure for the demanding end user while still retaining their aesthetic and comfort characteristics. Finally, since clothing is always "on" the individual, it provides real-time data (both structured and unstructured) that can be traced back to the source, thus effectively addressing the four dimensions of big data—velocity, volume, variety, and veracity, shown in Figure 5.6.

Thus, the metawearable of textiles presents the ideal technology for the successful practice of *personalized* mHealth and enhances the quality of life.

5.5 The Wearable Motherboard or Smart Shirt

The Wearable Motherboard or smart shirt is the first such metawearable that has been successfully developed [40]. The comfort or base fabric provides the necessary physical infrastructure for the Wearable Motherboard shown in Figure 5.9. The base fabric is made from typical textile fibers (e.g., cotton and polyester), where the choice of fibers is dictated by the intended application. The conducting yarns integrated into the fabric serve as data buses and constitute the information infrastructure. An interconnection technology has been developed and used to route the information (signals) through desired paths in the fabric, thereby creating a motherboard that serves as a flexible and wearable framework into which sensors and devices can be plugged. For instance, when sensors for vital signs such as heart rate, electrocardiogram, and body temperature are plugged in, the wearer's physical condition is monitored. The plug-and-play capability greatly expands the scope of applications and range of uses of this technology, both of which are critical for mHealth.

5.5.1 The Wearable Motherboard Architecture

The Wearable Motherboard architecture is shown in Figure 5.10 along with its schematic representation in the garment. The signals from the sensors flow through the flexible data bus integrated into the structure to the multifunction processor/controller. This controller, in turn, processes the signals and transmits them wirelessly (using the appropriate communications protocol) to desired locations (e.g., doctor's office, hospital, and care

FIGURE 5.9
The Wearable Motherboard or smart shirt: adult and infant versions.

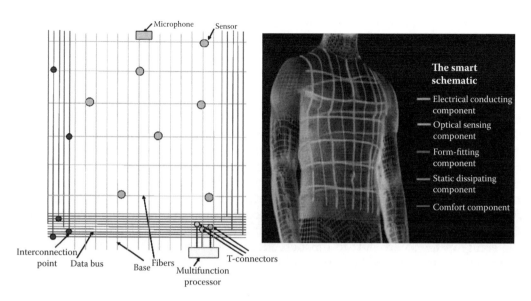

FIGURE 5.10
The Wearable Motherboard architecture.

coordinator). The bus also serves to transmit information *to* the sensors (and, hence, the wearer) from external sources, thus making the smart shirt a valuable bidirectional information infrastructure. The controller provides the required power (energy) to the Wearable Motherboard. With the ubiquity of the smartphone, all the processing and communication can be shifted to it, thereby obviating the need for a stand-alone controller.

Several versions of the smart shirt have been produced and with each succeeding version, the garment has been continually enhanced from all perspectives: functionality, capabilities, comfort, ease of use, and aesthetics. It also successfully withstood the series of industry-standard launderability tests (washing and drying) typically carried out on textiles and apparel.

5.5.2 Testing of the Smart Shirt

The smart shirt has been successfully tested in a variety of applications. The vital signs of individuals, namely, heart rate, respiration rate, ECG, and body temperature have been measured using commercial off-the-shelf sensors that plugged into the smart shirt worn by the individual. Initial testing was done at Crawford Long Hospital in Atlanta and followed by another set of tests in the Department of Physiology at Emory University to prove the concept and feasibility of vital-signs monitoring using the smart shirt. A physician reviewing the data both from the smart shirt and directly from the body (the control) concluded that the waveforms and heart-rate data were identical and confirmed the functionality of the smart shirt.

The medical version of the smart shirt was then tested in the researchers' laboratory by using a Nihon Kohden ECG hospital monitor. A subject wearing the smart shirt moved around freely in the laboratory and in the building, with the vital-signs data (heart rate, respiration rate, and ECG waveform) being remotely (wirelessly) recorded on the machine. Each test typically lasted 30 min and 10 tests were conducted on several different days.

The consistent quality of the signals and waveforms during the tests clearly demonstrated that the smart shirt successfully monitored the vital signs of the subject and also interfaced with a commercial system that is used in hospitals.

Then the smart shirt was tested in an area hospital that deployed the Nihon Kohden system. A subject wore the garment and the data were transmitted wirelessly by the controller on the garment to the hospital monitor. Initially, the subject was stationary and the data (heart rate, respiration rate, and ECG waveform) were recorded in the monitor. Simultaneously, the vital signs were recorded directly from the subject and these data served as the *control* for the experiment. Then the subject walked at a measured pace in the hospital and the data were wirelessly recorded by the monitor (Figure 5.11). The subject then walked very fast in the hospital and the quality of the data was similar to that shown in Figure 5.11, with both the control and readings from the smart shirt being identical. Finally, the subject ran in the hospital and the vital signs and waveform were recorded; the traces from the smart shirt and the control were again identical, thus demonstrating the ability of the smart shirt to function effectively even when the user was mobile, a key requirement for mHealth applications since the user would be mobile with the corresponding changes in the ambient environment.

To demonstrate the simultaneous monitoring of multiple parameters, the user's body temperature, heart rate, respiration rate, and ECG were monitored concurrently while the user was listening to music from an MP3 player plugged into the shirt through a pair of headphones that was also plugged into the shirt. To demonstrate its successful operation at high speeds, it was tested on a race-car driver on the Daytona 500 track. At speeds exceeding 180 mi/h, the driver's ECG signals were recorded on the monitor placed in the

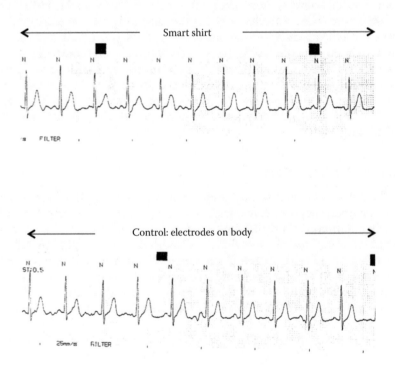

FIGURE 5.11
Wireless transmission of vital signs from the smart shirt. *Top trace*: through the smart shirt; *bottom trace*: control.

pit on the sidelines. An infant version of the smart shirt for monitoring vital signs was also successfully tested on two healthy babies following institutional review board (IRB)–approved protocols at Egleston Hospital Children's Healthcare of Atlanta.

The garment is comfortable and easy to wear and take off, similar to a typical undershirt. All these tests conclusively demonstrated the ability of the smart shirt to unobtrusively monitor the vital signs of individuals (from infants to adults) engaged in normal activities in an easy-to-use form factor with the convenience and familiarity associated with a garment.

5.5.3 Realizing mHealth through the Wearable Motherboard

We will now illustrate the value of the motherboard architecture for mHealth as it enables the *same garment* to be quickly reconfigured for a different application by changing the suite of sensors.

Consider the daily life of a metalworker leaving home with a garment that is measuring his vital signs (e.g., heart rate, body temperature, and SpO_2) through the plugged-in sensors; optionally, the conducting fibers in the garment can themselves act as sensors to capture the wearer's heart rate and ECG [41]. Upon reaching the factory, the worker can plug in special-purpose sensors into the garment to monitor particles and gases in the factory environment (in addition to the vital signs that are already being monitored). The worker can also plug in a microphone into the garment to record his physical condition as the day progresses, especially the impact of the environment on his respiratory health. Upon leaving, the factory environmental sensors can be replaced with those for measuring the pollution in the air (e.g., pollen count), so that the impact of the external environment on the individual's health can be tracked. These sensors can then be detached upon reaching home and a different garment can be used to record the vital signs (if they are needed as part of the care protocol) or to listen to music through the headphones plugged into the garment. That is the power of the Wearable Motherboard paradigm: it is an effective metawearable and the structure has the *look* and *feel* of traditional textiles, with the fabric serving as a comfortable information infrastructure for mHealth.

 Multifunctional

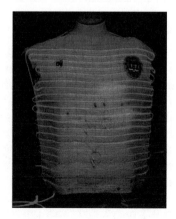

FIGURE 5.12
The military version of the smart shirt.

5.5.3.1 Another Example

The military version of the smart shirt worn by the soldier in battlefield shown in Figure 5.12 uses optical fibers to detect the penetration of a projectile (e.g., bullets and shrapnel); at the same time, the other sensors monitor the soldier's vital signs. In fact, the motivation for the Wearable Motherboard came from the need to enhance the quality of life for soldiers on the battlefield; since medical resources are limited in a combat zone, a technology that could in real time determine the condition of the wounded soldier would help the medic save valuable time in attending to the soldiers who could be saved during the golden hour. A soldier always wears his uniform; so it becomes the infrastructure for the sensors and processors and the Wearable Motherboard paradigm is born. When a soldier is shot, the optical fibers in the smart shirt are penetrated; the extent of the injury along with soldier's vital signs is sent to the medics for prompt intervention that could potentially save the soldier's life.

Upon returning from combat, the regular version of the garment can be used to continuously monitor the soldier and facilitate easy integration into society, yet another critical facet of mHealth. The Wearable Motherboard can be tailored to be a head cap so that the soldier's brain activity can be tracked by recording the electroencephalogram and, thus, serve as a viable and easy-to-use unobtrusive platform for issues related to post-traumatic stress disorders.

These two examples illustrate the value of wearable technology in the form of the Wearable Motherboard, a metawearable for the successful practice of mHealth, i.e., coordinating and providing holistic care to individuals. It will have a significant impact on the practice of medicine since it fulfills the critical need for a viable and easy-to-use technology that can enhance the quality of life while reducing healthcare costs across the continuum of life, namely, from newborns to senior citizens, and across the continuum of medical care, namely, from homes to hospitals and everywhere in between. By having a technology that not only is ubiquitous but also has the ability and intelligence to respond to the changes in the needs of the wearer, the quality of preventive care can be significantly enhanced, thus reinforcing the paradigm that "investment in prevention is significantly less than the cost of treatment."

5.6 Looking Ahead: WOW and the Future of mHealth

Let us project the metalworker scenario presented earlier into the future. The Wearable Motherboard provides the platform for the unobtrusive collection of the worker's data in real time and at all times. Imagine if these data were automatically fed to the individual's EHR, which is already augmented with the worker's occupational information using the Standard Occupational Classification (SOC) code system developed by the US Bureau of Labor Statistics. Further imagine that this EHR is linked to O*Net OnLine, an online web portal sponsored by the US Department of Labor, which provides occupation-specific information [42]. Figure 5.13 shows a snapshot of information for "grinding and polishing workers, hand" corresponding to the worker's SOC code in the EHR.

Since the worker is suffering from asthma, a chronic disease, which affects nearly 8% of the US adult population [43], the primary care physician can examine the data collected over time from the EHR and determine if the worker is taking adequate precautions by using personal protective equipment in the workplace so that the occupational hazards (known

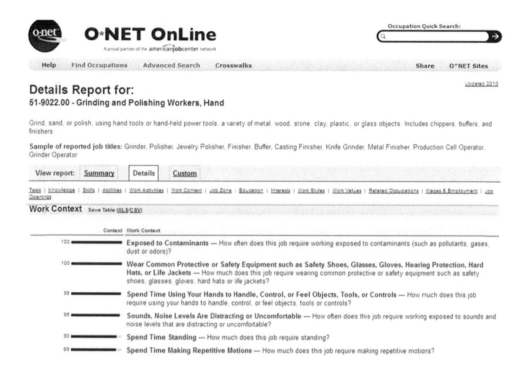

FIGURE 5.13
A snapshot of occupation-specific information from O*Net OnLine.

easily to the physician from the link in the EHR to the O*Net database, Figure 5.13) do not exacerbate the worker's asthma. This big-data mining will add significant value to the physician's clinical examination during the visit and, thereby, lead to better counseling, treatment, and care of the worker. For instance, if the respiratory function is found to be significantly better on nonworking days, the impact of the workplace on the worker's condition can be effectively gauged (assuming, of course, that all other factors, e.g., social and food, are same or similar) and the worker can be advised to seek other job opportunities to have a better overall quality of life. This is indeed the future of mHealth powered by wearables—providing predictive personalized treatment and counseling anytime and anywhere.

5.6.1 The Chain Reaction

Taking this scenario one step further, assume that the worker suddenly becomes ill in the workplace and has breathing difficulties. The data fed from the garment can be sent directly to the ambulance personnel and to the nearby hospital and, even as the worker is being rushed in the ambulance, preparations can be made in the hospital to receive and treat the worker immediately upon arrival. Since the hospital can access all the information through the EHR (including the occupational information), the *right* care can be given at the *right* time at the *right* cost and a life can possibly be saved. Upon release from the hospital, the individual's recovery can be monitored and the care coordinated by taking into account the employment and social aspects, among other factors, thus providing value-based care that is tailored to the individual, the goal of mHealth, once again realized through the technology of wearables.

The value of having access to all the patient information is also realized by the hospital in a chain reaction. By providing the right treatment, the hospital can reduce preventable medical errors. By integrating radio-frequency identification (RFID) tags into the Wearable Motherboard, it can better track and identify patients correctly so that never events such as amputating the incorrect body part or transfusing the incorrect blood type can be avoided. By integrating pressure sensors and accelerometers into the garment, the patient on the hospital bed can be continuously monitored and the patient's position shifted to avoid the occurrence of bedsores (another never event); the accelerometer data can be used to detect and prevent patient falls, another never event. In all instances, the metrics of cost, quality, and access to care will be positively impacted, along with the patient outcomes. The result will be the delivery of patient-centric value-based care. This chain reaction will ripple through, resulting in a larger client base because of better performance and enhanced reputation of the hospital. Thus, the desired transformation of the healthcare system can be realized and the practice of mHealth will become an everyday reality.

5.7 Conclusions

In conclusion, smart wearables—in the form of the metawearable clothing—are increasingly becoming an integral part of our digital lives and are critical for the successful practice of mHealth. They can also transform the current healthcare delivery system by enabling the harnessing of big data to create the patient-centric value-based system of tomorrow.

Acknowledgments

The patient's bill of rights and the provider's bill of rights started as class exercises for students in Professor Jayaraman's healthcare management courses. They have evolved over time with refinements from batches of students over the years. Funding for the development of the Wearable Motherboard was provided by the Defense Advanced Research Projects Agency (DARPA) through the US Department of the Navy under Contract No. N66001-96-C-8639. Professor Jayaraman would also like to acknowledge funding from Kolon Corporation for his research in technology and innovation focusing on wearables and healthcare. Finally, the authors would like to thank the former members of the Wearable Motherboard team for their contributions.

References

1. "Crossing the Quality Chasm: A New Health System for the 21st Century," Committee on Quality of Healthcare in America, Institute of Medicine, Washington, DC, 2001.
2. HealthCast 2020: Creating a Sustainable Future, PricewaterhouseCoopers, 2005.

3. National Health Expenditure Projections 2012–2022, http://www.cms.gov/Research-Statistics -Data-and-Systems/Statistics-Trends-and-Reports/NationalHealthExpendData/National HealthAccountsProjected.html, Last accessed: December 25, 2014.

4. The World Bank, Health Expenditure, Total (% of GDP), http://data.worldbank.org/indicator /SH.XPD.TOTL.ZS/countries, Last accessed: December 15, 2014.

5. Living Well with Chronic Illness: A Call for Public Health Action, Institute of Medicine Report, National Academies Press, Washington, DC, January 2012.

6. Health Insurance Marketplace Summary Report, ASPE Issue Brief, US Department of Health and Human Services, May 1, 2014, http://aspe.hhs.gov, Last accessed: May 27, 2014.

7. "To Err is Human: Building a Safer Health System," Committee on Quality of Health Care in America, Institute of Medicine, Washington, DC, 2000.

8. Provider Preventable Conditions, http://www.medicaid.gov/Medicaid-CHIP-Program-Infor mation/By-Topics/Financing-and-Reimbursement/Provider-Preventable-Conditions.html, Last accessed: May 26, 2014.

9. Hsia, R. Y., Kellermann, A. L., and Shen, Y., "Factors Associated with Closures of Emergency Departments in the United States," *JAMA* 2011; 305(19):1978–1985. doi:10.1001/jama.2011.620.

10. Retail Medical Clinics: From Foe to Friend?, Accenture, http://www.accenture.com, Last accessed: May 26, 2014.

11. Meir, B., McGinty, C., and Creswell, J., "Hospital Billing Varies Wildly, Government Data Shows," *The New York Times*, May 8, 2013.

12. "1 Billion Smartphones Shipped Worldwide in 2013," http://www.pcworld.com/article /2091940/global-smartphone-shipments-topped-1-billion-in-2013.html, Last accessed: May 25, 2014.

13. Shroeder, S. A., "We Can Do Better—Improving the Health of the American People," *N Engl J Med* 2007; 357:1221–1228, September 20, 2007.

14. PwC Health Research Institute Top Issues Consumer Survey, PricewaterhouseCoopers, 2013.

15. "Emerging mhealth: Paths for Growth," PricewaterhouseCoopers, June 2012.

16. Park, S., Chung, K., and Jayaraman, S., "Wearables: Fundamentals, Advancements and Roadmap for the Future," in *Wearable Sensors, in Print,* Sazonov, E., and Newman, M. (eds.), Elsevier, Amsterdam, Netherlands, 2014.

17. Park, S., and Jayaraman, S., "On Innovation, Quality of Life and Technology of BodyNets," Proceedings of the Third International Conference on Body Area Networks, March 13–15, 2008, Tempe, Arizona 2008.

18. DeNavas-Walt, C., Proctor, B. D., and Smith, J. C., "Income, Poverty, and Health Insurance Coverage in the United States: 2011," *Current Population Reports*, pp. 60–243, U.S. Census Bureau, Washington, DC, 2012.

19. 50 Years after Thalidomide: Why Regulation Matters, http://blogs.fda.gov/fdavoice/index .php/2012/02/50-years-after-thalidomide-why-regulation-matters/, Last accessed: May 26, 2014.

20. Marcus, A. D., "Patients as Partners," *The Wall Street Journal*, April 16, 2012, p. R2.

21. McKinsey Global Institute, Big Data: The Next Frontier for Innovation, Competition and Productivity, June 2011.

22. Accountable Care Organizations, http://www.cms.gov/Medicare/Medicare-Fee-for-Service -Payment/ACO/, Last accessed: May 28, 2014.

23. Bundled Payments for Care Improvement (BPCI) Initiative: General Information, http://innovation .cms.gov/initiatives/bundled-payments/, Last accessed: May 28, 2014.

24. Hughes, G. D., "Evidence-Based Medicine in Health Care Reform," Otolaryngol Head Neck Surg 2011; 145(4):526–529, October 2011.

25. Sackett, D. L., Rosenberg, W. M. C., Gray, J. A. M., Haynes, R. B., and Richardson, W. S., "Evidence Based Medicine: What It Is and What It Isn't," *British Medical Journal* 1996; 312(7023):71–72, January 13, 1996.

26. Incorporating Occupational Information in Electronic Health Records, Institute of Medicine Letter Report, National Academies Press, September 2011.

27. Torén, K., and Blanc, P. D., "Asthma Caused by Occupational Exposures is Common—A Systematic Analysis of Estimates of The Population-Attributable Fraction," *BMC Pulm Med* 2009; 9:7.

28. Steenland, K., Burnett, C., Lalich, N., Ward, E., and Hurrell, J., "Dying for Work: the Magnitude of U.S. Mortality from Selected Causes of Death Associated with Occupation," *Am J Ind Med* 2003; 43:461–482.

29. Emergency Department Utilization: Update on Assumed Savings from Best Practices Implementation, Washington State Healthcare Authority, Olympia, Washington, March 20, 2014.

30. Park, S., and Jayaraman, S., "The Wearables Revolution and Big Data: The Textile Lineage," Proceedings of the 1st International Conference on Digital Technologies for the Textile Industries, Manchester, United Kingdom, September 5–6, 2013.

31. Fitbit, http://www.fitbit.com/, Last accessed: May 20, 2014.

32. Nike FuelBand, http://store.nike.com/us/en_us/pd/fuelband-se/pid-924485/pgid-924484?cp=usns_kw_AL!1778!3!30651044462!e!!g!nike%20fuelband, Last accessed: May 21, 2014.

33. Jawbone, https://jawbone.com/up, Last accessed: March 20, 2014.

34. Park, S., and Jayaraman, S., "Sensor Networks and the i-Textiles Paradigm," Proceedings of the Next Generation PC 2005 International Conference, COEX, Seoul, South Korea, pp. 163–167, November 3–4, 2005.

35. Jayaraman, S., "Fabric Is the Computer: Fact or Fiction?" Keynote Talk at Workshop on Modeling, Analysis and Middleware Support for Electronic Textiles (MAMSET) at ASPLOS-X (Tenth International Conference on Architectural Support for Programming Languages and Operating Systems), San Jose, California, October 6, 2002.

36. Park, S., and Jayaraman, S., "Smart Textiles: Wearable Electronic Systems," *MRS Bull* 2003; 28(8):586–591, August 2003.

37. Park, S., and Jayaraman, S., "Wearable Sensor Network: A Framework for Harnessing Ambient Intelligence," *J Amb Intel Smart En* 2009; 1(2):117–128.

38. Gopalsamy, C., Park, S., Rajamanickam, R., and Jayaraman, S., "The Wearable Motherboard™: The First Generation of Adaptive and Responsive Textile Structures (ARTS) for Medical Applications," *Virtual Real* 1999; 4:152–168.

39. S. Park, and S. Jayaraman, "Adaptive and Responsive Textile Structures," in *Smart Fibers, Fabrics and Clothing: Fundamentals and Applications,* Tao, X. (ed.), pp. 226–245, Woodhead Publishing Limited, Cambridge, United Kingdom, 2001.

40. Rajamanickam, R., Park, S., and Jayaraman, S., "A Structured Methodology for the Design and Development of Textile Structures in a Concurrent Engineering Environment," *J Text I* 1998; 89(3):44–62.

41. Jayaraman, S., and Park, S., A Novel Fabric-Based Sensor for Monitoring Vital Signs, US Patent No. 6,970,731, November 29, 2005.

42. O*Net Online, http://www.onetonline.org/link/summary/51-9022.00, Last accessed: May 29, 2014.

43. Asthma, FastStats, Centers for Diseases Control and Prevention, http://www.cdc.gov/nchs/fastats/asthma.htm, Last accessed: May 29, 2014.

6

Wireless Instrumentation and Biomedical Applications

João Paulo Carmo and José Higino Correia

CONTENTS

6.1 Introduction

Modern telemedicine, telehealth, and mobile health systems are almost all dependent on the appropriately designed and implemented sensors, wireless instruments and instrumentation systems, and networks for gathering the relevant information and transmitting it to target databases for further processing. Therefore, it is important to have an appreciation of the basic technology and operational principles of wireless instruments and instrumentation systems. This chapter describes such technologies and gives examples as they are applied in biomedicine. Many other examples can be found throughout the book.

Almost anyone knows that a measurement is the process of comparing a quantity with another one of the same species (e.g., length, volume, and area) whose result is a number. Such a measurement can be done without ambiguities in a straightforward manner with the help of a measurement system, which requires a specific instrument for achieving such a task [1]. This instrument can be a simple instrument for directly measuring a physical quantity (e.g., a voltmeter for measuring an electric potential difference between two points of a circuit, an ammeter for measuring a current flowing in a branch of a circuit, and a thermometer for measuring temperatures) and can take one of these following forms regarding their internal working and signal processing: analog or digital [2]. The ability to connect and communicate with external devices [3] (using dedicated cables and/or communication networks) as well as the inherent flexibility [4] (the ease of adding new functions and/or reconfiguration of the existing ones) makes digital instruments undoubtedly

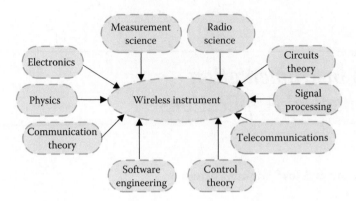

FIGURE 6.1
The multidisciplinary areas to take into account when designing a wireless instrument.

those with major potential for use in several fields of human activity (heavy industry, medicine, and transportation systems are some application examples). The current evolutionary step of measurement instruments is integrating functions to provide wireless transfer of data. In this sequence of ideas, the developments of the microelectronics and microsystems industry allowed the engineers to successfully develop this new measuring paradigm [5]. This resulted in new possibilities for measuring, acquiring, transferring, and storing and for analyzing the physical world: the embedded systems [6] and wireless sensors networks [7] are two new ways for achieving such goals, with wireless being the leader. This lead to the wireless instrument concept, which by its nature requires multidisciplinary concepts such as measurement science, electronic circuits design, microelectronics and microsystems fabrication, wireless communication systems, and networking [8]. Figure 6.1 reinforces this idea by illustrating how many disciplines must be employed for designing a wireless instrument. The primary focus of this chapter is the presentation and integration of these concepts. This chapter also presents biomedical applications based on wireless instruments and the new application concepts.

6.2 Measurement Systems

Figure 6.2 shows a block diagram of a generic measurement instrument. The blocks of a measurement system can be grouped into three major types: the real world (representing

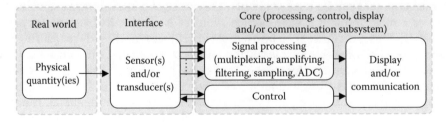

FIGURE 6.2
Block diagram of a generic measurement instrument.

the physical quantity to be acquired), the interface block (with the sensor), and the core (e.g., the instrument itself).

There are situations where the interface block can be part of the core; e.g., a voltmeter does not require any external sensor because this is already embedded inside the measuring instrument; thus, the sensing tips can directly touch the electrical potentials. In this context, it must be clarified that a sensor should not be confused with an instrument, because the latter can perform the same function of the former; but if the former is passive (for example, a physical quantity–dependent resistor mounted in a Whitestone bridge), then additional circuits must be provided for obtaining the signal from the sensor. This means that the set composed by sensor and powering system makes an instrument.

The core blocks can include electronics of control for acquisition from the sensor. The core also provides signal-processing functions for signal-conditioning purposes. These latter functions include amplification (with the possibility of adjusting the gain), filtering (either low-pass or band-pass or even high-pass filtering), and analog-to-digital conversion (ADC). Then the user can read the acquired values in a dedicated display. A more sophisticated core system can interface with the external world either to connect several measurement instruments or to send data to a central unit for further processing. These communications can be done using wired buses (e.g., I²E bus [9], general-purpose interface bus [GPIB] [10], RS-232 [11–12], parallel ports, [13] or even universal serial bus [USB] [14]) or wirelessly (e.g., IEEE 802.11 [15], ZigBee [16], or Bluetooth [17] or using a customized solution [18]).

The core blocks of measurement instruments can be analog or digital. Analog is the less versatile core because it requires the presence of a person to record the measurements. This type of instrument is very limited and very difficult to be adapted to wide disparities of signals to measure. Furthermore, it is not possible to send wirelessly the physical quantities unless a specific interface with an analog modulation scheme is provided.

A digital core can be used for connecting sensors (whose output can provide signals in the analog or in the digital domain). The difference from their analog counterparts resides in the conversion component used in the final processing stage, e.g., the sampler and the analog-to-digital converter block. The inclusion of multiplexers enables the acquisition of multiple channels with a single measurement instrument. This topic is the focus of discussion in the following subsection. Then and after the analog-to-digital conversion, the acquired measurements can be presented in a numerical display. These cores can also be built with internal memory for storing the ADC-converted samples for rendering in a more complete display system (e.g., a planar screen) or for remote transmitting through a communication interface. This core also allows for changing and/or for programming the amplifier's gain, thus allowing for adapting to wide variations in physical quantities. Finally and thanks to the latest developments in microelectronics, by making available sensors with digital outputs (for example, integrated monolithic temperature sensors [19], Hall-effect magnetometers [20], and accelerometers [21], among others), it is possible to have full digital and reusable cores.

The judicious selection of sensors and cores can be decisive points for fabricating wireless instruments with low power, reduced sizes, and low prices. This statement is especially evident for measurement instruments composed of reusable cores (for controlling and displaying/communicating), monolithic sensors (for signal acquisition), and on-chip signal-conditioning circuits [22] (for signal processing).

6.2.1 Multiplexing Structures

The multiplexing operation is required in measurement instruments for simultaneously acquiring more than one signal at once. Signal multiplexing is not as simple a matter to treat at first sight as it seems to be. The first issue is concerned with the availability of a variety of multiplexing structures and the decision to select the most suitable. This poses trade-off problems related to implementation costs (e.g., the more complex the multiplexing system, the higher will be the cost) and specifications (e.g., the more general purpose the device, the higher will be the complexity and the cost). Secondly, it is mandatory to have a clear knowledge of the input characteristics. The most important issues are the input impedance, the dynamic range, the bandwidth, the balancing type of signals (e.g., single ended or differential), power-supply interference rejection, and interference between input channels due to the multiplexer and noise.

The multiplexing configuration can assume one of the following classifications: either low level or high level for analog or digital multiplexers, respectively.

Figure 6.3 shows the simplest structure. The signals at the outputs of sensors connect to an analog multiplexer (with single-ended or differential inputs). It is important to have a variable-gain amplifier when sensors with different signals are used, in order to provide signals within the full dynamic range of inputs of the ADC. This structure poses significant restrictions for solving speed bottlenecks: first, the multiplexer must be fast enough for switching the different analog channels; second, the bandwidth of the ADC and sample-and-hold (S/H) (which is actually a part of the ADC) circuits must be high enough to avoid distortion of the analog signals for conversion.

The low-level shared-amplifier configuration is less flexible in terms of plurality of physical quantities to measure; the ADC must be compatible with the sensor with the highest bandwidth; the direct acquisition of multiple signals can pose shielding problems because the sensors can be located very far away from each other. However, the low-level shared-amplifier configuration is the configuration with the highest potential for fabricating small electronic modules with a high degree of integration (e.g., one multiplexer, one amplifier, one analog filter, one S/H plus ADC, and control electronics).

Figure 6.4 structure is very similar to that illustrated in Figure 6.3. The only exception is the use of sensors with associated amplifiers. These dedicated amplifiers guarantee signals with equal excursions for use in the full dynamic range of the ADC. The bandwidth considerations are the same as those made in the previous multiplexing configuration. A high degree of integration is still possible to achieve with this configuration, but by sacrificing the compactness and small size due to the use of multiple amplifiers.

The low-level multiplexing configurations have a high integration potential, allowing the fabrication of wireless instruments with small sizes and low-power consumption. However, the shared nature of the analog multiplexer can cause problems, such as finite

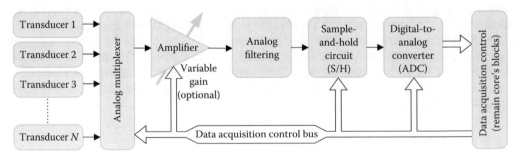

FIGURE 6.3
Low-level multiplexing configurations with shared amplifier.

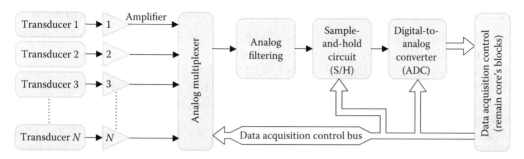

FIGURE 6.4
Low-level multiplexing configurations with dedicated amplifiers.

impedance of nonselected analog channels, cross talk between channels, nonzero switching times between channels, and handling of different bandwidths between channels.

Figure 6.5 shows a high-level multiplexing configuration, providing a dedicated set of amplifiers, analog filters, S/H circuits, and ADCs for each sensor. After conversion to the digital domain, the acquired physical measures are digitally multiplexed, avoiding the problems of the analog multiplexers. This configuration can pose integration restrictions by requiring circuits with increased sizes because each analog channel requires a dedicated signal processing chain. However and despite these drawbacks, a high-level configuration is the most flexible of those analyzed because different amplifiers and filters can be provided for an extensive set of sensors. Moreover, this configuration offers the possibility to select the most suitable ADC for the respective analog chain. More important is that it can accommodate channels with different sampling frequencies, since the switching speed of the analog multiplexer is not exceeded. This configuration also allows a variety of channel selecting policies for desired channels with desired sampling frequencies.

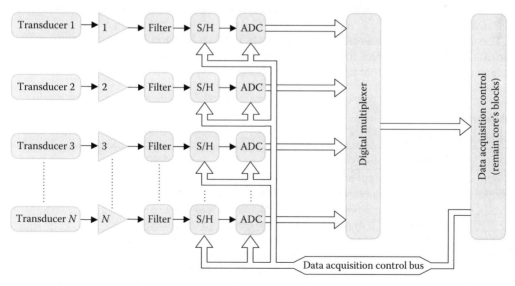

FIGURE 6.5
High-level multiplexing configuration.

6.2.2 Wireless Instruments Seen from the Communication Protocol Point of View

The design of a wireless instrument (as well as a generic measurement instrument) cannot be done without taking into account the communication protocol. Table 6.1 shows the layer structure proposed by the Open Systems Interconnection (OSI) model [23].

Basically, layer 1 specifies the modulation in conjunction with the line coding m (coding and modulation = codulation [24]), the direct-current (DC) balancing, and single-ended/differential-ended balancing.

Normally, layer 2 is divided in two sublayers, e.g., the MAC and the LLC. The MAC sublayer is on the bottom and defines the mechanisms (or rules) in which a given transmitter is allowed (or not!) to access the physical medium for signal injection (corresponding to the data that are intended to be sent toward the target receiver). The top LLC sublayer specifies the types of frames: data or control frames. This sublayer also provides the clear definition of frame formats in terms of their contents (their fields). The data frames are used to transport the useful information in a field known as payload, while the other frames (e.g., the control frames) are used to inform the transmitter if the previous transmissions were received and processed by the receiver with or without errors. The control frames can also be used for doing the flow control, in order to avoid data congestion in the receiver or across the network (with the consequence that loss of data occurs).

The *network layer* specifies a set of procedures for guaranteeing a reliable transmission between consecutive nodes along the network. Examples of procedures include, for example, the detection/correction of transmission errors and flow controlling. This layer also establishes the routing paths for the messages.

The *transport layer* ensures a reliable communication between terminals, e.g., between end-to-end users. This layer also provides error-control procedures to verify the correct reception of all packets that form the messages (e.g., error-free packets). Another important procedure provided by the transport layer is grouping the packets in the correct sequence order for obtaining a correct reassembled segment.

The *session layer* provides mechanisms for allowing the hosts to establish a communication. This layer also provides recovery mechanisms when an interruption occurs during the communication.

The definition and conversion between data formats is done in the presentation layer. Normally, it is in this layer that the data are encoded and/or converted to (and obviously converted from) the format used by the application.

TABLE 6.1

Layers in the OSI Model

	Data Unity	Layer	Function
Host layers	Data	Application	Final communicating application
		Presentation	Data formatting and cryptography
		Session	Communication between hosts
	Segments	Transport	End-to-end communication reliability
Medium layers	Packets	Network	Node-to-node routing and communication reliability
	Frames	Data link	Physical identification (medium access control [MAC] and logical link control [LLC] sublayers)
	Bits	Physical	Signal transmission through the communication channel (codulation; other physical aspects of signals)

The last layer (e.g., the *application layer*) provides interfaces between the application itself and between the protocol stack on the bottom.

Figure 6.6 provides a better understanding of how these concepts can be applied when designing a wireless instrument. This example helps to identify the blocks inside the wireless instrument that implement functions defined by the OSI model.

The application is simply the measurement part of the wireless instrument (e.g., the measurement instrument), which is composed of the core measurement subset (this subset includes the signal-acquisition block, the signal-processing block, the control block, the memory block—a part of the control block—and the display control—actually, this block is optional in wireless instruments). Additionally and as previously stated in Subsection 6.2.1, the interface block can be a part of the measurement instrument. The remaining part of the wireless instrument is implemented by the core communication subset. This subset is responsible for communicating with external devices (other wireless instruments or with a central processing unit or even a remote measurement unit controlled by a remote user).

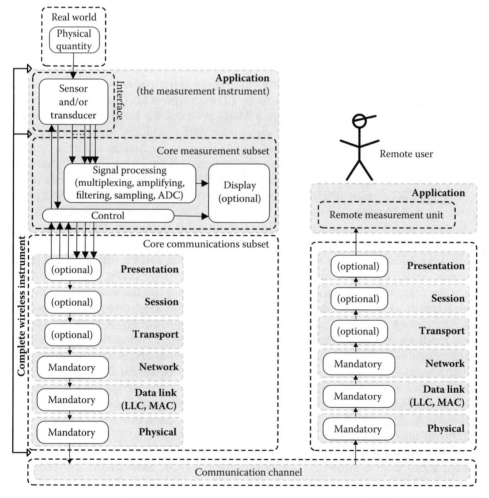

FIGURE 6.6
Relation between wireless instruments with the stacks of the OSI model.

Normally and as shown in Figure 6.6, there are three optional functions inside a wireless instrument. This results from the need to simplify the transmission procedures in order to minimize the latencies (this is the case of the IEEE 802.15.4 that supports the low-level layers of the ZigBee protocol stack) and maximize the data throughput.

The most common function of the network layer is the cyclic-redundancy check (CRC) generation (by the wireless instrument in the transmitter's side) and decoding (in the receiver's side) for detecting transmission errors. It is possible to use forward error correction (FEC) schemes for error correcting but with the cost of sending redundant bits for each bit of information (e.g., each bit of acquired data) and larger bit rates. In the case of auto-repeat reQuest (ARQ) schemes, the CRC generation and decoding can be used by the receiver to do a request to the transmitter for retransmission of an erroneous packet.

The LLC sublayer functions cannot be bypassed (or at least, a simple implementation must be provided) because at least a frame-alignment word (or synchronization character) must be provided in the transmitter in order to allow the receiver to start the bits reception. The LLC implemented by Carmo et al. [18] uses a header composed of an alternate sequence (with a length equal to an integer multiple of 8) 1s and eight 0s for DC balance, while the remaining header contains a synchronization character with 8 bits (this complete header is used by both control and data frames).

The simpler way for implementing MAC sublayer functions is in point-to-point communications with only one transmitter and receiver. In this situation, the transmitter can send data without restrictions because the communication channel is always available and ready for use. Sophisticated MAC protocols are required for managing the medium access when the general scenario (with a multiplicity of transmitters and a multiplicity of receivers) is present. Silva et al. [25] proposes a MAC protocol for transmitting signals from transmitters placed on a multiplicity of paraplegic patients (doing hydrotherapy inside a swimming pool) into a base station, which stores the data for further analysis by a health professional. Note that the most common applications are based on simple point-to-point configurations as it is the case of the work done by Dias et al. [26], where a wireless instrument acquires EEG signals and transmits them into a base station connected to a personal computer for data logging. This solution uses wireless modules based on IEEE 802.15.4, but a solution based on a microcontroller with a simple radio-frequency (RF) module could be used for reducing the latencies.

Finally and taking into account the wired case (which can be directly transposed to the wireless case), the physical layer can be implemented by doing a direct connection between the transmitter and the receiver (e.g., connecting the transmitter's output to the receiver's input). A wireless instrument must provide at least one modulation scheme in order to reduce the errors introduced during communication. A huge set of flexible wireless modules from third-party manufacturers ready for connection into the wireless instrument are available at low cost [27–30]. Modulation is an important issue in wireless instruments, but a coding scheme is recommended either for DC balancing or even for error control and for synchronization. In this context, the RF complementary metal–oxide–semiconductor (CMOS) receiver at 433 MHz for integration on implantable devices by Carmo et al. [31] was designed taking into account the following *codulation* scheme: simultaneous on/off keying (OOK) modulation and biphase code [31]. Another example of physical layer definition is the RF transmitter proposed by Morais et al. [32] for operation at 433 MHz but with a different codulation scheme: OOK modulation and pulse-width modulation (PWM) code. This RF transmitter is compatible with the commercial receiver unit model LM-RXAM2433 (from the manufacturer Low Power Radio Solutions [LPRS] Inc.) and was tested with success for soil moisture measurements [33].

6.3 Technology for Wireless Systems

6.3.1 Operational Issues

The selection of the operation frequency is important. In the first place, the dimensions of the antennas are imposed by the frequency. For an acceptable efficiency, the antenna size must be the same order as one-quarter of the wavelength, λ [m], which is given by $\lambda = c/f$, with $c = 3 \times 10^8$ m·s⁻¹ being the speed of light in a vacuum and f [Hz] the frequency of operation. Decreasing the dimension of an antenna implies the use of high frequencies for achieving such size. This issue can be solved by modulation. The most used modulations in wireless instruments (especially in laboratory environments, e.g., closed and relatively absent of common band interferences) are *amplitude-shift keying* (ASK), *phase-shift keying* (PSK), and *frequency-shift keying* (FSK). Both ASK and PSK need the same bandwidth given by BW = $2R_b$ [Hz], where R_b [bps] is the bit rate per second. The bandwidth required by FSK modulation is slightly higher and is BW = $2R_b + |f_1 - f_2|$, where $|f_1 - f_2|$ is the frequency shift between the two carriers f_1 [Hz] and f_2 [Hz]. The bit error probability (BEP) of each modulation is in Table 6.2 [34], as well as the plots of their respective values as function of the ratio E_b/N_0 shown in Figure 6.7. The ratio E_b/N_0 is the energy per bit, E_b [J] divided by the spectral density of the noise (additive white Gaussian noise) N_0 [W] given by Couch [34]:

$$\frac{E_b}{N_0} = \frac{S/r_b}{N/\text{BW}} = \left(\frac{S}{N}\right) \times \frac{\text{BW}}{r_b}, \tag{6.1}$$

where $N = N_0\text{BW}$ [W] is the filtered noise at the output of a band-pass filter BW [Hz].

The selection of the frequency considering only the bit rate and antenna size is not enough because as is general knowledge, the antenna is perhaps the most critical subsystem in wireless communications. This requires an antenna small enough for integration with the transmitter but not so small as to compromise this same miniaturization. The size reduction can be a problem because the antenna must be designed for transferring the highest possible power to the receiver. In this context, the small size of antennas can

TABLE 6.2

BEPs for the Modulations ASK, PSK, and FSK

Modulation	BEP	
ASK with coherent detection	$Q\left(\sqrt{\dfrac{E_b}{N_0}}\right)$	$Q\left(\sqrt{\dfrac{S}{N}}\right)$
ASK with noncoherent detection	$\dfrac{1}{2}e^{-\left(\frac{1}{2}\right)\left(\frac{E_b}{N_0}\right)}, \left(\dfrac{E_b}{N_0}\right) > \dfrac{1}{4}$	$\dfrac{1}{2}e^{-\left(\frac{S}{N}\right)}, \left(\dfrac{S}{N}\right) > \dfrac{1}{8}$
PSK	$Q\left(\sqrt{\dfrac{2 \times E_b}{N_0}}\right)$	$Q\left(\sqrt{2} \times \sqrt{\dfrac{S}{N}}\right)$
FSK	$Q\left(\sqrt{\dfrac{E_b}{N_0}}\right)$	$Q\left[\sqrt{1 + \left(\dfrac{|f_1 - f_2|}{4r_b}\right)} \times \sqrt{\dfrac{S}{N}}\right]$

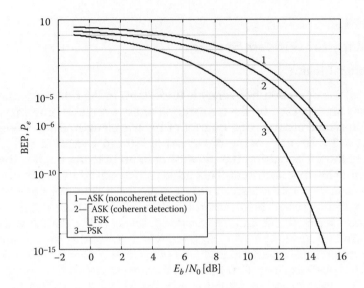

FIGURE 6.7
BEP versus the E_b/N_0 [dB] ratio for the modulations ASK, PSK, and FSK.

introduce additional problems of impedance matching [35] which must be solved. As seen later, wafer-level-packaging (WLP) techniques can be used for overcoming these problems [36].

The communication range is another issue to take into account when selecting the frequency. This is related to the attenuation of RF signals, whose free-space value increases with transmitter/receiver distance d [m] and frequency f [Hz] [37]:

$$L_f(d, f) = -20 \log_{10}(d) - 20 \log_{10}(f) + K_f \text{ [dB]}, \tag{6.2}$$

with $K_f = -20 \log_{10}[c/(4\pi)]$ [dB]. This means that for a simultaneously given transmitted power P_t [dB] and receiver's sensitivity S_r [dB], the frequency is limited by the range $d_{\max}$ [m]:

$$f \leq 10^{\left[(P_t-S_r)-20\log_{10}(4\pi d_{\max})\right]/20} \text{ [Hz]}. \tag{6.3}$$

The free-space model is the most optimistic approach for calculating the link budget because the additional losses due to the surrounding environment (lossy propagation mediums, buildings, terrain conditions, vehicles, persons, shadowing, and systemic implementations, among other factors) is not taken into account. Therefore, a general loss model must be used [37]:

$$L(d) = \alpha c^{-n} + \chi, \, n \geq 2, \text{ and } \alpha < 1. \tag{6.4}$$

Alternatively, $L(d) = -10n \log_{10}(d) + 10 \log_{10}(\alpha) + \chi_{\text{dB}} = A \log_{10}(d) + B + \chi_{\text{dB}}$ [dB]. The factor $A = -10n$ is very important to analyze because it justifies why the distance-dependence loss is higher than that observed in free space. The signal fluctuations (also known as fading) do not contribute statically to the loss, but dynamically. This dynamic behavior can impose severe restrictions when designing a wireless link because a superdimensioned RF receiver must be provided to overcome the temporary losses of signal power. The distance-dependent loss model is normally enough to predict the link budget especially for short distances (typically under 20 m) and closed spaces (laboratories, hospitals, residences, and trains, among others). Very good references can be found in Pätzold [38], Blaunstein and Andersen [39], and Bertoni [40] to deal with the fading.

Figure 6.8 shows the available frequency bands for the different technologies used in wireless communications. Suitable frequencies for possible use in wireless instruments are those that belong to the so-called industrial, scientific, and medical (ISM) band, due to its unregulated usage. These frequencies can be freely used without being subject to standardization but keeping the emission powers below the maximum levels imposed by regulations. This usage flexibility leads to the widespread use of new applications as will be discussed further.

6.3.2 RF Interfaces

A wireless instrument communicates with the external world by RF. Thus, a wireless interface must be provided for allowing RF communications. Figure 6.9 shows a generic schematic block of a wireless microsystem performing functions of a stand-alone wireless instrument. These microsystems are composed of sensors and electronics for control and signal processing, by memory and by an RF interface (the RF transceiver) for connecting to an associated antenna. The dimensions of the RF transceiver must be comparable with other elements integrated in the microsystem (e.g., the sensors and remaining electronics). The miniaturization of electronics and the spreading of fabrication processes for integrating heterogeneous technologies (e.g., CMOS, SiGe, III/V technologies, microelectromechanical

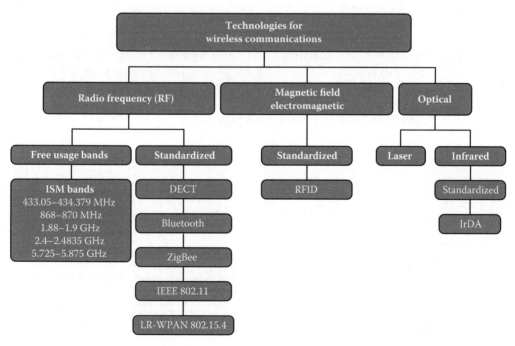

FIGURE 6.8
Currently available frequencies for wireless applications.

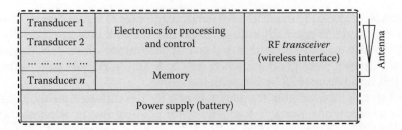

FIGURE 6.9
Generic microsystem architecture that connects to an associated antenna.

systems [MEMSs], among others) results in the mass production of wireless microsystems at low prices. All these issues combined with the flexibility to select which and the number of sensors for integrating together with the RF transceiver and remaining electronics allow engineers to design a wide number of devices for a wide number of applications. This last goal can be easily achieved with multichip-module (MCM) techniques applied to a limited number of components (which can be of different technologies). In conclusion, the technology is also a major point of allowing the fabrication of wireless microsystems for use in wireless instruments. In this section, a few examples for each of the ISM bands in Figure 6.8 are presented for a better view of wireless instruments potential.

Figure 6.10 shows the block diagram of a wireless instrument for monitoring the body movements of individuals doing hydrotherapy [41]. This wireless instrument is modular and is composed as follows: a module with two MEMS accelerometers with three degrees of freedom (with three axes), a module with low-level multiplexing for signal processing and analog-to-digital conversion, and a third-party RF module (with an RF transceiver at 2.4 GHz and control electronics). The analog electronics in the core measurement subset is controlled by the core communications subset. The core communications subset is a MICAz RF module at 2.4 GHz (fabricated by the Crossbow Company) for communicating with external devices and for controlling and managing the data acquisition process [42]. The accelerometers module was designed for measuring the movements of the individuals by obtaining information about the instantaneous roll, yaw, and pitch. This wireless instrument was designed for low-power and high-throughput communications using a specific MAC protocol for achieving such goals [43].

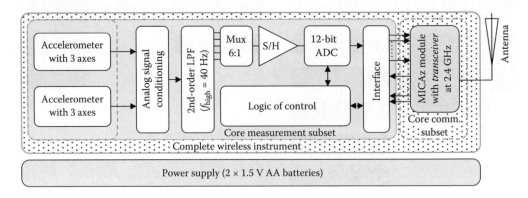

FIGURE 6.10
Block diagram of a wireless instrument at 2.4 GHz ISM band for monitoring the body movements of individuals.

Figure 6.11 shows the block diagram of a sensorial RF interface for operation at 433 MHz [32]. This RF interface has a differential (instrumentation) analog input for interference reduction purposes and allows the connection of other types of sensors. The analog-to-digital conversion is done by a $\Sigma\Delta$ modulator for coding the input analog signals, whose result is a bit stream for encapsulation in a frame for RF transmission. As stated by Morais et al. [32], this interface was especially designed for connecting into the soil moisture sensor developed by Valente et al. [33]. The wireless instrument illustrated in Figure 6.11b was developed for measuring the soil moisture of greenhouses and uses the latter RF interface to send the acquired data into an external storing and/or analyzing unit.

Figure 6.12a shows the block diagram of a receiver for operation in the 433 MHz ISM band that was developed for use in implantable microsystems [31]. The selected architecture explores the super-regeneration phenomena to achieve a high sensitivity. This receiver can be supplied with a voltage of only 3 V for demodulating signals with powers in the range [−100, −40] dB. The combination of modulation and coding scheme is OOK modulation combined with a variation of the Manchester code (e.g., a biphase code). The AMIS 0.7 µm CMOS process was selected for targeting the requirement to fabricate a low-cost receiver. Figure 6.12b shows a photograph of the first prototype (shaded area) which was integrated in a die with an area of 5×5 mm^2. An advantage of this receiver is being fully compatible with commercial transmitters and the transmitter fabricated by Morais et al. [32], for the same coding scheme (the variation of the Manchester code) in the transmitter.

The power consumption of a wireless instrument limits its working time, especially when functioning with batteries. In this context, the selection (or even further, the design) of RF transceivers cannot neglect this issue because this is the block with major impact in the total power consumption, when compared with the entire electronics in the instrument [44]. Furthermore and despite the spreading of microelectronics fabrication processes with the potential to achieve smaller power consumptions, the RF transceiver is irremediably the subsystem of higher power consumption [45–46]. Hence, it demands the integrated definition of architectures and methods of control, as well as providing the means to predict the power consumption of the RF system. Figure 6.13a shows a photograph of an RF CMOS transceiver at 2.4 GHz that allows the implementation of control actions for optimizing the power consumption [23,47]. This RF CMOS transceiver was fabricated in a standard 0.18 µm CMOS process for achieving low-power consumption with a low-voltage supply. As shown in Figure 6.13b, the design of this RF CMOS transceiver predicted the use of the control signal to either select the transmitter or the receiver in order to allow its integration with electronics to perform custom control (Figure 6.13c).

It is possible to explore the band located between 5.7 and 5.89 GHz for implementing wireless instruments [48]. This band permits the fabrication of antennas, whose small dimensions allows their integration with the electronics by using WLP techniques [36]. The integration of antennas and electronics in the same microsystem results in fewer impedance mismatching problems. Moreover, the antenna and electronics cointegration systematizes the fabrication process and, at the same time, results in microsystems with a small cost per unit. The work by Dias et al. [49] takes all of this into account to provide a low-power/low-voltage wireless interface at 5.7 GHz with dry electrodes for implementing wireless instruments as parts of cognitive networks. Figure 6.14a shows a photograph of the wireless interface measuring 1.5×1.5 mm^2. The schematic in Figure 6.14b shows the block diagram of the RF part at 5.7 GHz. The digital signals $\{S_0, S_1, S_2, S_3\}$ select the target frequency in the range $f_{out} = f_{ref} \times (400 + 2S) = f_{ref} \times [400 + 2(S_0 + 2S_1 + 4S_2 + 8S_3)]$, whose range is located between 5.42 and 5.83 GHz for a reference frequency $f_{ref} = 13.56$ MHz.

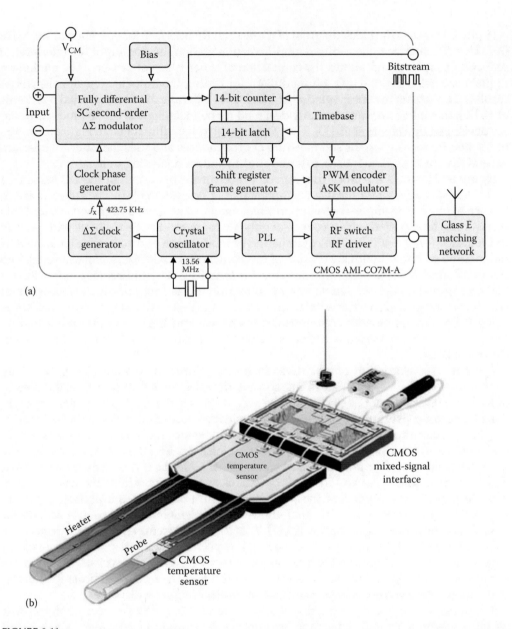

(a)

(b)

FIGURE 6.11
(a) The block diagram of a sensorial RF interface for operation in the 433 MHz ISM band. (Reprinted from *Journal Sensors and Actuators A: Elsevier Science Direct*, 115, Morais, R., Valente, A., Couto, C., and Correia, J. H., A wireless RF CMOS mixed-signal interface for soil moisture measurements, 376–384, Copyright (2004), with permission from Elsevier.) (b) A wireless instrument composed of the RF interface mounted in a soil moisture sensor for utilization on greenhouse environments. (Reprinted from *Journal Sensors and Actuators A: Elsevier Science Direct*, 115, Valente, A., Morais, R., Couto, C., and Correia, J. H., Modeling, simulation and testing of a silicon soil moisture sensor based on the dual-probe heat-pulse method, 434–439, Copyright (2004), with permission from Elsevier.)

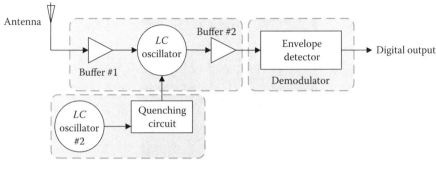

(a)

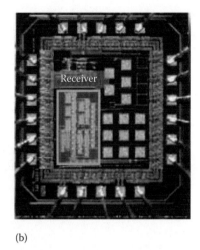

(b)

FIGURE 6.12
(a) The block diagram of the super-regenerative receiver at 433 MHz and (b) a die photograph containing the first prototype of the super-regenerative receiver. (Reprinted from *Microelectronics Journal: Elsevier Science Direct*, 42, Carmo, J. P., Ribeiro, J. C., Mendes, P. M., and Correia, J. H., Super regenerative receiver at 433 MHz, 681–687, Copyright (2011), with permission from Elsevier.)

An integrated low-cost solution for wireless instruments based on a microdevice fabricated with low-power consumption 0.18 μm CMOS process is presented by El-Hoiydi et al. [50]. This microdevice is naturally composed of an RF transceiver, a reduced instruction set computer (RISC) microcontroller, random-access memory (RAM), a power-supply management circuit, analog electronics of signal conditioning and analog-to-digital conversion, and circuits for providing communication based on serial peripheral interface (SPI) and inter-integrated circuit (I²C) buses. The control electronics was developed for implementing a specific communication protocol for use with multiple wireless instruments and low-power consumption, e.g., the WiseMAC protocol. According to the authors, this protocol working together with their RF transceiver achieves power consumptions 30 times smaller than those obtained with the IEEE 802.15.4 (which defines OSI layers 1 and 2 functions). Furthermore, the operation frequency can be selected from 433 and 868 MHz, as well as with either OOK modulation or FSK modulation. According to El-Hoiydi et al. [50], their RF transceiver presents a power consumption of either 2.5 or 39 mW, when either the receiver or the transmit operation mode is selected.

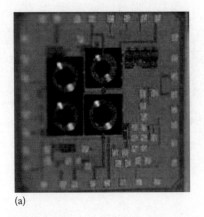

(a)

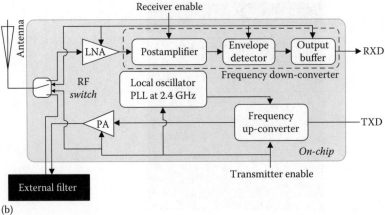

(b)

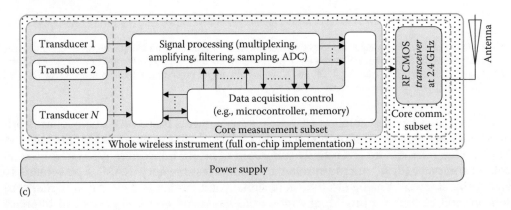

(c)

FIGURE 6.13
(a) Photograph and (b) block schematic of an RF CMOS transceiver at 2.4 GHz specially designed for stand-alone wireless instruments in biomedical applications; (c) a schematic showing the integration concept of the RF CMOS transceiver, sensors, and electronics in the same microsystem.

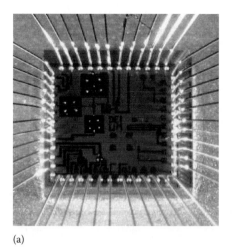

(a)

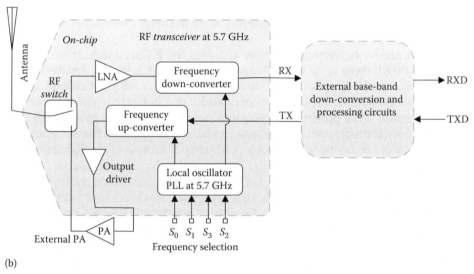

(b)

FIGURE 6.14
For the frequency of 5.7 GHz: (a) a photograph and (b) the schematic block of the RF part of a wireless interface at 5.7 GHz.

6.4 Networks of Wireless Instruments

A wireless sensor network can be considered (in a *latum sense*) a network of wireless instruments whose sensorial nodes are the wireless instruments themselves. The stand-alone operation without the need of a human operator for doing maintenance and/or for replacing the batteries (that provide the supply of power) are the main differences of this kind of wireless instrument from the most conventional available. In this sequence and as shown in Figure 6.15, a wireless sensors network can be considered a distributed sensor network constituted by a high density of nodes. It is expected for each node to run simple protocols and provide low data rates in order to keep power consumptions below

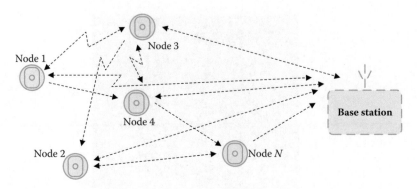

FIGURE 6.15
A schematic illustrating the wireless sensors network concept.

reasonable levels. These issues are of major interest, especially for nodes powered by batteries and without energy-harvesting capabilities because their useful life will be limited by the remaining charge.

In a wireless sensor network, each node acquires the physical data from the sensors and transmits by RF these same data toward a base station for storing and further analysis. The adjacent sensorial nodes can be used for storing the information and relaying it into the base station when the establishment of an RF link between a given sensorial node and the base station is not possible. Upon resuming, the information in a hidden node (in relation to the base station) is not lost because it can be relayed across the distributed infrastructure that forms the network of wireless sensors. However, their distributed nature and the nonexistence of a central controller imply the development of complex algorithms for dealing with the multiplicity of nodes. This is especially true when the topology of the network changes due to the malfunction of a node and/or when a new node is placed in the network or even when the existent nodes occupy new physical positions [51]. Moreover and contrary to what happens in wired networks, this type of network brings new problems: propagation aspects of RF signals and power-supply issues. The shared nature of the propagation medium is another problem because this makes the signals vulnerable to interference and multiple paths (fading), especially for mobile nodes and when a wide number of surrounding obstacles are present [52]. The shared mediums also introduce problems related to the security and confidentiality of the data.

The increased application potential of wireless sensors networks in several fields of human society (such as the industrial, biomedical, transportation, domestic, and energy fields, among others) resulted in the need for their standardization as well as for their wide acceptance. Historically, the first wireless networks were mere technologic extensions of IEEE 802 local networks. Basically, the target of the local wireless networks was the interconnection of computers (as it was a common wired network). With time, other wireless networks appeared, such as Bluetooth for connecting computers to their peripherals and IEEE 802.15.4 for wireless sensors networks. IEEE 802.11 and Bluetooth protocols are very heavy and complex and, thus, have the potential to require devices with high power consumption. These are the reasons that made these protocols not suitable for wireless sensors networks but only for point-to-point connections. In this context, the need for protocols with low power consumption and simple procedures resulted in the establishment of the IEEE 802.15 work group. The joint actions developed by this work group resulted in the proposal for three different classes of wireless operation. The focus of the first class of operation prioritized the bit rates, whereas the second was targeted at power consumption, while the third class was more concerned with the quality of service (QoS). The need for protocols for low-power devices resulted in the proposal of IEEE 802.15.4 as a basic set of rules for application in wireless sensor networks. The IEEE 802.15.4 protocol was developed for low-complexity applications and distances of up to 10 m, allowing bit rates of up to 250 kbps. Furthermore, the IEEE 802.15.4 protocol was proposed for a wide range of uses, ranging from consumer electronics, industrial and domestic automation, personal healthcare, and interconnection of computer peripherals. To resume, the IEEE 802.15.4 protocol defines the two lower OSI layer functions. Note in that Figure 6.16, two versions of the physical layer of the IEEE 802.15.4 protocol can be provided [53].

The first version of the physical layer uses either 868 or 915 MHz in Europe or in the United States, respectively. The European version uses only one RF channel for transmitting the maximum bit rates of 20 kbps, whereas the US version allows the use of 10 simultaneous channels spaced 2 MHz apart and maximum bit rates of 40 kbps per channel. The second version of physical layers uses the 2.4 GHz band and supports the use of 16 simultaneous channels spaced 5 MHz apart and maximum bit rates of 250 kbps per channel. Table 6.3 shows that the IEEE 802.15.4 protocol uses spread-spectrum techniques for increasing the resilience against a variety of factors that include interference from other radio stations and the fading resulting from a multiplicity of radio-wave paths. The spread-spectrum techniques also make easy the clock-synchronization task in the receiver. These modulations belong to the constant-amplitude modulations group and are very complex to implement because analog products in four quadrants are required. Fortunately, the RF *transceiver* CC2420 from the Chipcon Company [54] is commercially available. This RF transceiver contains internally a core ready for implementing all the IEEE 802.15.4 functions (naturally, the second version of the physical layer) and consumes only 19.7 mW when operating in the receiving mode, as well as 17.4 mW when operating in the transmitting mode.

The IEEE 802.15.4 also defines data-link layer standardized protocols (e.g., MAC and LLC). In this context, the LLC sublayer of the IEEE 802.15.4 uses the same type I LLC

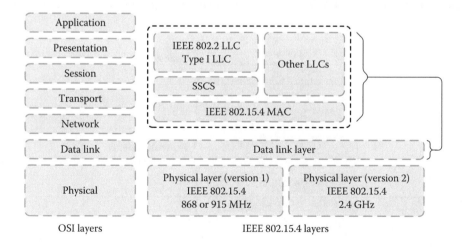

FIGURE 6.16
The OSI layer functions in the IEEE 802.15.4 protocol.

TABLE 6.3

The Main Characteristics of the Physical Layer of the IEEE 802.15.4 Protocol

Band	Frequencies	Bit Rate	Symbol Rate	Spreading Parameters	
				Modulation	Chip Rate
868 MHz	868–868.6 MHz	20 kbps	20 ksymbols/s	Binary phase-shift keying (BPSK)	300 kchips/s
915 MHz	902–928 MHz	40 kbps	40 ksymbols/s	BPSK	600 kchips/s
2.4 GHz	2.4–2.4835 GHz	250 kbps	62.5 ksymbols/s	Offset quadrature phase-shift keying (O-QPSK)	2 Mchips/s

frame formats and procedures specified by the standard IEEE 802.2. The main difference between those used by the local area networks and those used by the wireless sensor networks is the adopted MAC sublayer. The MAC sublayer adopted by the IEEE 802.15.4 (e.g., the IEEE 802.15.4 MAC) is closer to the hardware than the ordinary MACs adopted for local area networks. The service-specific convergence sublayer (SSCS) allows the adoption of other proprietary LLCs as an alternative to this one defined by the IEEE 802.2 (e.g., the type I LLC). The purpose of this model is to allow the IEEE 802.15.4 MAC to implement medium access mechanisms not defined in the IEEE 802.2 [53]. The structure of the MAC sublayer frames is flexible enough to allow the deployment of networks with a wide range of topologies and applications. Typically, an IEEE 802.15.4 MAC frame contains the following fields: a control field to indicate its type; a sequence field to indicate the number of frames for transmission; two fields with receiver and sender addresses information; a field with the information itself (designated as payload); and a field for data integrity check (e.g., CRC for transmission errors verification).

Figure 6.17 shows that the ZigBee protocol is an extension of the IEEE 802.15.4 protocol. The ZigBee uses the IEEE 802.15.4 protocol to implement the physical and data-link layer functions. Furthermore, the ZigBee supports a wider range of high-level functionalities (not present in the IEEE 802.15.4, which is closer to the hardware) as in the case of the cryptography and management policies in environments with multiple users, as well as error

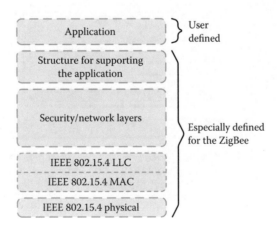

FIGURE 6.17
Adopted layers from the OSI model for use by ZigBee.

control [53]. The CC2430 is another integrated circuit fabricated by the Chipcon Company [55] that simplifies the task of implementing ZigBee networks. The CC2430 contains internally an RF transceiver and an additional core for implementing the ZigBee high-level functions. The RF part of the CC2430 consumes 21 mW in the receiving mode and 30 mW in the transmitting mode. The CC2430 is at this date the first integrated circuit to offer a full ZigBee solution for the market. This integrated circuit is also composed of flash memory up to 128 kB, 8 kB of RAM, an SPI interface, and a few pins for connecting analog and digital signals.

The ZigBee protocol was defined in response to the increased demand for wireless sensors by the industry and the need for new personal applications. Thus, ZigBee was developed to allow the fast prototyping of wireless sensor networks. In this context, it is possible to find a wide number of solutions to implement wireless sensor networks. A few companies (that include Crossbow [42], Dust Networks [56], and Sensicast Systems [57]) offer products such as radios (motes) and sensor interfaces. The motes are battery-powered devices that run specific software. In addition to running the software networking stack, each mote can be easily customized and programmed, since it runs open-source operating systems which provide low-level event and task management. Mote processor/radio module families working in the 2.4 GHz ISM band that support IEEE 802.15.4 and ZigBee are available from the Crossbow Company.

However and despite the ease inherent in the solutions based on motes, they can be very expensive when full custom network prototypes are required. The wireless sensors network solution by Carmo et al. [18] uses peripheral interface controller (PIC) microcontrollers from the Microchip Company to meet a wide range of small-volume applications with a low cost and in a ready-to-use fashion. Their solution uses a PIC microcontroller to provide the basic services of communication and control. Thanks to the serial connection of ADC chains, this solution is scalable in the sense that it is possible to expand the number of sensors to be attached. The main drawback is that the maximum sampling frequency is limited by the number of sensors: the maximum sampling frequency per sensor is limited to f_s/N [Hz], where f_s [Hz] is the maximum sampling frequency when only one sensor is present and N is the number of sensors. To finish, the reprogramming of the microcontroller increases the functionalities with new services of nodes.

6.5 Examples of Wireless Instruments in Biomedical Applications

6.5.1 Commercial Off-the-Shelf (COTS) and Customized Applications

Biomedical applications have a high potential for using wireless instruments. An example that confirms this statement is the wireless monitoring systems of human body information as a growing field. Body area networks comprise smart sensors able to communicate wirelessly to a base station.

Examples of applications are a wireless EEG, which is expected to provide a breakthrough in the monitoring, diagnostics, and treatment of patients with neural diseases. Wireless EEG modules composed of the neural electrodes, processing electronics, and an RF transceiver with an associated antenna will be an important breakthrough in EEG diagnostics. Two approaches can be used for implementing wireless EEG systems:

the COTS and the customized solutions. A commercial off-the-shelf solution uses discrete integrated circuits and passive components for making the wireless instrument, whereas a customized solution is designed from scratch and further integrated on a single microdevice in order to optimize the size and power consumption and allow a power supply with small batteries (for example, class AA, coin-sized batteries). The system proposed by Dias et al. [58] is an example of a COTS system for acquiring EEG signals and transmission by RF. Basically, this wireless EEG system uses a MICAz module [42] at 2.4 GHz for RF transmission and for controlling and converting the physical data. This system uses two 1.5 V class AA batteries for a power supply and achieves maximum bit rates of 120 kbps. Other features of this system include resolution of about 4 µV and power consumption of 15 mW and acquiring of signals with five single-ended channels. This wireless acquisition system fits approximately in 5.7 × 4.8 × 2.0 cm³. Figure 6.18 shows the block diagram of the acquisition part of this wireless EEG system, and note that a reference voltage must be added to the acquired signals before the analog-to-digital conversion can be done because negative potentials are not provided by the power-supply system (the batteries can provide the following electrical potentials: ground V_{dd} and $V_{dd}/2$). Further explanations of this analog circuit (especially for the requirement of neutral and signal ground [SGND] electrodes) are provided by van Rijn et al. [59].

Customized solutions require the development of dedicated microelectronic systems or at least dedicated application-specific integrated circuits (ASICs). The wireless EEG system proposed by Yazicioglu et al. [60] pushes further the concept of wireless EEG, by using the heat of human body for powering the whole wireless instrument itself. Their wireless EEG system is fully autonomous in terms of power supply and uses a

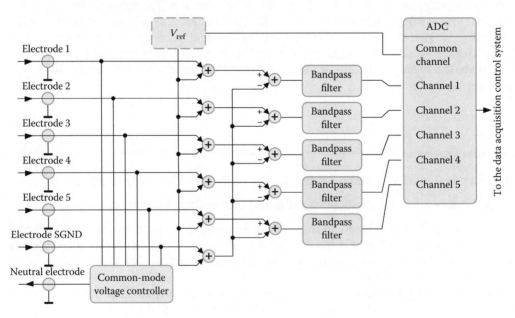

FIGURE 6.18
Block diagram of the EEG analog signal-processing part found by Dias et al. [58].

thermoelectric generator to convert the temperatures differences between the environment (the coldest side, at the temperature T_c [K]) and the forehead of the subject (the hottest side, at the temperature T_h [K]) that uses the wireless EEG system [60]. The output voltage depends on the temperature difference $\Delta T = T_h - T_c$ [K], explained by the Seebeck effect [61]. This wireless EEG system can acquire signals from eight EEG channels, whose inputs are differential (instrumentation) for noise and interference reduction. Each channel uses a new concept developed by Yazicioglu et al. [60] that is known as alternating current–coupled chopper-stabilized instrumentation amplification (ACCIA) for achieving a high common-mode rejection ratio (CMRR) and at the same time eliminating the flicker noise of the transistors as well as for filtering the differential DC voltage generated between two EEG electrodes [60]. A dedicated ASIC was developed for achieving a complete readout front-end for the eight EEG channels and, thanks to it, the complete wireless EEG module mounted with the RF front-end and with a backup lithium battery occupies a volume of 1 cm^3.

Another example of a customized solution is the Sensium TZ1030 sensor interface shown in Figure 6.19a [62]. The Sensium TZ1030 sensor interface was developed by Toumaz Technology Limited for operation in the following bands: 863–870 and 902–928 MHz in Europe and the United States, respectively. Internally, the TZ1030 is composed of analog and digital electronics for interfacing and calibrating the sensors. The sensors are external to the TZ1030 and can attach directly to it. An RF transceiver, an 8051-compatible microcontroller, RAM, and flash memories are also provided and make the TZ1030 a compact solution for an easy placement on the subject's body. The control software contains procedures for local processing of the information, in order to reduce the amount of information and the total transmission times. These features make the TZ1030 a low-power consumption solution. Together with an appropriate external sensor, the TZ1030 is ready for acquiring ECGs, temperature, glucose levels, and oxygen levels in the blood. Figure 6.19b shows a possible architecture for rapid development of wireless BANs.

6.5.2 Active Concepts for Biomedical Wireless Instruments

New techniques for implementing wireless instruments can be found in the literature. These techniques are extremely innovative due to the breakthrough introduced in the way the measurements are done. The work proposed by Karacolak et al. [63] takes into account the variation in the electric parameters for continuously measuring the concentration of glucose (the electric parameters vary with the sugar concentrations).

Alternatively, the research group of Chow et al. [64] explores an uncommon (but still very innovative) methodology that uses cardiovascular stents to receive RF signals inside the human body. In this work, the stents are used as radiating structures for transmitting the measurements across the tissues of the human body.

Finally, the work proposed by Rodrigues et al. [65] uses a MEMS antenna with a U-shaped cantilever structure. Basically, this cantilever is sensitive to the magnetic field component of electromagnetic waves and will oscillate. A piezoelectric material layer of polyvinylidene fluoride (PVDF) is used to convert the magnetic field into a voltage useful enough to be understood by the reading circuit. The major innovation of this technique allows the integration of antennas with implantable devices by way of WLP techniques for achieving the fabrication of small-sized devices. Their antenna occupies an area of only 1.5×1.5 mm^2 [65].

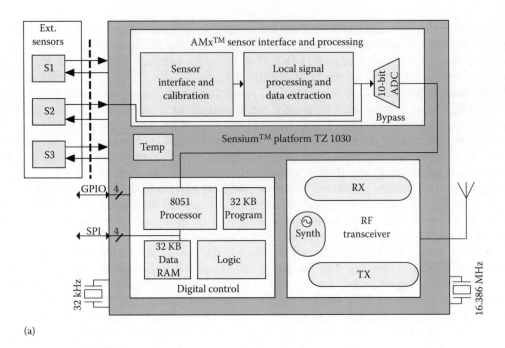

(a)

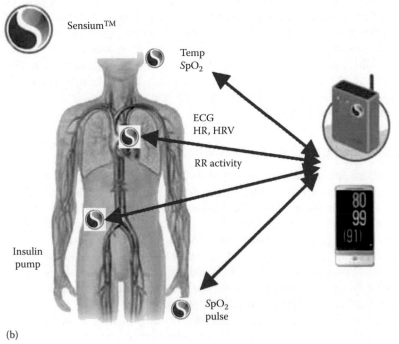

(b)

FIGURE 6.19

(a) Sensium TZ1030 sensor interface for transmission at 863–870 MHz and (b) system architecture for biomedical applications using the TZ1030. (Reproduced from Sensium TZ1030, Ultra low power smart sensor interface and transceiver platform, Toumaz Technology Limited, online [August 27, 2014]: http://www.toumaz.com. With permission.)

References

1. D. Buchla and W. McLachlan, *Applied Electronic Instrumentation and Measurement*, First edition, Prentice Hall, 1991.
2. W. D. Cooper and A. D. Helfrick, *Electronic Instrumentation and Measurement Techniques*, Third edition, Prentice Hall, 1985.
3. J. H. Correia, G. de Graaf, M. Bartek, and R. F. Wolffenbuttel, "A single-chip CMOS optical microspectrometer with light-to-frequency converter and bus interface," *IEEE Journal Solid-State Circuits*, Vol. 37, No. 10, pp. 1344–1347, October 2002.
4. G. R. Tsai and M. C. Lin, "FPGA-based reconfigurable measurement instruments with functionality defined by user," *EURASIP Journal on Applied Signal Processing*, pp. 1–14, January 2006.
5. I. F. Akyildiz, W. Su, Y. Sankarasubramaniam, and E. Cayirci, "Wireless sensor networks: A survey," *Computer Networks*, Vol. 38, No. 4, pp. 393–422, March 2002.
6. V. Raghunathan, C. L. Pereira, M. B. Srivastava, and R. K. Gupta, "Energy-aware wireless systems with adaptive power-fidelity tradeoffs," *IEEE Transactions on Very Large Scale Integrated (VLSI) Systems*, Vol. 13, No. 2, pp. 211–225, February 2005.
7. W. Wu, M. A. T. Sanduleanu, X. Li, and J. R. Long, "17 GHz RF front-ends for low-power wireless sensor networks," *IEEE Journal of Solid State Circuits*, Vol. 43, No. 9, pp. 1909–1919, September 2008.
8. H. Eren, *Wireless Sensors and Instruments: Networks, Design and Applications*, CRC Press, 2006.
9. J. H. Correia, G. de Graaf, M. Bartek, and R. F. Wolffenbuttel, "A CMOS optical microspectrometer with light-to-frequency converter, bus interface and stray-light compensation," *IEEE Transactions on Instrumentation & Measurement*, Vol. 50, No. 6, pp. 1530–1537, December 2001.
10. F. J. Naivar, "CAMAC to GPIB interface," *IEEE Transactions on Nuclear Science*, Vol. 25, No. 1, pp. 515–519, February 1978.
11. L. Korba, S. Elgazzar, and T. Welch, "Active infrared sensors for mobile robots," *IEEE Transactions on Instrumentation & Measurement*, Vol. 43, No. 2, pp. 283–287, April 1994.
12. R. Mukaro and X. F. Carelse, "A microcontroller-based data acquisition system for solar radiation and environmental monitoring," *IEEE Transactions on Instrumentation & Measurement*, Vol. 48, No. 6, pp. 1232–1238, December 1998.
13. D. R. Muñoz, D. M. Pérez, J. S. Moreno, S. C. Berga, and E. C. Montero, "Design and experimental verification of a smart sensor to measure the energy and power consumption in a one-phase AC line," *Measurement: Elsevier Science Direct*, Vol. 42, No. 3, pp. 412–419, April 2009.
14. A. Depari, A. Flammini, D. Marioli, and A. Taroni, "USB sensor network for industrial applications," *IEEE Transactions on Instrumentation & Measurement*, Vol. 57, No. 7, pp. 1344–1349, July 2008.
15. G. Bucci, E. Fiorucci, C. Landi, and G. Ocera, "Architecture of a digital wireless data communication network for distributed sensor applications," *Measurement: Elsevier Science Direct*, Vol. 35, No. 1, pp. 33–45, January 2004.
16. A. Wheeler, "Commercial applications of wireless sensor networks using ZigBee," *IEEE Communications Magazine*, Vol. 45, No. 4, pp. 70–77, April 2007.
17. L. Ferrigno, V. Paciello, and A. Pietrosanto, "Performance characterization of a wireless instrumentation bus," *IEEE Transactions on Instrumentation and Measurement*, Vol. 59, No. 12, pp. 3253–3261, December 2010.
18. J. P. Carmo, P. M. Mendes, C. Couto, and J. H. Correia, "A low-cost wireless sensor network for industrial applications," Proceedings of Wireless Telecommunications Symposium 2009, Praha, Czech Republic, Session D-2, pp. 1–4, April 22–24, 2009.
19. A. Bakker and J. H. Huijsing, "Micropower CMOS temperature sensor with digital output," *IEEE Journal of Solid-State Circuits*, Vol. 31, No. 7, pp. 933–937, July 1996.

20. M. Motz, D. Draxelmayr, T. Werth, and B. Forster, "A chopped hall sensor with small jitter and programmable 'true power-on' function," *IEEE Journal of Solid-State Circuits*, Vol. 40, No. 7, pp. 1533–1540, July 2005.
21. J. Chae, H. Kulah, and K. Najafi, "A monolithic three-axis micro-g micromachined silicon capacitive accelerometer," *IEEE Journal of Microelectromechanical Systems*, Vol. 14, No. 2, pp. 235–244, April 2005.
22. A. Arnaud and C. Galup-Montoro, "Fully integrated signal conditioning of an accelerometer for implantable pacemakers," *Analog Integrated Circuits and Signal Processing*, Vol. 49, No. 3, pp. 313–321, 2006.
23. W. Stallings, *Data and Computer Communications*, Prentice Hall, 2003.
24. B. Pattan, *Robust Modulation Methods and Smart Antennas in Wireless Communications*, Prentice Hall, 1999.
25. H. R. Silva, L. A. Rocha, J. A. Afonso, P. C. Morim, P. M. Oliveira, and J. H. Correia, "Wireless hydrotherapy smart-suit network for posture monitoring," Proceedings of IEEE International Symposium on Industrial Electronics—ISIE 2007, Vigo, Spain, pp. 2713–2717, June 2007.
26. N. S. Dias, J. P. Carmo, P. M. Mendes, and J. H. Correia, "Wireless instrumentation system based on dry electrodes for acquiring EEG signals," *Medical Engineering & Physics: Elsevier Science Direct*, Vol. 34, No. 7, pp. 972–981, 2012.
27. Taiyo Yuden Functional Modules, Taiyo Yuden Co. Online (August 27, 2014): http://www.yuden.co.jp/ut/product/category/module/.
28. Linx RF modules, Linx Technologies Inc. Online (August 27, 2014): http://www.linxtechnologies.com/.
29. Radiometrix Wireless Data Transmission, Radiometrix Ltd. Online (August 27, 2014): http://www.radiometrix.com/.
30. Wireless Solutions for a Connected World, Low Power Radio Solutions (LPRS Ltd.). Online (August 27, 2014): http://www.lprs.co.uk/.
31. J. P. Carmo, J. C. Ribeiro, P. M. Mendes, and J. H. Correia, "Super regenerative receiver at 433 MHz," *Microelectronics Journal: Elsevier Science Direct*, Vol. 42, pp. 681–687, 2011.
32. R. Morais, A. Valente, C. Couto, and J. H. Correia, "A wireless RF CMOS mixed-signal interface for soil moisture measurements," *Journal Sensors and Actuators A: Elsevier Science Direct*, Vol. 115, pp. 376–384, September 2004.
33. A. Valente, R. Morais, C. Couto, and J. H. Correia, "Modeling, simulation and testing of a silicon soil moisture sensor based on the dual-probe heat-pulse method," *Journal Sensors and Actuators A: Elsevier Science Direct*, Vol. 115, pp. 434–439, September 2004.
34. L. W. Couch II, *Digital and Analog Communication Systems*, Fifth edition, Prentice Hall, 1996.
35. M. D. Weiss, J. L. Smith, and J. Bach, "RF coupling in a 433-MHz biotelemetry system for an artificial hip," *IEEE Antennas and Wireless Propagation Letters*, Vol. 8, pp. 916–919, 2009.
36. P. M. Mendes, J. H. Correia, M. Bartek, and J. Burghartz, "Analysis of chip—Size antennas on lossy substrates for short-range wireless microsystems," Proceedings SAFE 2002, Veldhoven, The Netherlands, pp. 51–54, November 27–28, 2002.
37. J. D. Parsons, *The Mobile Radio Propagation Channel*, Pentech Press, 1992.
38. M. Pätzold, *Mobile Fading Channels*, Wiley-Blackwell, 2002.
39. M. Blaunstein and J. B. Andersen, *Multipath Phenomena in Cellular Networks*, Artech House Publishers, 2002.
40. H. L. Bertoni, *Radio Propagation for Modern Wireless Systems*, Prentice Hall, 2000.
41. L. A. Rocha, J. A. Afonso, P. M. Mendes, and J. H. Correia, "A body sensor network for e-textiles integration," Proceedings of Eurosensors XX, Gothenburg, Sweden, September 2006.
42. Crossbow (2009). Wireless measurement systems, Crossbow Inc. Online (August 27, 2014): http://www.xbow.com.
43. J. A. Afonso, L. A. Rocha, H. R. Silva, and J. H. Correia, "MAC protocol for low-power real-time wireless sensing and actuation," Proceedings of IEEE International Conference on Electronics, Circuits and Systems—ICECS 2006, Nice, France, December 2006.

44. J. A. Gutierrez, M. Naeve, E. Callaway, M. Bourgeois, V. Mitter, and B. Heile, "IEEE 802.15.4: Developing standards for low-power low-cost wireless personal area networks," *IEEE Network*, Vol. 5, No. 15, pp. 12–19, September/October 2001.

45. C. Enz, N. Scolari, and U. Yodprasit, *"Ultra low-power radio design for wireless sensor networks,"* Invited Paper, Proceedings of the IEEE International Workshop on Radio Frequency Integration Technology: Integrated Circuits for Wideband Communication and Wireless Sensor Networks, Singapore, December 2005.

46. C. C. Enz, A. El-Hoiydi, J. D. Decotignie, and V. Peiris, "WiseNET: An ultralow—Power wireless sensor network solution," *IEEE Computer*, Vol. 37, No. 8, pp. 62–70, August 2004.

47. J. P. Carmo and J. H. Correia, "Low-power/low-voltage RF microsystems for wireless sensors networks," *Microelectronics Journal: Elsevier Science Direct*, Vol. 40, No. 12, pp. 1746–1754, December 2009.

48. Callaway, Jr., E. H., Chapter 3: The physical layer, *Wireless Sensor Networks, Architectures and Protocols*, CRC Press, 2004.

49. N. S. Dias, J. P. Carmo, P. M. Mendes, and J. H. Correia, "A low-power/low-voltage CMOS wireless interface at 5.7 GHz with dry electrodes for cognitive networks," *IEEE Sensors Journal*, Vol. 11, No. 3, pp. 755–762, March 2011.

50. A. El-Hoiydi, C. Arm, R. Caseiro, S. Cserveny, J. D. Decotignie, C. Enz, F. Giroud, S. Gyger, E. Leroux, T. Melly, V. Peiris, F. Pengg, P. D. Pfister, N. Raemy, A. Ribordy, D. Ruffieux, and P. Volet, *"The ultra low-power WiseNET system,"* Proceedings Design, Automation and Test in Europe, DATE '06, Munich, Germany, pp. 1–5, March 6–10, 2006.

51. J. A. Afonso, H. D. Silva, P. Macedo, and L. A. Rocha, "An enhanced reservation-based MAC protocol for IEEE 802.15.4 networks," *Sensors*, Vol. 11, No. 4, April 2011.

52. W. Y. Lee, *Wireless and Cellular Communications*, Second edition, McGraw-Hill, 1998.

53. E. Callaway, P. Gorday, L. Hester, J. A. Gutierrez, M. Naeve, B. Heile, and V. Bahl, "Home networking with IEEE 802.15.4: A developing standard for low-rate wireless personal area networks," *IEEE Communications Magazine*, Vol. 40, No. 8, pp. 2–9, August 2002.

54. Smart RF CC2420, 2.4 GHz IEEE 802.15.4/ZigBee-ready RF transceiver, Texas Instruments Incorporated. Online (August 27, 2014): http://www.ti.com/.

55. Smart RF CC2430, A true system-on-chip solution for 2.4 GHz IEEE 802.15.4/ZigBee, Texas Instruments Incorporated. Online (August 27, 2014): http://www.ti.com/.

56. Dust, Dust Networks Inc. Online (February 27, 2012): http://www.dust-inc.com/.

57. Sensicast, Sensicast Systems. Online (August 27, 2014): http://www.sensicast.com/.

58. N. S. Dias, J. F. Ferreira, C. P. Figueiredo, and J. H. Correia, *"A wireless system for biopotential acquisition: An approach for non-invasive brain-computer interface,"* Proceedings of IEEE International Symposium on Industrial Electronics—ISIE 2007, Vigo, Spain, pp. 2709–2712, June 4–7, 2007.

59. A. C. M. van Rijn, A. Peper, and C. A. Grimbergen, "High-quality recording of bioelectric events, Part 1: Interference reduction, theory and practice," *Medical & Biological Engineering & Computing*, Vol. 28, No. 5, pp. 389–397, September 1990.

60. R. F. Yazicioglu, T. Torfs, P. Merken, J. Penders, V. Leonov, R. Puers, B. Gyselinckx, and C. V. Hoof, "Ultra low-power biopotential interfaces and their applications in wearable and implantable systems," *Microelectronics Journal: Elsevier Science Direct*, Vol. 40, No. 9, pp. 1313–1321, September 2009.

61. J. P. Carmo, L. M. Goncalves, and J. H. Correia, "Thermoelectric microconverter for energy harvesting systems," *IEEE Transactions on Industrial Electronics*, Vol. 57, No. 3, pp. 861–867, March 2010.

62. Sensium TZ1030, Ultra low power smart sensor interface and transceiver platform, Toumaz Technology Limited. Online (August 27, 2014): http://www.toumaz.com.

63. T. Karacolak, A. Z. Hood, and E. Topsakal, "Design of a dual band implantable antenna and development of skin mimicking gels for continuous glucose monitoring," *IEEE Transactions on Microwave Theory and Techniques*, Vol. 54, No. 4, pp. 1001–1008, April 2008.

64. E. Y. Chow, Y. Ouyang, B. Beier, W. J. Chappell, and P. P. Irazoqui, "Evaluation of cardiovascular stents as antennas for implantable wireless applications," *IEEE Transactions on Microwave Theory and Techniques*, Vol. 57, No. 10, pp. 2523–2532, October 2009.
65. F. J. O. Rodrigues, J. H. Correia, and P. M. Mendes, *"Modeling of a neural electrode with MEMS magnetic sensor for telemetry at low frequencies,"* Proceedings MicroMechanics Europe, MME 2009, Toulouse, France, pp. D19/1–D19/4, September 20–22, 2009.

7

Context-Aware Biomedical Smart Systems

François Philipp and Manfred Glesner

CONTENTS

7.1 Introduction

The technological advances in microelectronic systems design enabled the implementation of smaller, low-cost, low-power, and higher-resourced embedded devices. Vanishing into our environment as defined by the paradigm of ubiquitous computing, each of these electronic gadgets includes intelligence to sense, analyze, actuate, and communicate with each other and the external world. This intelligence makes them aware of the application context and able to give customized feedback to the user in an adaptive manner.

In the healthcare domain, these systems can be commonly found in applications such as remote diagnosis, automatic treatment, intelligent implants, patient monitoring, or fitness coaching. Accompanying patients and doctors in their daily routine, they mainly target improvement of diagnosis by offering direct feedback and noninvasive continuous monitoring of vital signs.

However, autonomous smart devices have limited computational capability and specific system-level approaches must be introduced to guarantee proper operation of the system. Combining the different enabling technologies into a reliably working system is a challenging and time-consuming task which requires the design of new development tools.

In this chapter, we first describe the main features of biomedical smart systems and briefly introduce their enabling technologies and modes of operation. An application example then illustrates how environmental information can be used in a smart way to develop an activity recognition system. Considerations on development tools are explained in Section 7.4.

7.2 Biomedical Smart Systems

7.2.1 Architecture

The concept of smart systems covers low-cost, miniaturized sensing devices to high-end modern diagnosis equipment. For healthcare, this includes smart lenses measuring glucose levels and body temperature sensors to three-dimensional (3D) medical devices and wireless EEG systems. These systems have in common an architecture whose functionality goes beyond simple sensing and processing. Combined with communication interfaces, energy-scavenging circuits, and reliable embedded software, biomedical sensors become more intelligent and deliver information which can be processed in a more efficient way. Through this improved connectivity, vital signs can be directly visualized on a smartphone or transmitted over the Internet. Autonomous operation is enabled by processors with lower power consumption and higher performance, collecting their energy from environmental sources. A generic modular representation of such smart systems is depicted in Figure 7.1.

The central intelligence of the digital processing core is boosted by the functionalities enabled by the following surrounding blocks:

- **Sensors:** Typical sensors used in smart wearable patient monitoring include ECG sensor, EEG sensor, EMG sensor, pulse oximeter, inertial sensor for activity or posture recognition, airflow sensor, temperature sensor, glucose sensor, perspiration sensor, blood pressure sensor, image sensor, etc. Among the most relevant sensor technology enabling biomedical smart systems, one can report MEMS sensors such as accelerometers, gyroscopes, or pressure sensors. MEMSs can also be used in an insulin pump for diabetes or in an artificial retina. Smart systems sensors are usually low cost and easy to integrate in a small device.

- **Energy harvesting:** Operation without a battery is enabled by extracting energy from the surrounding environment. The most common harvesting circuits collect energy from photovoltaic cells, ambient radiations, thermoelectric generators, piezoelectric generators, and other electromechanical systems or even chemical processes. Used in complement with or in replacement of a battery, these energy

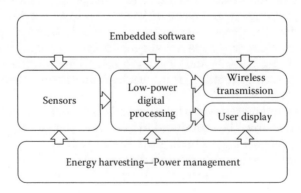

FIGURE 7.1
Building blocks of a smart system device.

sources are renewable but have limited efficiency, which requires particular care in the system design. Energy harvested from body heat or body movement is sufficient to regularly send measurements to a base station over a wireless link.

- **Wireless connectivity:** Ranging from a few centimeters to several hundreds of meters, the wireless connectivity of smart systems allows exchanging data with surrounding devices. This can be a standardized wireless connection such as Bluetooth low energy, ZigBee, ANT, Wi-Fi, or any low-power wireless personal area network (PAN), but also a simple RFID tag. This feature makes the device very portable and comfortable to use. For some systems, wired connection such as Ethernet is used.
- **Actuators:** The local intelligence is commonly used to give direct feedback. This can be implemented by a simple visual or audio system but also by giving mechanical or electrical impulses. In vivo actuators can be controlled to optimize drug delivery.

In practice, smart systems are built as any combination of these submodules [1]. Design requirements may vary greatly according to the type of module used. At a higher level of abstraction, a smart system device may be used as stand-alone node or be part of a network and cooperate with other neighboring devices. We can thus distinguish these two types of smart systems infrastructure:

- **Stand-alone:** A single device is operating autonomously and does not need any permanent connection with a central unit. External connectivity is still available to transfer data or reconfigure the functionality.
- **Network:** Several devices are spatially distributed and cooperate by exchanging sensor data and feedback information. A central node or sink is usually coordinating the network and interfacing other networks (gateway). For healthcare, wireless body area networks are typical examples of such infrastructure.

Following this definition, devices such as smartphones can be considered as smart biomedical devices. Embedded sensors such as accelerometer, microphone, or image sensor can be used to monitor vital signs of a patient by using embedded software applications. Smart systems are usually based on "normal" systems that have been made intelligent by integrating processing, sensing, and actuating capability. For instance, smart watches are wristwatches integrating a highly resourced processor, wireless connectivity, a user display, and a set of embedded sensors.

7.2.2 Applications of Smart Systems in Healthcare

When correctly used in our daily life, smart systems can greatly contribute to monitoring and improving health conditions. In the classical patient–doctor interaction, the measure of vital signs is performed with high-end equipment under the supervision of an expert. Figure 7.2 illustrates the relationships in this traditional approach. This process can, however, be realized over only a short amount of time, in the environmental conditions of the diagnosis location (in the medical office, for example). In certain cases, the usage of these devices can be expensive but they deliver highly reliable data. With this approach, the diagnosis can be biased by the short duration of the measure and the environmental conditions. Drug delivery, reeducation, and physical training are similar examples where the

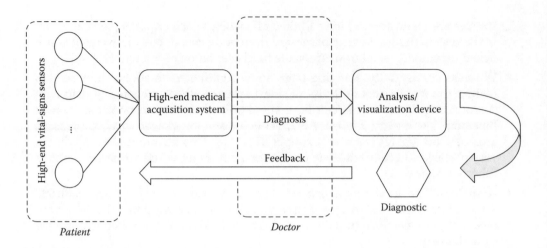

FIGURE 7.2
Traditional doctor–patient environment. The doctor supervises short-term data acquisition with high-end sensors. The patient and the diagnosis process are tightly coupled: the process can be realized only in a specific location under the supervision of a specialist.

supervision of an expert is required to achieve higher efficiency. But specialized medical staff are not continually available to monitor these repetitive operations. Smart systems are in this case a good and low-cost alternative solution to automate these applications. Integrated into our environments or in a familiar device in an unobtrusive way, they are not seen as medical equipment by the patient and are more likely to be accepted.

Continuous remote monitoring of patients is a very large source of data for the medical community. Using accumulated sensor data, the pattern responsible for specific diseases or conditions can be identified and an improved prevention strategy can be organized. Doctors having access to a complete and comprehensive database of their patient's vital signs are able to improve the quality and the speed of their diagnosis. These continuous records can also be used to identify specific behaviors, for example, for patients suffering from mental disabilities.

Based on the real-time health condition of the patient, drug delivery and treatment can be optimized to guarantee an optimal recovery. Emergency cases such as fall or loss of consciousness can be immediately detected by wearable sensors (accelerometers and ECG) or sensors deployed in the living environment (movement sensor, floor vibrations, etc.). In fitness applications, the device can give feedback to the user about performance and the quality of exercises in order to improve training (see Figure 7.3). Actuation can be based on vibration indicating erroneous movements and optical or spoken indicators of current performance. This self-awareness of vital signs, also known as personal informatics, encourages healthier behavior, performance analysis for athletes, or prevention of attacks.

A major feature of smart systems is to use the computational power of the central processing unit to implement an intelligent closed control loop. In the critical context of medical applications, such control systems must be highly robust, requiring a high level of accuracy and stability. Glucose level control, pain management, and arterial oxygen tensions are some typical examples requiring particular care. A nonexhaustive list of devices based on adaptive control loops and artificial intelligence used in medical applications can be found in an article written by Abbod et al. [2].

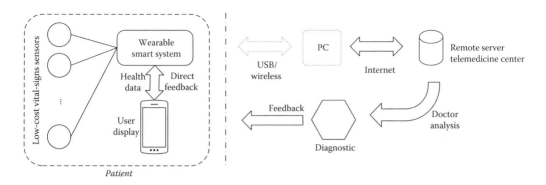

FIGURE 7.3
Smart system diagnosis. The patient can get direct feedback from the wearable device. The doctor can access log data from a remote location and send detailed analysis from long-term measurements.

In general, the users of connected smart systems for healthcare can be categorized in three main categories, according to the severity of their condition:

- Chronically monitored patients requiring constant monitoring of vital signs: This is, for example, the case of persons suffering attention deficit hyperactivity disorder or persons requiring frequent in vivo drug delivery.

- Information seekers or persons with a medical condition looking for data to improve their condition: We can, for example, cite overweight persons tracking their calorie balance or glucose sensing for persons suffering from diabetes.

- Motivated healthy persons or athletes interested in their biometric data to improve or track their performance: This includes all fitness and sports-related applications.

A class of smart systems corresponds to each of these categories. Some devices are mainly targeted to mainstream consumers and will focus on functionalities such as connectivity to social networks, external appearance, or low cost. On the other hand, devices used in a medical context will rather concentrate on high reliability and accuracy.

7.2.3 Microelectronic Technologies Enabling Smart Systems

One of the main requirements of smart biomedical systems is to be noninvasive. As they might be worn for long time periods, they should not interfere with the daily routines of the users. The device should be small and user friendly. A miniaturized implementation of the different building blocks defined by Figure 7.1 is, therefore, necessary to fit all the necessary intelligence into a seamlessly operating device. This small size notably triggers in vivo applications and smart implants. As a rule of thumb, the circuit size of an average smart system that can be used for biomedical applications is under 2 cm². With ultralow power consumption and energy-harvesting techniques, the device is able to operate autonomously for long time periods.

This feature has notably been enabled by recent progresses in the development of integrated electronic systems. The enabling technologies can be classified in two major trends known as "More Moore" and "More than Moore." Both trends are referring to Moore's law, which defines the growth of the numbers of transistors in integrated circuits: the first is concerned with the increasing amount of digital logic available in complex system-on-chip

devices while the latter describes the technological diversification of integrated components. Three main innovations are particularly relevant for smart systems in healthcare.

- Modern chips combine MEMS technology with analog/RF circuitry or energy-
  harvesting components into a same package. Heterogeneous types of integrated
  circuits stacked or tiled into the same physical container are known as *systems-in-
  package* (SiPs). Here, individual dies are connected to each other by using wires.
  This technology significantly reduces the board size since the functionalities of
  several chips are gathered in the same module. Biosensors can, thus, be mixed
  with powerful processors and communication interfaces in minimal spaces.

- Another approach to save board space is *three-dimensional stacking*. In this case, the
  circuits are not connected by wires but directly by through-silicon vias (TSVs).
  Each circuit is a layer of a complete heterogeneous stack connected to each other
  in a three-dimensional fashion. Sensors can, thus, be placed directly on top of
  processing modules and communication interfaces for a very deep integration.

- Another major enabling technology for seamless integration of circuits is *flexible
  integrated circuits (ICs)*. Using novel types of material and fabrication techniques,
  circuits can be integrated on a bendable or even stretchable support. Used in combination with flexible boards, the circuit can, thus, be implemented as a patch
  fitting to the surface where it is installed. This flexibility is particularly useful
  for applications such as smart lenses, arm-worn devices, intelligent clothes, or
  implants. Sensors, RFID interfaces, or displays are examples of smart systems
  components that can be created using this technology.

7.2.4 Considerations on Hardware–Software Codesign

Enabling intelligent behavior of smart systems relies on the computational power, the energy
efficiency, and the memory available on the device. Highest performance and energy efficiency is usually achieved by implementing dedicated circuits, also known as application-
specific integrated circuits (ASICs). On the other hand, software offers full flexibility and
requires significantly lower development time. An intermediate solution combining the benefits from hardware acceleration and software reprogrammability is reconfigurable hardware.
Modern field-programmable gate arrays (FPGAs) mix processor cores with programmable
logic, enabling low-cost fully configurable system-on-chips. Depending on individual performance and cost requirements, smart systems processing units are implemented as combination of these approaches for true hardware–software codesign. For healthcare applications,
the following features have been identified as the most critical for the processing core:

- Safe and reliable operation is the core of any system including actuation. A malfunction of actuators may damage the system itself or even damage the environment where it is located. Accuracy improvement and fault identification are
  examples of techniques to achieve these goals.

- The system must be reactive and meet real-time constraints. Wireless communication stacks follow elaborated control flows requiring accurate timing and large
  buffering capability.

- The system must work autonomously and be self-healing. Minimal human intervention is necessary for proper operation.

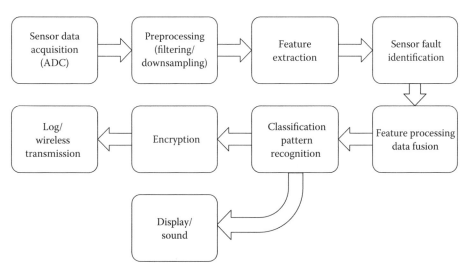

FIGURE 7.4
Generic flowchart for smart biomedical signal processing. Simple data analysis is performed on the device itself. The implementation of each block can be allocated to specific accelerators to improve their individual performance or efficiency.

Achieving such dependability implies employing complex algorithms and test techniques which may overload the processor or largely increase the power consumption. Algorithms running on the device must, therefore, be simultaneously robust, lightweight, and energy efficient. Solutions based on reconfigurable hardware accelerators have been shown to be one of the most suitable for facing such challenges [3]. A typical sensor data-processing flow on smart systems is depicted in Figure 7.4. Most of these tasks directly depend on results from previous computations. Once the most critical computation costs have been identified, each task can be sequentially accelerated by mapping it to a generic reconfigurable hardware accelerator. An example of such a setup will be introduced in the following section.

7.3 Building a Smart System for Activity Tracking

To illustrate the features of smart system used in medical applications, we take the example of an activity-tracking device. Such systems are used to constantly monitor the activity of a patient and detect abnormal behavior or specific activity patterns. They are most commonly used in detecting falls of elderly persons, sleep analysis, or monitoring persons suffering from behavioral disorders. They can also be used for fitness purposes to automatically detect and evaluate a type of exercise. Multiple academic prototypes and commercially available products are implementing this system. Most of them are based on wrist-worn devices, similar to watches. Some projects include devices worn around the leg or the waist.

The basic operation of such systems is to record the user's vital signs and general movements, extract relevant features to label the activity, and log the result. For computationally limited systems, raw sensor data are saved in a local nonvolatile memory and analyzed offline when the data have been transmitted to a PC or a remote server. A smart system will be able to perform the data analysis online, which may require a significant amount of memory and computation power if the classification task is complex. A complete survey on methods for activity tracking can be found in an article from Lara and Labrador [4]. A short introduction on the machine learning processes involved in this application is given thereafter.

During consecutive time intervals I_j, a user is performing a given activity A_j^k from a group of considered activities A. During each interval, sets of n time-series data $X_j = \{X_j^1, X_j^2, ..., X_j^n\}$ are recorded, where X_j^i is the time series of a given observation (for example, acceleration along one axis of an accelerometer and pulse rate). A specific feature set F_j can be extracted from each X_j. A classifier C will then label each interval with the most probable activity $\hat{A}_j$ corresponding to this feature set based on comparison with previously recorded training feature sets:

$$C(F_j) = \hat{A}_j = \arg\max_i p(A^i \mid F_j). \tag{7.1}$$

The precision of the system is then defined by the ratio of correctly recognized classes to the number of considered time intervals. A robust system would include label for unrecognized activities in order to take classes that are not part of A into account.

Thanks to simplified integration of multiple sensors through smart systems, a higher number of types of observations are available, resulting in a larger feature set. Obtaining more features give additional information which may improve the accuracy of the classifier. However, as the size of the feature set increases, the computational complexity of both the extraction and classification processes increases. In addition, increasing the number of features does not necessarily imply an improvement of the classifier accuracy. This phenomenon is well known as the curse of dimensionality. An example of this phenomenon is given by Figure 7.5, where the recognition ratio drops for higher numbers of features. Some of the features may be irrelevant or redundant for the considered activities and induce unnecessary complexity. Increasing the feature spaces also requires an extension of the reference training database, which causes exponential data growth. In general, note that the recognition rate can vary greatly according to the selected features. However, optimal sets are resulting in rates close to 90%.

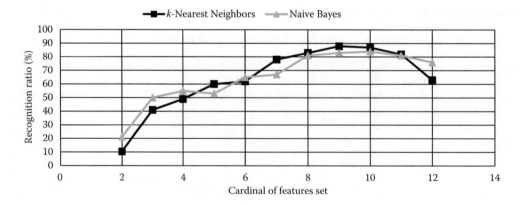

FIGURE 7.5
Recognition ratio for an activity-recognition application as a function of the feature set cardinal. Precision drops for higher numbers of features due to overfitting.

Acquiring statistically significant training data sets is also a difficult and costly task. However, the Internet connectivity of smart systems could be used to automatically collect data from users on a voluntary basis. This trend is known as participatory sensing. Considerations on privacy and security put apart, very large data sets can be built and exploited for classification purposes.

For a specific set of classes, it is possible to select a feature subset F' which maximizes the classifier accuracy or minimizes the computation effort, depending on the evaluation criteria and application requirements.

In order to restrict the search space, a subset of classes can be created on the basis of environmental context. Each class subset corresponding to a context is associated with a feature subset F' giving the best results on the test data. In terms of training, this solution is particularly expensive since it requires an extensive search in a large multidimensional space. The complexity of finding F' will grow exponentially with the number of classes, considered features, and context.

7.3.1 Context Sensing

For activity tracking, context is commonly associated with a spatial location. But additional info such as time of the day, presence of other people, or user feedback can be used to define subsets [5]. For example, outdoor sport activities such as jogging or cycling are unlikely to take place indoors in a bathroom, and, conversely, toothbrushing or vacuuming is unlikely to be done outside. These simple suppositions allow enlarging the class set with a low impact on the global accuracy or computation costs. This paradigm of using environmental context to adapt the operation of a system is known as ambient intelligence (AmI) [6].

Context sensing is an aspect which has been largely facilitated by smart systems (see Figure 7.6). The availability of a larger number of sensors and improved connectivity makes it easy to retrieve information about the surrounding environment. We take the example of indoor positioning systems (IPSs), which can give very accurate information about the location of the person. For instance localization based on fingerprinting will use an existing wireless communication infrastructure to build a reference database (radio map). Discrete reference positions are established by estimating the received signal strength indicator of wireless packets exchanged with the nearby access points. After this

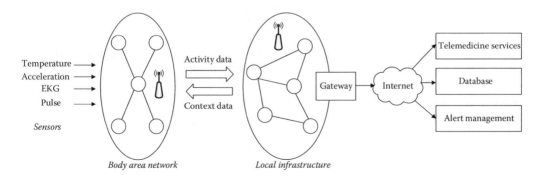

FIGURE 7.6
Ambient intelligence infrastructure based on smart systems for activity tracking. Local infrastructure (smart home, desktop, etc.) connects the wearable devices with the remote services. The BAN interacts with the infrastructure to obtain context information.

training phase, a mobile wireless system can be localized by applying pattern-recognition techniques. Symbolic localization such as rooms or floors can be estimated with very high success rates. Other techniques based on geometrical considerations (triangulation, multilateration, etc.) demonstrate in general lower accuracy for indoor positioning because of their high sensitivity to environmental disturbances (reflections, noise, etc.).

Additional context information can be delivered by intelligent devices installed to form smart homes [7]. Besides the sensors worn by the user, modern homes collect useful information which can be used to track activities. Passive infrared sensors are typically used to detect motion in their coverage area. Detected intrusion can be shared to give a first estimate of the user location. In the context of smart grids, intelligent household devices integrate sensors to precisely measure their electricity consumption or periods of operation. Times when the television is switched on or when water is flowing in the bath can thus be easily detected. This information is used to define user–object interaction and refine the context identification.

7.3.2 Implementation

This scenario is typical for smart systems since it requires the combination of local sensor data with environmental data retrieved from communication with surrounding smart objects. The application requires a significant amount of processing power to handle the feature extraction and classification problem. In addition, the system must detect and adapt to changing environmental conditions.

A wearable sensor platform for general purpose biomedical signal processing was implemented by the microelectronics research group at the University if Darmstadt [8]. A low-power FPGA used for signal processing is combined with an RF system-on-chip for control and communication. Motion and simple body vital sensors (accelerometer and temperature and pulse sensors, respectively) are connected to custom interfaces implemented in the FPGA. In particular, the feature extraction and classification algorithms are accelerated by special digital signal processing (DSP) hardware blocks. In general, hardware acceleration on FPGA is two to three orders of magnitude faster than software implementation for these types of algorithms. As a consequence, energy consumption is also significantly lower. Saving the raw data in a nonvolatile memory or transferring it over the wireless is in this case always slower and less energy efficient than directly processing the data.

The system is running the operating system Contiki supporting Internet protocol version 6 over low-power wireless personal area network (6LowPAN), a protocol stack based on Internet protocol version 6 (IPv6) for devices with low processing resources [9]. This standard is now commonly used for smart systems and Internet of Things applications. This connectivity makes the device easy to integrate with a surrounding smart infrastructure.

The hardware accelerator implemented on the embedded FPGA supports dynamic reconfiguration. Preprogrammed function units can be loaded into a hardware coprocessor and loaded at runtime according to user-defined rules. The template architecture promotes resources sharing by modifying its operation according to external events. This feature fits well to the context awareness required by the application. The feature-extraction algorithms can be dynamically reprogrammed so that only the context-specific optimal feature subset F' is computed. This process is illustrated by Figure 7.7. Reducing the feature set has a significant impact on the performance of the system: from 47 possible characteristics, only 12 to 17 are actually used to perform the classification, efficiently reducing the computational overhead.

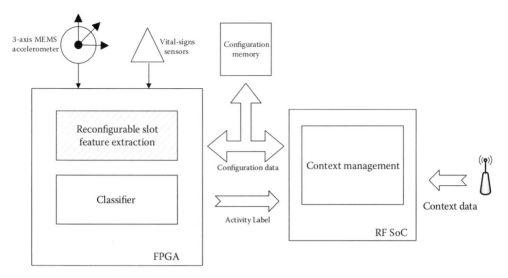

FIGURE 7.7
Smart sensor node for activity recognition. Hardware accelerators for feature extraction are dynamically reconfigured according to context data in order to maximize the classifier accuracy and minimize the computation overhead.

This example was selected to show that the development of smart systems implies multidisciplinary integration. The final implementation is based on novel concepts issued from diverse topics such as pattern-recognition mechanisms, wireless communication, digital signal processing, and dynamic hardware reconfiguration. Integrating all this intelligence in a small low-power embedded system remains the main challenge for smart systems engineers. Therefore, the implementation success mainly relies on the availability of appropriate design tools.

7.4 Development Tools for Smart Systems

As a combination of subsystems which are usually considered independently, smart systems do not have a standard development flow. Combining sensor data acquisition with wireless communication and power management requires a deep knowledge of the diverse implementation aspects. Smart systems usually have stringent requirements in terms of energy consumption, performance and reliability that should be met as well. High-level application constraints must be mapped to the different subcomponents in order to make the right design choices. Information such as sensor sampling rate and wireless range can be used to scale the energy scavenger and the processing power at system level. Programmability is, therefore, a critical issue for smart systems. The increasing complexity coming for heterogeneity and constrained resources must be tackled by novel design methods. These development flows must also be scalable and portable for the variety of smart systems applications.

Prototyping smart systems have been largely facilitated by the availability of modular development platforms. Instead of applying a highly integrated approach (system-on-chip or system-in-package), board-level design is adopted. Around a central microcontroller or FPGA board, sensor, communication, or power-supply boards can be connected in a plug-and-play fashion by using standardized interfaces such as serial peripheral interface (SPI) or interintegrated circuit (I^2C) buses. In terms of software, module-specific libraries can be easily integrated in the operating system and imported in the project. Popular microcontroller-based development platforms such as Arduino or Raspberry Pi development boards [10] are excellent solutions for such rapid system prototyping approaches [11]. The Cooking Hacks e-Health Sensor Platform v2.0 is a solution combining nine biomedical sensors which is compatible with Arduino and Raspberry Pi development boards. Algorithms and functionalities of a reliable portable medical station can be tested and validated at low cost and in a very short development time by using this equipment, making it also suitable for educational purposes. Texas Instruments MSP430 LaunchPad is another example of development kit used for biomedical applications.

Arduino and other open-source hardware projects largely stimulated the design of low-cost smart systems. Low cost, a simple programming model, and a large number of external libraries were the main triggers of this success, making the development very affordable. The Arduino integrated development environment (IDE) is a combination of multiple open-source projects and mainstream programming languages such as Processing, Wiring, C++, and Java into an easy-to-use interface [12]. Complex applications combining sensing, data processing, communication, and actuating can be deployed with only a few lines of code. On top of the hardware innovations, smart systems benefited from high-level software infrastructures customized for rapid development. Low-level hardware interfaces and resource management routines are no longer the main concern of the smart systems designer. Hardware abstraction layers provide the necessary interfaces to integrate heterogeneous components in the design in a flexible way.

If the Arduino programming model is suitable for microcontroller-based systems, a similar approach can be adopted for system-on-chip or FPGA designs. Xilinx, one of the worldwide leading companies for FPGAs, replaced its old programming environment where hardware and software are loosely coupled by a new tool promoting hardware–software codesign and rapid prototyping [11]. Transactions between the operating system and customized or IP hardware accelerators are automatically generated by the tool and abstracted to the user.

For low-power systems, the $(GECO)^2$ environment accelerates the deployment of sensor data-processing hardware accelerators by providing a complete design interface [13]. The target platform architecture template is the one described previously in Subsection 7.3.2. Using the design flow as depicted by Figure 7.8, a platform for smart systems can be simply programmed without writing code. Each design step can be completed by setting parameters and configure operations through an intuitive graphical interface. Libraries of preprogrammed components and tasks can be imported and inserted in the generic architecture template. Finally, designs can be loaded and activated on remote platforms at run time. The operation of the platform is defined within an event-based framework integrated into an open-source operating system. Altogether, the design framework is largely reducing development time and offers support to reconfigurable hardware acceleration.

The drawback of these approaches is a loss of flexibility and performance. High-level solutions offer generality at the price of additional resources and computation time which are critical in most smart systems. The design tools should therefore optimize the low-level template to minimize this overhead.

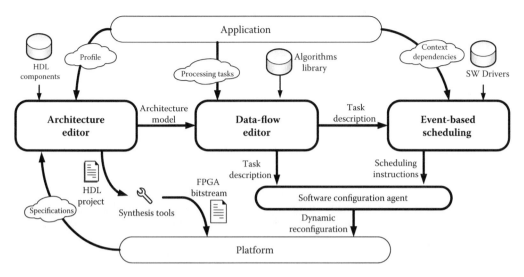

FIGURE 7.8
(GECO)² design flow for smart sensors. (1) The processing unit is customized for the target platform with application-specific function units (architecture editor), (2) tasks and data processing algorithms are mapped on the architecture model (data flow editor), and (3) tasks are deployed and scheduled on remote platforms at runtime (event-based scheduling).

7.5 Conclusion

Following the idea of ubiquitous computing developed by Marc Weiser, intelligent, connected objects are vanishing into our environment. These smart items are deployed everywhere in our home or in our clothes in order to improve our well-being and monitor our vital signs. Boosted by technological improvements in domains such as microelectronic integration and wireless connectivity, the development of smart systems has also been accelerated by the availability of low-cost components and high-level design tools. The increasing popularity and accessibility of these devices will contribute to the large-scale collection of health and vital signs data for medical purposes which can be used to constantly improve the quality of healthcare services.

References

1. Crepaldi, M. et al., A Top-Down Constraint-Driven Methodology for Smart System Design. *IEEE Circuits and Systems Magazine,* 14(1), pp. 37–57, 2014.
2. Abbod, M., Linkens, D., Mahfouf, M. & Dounias, G., Survey on the Use of Smart and Adaptive Engineering Systems in Medicine. *Artificial Intelligence in Medicine,* 26, pp. 179–209, 2002.
3. Kim, C., Cho, Y., Konijnenburg, M., Ryu, S. & Kim. J., ULP-SRP: Ultra Low Power Samsung Reconfigurable Processor for Biomedical Applications. *2012 International Conference on Field-Programmable Technology (FPT),* December 10–12, 2012, pp. 329–334, 2012.
4. Lara, O. D. & Labrador, M. A., A Survey on Human Activity Recognition Using Wearable Sensors. *IEEE Communications Surveys & Tutorials,* 15(3), pp. 1192–1209, 2013.

5. Riboni, D. & Bettini, C., Context-Aware Activity Recognition through a Combination of Ontological and Statistical Reasoning. *Proceeding UIC '09 Proceedings of the 6th International Conference of Ubiquitous Intelligence and Computing*, pp. 39–53, 2009.
6. Acampora, G., Cook, D. J., Rashidi, P. & Vasilakos, A. V., A Survey on Ambient Intelligence in Healthcare. *Proceedings of the IEEE*, 101(12), pp. 2470–2494, 2013.
7. Chen, L., Nugent, C. D. & Wang, H., A Knowledge-Driven Approach to Activity Recognition in Smart Homes. *IEEE Transactions on Knowledge and Data Engineering*, 24(6), pp. 961–974, 2012.
8. Philipp, F. & Glesner, M., Low Power Reconfigurable Computing for Biomedical Signal Processing. *International Journal of Applied Biomedical Engineering*, 6, pp. 47–55, 2013.
9. Dunkels, A., Gronvall, B. & Voigt, T., Contiki—A Lightweight and Flexible Operating System for Tiny Networked Sensors. *29th Annual IEEE International Conference on Local Computer Networks, 2004*, November 16–18, 2004, pp. 455–462, 2004.
10. Cooking Hacks, 2013. *e-Health Sensor Platform V2.0 for Arduino and Raspberry Pi [Biometric/Medical Applications]*. [Online] Available at: http://www.cooking-hacks.com/documentation/tutorials /ehealth-biometric-sensor-platform-arduino-raspberry-pi-medical#step8_2. Accessed on June 12, 2015.
11. Xilinx, 2015. UltraFast Design Methodology Guide for the Vivado Design Suite, Available at http://www.xilinx.com/support/documentation/sw_manuals/ug1046-ultrafast-design-meth odology-guide.pdf. Accessed on June 12, 2015.
12. Anon., n.d. *Arduino Home Page*. [Online] Available at: http://www.arduino.cc. Accessed on June 12, 2015.
13. Philipp, F. & Glesner, M., (GECO)², A Graphical Tool for the Generation of Configuration Bitstreams for a Smart Sensor Interface Based on a Coarse-Grained Dynamically Reconfigurable Architecture. *22nd International Conference on Field Programmable Logic and Applications*, August 29–31, 2012.

Further Reading

Encarnação, J. L. (Ed.), Ambient Intelligence—The New Paradigm for Computer Science and for Information Technology, *IT—Information Technology*, 50(1), pp. 5–6, doi: 10.1524/itit.2008.9048, September 2009.
Krohn, R. & Metcalf, D. (Eds.), *mHealth: From Smartphones to Smart Systems*, HIMSS, Chicago, 2012.
Nakashima, H., Aghajan, H. & Augusto, J. C. (Eds.), *Handbook of Ambient Intelligence and Smart Environments*, Springer, New York, 2010.
Poslad, S., *Ubiquitous Computing: Smart Devices, Environments and Interactions*, Wiley, Hoboken, 2009.

Section II

Wireless Technologies and Networks

8

Technologies for mHealth

Jinman Kim, Christopher Lemon, Tanya Baldacchino,
Mohamed Khadra, and Dagan (David) Feng

CONTENTS

8.1 Introduction to mHealth

There is now a massive amount of digital health data in the modern healthcare environment including, for example, numerous hospital information system databases, medical images, videos, dictation, and medical reports (text and lab reports). These data are stored and exchanged, typically via electronic medical records (EMRs) [1], and are accessed by a heterogeneous medical workforce including doctors and nurses and technical, scientific, administrative, and support staff. Patients also have access to their health data in the form of patient medical records (PMRs) [2]. In recent years, we have already witnessed a revolution in mobile device technologies, in the form of smartphones, iPads, and tablets that can assimilate all types of digital health data. These devices are readily affordable ($200–$500), practical (high-capacity battery sufficient for a whole-day usage and high-resolution screens) and have powerful computation and intuitive user interfaces (UIs) (touch screens). The majority of these devices are also equipped with rich communication capabilities (Internet and other networks) that enable videoconferencing, data sharing, remote desktop, e-mails, and real-time alerts, as well as tracking capabilities, e.g., with global positioning system (GPS) and Wi-Fi triangulation [3]. Further, these mobile devices have myriad applications, or apps, that are designed with user-friendly and intuitive interfaces, thus contributing to the acceptance of these mobile device technologies by the general population. This has allowed their introduction into the healthcare environment with the promise of improving access and care for both patients at home and medical workforces at clinical institutes (hospitals, clinics, and research institutes). Wearable computing technology such as smart glasses and smart watches, another variation of mobile device technologies that are small and designed to be worn by the users at all times, are beginning to find potential in clinical environments [4]. We are in the midst of an era of mobile health, here on referred to as mHealth, where it is possible for desktop-based applications and systems to be replaced by mobile device equivalents, in addition to new capabilities and functions being introduced that are specifically designed to exploit the properties of mobile devices in providing pervasive and distributed computing. As an example, in a recent study by Sclafani et al. [5], they found that 40% of survey respondents among academic physicians and trainees used a tablet device for various functions in a clinical setting such as accessing EMRs. The application of mHealth technologies is rapidly reshaping the healthcare practices in the provision of better healthcare.

The definition of *mHealth* is widely adopted as the application of networked mobile device technologies in healthcare environments. mHealth broadly encompasses the use of mobile telecommunication and multimedia technologies as they are integrated within increasingly mobile and wireless healthcare system. These technologies cover areas of networking, mobile computing, health sensors such as vital-signs monitoring [6], and other communication technologies. mHealth has enjoyed rapid developments in recent years through the provision of new and expansion of existing healthcare services [7]. The intention behind designing and implementing mHealth solutions into health systems is to improve health processes and outcomes. Major goals include decreasing costs and improving equity and efficiency, as well as enhancing care [8–9]. The use of mobile technologies to support the achievement of health objectives, such as accessibility of healthcare to rural populations and improved ability to diagnose and track diseases and epidemics, has the potential to transform the face of health service delivery and introduce efficiency as well as better patient care.

In this chapter, we introduce technologies for mHealth that are instrumental to modern healthcare adoption, including networking, mobile devices, and data exchange. A key requirement for success of mHealth ties strongly with user perspective and acceptance (both patients and healthcare staff) of the technology, which we discuss in terms of usability evaluation. The requirements of institutional support for mHealth implementation are then discussed followed by case studies in mHealth applications and systems.

8.2 mHealth Technologies

This section will discuss new and emerging technologies that are fundamental to mHealth systems and applications, discussed in terms of healthcare adoption and requirements.

8.2.1 Mobile Devices

A mobile device is a portable, handheld computing device, comprising a display screen with touch input and/or a miniature keyboard. There are various types of mobile devices including PDAs, mobile phones ("feature phones") that are low cost and have minimal functions of making phone calls and sending short message service (SMS) messages, and basic pagers with their sole function of receiving alerts. In this chapter, we define mobile devices as modern variations comprising smartphones, tablets, and other wearable devices that are equipped with rich communication and power-computing capabilities. These devices typically weigh less than 200 g for smartphones (4 in or 10.2 cm; the screen-size Google Nexus device is 139 g [10]) and between 300 and 600 g for tablets (7 in or 17.8 cm; Nexus is 340 g and 9.7 in or 24.6 cm, and iPad Air is 478 g [11]). There are numerous manufacturers of mobile devices, with current dominant companies being Apple, Samsung, Huawei, Sony, Microsoft (acquisition of Nokia), HTC, LG, Google (acquisition of Motorola), and ZTE. These mobile devices are operated by an operating system (OS) and can run various types of application software, commonly known as apps. Most devices are also equipped with extensive networking functions including Wi-Fi and third-/fourth-generation mobile telecommunications (3G/4G) data that enable Internet connections, as well as other short-distance networking standards (see Subsection 8.2.2). A camera that is capable of taking high-resolution photos and videos is included on these devices along with a high-capacity battery source such as a lithium battery that is sufficient to power the device for a whole day with a single charge. Figure 8.1 shows examples of different types of mobile devices from major manufacturers with different OSs. On the left is an Apple iPad (third generation) with Apple iOS 7, running an image viewer [12] with whole-body positron emission tomography–computed tomography (PET-CT) medical images in digital imaging and communication in medicine (DICOM) format (OsiriX public data set [13]); in the middle is a Nexus device with Google's Android 4.0 OS, running a dialysis logbook for real-time data entry; and on the right is a Nokia Lumia 1320 operated by Microsoft Windows Phone 8.1 OS, demonstrating its use in running a PMR that is integrated to Microsoft's HealthVault [14]—an online PMR repository.

8.2.2 Networks

The appeal of networked mobile technologies is that they enable communication and sharing of data among the users (patients and healthcare staff), irrespective of location

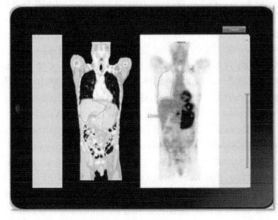

FIGURE 8.1
Selection of mobile devices commonly used in mHealth applications and systems. *Left*: an Apple iPad (9.7 in screen tablet, iOS 7); *middle*: Nexus (4 in smartphone, Android OS); *right*: Nokia Lumia (7 in, Windows Phone OS).

and time. The networking and communication requirements for mHealth applications and systems are broad; for basic messaging and communication services, a simple phone capable of SMS and voice [second-generation mobile telecommunications (2G)] is sufficient [15]; whereas for video consultations (patients to doctors) and data distribution (retrieving medical images from a hospital imaging repository), a high-capacity data network of 3G/4G is necessary [16–17].

As a result of rapid network advances, the capacity for improved access to information and two-way communication has become more available to the population. mHealth involves the use of a mobile device's networking/communication functions comprising 2G, 3G, and 4G networks for communications and data distributions. Wi-Fi is also typically available for network data connection. GPS is also a core part of mobile devices for its geopositioning capabilities; it has a wide variety of uses in mHealth as it provides location that can track the user of the mobile device, e.g., to track patients with dementia suspected of wandering [18], as well as location indexing for mobile outreach nurses [19]. For networking between short distances, two main technologies are readily available in mobile devices: Bluetooth [20] and near-field communication (NFC) [21] technologies. These are wireless data-exchange technologies over short distances and are primarily used for extending, e.g., a smartphone's function with a Bluetooth-enabled hands-free headphones or medical devices such as glucose monitoring devices [6]. All these network capabilities are found in the example devices in Figure 8.1. iPad (third generation) is equipped with Wi-Fi, 3G/4G, GPS, and Bluetooth; whereas Nexus and Lumia further has NFC.

8.2.3 Health Information (Data) Exchange

Health information (data) exchange is a key enabler for interoperability between different health data sources that is essential to facilitate improvements in healthcare quality and efficiency [22]. The importance of health data exchange is to ensure that the different organizations can share and use the data regardless of the application and system that has created/stores the data [23].

With mHealth, because mobile technologies are being used in conjunction with existing systems, these systems also adopt the widely supported healthcare standard data types of health level 7 (HL7)—a comprehensive framework and related standards for the exchange, integration, sharing, and retrieval of electronic health information that supports clinical practice and the management, delivery, and evaluation of health services [24]—and DICOM—an international standard format for medical images and related information for exchange of data and its quality necessary for clinical use [25]. As an example, there is medical image viewing software available for mobile devices that can connect to a picture archiving and communication system (PACS) and support DICOM standards, such as OsiriX MD viewer for iPad [13] and ResolutionMD for Android as well as iPhone/iPad devices [16]. In another example, in a recent study by Lee et al. [26], a mobile vital-signs measurement and data-collection system for chronic disease management was presented using HL7 messaging interface for interoperability of clinical data exchange.

For mHealth applications that are specific to mobile devices, such as "shared-use cases" between a healthcare device and a mobile device, specific functional requirements and a standard are needed. As a case study of such an application, a glucose meter device can be paired to a mobile device (via Bluetooth or data cable connection) to facilitate exchange of data and journal with a web-based patient portal where glucose data and eating habits are recorded and shared with the care team [6]. Realizing this need, the HL7 mobile health work group was developed with the objective of creating and promoting health information technology standards and frameworks for mHealth [27]. The work group identifies and promotes mobile health concepts for interoperability as adopted and adapted for use in the mobile environment. They also coordinate and cooperate with other groups interested in using mobile health to promote health, wellness, public health, clinical, social media, and other settings.

8.3 User Perspective and Usability

There is a large body of literature identifying that using mHealth technologies can help improve health processes and outcomes. However, the history of uptake of mHealth has not been consistent across all health systems. This seems to be at least in part due to the fact that there is no clear method of evaluating the usefulness to the user of mHealth technologies [28–31]. The user perspective and usability are critical in establishing how mHealth technologies will affect the functioning of a health system.

8.3.1 User Perspective of mHealth

User perspective has been considered in many studies related to mHealth. There is evidence that a variety of patients, including those living locally with chronic diseases, as well as those living remotely such as in rural towns, see value in using mHealth [32–34]. There is also evidence that providers see value in using mHealth technologies. However, in cases where levels of commitment increased, such as when technical difficulties were experienced, these attitudes sometimes changed back to neutral [35]. As expected, users who were already using such technologies were more optimistic compared to users who had not used them previously, suggesting that teaching and awareness of technology is a crucial requirement. Overall, other studies have shown that providers find mHealth to be a useful tool in maintaining care away from the clinical environment and ensuring that patients are adjusting to their normal lives after treatment [36–37]. Importantly, some research has indicated that mHealth was not always adopted with optimism. Vuononvirta et al. [35] showed that there was variability in attitude among nurses. Another study used semistructured interviews to show that mobile nurses using mHealth to manage blood pressure and blood-sugar levels of elderly patients were concerned about sustainability, additional workload, and reliance on others to also use the technology effectively [38]. While these results show an initial interest in user perspectives, the measures used in these studies were not validated. This is problematic for health systems wishing to implement or build their own mHealth devices similar to those mentioned. Other literature has been more convincing in assessing the user perspective. In one study on using mHealth in mobile nursing services, a 39-item questionnaire was built using consideration of existing literature and expert review. This resulted in high content validity. Conclusions regarding satisfaction were, thus, more robust. In addition, it was also found that using mHealth in mobile nursing settings can help improve information acquisition, identification, integration, and interpretation and, therefore, overall nursing performance [39]. Another study evaluated the satisfaction and acceptability of a transnational telephonic electrocardiogram system by using a validated medical system satisfaction questionnaire. The questionnaire features five aspects of telemedicine use: general functionality, technical quality, acceptance, data fidelity and security, and satisfaction. Researchers found high acceptance and a perception of significant value in using the system for remote management of cardiac patients [40].

There have been studies that integrated broader theories to help strengthen claims about user perspectives. Evans et al. [41] evaluated the use of mHealth for health promotion by using behavioral change theories. The study conducted an evaluation of a text messaging service that provides regular updates to pregnant women about how to improve their baby's health and their own health. Researchers examined how the technology affected health behaviors and medical outcomes according to the social cognitive theory and health belief model. Participants completed a questionnaire designed by the researchers to measure health attitudes and behaviors. Behavioral outcomes were obtained from validated measures such as the Behavioral Risk Factor Survey. On a broader scale, Brown and Shaw, [42] introduced an evaluation guide for any health technology called the clinical, human and organizational, education, administrative, and social (CHEATS) framework. Each of these domains contains indicators of user experience, such as human and organizational, which contains interface considerations, while administrative contains considerations of convenience. They all relate to patient experience. These domains can be used to organize existing studies and help guide novel studies in ensuring their approach to evaluation is comprehensive.

8.3.2 mHealth Usability

In the preceding section on user perspective, several conceptual issues in the studies were discussed; however, it is unclear what satisfaction or other ideas such as acceptability mean in relation to evaluating mHealth technologies. In addition, these concepts are discrete. They do not provide a clear understanding of which technologies may be better than others. Such evaluations are limited in their practical use. A well-established, robust aspect of the user experience that is beginning to emerge in mHealth literature is usability. Usability research focuses on optimizing products for specific human interactions. According to the International Organization for Standardization (ISO) [43], usability may be defined as the extent to which a product can enable a user to complete a defined goal with effectiveness, efficiency, and satisfaction. Usability in healthcare technology design has been argued to be essential in ensuring that nuances of devices are effectively and easily integrated into medical work flows [44]. Studies designed around this concept can provide a clear scale about the user experience, far beyond discrete conclusions. There are multiple ways to measure usability. These include using representative focus groups for observation during controlled tests or within intended environments; the think-aloud method, requiring users to elicit their thought processes and feedback during testing; and questionnaires and quantitative measures such as time taken to complete tasks and the number of errors [45–47]. These have all been shown to be effective in understanding and utilizing the user perspective of health technologies.

Some of the first mHealth studies to use usability methodologies to assess the user experience obtained mixed results. Many of the methods used are qualitative. A study by Luxton et al. [48] used observations, interviews, and self-report questionnaires to assess the use of a smartphone-based (Apple iPhone 4) videoconferencing system for remote care. Users' experiences with technology were considered. They were then observed using the smartphone in a controlled environment. The questionnaire was specifically designed to assess ease of use, comfort, and preferences. They were also asked to give verbal and free-form feedback while completing the tests. Researchers found that users were overall content with the size of the device and were optimistic about using the device as an adequate healthcare tool among patients and providers. However, they also found that the quality of video was not satisfactory when the connection was poor. This lowered the quality of the users' experience. In another study, Zargaran et al. [49] built an electronic trauma medical record system for mHealth devices. The system was built using observations of trauma clinical work flow, interviews with various emergency health staff, and examination of paper-based methods. The record was designed for iPads and used to capture basic demographic and health data of presenting patients, a record of any operations performed, and details of the discharge summary. The system's usability was examined using the think-aloud method by a range of trauma clinicians from high- and low-resourced settings as well as urban and rural environments. Testing was conducted in a controlled environment, focusing on obtaining data on interface, operation, and interaction quality. Results indicated that users were able to quickly learn to use the device and easily overcome errors and overall reductions in time to record information about patients. It was also revealed that due to unreliable connections in some areas, the device needed better off-line capabilities. The device was also field-tested. Results indicated that many health staff found the system to be user friendly and intuitive. There were some challenges with network strength and concerns about possible theft.

8.4 mHealth Implementation

The potential benefits of mHealth implementation in a clinical environment, including hospitals, are well recognized for improvements in access to healthcare, development of healthcare professionals, and the potential reduction of costs associated with patient diagnosis, treatment, and follow-up care [5,8–9,16,33,42]. However, implementation of mHealth technologies encompasses many challenges. This is particularly true for hospitals which have a detailed set of compliances and requirements. mHealth brings with it new technologies that have not been evaluated in their operational environment. The application of mHealth in a clinical setting requires new work-flow processes, policies, and maintenance and patient privacy and safety. The challenges of implementing mHealth include attaining support of essential units in health, addressing the barrier of medical record privacy, introducing technology to the healthcare setting (to staff and also the interoperability with existing HISs), and ensuring sustainable implementation. Fortunately, mHealth has been able to leverage the policy and procedures that may exist for more established and widely implemented telehealth applications and systems in the hospital environment such as video-conferencing and remote diagnosis [50]. In the context of this section, we define mHealth as a technology that is providing a higher level of mobility and accessibility while retaining relatively similar health services to telehealth.

The endorsement of hospital implementation of mobile devices, such as iPads and iPhones, has been a slow and arduous journey. Resources that are currently available in the form of medical and health apps support the idea of "bring your own device" (BYOD) [51]. BYOD allows access to health information (patient data, clinical references, etc.) in the user's own mobile device and it has quickly become an important part of health professionals; however, the use of such mobile devices is often unsupported. The implementation of hospital policy that supports BYOD will promote change in the work culture, resulting in an era of technologically savvy health professionals. The goal is not to be limiting—but accommodating and safe [9]. Accessing patient information by using mobile devices, however, introduces complications in the implementation of mHealth. The risk of breaching the confidentiality of a patient's medical record is a barrier many health districts across Australia have not overcome. BYOD adds a further complexity on questioning how a Patient's medical records may be deleted while maintaining the owner's personal information. Some hospitals have adopted a practice of issuing mobile devices to selected employees. The equipment is the property of the hospital; therefore, it is ensured that correct security and approved apps are in use.

The implementation of new technologies such as BYOD must be governed and guided to meet the requirements of the hospital's clinical needs while retaining complete patient privacy and safety. Governance over the implementation of telehealth is essential to its success where a lack of support or ownership by senior management introduces risks of failure when uncertainty and complexities that were not initially planned arise. Establishing a governance committee or a telehealth steering committee will assist in ensuring that all affected sectors are considered and engaged in the process. It also ensures there is a level of readiness for the organization to change to allow the implementation of telehealth. In recent times telehealth readiness has been identified as vital to the successful implementation of telehealth [52–53]. A successful committee will be led by a senior organizational manager or director and committed telehealth advocates. The governance committee will be responsible for the evaluation of telehealth project proposals and the approval of implementation of proposed projects including mHealth. All projects should appoint a "clinical champion" (usually a department head) to drive the project from a clinical perspective and they should be invited to participate in the committee meetings and contribute by providing status and evaluation of their ongoing projects. mHealth implementation may require change of management, restructuring, cultural change, and, in some situations, an increased workload. With mHealth, the possibility of frequent use of mobile devices outside of the hospital environment adds further complexity in the security and the confidentiality of information flow, thus requiring special attention and assurances in the quality of data exchange.

The committee must also include a representative from the organization's information technology services (ITSs) to ensure early planning of integration and support of the existing information systems and hardware/network infrastructure in the hospital. This includes, for example, software compatibility, use of hospital data exchange (file format standards), security needs, technical support (on/after hours), and availability of networks [54]. This relationship will be necessary for the success of telehealth implementation, growth, and sustainability within the hospital. As a group, the committee will ensure that the technology is integrated or supported within the organization's current service provisions and provide ongoing maintenance. The use of mobile devices and the development of guidelines around their use should be addressed within this committee in consultations [55]. A representative from the clinical governance department is also essential to the committee, with a responsibility to ensure patient privacy and that safety is maintained. The benefits and risks to the patient must be identified and carefully calculated for all proposed projects including the use of mHealth.

As with any new technology introduced at the hospital, it is essential to have a clear cost–benefit and long-term sustainability plan. These benefits should be weighed against the relatively low costs of implementation of mHealth technology and the services it provides. The risk of not collaborating in this effort will result in the underuse of modern technology in the health sector and a lack of progress in the holistic care of patients. A large number of mHealth, or more generally telehealth, projects are operational during the initial funded stage, but once the funds are exhausted, the cost of maintenance and operations is often too high for continuous operation. Current evidence is sparse for efficacy of mHealth. Although these technologies may be appealing and seemingly innocuous, research is needed to assess when, where, and for whom mHealth devices, apps, and systems are effective [56]. Evaluation is part of a process that can determine cost-effectiveness and involves educating the public about the benefits of technology. These were reported to be among the most important barriers to mHealth adoption [8]. As part of the evaluation of mHealth systems, prior to implementation, a sustainability model must be analyzed and its benefits in terms of costs, as well as patient care/experience, must be measured and evaluated.

The implementation of new technology within a healthcare setting can be overwhelming to those people who are required to operate the equipment. It is, thus, essential that mHealth solutions replicate the current provision of a service in the clinical work flow as closely as possible. This will reduce the impact on existing staff and encourage effective adaptation to the new method of service delivery. Fortunately, modern technologies are enabling the applications and systems to be more user friendly and simplified [34,39]. To increase the acceptance of technology, it is important to demonstrate the mHealth device and create a comfortable environment for users to familiarize themselves with it. Basic troubleshooting guidelines and experienced support personnel are essential in assisting the use and encouraging the uptake of mHealth. Training and education are vital to the successful implementation of any telehealth initiative, including mHealth [57]. Training should be initiated by the members of staff that are most proficient in utilizing the equipment and developing a program to train the trainer is vital to ensuring a sustainable training plan. An example of this is seen in the use of an iPad with an app created specifically for health assessment in the community [58]. The developer of the app carried out the education in stages, focusing initially on a group workshop session, introducing the app to staff, and allowing familiarization. The next stage was to support staff members and patients in the utilization of the app in the clinical setting. The final stage was to allow patients to use the iPad and the app independently. Each stage was followed by an evaluation and adjustments made to the app as required.

The implementation of mHealth technology and specific projects must be supported by organizational policies and procedures. These official documents must specify who may use the equipment, what the equipment can be used for, and what process to follow if there is a potential security breach. The impact on the patient must also be considered; therefore, guidelines must be established to ensure patient safety and that privacy will not be jeopardized. Documentation of the experiences and lessons learned during telehealth should be generated for future reference and for use during changes to the existing implementations with new technologies that are constantly being introduced.

8.5 mHealth Case Studies

mHealth has introduced improvements to a wide variety of existing healthcare processes, and in some situations, provided new enabling technologies for healthcare services. Many mHealth systems have been developed within both research and clinical environments with evidence of improvements to healthcare services. Specifically for the hospital environment, in an attempt to assist in the work-flow processes, mobile device apps have been widely deployed for, e.g., medical image viewing [59–60], patient report viewers [61–62], remote database accessibility [63], and video streams of an operating theater [64]. The adoption of mobile devices has also produced many patient-centric applications. A "smart monitor" product leveraged the pervasiveness of a mobile device via its location tracking capabilities for an epilepsy patient, where a watch is equipped with a GPS module and a sensor to continuously monitor someone for abnormal shaking motions that may be indicative of a seizure [65]. Similarly, to monitor daily activities of dementia patients, a portable and wearable monitoring device (smart watch with a three-axis acceleration sensor) is being introduced with various sensor technologies to monitor emergency situations such as falling down and wandering activities as a result of memory and cognitive impairment [66]. In this section, we present selected mHealth research and implementation projects that has been conducted at the Institute of Biomedical Engineering and Technology, at the University of Sydney, Australia, in collaboration with the Nepean Telehealth Technology Centre (NTTC), Nepean Hospital, and the Royal Prince Alfred (RPA) Hospital.

8.5.1 Outreach Mobile Nursing

Multidisciplinary healthcare refers to multiple professionals contributing to a single holistic form of care. Contributors may include medical, nursing, and allied health professionals [36,39,67–68]. The benefits of multidisciplinary care have been well documented. These include enhancing patients' understanding of their healthcare experience, increased patient satisfaction, and improved management of complex clinical problems [69–70]. In order to provide multidisciplinary care, all staff members of a health service must be able to maintain continuity of care, which may be defined as the perception that medical providers are sharing and using sufficient knowledge of the patient's past and present to shape and execute their treatment plan [71–72]. Continuity of care in multidisciplinary healthcare is dependent upon effective exchange of information. This exchange mostly occurs via interpretation of PMRs. Thus, continuity of care is contingent upon the quality of PMRs, which can be measured by examining their comprehensiveness, accuracy, and accessibility [34,68].

This subsection summarizes the outcomes from the development of an mHealth app for helping the staff of a multidisciplinary medicine nursing clinical service called Nepean Outreach Service (NOS), based at Nepean Hospital, in order to better maintain continuity of care [19,58]. NOS staff attends to outpatients and discharged patients needing ongoing care. In this process, mobile nursing staff visits patients' homes to record clinical notes in paper-based PMRs and administer treatments. PMRs are then used by medical and specialist staff at Nepean to conduct weekly patient reviews, adjust treatments, and escalate treatment and care. In multidisciplinary services like NOS, PMRs are primarily maintained and produced by the nurses. There are many methods of developing PMRs. NOS nurses currently use conventional methods comprising handwriting notes, building databases, and having verbal consultations. Constructing notes and databases can be overly time consuming. Conversations between colleagues are easily forgotten. Data can be only physically disseminated. These methods develop brief, potentially inaccurate, difficult-to-access, and, hence, low-quality PMRs. In services like NOS, this can result in poor continuity of care. Existing literature indicates that continuity of care is significantly affected by the degree of comprehensiveness, accuracy, and accessibility of PMRs. The NOS project examined problems in existing methods of developing paper-based PMRs and addressed them in the new mHealth Outreach App framework depicted in Figure 8.2. The framework, comprising an app (front-end) and a centralized database (back-end) system, provides functions for consistently taking PMRs by capturing written and visual data at a patient's home, as well as the ability to share the data via e-mails

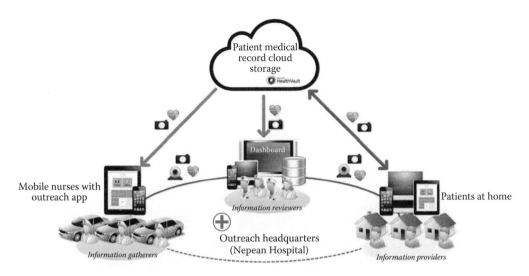

FIGURE 8.2
Continuity-of-care-based telehealth framework for NOS.

and facilitate videoconferencing communications between the NOS staff. The framework further enables the PMRs to be uploaded to a cloud-based online patient record repository (based on Microsoft HealthVault [14]) to share the PMRs between all the NOS staff including clinicians, nurses, and patients.

In our pilot study, NOS mobile nurses used the app to develop PMRs for patients requiring wound care or medications in their home environments. The app enabled NOS mobile nurses to develop better-quality PMRs compared to traditional paper-based note approach. With the app, the NOS staff was able to improve the maintenance of continuity of care. The use of mHealth technologies to develop PMRs, to capture, analyze, and share health information in a digital format, was shown to be able to create better quality PMRs. In the design of Outreach App, findings from requirement analyses, involving observation of a mobile nursing routine; examinations of documentation procedures; and interviews with key nursing, administrative, medical, and specialist staff, guided the creation of functions and user interfaces of Outreach App. Further, field tests in real mobile nursing routines gave clear representations of how the app affected the development of PMRs.

Figure 8.3 presents the main user interface of Outreach App, which features three key functions. The Patient Notes function enabled users to record and review time-stamped notes in lists while creating new notes. The Camera function allowed users to record and review the progress of lesions in the same manner as the Patient Notes function. Users could add captions to images. They could also annotate them by drawing on the screen with their finger. The Video Call function could be used to transfer the user to Skype, where a videoconference consult could be had with other users or health professionals at any time deemed necessary. For our experiments, we used Skype but this can be swapped out for any videoconferencing software to fit the need of the organization.

An Outreach App database was also created for a desktop computer at the Outreach headquarters, with the interface shown in Figure 8.4. The database was used for local secure storage of PMRs not in use. The interface design is consistent with the mobile app. The desktop user interface allows users to view PMRs in a single pop-up window. All the notes and pictures are sorted by date and each record is associated with

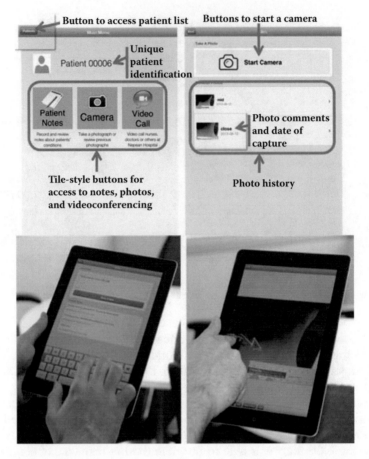

FIGURE 8.3
Upper left: user interface of Outreach App showing the main menu; *lower left*: the Camera function; *upper right*: inserting patient notes by using the app; *lower right*: annotating the pictures taken of the patient.

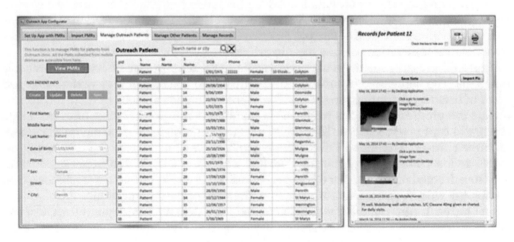

FIGURE 8.4
Left: the Outreach database back end for storing all the PMRs; *right*: for each patient, a summary page that comprises all the notes and pictures from patient's PMR may be rendered.

a nurse's name. A summary page can be printed if a paper PMR is required. It can be used to synchronize data between multiple devices across the medical and nursing team. Therefore, when medical or nursing staff begin their rounds, they can instantly download PMRs for the patients to be seen. Upon the end of their rounds, they can automatically upload the new data back onto the database. To connect an Apple device to the back-end system, Apple Mobile Device Service (AMDS) was used. Comparative to the simplicity of iTunes, this enabled automatic transfer of data between an Apple device and a computer.

Outreach App was designed and developed for iPad 2 and iPad 3 with iOS 6. iPad was chosen for its screen size, comfortable form factor, long battery life, and high-quality camera (1080p) for capturing and viewing pictures, in addition to text notes. Apple's iOS platform was also deemed to be the most familiar among providers and patients [73]. A 3G network provided Internet access and a virtual private network (VPN) were used for security. Outreach App features multiple layers of security. The iPad device is locked with a pass code. A password is required to access PMRs in the app, which are encrypted using the Advanced Encryption Standard (AES) 256-bit cryptoengine, with a key derived from the pass code and the device's unique identifier. For PMRs uploaded to HealthVault, these data are transmitted using encryption with secure sockets layer (SSL). SSL is a cryptographic communication protocol that prevents third parties from viewing information. Additionally, if an iPad is misplaced, the device can be located using the Find My iPad feature in an iOS that identifies the location of the iPad with GPS sensor built-in to the device. Data can also be erased remotely [74]. The synchronization process at the beginning and end of shifts ensures that the device holds information only for a single day and of selected patients. This further improves security where the potential data loss is minimized, in case of misplaced devices.

The computational performance of Outreach App was measured using an iPad 3. During picture taking and sharing, the peak RAM usage was ~15.0%. The peak central processing unit (CPU) usage was 47.6%, which occurred during image uploading. In our environment using 3G networking, the time taken to upload a picture (1024 × 1024 pixels resolution) was an average of 11.1 s per image (100 uploads). Upload time for notes were negligible at <1 s as the data size was only an average of 8 KB (100 uploads).

8.5.2 mHealth Imaging

Over the past decade, the adoption of PACS in hospitals has dramatically improved the ability to digitally share medical image studies via portable storage, mobile devices, and the Internet [59–60]. This has led, in turn, to increased productivity, greater flexibility, and improved communication between hospital staff, referring physicians, and outpatients. However, many of these sharing and viewing capabilities are limited to proprietary vendor-specific applications. Furthermore, there are still interoperability and deployment issues which reduce the rate of adoption of such technologies, thus leaving many stakeholders, particularly outpatients and referring physicians, with access to only traditional still images with no ability to view or interpret the data in full.

With the advent of mobile technology, there have been many remote-imaging solutions that are enabling the access to PACS by using mobile devices, in particular those for iPad and iPhones. With modern devices boosting high resolution and rich color display (as an example, Retina display boosting 326 pixels per inch) and fast and secured network connectivity, the remote access of PACS has been widely deployed. These remote-imaging applications typically work by streaming the images from PACS to mobile devices (image distribution) [60]. There are several pieces of evidence of their benefits, including those for use in telemedicine as remote consultation, e.g., telestroke [16] and teleradiology [75], where timely access to medical images is an important requirement.

Another application area that is becoming mobile centric is distributing images for patients and referring doctors who do not have access to PACS and also have different needs of the imaging data. The push for patient-centric and participatory healthcare requires that patients should be active participants in their ongoing care [76]; and this necessitates that they be given direct access to, and an understanding of, the imaging data that underlie their physician's decision making process [77].

Many referring physicians show a strong preference for direct access to PACS data and to hospital colleagues: when ordering a radiological study, they would prefer to be able to "call up" the images immediately and have the chance to discuss the case collaboratively with the radiologist, rather than receiving a simple textual report and a stand-alone DICOM viewer on a disc [78]. Similarly, in intrahospital or emergency cases, delivering radiology studies to the right person on time can be critical, necessitating an informatics-based distribution approach [79]. In current practice, it is common for patients to be given a digital video disc (DVD) or compact disc (CD) of their images bundled together with image viewing software. Proprietary systems such as Codonics Virtua [80] or Medigration's MediImage [81] are a good baseline, producing (often platform-specific) stand-alone CDs/DVDs. Clinicians need only transfer a study via DICOM and then physically hand the resulting disc to the patient prior to the end of their visit. Relying entirely on physical media, however, this approach neglects many of the benefits of digital imaging, offering few advantages over paper/film records when it comes to distribution. More full-featured proprietary systems such as MIM, Mobile MIM, and MIMcloud [82] and Siemens's syngo WebSpace [83] offer a much wider range of distribution and sharing features but centralize their offering around their vendor's own system. This is easier for vendors to implement, as there is no need for compatibility with or integration into any work flow but the company's own, but the transition cost for hospitals is high. Furthermore, outpatients and referrers usually cannot access the full benefits of the system, which is located within the hospital.

In a recent study, Constantinescu et al. [12] presented the INVOLVE2 distribution system for medical image display across numerous devices and media, which uses a preprocessor and a built-in networking framework to improve compatibility and promote greater accessibility of medical data. The core functionality of INVOLVE2 is illustrated in Figure 8.5. Its primary objectives were to develop a full-featured medical image deployment platform that supports physical sharing but is optimized to also operate effectively across the Internet or in a browser, usable on the widest possible range of consumer hardware, designed for operation by untrained users, and capable of networked distribution of medical imagery when necessary. This system consisted of three main software modules: (1) a preprocessor, which collates and converts imaging studies into a compressed and distributable format; (2) a PACS-compatible work flow for self-managing distribution of medical data, e.g., via CD, USB, network, etc.; and (3) support for potential mobile and web-based data access. The image viewing software included in our cross platform CDs was designed with a simple and intuitive UI for use by outpatients and referring physicians. Furthermore, digital image access via mobile devices or web-based access enables users to engage with their data in a convenient and user friendly way.

The described INVOLVE2 medical image distribution system was demonstrated for its use inside and outside a simulated hospital setting, and evaluated for its ability to run quickly and effectively on a wide range of mobile devices and across the Internet. The

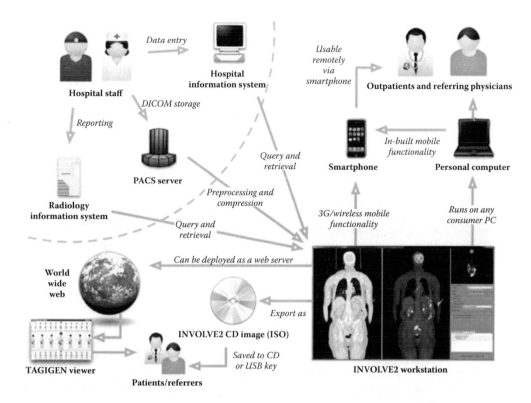

FIGURE 8.5
An overview of the INVOLVE2 medical image distribution system highlighting its main capabilities and the users from inside and outside of the simulated hospital setting who may interact with the system. (With kind permission from Springer Science+Business Media: *Health Information Science and Systems*, A patient-centric distribution architecture for medical image sharing, 1, 2013, 3, Constantinescu, L. et al.)

results demonstrate that the described distribution system meets performance targets, supports a wide variety of consumer devices, and runs effectively across the network or from a CD or USB key. The system was developed to a high standard using powerful, nonproprietary technologies, thus potentially enabling patient participation by granting easy access to complex data, and its unique distribution work flow enables fast, effective sharing.

8.5.3 mHealth Applications

One of the most frequent uses of mobile technology in healthcare is with apps developed for mobile devices. Apps are software applications specifically designed for mobile devices, initially popularized by iPhone smartphones and the App Store, introduced in 2008. Since then, there has been an accelerated development of applications, resulting in huge number of apps available to consumers. The rapid rise of tablets, including iPad, Galaxy Tab, and Windows Surface, in recent years has also contributed to the growth in the availability of apps. As of November 2013, there were over 900,000 apps available for download from the Apple App Store and over a million from the Google Play store [84]. These apps are increasingly becoming relevant to mHealth, with intended users being healthcare practitioners and patients. There are currently over 40,000 medical apps available for smartphones alone with an annual ever-increasing number of downloads, which reached 247 million in 2012 [85]. These apps often make use of the properties of the mobile devices that are useful for presenting and consuming health-related data such as its multimedia capabilities and the built-in sensors. For example, a GPS sensor is used in apps such as RunKeeper and PEAR Sports to monitor a patient's own health [86]. These apps also utilize a mobile device's communication capabilities for videoconferencing and sharing of data [87], as well as portability and accessibility, such as the case of a reference dictionary that is now specifically formatted for mobile devices and available as an app [88]. Furthermore, a recent literature review by Calvillo et al. [89] on how technology is empowering patients concluded that health literacy of patients, remote access to health services, and self-care mechanisms are the most valued ways to accomplish patient empowerment.

With the rise of medical apps, there are a growing number of studies looking into the review of mHealth apps in large variety of medical conditions and diseases, including diabetes [6], asthma [90], smoking [91], and pregnancy [92]. In a study by Demidowich et al. [6], the authors presented usability evaluation among 42 diabetes management apps available for Android smartphones and found that there are wide variety of apps with varying degrees of usefulness, the higher-rated apps generally providing greater level of diabetes management features such as a tool to track insulin and dose calculators. In another study, Tripp et al. [92] described the diverse nature of pregnancy-related mHealth apps and their potential impact on maternity care. They concluded that the popularity of pregnancy-related apps could indicate a shift toward patient empowerment within maternity care provision. The traditional model of shared maternity care needs to accommodate electronic devices into its functioning. This, combined with the fact that smartphones are widely used by many women of childbearing age, is modifying maternity care and experiences of pregnancy.

Despite this massive growth in the number of mHealth apps, most of the reviews are based on a particular medical condition. These reviews are most often conducted manually, which becomes problematic when a broader medical condition is considered, such as diet or all the apps related to well-being, which results in the excess of 10,000 s of apps in either Apple's App Store or the Google Play store. With such large number of apps, manual analysis is not feasible; however, there is very limited capability to analyze the apps, where the ability to query the app store is limited to key words and predefined categories. Fortunately there are several tools that are specifically designed for use in analyzing the apps. uQuery [93] is an App Store search engine that allows filtering of available medical apps by using, for example, ratings, number of downloads, and price, which can be further sorted by relevance or popularity. These filters are compiled by uQuery by indexing all the applications currently available in the United States iTunes App Store. AppMonsta [94] is a paid service that provides a spreadsheet filled with proprietary information about the apps that can be used to infer "insights" via their sentiment index that is a prediction of "virality" (measuring the word-of-mouth sharing of the app). Medical App Journal [95] is another app search engine specifically made for medical apps. By performing a search with key words, the result containing apps that have been referred to in a peer-reviewed journal is retrieved. Although the inclusion of journal citation is useful, this limits the availability of apps and also newer apps that are yet to appear in a peer-reviewed journal. All these app search engines demonstrate the importance and value of analyzing all the available apps. The analysis of apps can be further enhanced by using information visualization tools such as Microsoft PivotViewer [96], as shown in Figure 8.6. PivotViewer enables interactive browsing of the app categories and making sense of interconnectedness between the apps in a visually driven user interface.

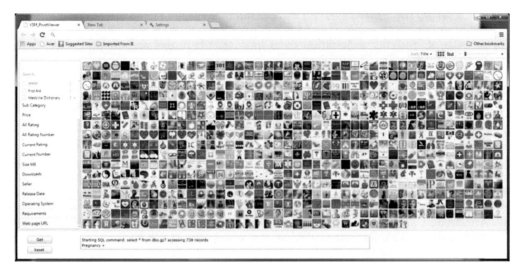

FIGURE 8.6
An example of a PivotViewer information visualization showing several hundreds of health-related apps available in the Google Play store. The left column comprises the menus that can be used to search among the apps based on their attributes, such as prices and ratings.

Acknowledgments

We would like to thank all the staff and patients involved in our mHealth projects at the Nepean Telehealth Technology Centre, Nepean Hospital.

References

1. Fichman, R. et al., The role of information systems in healthcare: Current research and future trends, *Inform Syst Res,* 22, pp. 419–428, 2011.
2. Archer, N. et al., Personal health records: A scoping review, *J Am Med Inform Assoc,* 18, pp. 515–522, 2011.
3. Bahl, P., and Padmanabhan, V., RADAR: An in-building RF-based user location and tracking system, *19th Annual Joint Conference of the IEEE Computer and Communications Societies* (INFOCOM '00), 2, pp. 775–784, Tel Aviv, Israel, March 2000.
4. UCI School of Medicine first to integrate Google Glass into curriculum, http://news.uci .edu/press-releases/uci-school-of-medicine-first-to-integrate-google-glass-into-curriculum/. Accessed on May 18, 2014.
5. Sclafani, J. et al., Mobile tablet use among academic physicians and trainees, *J Med Syst,* 37, 9903, 2013.
6. Demidowich, A. P. et al., An evaluation of diabetes self-management applications for Android smartphones, *J Telemed Telecare,* 18(4), pp. 235–238, 2012.
7. Istepanian, R. et al., Guest editorial introduction to special issue on mHealth: Beyond seamless mobility and global wireless healthcare connectivity, *IEEE T Info Tech Biomed,* 8(4), pp. 405–414, 2004.

8. Free, C. et al., The effectiveness of mobile-health technologies to improve health care service delivery processes: A systematic review and meta-analysis, *PLOS Med*, 10(1), e1001363, 2013.

9. Shrestha, R., Mobility in healthcare and imaging: Challenges and opportunities, *Applied Radiology*, 41(9), pp. 25–28, 2012.

10. Google Nexus, http://www.google.com/nexus/. Accessed on May 18, 2014.

11. Apple iPad, https://www.apple.com/au/ipad/compare/. Accessed on May 18, 2014.

12. Constantinescu, L. et al., A patient-centric distribution architecture for medical image sharing, *Health Information Science and Systems*, 1(3), 2013.

13. OsiriX, http://www.osirix-viewer.com/. Accessed on May 18, 2014.

14. Microsoft HealthVault, https://www.healthvault.com/au/en. Accessed on May 18, 2014.

15. Cormick, G. et al., Interest in pregnant women in the use of SMS (short message service) text messages for the improvement of perinatal and postnatal care, *Reprod Health*, 9(9), 2012.

16. Demaerschalk, B. M. et al., Smartphone teleradiology application is successfully incorporated into a telestroke network environment, *Stroke*, 43, pp. 3098–3101, 2012.

17. Armstrong, G., FaceTime for physicians: Using real time mobile phone–based videoconferencing to augment diagnosis and care in telemedicine, *Eplasty*, 11, e23, 2011.

18. Lin, C.-C. et al., Wireless health care service system for elderly with dementia, *IEEE T Info Tech Biomed*, 10(4), pp. 696–704, 2006.

19. Haraguchi, D. et al., A telehealth framework for mobile nursing: Improving patient medical record management and staff communications, *IFMBE, Springer*, pp. 67–70, 2014.

20. Bluetooth, http://www.bluetooth.com/. Accessed on May 18, 2014.

21. NFC Forum, http://nfc-forum.org/. Accessed on May 18, 2014.

22. Kuperman, G. J., Health-information exchange: Why are we doing it, and what are we doing? *J Am Med Inform Assoc*, 18, pp. 768–782, 2011.

23. Adler-Milstein, J. et al., A survey of health information exchange organizations in the United States: Implications for meaningful use, *Ann Intern Med*, 154(10), pp. 666–671, 2011.

24. Health level 7, http://www.hl7.org/. Accessed on May 18, 2014.

25. Digital imaging and communications in medicine, http://medical.nema.org/Dicom/. Accessed on May 18, 2014.

26. Lee, S.-B. et al., Improving chronic disease management with mobile health platform, *IEEE EMBC, 35th Annual International Conference of the IEEE*, pp. 2275–2278, 2013.

27. HL7 mobile health work group, http://www.hl7.org/Special/committees/mobile/. Accessed on May 22, 2014.

28. Kaplan, B., Evaluating informatics applications—Some alternative approaches: Theory, social interactionism, and call for methodological pluralism, *Int J Med Inform*, 64, pp. 39–56, 2001.

29. Pan, E. et al., The value of provider-to-provider telehealth, *Telemed J e-Health*, 14(5), pp. 446–453, 2008.

30. Whetton, S., Successes and failures: What are we measuring? *J Telemed Telecare*, 11, pp. 98–100, 2005.

31. Becevic, M. et al., TeleMDID: Mobile technology applications for interactive diagnoses in tele-dermatology clinics, *2013 IEEE 15th International Conference on e-Health Networking, Applications & Services (Healthcom)*, Lisbon, Portugal, pp. 429–433, 2013.

32. Dehours, E. et al., User satisfaction with maritime telemedicine, *J Telemed Telecare*, 18, pp. 189–192, 2012.

33. Kobb, R. et al., Enhancing elder chronic care through technology and care coordination: Report from a pilot, *Telemed J e-Health*, 9(2), pp. 189–195, 2003.

34. Young, L. B. et al., Home telehealth: Patient satisfaction, program functions and challenges for the care coordinator, *J Gerontol Nurs*, 37(11), 38–46, 2011.

35. Vuononvirta, T. et al., The compatibility of telehealth with health-care delivery, *J Telemed Telecare*, 17, pp. 190–194, 2011.

36. Hong, H. S., Adoption of a PDA-based home hospice care system for cancer patients, *CIN: Comput Inform Nu*, 27(6), pp. 365–371, 2009.

37. Williams, K. et al., In-home monitoring support for dementia caregivers: A feasibility study, *Clin Nurs Res*, 22(2), pp. 139–150, 2012.

38. Chang, C. et al., Telecare for the elderly—Community nurses' experiences in Taiwan, *CIN: Comput Inform Nu*, 31(1), pp. 29–35, 2013.
39. Hsiao, J., and Chen, R., An investigation on task-technology fit of mobile nursing information systems for nursing performance, *CIN: Comput Inform Nu*, 30(5), pp. 265–273, 2012.
40. Vanagas, G. et al., Clinical-technical performance and physician satisfaction with a transnational telephonic ECG system, *Telemedicine and e-Health*, 14(7), pp. 695–700, 2008.
41. Evans, W. D. et al., Mobile health evaluation methods: The Text4baby case study, *J Health Commun*, 17(Suppl 1), pp. 22–29, 2012.
42. Brown, M., and Shaw, N., Evaluation practices of a major Canadian telehealth provider: Lessons and future directions for the field, *Telemed J e-Health*, 14(8), pp. 769–774, 2008.
43. Ergonomic requirements for office work with visual display terminals (VDTs), ISO 9241-1:1997, http://www.iso.org/. Accessed on May 18, 2014.
44. Bates, D. W. et al., Ten commandments for effective clinical decision support: Making the practice of evidence-based medicine a reality, *J Am Med Inform Assn*, 10(6), pp. 523–530, 2003.
45. Hornbaek, K., Current practice in measuring usability: Challenges to usability studies and research, *Int J Hum-Comput St*, 64, pp. 79–102, 2006.
46. Svanœs, D. et al., Usability testing of mobile ICT for clinical settings: Methodological and practical challenges, *Int J Med Inform*, 79, pp. 24–34, 2010.
47. Lin, C. A. et al., Usability testing by older adults of a computer-mediated health communication program, *J Health Commun*, 14(2), pp. 102–118, 2009.
48. Luxton, D. et al., Usability and feasibility of smartphone video capabilities for telehealth care in the U.S. military, *Telemed J e-Health*, 18(6), pp. 409–412, 2012.
49. Zargaran, E., The electronic trauma health record: Design and usability of a novel tablet-based tool for trauma care and injury surveillance in low resource settings, *J Am Coll Surgeons*, 218(1), pp. 41–50, 2014.
50. Hunter New England (HNE), Health Telehealth Clinical Strategy 2010–2014, http://www.hnehealth.nsw.gov.au/telehealth. Accessed on May 18, 2014.
51. Moyer, J., Managing mobile devices in hospitals: A literature review of BYOD policies and usage, *J Hosp Librarianship*, 13(3), pp. 197–208, 2013.
52. Jennett, P. A. et al., Preparing for success: Readiness models for rural telehealth, *Health Telematics Unit*, 51(4), pp. 279–285, 2005.
53. Weiner, B., A theory of organizational readiness for change, *Implement Sci*, 4(67), 2009.
54. Wade, V. A., and Hamlyn, J. S., The relationship between telehealth and information technology ranges from that of uneasy bedfellows to creative partnerships, *J Telemed Telecare*, 19(7), pp. 401–404, 2013.
55. Wirth, A., Enabling mHealth while assuring compliance: Reliable and secure information access in a mobile world, *Biomed Instrum Technol*, 46(2), pp. 91–96, 2012.
56. Kumar, S. et al., Mobile health technology evaluation, *Am J Prev Med*, 45(2), pp. 228–236, 2013.
57. Krupinski, E. et al., Successful models for telehealth, *Otolaryng Clin N Am*, 44(6), pp. 1275–1288, 2011.
58. Lemon, C. et al., Maintaining continuity of care in a multidisciplinary health service by using m-health technologies to develop patient medical records, *IFMBE, Springer*, 42, pp. 84–87, 2014.
59. Choudhri, A., and Radvany, M., Initial experience with a handheld device digital imaging and communications in medicine viewer: OsiriX mobile on the iPhone, *J Digital Imaging*, 24(2), pp. 184–189, 2011.
60. Constantinescu, L. et al., SparkMed: A framework for dynamic integration of multimedia medical data into distributed m-health systems, *IEEE T Inf Technol Biomed*, 16(1), pp. 40–52, 2012.
61. Holzinger, A. et al., Design and development of a mobile computer application to reengineer workflows in the hospital and the methodology to evaluate its effectiveness, *J Biomedical Informatics*, 44(6), pp. 968–977, 2011.
62. Gomes, P., and Antunes, M., Mobile edoclink: A mobile workflow and document management application for healthcare institutions, *Procedia Technology*, 5, pp. 932–940, 2012.

63. Gomex-Iturriaga, A. et al., Smartphones and tablets: Reshaping radiation oncologists' lives, *Reports of Practical Oncology and Radiotherapy*, 17(5), pp. 286–280, 2012.
64. Lane, J. S. et al., Development and implementation of an integrated mobile situational awareness iPhone application BigiVUTM at an academic medical center, *Int J CARS*, 7, pp. 721–735, 2012.
65. Smart Monitor, http://www.smart-monitor.com/. Accessed on May 18, 2014.
66. Shin, D., Smart watch and monitoring system for dementia patients, *GPC 2013 LNCS 7861*, pp. 577–584, 2013.
67. Caplan, G. A. et al., A randomized, controlled trial of comprehensive geriatric assessment and multidisciplinary interventions after discharge of elderly from the emergency department, *J Am Geriatr Soc*, 52, pp. 14–23, 2004.
68. Funk, L., and Stajduhar, K., Analysis and proposed model of family caregivers' relationships with home health providers and perceptions of the quality of formal services, *J Appl Gerontol*, 32(2), pp. 188–206, 2011.
69. Aizer, A. A. et al., Multidisciplinary care and management selection in prostate cancer, *Semin Radiat Oncol*, 23(3), pp. 157–164, 2013.
70. De leso, P. B. et al., A study of the decision outcomes and financial costs of multidisciplinary team meetings (MDMs) in oncology, *Brit J Cancer*, 109, pp. 2295–2300, 2013.
71. Gjevjon, E. R. et al., Continuity of care in home health-care practice: Two management paradoxes, *J Nurs Manage*, 21, pp. 182–190, 2013.
72. Haggerty, J. L. et al., Continuity of care: A multidisciplinary review, *Brit Med J*, 327, pp. 1219–1221, 2003.
73. IDC, Worldwide tablet market surges ahead on strong first quarter sales, 2013, http://www.idc.com/getdoc.jsp?containerId=prUS24093213. Accessed on May 18, 2014.
74. Hoog, A., and Strzempka, K., iPhone and iOS forensics: Investigation, analysis and mobile security for Apple iPhone, iPad and iOS devices, *Syngress Press*, 2011.
75. Benjamina, M. et al., From shared data to sharing workflow: Merging PACS and teleradiology, *Eur J Radiol*, 73(1), pp. 3–9, 2010.
76. Kvedar, J. C. et al., Up from crisis: Overhauling healthcare information, payment, and delivery in extraordinary times, *Telemed J e-Health*, 15(7), pp. 634–641, 2009.
77. Weitzman, E. et al., Acceptability of a personally controlled health record in a community-based setting: Implications for policy and design, *J Med Internet Res*, 11(2), 14, 2009.
78. Orenstein, B. W., PACS: It's not just for radiology anymore, *Radiol Today*, 9(22), 10, 2008.
79. Nagy, P. G., Using informatics to improve the quality of radiology, *Appl Radiol*, pp. 9–14, 2008.
80. Codonics Virtua, http://www.codonics.com/Products/Virtua/virtua.php. Accessed on May 18, 2014.
81. MediImage, http://www.medigration.de/homeEn/patientencd.html. Accessed on May 18, 2014.
82. Mobile MIM, http://www.mimsoftware.com/markets/mobile/. Accessed on May 18, 2014.
83. Siemens Medical Solutions syngo WebSpace, https://www.smed.com/webspace/. Accessed on May 18, 2014.
84. Mashable, "Google Play Hits 1 Million Apps," http://mashable.com/2013/07/24/google-play-1-million/. Accessed on May 18, 2014.
85. Allied World Health, http://www.alliedhealthworld.com/visuals/smartphone-healthcare.html. Accessed on May 18, 2014.
86. Levine, B. A., and Goldschlag, D., Apps and monitors for patient health, *Contemporary OB/GYN*, 2013.
87. VSee, http://vsee.com/telemedicine. Accessed on May 18, 2014.
88. *Mosby's Mobile Dictionary of Medicine, Nursing & Health Professions*, Seventh Edition, iTunes, Accessed on May 18, 2014.
89. Calvillo, J., How technology is empowering patients? A literature review, *Health Expect*, epub ahead of print, 2013.
90. Huckvale, K. et al., Apps for asthma self-management: A systematic assessment of content and tools, *BMC Medicine*, 10:144, 2012.

91. Abroms, L. C. et al., iPhone apps for smoking cessation, *Am J Prev Med*, 40(3), pp. 279–285, 2011.
92. Tripp, N. et al., An emerging model of maternity care: Smartphone, midwife, doctor?, *Women and Birth*, 27, pp. 64–67, 2014.
93. uQuery, http://www.uquery.com/. Accessed on May 18, 2014.
94. AppMonsta, http://appmonsta.com/. Accessed on May 18, 2014.
95. Medical App Journal, http://medicalappjournal.com/. Accessed on May 18, 2014.
96. Microsoft PivotViewer, http://www.microsoft.com/silverlight/pivotviewer/. Accessed on May 18, 2014.

9

Wireless Body Area Network Protocols

Majid Nabi, Twan Basten, and Marc Geilen

CONTENTS

9.1 Introduction

Wireless body area networks (WBANs) differ from other types of wireless networks in some important aspects, related to, for example, dynamics and energy constraints. This raises the need for dedicated architectures and protocol stacks. This chapter focuses on communication of sensor nodes deployed on or in a human body with a particular wireless device on the body (gateway) to deliver the sensed data. Architectures and protocols for such communication, considering the requirements, constraints, and challenges associated with these networks, are discussed.

9.2 WBAN Characteristics

WBANs have specific characteristics that differentiate them from other wireless sensor networks (WSNs). Special attention is required for designing appropriate communication protocols for WBANs. In this section we point out such special characteristics of WBANs.

9.2.1 Body Sensor Devices

Body sensor devices are expected to develop into very tiny, lightweight, and even flexible nodes to be conveniently used in clinical measurements or to be wearable during normal daily life. Besides strict form factor requirements, these devices should have proper sensing, processing, (wireless) communication, and energy supply. Nodes need to be placed strategically, according to the sensor type, on or in the human body or they need to be embedded in clothing. The use of redundant nodes to deal with different kinds of failures is often not possible in WBANs. These characterizing properties of the body sensor devices are the source of several challenging issues in designing reliable communication protocols.

9.2.2 Energy Consumption Constraints

The physical constraints of on-body devices limit the energy source of the wireless body nodes. The battery of on-body sensor nodes needs to be very small to satisfy the size requirements of the nodes. Moreover, some sensor nodes such as certain implantable sensors rely on energy scavenging from body sources [1] (e.g., producing energy from body temperature, body motion, or heartbeats [2]). Consequently, the total amount of consumable energy for both wireless communication and computation is severely limited. Therefore, the energy consumption constraints of body sensor nodes are tighter than in many other wireless sensor applications. In all WBAN design steps, from designing the physical layer of sensor devices to communication protocols and assigning computation tasks, energy consumption should be carefully considered and minimized.

9.2.3 Quality of Wireless Links

Experimental observations show that the quality of wireless links between on-body sensor nodes is low and time variant. Studies by Natarajan et al. [3–4] show that between some on-body nodes the packet reception ratio over wireless links is lower than 60% for more than 40% of the experiments. The status of links varies with different postures [5]. Moreover, even within a given posture, link quality may vary and have intermittent connections or failures. For instance, while walking, a periodic variation between low and high quality is observed for the link between two nodes placed on the wrist and stomach of the person wearing the nodes [6]. In the following, we summarize some of the prominent causes for the time variability and low quality of links in WBANs.

- **Low-power RF radio:** The main energy-consuming component of wireless sensor devices is their RF radio transceiver. To satisfy very tight energy consumption constraints and to enlarge sensor-node lifetime, ultralow-power RF devices are used in WBANs. Moreover, to provide very small-sized features and to satisfy convenience requirements, a big RF antenna is not affordable. Very small onboard or even on-chip antennas are used for body sensors. A low-power radio transceiver with a tiny antenna leads to a very short transmission range for these nodes (sometimes below 1 m [7–9]).

- **Lossy communication channel:** Water severely absorbs some frequency bands in the spectrum of the RF waves. The 2.4 GHz industrial, scientific, and medical (ISM) band is one of those frequency bands. This frequency band is one of the carrier frequencies suggested for WBANs by IEEE 802.15.6 [10]. Since 50% to 65% of the human body consists of water, the human body can severely shadow the RF signal and present a highly attenuating channel. Empirical investigations show that the propagation loss around and in a human body is high [11–12]. This makes the links between body nodes unreliable and further limits the transmission range. Moreover, the radio channels in WBANs are shown to be subject specific [13] and its characteristics vary over time due to mobility and posture changes.

- **External interference:** The 2.4 GHz ISM band is an unlicensed frequency band that may be simultaneously used by other commercial devices in the environment of a WBAN. IEEE 802.11 (wireless local area network [WLAN]) and IEEE 802.15.1 (Bluetooth) are examples of coexistent users of this frequency band. Microwave ovens also use waves in this frequency range. This leads to likely external interference on WBAN links. An experimental investigation [14] on the impact of WLAN interference on WBANs concludes that when body nodes use a very low transmission power, the external interference can cause substantial packet losses in WBANs (more than 50% transmission failures during the experiments when a transmission power of –25 dBm is used by the body nodes).

9.2.4 Mobility

The human body is mobile and can be in several postures. The movements of the sensor nodes deployed on different positions of the body change the distances between the nodes. An important effect of such movements is that the channel coefficient substantially changes, depending on the fraction of the distance between two nodes that the RF waves should pass through or along the body. The wireless link between two nodes that are far away from each other, but can see each other through a direct line of sight, may be of good quality, whereas two close nodes cannot reach each other because of the positioning of the body between them.

The mobility of nodes causes frequent changes in the quality of links in WBANs. Therefore, postural changes of the human body may frequently change the network topology. Thus, assumptions about the node neighborhood based on only the relative node positions are not realistic. The network protocol should be robust against such frequent link quality fluctuations and topology changes.

9.2.5 Heterogeneity

Several types of biosensors may be used in a WBAN, based on the application requirements and the conditions of the person wearing the WBAN. Different sensors are likely to have different data sampling rates and precisions. These differences result in a broad range of packet transmission rates for different body nodes. An ECG sensor requires around 250 samples per second, while 1 sample per minute may be enough for a temperature sensor. Even a single sensor may require different wireless data transmission rates based on the application scenario and the body status. For instance, a sensor node equipped with an ECG sensor may transmit all data samples (around 250 samples per second) to be processed or recorded by a central station. In another scenario, only the heartbeat rate might be enough. So the sensor node needs to perform some processing, calculate the heart rate, and transmit it once per few seconds. In the latter case, the transmission data rate is lower than that in the former case. In a health-monitoring application, the sensor node may decide to use one of the two mentioned scenarios based on the health status of the person. For instance, in an emergency situation, it sends all data samples, while it transmits only heartbeat rate in normal conditions.

Besides heterogeneity in terms of data sampling specifications, a health application may also have different quality-of-service (QoS) requirements for different body sensors. Data samples from some sensors may be of higher importance than others; losing more important data imposes a greater risk of application failure. Therefore, the application has different tolerances for data losses from different sensor nodes. Such spatial and temporal heterogeneity in sampling and QoS requirements in WBANs requires appropriate support from the protocol stack in order to provide efficient and reliable data communication.

9.2.6 Network Scale

Among the above challenging properties of WBANs, one relaxing property of WBANs is that the number of sensor nodes on a body is limited. Deploying many nodes on the body may cause inconvenience in daily life. So the problem of scalability of protocols is not a challenging issue in WBANs. This also allows protocol designers to make some simplifying assumptions. For instance, one may assume that a central node on the body that plays the role of a hub (the gateway node) has information about sampling specifications and wireless transmission rates of the body nodes within the WBAN.

9.3 WBAN Architectures

Literature on WBANs describes several medium access control (MAC) and data propagation protocols for communication within WBANs. Although differing in the details of the protocols, they all assume either a star topology or a multihop architecture for delivering the sensed data to the gateway node. These two design approaches are discussed in this section. We first introduce some notation and terminology.

9.3.1 Preliminaries

Consider that $S = \{s_1, s_2,...,s_N\}$ is the set of N sensor nodes deployed at different positions on a human body. These nodes are equipped with desired biosensors to sense certain biological signals of the body. The sensed signal may be processed by the node's embedded processor. The raw data or the processed data are communicated with other nodes or some base stations to be further processed, stored, or transmitted again to further stations. We distinguish on-body nodes, part of the WBAN, from nodes and stations that are not positioned on the body; the latter form the *ambient* network. There are two general schemes for the wireless communication between WBAN and the ambient network. In the first scheme, the individual body sensor nodes directly communicate with the ambient network. As an example, an efficient communication protocol is presented by Nabi et al. [15], which assumes direct wireless links between body nodes and the ambient sensor nodes in the neighborhood. In the second scheme, which is the focus of this chapter, all nodes on the body communicate with a specific wireless device on the body. This node, called the *gateway*, is a device such as a smartphone and has more powerful features than the body nodes. This can include more battery capacity, more powerful computation capability, and a higher transmission range. The gateway gathers information from all nodes in the WBAN and then sends the collected (or aggregated) data to a base station in the ambient network. Depending on the application scenario, the gateway may receive some information from the base station and forward this to the body sensor nodes. We assume $s_N \in S$ is the gateway node, which is a part of the WBAN.

9.3.2 Star Architecture

The first choice for designing a WBAN is a star (one-hop) architecture illustrated in the left side of Figure 9.1. In this architecture, the sensor nodes are supposed to send sensed data directly to the gateway. Considering the short distances between nodes in a WBAN, a star architecture may be a reasonable option that in general is simple to implement with an overall low latency because of direct links to the gateway. Moreover, nodes in a star architecture are independent. For instance, in the case of failure of a node other than the gateway, other nodes continue to function as normal.

The power constraint and short transmission range of on-body sensor nodes in some WBAN applications and the severe wave propagation loss through and around the body lead to a low and time-variant quality of wireless links. In such circumstances, a star architecture may suffer from insufficient reliability and low data delivery ratio (DDR). For instance, in the experimental setup used by Nabi et al. [16], some body nodes are disconnected from the gateway in certain postures of the body, even if they use their highest transmission power.

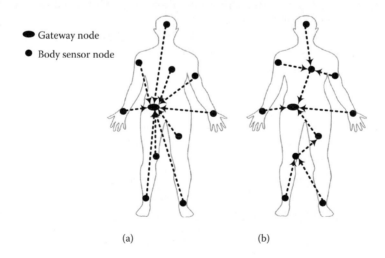

(a) (b)

FIGURE 9.1
Common approaches for communication in WBANs. (a) A star architecture in which all body nodes directly communicate with the gateway; (b) a multihop architecture. (From Nabi, M. et al., On-demand data forwarding for automatic adaptation of data propagation in WBANs, in Proc. IEEE Conference on Sensor, Mesh and Ad Hoc Communications and Networks [SECON], pp. 326–334, IEEE, 2012. With permission.)

9.3.3 Multihop Communication

The alternative architecture for WBANs is multihop communication, shown in the right side of Figure 9.1. In such networks, nodes are not required to necessarily send their data directly to the gateway. Instead, they send their data to certain neighbors in range according to the underlying routing structure, hoping that their data reach the gateway in one or more hops.

Multihop protocols are shown to have in general a high end-to-end DDR [16]. The other advantage is that the body nodes are released from the requirement of reaching the gateway. Thus, they may be equipped with very low-power radios. Considering the low transmission range of typical body sensor nodes and the low wireless channel quality, a multihop architecture can be beneficial, and in some cases it is the only possible option. On the other hand, multihop protocols are more complicated and the network may suffer from longer latencies. No solution will be optimal for all applications of WBANs because of different constraints and requirements.

9.3.4 Protocol Stacks for WBANs

The characteristics of WBANs, pointed out in Section 9.2, demand a lightweight and energy-efficient protocol stack for communication between body sensor nodes (intra-WBAN communication). The mission is to deliver sensed data sampled by the body sensor nodes to the gateway node on the body, satisfying data delivery and energy consumption requirements of the application. Compared to the conventional wireless networks, a simpler protocol stack is considered for WBANs; in many cases it consists of only the physical and the MAC layers. The application layer is developed directly on top of the MAC layer. This is, in particular, the case for star architecture networks because no routing mechanism is required. However, in multihop architectures, a lightweight network layer is used on top of the MAC layer to manage multihop data routing toward the gateway node. A close relation usually exists between these layers in order to better optimize communications. This means that in many cases, for designing a layer of the protocol stack, some assumption may be made about the support of other layers. Cross layer protocol design is also considered to further optimize WBANs.

In the rest of this chapter, we present the MAC layer mechanisms and standards that are developed for intra-WBAN communications. Data delivery requirements and data routing mechanisms for WBANs are discussed in Chapter 10.

9.4 MAC Mechanisms

The MAC mechanism is responsible for managing the access of different nodes to the shared medium (wireless channel) and resolving collisions. MAC protocol design for WBANs has minimized the energy consumption as a main objective. To this aim, radio activities such as listening to the wireless channel, in particular, should be optimized. There are various MAC protocols developed for WSNs that are low power compared to the MAC layers for conventional wireless networks such as WLANs. However, the characteristic properties of WBANs require further consideration to make them even more power efficient, simple, and optimized for WBAN dynamics. In this section, we discuss the MAC layers that are developed and used specifically for WBANs.

9.4.1 MAC Paradigms

In general, MAC protocols for short-range wireless communications, such as in WSNs, exploit one of two major paradigms: *contention-based* or *schedule-based* access. Carrier sense multiple access (CSMA) and ALOHA are the base mechanisms for designing many contention-based protocols. In CSMA, a node senses the channel when it has a packet for transmission. If the channel is found to be clear, the node turns on its transmitter and sends its packet. Otherwise, it repeats this procedure some time later according to a specific back-off mechanism. In the ALOHA mechanism, no carrier sensing is performed. It means that the node immediately accesses the medium when it has a packet for transmission. In a variation of ALOHA, called slotted ALOHA, the time is divided into synchronized time slots. Nodes having a packet can start transmission only at the beginning of time slots. This reduces the probability of interrupting the packets that are currently being transmitted by other nodes in the neighborhood, in particular for networks with higher traffic loads.

The contention-based protocols are considered well suited for WSNs due to their simplicity, flexibility, and the fact that there is no need for much network infrastructure support. There are no constraints on network topology and node neighborhoods. However, these protocols suffer from inefficient use of the bandwidth due to collisions, especially when they are not synchronized and cannot use a slotted approach. Therefore, this paradigm is not well suited for dense networks with high data traffic. Although CSMA uses carrier sensing to avoid collisions, they can still happen due to the hidden terminal problem. This happens when two nodes are not in the interference range of each other, but their interference ranges overlap. Thus, they cannot detect each other's transmission by carrier sensing, but they collide at the location of a third node. Carrier sense multiple access with collision avoidance (CSMA/CA) methods are used to alleviate this problem. The sender nodes check the channel clearance at the place of the receiver by asking a clear channel confirmation from the receiver. However, such a solution is power costly for the rather short data packets in WBAN applications.

In schedule-based protocols, the time division multiple access (TDMA) mechanism is used by wireless nodes to share the medium without collisions. In TDMA-based protocols, sensor nodes are synchronized and time slots that are unique in the neighborhood are reserved for them for communication. Therefore, the medium access is collision free, which makes the mechanism more efficient in terms of bandwidth usage compared to the contention-based mechanisms. Synchronization of nodes is an important requirement to provide efficient schedule-based communication, which is a complicating requirement for these protocols. However, compared to large-scale WSNs, synchronization is less challenging in WBANs due to the small size of the network and the role of the gateway in coordinating the body nodes in a centralized synchronization scheme. Body nodes can align their schedule by receiving beacon packets from the gateway node. That is the reason, together with efficiency and low-power requirements, that TDMA-based mechanisms are widely considered as a proper MAC mechanism for WBANs.

The energy efficiency of the MAC mechanism is a vital requirement for any MAC protocol for WBANs. This goal is mainly achieved by introducing very low-duty-cycle communication in which the wireless transceivers of the body nodes periodically activate for a very short duration to communicate sensed data and are deactivated most of the time. Using both contention-based and schedule-based approaches, it is important to reduce the duty cycle, avoiding unnecessary listening or transmission of packets. However, duty cycling affects the latency of data delivery, as nodes need to wait for the communication opportunity before they can transmit their data. This limits the maximum time interval between active times of the nodes in the duty cycling scheme.

Designing appropriate MAC protocols for WBANs in various medical and health-monitoring applications is an ongoing research field. In the following, we present various ideas that are proposed for designing efficient MAC protocols for WBANs.

9.4.2 Low-Duty-Cycle TDMA-Based MAC

A TDMA-based MAC protocol for WBANs is designed by Marinkovic et al. [17]. It aims for energy efficiency and communication reliability in WBANs with a star topology. In the TDMA mechanism, the communication activities of the body nodes are performed in a periodic manner using fixed-size TDMA frames with length T_{frame}. The frame includes an active part with a fixed number of time slots which are used for communication of sensor nodes with the gateway. The channel is silent during the rest of the time frame and all body nodes are in the idle (low-power) mode, which leads to low average energy consumption for nodes. Figure 9.2 shows the structure of TDMA frames and time slots of this MAC protocol. The gateway may use the inactive part of the frame for communication with a base station in the ambient network.

The gateway assigns a unique range of transmission slot (SL_1–SL_{N-1}) from the active part of the frame to each body node. Some k extra range of slots (RSL_1–RSL_k) are reserved for packet retransmission in the case that the communication in the dedicated slot to some nodes fails. The number of the reserved slots in each TDMA frame depends on expected packet error rate (PER) of communication over the wireless links in the WBAN and the application DDR requirement. Simulation results by Marinkovic et al. [17] show that a higher number of reserved slots leads to a higher DDR, especially for links with higher PER.

A time slot starts with packet transmission by the sensor node and follows with an acknowledgment by the gateway if the packet is successfully received. There is a guard time (T_{guard}) before starting the next slot to compensate for small drifts of the nodes. If the gateway does not receive a packet of a node, it replies with a negative-acknowledge character (NACK) packet that includes information about the reserved slot that the node can use for packet retransmission. Communication in all time slots is scheduled by the gateway and is assumed to be collision free. If the node does not receive an acknowledge (ACK)

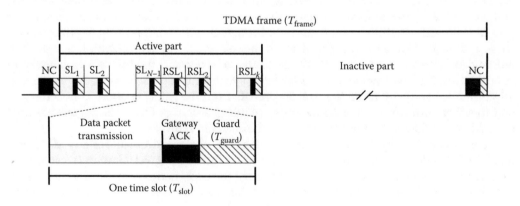

FIGURE 9.2
The structure of TDMA frames and data packet transmission time slots in the MAC layer designed by Marinkovic et al. [17].

character or NACK packet from the gateway, it does not retransmit the packet, leading to a packet drop.

Using such a TDMA-based medium access requires a good synchronization of all nodes in order to avoid collisions and provide efficient and reliable communication. A centralized synchronization mechanism with gateway beacons is used. The gateway transmits a network control (NC) packet at the end of each TDMA frame. Besides its use for synchronization, the gateway includes slot schedule information in the NC packets. Sensor nodes align their frames with the NC packets that they receive from the gateway. To reduce energy consumption overhead imposed on sensor nodes for receiving NC packets, the mechanism allows the nodes to synchronize once after a few TDMA frames. Assuming that the accuracy of the nodes' clock is θ ppm (the node's clock may be at most θ μs ahead or behind in 1 s with respect to the perfect clock), at most $N_R = T_{\text{guard}}/(2\theta \times T_{\text{frame}})$ time units can pass before the sensor needs to resynchronize. This is because the drift between the clock of a sensor node and the clock of the gateway should not exceed the guard time T_{guard}. The clocks may in the worst case have $2\theta \times T_{\text{frame}}$ drift in one TDMA frame.

The protocol achieves energy efficiency with collision-free communication and low duty cycling. Assume that a sensor node listens to the NC packet for synchronization only once in N_R TDMA frames, and the slot length of NC packet transmission is equal to the length of data packet transmission slots (T_{slot}). In this case, the average duty cycle (D) of communication for a sensor node is given by Equation 9.1:

$$D = \frac{T_{\text{slot}}}{T_{\text{frame}}}\left(1 + \frac{1}{N_R} + \text{PER}\right), \tag{9.1}$$

where PER stands for the packet error rate. In each TDMA frame, a body node transmits a packet in its slot with length T_{slot}.

With the probability of PER, the packet drops and the node needs to retransmit it in a reserved slot. Moreover, once per N_R frame, the body node needs to listen to the gateway to receive an NC packet. A longer frame results in a lower duty cycle, which reduces the energy consumption. However, this increases the latency of delivering sensed data to the gateway and reduces the throughput. Moreover, the data sampling rate of the sensor and the amount of data that can be included in one data packet should also be taken into account.

The strong points of this protocol are its simple mechanism and energy efficiency. Considering the acknowledgment mechanism of this protocol, only a star topology is assumed to be used. All body nodes are assumed to have similar data sampling rates and, thus, equal communication traffic loads. Equal size slots are assigned to all sensors in the WBAN. This may not be a realistic scenario in some health applications in which there are different biosensors with a variety of sampling rate requirements.

9.4.3 Gossiping TDMA-Based MAC

The WBAN protocol stack presented by Nabi et al. [16] uses a TDMA-based MAC mechanism that supports both star and multihop WBAN architectures. This MAC protocol aims to prepare a platform suitable for running a gossiping data dissemination algorithm on top. The basic mechanism of this MAC layer is taken from the gossip medium access control (gMAC) [18] protocol. However, it is optimized and simplified for WBANs. For instance, in the original gMAC, unique time slots in the two-hop neighborhood are dynamically assigned to sensor nodes for packet transmission; it provides spatial RF channel reuse. In the MAC layer used by Nabi et al. [16], fixed transmission slots are assigned to body nodes at design time. Considering the size of typical WBANs, and the frequent topology changes due to posture changes, such a simple fixed slot allocation is more efficient and reliable.

Figure 9.3 shows the layout of the TDMA frames and time slots in the gossiping TDMA-based MAC layer by Nabi et al. [16]. A frame consists of an active part and an inactive part. The active part of the frame includes a fixed number of time slots that are used for communications. The channel is silent during the inactive part of the frame because none of the nodes transmit in that part. Therefore, all nodes go to the idle mode in the inactive part of the frame, leading to low average energy consumption for nodes. A guard time, with length T_{guard}, is inserted at the beginning and the end of every time slot to avoid problems due to small phase differences. Assume that time slot SL_i is assigned to body node s_i. Relative to its own clock, node s_i waits for one guard time and then starts packet transmission. Consecutive transmissions are, therefore, $2 \times T_{\text{guard}}$ apart from each other. The body nodes that want to receive packets from s_i listen to the channel from the beginning of the slot.

The upper layer protocol determines whether a body node listens to other nodes. The list of slots that a node should listen to in a frame is given by the upper layer (e.g., the routing protocol) at the beginning of each frame. However, for maintaining the synchronization between nodes, as well as for receiving application layer and control packets from the gateway, every node listens to the gateway slot at every frame. Therefore, minimum radio activity of a body node in a TDMA frame consists of listening to the time slot of the gateway and a packet transmission in its own time slot. Assuming that a node listens in $1 \leq N_l \leq N - 1$ time slots (including the gateway's slot) in each frame, its radio duty cycle (D) is given by Equation 9.2:

$$D = \frac{T_{\text{slot}} \times N_l + T_{\text{tx}}}{T_{\text{frame}}}, \tag{9.2}$$

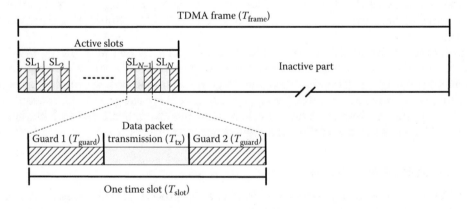

FIGURE 9.3
The structure of TDMA frames and time slots in the gossiping low-duty-cycle TDMA-based MAC protocol. (From M. Nabi et al., On-demand data forwarding for automatic adaptation of data propagation in WBANs. In Proc. IEEE Conference on Sensor, Mesh and Ad Hoc Communications and Networks [SECON], pp. 326–334. IEEE, 2012.)

where T_{slot} and T_{tx} stand for a time slot duration and the length of real packet transmission, respectively ($T_{slot} = T_{tx} + 2 \times T_{guard}$).

If the upper layer protocol relies on a star architecture, the node listens only to the gateway ($N_l = 1$), which provides the minimum duty cycle using this protocol with a given frame length (T_{frame}). If the upper layer protocol supports a multihop architecture, the node may listen in more than one slot, which leads to a greater duty cycle and higher energy consumption.

There is no acknowledgment mechanism in these protocols and all packets are broadcast to the all listening nodes. There is no field for a destination address in the MAC header of the packets. Each transmission packet includes several data items which are either sensed by the sender node or received from other nodes to be forwarded (in a multihop architecture). As mentioned, every node transmits exactly one packet in each TDMA frame. If the data sampling rate of the node is higher than the transmission rate, determined by the frame length, the node may buffer its data and send several data items in one transmission packet. Since there is no acknowledgment in this protocol, a probabilistic packet retransmission may be used to deal with possible packet transmission failures [19]. The data items that are put in a transmission packet are given by the upper layer; thus, the retransmission strategy is also a part of the upper layer protocol.

9.4.4 Battery-Aware MAC

A battery-aware TDMA scheduling mechanism for WBANs is proposed by Su and Zhang [20]. It tries to maximize the network lifetime. The main idea of this mechanism is to store sampled data items in a transmission queue in the sensor node and transmit them together at the right time. Three criteria are considered for deciding about the right time for data transmission: (1) According to the electrochemical properties of batteries, the batteries are capable of recovering the initial voltage value after they remain idle for a sufficiently long period of time. It means that if the MAC mechanism increases the idle time between consequent battery discharges (radio activities), the battery has more time for self-recharge, which leads to longer lifetime of the sensor node. This mechanism tries to maximize the idle periods while satisfying latency requirements by putting sampled data items in a queue and transmitting big data packets instead of frequent transmission of small data packets. (2) The second criterion for packet transmission is the quality of wireless links. The protocol uses gateway beacons to estimate the quality of links and transmit the packet when the link is in a good state. (3) The last criterion is the number of data items in the queue. If the queue is full, new data items will be discarded. Therefore, the mechanism takes into account the number of data samples that it can send in the next packet transmission, and the current number of samples in the queue to decide about the right time for a packet transmission. Long data packets may lead to degradation of DDR because a single packet drop leads to several data losses. There is also a higher chance of packet drop for longer packets. This is a negative factor of the larger data packets that are considered in this MAC protocol.

9.4.5 H-MAC: Heartbeat-Driven Medium Access Control

H-MAC [21] is a TDMA-based protocol for body sensor nodes that aims to reduce energy consumed for exchanging periodic synchronization beacons. The idea is to use the body heartbeat rhythm as a reference for synchronizing sensor nodes on the body. Through the circulatory system of the human body, the blood propagates the heartbeat rhythm, generated by the cardiac muscles, throughout the body. Such heartbeat pulsation affects various physiological parameters of the body. Because of that, the rhythm can be observed in different biological signals of the body and can be detected by various biosensors. The ECG signal records the electrical changes that are caused by the contraction of heart muscles. The heart blood pumping changes the blood volumes of different vessels and can be detected by ambulatory blood pressure (ABP) sensors. A phonocardiography (PCG) sensor can detect heart rhythm from its sound. These are just examples of body sensors that have already information about the heartbeat.

H-MAC proposes to use the heartbeat rhythm as the reference for scheduling radio packet transmissions. Every node uses its own sensory data for heartbeat peak detection. Some points motivate this idea. First, many biosensors already have such peak extraction ready in one way or another. Thus, it does not impose a heavy overhead for peak detection in those cases. Second, it is relatively robust against interference and peak detection algorithms are sufficiently reliable. Therefore, using that as a synchronization reference increases the robustness of the MAC mechanism. Third, the peak interval of the heartbeat is in the range of 300 to 1500 ms (assuming the heartbeat rate of 40–200 beats/min). This is an acceptable TDMA time slot for many health applications. Thus, it can be directly used to schedule time slots.

TDMA time slots are aligned with heartbeat peaks. However, the gateway still plays a key role in coordinating slot usage by the body nodes. Like other nodes, the gateway is also assumed to detect peaks by its sensors. It determines the length of a TDMA frame cycle and assigns exclusive time slots to body nodes, taking into account their data rate requirements and the number of nodes in the WBAN. This information is included in a control message that is transmitted with the first peak detection. Every node has a peak counter that is used to recognize the frame cycles and the dedicated transmission slots. In this way, a TDMA frame cycle is composed of a number of heartbeat peaks that are equivalent to the number of transmission slots per frame. Because the peaks detected at different parts of the body are not exactly aligned (due to the propagation delay of the circulatory system of the body), guard periods are used. The interval between two consecutive peaks is considered large enough to accommodate a packet transmission as well as a sufficiently large guard to compensate for peak propagation delay.

In the case that a node misses some peaks or counts false peaks, it loses its frame synchronization. This can lead to collisions, as the node may wrongly transmit in slots of other nodes. The H-MAC protocol has an embedded synchronization recovery when such a case is detected. A node detects this by observing abrupt changes in the peak intervals. The heart rate and accordingly the peak intervals change smoothly and continuous in normal situations. Therefore, abrupt changes are considered as a mistake in counting the peaks, which triggers resynchronization.

To make this technique applicable for more WBAN applications, where some nodes do not have access to heartbeat rhythm information, an integration of H-MAC and radio beacon–based scheduling mechanism is suggested by Li and Tan [21]. In this scenario, the gateway has access to heartbeat rhythm information and sends beacons to those nodes that do not have access to such information. However, this integration has not been sufficiently developed and evaluated and needs further investigation and development. Using a periodic biological event as a reference for time synchronization in WBANs is an interesting idea. The applicability, efficiency, and reliability of such an approach in a real-world WBAN needs to be practically investigated.

9.4.6 BANMAC: Body Area Network Medium Access Control

The quality of wireless links in WBANs changes due to the movements of the human body. In particular, the link quality between some nodes on the body fluctuates periodically in some postures. For instance, experiments [22] show periodic oscillations of the value of the received signal strength indicator (RSSI) for the link between two nodes deployed on the right wrist and stomach during walking. The measured RSSI value fluctuates between −80 and −60 dBm with a frequency that matches the walking-step frequency of the person wearing the nodes (i.e., 1.2 steps per second). To realize an efficient communication, BANMAC [22] tries to schedule packet transmissions when the quality of a link is predicted to be in its good state. This way, the chance of successful packet delivery is increased, which leads to better DDR and less energy consumption for retransmissions.

BANMAC proposes an algorithm using fast Fourier transformation (FFT) to predict the opportune transmission window (OTW) during which the link quality is predicted to be good. Some probe packets are transmitted by the gateway between the data packets. The receiving body nodes include their measured RSSI of the probe packet in their transmission data packets. The gateway applies an FFT on the received RSSI to find the dominant frequency. Using these calculations, the gateway makes a prediction of the OTW. It is then used to schedule the ongoing communication, hoping that it increases the chance of successful delivery.

The idea of using link quality prediction for scheduling communication is an attractive approach. However, the performance of such a MAC mechanism strongly depends on the accuracy of the predictions. Moreover, the complexity of the prediction algorithms and packet exchange overhead required for RSSI measurements are limiting factors for applicability of such mechanisms.

9.5 Communication Standards

Several wireless communication standards are suitable for WBANs or even specifically developed for this purpose. In particular, the IEEE 802.15.4 [23] standard that is designed for low-data-rate wireless personal area networks (LR-WPANs) is widely considered as a solution or the base for other WBAN protocols. Moreover, the IEEE 802.15.6 [10] is a newer standard specifically designed for ultralow-power WBAN communications. Ultralow-power radio transceivers supporting these standards are now available, which can be used in developing low-power wireless body sensor devices. For instance, Liu et al. [24] presents a multistandard (i.e., IEEE 802.15.4, IEEE 802.15.6, and Bluetooth low energy) radio transceiver that consumes very little energy for radio communication (3.8 and 4.6 mW DC power from a 1.2 V supply in receive mode and transmission mode, respectively). Availability of such technologies paves the way for realizing many interesting WBAN applications. In this section, we review these standards and their mechanisms of managing the access of body nodes to the shared radio channels.

9.5.1 IEEE 802.15.4 LR-WPAN Standard

IEEE 802.15.4 LR-WPAN [23] is a low-cost communication standard that fits applications and devices with power constraints and relaxed throughput requirements. The main objectives are reliability, power efficiency, ease of installation, and flexibility with a simple implementation. These features match protocol requirements for WBANs, therefore making this standard a proper candidate for these networks. LR-WPAN offers three frequency bands with different data rates. The data rate is 250 kbps in the 2.4 GHz ISM band, 40 kbps at 915 MHz, and 20 kbps at 868 MHz. In total 27 subchannels can be used in LR-WPAN (i.e., 16 subchannels in the 2.4 GHz band, 10 subchannels in the 915 MHz band, and 1 in the 868 MHz band).

Both star and multihop WBAN architectures can be developed based on the IEEE 802.15.4 standard, as it provides two different topologies, namely, the star topology and the peer-to-peer topology. Figure 9.4 illustrates the star and peer-to-peer topologies in LR-WPAN by two example networks. The PAN coordinator is the main destination (sink) of data packets generated in the network; it plays the main coordination role in the network. In a WPAN star topology, all wireless nodes communicate with the PAN coordinator. This is what we require in a star WBAN architecture. In the peer-to-peer topology of LR-WPAN, a PAN coordinator also exists. However, any node is able to communicate with any other node as long as the nodes are in the radio range of one another. A peer-to-peer topology of LR-WPAN can be used in a multihop WBAN architecture that allows body nodes to deliver their sensed data to the gateway (PAN coordinator) by traversing multiple hops. Note that the protocol for multihop packet routing is not a part of this standard and can be developed on top of that.

LR-WPAN provides two operational modes, beacon-enabled mode and non-beacon-enabled mode. In the non-beacon-enabled mode, every node uses an unslotted CSMA/CA mechanism or an ALOHA scheme to access the medium and transmit its packet. This mode provides a very simple behavior. In CSMA/CA, each node first senses the channel and starts transmitting its packet if the channel is clear. Otherwise, it tries again later using a particular exponential back-off mechanism. In the ALOHA scheme, a node transmits without sensing the channel, which is appropriate for lightly loaded networks because of lower collision probability. In any way, in a non-beacon-enabled network, the nodes that want to receive packets from their neighbors need to stay awake and listen to the channel because they are not synchronized with the other nodes. In a star topology, only the coordinator keeps listening to the channel. In a peer-to-peer architecture, every node needs to perform this energy costly task. This makes this mode typically not a suitable option for WBANs.

In the beacon-enabled mode, a superframe structure is used; its exact format is defined by the coordinator node. Each superframe is started by a beacon transmission by the coordinator. The beacons are used to identify the PAN coordinator, to describe the structure of superframes, and to synchronize the nodes. The superframe may have active and inactive parts, which provides the possibility of gaining energy efficiency by duty cycling communications. The active part of the superframes is divided into 16 equally sized slots. The beacon frame is sent in the first slot of each superframe. The active part of the superframe

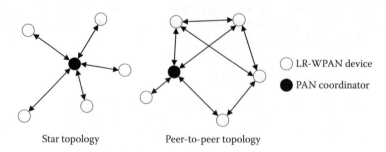

Star topology   Peer-to-peer topology

○ LR-WPAN device

● PAN coordinator

FIGURE 9.4
Examples of topologies supported by the IEEE 802.15.4 LR-WPAN [23] standard.

consists of a contention access period (CAP) and a contention-free period (CFP). Figure 9.5 depicts the structure of an IEEE 802.15.4 superframe. Nodes that have data packets for transmission may compete with other transmitting nodes during the CAP by using a slotted CSMA/CA mechanism. The CFP starts immediately after the CAP, and contains guaranteed time slots (GTSs) which are exclusively dedicated to some nodes by the PAN coordinator. At most seven GTSs can be allocated in each superframe. One GTS can occupy more than one time slot, based on the data transmission requirements of the node that uses the GTS. A node that needs a reliable data transmission requests the PAN coordinator for GTS allocation, describing its requirements. The PAN coordinator decides whether to allocate a GTS to the requesting node, based on the requirements of the node and the current available capacity in the superframe. A first-come-first-served scheme may be used by the coordinator to assign GTS slots. Note that a node that has been allocated a GTS may also operate in the CAP.

The duration of active and inactive parts of the superframe is given by two parameter values, the beacon order (BO) and the superframe order (SO), where $0 \leq SO \leq BO \leq 14$. The parameter BO determines the beacon interval (BI) in terms of the number of symbols; this is the time interval between two subsequent beacon transmissions by the coordinator. The length of the active part, superframe duration (SD), is determined by the value of SO. The formulas for calculating the length of each part are given in Figure 9.5. The coordinator determines these values based on the requirements of the application. A higher value of BO leads to a lower duty cycle at the cost of a higher latency for data delivery.

An optional acknowledgment mechanism can also be used. A successful packet reception is optionally confirmed with an acknowledgment packet by the receiver. If the sender does not receive an ACK packet after a certain time period, it assumes that the transmission failed and retransmits the packet. There is a limit to the number of retransmissions for each packet. When the acknowledgment is not enabled, the sender always assumes that the transmissions succeed.

The beacon-enabled mode of the IEEE 802.15.4 standard offers a synchronized, duty-cycled communication in combination with contention-based and contention-free medium access. This fits many WBAN applications. Heterogeneous WBANs in which some nodes may have real-time data with high sampling rate while others may have relaxed requirements can benefit from different medium access schemes. Since this protocol is a

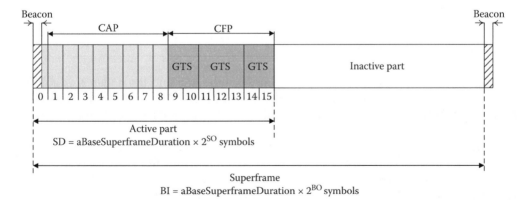

FIGURE 9.5
The structure of a superframe in the beacon-enabled mode of the IEEE 802.15.4 LR-WPAN standard [23]. The parts indicated by numbers 0 to 15 are time slots.

communication standard, it provides a detailed communication protocol description that determines exact packet framing, wireless channel scan and selection mechanisms, a network initialization mechanism, joining and leaving schemes, security, etc. The standard covers the physical and MAC layers of the protocol stack. Proper upper layers such as data routing need to be built on top of the IEEE 802.15.4 standard.

9.5.2 IEEE 802.15.6 WBAN Standard

The IEEE standard association formed the IEEE 802.15 Task Group 6 in 2007 in order to develop a communication standard optimized for ultralow-power devices operating on, in, or around the human body, serving a variety of applications. Characteristics of WBANs such as heterogeneity and power constraints have been taken into consideration to specify different mechanisms and protocol layers in this standard. For instance, the standard specifies four priority levels for WBAN services as nonmedical services, mixed medical and nonmedical services, general health services, and highest-priority medical services. Moreover, eight user priority (UP) levels are available to tag the data packets, which are set based on the nature of the sensor type. Higher priority gives higher chance to the transmitter of the packet to win the channel access competition. In the following, the priorities are discussed in detail.

The current version of the IEEE 802.15.6 [10] standard, released in February 2012, defines three physical layers, narrowband (NB), ultrawide band (UWB), and human body communications (HBC). The NB physical layer operates in seven different frequency bands with a variety of channels, bit rates, and modulation schemes for each frequency band. The UWB physical layer is included in the IEEE 802.15.6 standard to offer ultralow-power communication, causing less interference to existing NB wireless devices. The HBC physical layer uses the electric field communication (EFC) technology. The band of operation is centered at 21 MHz. The standard specifies a common MAC for all the supported physical layers.

In the terminology used in this standard, the gateway is referred to as the hub. The main topology considered is the one-hop star architecture. However, a two-hop extended star WBAN is also supported in which the hub and a node may use a relay-capable node to exchange packets. In these topologies, the hub is responsible for coordinating medium access. Three channel access modes are defined in the standard:

- Beacon mode with superframes
- Nonbeacon mode with superframes
- Nonbeacon mode without superframes

In the first two modes, time referencing between the hub and the nodes is established. The hub determines the superframe boundaries and defines the number of allocation slots in the superframes. In the last mode, the hub does not provide any time reference in its WBAN.

In the first mode (beacon mode with superframes), time is divided into equal-length superframes that are determined by the hub through transmitting beacons. A superframe may be active or inactive. The hub may maintain a number of inactive superframes after an active superframe, which leads to duty-cycled communication. Figure 9.6 shows the layout of an active superframe in this mode. In the inactive superframes, the hub does not transmit any beacon and, therefore, no access to the medium is expected. The hub organizes an active superframe by allowing several medium access phases, which are two

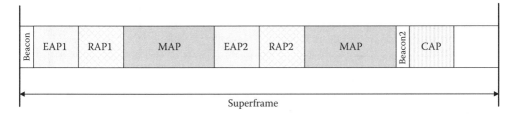

FIGURE 9.6
The layout of an active superframe in the beacon mode with superframes of the IEEE 802.15.6 standard [10].

exclusive access phases (EAP1 and EAP2), two random access phases (RAP1 and RAP2), two managed access phases (MAPs), and a contention access phase (CAP). The size of each phase is determined by the hub and can also be zero. If the length of the CAP is not zero, a second beacon is transmitted before the start of the CAP.

In the MAPs, uplink and downlink allocation intervals may be scheduled by the hub. Two channel access schemes are provided in this phase. The first scheme is improvised unscheduled access, in which the hub initiates exchange of particular types of management or data frames with a node without advance notice or previously announced times. This scheme is called polling communication. Nodes must stay awake and wait for a frame from the hub, indicating the start of data or management frame exchange. The other scheme is scheduled access that is contention free. A node may exchange frames with the hub in every m superframe (m periodic), for some nonnegative integer m. This allows the device to sleep between transfers. Using this scheme, the node transmits its packet in its reserved time slots.

A contention-based mechanism access is used in all other phases (EAPs, RAPs, and CAP). CSMA/CA or slotted ALOHA is considered for the narrowband and ultrawide band physical layers, respectively. Only the nodes that want to send data-type frames of the highest UP are allowed to use EAP1 and EAP2. This UP is only used for data containing emergency information. Table 9.1 presents the list of eight different UPs that are supported in the standards to be assigned to data packets [10]. The nodes sending data of all priorities other than UP_7 use either a RAP phase or the CAP for transmitting their data packets. In order to improve channel utilization using the CSMA/CA mechanism, a node with UP_7

TABLE 9.1

UP Levels and Channel Access Parameter Values in the IEEE 802.15.6 Standard

Priority	UP	Traffic Designation	Minimum Contention Window	Maximum Contention Window
Lowest	UP_0	Background	16	64
	UP_1	Best effort	16	32
	UP_2	Excellent effort	8	32
	UP_3	Video	8	16
	UP_4	Voice	4	16
	UP_5	Medical data or network control	4	8
	UP_6	High-priority medical data or network control	2	8
Highest	UP_7	Emergency or medical implant event report	1	4

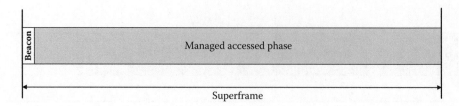

FIGURE 9.7
The layout of an active superframe in the nonbeacon mode with superframes of the IEEE 802.15.6 standard [10].

may combine EAP1 and RAP1. To prioritize data traffic of other user priorities (UP_0–UP_6), different values are used for the minimum and maximum lengths of the contention window in the CSMA/CA. A shorter contention window allows the node to back off shorter after a channel access failure, which eventually leads to a higher chance of winning the channel. A similar mechanism is also applied for channel access prioritization when the slotted ALOHA is used.

In the second operation mode (nonbeacon mode with superframes), the hub does not send beacon packets. However, the hub provides timing reference by including required timing information in its particular transmission frames (called T-Poll frames). There is only a MAP in each superframe. It means that in this mode, the packet exchanges are either polling or scheduled in allocation slots that are managed by the hub. No contention-based access mechanism is performed in this mode. Figure 9.7 depicts the structure of a superframe in this mode. In the third operation mode of the standard (nonbeacon mode without superframes), no beacon is transmitted; neither superframe nor allocation slots are established. In this mode, all nodes treat any time interval as an EAP1 or RAP1 and use CSMA/CA-based random access for packet transmission.

Although the medium access coordination between BANs at the MAC sublayer is not specified in the IEEE 802.15.6 standard, two optional mechanisms for coexistence and interference mitigation between adjacent or overlapping WBANs are provided. The first one is beacon shifting, in which the hub transmits its beacons at different time offsets, according to a particular beacon shifting sequence, not being used by other neighbor hubs (WBANs). The second suggested mechanism is channel hopping. In this mechanism, the hub changes its operating channel periodically according to a particular channel hopping sequence. The body nodes should hop accordingly to the same channels to remain connected with their hub.

9.6 Conclusions

This chapter discusses protocols for communication in WBANs. The focus is on networks in which all body nodes are to deliver their sensed data samples to a gateway node on the body. This gateway node plays the role of a hub to further transfer body data to an ambient network or to a central base station. The gateway may also perform some data processing or aggregation. This chapter first points out the characteristics of WBANs that should be considered in designing proper communication protocols. Afterward, prominent network architectures, MAC mechanisms, and standards for communication in WBANs are discussed.

Strict energy consumption constraints, very short-range RF radio, time-varying link quality, heterogeneity in terms of sensing parameters, and application data delivery requirements, and high mobility are the key characteristics of WBANs. Research on communication in WBANs has resulted in several efficient communication protocols and mechanisms of achieving required performance in such networks. For accessing the shared wireless medium, the TDMA mechanism has been the most prominently considered approach because of its efficiency in bandwidth usage and good performance predictability. In some mechanisms that use for simplicity a collision-based MAC mechanism, collision-free scheduled time slots are dedicated to the body sensors with strict performance requirements.

WBANs have gone beyond their early research phase and have already entered the standardization and industrial realization. Several communication standards are well suited for WBAN communication. The IEEE 802.15.4 LR-WPAN standard is widely used for communication in WBANs and their surrounding networks. Moreover, the IEEE 802.15 Task Group 6 has provided a communication standard, especially for WBANs taking special needs of these networks. These standards draw inspiration from the research results on WBAN protocols. There are low-power radio transceivers that support the standards and that can be used for developing low-power wireless body nodes. As an example, a multistandard ultralow-power radio chip has been developed by Imec-NL that supports the IEEE 802.15.4, IEEE 802.15.6, and Bluetooth low-energy standards. Availability of such devices is a sign of the advances in development and use of WBANs in various health and wellness applications.

References

1. E. Romero, R. Warrington, and M. R. Neuman. Energy scavenging sources for biomedical sensors. *Physiological Measurement*, 30(9), 2009.
2. A. Zurbuchen, A. Pfenniger, A. Stahel, C. T. Stoeck, S. Vandenberghe, V. M. Koch, and R. Vogel. Energy harvesting from the beating heart by a mass imbalance oscillation generator. *Annals of Biomedical Engineering*, 41(1):131–141, 2013.
3. A. Natarajan, B. de Silva, K.-K. Yap, and M. Motani. To hop or not to hop: Network architecture for body sensor networks. In Proc. 6th annual IEEE Communications Society Conference on Sensor, Mesh and Ad Hoc Communications and Networks (SECON), pp. 682–690. IEEE, 2009.
4. A. Natarajan, M. Motani, B. de Silva, K.-K. Yap, and K. C. Chua. Investigating network architectures for body sensor networks. In Proc. 1st ACM SIGMOBILE International workshop on Systems and Networking Support for Healthcare and Assisted Living Environments (HealthNet), pp. 19–24. ACM, 2007.
5. M. Quwaider and S. Biswas. Probabilistic routing in on-body sensor networks with postural disconnections. In Proc. 7th ACM International Symposium on Mobility Management and Wireless Access (MobiWAC), pp. 149–158. ACM, 2009.
6. J. Cai, S. Cheng, and C. Huang. MAC channel model for WBAN. Technical report, 15-09-0562-00-0006, IEEE P802.15 Working Group for Wireless Personal Area Networks (WPANs), 2009.
7. T. Falck, H. Baldus, J. Espina, and K. Klabunde. Plug 'n play simplicity for wireless medical body sensors. *Mobile Networks and Applications*, 12(2–3):143–153, 2007.
8. D. Sagan. RF integrated circuits for medical applications: Meeting the challenge of ultra low power communication. Ultra-Low-Power Communications Division, Zarlink Semiconductor, 2005.

9. E. Strömmer, M. Hillukkala, and A. Ylisaukkooja. Ultra-low power sensors with near field communication for mobile applications. In Proc. International Conference on Wireless Sensor and Actor Networks (WSAN), pp. 131–142. Springer, 2007.

10. IEEE standard for local and metropolitan area networks—Part 15.6: Wireless body area networks (WBANs). In IEEE Std 802.15.6-2012, pp. 1–271, 2012.

11. A. Sani. Modelling and Characterisation of Antennas and Propagation for Body-Centric Wireless Communication. PhD thesis, Queen Mary University of London, London, United Kingdom, 2010.

12. E. Reusens, W. Joseph, B. Latre, B. Braem, G. Vermeeren, E. Tanghe, L. Martens, I. Moerman, and C. Blondia. Characterization of on-body communication channel and energy efficient topology design for wireless body area networks. *IEEE Transactions on Information Technology in Biomedicine (TITB)*, 13(6):933–945, 2009.

13. Y. Zhao, A. Sani, Y. Hao, S. Lee, and G.-Z. Yang. A subject-specific radio propagation study in wireless body area networks. In IEEE Conf. Loughborough Antennas and Propagation (LAPC), pp. 80–83. IEEE, 2009.

14. J.-H. Hauer, V. Handziski, and A. Wolisz. Experimental study of the impact of WLAN interference on IEEE 802.15.4 body area networks. In Proc. 6th European Conference on Wireless Sensor Networks (EWSN), pp. 17–32. Springer-Verlag, 2009.

15. M. Nabi, M. Blagojevic, M. C. W. Geilen, T. Basten, and T. Hendriks. MCMAC: An optimized medium access control protocol for mobile clusters in wireless sensor networks. In Proc. IEEE Conference on Sensor, Mesh and Ad Hoc Communications and Networks (SECON), pp. 28–36. IEEE, 2010.

16. M. Nabi, M. Geilen, T. Basten. On-demand data forwarding for automatic adaptation of data propagation in WBANs. In Proc. IEEE Conference on Sensor, Mesh and Ad Hoc Communications and Networks (SECON), pp. 326–334. IEEE, 2012.

17. S. J. Marinkovic, E. M. Popovici, C. Spagnol, S. Faul, and W. P. Marnane. Energy efficient low duty cycle MAC protocol for wireless body area networks. *IEEE Transactions on Information Technology in Biomedicine (TITB)*, 13(6):915–925, 2009.

18. F. van der Wateren. MyriaCore implementation details, the inside of MyriaCore and gMac. Technical report, Chess Company, the Netherlands, 2010.

19. M. Blagojevic, M. Nabi, M. Geilen, T. Basten, T. Hendriks, and M. Steine. A probabilistic acknowledgment mechanism for wireless sensor networks. In Proc. 6th IEEE Conference on Networking, Architecture, and Storage (NAS), pp. 63–72. IEEE, 2011.

20. H. Su and X. Zhang. Battery-dynamics driven TDMA MAC protocols for wireless body-area monitoring networks in healthcare applications, *IEEE Journal on Selected Areas in Communications*, 27(4):424–434, 2009.

21. H. Li and J. Tan. Heartbeat-driven medium-access control for body sensor networks. *IEEE Transactions on Information Technology in Biomedicine (TITB)*, 14(1):44–51, 2010.

22. K. S. Prabh and J.-H. Hauer. Opportunistic packet scheduling in body area networks. In Proc. 8th European Conference on Wireless Sensor Networks (EWSN), pp. 114–129. Springer-Verlag, 2011.

23. IEEE standard for local and metropolitan area networks—Part 15.4: Low-rate wireless personal area networks (LR-WPANs). In IEEE Std 802.15.4-2011, pp. 1–314, 2011.

24. Y.-H. Liu, X. Huang, M. Vidojkovic, A. Ba, P. Harpe, G. Dolmans, and H. de Groot. A 1.9nJ/b 2.4 GHz multistandard (Bluetooth low energy/ZigBee/IEEE 802.15.6) transceiver for personal/body-area networks. In Proc. IEEE Solid-State Circuits Conference Digest of Technical Papers (ISSCC), pp. 446–447. IEEE, 2013.

10

Wireless Body Area Network Data Delivery

Majid Nabi, Marc Geilen, and Twan Basten

CONTENTS

10.1 Introduction

In typical applications of WBANs, the data sensed by the sensor nodes on the body need to be delivered to some stations for processing, analysis, and storage. This chapter discusses the requirements and mechanisms for efficient and reliable data delivery in WBANs. The main focus of the chapter is on data delivery to a gateway node within a WBAN (intra-WBAN data delivery); it furthermore briefly covers data delivery to a central station in a heterogeneous ambient network.

10.2 Data Delivery Requirements in WBANs

Different health or wellness applications may require different network organizations for data delivery as the requirements for data sampling and data delivery may also vary from one application to another. In this section, the requirements of data delivery in typical WBAN applications are discussed.

10.2.1 Network Organization for Data Delivery

Health- and well-being–monitoring applications may use specific network organizations including WBANs. Figure 10.1 shows an example. A WBAN consists of several low-power wireless biosensor devices worn by a human. WBANs are considered as the first-level (level 1) networks in the network organization. Typically, a more powerful node, called the *gateway*, is also a part of the WBAN; it plays the role of a hub for gathering data and forwarding them to higher-level networks. The body data are to be delivered to a local central station. In one scenario, this data delivery may be done through a direct communication of the gateway node with the central station. In another scenario, the body data may be delivered to an ambient WSN to be forwarded toward the central station. The ambient network may consist of a variety of wireless nodes deployed in the building or on the furniture. These nodes may be used only for relaying body data, or they may be equipped with sensors themselves to sense ambient parameters such as motion, humidity, and temperature. Processing ambient data can provide awareness of the activities of the monitored person.

The combination of WBANs and the ambient network forms a PAN at the second level (level 2) in the network organization. Combined body data and ambient information need to be delivered to a central station in the building for processing, storage, and further communication. Based on the application scenario, the central station may communicate the gathered information to a medical center to be observed by care workers and physicians (level 3). This communication may be realized with a wireless or wired network such as a cellular network or the Internet. Feedback may also be sent back toward the monitored persons from the medical center. In this chapter, we discuss requirements and mechanisms for delivery of body data sensed by the WBAN nodes to the gateway node on the body and subsequently to the local central station. Communication to the third level is subject to different principles and constraints outside the scope of WBANs.

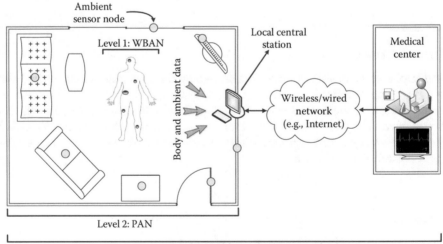

FIGURE 10.1
An example of a network organization for health applications. The combination of WBANs and the ambient network forms a local network for monitored persons. The local central station may communicate with medical centers through a wireless or wired network.

10.2.2 Data Generation and Transmission

Biosensors are the main source of data in WBANs. Three aspects in the sensing process of a sensor node can be distinguished, namely, *signal sampling*, *data sample generation*, and *data transmission*. Each sensor samples a particular biological signal. The main parameters of the signal sampling process are the resolution and frequency of sampling, the right values of which depend on the nature of the signal being sampled and the application requirements. There can be a wide range of sampling frequency requirements for different biosensors. A body sensor node samples the signal with the specified sampling frequency and then communicates either the raw signal samples or the results of processing or interpreting the signal samples (e.g., heartbeat extraction from ECG samples). The data generation frequency is then the number of data samples per second that the sensor node prepares for transmission. The node may buffer data samples for a period of time and then transmit multiple data samples at once. A *data packet* is the transmission unit that a node sends in one transmission; it may include several data samples as well as control information. Thus, the data packet transmission frequency of a node may differ from the data generation frequency of the node. In some applications, there may be sensors that require event-based data generation and transmission. This means that data transmission is triggered by the occurrence of particular events. For instance, observing specific patterns or abrupt changes in signal sample values may be considered events that trigger data transmission.

Besides the sampling frequency and resolution, three different scenarios may apply for gathering sensor data, namely, *status information, events*, and *continuous waveform* signals. In the first scenario, although periodic data sampling and transmission may be performed, the application is interested only in the most recent update of the current signal value and any variations in the signal prior to that update do not provide additional information. In this case, when a more recent data sample reaches the gateway (or the central station), the existing, older data samples are considered outdated and are discarded or ignored. For example, a health application may be interested to have the latest updated value of temperature of the monitored person's body (for detecting fever). In the second scenario, no periodic data transmission is performed. The sensor nodes may transmit event data only when they detect that the corresponding event occurs. As examples, detection of a fall and a heart attack may be considered as events in a health application. The important aspect of delivery of such event data is the speed of data delivery. In the last scenario, the continuous waveform signal, the temporal changes in the signal are important for the application. Therefore, even if new data samples arrive at the destination, the previous data samples are still of interest. Monitoring and processing ECG or EEG waveforms are examples of this kind of data scenario.

10.2.3 Data Delivery Requirements

In this chapter, we use the same notation as in Chapter 6. We consider $S = \{s_1, s_2,\ldots,s_N\}$ as the set of N sensor nodes deployed on a body, forming a WBAN; the last node (s_N) is the gateway node and plays the role of a hub for gathering sensed data from all body sensor nodes. As the natures of various body signals being sensed are different, there is usually a high diversity in sampling and also data delivery requirements for different body sensor nodes. This implies that each sensor needs to be treated considering its own specifications. In this section, we discuss metrics that reflect the end-to-end performance of data delivery in a WBAN application, from individual sensor nodes to their destination (the gateway node or the central base station). The relevant data delivery performance metrics for a particular sensor depend on the data scenario (status information, events, or continuous waveform) that is considered for that node.

We define $\text{DDR}_i^H(t)$, the *data delivery ratio* (DDR) of node s_i at time t, as the fraction of the H last data samples of node s_i before time t that reaches the destination node. Equation 10.1 calculates the DDR value for node s_i at time t for a window size H:

$$\text{DDR}_i^H(t) = \frac{1}{H}\sum_{k=1}^{H} D_i(t,k), \tag{10.1}$$

where $D_i(t, k)$ is a 0- or 1-valued function that shows whether the kth data sample of node s_i up to time t (starting to count backward from time t) reaches the destination. A value of 1 indicates a successful delivery; 0, a failure. This metric is most relevant for the continuous waveform data scenario. In order to be able to reconstruct the corresponding signal waveform at the destination, a certain percentage of data samples need to be received by the destination (assuming a uniform distribution of data losses). According to the characteristics of the continuous waveform (e.g., the rate of changes in the waveform of the signal being sampled), the WBAN application may specify a data delivery constraint DC_i for the DDR for a sensor node s_i. It means that the application expects the value of $\text{DDR}_i^H(t)$ to remain above this threshold $\left[\text{DDR}_i^H(t) \geq \text{DC}_i \text{ for all } i, t\right]$. However, because of the stochastic nature of the wireless network, this may not be 100% satisfiable during network operation, especially for short averaging windows (small H). Therefore, the application needs to have some tolerance for a certain percentage of violations of this constraint. As an example, an application may require the DDR for a node to be higher than 85% for 90% of the time. Such a constraint ensures that the original signal can be reconstructed at the destination with sufficient accuracy.

Besides the fraction of the delivered data samples, the distribution of data losses over time is also an important aspect of data delivery. In the discussion about DDR requirements, a uniform distribution of data losses throughout the averaging window is assumed. In case of *burst data losses*, the destination may not be able to reconstruct the waveform, although the DDR requirement is satisfied. For any time t such that $D_i(t, 1) = 1$, $\text{BL}_i(t)$ is defined as the length of a burst data loss that happened right before time t; it is calculated by Equation 10.2. $\text{BL}_i(t) = l - 1$, where l is such that

$$D_i(t,k) = 0 \text{ for } k = 2,\ldots,l-1, \quad \wedge \quad D_i(t,l) = D_i(t,1) = 1. \tag{10.2}$$

This equation states that the lth data sample generated before time t is the last received data sample before the burst of losses. Long burst data losses result in failure of reconstructing the signal for the continuous waveform data scenario; it may lead to an application failure. Therefore, WBAN applications may specify constraints on the value of $\text{BL}_i(t)$. For instance, it may be necessary for the length of burst data losses to be shorter than a certain limit for 90% of the times that losses happen.

The speed of data delivery to the destination is an important aspect for many applications. Assume that $\Gamma_i^{t_g}$ is a data sample generated by node s_i at time t_g. The *latency* $L\left(\Gamma_i^{t_g}\right)$ of this single data sample $\Gamma_i^{t_g}$ is the time between the moment of generation of that data sample (t_g) at the source node until the time $t_a\left(\Gamma_i^{t_g}\right)$ that it arrives at the destination for the first time (if it reaches the destination), given by Equation 10.3:

$$L\left(\Gamma_i^{t_g}\right) = t_a\left(\Gamma_i^{t_g}\right) - t_g. \tag{10.3}$$

The latency value for a data sample consists of the time that it is traversing a path toward the destination, which also includes the time that the data are suspended in forward queues of the relaying nodes on the routing path. It is clear that a latency value is undefined for a lost data sample. This metric can be relevant for both event and continuous waveform data scenarios. For event data, it is usually of importance that the occurred event is reported to the destination as soon as possible within a certain time limit. For continuous waveforms, latency determines the delay between the occurrence of the real biological signal and the observation of its reconstructed version at the destination. However, if the DDR requirements are satisfied, the latency does not affect the reconstruction of the signal being sampled. As discussed for DDR, there is in any case a positive probability of violation of latency constraints for individual data samples because of the stochastic behavior of the network. A common specification is to require that for a certain ratio of the received data samples at the sink nodes generated by node s_i, the achieved latency be lower than the constraint.

Although the DDR, burst length, and latency metrics reflect the performance of data delivery for event and continuous waveform data scenarios, none of them can directly show how up to date the gathered information is when a status information scenario is considered. Some WBAN applications may be interested to keep their observations of the status of the monitored person as up to date as possible. The latency and DDR of the data delivery to the destination play a role in this. However, the *age* of the gathered data samples at the destination reflects this type of requirement of the application precisely. The age $Age_i(t)$ of data samples from node s_i at the destination at time t is the time difference between the current time t and the sample generation time of the last received data sample generated by node s_i at the destination, given by Equation 10.4:

$$Age_i(t) = t - \max\left\{t_g \mid t_a\left(\Gamma_i^{t_g}\right) \le t\right\}. \tag{10.4}$$

It is clear that data losses will increase the age; burst data losses, in particular, result in higher value of the age. Accordingly, the latency of the last received data item affects the age value. When a status information scenario applies to data delivery in an application, age can properly be used as a performance metric. Thus, the application may specify a certain constraint on the maximum or the average of ages for different body sensor nodes at the destination.

The metrics mentioned in this section are examples of data delivery metrics for various WBAN scenarios. Depending on the application scenario, variants of these metrics may be considered. Besides the metrics that reflect data delivery performance, other types of QoS metrics may be required by the application. Energy consumption, network or node lifetime, and fault tolerance are examples of such metrics.

10.3 WBAN Adaptation for Efficient Data Delivery

To cope with dynamics, WBANs require proper network adaptation mechanisms in order to achieve acceptable data delivery performance. The objective of WBAN adaptation is to follow and adapt to the changes in the network connectivity in order to continuously satisfy the data delivery requirements and optimize the power consumption. Fixed network architecture may lead to unnecessary power consumption of highly constrained body sensor nodes or to violating the data delivery constraints in some time periods. This section discusses general WBAN adaptation approaches and methods for estimating the quality of wireless links in WBANs, which is a prerequisite for typical WBAN adaptation mechanisms. In the following subsection, we go into the details of some adaptation mechanisms for intra-WBAN data delivery.

10.3.1 WBAN Adaptation

Data packet transmission from a body node to the gateway may fail because of insufficient quality of the wireless link, collisions, and interference. There are several ways that a node can compensate for such packet drops. The first strategy is to retransmit the packet within a limited retransmission budget, assuming a star topology. If the communication protocol uses gateway acknowledgments, the body node retransmits the data packet if it does not receive any acknowledgment packet from the gateway within a certain time period. However, if the quality of the link is not good, the retransmissions may fail as well. The sender may give up after a limited number of retransmissions; this means the loss of data samples included in that data packet. The application requires a certain DDR. This may not be satisfied by only retransmission of the dropped packet if the quality of the wireless link is not sufficiently good. A body sensor node s_i is said to be *disconnected* from the gateway if it cannot satisfy the data delivery requirements by sending its packets directly to the gateway node within the available transmission budget. Considering the definition of disconnection from the gateway, the connection status of a body node depends on the quality of its wireless link to the gateway, the data delivery requirements for that node, and the packet retransmission budget. This means that we may have two nodes with the same link quality where one is considered connected and the other is considered disconnected, because of different packet transmission rates and DDR requirements.

When a body node goes to a disconnected state, the network protocol should adapt itself to improve data delivery for the disconnected node and to bring the DDR of that node within the acceptable range. The protocol should also adapt when a node reconnects to the gateway to optimize the consumption of network resources (e.g., energy). There are two common mechanisms that may be used to deal with this disconnection problem. The first one is to adapt the transmit power of the node (e.g., as shown by Smith et al. [1]), hoping to improve the link quality to the gateway and getting out of the disconnected state. This method resolves the issue as far as a higher transmit power is available for the body sensor node. The second approach is to provide multihop data routing, by getting the help of other body nodes that currently have good links to the gateway (e.g., on-demand listening and forwarding [ODLF] as discussed in the next section and by Nabi et al. [2]). Also a combination of these two approaches may be applied.

A prerequisite step for the protocol for adapting to connection changes is to reliably and timely detect such changes. Given the data delivery requirements for a particular node, we need to measure the quality of the wireless link from that node to the gateway in order to detect the connection state. The link quality states the probability of successful packet transmission over that link. Figure 10.2 shows a typical procedure for WBAN adaptation mechanisms and the role of link quality estimation (LQE). Because of frequent topology and link quality changes in WBANs, the LQE method should react fast enough to follow the changes in the quality of links. Late disconnection detection may lead to data loss or even violation of DDR requirements for the body nodes. Late reconnection detection is also expensive,

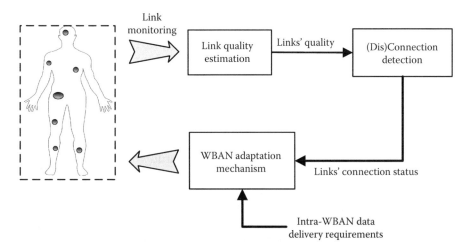

FIGURE 10.2
The structure and components of a typical WBAN adaptation mechanism.

because the protocol may be performing unnecessary data forwarding or the body node uses unnecessarily high transmit power. The method should not be very sensitive to incidental link failures or connections. It needs to be stable enough to avoid frequent fluctuations in order to keep the overhead of the adaptation mechanism low and to make it robust.

To decide if node s_i is disconnected from the gateway, a threshold l_i is applied to the estimated quality of the link. This threshold is given individually for each node based on data delivery requirements. Nabi et al. [2] presented a method for deciding on the appropriate threshold based on the DDR requirement RC_i, the transmission and sampling rate of node s_i. For disconnection detection using any LQE method, hysteresis is applied to prevent frequent switches when the value of the LQE is around the threshold. A link is concluded to have changed from a connected to a disconnected state when the LQE drops below the given threshold $l_i - \Delta l$, where Δl determines the amount of hysteresis applied. Accordingly, for changing from disconnected state to connected state, the quality estimate should go higher than $l_i + \Delta l$. In the following, relevant LQE techniques for WBANs are reviewed.

10.3.2 Link Quality Estimation for WBANs

There are several techniques for estimating the quality of wireless radio links. These techniques are firstly categorized as hardware-based and software-based methods. Link quality indicator (LQI), received signal strength indicator (RSSI), and signal-to-noise ratio (SNR) are hardware-based LQE methods provided by many radio devices. The hardware-based link quality values are directly given by the radio device and no further computation is necessary in software. Several WBAN mechanisms use hardware-based LQE methods as their base; examples are those by Smith et al. [1] and Prabh and Hauer [3]. However, these techniques suffer from some limitations. First, the provided values are only for successfully received packets and the signals that are not detected by the receiver are not taken into consideration. This may lead to overestimating the link quality. Second, these estimators are suitable only for classifying the links as very good or very bad and are not accurate enough to provide a fine-grained estimation of the link quality [4]. Some existing low-power radio chips (e.g., Nordic nRF24L01 [5]) are not equipped with an LQE.

The software-based LQE methods are done in software based on the ability of the links to deliver the whole packets (data packets or beacons). A survey on various LQE techniques for WSNs is presented by Baccour et al. [6]. Software-based LQE may be done at the sender or receiver side of the wireless link. In WBANs, receiver-side LQE methods are more suited for estimating the quality of links toward the gateway node. The main flow of data in typical WBANs is from body nodes to the gateway. So the gateway can use the existing data traffic for LQE without the need for many extra packet (beacon) transmissions by the body sensor nodes. Here, we focus on the methods that have been presented or used for WBANs. The overview presented in the remainder is taken from Nabi et al. [7], who also present an experimental study of the performance of these estimators.

Assume that each body node in a WBAN periodically transmits (data or control) packets to the gateway node. $L_i(\tau)$ is a logical value (0 or 1), which is 1 if the packet transmitted by node s_i with integer sequence number τ is directly received by the gateway; otherwise, it is 0. In fact, $L_i(\tau)$ shows the packet reception status over the link from s_i to the gateway. Based on this knowledge, the gateway is able to quantify the quality of its incoming links.

The packet reception ratio (PRR) at any time is the fraction of the last w transmitted packets that are successfully received by the gateway over the link from node s_i, given by Equation 10.5:

$$\mathrm{PRR}_i^w(\tau) = \frac{1}{w} \sum_{k=0}^{w-1} L_i(\tau - k). \tag{10.5}$$

$\mathrm{PRR}_i^w(\tau)$ presents a uniform moving average of the packet reception over the link from node s_i to the gateway over the last w packet transmissions. A shorter window leads to a more agile estimation at the cost of less stability (more fluctuations). Especially when the quality of the link is not on the extreme good or bad sides, a greater window size is required to accurately quantify the link quality. PRR can be used as an LQE itself, but it is also the basis for some other LQE techniques. A nonuniform averaging may also be used to put higher weight on the recent link receptions. For instance, time-weighted moving average (TWMA) is a moving average over a window of size w using a linear weighting function. It places linearly heavier weight on more recent samples to make the quality estimation more reactive to temporal dynamics compared to PRR. Equation 10.6 calculates this estimate:

$$\mathrm{TWMA}_i^w(\tau) = \frac{1}{\sum\limits_{k=0}^{w-1}(w-k)} \left[\sum_{k=0}^{w-1}(w-k) \times L_i(\tau - k) \right]. \tag{10.6}$$

The exponentially weighted moving average (EWMA) is another example that uses an infinite window with exponential weights. Calculating this estimate is simple and memory efficient as its recursive calculation, given in Equation 10.7, requires only the storage of the previous estimate:

$$\mathrm{EWMA}_i^w(\tau) = \alpha \times \mathrm{EWMA}_i^w(\tau - 1) + (1 - \alpha) \times L_i(\tau), \tag{10.7}$$

where $0 \leq \alpha < 1$ controls the influence of the history of the link packet reception. This estimator with low values for α (not close to 1) is very agile because it discounts previous link states very fast.

In some WBAN applications, a very stable estimation is required. Window mean with exponentially weighted moving average (WMEWMA) is proposed by Woo and Culler [8] to make the PRR estimation more stable by using an EWMA filter. The recursive calculation of the estimation is given by Equation 10.8:

$$\text{WMEWMA}_i^{w,\alpha}(\tau) = \alpha \times \text{WMEWMA}_i^{w,\alpha}(\tau-1) + (1-\alpha) \times \text{PRR}_i^w(\tau). \tag{10.8}$$

WMEWMA has two control parameters, which are the window size w in calculating PRR and the history control coefficient α for the EWMA filter. A high value of the coefficient α, close to 1, is used to make a filter that passes only low-frequency fluctuations on PRR, which results in a more stable estimate than PRR. Setting $\alpha = 0$ converts WMEWMA to the base PRR estimate.

Link likelihood factor (LLF) is also a PRR-based LQE method, which is proposed by Quwaider and Biswas [9], specifically for estimating the quality of wireless links in WBANs. This estimator tries to represent the likelihood for the link to be connected at any given time. The main point is to differentiate incidental link status from long-term behavior of the link by taking the history of the link into account. Equation 10.9 gives the recursive calculation of LLF estimation. The only control parameter of the estimator is the window size w for calculating PRR on which the LLF estimate is built:

$$\text{LLF}_i^w(\tau) = \begin{cases} \text{LLF}_i^w(\tau-1) + \left(1 - \text{LLF}_i^w(\tau-1)\right) \times \text{PRR}_i^w(\tau) & L_i(\tau) = 1 \\ \text{LLF}_i^w(\tau-1) \times \text{PRR}_i^w(\tau) & L_i(\tau) = 0 \end{cases} \tag{10.9}$$

Figure 10.3 illustrates the behavior of this metric compared to its base PRR for an example wireless link. If the receiver (gateway) successfully received the last transmitted packet by node s_i [$L_i(\tau) = 1$], the LLF value increases toward 1 with a rate determined by the history of this link (PRR) and the deviation of the previous LLF value $\text{LLF}_i^w(\tau-1)$ from its maximum value (i.e., 1). Notice that if the link has shown a good connection in the recent history of length w, LLF will converge to 1 very fast. Around $\tau = 30$ in Figure 10.3, the LLF value starts increasing later than PRR because of the poor quality record of this link. However, it reaches 1 faster than PRR, when it observes subsequent packet receptions,

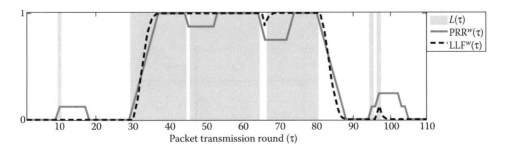

FIGURE 10.3
An illustrative example of the behavior of LLF estimation compared to the base PRR link quality estimation.

which is a sign of long-term improvement of the link. On the other hand, if the link has a poor PRR, the last successful reception is considered as a likely temporary connection and so LLF increases with a lower rate (the case around $\tau = 10$ in Figure 10.3). In the second case, in which the last packet is lost [$L_i(\tau) = 0$], the metric is decreased, again with a rate proportional to the history of the link (PRR). If PRR is high, the last link failure is supposed to be an intermittent disconnection and the metric decreases with a low rate (e.g., $\tau = 45$ in Figure 10.3). The main idea behind this is that if the link has shown a good record, incidental disconnections do not decrease the metric too much and vice versa. Based on this, the LLF estimator is expected to respond fast to real dynamics of the link quality and tolerate intermittent link states. However, experiments show that for medium-quality links, this estimator suffers from frequent fluctuations.

The smoothed link likelihood factor (SLLF) is proposed by Nabi et al. [7] to alleviate the stability problem of the LLF method. The LLF metric is very agile in detecting connection changes in WBANs, but it suffers from frequent fluctuations especially when the quality of a link is moderate [7]. SLLF uses an EWMA filter to smooth the LLF estimate, similar to the filter which is used to smooth the PRR in the WMEWMA estimator. This filter is able to improve the stability of LLF at the cost of low degradation of the speed of detections. Equation 10.10 shows the calculation for SLLF:

$$\mathrm{SLLF}_i^{w,\alpha}(\tau) = \alpha \times \mathrm{SLLF}_i^{w,\alpha}(\tau-1) + (1-\alpha) \times \mathrm{LLF}_i^w(\tau), \tag{10.10}$$

where $0 \leq \alpha < 1$ is the history control coefficient. Experiments by Nabi et al. [7] show that this technique trades off the high agility of LLF for acquiring more stable estimations. With $\alpha = 0$, this estimator turns to the base LLF estimation. A higher value of α, closer to 1, provides more stable estimation.

Experimental investigations were performed by Nabi et al. [7] to study and compare the performances of the aforementioned LQE techniques for connection detection in real-world WBANs in terms of detection agility, stability, and reliability. The experiments show that there is a trade-off between agility and stability of detection. The parameter settings of individual estimators also affect their performance properties. Figure 10.4 provides an overview of stability versus agility of different software-based estimators. The WMEWMA provides the most stable but the least agile estimator. The EWMA, on the other hand, provides the most agile but the least stable estimator. The experiments show that SLLF can make a good trade-off between stability and agility. Detailed analysis of the performance of the estimators as well as their effect on the application-level QoS in an adaptive WBAN is presented by Nabi et al. [7].

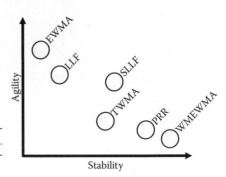

FIGURE 10.4
Qualitative visualization of the relative overall agility and stability of LQE methods for (dis)connection detection in WBANs. The position of each estimator may move depending on its specific configuration of parameters.

10.4 Mechanisms for Intra-WBAN Data Delivery

In this section, we focus on data delivery from all body sensor nodes within a WBAN to the gateway node on the body. In Chapter 9, two general network architectures (star and multihop) for intra-WBAN communication are discussed. In the WBANs with a star architecture, the data delivery to the gateway is simple and no further packet routing is necessary. On the other hand, multihop architectures are considered to improve the robustness and reliability of communications in a WBAN. In some applications, a multihop approach may even be the only design option for ensuring a continuous connectedness of all body nodes to the gateway. However, for multihop data forwarding, a routing algorithm is required to determine the data routing paths that body nodes should use to send their data packets. Because of the small scale of WBANs, the high dynamics, and the simplicity and low overhead requirements of WBANs, many routing mechanisms that are invented for wireless sensor and ad hoc networks are too heavy, suboptimal, and inefficient for WBANs. In the following, the approaches that have been considered specifically for data forwarding in WBANs are discussed. Figure 10.5 provides an overview of these mechanisms.

10.4.1 Location-Based Data Forwarding

A simple way to organize a multihop architecture for data routing in WBANs is to consider the position of the nodes on the body at design time, in order to fix the next hop neighbors for each node. A location-based forwarding mechanism is developed by Quwaider and Biswas [10]. Every node is programmed with an ordered list of all nodes in the WBAN, which is sorted based on the nodes' manually measured distance to the gateway. Assume DL = $[s_k | d(s_k) \leq d(s_{k+1}), k = 1, 2,..., N - 1]$ is the ordered list where $d(s_k)$ stands for the physical distance of node s_k to the gateway. When a node s_i has a data packet, it selects from its neighbors the closest node to the gateway and forwards its packet to that node. In other words, it selects the first node in the list DL that it is currently connected to. Therefore, a neighborhood discovery and LQE mechanism is also necessary, which is not discussed by Quwaider and Biswas [10]. The data packet may include received data samples from the other nodes to be forwarded or the node's own sensed data samples. This way, data samples are moved closer to the gateway hop by hop.

Considering the postural changes of the human body, the relative distance of the body nodes to the gateway substantially changes over time. Moreover, a major factor influencing

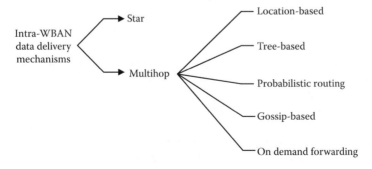

FIGURE 10.5
An overview of mechanisms for intra-WBAN data delivery.

the quality of wireless links in WBANs is the body shadowing effect, and the distance between nodes may have a less important influence on the link behavior. These issues suggest that a position-based mechanism may not be the best option for these networks.

10.4.2 Tree-Based Routing

Tree structures are commonly used for data routing in WSNs. A tree-based multihop protocol for WBANs has been developed by Latr et al. [11] in which the routing tree is set up autonomously to route data from the sensor nodes to the gateway. The protocol is cross layer (MAC and routing); it sets up the spanning tree and at the same time schedules time slots to reduce collisions. Communication between nodes is performed in cycles. Each cycle consists of some slots that are allocated to nodes by their parent node in the tree structure. A cycle starts with a control subcycle that is used for propagation of scheduling schemes from parents to their children. This starts by the root node (the gateway) down to the leaf nodes in the tree (supposed that a tree is already set up). After the control subcycle, data packet transmission is started by the leaf nodes up to the root by using the schedules established in the control phase of the cycle.

There is a contention slot in each cycle that can be used by new nodes to join the network. A new node listens to the packet exchanges in the network and selects a node as its possible parent. Then it sends a request to that node in the contention slot. This request includes the number of slots the node needs for transmitting its data, which depends on the data generation rate of the node. The parent that receives the request will include it in the control packets of the next cycle. Each node must transmit a packet in every cycle, either a data packet or a hello message if there is no data to be transmitted. In this way, parents and children can keep track of each other's existence.

A common problem of every tree structure in WSNs is that in the case of a node failure or node movement, the tree has to be reconstructed. Tree reconstruction may be an option only with low node mobility [11]. A posture change in a human body will likely change the body network topology. Moreover, frequent posture changes demand frequent tree reconstructions, which are energy and time costly. The complexity of the tree (re)construction algorithm is also a drawback of using this mechanism for computation- and energy-constrained nodes in WBANs.

10.4.3 Probabilistic Routing

To get rid of fixed or complex routing paths and to take into account the link quality, a probabilistic routing mechanism is proposed by Quwaider and Biswas [9]. The main idea is to use the forwarding nodes that have a better link to the gateway node. Each node uses the LLF metric (explained in Section 10.2) to continuously measure the quality of its link from other nodes in the WBAN (including the gateway) by exchanging periodic hello messages. All nodes broadcast their estimation of the quality of their link to the gateway. Using this information, body nodes try to estimate the overall quality (or cost) of their possible paths to the gateway. Each node then sends its data packets to a neighbor with the best quality link to the gateway. If a node itself does not have a good link to the gateway and cannot find any neighbor that has a better link to the gateway, it keeps buffering its data until it finds a good option for data packet transmission.

A drawback of this protocol is the reliance on symmetric links. Node s_i uses the reception rate of the hello messages transmitted periodically by another body node s_j to estimate the quality of its outgoing link toward node s_j. Assuming symmetric links for low-power wireless links is typically unrealistic. In particular, in many WBAN applications, the gateway node is supposed to be more powerful with a higher transmission power than the body nodes, which makes the link between the gateway and body nodes more likely to be asymmetric. Another drawback of this approach is the data-exchange overhead imposed for detecting the quality of links between all nodes. Because of frequent topology changes and high link quality variations, the estimated best forwarding paths may not remain valid for a long time and frequent adaptation is necessary. It also means that broadcasting the estimated link quality values should be done often enough to follow dynamics in the network. These frequent costly tasks need to be done by the energy-constrained body nodes.

10.4.4 Gossip-Based Data Forwarding

A multihop data dissemination scheme that relies on a gossiping strategy to realize a robust and simple data delivery for WBANs is proposed by Nabi et al. [12]. The key idea is to avoid making any assumption about the position of the nodes, establishing and maintaining complex routes, or imposing overhead for measuring the quality of wireless links. Every node in the WBAN participates in data forwarding without making specific routes toward the gateway. This mechanism assumes a low-duty-cycle TDMA-based MAC protocol (without any acknowledgment mechanism) underneath that allows the node to listen to all active slots in each TDMA frame.

Figure 10.6 illustrates the behaviors of this mechanism. A body node inserts the received data samples from the other nodes in the WBAN in a local data pool. The existing data samples in the pool, already forwarded data samples, and the node's own data samples are filtered out and do not enter the data pool. At each TDMA frame, it has a fixed time slot to use for transmission of a data packet including its own data samples as well as some data samples from its data pool. The number of data samples that a node can forward at each TDMA frame depends on the length of the data packets and the time slot duration that is assigned to the node. In Nabi et al.'s scheme [12], data samples are selected for transmission uniformly and randomly from the data pool. This way, all data samples are statistically forwarded equally often. A particular prioritization scheme can be applied for data sample selection from the pool in order to give more chances for the more important data to be forwarded; this is discussed in Section 10.4. Data samples are removed from the data pool if there is no room in the pool for arriving data samples. According to the

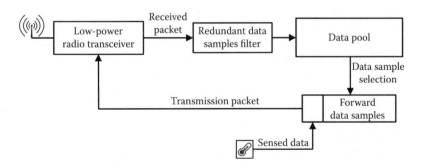

FIGURE 10.6
The block diagram of the gossip-based data forwarding mechanism used in [12] for multihop intra-WBAN data delivery.

application requirement, a particular strategy may be taken for selecting the data samples that are overwritten. A strategy may dedicate a fixed memory space to every node in the WBAN. Overwriting already forwarded data samples can also be an option.

The gossip-based mechanism is robust against frequent topology changes in WBANs because it does not rely on any specific routes. It does not need to adapt to the dynamics of the network. It also allows a simple implementation that is suitable for the body nodes. However, there are some concerns. In this mechanism, listening and data forwarding is done for all nodes, even for the nodes that themselves have a good connection directly to the gateway. This may lead to inefficient behavior of the mechanism due to unnecessary listening and data forwarding. It also unnecessarily increases the network traffic load, which may cause high latency of data delivery to the gateway.

10.4.5 On-Demand Data Forwarding

Relying on the location of nodes in a network with highly dynamic topology, establishing and maintaining tree-based routing structures, high packet-exchange overhead for LQE by individual nodes, and unnecessary listening and data forwarding are the main concerns for the previously described methods. Moreover, they do not exploit the features and the role of the more powerful node in the WBAN, the gateway.

The ODLF [2] mechanism creates a hybrid star/multihop network architecture for WBANs with a robust, power-efficient, and simple data delivery protocol. When all nodes have a good connection to the gateway, the body sensor nodes send their data directly to the gateway node. Thus, the network will have a star topology. If some nodes do not have a proper link to the gateway, the topology will automatically adapt the required level of multihop communication. However, the nodes do not maintain any routing path and do not exchange packets for LQE. Instead, the gateway plays a key role in managing the connection and data delivery of all nodes. The gateway node continuously estimates the quality of its entire incoming links. Using this estimation, the gateway detects the body nodes that do not have a sufficiently good connection. The gateway then distributes a short summary of the connection information in the form of a bit pattern to inform body nodes about the nodes that cannot properly reach the gateway. The body nodes receive this information and start listening to the nodes disconnected from the gateway. If they receive any packet from disconnected nodes, they use a store-and-forward scheme to propagate the received data samples, aiming to deliver them to the gateway. This results in on-demand multihop data dissemination, for which no specific routing structure is required to be established and maintained. This keeps the network protocol simple and robust.

Nodes in WBANs may experience two kinds of disconnection. The first kinds of outages are rather short term and caused by fading due to movement of the nodes or a temporary interference. This frequently happens in mobile postures such as walking. The transmission of a node may happen while the RSSI has its high peak and so it succeeds. On the other hand, the packet may be transmitted while the link is in its weak state and so it fails. Therefore, we may see some transient failure or success in packet transmission of a node. If a node has a good link to the gateway, this kind of short outages can be combated using retransmission of the packets by the source node itself. The second kinds of outages, which are especially the target of the ODLF mechanism, are longer-term outages. Shadowing caused by posture changes and also movements can bring a link to a situation in which the node cannot reach the gateway for a long period of time. Experiments by Nabi et al. [2] show that some nodes cannot send any packet to the gateway for the whole time during which the same type of posture was kept. This kind of outage is especially very problematic since there may be no data reception from a node for minutes or even longer. This, in turn, may lead to a serious failure of the application. As the core of the ODLF mechanism, the gateway uses the SLLF metric (originally LLF) to measure the quality of links based on which it reports the nodes that are disconnected from it. This is done by comparing the link quality with thresholds l_i given for individual sensor nodes. Based on the connection status of the node, the gateway determines whether node s_i needs data forwarding help.

Each node transmits its data packets hoping that the gateway receives the packets. On the other hand, nodes receive the requested set $\varphi^t = \{s_i \in S \mid s_i \text{ is disconnected}\}$ from the gateway, which represents the disconnected nodes from the gateway at time t. The gateway node usually uses a more powerful radio. Thus, we assume that the links from the gateway to the body nodes are of good quality. Intermittent failures of these links do not lead to a failure of the mechanism. Each node listens to the nodes in the latest received φ^t. Subsequently, if it receives any data packet from those nodes, it participates in forwarding data samples included in that data packet. Note that a node s_i tries to participate in data forwarding for the nodes in the requested set regardless of whether node s_i itself has a direct link to the gateway. This provides a multihop structure (beyond two hops) for data propagation within the WBAN. For example, suppose that node s_j cannot reach the gateway due to a large distance or low transmission range while node s_i is in its radio range. So if s_i listens to s_j, it receives and forwards the data of node s_j. If node s_i itself has a direct link to the gateway, then the packets of s_j reach the gateway in two hops (Figure 10.7a). Otherwise, the same procedure happens for all propagated information from node s_i by another node, say s_k (Figure 10.7b).

In ODLF, the amount of listening activities of the nodes depends on the number of nodes in the requested set. When all nodes have good connection to the gateway, the requested set is empty and nodes do not listen to any other nodes in the WBAN (except the gateway). If some nodes do not have good enough links to the gateway, other nodes listen to them to receive and forward their data. Therefore, the actual energy consumption of nodes in the WBAN that use ODLF depends on the quality of links in the network. More energy is consumed only in the case that it is necessary to ensure data delivery for the disconnected nodes.

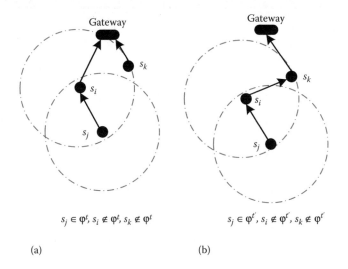

$$s_j \in \varphi^t, s_i \notin \varphi^t, s_k \notin \varphi^t \qquad s_j \in \varphi^t, s_i \notin \varphi^t, s_k \notin \varphi^t$$

(a)   (b)

FIGURE 10.7
Two examples of (a) two-hop and (b) three-hop data forwarding in the ODLF mechanism. (From M. Nabi et al., On-demand data forwarding for automatic adaptation of data propagation in WBANs. In Proc. IEEE Conference on Sensor, Mesh and Ad Hoc Communications and Networks [SECON], pp. 326–334. IEEE, 2012.)

In order to study the performance of ODLF in terms of energy consumption and data delivery, a real WBAN was tested [2], using 11 low-power wireless nodes deployed on a human body. A TDMA-based MAC mechanism was used underneath (see Section 9.4.3), which provides a collision-free communication for nodes. Figure 10.8 shows online energy consumption per TDMA frame averaged over all body nodes in two postures (sitting and standing). During each posture, a pure star architecture, gossip-based data forwarding (Section 10.4.4 [12]) and ODLF were separately used. The nodes used −18 dBm as their transmit power, which is the lowest transmit power level of the nodes. In using a star architecture, the energy consumption is a constant low value as all nodes send only their own data packets to the gateway. In the gossip-based mechanism, the energy consumption is again a constant value because nodes listen to all other nodes at each TDMA frame. Note that it is assumed that the transmission activities in the gossip-based mechanism are fixed and constant. In the ODLF mechanism, the energy consumption varies according to the connection of nodes to the gateway, and on average it is between the energy consumption with the star and gossip-based mechanisms. In the sitting posture, the ODLF energy consumption is slightly higher than in the standing posture, which is an indication of worse link conditions while sitting when compared to standing.

The ODLF mechanism consumes more energy to ensure data delivery if there is a route from the source node to the gateway but no direct connection. Figure 10.9 presents the achieved DDRs of three nodes as examples to show the impact of using ODLF compared to using a star architecture. Node s_1 is a node that always has a good connection to the gateway. Thus, it can deliver its sensed data directly to the gateway and the ODLF mechanism has no impact on its DDR. The story is quite different for s_2 and s_3, because of their physical positions on the body with respect to the gateway (arm and back, respectively). Relying on a direct link to the gateways will result in a very poor DDR for these two nodes. The

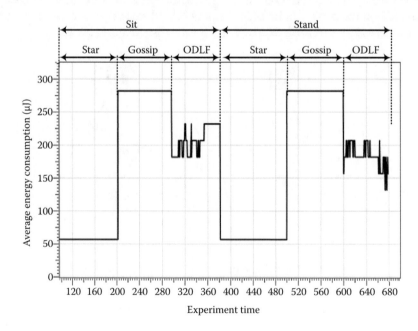

FIGURE 10.8
Online energy consumption per TDMA frame, averaged over 10 body nodes (the gateway is excluded) in a real WBAN deployment in two different postures, using different data forwarding schemes.

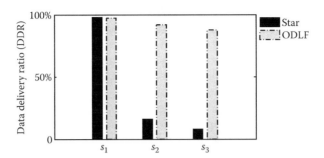

FIGURE 10.9
Achieved data delivery ratios of three nodes in a WBAN with 11 wireless nodes using a star topology and using the ODLF mechanism for data delivery. The gateway is placed on the belt (right side of the stomach). Nodes s_1, s_2, and s_3 are placed on the right thigh, left upper arm, and back, respectively.

ODLF mechanism provides a data forwarding support for these two nodes and provides substantially better DDR values.

10.5 Prioritized WBAN Data Delivery

Health-monitoring applications show a considerable heterogeneity in terms of data delivery requirements and sensing specifications. Body sensor nodes sense and transmit biological signals of patients. The wireless nodes in the surrounding ambient network (the level 2 network in Figure 10.1) may generate data and contribute in a multihop delivery of both WBAN and ambient data to the local central station. Data samples from body sensor nodes are often of higher importance than ambient information. This means applications may have tighter data delivery latency and DDR constraints for WBAN data delivery than for the ambient information. For example, losing important ECG data or detected falls or very late delivery of such information may lead to severe danger for the person being monitored. Also, there may be substantial differences in importance and data delivery requirements between different biosensors within a WBAN. Movement of the person wearing the WBAN causes the distance of the WBAN to the central station to vary over time. This changes the performance of WBAN data delivery over time. The data delivery requirements may even change according to the health status of the monitored person (e.g., normal versus emergency situations). The considerable temporal and spatial diversity in data importance and delivery requirements in health applications requires data prioritization. Without data prioritization, unnecessary service may be provided for data of lower importance while limiting the nodes with more important demands.

Data prioritization may be applied in different layers of the networking protocol stack. The MAC layer may contribute in prioritization by giving a better chance to the nodes with important data to win the channel access competitions. For instance, the IEEE 802.15.6 standard [13] defines seven user priorities (UPs). The MAC layer of this standard presents several ways of giving higher priority to data packets with higher priority levels. Collision-free medium access is provided for body nodes with data packets of the highest priority (UP_7). For other priority levels, shorter back-off periods are considered for transmission of data packets with higher priorities to increase their chance of successful transmission.

Although this standard considers intra-WBAN communication only, the same approach can be exploited for the MAC layer of the nodes in the ambient network. More details about the IEEE 802.15.6 MAC can be found in Chapter 9.

In multihop data delivery, the network layer may also effectively participate in prioritization. Kim et al. [14] propose the priority-based hybrid routing (PHR) mechanism, in which data samples are classified by the source node into primary or secondary priority classes. Different data dissemination schemes are used for dissemination of data samples in different priority classes. A multipath diffusion-based mechanism is used for forwarding the packets in the primary priority class to provide a more reliable and faster data delivery. For data samples in the secondary class, a single-path routing mechanism based on the ad hoc on-demand distance vector (AODV) [15] approach is exploited. Compared to a multipath routing, a single-path approach is more prone to data loss.

Another approach for data prioritization in multihop routing is to prioritize the data sample selection by the relay nodes in the routing path while forwarding the received data packets. The priority-based dynamic adaptive routing (PDAR) protocol [16] categorizes data packets into two classes of vital and common packets. Accordingly, every node on the routing path maintains two separate data queues, each dedicated to a certain class of packets. The packets in the higher-priority queue (vital packets) are always sent before packets in the lower-priority queue (common packets). PDAR uses a first-in-first-out (FIFO) scheme to forward data samples in each priority class. In the case that there are multiple data samples in the vital class, for instance, they are forwarded using the order of their arrival.

The prioritization mechanisms mentioned so far have a specific criterion for assigning the priorities to data samples and a specific means for providing proper services according to the priorities. They all use a fixed priority assignment by the source node. Considering the temporal and spatial heterogeneity in the networks including WBANs, Nabi et al. [17] proposes a mechanism to dynamically assign priorities to data samples waiting to be forwarded at any relaying node. The mechanism aims to provide differentiated data delivery services for data samples according to their requirements. It also considers scenarios in which the data delivery requirements change over time. To provide such flexibility, instead of attaching priority values to the data packets, relative data delivery requirements (such as deadlines) are attached to the individual data samples by the source nodes. Then the priorities are calculated at each relay node on the routing path, taking the attached data delivery requirements and the history of the data sample into account. In this way, a source node can change the requirements for its data samples at any time. In addition, as the history of the data sample (for instance, the time it spent on the path) is taken into account, dynamic priority calculation provides appropriate services for nodes farther away from the central station. This is specifically interesting for WBANs for which the hop distance to the central station varies over time due to the WBAN mobility.

10.6 Conclusions

Health-monitoring applications expect certain performance of WBAN data delivery. Depending on the nature of the biological signals being sensed and the application scenario, several data scenarios can be distinguished in WBAN applications as, namely, status information, events, and continuous waveform signals. Consequently, various data delivery performance metrics can be considered for which the application may have particular constraints. For instance, the DDR metric fits the data scenario of continuous waveform signals, latency is important for events, and the age of data at destination is the best metric for status information.

Network adaptation is a key requirement in achieving reliable and efficient data delivery in WBANs. LQE is a prerequisite for the adaptive WBAN mechanisms; it is used to trigger adaptations in response to topology and link quality changes. An LQE technique should be agile enough to follow the changes in the quality of wireless links and, at the same time, it should not be very sensitive to incidental link failures or connections. The method needs to be stable enough to avoid frequent fluctuations in order to keep the overhead of the adaptation mechanism low and to make it robust. The agility and stability of the estimators are in conflict and making a good trade-off between these metrics is a key characteristic of an appropriate LQE method for WBANs. As an example, SLLF is an LQE method that is specifically designed for WBANs; it can make a good trade-off between agility and stability of the link quality estimation. However, any decision about the best estimator and its configuration for a WBAN application depends on the type of adaptation mechanism and the application constraints.

A star topology is the simplest way for data delivery from the body nodes to the gateway node. However, in some applications, it is not sufficiently reliable, and some form of multihop intra-WBAN data delivery is required. Frequent changes in the WBAN topology (due to posture changes and mobility) suggest that a static or location-based routing mechanism is not a proper option. Exploiting data forwarding only when direct links to the gateway cannot satisfy the required data delivery performance is considered an efficient and reliable approach.

Considering the high temporal and spatial heterogeneity of WBAN applications, prioritized data dissemination is necessary to properly distribute the limited network resources such as bandwidth, energy, and storage. Prioritization may be used for intra-WBAN data delivery as well as data dissemination in the surrounding ambient network toward a central station. There are various approaches to assigning priorities to data samples (for instance, static or dynamic), levels of priorities, and the way priorities influence protocol behavior. In any case, the prioritization mechanism needs to be designed taking into account the WBAN characteristics such as high dynamism, mobility, and heterogeneity. Moreover, it may need to support multiscenario applications in which the source nodes may change their data delivery requirements over time. Because of these needs, a dynamic priority assignment mechanism taking individual delivery requirements of data samples into account is a proper approach for WBAN data delivery.

References

1. D. Smith, L. Hanlen, and D. Miniutti. Transmit power control for wireless body area networks using novel channel prediction. In Proc. IEEE Wireless Communications and Networking Conference (WCNC), pp. 684–688. IEEE, 2012.
2. M. Nabi, M. Geilen, and T. Basten. On-demand data forwarding for automatic adaptation of data propagation in WBANs. In Proc. IEEE Conference on Sensor, Mesh and Ad Hoc Communications and Networks (SECON), pp. 326–334. IEEE, 2012.

3. K. S. Prabh and J.-H. Hauer. Opportunistic packet scheduling in body area networks. In Proc. 8th European Conference on Wireless Sensor Networks (EWSN), pp. 114–129. Springer-Verlag, 2011.

4. R. Fonseca, O. Gnawali, K. Jamieson, and P. Levis. Four bit wireless link estimation. In Proc. 6th Workshop on Hot Topics in Networks (HotNets-VI). ACM, 2007.

5. NORDIC Semiconductor nRF24L01 data sheet, 2007.

6. N. Baccour, A. Koubaa, L. Mottola, M. A. Zuniga, H. Youssef, C. A. Boano, and M. Alves. Radio link quality estimation in wireless sensor networks: A survey. *ACM Transactions on Sensor Networks (TOSN)*, 8(4):1–33, 2012.

7. M. Nabi, M. Geilen, and T. Basten. An empirical study of link quality estimation techniques for disconnection detection in WBANs. In Proc. ACM Conference on Modeling, Analysis and Simulation of Wireless and Mobile Systems (MSWiM), pp. 219–228. ACM, 2013.

8. A. Woo and D. Culler. Evaluation of efficient link reliability estimators for low-power wireless networks. Technical Report UCB/CSD-03-1270, EECS Department, University of California, Berkeley, 2003.

9. M. Quwaider and S. Biswas. Probabilistic routing in on-body sensor networks with postural disconnections. In Proc. 7th ACM International Symposium on Mobility Management and Wireless Access (MobiWAC), pp. 149–158. ACM, 2009.

10. M. Quwaider and S. Biswas. On-body packet routing algorithms for body sensor networks. In Proc. International Conference on Networks and Communications (NetCoM), pp. 171–177. IEEE, 2009.

11. B. Latr, B. Braem, C. Blondia, I. Moerman, E. Reusens, W. Joseph, and P. Demeester. A low-delay protocol for multihop wireless body area networks. In Proc. 4th International Conference on Mobile and Ubiquitous Systems (MobiQuitous), pp. 1–8. IEEE, 2007.

12. M. Nabi, T. Basten, M. Geilen, M. Blagojevic, and T. Hendriks. A robust protocol stack for multi-hop wireless body area networks with transmit power adaptation. In Proc. 5th International Conference on Body Area Networks (BodyNets). ICST, 2010.

13. IEEE standard for local and metropolitan area networks—Part 15.6: Wireless body area networks (WBANs). In IEEE Std 802.15.6-2012, pp. 1–271, 2012.

14. S. Kim, S. Lee, H.-J. Ju, D. Ko, and S. An. Priority-based hybrid routing in wireless sensor networks. In Proc. IEEE Wireless Communications and Networking Conference (WCNC), pp. 1–6. IEEE, 2010.

15. C. E. Perkins and E. M. Royer. Ad hoc on-demand distance vector routing. In Proc. 2nd IEEE Workshop on Mobile Computing Systems and Applications (WMCSA), pp. 90–100. IEEE, 1999.

16. J. Chen, M. Zhou, D. Li, and T. Sun. A priority based dynamic adaptive routing protocol for wireless sensor networks. In Proc. International Conference on Intelligent Networks and Intelligent Systems (ICINIS), pp. 160–164. IEEE, 2008.

17. M. Nabi, M. Blagojevic, M. C. W. Geilen, and T. Basten. Dynamic data prioritization for quality-of-service differentiation in heterogeneous wireless sensor networks. In Proc. 8th IEEE Communications Society Conference on Sensor, Mesh and Ad Hoc Communications and Networks (SECON), pp. 296–304. IEEE, 2011.

11

Use of Small-Cell Technologies for Telemedicine

Edward Mutafungwa and Jyri Hämäläinen

CONTENTS

11.1 Introduction

11.1.1 Background and Motivation

The World Health Organization (WHO) has recently noted that lifestyle-related chronic noncommunicable diseases, such as cardiovascular diseases, diabetes, chronic obstructive pulmonary disease, and cancers, are the leading cause of death globally [1]. For instance, in 2008, chronic diseases accounted for almost two-thirds of the 57 million deaths globally, with the fraction of chronic disease–related deaths projected to increase significantly in the coming years. Furthermore, findings from a separate study commissioned by the World Economic Forum (WEF) projected that the burden of chronic noncommunicable diseases up to year 2030 will generate cumulative losses equivalent to 75% of global gross domestic product for the year 2010 [2].

Findings from the aforementioned and similar studies underline the need for interventions, particularly those centered on prevention of chronic noncommunicable diseases through lifestyle changes. In the WHO study, it was noted that most of the chronic diseases are preventable through reduction of the four primary behavioral risk factors, namely, tobacco smoking, physical inactivity, excessive alcohol consumption, and unhealthy diet [1]. There is strong evidence that adoption in these four key health behaviors may add up to 14 years to one's life [3]. Policy makers worldwide are implementing comprehensive approaches to tackling chronic diseases through use of reliable national statistics for effective policy making; support of public-awareness campaigns on prevention of chronic diseases; and reductions of inequalities in health. While such preventive care interventions are typically implemented at a societal, organizational, or community level, ultimately the success of chronic disease prevention programs depends on behavioral changes and lifestyle choices made by individuals continuously on a daily basis.

This calls for more integrated person-centered health and wellness solutions that enable implementation of new care models that narrow the gap between hospital-based and home-based care. The underlying driver for this care modality is the fact that demand for healthcare services is fast exceeding the capacity of traditional hospital-based care due to a rapidly aging demographic and prevalence of chronic diseases [4]. Therefore, there is a critical need for innovative ways for cost-effective delivery of healthcare services while simultaneously exceeding (if not meeting) the populations' expectations on the quality of care [5]. In addition to cost savings there is now a general acknowledgment that long-term out-of-hospital care provides a humane and comfortable setting while also minimizing some of the common challenges encountered in a hospital setting (e.g., hospital-acquired infections [6] and prolonged waiting times [7]).

11.1.2 Toward Connected Personal Health Systems

The vision of ubiquitous and personalized delivery of healthcare services is now rapidly becoming a reality through advances in telemedicine that challenge the legacy care models that rely on direct physical presence and contact between care providers and patients for healthcare delivery [4–5,8–10]. To that end, personalized telemedicine or telehealth system implementations (generically referred to as personal health systems [4]) offer an attractive solution as they enable proactive person-centered care at fraction of the cost of hospital-based care, resulting in significant economic benefits at the national scale [8–10]. Specifically, personal health systems improve the quality of care by enabling

- Continuity of care beyond the traditional hospital domain
- Improved personal safety by early identification of medical risks and rapid initiation of intervention measures
- Individual empowerment through increased involvement and accountability for managing health risks or chronic diseases at a personal level, and
- Improved capabilities in engaging and receiving support from professional care providers, family members, friends, online communities, and other relevant social groupings.

The integrated personal health system implementations typically constitute wirelessly connected personal health system devices, such as glucose meters, activity monitors, and heart-rate monitors, that are implanted within or worn by an end user (patient or monitored subject), or deployed around their home environment [10–13]. Furthermore, personal health system hubs collect data from the heterogeneous personal health devices and may provide an interface for presenting the generated contextual health information to the end user. Moreover, the hub converts the collected health data for onward transfer so as to enable personal health system information sharing and expert analysis among relevant actors (physicians, wellness mentors, etc.) and remote archiving in health data repositories (e.g., electronic patient records).

11.1.3 Mobile Technologies for Personal Health Systems

The wide-scale adoption of personal health systems by end users and care providers requires a standards-based approach so as to guarantee successful development of interoperable systems and devices [14]. Another key success factor for significant adoption of personal health system solutions is the increased integration of mobile network operators (and communications service providers) in the personal health system value chain, thus enabling the deployment of relatively robust, flexible, and manageable end-to-end personal health system solutions. This stems from the fact that mobile operators continuously enhance and optimize their network assets and processes (billing, subscription management, security management, etc.), to enable themselves to provide services reliably for millions of subscribers and devices in their networks [13]. Furthermore, the mobile sector, which has benefited significantly from adoption of an open standards-based approach, is now in intensive standardization activities to specify network enhancements for the ongoing paradigm shift from human-to-human communications to the traffic being dominated by the Internet of Things (IoT) and machine-to-machine (M2M) communications [15–17]. The commonly quoted projection is of over 50 billion connected devices by 2020 through embedding of mobile wireless modules in things or objects ("machines" in M2M systems), such as cars, domestic white goods, buildings, factory equipment, and personal health devices [16].

In the case of personal health systems, the privacy and convenience of the home environment represents the most common setting for connected personal health system use cases [14]. This is because the home environment provides facilities for placement and usage of health devices, other than those that are either implanted in or worn by the end user.

The fact that humans spend on the average 58%–70% of their daily life in their indoor residential environment further supports this observation [18]. The fraction of time spent in one's home is even higher (75%–95%) for less mobile groups, such as the elderly (>64 years), home care patients, and infants (<1 year) [18]. These statistics emphasize the need for intelligent solutions that enable a smart and efficient home environment not just for personal health services but also for home automation, security, energy efficiency, entertainment, and so on [16]. To that end, there is crucial need for continued development of mobile technologies to support M2M systems operating from within and beyond the home environment [17].

Notably among the mobile technology developments are miniature base stations, commonly referred to as femtocells or residential small cells, intended for deployment in indoor environments, such as homes, enterprise spaces, shopping malls, and hospitals [19–20]. Development of small cells was initially driven by the fact that over 80% of mobile data traffic has been originating from those indoor environments [20]. Residential small cells not only improve mobile network coverage and capacity in the home environment, but also provide a platform for deployment of home-based mobile services and interworking with non-mobile-based home networks [21]. In this chapter we present the small-cell concept and outline the benefits of leveraging small cells within connected personal health implementations, with a particular focus on personal health systems in the home environment.

11.1.4 Scope and Organization of the Chapter

The remainder of the chapter is organized as follows. Section 11.2 provides a more detailed background on personal health system solutions and their current developments in various standardization bodies. Section 11.3 provides an overview of small cells and their architectural features while Section 11.4 describes how small cells could be used as a gateway for personal health systems and the benefits that could be attained from that configuration. These benefits are explored further in a simulation case study in Section 11.5 and the general conclusions are presented in Section 11.6.

11.2 Background on Personal Health Systems

11.2.1 Revisiting Connectivity Needs for Personal Health Systems

Personal health systems are implemented end to end so as to interlink the end-user domain (patient or monitored individual) and the care-provider domain (e.g., hospital and fitness center). The end-user domain essentially covers the end user's personal space and their immediate surroundings, which could usually be their place of abode (home environment) or physical work space (e.g., office). In telemedicine the end-user personal space is commonly classified according to the spatial connectivity range of the personal health devices and other relevant sensing entities in their environment (see Figure 11.1). It is notable that some parallels can be drawn between this concept of connected personal space and areas of proxemics theory [22] and neuropsychology [23]. Proxemics theory was pioneered by Edward Hall, who spatially classified the interpersonal space around an individual in the following zones: intimate (15 to 46 cm), personal (46 cm to 1.2 m), social (1.2 to 3.6 m), and public (over 3.6 m) [22]. On the other hand, spatial representation in neuropsychology

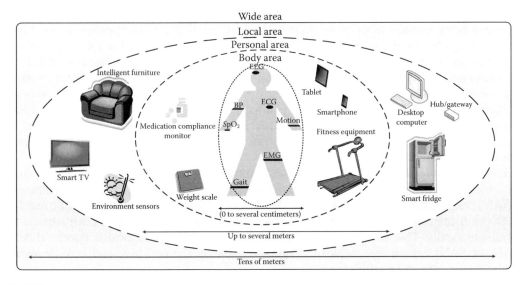

FIGURE 11.1
Typical classification of local connectivity regions for personal health systems. BP, blood pressure; ECG, electrocardiography; EEG, electroencephalography; EMG, electromyography; SpO_2, peripheral capillary oxygen saturation.

draws a distinction between peripersonal space (space within arm's reach) and extrapersonal space (space beyond arms reach) [23].

Common spatial classification for connected personal health begins from the body area that includes devices attached to, worn by, or embedded within the end user (for monitoring physiological signals, posture, gestures, emotions, etc.) and the interconnection of these devices forms a BAN (see Johny and Anpalagan's paper [24] and references quoted therein). This area could be considered to be analogous to the intimate zone in proxemics theory. The BAN usage in connected personal health scenarios allows for continuous monitoring of the end user without significant or noticeable distraction to their daily routines. The BAN devices are usually interconnected for communications in, on, or around the end user's body using short-range wireless technologies, such as Bluetooth low energy, IEEE 802.15.6 medical body area network (MBAN), near-field communication (NFC) (ISO 18092), and Zarlink ZL70101 [24].

The personal area then covers the space that could be considered to be the personal and social zones (proxemics theory), or the peripersonal and initial parts of the extrapersonal regions (neuropsychology). The PAN provides connectivity for PAN devices deployed in this space that goes beyond the BAN region (see Figure 11.1). These PAN devices are usually portable and personal (smartphone, tablets, etc.) or are exclusively dedicated to the end user during their brief period of use (e.g., weight scale and fitness equipment). Short-range wireless technologies, such as Bluetooth classic, Bluetooth low energy, ANT+, and ZigBee, enable connectivity between PAN devices or between BAN and PAN devices. The PAN device connectivity to remote entities in the care-provider domain is then provided by pairing with end-user communication devices or PAN devices with embedded cellular modems. The utilization of PAN devices as a gateway device is also useful for BAN devices equipped with only short-range radios via the BAN-to-PAN link.

The local area then covers the region considered the public zone (proxemics theory), or outer extrapersonal space (neuropsychology). Again using contemporary networking terminology, the local area network (LAN) covers the areas beyond the PAN (e.g., entire home or office environment), typically in a range of few tens of meters, and uses short- to medium-range wireless technologies, such as Wi-Fi and ZigBee. Usually this local area is a wider shared space with other occupants (e.g., household members), and LAN devices (smart television [TV], furniture, home gateway, etc.) are devices deployed for simultaneous shared use with co-occupants. The LAN device connectivity to remote entities is similar to that of PAN devices and in some cases may even utilize local broadband wired connections depending on device type.

Finally, the wide area is the broad region beyond the local area where the remote care–provider entities would normally be located (physically or in the cloud). Wide-area-network (WAN) connectivity for LAN, PAN, or BAN devices could be wireless using public mobile network infrastructure or through other fixed broadband Internet access infrastructure available locally. This end-to-end connectivity via the WAN and local networks (LAN, PAN, or BAN) essentially enables the linkage between the end-user and care-provider domains for personal health systems.

11.2.2 General System Requirements for Connectivity Providers

The connected personal health services and their usage scenarios impose a number of requirements on the underlying network providing connectivity for the personal health system. The notable high-level system requirements for the network(s) include the following:

- **Accessibility:** The authorized personal health service connection requests shall have uninhibited access to the network services, preferably with high access success rates and low access delays.

- **Availability:** The personal health service connectivity shall be maintained regardless of user location within the intended area of coverage (e.g., home environment).

- **QoS guarantees and prioritization:** The network shall maintain the necessary performance levels (achievable throughput, delay, jitter, loss rates, etc.) in order to guarantee the QoS level requested by a personal health service. Moreover, the most preferential QoS class shall be assigned to personal health service connections and traffic flows, particularly during critical episodic or medically critical instances.

- **Service retainability:** The network shall ensure the maximum possible service retainability for personal health service connections by minimizing call-drop rates, dropped data connections, connection timeouts, and so on.

- **Seamless mobility and roaming:** The network(s) shall provide service continuity for personal health service connections, so as to ensure that the portable personal health services are uninterrupted when the end user traverses across different networks or locations (e.g., from home to bus stop and onward).

- **Service usability:** The required intervention by the personal health service end user in obtaining the network connectivity service should be minimized. This could be facilitated by the network providing network-initiated service announcements, automated network service discovery, network device discovery, and automated configuration of best available service for personal health services.

- **Security:** Connected personal health services in most cases involve the transfer of confidential end users' personal, health, activity, and medical data. In practice, the transfer of these data has to comply with stringent guidelines applicable within a particular region, such as the European General Data Protection Regulation (successor of the Data Protection Directive) [25] and the United States Health Insurance Portability and Accountability Act (HIPAA) Privacy Rule [26]. Therefore, personal health service signaling and data flows shall be protected within the network from all potential security breaches, such as data corruption or exposure, denial-of-service attacks, unauthorized access, eavesdropping, man-in-the-middle attacks, and data exposure.

- **Interoperability:** The network should provide standard interfaces that allow for seamless interconnectivity of personal health devices and subsystems. The issue of interoperability has been one of the most critical factors in the implementation of personal health systems [14] and is addressed further in the in the next subsection.

11.2.3 Achieving Interoperability for Personal Health Systems

Personal health system integration has been greatly hampered by closed incompatible proprietary solutions by competing vendors, each with their own data formats, data-exchange protocols, and so on [14]. In a bid to realize economies of scale and drive up adoption of personal health system solutions, there have been efforts to address the aforementioned fragmentation by pushing for personal health system designs based on commonly agreed standards. This subsection provides an overview of the most widely accepted standards for implementing interoperable personal health system within the mobile framework.

11.2.3.1 Continua Reference Architecture

The Continua Health Alliance is arguably the leading open-industry group providing interoperability guidelines, testing, and certification programs for the personal health system device implementers and system integrators [27]. The current Continua guidelines specify an end-to-end harmonized Continua reference architecture (see Figure 11.2) based on a suite of standards that enable implementation of interoperable personal health system devices and system interfaces [27]. The Continua design guidelines have recently been adopted by the International Telecommunication Union (ITU) as the first global standard for personal connected health devices and systems [28]. The end user–owned devices in the defined Continua reference architecture are the touch area network (TAN), PAN, and LAN health-monitoring devices and the application-hosting devices (AHDs), which in most cases are physically located in the end user's domain (see Figure 11.2).

The TAN, PAN, and LAN monitoring devices are essentially health sensors or actuators, which may be worn, implanted, or carried by the monitored individual or alternatively deployed within their home environment. The Continua PAN device characteristics are essentially defined to be similar to the general PAN and some BAN devices introduced in the classification in Subsection 11.2.1. The same is the case for the similarities between Continua LAN devices and LAN devices in Section 11.1. The TAN devices were introduced in later revisions of Continua guidelines to cover NFC-based health devices.

The AHDs are devices such as smartphones and dedicated personal health system hubs that aggregate data from monitoring devices and provide a connectivity gateway toward

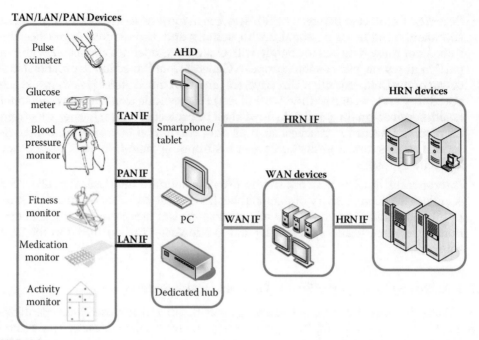

FIGURE 11.2

Continua reference architecture. AHD, application hosting devices; HRN, health reporting network; IF, interface; LAN, local area network; PAN, personal area network; TAN, touch area network; WAN, wide area network.

remote WAN or health-reporting network (HRN) devices. Moreover, the Continua reference architecture also allows for the Continua AHD to be integrated in the same unit with some health sensing or actuation functionality. For instance, smartphones may have integrated sensors and health-monitoring apps while also carrying out the function of interfacing to PAN devices as well as sending health data to WAN devices via the mobile network.

The WAN and HRN devices are deployed and managed in the care-provider domain. For instance, a WAN device could be a web server platform that receives aggregated health data from an end user's smartphone (AHD device) and uses it to provision a personalized health service (e.g., medicine reminder service). The HRN devices sit on the edge of the personal health system and provide long-term information storage (e.g., electronic medical records in hospital data servers).

11.2.3.2 IEEE 11073 Device Specializations

The Continua guidelines also specify network interfaces between two or more Continua-certified devices (see Figure 11.2). The TAN, PAN, and LAN interfaces use the baseline IEEE 11073-20601 for higher-layer functions, such as connection management, device configuration information exchange, and abstract-to-transmission data format conversion. Each TAN, PAN, and LAN monitoring device type has a specialized IEEE 11073-104xx standard (e.g., IEEE 11073-10417 for glucose meters) that specifies how it utilizes the IEEE 11073-20601 standard to fulfill its function (see Figure 11.3). The IEEE 11073–compliant data transport over the TAN, PAN, and LAN interfaces is currently specified for Bluetooth, USB, ZigBee (IEEE 802.15.4), and NFC [28]. Additionally, Wi-Fi (IEEE 802.11x) and mainstream mobile

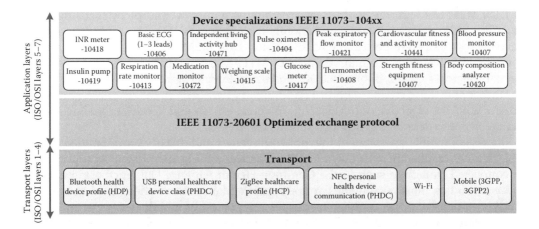

FIGURE 11.3
IEEE 11073 personal health device conceptual framework. 3GPP, third generation partnership project; 3GPP2, third generation partnership project 2; ECG, electrocardiography; INR, international normalized ration.

technologies standardized by the Third Generation Partnership Project (3GPP) or Third Generation Partnership Project 2 (3GPP2) may be used for health data transport in the case that AHDs are physically deployed at long distances away from the monitoring devices.

The WAN and HRN interfaces use modifications or transformations of the IEEE 11073 semantics into the health level 7 (HL7) formats widely used in the care-provider domain [29]. The data-exchange protocols for WAN and HRN interfaces are transport independent and may utilize Wi-Fi, mobile, or fixed technology standards for connectivity between devices in the end-user domain (monitoring devices and AHDs) and WAN or HRN devices [28].

11.2.4 mHealth Developments within Mobile Technology Standardization

Mobile networks provide the kind of ubiquitous coverage, service manageability, and security features that add significant value in the implementation of personal health systems [11]. This has seen the increased commercialization and rapid adoption of so-called mobile health or mHealth solutions [12]. The potential of this synergy is recognized by standardization bodies in both the mobile and health domains. To that end, the Continua Health Alliance has produced a supplementary set of implementation guidelines for Continua-certified personal health device vendors with embedded 3GPP or 3GPP2 communication modules for operation in the mobile environment [30]. These guidelines include essential design and operating considerations for aspects such as remote device management, device efficiency in the consumption of power and network resources, network performance levels (throughput, latency, etc.), and device certification. These Continua Health Alliance efforts are reciprocated by the CDMA (code-division multiple access) Certification Forum (CCF) and Global Certification Forum (GCF) [31], the bodies which certify 3GPP and 3GPP2 mobile devices, respectively.

The considerations for mobile health use cases in mobile technology standardization bodies are being incorporated within the general frameworks of M2M and machine-type communication (MTC) [17]. The MTC/M2M communication standards are currently being developed by, among others, the 3GPP and the European Telecommunications Standards Institute (ETSI). The MTC standardization activities in 3GPP are closely aligned to those of ETSI M2M but with the 3GPP standards having a narrower scope by focusing mostly on 3GPP radio access networks and mobile core networks [17]. The mHealth use cases have been described in both 3GPP MTC [32] and ETSI M2M working groups [33], and are considered to be one of the main drivers for MTC/M2M in addition to smart grids, intelligent transport systems, public security, and so on. Furthermore, ETSI has teamed up with a number of leading standard bodies and industry consortia in the oneM2M partnership project with the objective of developing universal end-to-end specifications for an M2M management system for applications such as health [34].

The role of mobile operators in the mHealth ecosystem is clearly central and significant [13]. This has resulted in most major mobile operators joining and playing a prominent role in connected initiatives such as the Continua Health Alliance [30]. Furthermore, mobile industry bodies, most notably the Groupe Speciale Mobile Association (GSMA), have been studying potential opportunities for deeper integration of mobile networks and health-care systems. To that end, the GSMA has proposed a high-level reference architecture for mobile health, which defines in detail how existing capabilities in mobile networks (connectivity, security management, identity management, billing platforms, etc.) can be effectively leveraged to add value to mobile health systems [35]. The GSMA reference architecture identifies both defined and missing logical architectural components from existing open standards. The main aim includes the use of the GSMA reference architecture to develop mobile health use cases and system requirements for detailed consideration in future 3GPP standardization.

11.3 Overview of Small-Cell Technologies

11.3.1 Drivers for Small-Cell Deployment

Ever since mobile networks first supported data services, the upward trend in growth of mobile data traffic has shown no sign of abating [36–38]. On the contrary, the future rates of growth in mobile data traffic are expected to be even more rapid due to the confluence of growth factors, namely, increase in number of users; increase in number of mobile devices per user, increase in average mobile data connection speeds, and increase in volume of traffic generated per user. The evidence of recent growth trends underlines this fact. For instance, the Amdocs late annual study "The State of the Radio Access Network (RAN)," which is on eight mature mobile markets in Europe and North America, noted that mobile data traffic grew on average by 111% in the year 2013 (compared to 16% growth in voice traffic) [36]. In the same line of study, Cisco observed an 81% global mobile data traffic growth for the year 2013 with projections for 11-fold traffic growth between the years 2013 and 2018 [37]. The commonly quoted projection in the mobile industry has been 1000-fold increase in traffic from the years 2010 to 2020 [38].

While there may be continued debate on the accuracy of some of the traffic growth projections [38], the general consensus is that mobile operators face a formidable challenge in evolving their networks to sustain or exceed end-user quality of experience expectations. This objective is particularly challenging when the cost and energy consumption of the evolved networks has to be reduced or maintained at current levels, when considered against the backdrop of flat or declining revenues [38]. To that end, the mobile industry and research community are actively investigating a number of solutions for cost- and energy-efficient scaling of network capacity (in bits per second per square kilometer) to accommodate the increased data traffic [38–41].

The availability of radio spectrum sets an upper limit on the achievable capacity gains, but spectrum is a both very costly and limited resource [42]. Therefore, network capacity scaling solutions should target not only the increase in available spectrum (hertz) but also improvements in the spectral efficiency (bits per second per hertz per cell) and spatial reuse of spectrum (cells per kilometer) [39]. The latter approach of reusing spectrum through dense spatial deployment of smaller cells is the most promising, with an estimate of up to 1600-fold theoretical capacity gain network [19,38–39]. These potential gains have enforced a paradigm shift from traditional homogeneous macrocell networks to a heterogeneous deployment of large macrocells and small cells [19–20,38–40].

Small cells is an umbrella term for reduced form-factor, low-powered short-range base stations with a range of a few tens to several hundred meters, deployed to provide improved network coverage and capacity in homes, enterprise environments, underserved areas, and public spaces [19]. Operator-deployed small cells include extensions to macrocells like remote radio heads, metrocells, microcells, picocells, and open-access femtocells. On the other hand, indoor closed-access femtocells are autonomously deployed and operated by end users in their place of residence, analogous to traditional Wi-Fi access point (AP) or digital subscriber line (DSL) modem. These femtocells are also referred to as residential small cells (note that in this chapter the terms *femtocells* and *residential small cells* may be used interchangeably) [19,40]. The restricted access to residential small cells allows for improvement throughputs for mobile users who would otherwise be connected to congested macrocells. This is significant as over 50% of mobile data traffic is generated from indoor residential environments [40]. Furthermore, residential small cells eliminate coverage holes (in the attic, basement, etc.) and provide improved indoor coverage compared to the coverage from the operators' outdoor cells. In addition to aforementioned quality of service improvements, small cells can provide an intelligent gateway and management interface for multiple M2M services, such as connected health, that are active in residential environments [21,43].

11.3.2 Small-Cell Architectures

In this section we present the architectural enhancements applied to legacy mobile radio access network architectures. These architectural enhancements also have an impact on the features that can be utilized for applications such as personal health systems. As noted previously, the discussion is limited to 3GPP-standardized small-cell architectures but equally applicable to small cells specified by other standardization bodies.

11.3.2.1 Universal Terrestrial Radio Access Network (UTRAN) Architecture

The 3GPP maintains the 2G Global System for Mobile Communication (GSM) standards and evolutions thereof, namely, the 3G Universal Mobile Telecommunications System (UMTS) and subsequent Long Term Evolution (LTE) standards. UMTS as an umbrella term includes all 3G-specified radio access technologies, namely, wideband code-division multiple access (WCDMA) and high-speed packet access (HSPA). The legacy UTRAN architecture constituted up to hundreds of 3G base stations or node Bs attached to a radio network controller (RNC) via an Iub interface over fixed infrastructure owned or leased owned by the operator [44]. The UTRAN is connected to a user device (user equipment or UE) via the Uu interface, which could be a standard WCDMA or HSPA radio link. On the other side, the UTRAN uses the Iu interface to connect to the core network which is responsible for traffic switching, routing, and service control.

The deployment of femtocells in UTRAN necessitated architectural enhancements for the operator due to their relatively larger numbers, plug-and-play deployment by the user, and utilization of Internet protocol (IP) backhaul links not directly owned or leased by the operator [20]. The end-to-end femtocell system logical architecture for UTRAN environment is depicted in Figure 11.4 [45]. The femto–base station for UTRAN is formally known as the home node B (HNB), which terminates the Iu–home (Iuh) interface toward the home node B gateway (HNB-GW).

The HNB-GW plays a role similar to the one performed by RNC for node Bs, by aggregating multiple HNBs and terminating the Iu interface toward the core network. The UE may be connected to multiple HNBs (soft handover) directly via the Iurh interface or through the HNB-GW. The security gateway (SeGW) provides Internet protocol security (IPSec) tunnels for secure transport of mobile traffic over shared fixed broadband access links connecting the HNBs to the HNB-GW. In practical implementations, the SeGW may be colocated with the HNB-GW.

Apart from the newly introduced elements, the HNB architecture uses standard Uu interfaces (between HNB and UE) and Iu interfaces (between HNB-GW and core network), which implies that a HNB appears to the UE like a normal node B, while the HNB-GW appears like an RNC to the core network. The home node B management system (HMS) is an architectural entity implemented using TR-069 specifications and was introduced for enabling remote administration of HNBs by the HNB owner [45].

The Gi interface provides an IP-based gateway with authentication, authorization, and accounting (AAA) services for data connections between mobile core network and external IP networks (e.g., public Internet). The Gi interface can be present at the HNB for directly connecting IP-capable UEs in the residential network via the HNB without the need for the user traffic being routed to the core network, as would usually be the case for node Bs in legacy UTRAN. This architectural enhancement is referred to as local IP access (LIPA) [45]. The local gateway (L-GW) is logically colocated within a HNB (in LIPA mode) and provides the local traffic offloading point using the Gi interface to the local IP network. The

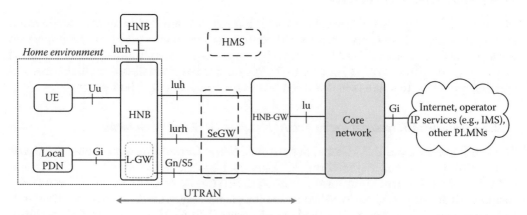

FIGURE 11.4
End-to-end logical architecture for UTRAN environment. HNB, home node B; HNB-GW, HNB gateway; HMS, HNB management system; IMS, IP multimedia subsystem; L-GW, local gateway; PLMNs, public land mobile networks; SeGW, security gateway; UE, user equipment; UMTS, universal mobile telecommunications systems; UTRAN, UMTS radio access network.

LIPA enhancement was introduced to direct some of the traffic away from the UTRAN and core network, by avoiding unnecessary detours to core network for data that originate and terminate in the same local network. Only the signaling traffic is sent to the core network via the Gn/S5 interface.

11.3.2.2 Evolved Universal Terrestrial Radio Access Network (E-UTRAN) Architecture

The development of LTE radio access technology in 3GPP was accompanied by the development of the UTRAN and core network under the system architecture evolution (SAE). The result of this standardization initiative was the E-UTRAN (or eUTRAN) and evolved packet core (EPC). The E-UTRAN is a simplified packet-based flat-architecture without the intermediate nodes (similar to an RNC in UTRAN) between the core network and base stations or evolved node Bs (eNBs) [46].

The femtocells have also been specified for E-UTRAN and these LTE femtocells are known as home evolved node Bs (HeNBs) in 3GPP vocabulary [46]. The end-to-end logical architecture for HeNB in E-UTRAN is shown in Figure 11.5. The specification of the E-UTRAN HeNB architecture followed the same considerations adapted for the development of the HNB architecture in UTRAN (described in Sub-subsection 11.3.2.1), such as securing backhaul traffic to/from the HeNB in IPSec tunnels, remote HeNB management with HMS, and local IP connectivity with LIPA. The notable differences between the E-UTRAN HeNB architecture and HNB architecture are the following:

- The HeNB terminates the user data traffic (user plane) and signaling traffic (control plane) via the S1 interface toward the EPC. This is done either directly to EPC or via the home evolved node B gateway (HeNB-GW). The S1 interface is the same as the corresponding interface used by eNBs for improved compatibility.

- The HeNB-GW is not mandatory and is considered only as an optional logical element and if deployed would aggregate control plane traffic.

The EPC constitutes a number of elements (e.g., mobility management entity) that carry out key management functions for mobile connectivity services provided by the network.

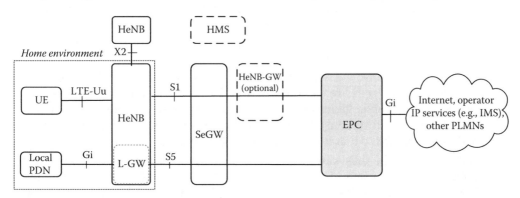

FIGURE 11.5
End-to-end logical architecture for E-UTRAN environment. EPC, evolved packet core; E-UTRAN, evolved UMTS RAN; HeNB, home eNode B; HeNB-GW, HeNB gateway; HMS, HeNB management system; IMS, IP multimedia subsystem; LTE, long term evolution; PDN, packet data network; P-GW, PDN gateway; PLMNs, public land mobile networks; RAN, radio access network; SeGW, security gateway; UE, user equipment; UMTS, universal mobile telecommunications system.

These include mobility management; AAA functions; subscription management; and management of connection quality, policies, and prioritizations. Furthermore, the EPC provides the interface toward the external IP networks, such as the public Internet and other cloud services. For a more detailed description of the EPC features, we refer the reader to relevant 3GPP standards (3GPP TS 23.401 [47]) and specialist text (e.g., that by Olsson et al. [48]).

11.3.3 Access Control Mechanisms

Although residential small cells or home node Bs/home evolved node Bs [HeNBs] are deployed autonomously by the small cell owner who would usually be a member of a household, it is essential to have access control mechanisms for controlled utilization of a small-cell resources both for the sake of the mobile network operator and for the small cell owner. This is based on the fact that the operator provides valuable radio spectrum bandwidth, backhaul capacity, and core network functionalities that are essential for the small-cell operation. The small cell owner is also a key stakeholder that procures the small cell and is responsible for recurring usage costs, such as subscription fees and powering costs.

The accessibility to mobile services provided over small cells varies according to small-cell ownership, intended coverage area for the small cell, and mobile operator's business model. Three access modes are commonly considered in practice [20,49]:

- *Closed-access* residential small cells, with access limited to closed subscriber group or CSG (e.g., members of a common household) exclusive access to the small-cell resources: The CSG members can be either temporary or permanent CSG members [49]. The membership of temporary CSG members is associated with an expiration time set by the small cell owner or operator, while the permanent CSG membership remains valid for as long the small-cell subscription is valid. In most cases, the small cell owner would have the administrative rights to add, remove, or modify details (UE identifiers, expiration time, etc.) of temporary members on a CSG membership list. The access control for closed-access small cells in UTRAN is usually carried out in HNB-GW or core network based on the closed subscriber group identity (CSG ID) initially reported by a UE requesting to camp on the small cell. In the case of E-UTRAN, access control is carried out only in the core network (EPC).

- *Open-access* operator-owned small cells, which are accessible to all subscribers: Open-access small cells are usually deployed by operator in outdoor hot spots or other spaces of public gathering (e.g., picocells deployed on avenue lampposts, stadiums, and train stations). In that regard the access control of an open-access small cell is similar to that for a conventional macrocell.

- *Hybrid-access* small cells, offering a trade-off between closed- and open-access modes, whereby for hybrid-access small cells, CSG members are prioritized and afforded unlimited access while non-CSG members are provided only best-effort services utilizing idle resources: The access control procedures for hybrid access small cells follow those similar to the HeNB closed-access case except that the mobility management entity (MME) performs membership verification based on the CSG ID and membership status (CSG member or nonmember) initially reported by the UE.

11.4 Small Cells for Personal Health Systems

11.4.1 Brief Review of Existing Personal Health Gateway Approaches

The reliable and continuous end-to-end connectivity of personal health systems depends greatly on the gateway that links the end-user and remote care-giver domains (as noted in Subsection 11.2.1). The smartphone has emerged as a leading platform for implementing the personal health gateway [12–13,15]. This is driven by a range of factors, including high adoption rates among target user group, increased computing power and storage space, integrated sensing capabilities (e.g., accelerometer and gyroscope), availability of high-resolution displays and cameras, and inclusion of mobile operating systems (e.g., Android and iOS) that support powerful third-party health apps. This positioning of smartphones as a key connected personal health gateway is evident in recent developments, such as the incorporation of the Bluetooth health device profile (HDP) in the popular Android operator system (version 4.0 onward), enabling Android phones to seamlessly interface to Bluetooth health devices [50].

The versatility of the smartphone as a platform and/or gateway for a range of personal applications is cementing its role not just for personal health but also for entertainment, home automation, education, and so on. However, the considerable computational demands placed on the smartphones by these applications have rapidly outpaced the growth in the smartphone battery capacity [51]. This need for frequent smartphone battery recharging has meant that the smartphone is limited to continuous operation as a gateway for personal health systems.

The limitation in smartphone-based gateways has prompted the development of custom gateways for personal health systems. The custom personal health gateway could function as stand-alone gateway in home environments and could take over the gateway function from the smartphone when the end user is located in the home. The portability of the smartphone allows for it to continue providing gateway functionality when the end user roams beyond the coverage area of the custom gateway. An example custom gateway is the Continua-certified Qualcomm 2net Hub that provides short-range radio interfaces (Bluetooth, ANT+, and Wi-Fi) for health devices deployed in the home environment and 2G/3G mobile radio interfaces for connectivity to cloud-based health servers [52]. The stand-alone gateway may also be implemented as a multipurpose gateway that provides gateway functionality for health, as well as other vertical applications in the home environment. For instance, the British project Hydra proposed the use of utility-provided smart meters equipped with ZigBee radios to simultaneously support local health device connectivity by using the ZigBee health care profile (HCP) in addition to the original function of monitoring energy consumption and gathering readings from home appliances [53].

This report describes how small cells present an additional attractive option for implementing an intelligent gateway for locally connected personal health devices in the home environment and their interconnection to remote elements in the care-provider domain.

11.4.2 General End-to-End Implementation

Figure 11.6 illustrates a generalized high-level view of an end-to-end personal health system deployment utilizing a residential small cell as the primary gateway. In this implementation the personal health devices could belong to one of the following categories:

- **Devices with an embedded mobile module:** These are devices with mobile radio interfaces (general packet radio service [GPRS], HSPA, LTE, etc.) that enable them to obtain direct wide area connectivity via the mobile radio access networks. This

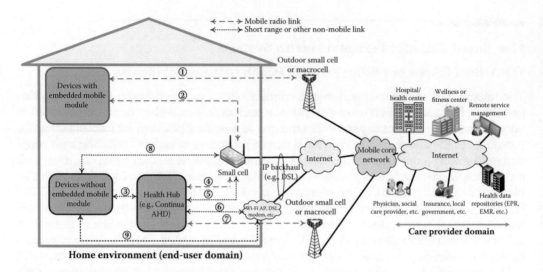

FIGURE 11.6
Personal health system implementation with a residential small-cell gateway option. 3GPP, third generation partnership project; AHD, application hosting device; AP, access point; DSL, digital subscriber line;
EMR, electronic medical record; EPR, electronic patient record; IP, internet protocol; OA&M, operations,
administration and management.

category includes not only traditional mobile devices (e.g., smartphone) but also
M2M/MTC devices discussed in Subsection 11.2.4.

- **Devices without an embedded mobile module:** These are devices equipped only
  with short-range radios (e.g., Continua TAN/PAN/LAN devices) that depend on a
  health hub or other gateway to obtain wide area connectivity via the mobile network or other public IP network.

- **Health hubs:** The health hub provides a wide area connectivity gateway for local
  health devices equipped only with short-range radios. Example devices in this
  category are the Continua AHDs described in Subsection 11.2.3. In the case that a
  small cell is deployed as the primary gateway, the health hub could serve as secondary gateway for personal health system.

The wide area connectivity to the remote care-provider domain is provided via a residential
small cell or legacy connectivity options as illustrated in Figure 11.6 and outlined below:

- **Small-cell connectivity:** The small cell provides mobile connectivity for health
  devices and hubs equipped mobile radio interfaces (links ② and ④ in Figure 11.6).
  Hubs provide an indirect bridge for small cells to health devices without mobile
  radio interfaces (link ③ on first hop and link ④ on second hop in Figure 11.6). In the
  case that the small cell is a multiradio small cell integrated with short-range radios
  (e.g., Bluetooth and ZigBee), it also provides a direct gateway for health devices
  and hubs having only short-range radios (links ⑤ and ⑧ in Figure 11.6).

- **Legacy connectivity:** For health devices and hubs with embedded mobile radios,
  the legacy option is connectivity via the outdoor macrocell (links ① and ⑦ in Figure
  11.6). In the case of health devices and hubs without mobile radio interfaces, the
  alternative would be to utilize a legacy gateway device, such as a Wi-Fi AP or DSL
  modem (links ⑥ and ⑨ in Figure 11.6).

11.4.3 Benefits of Small Cells for Personal Health Systems

The use of small cells as a gateway for personal health systems brings a range of benefits by leveraging mobile connectivity and well-established service management (e.g., subscription data management and remote device management) in mobile networks as noted previously in Subsection 11.2.4. Moreover, it allows personal health systems to gain the mobile service delivery advantages that small cells provide over traditional macrocell deployments. In this subsection we highlight and briefly outline some of the key benefits gained through use of small cells in this context.

11.4.3.1 QoS, Charging, and Policy Control

The evolving mobile core network is offering increasingly sophisticated ways for differentiation of both services and subscribers [48]. Data bearers provide end-to-end logical transport for IP traffic across mobile networks and each bearer is assigned specific QoS parameters (e.g., guaranteed bit rates, packet delay, and packet error rate). This enables differentiation in treatment of traffic flows from different services (e.g., health data flows versus nonhealth data flows) during admission control and resource allocation procedures for different service requests.

The introduction of policy control and charging (PCC) functionality in 3GPP Release 8 systems provides even more advanced QoS differentiation operating at the relatively fine-grained per-service session level rather than on a per-bearer level [48]. This could, for instance, enable differentiation among health service data flows according to their criticality [54]. For instance, ambulatory heart-rate data could be prioritized over other noncritical health data that may be produced from a common personal health system. The PCC also includes functionality for both real-time online charging and offline charging, which could result in many alternative billing mechanisms and new innovative personal health business models. For instance, from generated charging data records (CDRs), it is possible to identify which segments of the personal health service should be billed to the patient and which are covered by their health insurance provider or local authority.

The rapidly emerging concepts of software-defined networking (SDN) and network function virtualization (NFV) are adding further agility to the mobile service provisioning by allowing the networks to adapt to service demand instantaneously rather than rely on overprovisioning of network resources [55]. The SDN/NFV upgrades enable the mobile operator to create new and virtualized subnetworks that could be customized by third-party service providers, such as the personal health service considered in this study. Initial SDN/NFV implementations in mobile networks focus on the mobile core network (EPC) due to its centralized architecture [55]. However, the long-term vision is of SDN/NFV upgrades targeting also the radio access network, where network edge components, such as residential small cells, could be virtualized and better exposed for third-party service providers.

11.4.3.2 Carrier-Grade Security

The secure handling of transferred end-user data was noted as one of the major requirements for connected personal health systems (recall Subsection 11.2.2). Residential small cells or HeNBs are consumer devices autonomously deployed by an end user in the same way as Wi-Fi APs, and both device types face common security threats, such as compromise of credentials, physical attacks, configuration attacks, protocol attacks, attacks on radio resources, and user data and identity privacy attacks. However, residential small cells are actually managed by the mobile operator via the mobile core network elements. This provides residential small cells with benefits, such as carrier-grade security features [56], which are relatively more robust compared to Wi-Fi security features privately managed by the Wi-Fi AP owner. The security measures implemented for small cells include the following [56]:

- **Small-cell authentication:** The mandatory mutual authentication of the small cell's device and the mobile operator's network occurs because the identity of the small cell is authenticated by the network and the identity of the mobile operator's network is authenticated by the small cell. This mechanism mitigates against the threat of a masquerading small cell in open-access mode being deployed to sniff data traffic from nearby end-user devices (e.g., health devices) attached to a legitimate small cell.

- **Protection of backhaul traffic:** The integrity of the end-user health data could be compromised on parts of the networks that are not under operator control, specifically in the traffic backhauling between small cells and mobile core network via public IP networks. To that end, the IPsec tunnels are used to protect all end-user and signaling data traffic between the small cell and SeGW (recall the architectures in Figures 11.4 and 11.5).

- **CSG-based access control:** The CSG-based access control mechanisms for closed- and hybrid-access mode small cells were introduced in Subsection 11.3.3. This provides useful means for a small-cell administrator to manage access for personal health systems by using the small cell as a secure hub and gateway. Access list management functions include addition or deletion of end-user health devices from the CSG, assignment of temporary access for certain devices, and barring access for any connection requests from devices outside the CSG.

11.4.3.3 Indoor Coverage and Capacity Enhancements

Small cells enable relatively improved indoor coverage compared to conventional macro-cell coverage mostly by avoiding the wall attenuation encountered in outdoor-to-indoor propagation [20,38–40]. This coverage improvement ensures service availability in all areas of the house (even in basements, attic, etc.) which are usually macrocoverage dead spots. Furthermore, small cells offer relatively larger capacities in both uplink and down-link and avoid capacity drops due to fluctuating load conditions. This is attributed to the fact that, unlike macrocell usage, the residential small-cell usage is typically restricted to a relatively smaller number of CSG members. This scalable capacity could potentially enable wide area connectivity of personal health-monitoring devices, producing rich multimedia observation data (e.g., high-resolution image fall sensors).

11.4.3.4 Improved Device Energy Efficiency

A majority of personal health devices are powered off grid by using batteries with very limited capacity (limited battery lifetime). Therefore, energy-efficient operation of the devices is essential to eliminate the need for frequent recharging or battery replacement. Personal health devices using short-range radios, such as ZigBee and Bluetooth low energy, have managed to achieve significant energy-efficiency gains by the going into sleep mode during periods of inactivity [57]. Furthermore, the range of indoor BAN, PAN, or LAN radio links is usually in the range of up to a few tens of meters between paired devices or gateway. This reduces the required level of device transmission powers needed to compensate for distant-dependent path loss, resulting in more power saving.

By contrast, health devices used with embedded mobile radios would have shorter sleep cycles due to more frequent signaling operations with mobile core network, although it should be noted that this reduction in signaling load and frequency is one of the key objectives for updated mobile standards in anticipation of M2M-type traffic dominance [17]. Furthermore, the mobile radio access links from indoor mobile health devices usually terminate at an outdoor macro–base station typically located kilometers away. This means in order to achieve radio performance targets (e.g., signal-to-interference-plus-noise ratio [SINR]) higher device transmit powers are needed to overcome the higher distance-dependent path losses, shadowing (due to buildings, hills, etc.), and cochannel interference. By contrast, small cells are usually deployed locally in home environments (<100 m from health devices), leading to relatively low losses (propagation losses, wall attenuations, etc.), and, hence, lower device uplink transmit power requirements. This enables longer battery life or reduced grid power consumption compared to the macrocell case.

11.4.3.5 Small-Cell Services Development

The benefits of small cells in terms of hot spot capacity and coverage enhancements are widely known. However, small cells also have the potential to enable a range of value-added services for mobile operators (and, in turn, third-party service providers) by leveraging context, presence, and location information of devices and subscribers in small-cell coverage areas [21]. To that end, the Small Cell Forum has specified a set of APIs that exposes small-cell awareness information (presence, location, or context) to the developer community to spur the creation of small-cell apps [58].

Connected personal health service developers may utilize the small-cell application programming interfaces (APIs) to build innovated applications that add value to the use. For instance, a virtual-fridge note application may prompt the small cell to send medicine reminders only when the patient returns to the home environment (detected by inbound handover of patient's handset). Another example is a work-flow enhancement application, whereby a small cell, upon detecting a visiting caregiver's smartphone entering the patient's small-cell coverage area, would then initiate the patient's health context data to be automatically updated or activated on the caregiver's handset.

11.4.3.6 Mobile Traffic Offloading

Multimedia traffic generated continuously from services such as personal health services places a huge strain on mobile networks and frequently create congestion in the network. One of the solutions standardized by 3GPP is the use of LIPA (introduced in Subsection 11.3.2) to alleviate congestion by having the HeNBs (small cell) locally route traffic that is destined for local networks, rather than have the traffic flow via the small-cell IP backhaul and mobile core network (see example in Figure 11.7). Additional traffic offloading could be achieved for traffic destined beyond the local environment by using selective Internet protocol traffic offloading (SIPTO) [59]. In the case of SIPTO, the traffic breakout point (local packet data network gateway or L-P-GW) is at an offloading point logically located between the small cell and mobile core network, again the objective being to relieve traffic from the mobile core network. The SIPTO feature is specified for both small cells and macrocells (eNBs), unlike LIPA which is available only for HeNBs. However, it should be noted

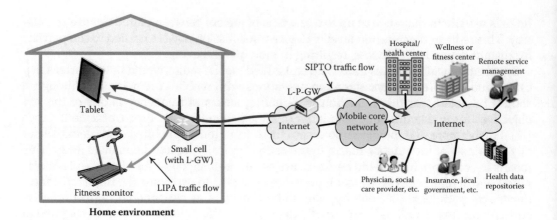

FIGURE 11.7
Example of mobile traffic offload with LIPA and SIPTO. IP, internet protocol; L-GW, local gateway; LIPA, local IP access; L-P-GW, local PDN gateway; PDN, packet data network; SIPTO, selective IP traffic offloading.

the LIPA and SIPTO offloading is carried out only for user data traffic while the lighter control plane traffic is routed via mobile core network as usual for purposes of authentication, billing, subscription management, and so on [59].

11.5 Simulation Case Study

The discussion on improvements in system performance and energy efficiency through use of small cells was introduced in Subsection 11.4.3. The objective of this simulation case study is to provide further insights on the possible performance gains achieved when using small cells instead of macrocells to provide wide area connectivity for personal health systems.

11.5.1 System Model

The RF uplink models are useful for the study of transmission power requirements (hence, power consumption) of wireless health devices. The typical building blocks for a 3GPP-compliant mobile wireless device include integrated circuits for RF, baseband, and mixed-signal processing functions [60–61]. Each of the functional blocks contributes to the overall budget for power consumption of the device. To that end, the majority of power consumption budget is attributed to the RF transceiver and modem circuitry [61]. The fraction of RF component's power consumption is even more significant in wireless embedded devices that would usually have relatively less or no user interface components (microphones, speakers, backlit displays, etc.) and a reduced set of integrated secondary radio interfaces (Bluetooth, GPS, NFC, etc.).

An LTE radio access environment is assumed for this simulation study, whereby comparisons are made for the case where the mobile health device (LTE UE) connected via a small cell (HeNB) versus the case where it is via a macrocell (eNB). In LTE networks the devices that are ON may be in either radio resource control (RRC) connected or RRC idle state depending on whether the UE context is registered at a serving cell (HeNB or eNB) or not [62]. In the RRC connected state, the UE transfers (or receives) data packets, monitors the shared signaling channels for any scheduled resource allocation, and provides measurement report feedback to the serving cell. By contrast, no transmission (or reception) of user data occurs in the RRC idle state except for periodic monitoring of common signaling and paging channels. This inactivity in RRC idle state allows for most of the device circuitry to be powered down (go into sleep mode), so as to prolong battery life. Battery-conserving opportunities can also be obtained by exploiting inactive periods in the RRC connected state (e.g., due to bursty traffic) [63]. Reduction of transmit power by reducing radio link distance provides even higher savings in power consumption in RRC connected state [64].

The LTE uplink adopts single-carrier frequency-division multiple access (SC-FDMA) as the radio interface. The SC-FDMA scheme maintains a low peak-to-average power ratio (PAPR), thus reducing UE power consumption, and also provides the benefit of high-frequency efficiency due to compact subcarrier spacing [62]. In LTE, the total system bandwidth W is divided into subcarriers with an intercarrier spacing of 15 kHz. Twelve subcarriers with 180 kHz bandwidth in total are grouped into a physical resource block (PRB) with 0.5 ms temporal duration, which is the basic radio resource unit allocated to a UE.

The radio propagation between UE and the serving cell experiences path losses mostly from building wall penetration losses, distance-dependent path loss, shadowing, and fast fading. For indoor-located UE, the penetration loss due to walls separating the apartments is explicitly modeled, while penetration loss due to internal walls within the apartment is taken into account by using a log-linear model that depends on the separation distance between walls. In this chapter, we adopt the 3GPP TR 36.814 channel models [65], to evaluate the path loss for all possible radio propagation scenarios, namely, indoor UE to HeNB, outdoor UE to eNB, indoor UE to eNB, and outdoor UE to HeNB propagation.

The UE uplink transmission power is determined by the 3GPP fractional power control method for the physical uplink shared channel (PUSCH) that bears the LTE uplink user data. We use a similar approach to that proposed by Castellanos et al. [66], which ignores the closed-loop corrections, resulting in the following open-loop power control scheme, where the UE transmit power P_t is expressed as

$$P_t = \min\{P_{\max}, P_0 + 10\log_{10} M + \alpha\text{PL}\}, \tag{11.1}$$

where $P_{\max}$ is the maximum UE transmit power, P_0 is a UE- or cell-specific parameter indicating the possible minimum UE transmit power, M is the number of PRBs assigned for a certain UE, α is the cell-specific path-loss compensation factor, and PL is downlink path loss estimated by the UE.

The uplink performance is quantified in terms of the UE uplink throughput by summing the data throughput over all the PRBs allocated to the UE. The throughput over PRB j is approximated using an attenuated and truncated form of the Shannon bound by mapping the SINR on PRB j to throughput S_j by using the following equation [67]:

$$S_j = \begin{cases} 0 & \text{SINR}_j \leq \text{SINR}_{\min} \\ \text{BW}_{\text{PRB}} \cdot \text{BW}_{\text{eff}} \sum_{j=1}^{N(j)} \log_2(1 + \text{SINR}_j/\text{SINR}_{\text{eff}}) & \text{SINR}_{\min} \leq \text{SINR}_j \leq \text{SINR}_{\max} \\ S_{\max} & \text{SINR}_j \geq \text{SINR}_{\max} \end{cases}$$

$$\tag{11.2}$$

where BW_{PRB} is the bandwidth of a single PRB and BW_{eff} and $SINR_{eff}$ are bandwidth and SINR efficiencies, respectively, that represent capacity loss due to system implementation and signal processing procedures.

11.5.2 System Simulation Assumptions

The simulated environment was an area covered by seven trisectored macrosites (total of 21 macrocells) in a wraparound layout (see Figure 11.8). Each macrosite covered three hexagonal sectors by using sectorized antennas and separate radio resource management (RRM) procedures. The small cells (HeNBs) were deployed randomly in a multiapartment building. The 100 m² apartments in the building were arranged in a 5 × 5 grid commonly used in 3GPP indoor simulation studies [68] (see Figure 11.8, inset). The building was located in a macrocell covered from a centrally located macrosite to obtain worst-case intercell interference scenarios. Furthermore, two locations were considered for the buildings (see Figure 11.8): at the cell center and the cell edge of the macrocell, representing locations where the macrocell coverage was the best and worst, respectively.

Comparative studies of the macrocell and small-cell systems were performed using a MATLAB® static simulator over a large number (10^4) of random snapshots (Figure 11.8 being an example snapshot). For each snapshot, the positions of the indoor and outdoor UEs and the indoor HeNBs were generated randomly and the performance was evaluated for the UEs of interest, that is, the indoor-located UEs. For each snapshot a constraint was defined that each apartment may have either one or no HeNB deployed. Furthermore, the HeNBs were deployed on the apartment walls, which is a practical assumption based on the need to plug in the small cell to a wall outlet for the fixed IP backhaul. The HeNBs were assumed to operate in same spectrum band as the macro-eNBs, creating a range

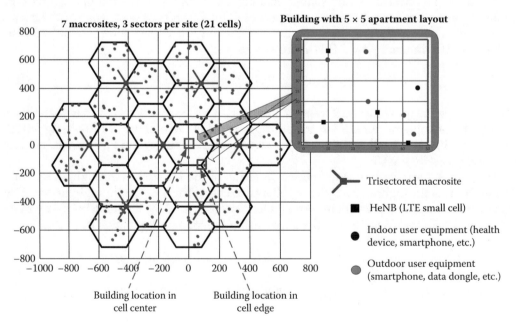

FIGURE 11.8
Layout of simulated environment.

of interference scenarios between the macrocell and small cells. The system simulation parameters are listed in Table 11.1 and are in line with commonly used 3GPP simulation guidelines outlined in 3GPP TR 36.942 [69].

11.5.3 Results and Discussions

The generated cumulative distribution function (CDF) plots of the average UE transmission power for the scenarios where the indoor UEs are connected via either the eNB or the HeNB are shown in Figure 11.9. For the HeNB case, simulations are run for different values of power control parameter P_0 so as to observe the range of average transmission power requirements for different power control scenarios. In the case of eNB, simulations are repeated for the scenarios where the building is in cell edge and at cell center. Considering same combination of power control parameters ($P_0 = -73$ dBm and $\alpha = 0.8$) for eNB and HeNB in Figure 11.4, the UE connection via the HeNB can provide 50th-percentile UE transmission power saving of 39 dB. To put that power saving into perspective by using a simple example, an alkaline AAA-size cell (with 1.41 Wh capacity) could potentially increase its lifetime by two to three orders of magnitude.

Figure 11.10 shows the CDFs for the average achievable UE throughputs for the same set of simulation runs. The results demonstrate at least a 10-fold 50th-percentile throughput gains for the HeNB connectivity case compared to the eNB case. For typical periodic and bursty

TABLE 11.1

System Simulation Parameters

Parameter	Value
System Parameters	
Intersite distance	3GPP macro case 1: 500 m [68]
Carrier frequency	2 GHz
Bandwidth	10 MHz (= 50 PRBs)
	48 PRBs for data and 2 PRBs for signaling
Bandwidth efficiency (BW_{eff})	0.4
SINR efficiency ($SINR_{eff}$)	1
Thermal noise power spectral density (PSD)	-174 dBm/Hz
(H)eNB Parameters	
eNB antenna gain	14 dBi
Receiver noise figure	7 dB
Antenna pattern	For eNB:
	$A(\varphi) = -\min[12(\varphi/\varphi_{3\,dB})^2, A_m]$
	$\varphi_{3\,dB} = 70°$ and $A_m = 20$ dB
	For HeNB:
	Omnidirectional
Traffic model/scheduler	Full buffer/round robin
UE Parameters	
Maximum transmit power	23 dBm
Maximum antenna gain	0 dBi
Building Parameters	
Apartment dimensions	10×10 m^2
Building layout	5×5 grid apartments [68]
Wall penetration losses	Inner walls = 5 dB, outer walls = 20 dB

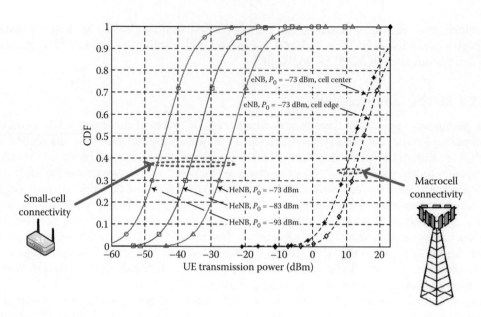

FIGURE 11.9
CDF of average UE transmission power.

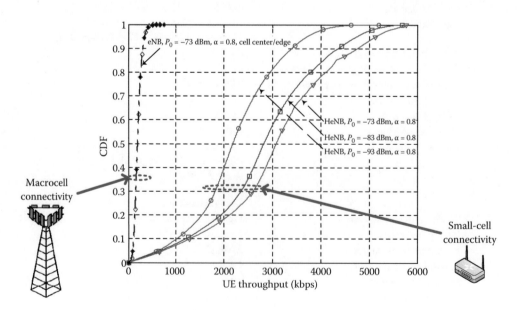

FIGURE 11.10
CDFs of average UE uplink throughputs.

telemetry traffic from monitoring devices, a higher throughput reduces the amount of time the UE remains in the connected state, which in turn reduces power consumption [70].

The energy-efficiency metric, defined as the number of joules required for transferring 1 bit (W/bit/s) [70], can be used here to provide a comparison of the UE energy-efficiency gains when connectivity is provided via an indoor HeNB rather than eNB. It is noted that for the same power settings ($P_0 = -73$ dBm, $\alpha = 0.8$) the energy-efficiency gains with HeNB versus eNB connectivity is three to five orders of magnitude in most throughput regions. In practice, performance gains may vary in different deployment scenarios depending on various factors, such as building design, operating frequency, traffic patterns, and UE type. However, evidence from numerous research and practical implementations has demonstrated consistent performance gains for heterogeneous networks with small cells compared to macro-only deployments [19–20,38–40].

11.6 Conclusions

This chapter presents the opportunity and benefits for leveraging residential small cells as gateways for personal health systems. To that end, the current trends in the development of personal health systems were reviewed and a brief overview of small-cell concept was provided. It is clear that small cells bring a range of benefits in terms of performance improvement, security, and service innovations that add value to personal health system implementations. While this study focused on personal health systems, the benefits of small cells could be equally useful for other telemedicine scenarios. For instance, in-hospital environments could gain advantages, such as integration of hospital worker mobile devices with hospital IT systems, high indoor capacity, and robust security, that are brought by small-cell deployments [71]. The mobile landscape is currently heading toward another phase of significant research and development toward future fifth-generation mobile telecommunications (5G) networks [72]. These developments will further enhance the benefits of small cell (for personal health systems and similar applications) through disruptive design changes, such as device-centric architectures, larger bandwidth in the millimeter wave region, and significantly improved support for M2M communications.

Abbreviations

2G	second-generation mobile telecommunications
3G	third-generation mobile telecommunications
3GPP	Third Generation Partnership Project
3GPP2	Third Generation Partnership Project 2
4G	fourth-generation mobile telecommunications
5G	fifth-generation mobile telecommunications
AAA	authentication, authorization, and accounting
AHD	application-hosting device
AP	access point

API	application programming interface
BAN	body area network
BP	blood pressure
CCF	CDMA Certification Forum
CDF	cumulative distribution function
CDMA	code-division multiple access
CDR	charging data record
CSG	closed subscriber group
DSL	digital subscriber line
ECG	electrocardiography
EEG	electroencephalography
EMG	electromyography
EMR	electronic medical record
EPC	evolved packet core
EPR	electronic patient record
EPS	evolved packet system
ETSI	European Telecommunications Standards Institute
E-UTRAN	evolved universal terrestrial radio access network
GCF	Global Certification Forum
GPRS	general packet radio service
GPS	global positioning system
GSM	Global System for Mobile Communication
GSMA	Groupe Speciale Mobile Association
HDP	health device profile
HeNB	home evolved node B
HeNB-GW	home evolved node B gateway
HIPAA	Health Insurance Portability and Accountability Act
HL7	health level 7
HMS	home node B management system
HNB	home node B
HNB-GW	home node B gateway
HRN	health-reporting network
HSPA	high-speed packet access
IEEE	Institute of Electrical and Electronics Engineering
IMS	Internet protocol multimedia subsystem
IMT	International Mobile Telecommunications
INR	international normalized ratio
IoT	Internet of Things
IP	Internet protocol
IPSec	Internet protocol security
LAN	local area network
L-GW	local gateway
LIPA	local Internet protocol access
L-P-GW	local packet data network gateway
LTE	Long Term Evolution
M2M	machine-to-machine
MBAN	medical body area network
MME	mobility management entity
MTC	machine-type communication

NFC	near-field communication
NFV	network function virtualization
OA&M	operations, administration, and management
PAN	personal area network
PAPR	peak-to-average power ratio
PDN	packet data network
P-GW	packet data network gateway
PLMN	public land mobile network
PRB	physical resource block
PUSCH	physical uplink shared channel
QoS	quality of service
RAN	radio access network
RF	radio frequency
RRC	radio resource control
SAE	system architecture evolution
SC-FDMA	single-carrier frequency-division multiple access
SDN	software-defined networking
SeGW	security gateway
S-GW	serving gateway
SIPTO	selective Internet protocol traffic offloading
SpO$_2$	peripheral capillary oxygen saturation
TAN	touch area network
UE	user equipment
UMTS	Universal Mobile Telecommunications System
UTRAN	universal terrestrial radio access network
WAN	wide area network
WCDMA	wideband code-division multiple access
WEF	World Economic Forum
WHO	World Health Organization

Acknowledgments

This work was partially supported by the Preventive Care Infrastructure Based on Ubiquitous Sensing (PRECIOUS) project, which is funded by the European Union's Seventh Framework Programme for research, technological development, and demonstration under Grant Agreement No. 611366.

References

1. World Health Organization (WHO), Global status report on noncommunicable diseases 2010, 2011.
2. Bloom, D. E. et al., The global economic burden of noncommunicable diseases. Report by the World Economic Forum and the Harvard School of Public Health, Geneva, 2011.

3. Khaw, K.-T., Wareham, N., Bingham, S., Welch, A., Luben, R., and Day, N., Combined impact of health behaviours and mortality in men and women: The EPIC-Norfolk prospective population study. *PLoS Medicine*, 5(1), 2008.

4. European Commission, Directorate-General for Economic and Financial Affairs, Projecting future health care expenditure at European level: Drivers, methodology and main result. Economic papers 417, Brussels, 2010.

5. Berry, L., and Mirabito, A. M., Innovative healthcare delivery. *Business Horizons*, 53, pp. 157–169, 2010.

6. Centre for Disease Control, HAI page, http://www.cdc.gov/hai/. Accessed April 15, 2014.

7. Iacobucci, G., Nearly 40% of hospitals missed emergency department waiting time target in last quarter, show figures. *British Medical Journal*, 346, 2013.

8. Bashshur, R., Shannon, G., Krupinski, E., and Grigsby J., The taxonomy of telemedicine. *Telemed J E Health*, 17(6), pp. 484–94, 2011.

9. Gartner: eHealth for a healthier Europe!—Opportunities for a better use of healthcare resources. Study on behalf of the Ministry of Health and Social Affairs in Sweden, 2009.

10. Piniewski, B., Muskens, J., Estevez, L., Carroll, R., and Cnossen, R., Empowering healthcare patients with smart technology. *IEEE Computer*, 43, pp. 27–34, 2010.

11. World Health Organization (WHO), mHealth: New horizons for health through mobile technologies. WHO Global Observatory for eHealth series, 3, 2011.

12. PricewaterhouseCoopers (PwC), Socio-economic impact of mHealth. Assessment report for the European Union commissioned by GSMA, 2013.

13. Foh, K.-L., Integrating healthcare: The role and value of mobile operators in eHealth. GSMA mHealth Programme paper, 2012.

14. Continua Health Alliance, Connected personal health in 2015: "Getting it right!"—Looking back on the emergence of integrated person-centered health. Continua Health Alliance vision paper, 2008.

15. Wu, G., Talwar, S., Johnsson, K., Himayat, N., and Johnson, K. D.: M2M: From mobile to embedded internet. *IEEE Communications Magazine*, 49(4), pp. 36–43, 2011.

16. Ericsson, More than 50 billion connected devices. Ericsson white paper, 2011.

17. Chen, K. C., and Lien, S. Y., Machine-to-machine communications: Technologies and challenges. *Ad Hoc Networks*, 18, pp. 3–23, 2013.

18. Institute of Medicine, *Climate Change, the Indoor Environment, and Health*. National Academies Press, Washington, DC, 2011.

19. Small Cell Forum (SCF), Small cells—What is the big idea? SCF Release 1, Doc. 030.01.01, 2012.

20. Zhang, J., and de la Roche, G., *Femtocells: Technologies and Deployment*. John Wiley & Sons Ltd, Chichester, 2009.

21. Small Cell Forum (SCF), Small cell services? SCF Release 1, Doc. 046.01.01, 2013.

22. Hall, E. T., *The Hidden Dimension*. Anchor Books, New York, 1966.

23. Gamberini, L., Seraglia, B., and Priftis, K., Processing of peripersonal and extrapersonal space using tools: Evidence from visual line bisection in real and virtual environments. *Neuropsychologia*, 46(5), pp. 1298–1304, 2008.

24. Johny, B., and Anpalagan, A., Body area sensor networks: Requirements, operations, and challenges. *IEEE Potentials*, 33(2), pp. 21–25, 2014.

25. European Commission's Directorate General for Justice, Protection of personal data, http://ec.europa.eu/justice/data-protection/index_en.htm. Accessed April 15, 2014.

26. U.S. Department of Health & Human Services, Health Information Privacy, http://www.hhs.gov/ocr/privacy/hipaa/understanding/coveredentities/De-identification/guidance.html. Accessed April 15, 2014.

27. Continua Health Alliance, http://www.continuaalliance.org/. Accessed April 15, 2014.

28. ITU-T Recommendation H.810, Interoperability design guidelines for personal health systems, 2013.

29. Health Level Seven International, https://www.hl7.org/. Accessed April 15, 2014.

30. Continua Health Alliance, Implementation guidelines for cellular modems embedded into medical devices. Continua Health Alliance white paper, 2012.
31. Global Certification Forum, Continua Health Alliance specifies GCF certification as requirement for personal health device certification. GCF press release, 2012, http://www.globalcertificationforum.org/news-events/press-releases/39-continua-oct-2012.html. Accessed April 15, 2014.
32. 3GPP TR 22.868, Study on facilitating machine to machine communication in 3GPP systems, 2007.
33. ETSI TR 102 732, Use cases of M2M applications for eHealth. Draft standard, 2010.
34. oneM2M Partnership Project, http://www.onem2m.org/. Accessed April 15, 2014.
35. GSMA, a high level reference architecture for mobile health. GSMA Mobile Health report, 2011.
36. Amdocs, The State of the Radio Access Network (RAN) 2014, 2014.
37. Cisco Systems, Cisco visual networking index: Global mobile data traffic forecast update 2013–2018, 2014.
38. Zander, J., and Mähönen, P., Riding the data tsunami in the cloud: Myths and challenges in future wireless access. *IEEE Communications Magazine*, 51(3), pp. 145–151, 2013.
39. Hwang, I., Song, B., and Soliman, S. S., A holistic view on hyper-dense heterogeneous and small cell networks. *IEEE Communications Magazine*, 51(6), pp. 20–27, 2013.
40. Weitzen, J., Mingzhe, L., Anderland, E., and Eyuboglu, V., Large-scale deployment of residential small cells. *Proceedings of the IEEE*, 101(11), pp. 2367–2380, 2013.
41. Bhat, P. et al., LTE-Advanced: An operator perspective. *IEEE Communications Magazine*, 50(2), pp. 104–114, 2012.
42. Lazarus, M., The great spectrum famine. *IEEE Spectrum*, 47(10), pp. 26–31, 2010.
43. Starsinic, M., System architecture challenges in the home M2M network, Long Island Systems Applications and Technology Conference (LISAT), New York, 2010.
44. 3GPP TS 25.401, UTRAN overall description, http://www.3gpp.org/DynaReport/25401.htm. Accessed April 15, 2014.
45. 3GPP TS 25.467, UTRAN architecture for 3G home node B (HNB); Stage 2, http://www.3gpp.org/DynaReport/25467.htm. Accessed April 15, 2014.
46. 3GPP TS 36.300, Evolved universal terrestrial radio access (E-UTRA) and evolved universal terrestrial radio access network (E-UTRAN); Overall description; Stage 2, http://www.3gpp.org/DynaReport/36300.htm. Accessed April 15, 2014.
47. 3GPP TS 23.401, General packet radio service (GPRS) enhancements for evolved universal terrestrial radio access network (E-UTRAN) access, http://www.3gpp.org/DynaReport/23401.htm. Accessed April 15, 2014.
48. Olsson, M., Sultana, S., Rommer, S., Frid, L., and Mulligan, C., *SAE and the Evolved Packet Core: Driving the Mobile Broadband Revolution*. Elsevier, New York, 2009.
49. 3GPP TR 23.830, Architectural aspects of home node B and home eNode B, http://www.3gpp.org/DynaReport/23830.htm. Accessed April 15, 2014.
50. Android BluetoothHealth class documentation, http://developer.android.com/reference/android/bluetooth/BluetoothHealth.html. Accessed April 15, 2014.
51. Palacin, M. R., Recent advances in rechargeable battery materials: A chemist's perspective, *Chemical Society Reviews*, 38(9), pp. 2565–2575, 2009.
52. Qualcomm Life, Introducing the award winning 2net Platform, http://www.qualcommlife.com/wireless-health. Accessed April 15, 2014.
53. Clarke, M., Palmer, C., and Jones, R., Using the smart meter infrastructure to support home based patient monitoring. *International Journal of Integrated Care*, 2011. http://www.ijic.org/index.php/ijic/article/view/729/1335. Accessed April 15, 2014.
54. Skorin-Kapov, L., and Matijasevic, M., Analysis of QoS requirements for e-health services and mapping to evolved packet system QoS classes. *International Journal of Telemedicine and Applications*, 2010. http://www.hindawi.com/journals/ijta/2010/628086/. Accessed April 15, 2014.

55. Informa Telecoms and Media, Mobile SDN: The future is virtual. Informa UK white paper, 2013.
56. 3GPP TR 33.820, Security of H(e)NB, http://www.3gpp.org/DynaReport/33820.htm. Accessed April 15, 2014.
57. Jin-Shyan, L., Yu-Wei, S., and Chung-Chou, S., A comparative study of wireless protocols: Bluetooth, UWB, ZigBee, and Wi-Fi, 33rd Annual Conference of the IEEE Industrial Electronics Society, IECON 2007, 5–8 November 2007, 2007.
58. Small Cell Forum developers site, http://www.smallcellforum.org/developers/. Accessed April 15, 2014.
59. 3GPP TR 23.829, Local IP access and selected IP traffic offload (LIPA-SIPTO), http://www.3gpp.org/DynaReport/23829.htm. Accessed April 15, 2014.
60. Shearer, F., *Power Management in Mobile Devices*. First ed. Newnes, Burlington, 2008.
61. Silven, O., and Jyrkkä, K., Observations on power-efficiency trends in mobile communication devices. *EURASIP Journal on Embedded Systems* 2007. http://jes.eurasipjournals.com/content/2007/1/056976. Accessed May 2, 2014.
62. Holma, H., and Toskala, A.(eds.), *LTE for UMTS: OFDMA and SC-FDMA Based Radio Access*. Wiley & Sons Ltd, Chichester, 2009.
63. Bontu, C., and Illidge, E., DRX mechanism for power saving in LTE, *IEEE Communications Magazine*, 47(6), pp. 48–55, 2009.
64. Haq Abbas, Z., and Li, F., Distance-related energy consumption analysis for mobile/relay stations in heterogeneous wireless networks. 7th International Symposium on Wireless Communication Systems (ISWCS), York, September 19–22, 2010.
65. 3GPP TR 36.814, Further advancements for E-UTRA physical layer aspects, http://www.3gpp.org/DynaReport/36814.htm. Accessed May 2, 2014.
66. Castellanos, C. U., Villa, D. L., Rosa, C., Pedersen, K. I., Calabrese, F. D., Michaelsen, P., and Michel, J., Performance of uplink fractional power control in UTRAN LTE. IEEE Vehicular Technology Conference (VTC Spring 2008), Singapore, May 11–14, 2008.
67. Mogensen, P. et al., LTE capacity compared to the Shannon bound. IEEE Vehicular Technology Conference (VTC Spring 2008), Dublin, April 22–25, 2007.
68. 3GPP R4-092042, Simulation assumptions and parameters for FDD HeNB RF requirements. 3GPP WG4, Meeting 51, 2009.
69. 3GPP TR 36.942, Evolved universal terrestrial radio access (E-UTRA); Radio frequency (RF) system scenarios, http://www.3gpp.org/DynaReport/36942.htm. Accessed May 2, 2014.
70. Lauridsen, M., Noël, L., Sørensen, T. B., and Mogensen, P., An empirical LTE smartphone power model with a view to energy efficiency evolution. *Intel Technology Journal*, 18(1), pp. 172–193, 2014.
71. Small Cell Forum (SCF), Release two—Enterprise, SCF Release 2, Doc. 102.02.01, 2013.
72. Boccardi, F., Heath, R. W., Lozano, A., Marzetta, T. L., and Popovski, P., Five disruptive technology directions for 5G, *IEEE Communications Magazine*, 52(2), pp. 74–80, 2014.

12

Dynamic Coexistence of Wireless Body Area Networks

Mohammad N. Deylami and Emil Jovanov

CONTENTS

12.1 Introduction

The widespread use of communication and information technologies is rapidly affecting conventional healthcare systems, which are designed to react to and manage illnesses. Recent developments in sensors, wearable computing, and ubiquitous communications have provided the medical experts and users with frameworks for gathering physiological data in real time over extended periods of time. Wearable sensor-based systems can transform the future of healthcare by enabling proactive personal health management and unobtrusive monitoring of a patient's health condition.

This evolution in healthcare systems has made possible the delivery of medical services at a distance, which is known as telemedicine [1]. Classical sensing devices such as Holter monitors are being transformed into miniature sensors that can be used for unobtrusive and continuous monitoring of physiological signs. A collection of such sensors can be organized in the form of a network around the user's body, which is known as a body sensor network (WSN) or body area network (BAN) [2–3].

With the advances in power-efficient wireless technologies and the maturity of WSNs, BANs can take advantage of wireless transmissions to make the usage of health-monitoring systems even more ubiquitous and unobtrusive. The resulting concept of a wireless body area network (WBAN) extends well beyond sensor connectivity as it can be a system-level approach to addressing issues related to biosensor design, interfacing, ultralow-power processing/communication, power scavenging, and data mining.

The first implementation of a WBAN for health monitoring as WPAN was proposed and implemented at the University of Alabama in Huntsville in 2000 by Jovanov et al. [2]. The term *WBAN* was first used by Van Damme et al. [4], who listed distinct features to differentiate WPANs and WBANs. Since then, the application of WBANs has inclined toward medical and healthcare applications and WBANs are becoming key elements in the infrastructure for patient-centered medical systems [5–7]. In addition to the application of WBANs for healthcare and athletic monitoring, workplace safety, consumer electronics, secure authentication, and safeguarding of uniformed personnel can also benefit from WBANs [7].

Ubiquitous, continuous, and pervasive monitoring of physiological signs without activity restriction and behavior modification is the primary motivation of WBAN implementation [6]. The notion of continuous monitoring is more emphasized by the fact that conventional monitoring approaches are generally limited to brief time periods and transient abnormalities cannot always be captured reliably (e.g., transient cardiac arrhythmias). Therefore, life-threatening disorders can go undetected if they occur infrequently. A collection of sensors can be integrated on a user's body by using wired channels, wireless channels, biochannels, or a combination of these. Wired sensor networks are typical for smart clothes applications, yet they are not applicable for ambient and implantable sensors, such as pacemakers. Design of low-power sensors and optimization of wireless communication channels have been active topics of research in this area [7–8].

Recent technological developments have enabled the design of power-efficient sensor miniaturization; yet a long-term and sustainable power supply remains a great challenge to WBANs. New trends in hardware design aim at decreasing processor power consumption up to 10–20 times and power-supply voltage to less than 500 mV [9]. In addition to the efforts directed to power-efficient designs, extensive studies have also been carried out in producing new materials and technologies for innovative sensors and continuous monitoring of vital signs. An example is the development of conductive fabric and weaving technologies to directly print patterned electrodes and circuits on fabric [10]. In addition, the availability of mobile devices with high-speed Internet connection and GPS has greatly extended the scope of WBANs.

Considering the increasing healthcare costs and the aging population [11–12], there is a need for cost-effective and noninvasive systems for monitoring the health or wellness status of users. WBANs can put forward new ways of enabling proactive personal health management and ubiquitous health monitoring. Continuous and nonintrusive monitoring of vital signs and keeping medical records updated in real time provide effective solutions to the challenges faced by the healthcare system. WBANs can introduce smart and affordable solutions for monitoring of diagnostic procedures, chronic conditions, recovery, and emergency events [13]. The launch of large commercial projects, such as MobiHealth [14], further accelerates the development of health-monitoring WBANs.

12.2 WBAN Architecture

WBANs need to incorporate low-power and miniature sensors, processors, and wireless transmitters for nonintrusive monitoring and long-term operation. Examples of the most common health-monitoring sensors are activity sensors (accelerometer and gyroscope),

ECG for monitoring heart activity, EEG for monitoring brain electrical activity, and EMG for monitoring muscle activity. Table 12.1 lists some of the most common health-monitoring sensors and their operational characteristics [1].

WBANs can be deployed as part of a multitier telemedicine system [8] as depicted in Figure 12.1. At the top of the hierarchy is a medical server (tier 3) that includes a network of interconnected services, medical personnel, and healthcare professionals. The medical server keeps electronic medical records of registered users in order to analyze the data patterns, recognize health conditions, detect emergency situations, and forward health-related instructions to the users. The medical server is connected to a personal server (tier 2) via the Internet by using a typical WLAN or WAN, such as the cell-phone network. The tasks of a personal server, which can be implemented on a mobile device, such as a smartphone, include the configuration and management of the WBAN, forwarding information to the medical server, and checking the health status of the user by processing the vital signs. Tier 1 is comprised of sensors that are capable of sensing, sampling, processing, and communicating the physiological signals with the personal server. The communication channel between the sensors and the personal server needs to be established using a power-efficient wireless technology such as IEEE 802.15.1 [15] or IEEE 802.15.4 [16].

12.2.1 Coexistence of WBANs

The shared nature of the wireless medium makes interference a key issue to be taken into account in the design of wireless systems. Since multiple technologies like Wi-Fi, Bluetooth, and ZigBee use the 2.4 GHz ISM bands; there may be multiple sources of interference for a

TABLE 12.1

Typical Health-Monitoring Sensors

Physiological Parameter	Sampling Rate (Hz)	Sample Size (bits)	Channels	Data Rate (kbps)
ECG	100–1000	12–24	1–3	1.2–72
EMG	125–1000	12–24	1–8	1.5–192
EEG	125–1000	12–24	1–8	1.5–192
Activity	25–100	12–24	1	0.3–2.4
Blood pressure	1–3	16–32	1	0.02–0.1
Temperature	<1 per minute	16–24	1	0.002

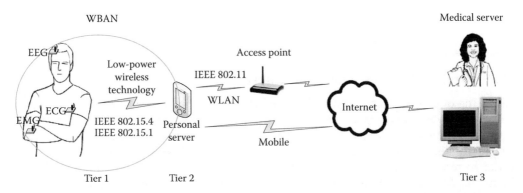

FIGURE 12.1
A multitier telemedicine system.

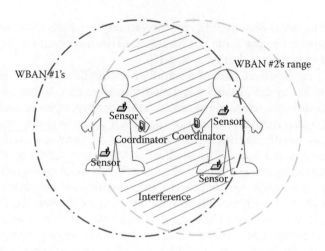

FIGURE 12.2
Coexistence of WBANs.

wireless transceiver at the same time. This problem is referred to as the spectral congestion problem, which includes homogeneous and nonhomogeneous interferences. Based on the different types of modulation and transmission powers, wireless transceivers may have different interference effects on each other.

The overlap of ranges between wireless networks is referred to as *coexistence*, which can be classified as *static* coexistence (for nonmobile network nodes) and *dynamic* coexistence (for mobile nodes) [17]. WBANs are as mobile as their users; therefore, they may dynamically coexist with a varying number of other WBANs. This is particularly important for medical environments, where multiple WBANs are likely to coexist in the same area, for example, when a group of people gather in the dining hall at an assisted living facility. Figure 12.2 depicts a case of dynamic coexistence.

At the time of coexistence, transmitted frames from different WBANs may collide, leading to lost messages, intermittent communication, and degradation of the quality of health monitoring. The criticality of health-monitoring data is a crucial factor in studying the effects of coexistence. For example, loss of communication in case of a heart attack may result in the death of a patient, in a dynamic coexistence situation.

12.3 Wireless Transmissions and Interference

Modeling of propagation and interference for radio waves can be an extremely complex task. Radio waves may travel from a source to a destination by multiple paths; therefore, multiple signals may arrive at a receiver with different powers. It may be impossible to exactly determine how signals reach a receiver and how they affect each other. In order to model interference in shared environments, various approximations have to be made. One may then estimate the range over which the communication link can be assumed reliable. We need to know that collision by no means has the same meaning as packet loss. The fact that two packets overlap in transmission time does not necessarily mean that they will be destroyed. The effect of collisions on transmissions is a function of numerous parameters, such as transmit power, type of modulation, and number of interference sources. For example, forward error correction (FEC) used in Bluetooth allows reliable operation in spite of some errors for a limited bit error rate (BER) in packets.

The maximum range of transmission determines how weak the signal gets before BER increases to an unacceptable level. When interference sources are present, this range is determined by the signal-to-noise ratio (SNR), defined as the ratio of signal power to the noise power, often expressed in decibels (dB) [18]. To find the effect of multiple transmissions on each other, we need to know the power of the received signals at a target receiver. We know that the ratio of the received signal power to the transmitted signal power is reversely proportional to the distance d between the transmitter and the receiver. The classic free-space link budget equation can be used to calculate the received signal powers [19]:

$$P_{r(dBm)} = P_{t(dBm)} + G_{t(dB)} + G_{r(dB)} + 20\log\left(\frac{\lambda}{4\pi}\right) + 10n\log\left(\frac{1}{d}\right),\tag{12.1}$$

where P_r is the received signal power, P_t is the transmitted power, G_t is the gain of the transmission (TX) antenna, and G_r is the gain of the receiving (RX) antenna. The carrier wavelength is given as λ and n is a factor for incorporating the effect of obstructions, called the *path-loss exponent*. In the simplified path-loss model (also known as the *distance-dependent model*), the antenna gains are assumed as 0 dB and the formula is changed to

$$PL = 20\log\left(\frac{4\pi}{\lambda}\right) + 10n\log(d),\tag{12.2}$$

where PL (path loss) is the difference between $P_{r(dBm)}$ and $P_{t(dBm)}$.

For several types of indoor transmission path analyses, IEEE uses a path-loss exponent of 2 for distances of less than 8 m and 3.3 for more than 8 m [19]. Figure 12.3 shows the extent of path loss on transmitted signals in the 2.4 GHz band. Note that increasing the transmit power of a transmitter by 3 dBm increases the indoor range of transmission roughly by a factor of 2 and that, in turn, increases the area under interference by a factor of 4 [20].

The IEEE 802.15 Task Group 2 (TG2) [21] has been established to study coexistence in WPANs. They have developed recommended practices on facilitating the coexistence of

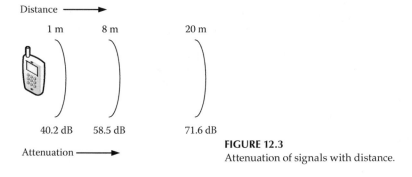

FIGURE 12.3
Attenuation of signals with distance.

WPANs and WLANs. They address the interference only between the IEEE 802.11b and IEEE 802.15.1 standards, yet the methods that they use for modeling interference can be used for modeling interference between other types of transceivers as well. To accurately model the effects of interference, the interaction between different types of modulation has to be taken into account. In order to have a complete interference model between multiple transceivers, we have to model the momentary state of the channel at each receiver. This necessitates knowing the exact location of the transmitters and receivers at every moment in order to have relative distances for calculating the received powers. Multipath effects in indoor transmissions can further complicate the interference modeling. We should also know that each transmitted packet may be affected by a varying number of interfering transmissions from different sources. Each of these transmissions may partially overlap with another; therefore, there are different levels of interference inflicted on each bit of a transmitted packet. As a result, we need the momentary value of the SNR in order to approximate the probability of a successful transmission for a single bit. These facts altogether leave us no choice but to consider simplifications in the modeling to make it feasible.

The first simplification that is often used in interference modeling is to approximate all types of interference as additive white Gaussian noise (AWGN) [19]. By using this simplification, standard equations that relate BER to SNR can be used. Figure 12.4 shows the steps needed to be taken in modeling interference in case that this simplification is made. The path loss has to be calculated for every interfering transmission, which is in reverse exponential relation to the distance. The spectral modification finds the fraction of the interfering signal that enters the receiver's frequency band. The power of the desired signal divided by the sum of the powers of the interfering signals will be the signal-to-interference ratio (SIR). The BER is then looked up from the standard tables that list the BERs of specific technologies based on the SIR.

The modeling can be further simplified if we assume that transmissions fail whenever packets collide. This is commonly done in studies of homogeneous interference, since the interfering signals are compliant and the probability of lost transmissions is higher. Assuming that collided packets are lost, there is no need for calculating BER anymore. Kim et al. [22] and Kim et al. [23] assume that the collided data packets and beacons are lost.

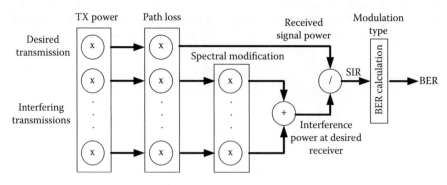

FIGURE 12.4
Steps in modeling interference.

12.4 Methods for Coexistence Management

Figure 12.5 presents the taxonomy of methods for coexistence management, where each category contains various methods. At the highest level, the problem may be classified as *homogeneous* or *heterogeneous* coexistence.

Multiple studies have focused on the mutual effect that different types of transmitters have on each other, which shall be referred to as *heterogeneous coexistence*. An example of this case is the coexistence of IEEE 802.11 and IEEE 802.15.4 networks. Multiple methods for mitigating the effects of this type of coexistence are reviewed by Yang et al. [24]. Knowing that the output power of IEEE 802.15.4 devices is typically as low as 0 dBm, whereas the output power of IEEE 802.11 devices is usually 15 dBm or above, IEEE 802.11 transmitters are considered as powerful sources of interference on IEEE 802.15.4 devices. The overlapping of the channels for these two technologies is shown in Figure 12.6. Channels 1, 6, and 11 are the default nonoverlapping channels for the operation of Wi-Fi devices in North America and that leaves us with only 4 nonoverlapped IEEE 802.15.4 channels from the 16 channels in the 2.4 GHz band.

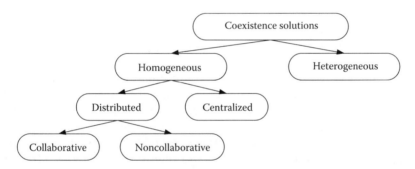

FIGURE 12.5
Taxonomy of the solutions for the coexistence problem.

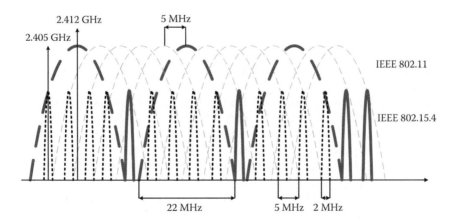

FIGURE 12.6
Overlap of channels between the IEEE 802.11b/g and IEEE 802.15.4 standards.

Many projects in the area of heterogeneous coexistence are based on experimental measurement of interference effects. It has been suggested that interference from IEEE 802.11 transceivers can result in substantial packet losses for IEEE 802.15.4 body area networks [25]. Using measurements of channel power, SIR, and packet loss ratio, it has been concluded that the effect of IEEE 802.15.4 on IEEE 802.11 transmissions is negligible, but up to 70% of IEEE 802.15.4–based transmissions may be lost due to interference from IEEE 802.11 [26]. Similar approaches have been studied by Guo et al. [27], Howitt and Shukla [28], Pollin et al. [29], and Shin et al. [30]. Some works have also focused on detecting interference from unknown sources and changing the working channel [31]. An analytical modeling of the effects of IEEE 802.11 on IEEE 802.15.4 in medical environments has been studied by De Francisco et al. [32], as well as a study on the cross channel interference issues in colocated IEEE 802.15.4–based industrial networks [33].

Another potential solution to the spectral congestion problem is the usage of cognitive radios, which operate based on the opportunistic usage of the frequency bands that are not heavily occupied by other transmissions. A radio is said to be cognitive if it is able to measure, sense, learn, and be aware of the parameters related to the radio channel characteristics, availability of spectrum and power, the operating environment, user requirements and applications, available network infrastructure and nodes, local policies, and other operating restrictions [34]. Cognitive radios are still in the proposal stage and have not been standardized.

Several wireless technologies take advantage of frequency hopping to counterinterference. The combination of frequency hopping and cognitive radios [15] results in adaptive frequency hopping [16], which has been proved to be an effective technique for further extending the ability of frequency hopping to counterinterference. WirelessHART [17] and ISA100 [18] are wireless communication standards specifically designed for control applications in industrial environments and both are based on IEEE 802.15.4. WirelessHART uses channel hopping and combines it with time-division multiple access (TDMA) to avoid interference and reduce multipath fading effects. ISA100 implements adaptive channel hopping to use the channels with minimum interference. Although frequency hopping may help with resolving interference, more effective methods are still needed for the detection and mitigation of dynamic coexistence.

Coexistence of multiple transceivers of the same type is another problem that has been extensively studied. This type of coexistence, which shall be called *homogeneous coexistence*, may be dealt by either *centralized* or *distributed* methods. In centralized methods, there is a central entity that regulates the usage of the time and frequency resources among the coexisting networks. An example of the centralized mechanisms is a resource allocator server, which receives requests from the WPAN coordinators and allocates transmission times to them [35].

In distributed methods, it is assumed that the coexisting networks are able to detect the coexistence situation and find a solution for it. They may do this in a *collaborative* or *noncollaborative* manner. In collaborative methods, networks cooperate to reach a regulation on their transmissions in order to minimize interference. The problem of assigning nonoverlapping time slots or channels for transmission can be approached based on graph-coloring algorithms. As an example, the coexisting WPANs may collaborate by exchanging information about their neighbors, in order to reach an arrangement, such that neighbors use different channels [36]. However, such algorithms are designed for static network topologies, where the high computational and communicational overheads can be justified. Interference can also be detected by observing packets from other networks, while interfering networks collaborate to determine which network should switch to a different channel [37]. Other methods have proposed finding and using multiple clean channels for transmissions in a single WBAN in order to reduce interference with other types of transceivers [38].

In noncollaborative methods, individual networks try to maximize their successful transmissions independently. Power control mechanisms are examples of this category, where each network tries to minimize interference by adjusting its transmission power according to the dynamic status of coexistence [39]. WPANs can also manage the coexistence situation based on an independent monitoring of the possible information that they can acquire from their coexisting networks. One solution that can be implemented as both a collaborative or a noncollaborative method is organizing the active and inactive periods of the coexisting WPANs such that the active period of each network overlaps with the inactive periods of the others. This method is well applicable in IEEE 802.15.4 WPANs that operate in the periodical active-inactive fashion.

12.5 Support for Coexistence in Different Wireless Standards

In this section, we review the built-in mechanisms in different standards that can help with mitigating the harmful effects of interference and coexistence.

12.5.1 The IEEE 802.11 (Wi-Fi)

Since IEEE 802.11 uses the carrier sense multiple access with collision avoidance (CSMA/CA) mechanism for transmissions, multiple coexisting BSSs can coexist in the same area with minimum collision. However, the throughput of the network is shared among them and transmissions may face long delays as a result of backoffs, as the number of coexisting BSSs increases. The fact that there are only three nonoverlapping channels for IEEE 802.11b/g that are used as default channels (channels 1, 6, and 11 in North America) increases the chance of interference. Channel agility is an optional method that lets access points switch channel when the current channel is under a high level of interference. However, it cannot be effective in coexistence between IEEE 802.15.1 and IEEE 802.11, because of the fast channel hopping of IEEE 802.15.1.

The level of transmit power is a key factor in the ability of wireless networks to coexist. The good-neighbor policy [19] indicates that the minimum transmit power for reliable communication should be used to minimize interference. Based on this, the IEEE 802.11 specification requires that transmit power control shall be implemented if a node is capable of transmitting with a power higher than 100 mW. Wi-Fi also supports several data rates that can be adapted based on the packet error rate (PER), signal-to-interference-plus-noise ratio (SINR), and RSSI. Lower bit rates result in higher energy per data bit and that would decrease the BER. If the interference is sporadic, faster data rates are more effective, since the packet transmission duration is shorter and that lowers the chance of collision.

12.5.2 The IEEE 802.15.1 (Bluetooth)

Bluetooth uses frequency hopping spread spectrum (FHSS) modulation at a nominal rate of 1600 hops/s. The usage of FHSS gives this technology a relatively high resilience to interference. Bluetooth is capable of mitigating interference by a number of ways using its basic operation and transmit power control. FEC, adaptive packet selection, and adaptive frequency hopping are the main methods. Bluetooth class 1 devices implement a feedback mechanism such that power control information can be exchanged between the master and slaves. The goal here is to set the transmit power such that the RSSI at the other end's receiver is optimal for reception. With adaptive packet selection the master can force the use of a specific type (size) of packets, such that a certain FEC can be provided.

The basic channel hopping sequence consists of 79 channels and adaptive channel hopping can use as few as 20 channels, since the Bluetooth specification requires the usage of at least 20 channels in the hopping sequence. Bluetooth devices are capable of categorizing channels as bad and good for use in the hopping sequence. Channel classification can be performed by all nodes in a piconet. The master of the piconet keeps a channel hopping map containing the collected information about the channels. The classification of channels into good and bad can be performed using multiple methods like RSSI, carrier sensing, and PER. Channel classification should be periodically updated to reflect the latest changes in the interference level of different channels.

12.5.3 The IEEE 802.15.3 (Ultrawideband or UWB)

UWB's close range of transmission (<10 m) and its low transmit power help to reduce interference with other systems. On the other hand, low transmit power means that BER can be high under interference. UWB devices are required to scan the candidate channels before they start a piconet; however, they are not required to be capable of detecting non-UWB signals. Devices are also able to scan for new channels when interference is detected using RSSI monitoring. A coordinator can ask a device in its piconet to scan for a free channel as well. Coordinators periodically monitor active channels for the presence of other networks and if it is detected they can change the operation of their piconet to another channel, reduce their transmit power, or even become a child piconet to a coexisting network for shared channel access [19]. The parent coordinator will allocate transmission time for the child piconet. There is also a neighbor piconet mechanism that allows two coexisting piconets to operate independently without one becoming another's child.

12.5.4 The IEEE 802.15.4 (ZigBee)

The capability of transmitting multiple packets at the same time is referred to as multiple access. With IEEE 802.15.4, direct sequence spread spectrum (DSSS) modulation can provide some degree of multiple access if simultaneous transmissions use orthogonal spreading sequences. Since the spreading sequences for ZigBee are not random and they are related to each other through cyclic shifts, IEEE 802.15.4 has basically no multiple-access capability.

Unlike Wi-Fi, Bluetooth, and UWB, ZigBee can operate in the 868 MHz European and 900 MHz ISM bands in North America. This might solve the problem of coexistence between ZigBee and other wireless technologies to a great extent. However, the data rates are lower than 250 kbps (40–100 kbps) in those bands and they are not available worldwide, which means that the 2.4 GHz ISM band is the main candidate. ZigBee devices are required to detect the channel being taken by other ZigBee devices by using clear channel assessment (CCA) and they may optionally support the detection of transmitted energy from other wireless transmissions. The combination of this mechanism and CSMA/CA would ease the interference problem. Channels occupied by Wi-Fi transmitters, which are one of the main sources of interference to ZigBee devices, may be blacklisted using scan mechanisms.

As stated previously, superframe arrangement can be an effective solution for the coexistence of IEEE 802.15.4–based WPANs. In a method based on this approach, it is assumed that an IEEE 802.15.4–based WPAN is trying to join a group of preexisting networks [22,40]. Based on the information carried in the beacons of the preexisting networks, the newly joining WPAN selects its superframe timings in a way that the overlap of the active periods is minimized. The first priority in this mechanism is avoiding the beacon collisions, since the transmission of the data frames depends on successful beacon transmissions. Then they try to find offsets between the transmissions of the beacons of the coexisting WBANs, in a way that the collisions caused by the overlap of the active parts of the superframes are minimized. This mechanism solves the problem of *static* (and not *dynamic*) coexistence, since the mobility of WPANs is not considered and superframes are arranged only when WPANs start their operation. Note that WBANs may coexist with a dynamic number of other WBANs as they move and enter each other's ranges. Therefore, finding a nonoverlapping period for the operation of a WBAN only when it is starting operation will not avoid collisions.

Multiple proposals have focused on the dynamic coexistence of IEEE 802.15.4–based medical WBANs. Examples of collaborative and noncollaborative methods in this area are collaborative superframe arrangement [41] and dynamic coexistence management [42]. In the first mechanism, the coexisting coordinators collaborate to find an optimal arrangement for their superframes. One of the coexisting coordinators finds this arrangement and the other coordinators adjust their superframes accordingly. In the second mechanism, each coordinator independently tries to adjust its superframe in relation to the superframes of the coexisting WPANs to minimize the overlap.

12.5.5 The IEEE 802.15.6

Resilience to interference is a key requirement in WBAN applications and the IEEE 802.15.6 standard attempts to maximize this resilience in a number of ways. When wireless channels become increasingly crowded, the interference from other networks may reduce the potential throughput and reliability. When the single-hop star topology is broken, the standard allows a single relay between the sensor and the coordinator. This can be useful when some of the single-hop links are unavailable because of attenuation or interference. Since transmission time is provided for the sensor nodes using beacons, successful beacon reception is crucial. Therefore, the IEEE 802.15.6 standard has a mechanism for shifting the beacon transmission time by an offset. The standard also provides a dynamic channel hopping mechanism that helps WBANs to avoid interference.

12.5.6 Proprietary Wireless Technologies

Most proprietary wireless technologies that are candidates for WBAN implementations use low-cost transceivers that target simplicity and low cost. Therefore, they do not include built-in mechanisms for interference mitigation.

12.6 Conclusion

WBANs have become the dominant implementation technology for ubiquitous health monitoring. The use of communication and information technologies is transforming healthcare systems and introduces a proactive approach and emphasis to wellness management, in addition to the management of diseases. One of the critical implementation issues is dynamic coexistence of WBANs. Dynamic coexistence is frequently ignored and may result in loss of critical health-monitoring data. The criticality of such information, in addition to the high probability of coexistence in healthcare environments, necessitates the development of mechanisms that enable the WBANs to detect and mitigate the effects of coexistence. In this chapter, we discussed WBAN architecture and applications and addressed the problem of dynamic coexistence for the most frequently used wireless technologies. Future studies may investigate the relation of coexistence and quality of service and propose methods for protecting the highly critical data in other types of wireless sensor networks from interference.

A more detailed and thorough study may take application specific patterns of traffic into account, such as dynamic coexistence in nursing homes and hospitals. Transmission of real data and traffic patterns should be considered in combination with dynamic coexistence, for an application specific optimal method.

References

1. D. Raskovic, A. Milenkovic, P. De Groen, and E. Jovanov, "From Telemedicine to Ubiquitous M-Health: The Evolution of e-Health Systems." *Biomedical Information Technology* (David Feng, ed.), Elsevier, Burlington, Massachusetts, 2007.
2. E. Jovanov, J. Price, D. Raskovic, K. Kavi, T. Martin, and R. Adhami, "Wireless Personal Area Networks in Telemedical Environment," *Proceedings. 2000 IEEE EMBS International Conference on Information Technology Applications in Biomedicine, 2000*, pp. 22–27, 2000.
3. A. Pantelopoulos and N. G. Bourbakis, "A Survey on Wearable Sensor-Based Systems for Health Monitoring and Prognosis," *IEEE Transactions on Systems, Man, and Cybernetics, Part C: Applications and Reviews*, vol. 40, no. 1, pp. 1–12, Jan. 2010.
4. K. Van Damme, S. Pitchers, and M. Barnard, "From PAN to BAN: Why Body Area Networks," presented at the Wireless World Res. Forum 2, Helsinki, Finland, 2001.
5. B. Latré, B. Braem, I. Moerman, C. Blondia, and P. Demeester, "A Survey on Wireless Body Area Networks," *Wireless Networks*, vol. 17, no. 1, pp. 1–18, 2011.
6. E. Jovanov, C. Poon, G. Yang, and Y. T. Zhang, "Guest Editorial Body Sensor Networks: From Theory to Emerging Applications," *IEEE Transactions on Information Technology in Biomedicine*, vol. 13, no. 6, pp. 859–863, Nov. 2009.
7. H. Cao, V. Leung, C. Chow, and H. Chan, "Enabling Technologies for Wireless Body Area Networks: A Survey and Outlook," *IEEE Communications Magazine*, vol. 47, no. 12, pp. 84–93, Dec. 2009.
8. C. Otto, A. Milenkovic, and E. Jovanov, "System Architecture of a Wireless Body Area Sensor Network for Ubiquitous Health Monitoring," *Journal of Mobile Multimedia*, vol. 1, pp. 307–326, 2006.
9. J. Kwong, Y. Ramadass, N. Verma, M. Koesler, K. Huber, H. Moormann, and A. Chandrakasan, "A 65nm Sub-Vt Microcontroller with Integrated SRAM and Switched-Capacitor DC-DC Converter," *Digest of Technical Papers. IEEE International Solid-State Circuits Conference, 2008. ISSCC 2008*, pp. 318–616, 2008.
10. R. Paradiso, G. Loriga, and N. Taccini, "A Wearable Health Care System Based on Knitted Integrated Sensors," *IEEE Transactions on Information Technology in Biomedicine*, vol. 9, no. 3, pp. 337–344, Sep. 2005.
11. M. Patel and J. Wang, "Applications, Challenges, and Prospective in Emerging Body Area Networking Technologies," *IEEE Wireless Communications*, vol. 17, no. 1, pp. 80–88, Feb. 2010.

12. S. Ullah, H. Higgins, B. Braem, B. Latré, and C. Blondia, "A Comprehensive Survey of Wireless Body Area Networks—On PHY, MAC, and Network Layers Solutions," *Journal of Medical Systems*, vol. 36, no. 3, pp. 1065–1094, 2012.

13. J. Xing and Y. Zhu, "A Survey on Body Area Network," presented at the Fifth International Conference on Wireless Communications, Networking and Mobile Computing (WiCom '09), pp. 1–4, 2009.

14. A. Van Halteren, R. Bults, K. Wac, N. Dokovsky, G. Koprinkov, I. Widya, D. Konstantas, V. Jones, and R. Herzog, "Wireless Body Area Networks for Healthcare: TheMobiHealth Project," *Wearable eHealth Systems for Personalised Health Management: State of the Art and Future Challenges, The Netherlands: IOS Press, Amsterdam*, vol. 108, pp. 23–34, 2004.

15. "IEEE 802.15 WPAN Task Group 1 (TG1)." [Online]. Available: http://www.ieee802.org/15/pub/TG1.html.

16. "IEEE 802.15 WPAN Task Group 4 (TG4)." [Online]. Available: http://www.ieee802.org/15/pub/TG4.html.

17. M. N. Deylami and E. Jovanov, "A Distributed Scheme to Manage the Dynamic Coexistence of IEEE 802.15.4–Based Health-Monitoring WBANs," *IEEE Journal of Biomedical and Health Informatics*, vol. 18, no. 1, pp. 327–334, Jan. 2014.

18. "Signal-to-Noise Ratio." [Online]. Available: http://en.wikipedia.org/wiki/Signal-to-noise_ratio.

19. R. Morrow, "Indoor RF Propagation and Diversity Techniques," in *Wireless Network Coexistence*, McGraw-Hill, 2004.

20. N. Golmie, "Interference Performance Evaluation." *Coexistence in Wireless Networks*, Cambridge University Press, Cambridge, United Kingdom, pp. 30–42, 2006.

21. "IEEE 802.15 WPAN Task Group 2 (TG2)." [Online]. Available: http://www.ieee802.org/15/pub/TG2.html.

22. T. H. Kim, J. Y. Ha, and S. Choi, "Improving Spectral and Temporal Efficiency of Collocated IEEE 802.15.4 LR-WPANs," *IEEE Transactions on Mobile Computing*, vol. 8, no. 12, pp. 1596–1609, 2009.

23. J. W. Kim, J. Kim, and D. Eom, "Multi-dimensional Channel Management Scheme to Avoid Beacon Collision in LR-WPAN," *IEEE Transactions on Consumer Electronics*, vol. 54, no. 2, pp. 396–404, 2008.

24. D. Yang, Y. Xu, and M. Gidlund, "Wireless Coexistence between IEEE 802.11– and IEEE 802.15.4–Based Networks: A Survey," *International Journal of Distributed Sensor Networks*, vol. 12, no. 2, pp. 301–318, 2011.

25. J. Hauer, V. Handziski, and A. Wolisz, "Experimental Study of the Impact of WLAN Interference on IEEE 802.15.4 Body Area Networks," presented at the Sixth European Conference on Wireless Sensor Networks (EWSN '09), 2009.

26. L. Angrisani, M. Bertocco, D. Fortin, and A. Sona, "Experimental Study of Coexistence Issues between IEEE 802.11b and IEEE 802.15.4 Wireless Networks," *IEEE Transactions on Instrumentation and Measurement*, vol. 57, no. 8, pp. 1514–1523, Aug. 2008.

27. W. Guo, W. M. Healy, and M. C. Zhou, "An Experimental Study of Interference Impacts on ZigBee-Based Wireless Communication inside Buildings," presented at the International Conference on Mechatronics and Automation (ICMA), 2010, pp. 1982–1987.

28. I. Howitt and A. Shukla, "Coexistence Empirical Study and Analytical Model for Low-Rate WPAN and IEEE 802.11b," *Wireless Communications and Networking Conference, 2008. WCNC 2008. IEEE*, pp. 900–905, Mar. 2008.

29. S. Pollin, M. Ergen, M. Timmers, A. Dejonghe, L. Van der Perre, F. Catthoor, I. Moerman, and A. Bahai, "Distributed Cognitive Coexistence of 802.15.4 with 802.11," *First International Conference on Cognitive Radio Oriented Wireless Networks and Communications*, pp. 1–5, 2006.

30. S. Y. Shin, H. S. Park, S. Choi, and W. H. Kwon, "Packet Error Rate Analysis of ZigBee under WLAN and Bluetooth Interferences," *IEEE Transactions on Wireless Communications*, vol. 6, no. 8, pp. 2825–2830, Aug. 2007.

31. M. Kang, J. Chong, H. Hyun, S. Kim, B. Jung, and D. Sung, "Adaptive Interference-Aware Multi-channel Clustering Algorithm in a ZigBee Network in the Presence of WLAN Interference," presented at the Second International Symposium on Wireless Pervasive Computing (ISWPC '07), 2007.

32. R. De Francisco, L. Huang, and G. Dolmans, "Coexistence of WBAN and WLAN in Medical Environments," *IEEE 70th Vehicular Technology Conference Fall (VTC 2009-Fall)*, pp. 1–5, 2009.

33. L. Lo Bello and E. Toscano, "Coexistence Issues of Multiple Co-located IEEE 802.15.4/ ZigBee Networks Running on Adjacent Radio Channels in Industrial Environments," *IEEE Transactions on Industrial Informatics*, vol. 5, no. 2, pp. 157–167, May 2009.

34. T. Yucek and H. Arslan, "A Survey of Spectrum Sensing Algorithms for Cognitive Radio Applications," *IEEE Communications Surveys & Tutorials*, vol. 11, no. 1, pp. 116–130, First Quarter 2009.

35. P. Ferrari, A. Flammini, D. Marioli, E. Sisinni, and A. Taroni, "Synchronized Wireless Sensor Networks for Coexistence," presented at the IEEE International Conference on Emerging Technologies and Factory Automation (ETFA), pp. 656–663, 2008.

36. K. Chowdhury, N. Nandiraju, P. Chanda, D. Agrawal, and Q. Zeng, "Channel Allocation and Medium Access Control for Wireless Sensor Networks," *Elsevier Ad Hoc Networks*, vol. 7, pp. 307–321, 2009.

37. R. C. Shah and L. Nachman, "Interference Detection and Mitigation in IEEE 802.15.4 Networks," presented at the International Conference on Information Processing in Sensor Networks, IPSN '08, pp. 553–554, 2008.

38. K. Hwang, S. Yeo, and J. Park, "Adaptive Multi-channel Utilization Scheme for Coexistence of IEEE802.15.4 LR-WPAN with Other Interfering Systems," *11th IEEE International Conference on High Performance Computing and Communications (HPCC '09)*, pp. 297–304, 2009.

39. G. Fang, E. Dutkiewicz, K. Yu, R. Vesilo, and Y. Yu, "Distributed Inter-network Interference Coordination for Wireless Body Area Networks," in *Global Telecommunications Conference (GLOBECOM 2010)*, pp. 1–5, 2010.

40. T. H. Kim, J. Y. Ha, S. Choi, and W. H. Kwon, "Virtual Channel Management for Densely Deployed IEEE 802.15.4 LR-WPANs," Fourth Annual IEEE International Conference on Pervasive Computing and Communications (PerCom), Pisa, Italy, pp. 103–115, 2006.

41. M. Deylami and E. Jovanov, "A Distributed and Collaborative Scheme for Mitigating Coexistence in IEEE 802.15.4 Based WBANs," presented at the 50th Annual Southeast Regional Conference (ACM-SE '12), 2012.

42. M. Deylami and E. Jovanov, "An Implementation of a Distributed Scheme for Managing the Dynamic Coexistence of Wireless Body Area Networks," *Southeastcon, 2013 Proceedings of IEEE*, pp. 1–6, 2013.

13

Mobile Health and Smartphone Platforms: A Case Study

M. B. Srinivas

CONTENTS

13.1 Introduction

The wider availability of mobile phones beginning late 1990s and their transformation into powerful smartphones over the past few years led to a large number of applications being developed in diverse areas such as agriculture, transportation, and e-governance, to name a few. Healthcare is one such area which has the potential to benefit from deploying smartphone platforms for both diagnostics and consultation. In this chapter, we focus on smartphone platforms and their role in providing mobile healthcare.

13.2 Mobile Health

Closely connected with mobile health is the term *telehealth*, which can be roughly defined as a healthcare system in which telephones (not necessarily a mobile) are used as a means of communication between doctors and patients. The communication can be verbal, visual, or sharing of content pertaining to patients' health records like medical reports, X-ray and ultrasound images, and MRIs. While the public switched telephone network (PSTN) was popular earlier, the current ubiquitous presence of mobile networks and the related infrastructure has made the use of mobile phones the center of telehealthcare.

In a typical mobile healthcare system shown in Figure 13.1, mobile phone, mobile network, remote-health kiosks, and primary/secondary/tertiary healthcare centers form the backbone of the entire system. While mobile phone and mobile networks provide necessary communications between hospitals and the kiosks, trained health workers at the kiosk are interacting with the patients (in person) and the physicians (remotely through mobile phone) in the hospitals for diagnosis and advice. Diagnostic devices themselves can now be designed around smartphones and, thus, can be made cheaper, portable, and to have the ability to communicate. Further, clinical parameters and diagnosis related to a given patient can be stored in EMRs or pushed on to a cloud (with appropriate data security) for later access that is independent of space and time. Thus a mobile-based healthcare system is expected to be truly pervasive.

13.2.1 Smartphone Platforms

One definition of *smartphone* is that "it is a mobile phone that is able to perform many of the functions of a computer, typically having a relatively large screen and an operating system (OS) capable of running general-purpose applications."

More recent models of smartphones are equipped with processors such as multicore central processing units (CPUs), digital signal processors (DSPs), and graphics processing units (GPUs), all on a single system-on-chip (SoC), which significantly enhance their computing/multimedia capabilities. For example, Apple's iPhone 5s has a custom SOC, Apple A6, which consists of a dual-core ARMv7-based CPU and a triple-core PowerVR SGX543MP3 GPU [1–2]. Samsung's Galaxy S4 has an eight-core Exynos 5410 processor made up of four ARM Cortex-A7 cores and four ARM Cortex-A15 cores with 2 GB of random-access memory (RAM) and tricore PowerVR SGX544 GPU [3–4]. Nvidia's recently released Tegra K1 SoC consists of 5 cores of ARM Cortex-A15 processor, 192 GPU cores, up to 8 GB of physical memory, and support for Nvidia Compute Unified Device Architecture (CUDA) [5]. Google's next version of the Nexus tablet, Nexus 9, may have Tegra K1 CPU [6]. The Snapdragon processor from Qualcomm is another powerful mobile processor, having a quad-core CPU, DSP, a GPU that can run at 550 MHz, and hardware accelerators, that can be found in several other models of smartphones. With this kind of raw processing

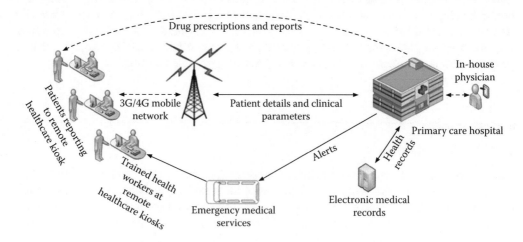

FIGURE 13.1
Architecture of modern telehealthcare system.

power and the availability of software tools that enable application development on these smartphone platforms, it is no wonder that even computationally intensive applications like ultrasound imaging are being developed to run on smartphones [7–8].

In addition to their raw computational power, all of the abovementioned smartphones are also equipped with sensors and sensor interfaces, such as digital camera, accelerometers, proximity sensors, and gyroscopes; GPS for location and navigation; and communication capabilities, such as Bluetooth, NFC, and Wi-Fi, that make them much more than a phone—practically a platform!

13.3 Mobile Operating Systems and Application Development

The choice of mobile operating system for smartphones is limited to four major players, namely, iOS from Apple, Android from Google, Windows from Microsoft, and BlackBerry OS from Research in Motion (RIM). According to a publication on "comparison of mobile operating systems" [9], with open access and more than 700,000 third-party applications, Android has a major share in mobile OS market, which is approximately 84%. Second is iOS, with about 12% market share. Windows and BlackBerry are in third and fourth positions, with 3% and 1% market shares, respectively [9]. In the most recent years, devices and applications in eHealth and mHealth have been more Android-based through the use of smartphones and tablets, whereas Windows-based applications are showing a trend of increase.

Android is the only OS which supports x86 architecture in addition to ARM architecture, while the rest support only ARM architecture (Windows RT, which is an OS for tablet, supports both ARM and x86 architectures). When it comes to languages used for building applications, iOS applications (apps) are developed using C, C++, and Objective-C. Similarly, C, C++, and Java for Android and C#, Visual Basic .NET (VB.NET), C, and C++ for Windows are the languages used. BlackBerry supports various languages such as C++/Qt, the fifth revision of hypertext markup language (HTML5)/JavaScript/cascading style sheet (CSS), ActionScript, and Java. One has to remember, however, that in all the abovementioned cases, depending on the choice of coding language, the tool set used for development also changes. There is no single tool set/development kit that supports all coding languages for that particular OS.

In mobile health applications, some of the fundamental features required of a smartphone are the ability to manage locally stored data through a file manager, support for multiple users, external storage support, and sharing of files through wireless means, over either Wi-Fi or Bluetooth. While Android supports all the abovementioned features (multiuser only on a tablet), iOS does not support any of the features (paid third-party application provides the user the file manager capability). When it comes to Windows, it lacks multiuser and file sharing over Wi-Fi, while BlackBerry provides all features mentioned except the multiuser [10].

An important functionality is the "USB on-the-go" or USB host, which enables applications to communicate with a peripheral device over USB. It also enables a mobile phone/tablet to act as a power source for the peripheral devices. Except for the Windows operating system, all other three operating systems support this functionality. In the case of Android, however, this support is provided only by OS 3.2 version and above.

It should be remembered that it also costs to develop an application on a smartphone. For example, Apple charges a fee of $99/year while others do not charge, but for publishing the app on the official store Google charges a one-time fee of $25.

13.3.1 Development of a Prototype Mobile Health Application

Mobile health (or mHealth) applications play a vital role as they provide the interface between a trained health worker and a hospital with telehealthcare facility. Thus, the application should be able to deal with various unforeseen situations like loss of data connectivity, low storage space, and loss of captured data. The application should also be able to collect various clinical parameters with different formats such as text, audio, and image files. The user interface should be capable of handling these formats without making the application look clumsy.

A basic mHealth application should mimic a doctor's prescription pad, in the sense that it will be able to collect patients' personal details like name, age, gender, and contact details as well as a prescription for his/her illness for the present visit. In a more advanced application, it should be able to communicate with external hardware to collect additional information necessary for diagnosis such as the temperature, heart rate, blood pressure, and ECG.

13.3.1.1 The Hardware

An in-house device is developed to measure the ECG, the blood pressure, and the heart rate of patients. In this device, a three-lead system records the electrocardiogram, pulse transit time (PTT) measurement records the blood pressure, and photoplethysmography (PPG) records the heart rate. A Cypress Programmable System-on-Chip 3 (PSoC3) is used as the main processing platform. The ECG signal is amplified by an instrumentation amplifier, passed through a passive high-pass filter, and buffered by an operational amplifier before being converted to digital form by using an analog-to-digital converter (ADC). The ADC is configured at 300 samples per second with a data width of 19 bits [11].

A noninvasive finger clip sensor is used for the PPG. This finger clip sensor measures blood volume through optical pulses. The passive-filtered PPG signal is fed into PSoC3; the input signal is buffered through the operational amplifier and digitized at a rate of 100 samples per second with a data width of 16 bits [11].

The PPG can also be used to calculate the heart rate of the patient. In this case, the analog PPG signals are fed to a comparator and the output of the comparator is used for driving

FIGURE 13.2
Application page for capturing measured parameters.

the interrupt of the timer block of PSoC3. The time periods between the pulses is calculated to determine the heart rate of the patient [11].

The samples of ECG and estimates of blood pressure/heart rate are transmitted to a smartphone/tablet which communicates with PSoC3 through a USB. The USB of PSoC3 is programmed to operate in a vendor-specific mode and the bulk transfer protocol of USB is used in communications [11].

Figure 13.2 shows an mHealth application developed on an Android tablet (works equally well on a smartphone) that is capable of communicating with the hardware device that measures pulse rate, blood pressure, and ECG as explained above. The user (health worker) will touch respective buttons on the touch screen for capturing and storing the parameters measured.

13.3.1.2 The Software

An application tried in here collects patients' personal information and the related diagnostic data. The data are stored in a remote server/cloud using wireless internet connectivity. A trained technician enters the registered patient ID if it exists; else, new patients' IDs, names, ages, genders, and addresses are registered by authorized personnel. A picture of each patient is also stored. Once a valid ID is entered, the details of the patient appear in the Android-based system, which also provides options to collect diagnostic data, view historical records, and make an audio/video call to the patient's doctor. The diagnostic data are collected by clicking on the appropriate soft buttons allocated for each parameter. The digitized ECG samples from PSoC3 hardware are reconstructed to form a graphical representation and stored as JPEG images. The blood pressure and heart-rate data are stored as string values. The historical diagnostic records of the patient are displayed in a chronologically ordered list when the View Historical Records option is selected. An option to view the historical and cumulative graphs of blood pressure and heart rate is also provided.

13.4 Results and Implementation

Figure 13.3 shows an ECG signal captured on a tablet. It is a digital ECG record; that is, an analog ECG signal has been captured from the hardware device mentioned above, digitized at 300 samples per second, and reconstructed on a digital ECG graph paper. It can be stored in local memory of the tablet, communicated to a remote doctor, or pushed on to a secure cloud for later reference.

The prototype application also has a feature of videoconferencing that enables the patient or the health worker to communicate with remote physician in case of need. Figure 13.4 shows a typical videoconferencing session between a doctor and a patient.

It is also possible to take advantage of smartphone's capabilities to process biomedical signals and use them to monitor vital signs referred to above. Figure 13.5a and b shows ECG and PPG signals captured, processed, and displayed on an inexpensive Windows Phone while Figure 13.5c shows the data related to pulse rate and oxygen saturation SpO_2. There are also other devices designed around a smartphone that make similar measurements and beyond [12–14].

Figures 13.6 and 13.7 show applications developed on an Android tablet for helping digitize rural health data. It is often important to collect and record data related to rural

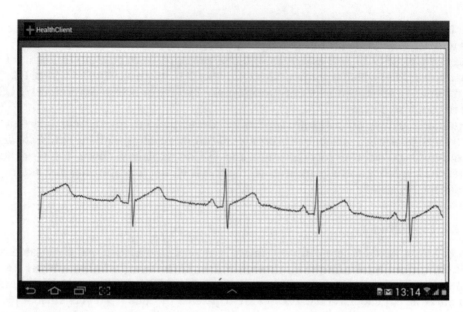

FIGURE 13.3
ECG signal being displayed by the application.

FIGURE 13.4
Mobile video call in progress.

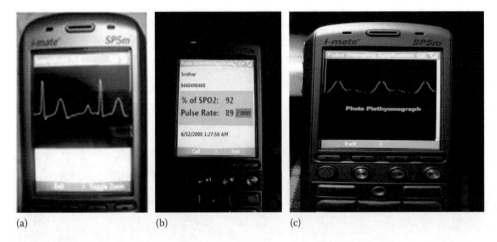

FIGURE 13.5
Windows Phone application showing (a) ECG, (b) photoplethysmograph, and (c) SpO_2 and pulse rate.

FIGURE 13.6
Application for selecting different categories.

FIGURE 13.7
Application page for entering necessary details for a selected category.

communities in developing countries for purposes such as maternal healthcare monitoring. While the current practice is to record such data on sheets of paper and then computerize these, such a practice may lead to errors while recording and reentering the data on a computer. The following applications demonstrate how data recording can be digitized and transmitted to remote servers for storage by using the mobile phone and mobile network. Since the data collection often involves monitoring a large number of parameters under different categories, the application interface can be programmed to change according to the category while keeping static data the same.

13.5 Conclusions

The influence of smartphones on healthcare is becoming significant with increasing sophistication of mobile technologies coupled with development of novel, affordable, and portable diagnostic devices. The day may not be too far away when the smartphone itself will become an all-in-one diagnostic device. The promise of high-speed and reliable data communication through mobile networks has the potential to make telehealthcare a more realistic proposition. With affordable, reliable, and secure technologies being developed, it can be safely said that very soon, mobile telehealthcare will be a part of mainstream healthcare system around the world.

Acknowledgment

The author gratefully acknowledges significant support provided by Microsoft Research, Redmond, for a major part of this work.

References

1. "iPhone 5 teardown," iFixit [Online]. Available: https://www.ifixit.com/Teardown/iPhone +5+Teardown/10525, accessed on September 3, 2014.
2. "Apple A6 teardown," iFixit [Online]. Available: https://www.ifixit.com/Teardown/Apple-A6 -Teardown/10528/, accessed on September 3, 2014.
3. "Samsung Galaxy S4 teardown: Board & chip shots and teardown images," TechInsights [Online]. Available: http://www.techinsights.com/inside-samsung-galaxy-s4/, accessed on September 3, 2014.
4. "Samsung Galaxy S4," Wikipedia [Online]. Available: http://en.wikipedia.org/wiki /Samsung_Galaxy_S4, accessed on September 3, 2014.
5. "Tegra K1," Nvidia [Online]. Available: http://www.nvidia.com/object/tegra-k1-processor .html, accessed on September 3, 2014.
6. R. Whitwam, "New Nexus 9 leak points to 64-bit Nvidia Tegra K1, 4 GB of RAM, and 2560 × 1600 screen," ExtremeTech [Online]. Available: http://www.extremetech.com/mobile/187752-new -nexus-9-leak-points-to-64-bit-nvidia-tegra-k1-4gb-of-ram-and-2560x1600-screen, accessed on September 3, 2014.
7. T. Fitzpatrick, "Ultrasound imaging now possible with a smartphone" [Online]. Available: http://news.wustl.edu/news/Pages/13928.aspx, accessed on September 3, 2014.
8. "Smartphone ultrasound: The MobiUS SP1 system," Mobisante [Online]. Available: http:// www.mobisante.com/products/product-overview/, accessed on September 3, 2014.
9. "Comparison of mobile operating systems" [Online]. Available: http://en.wikipedia.org/wiki /Comparison_of_mobile_operating_systems, accessed on September 3, 2014.
10. R. Hattersley, "What's the best mobile OS: iOS, Android, Windows Phone 8 or BlackBerry 10?," PC Advisor, May 1, 2013 [Online]. Available: http://www.pcadvisor.co.uk/buying-advice /mobile-phone/3445056/whats-best-mobile-os/, accessed on September 3, 2014.
11. S. Vaidya, M. B. Srinivas, P. Himabindu, and D. Jumaxanova, "A smartphone/tablet based mobile healthcare system for developing countries," *35th Annual International Conference of the IEEE Engineering in Medicine and Biology Society (EMBC), 2013*, Osaka, July 3–7, 2013.
12. D. Albert, B. R. Satchwell, and K. N. Barnett, "Wireless, ultrasonic personal health monitoring system." United States of America Patent US8301232 B2, October 30, 2012.
13. H. E. Reinhold Jr., S. Leibovitz, and R. Kazaz, "Electrocardiographic monitoring system and method." United States of America Patent 14/085165, March 20, 2014.
14. G. Lin, T. Nakajima, P. Rahul, and A. Hodge, "Seamlessly embedded heart rate monitor." United States of America Patent US8615290 B2, December 24, 2013.

Section III

Sensors, Devices, Implantables, and Signal Processing

14

Medical Sensors for Mobile Communication Devices

Jacob Fraden

CONTENTS

14.1 Introduction

A mobile communication device (MCD), such as a smartphone or a tablet, nowadays becomes a bionic extension of ourselves. No longer is a telephone just for remote (*tele*) transmission of sound (*phone*)—it has evolved into our personal cybervalet that can perform a multitude of services from composing and sending a letter to booking a restaurant to playing music. For doing any useful job, an MCD needs to receive information from the outside of its own shell by using a number of built-in sensors and transducers, some of which are used for the human–MCD interface, that is, for receiving the user commands (keypad, microphone, and accelerometer) and providing feedback. Other sensors are required for perceiving properties of the environment. Today MCDs contain a rather limited number of sensors, yet they support thousands of apps, some of which may be useful for medical purposes. MCD is further evolving from being a cybervalet to a personal cyberdoctor.

All too often an MCD is used as a mere attachment to an external medical device and its functions are limited to displaying data, human interface, and some computing and communication. In most cases, this does not make much sense as the display and human interface components are not expensive and are readily available and should be part of the medical device itself. A well-designed medical device really needs no MCD extension.

The real purpose of a medical MCD is to become an in-pocket personal health watchdog and provide an ability for a quick, easy, and accurate medical data collection and processing, something like a *tricorder* from the *Star Trek* science fiction series. It should never (at least in our lifetime) be a substitute for a professional-grade diagnostic device used by medical professionals. Take, for example, an MCD with dry medical electrodes built into a protective case (www.alivecor.com) that can collect a simple one-lead ECG signal between two points of the patient body (e.g., chest and left or right hand). One ECG lead is fine for measuring a few basic vital signs, such as a heart rate or arrhythmia, or for doing a quick ECG check. Yet, under no circumstances is it a replacement for a standard 12-lead ECG recording that is needed for a serious cardiac diagnostic.

A portable medical MCD by its very nature and definition has a clearly limited field of applications. Its main purpose is to work in the hands of a lay user and serve the lay user's health needs. It should not be developed for professional diagnostic use but rather adapted primarily for consumers with little or no medical training. This approach puts extra demands on automation of data collection and signal processing.

Medical uses of an MCD consist of two intimately connected major areas: (1) data collection (sensors) and (2) data storage and processing (internal or via cloud). Here, we address the first area—sensing the health-related signals. Some sensors are already part of a generic MCD and can be employed for certain medical purposes. A list of these general-purpose sensors is rather short:

1. Imaging camera (still photo and video).
2. Microphone: detects sound mostly in the audible frequency range.
3. Accelerometer: detects motion of the MCD and direction of gravity force.
4. Gyroscope: detects spatial orientation of the device.
5. Magnetometer (compass): detects strength and direction of magnetic fields.
6. GPS: an RF receiver and processor for identifying global coordinates.
7. Proximity detector: for indicating closeness of the MCD to the user's body.

All these sensors support thousands of apps and only a handful of the medical apps. Their uses are limited to detecting mostly images in the visible spectrum range and the mechanical and magnetic properties of objects. Other physical and chemical functions, as well as vital signs, cannot be detected by those sensors.

But what must one do if the currently embedded sensors cannot perform a particular job that requires medical signals? The answer is just too obvious—install specialized sensors. Then the questions are which, how, and where to install.

A medically tailored MCD should have capabilities for detecting a broad spectrum of signals including vital signs and environmental factors. These signals can be used for diagnostic purposes, as well as for behavior modification and tracking physical activity in real time, to detect symptoms of acute or chronic diseases, such as strokes, cardiac emergencies, depression, and attention deficit hyperactivity disorder (ADHD). Figure 14.1 shows a nonexhaustive list of the sensors and detectors that are desirable for use with the medical MCDs. Some sensors in the list at this time are just wishful thinking, but others are real possibilities if not a reality. Many of the listed devices either already exist or are being developed for use in a specialized medical device, both stationary and portable. The problem is that such sensors in their present configurations are not suitable for direct use with an MCD. To become practical, they need significant modifications in many respects: size, speed of response, easy of use, power consumption, etc.

14.2 Requirements for MCD Sensors

Perhaps the most important feature an MCD sensor should have in general, and for medical applications in particular, is a full integration with other supporting components, including, among others, data processing and communication circuits. The latter is for

1. Noncontact body core thermometer
2. Thermal imaging camera
3. Ultraviolet (UV) radiation detector
4. Detector of ionizing radiation
5. Sensor of electromagnetic pollution in various frequency ranges
6. Analyzer of gases (e-noses): carbon monoxide, radon, poison gases, etc.
7. Detector of bacteria and viruses
8. Arterial blood pressure or pressure trend
9. Blood oximeter
10. Blood factor (cholesterol, glucose, etc.) detectors
11. Pregnancy detector
12. Ovulation detector/predictor
13. Breathalyzer (alcohol, hydrogen sulfide, etc.)
14. Food composition detector (pH, fat, protein and carbohydrates, freshness)
15. Electrophysiological sensors (ECG, body impedance, etc.)
16. Heart rate, arrhythmia, and R wave–to–R wave (R-R) interval monitor
17. Fluid composition analyzer (saliva, urine, tears)
18. Radar for subcutaneous imaging (to replace X-ray, ultrasound)
19. Behavioral detectors (body orientation and motions)

FIGURE 14.1
List of medical sensors desirable for a mobile device.

interfacing with the MCD internal processor to communicate the sensed signals in forms suitable for interpretation. Thus, the most important requirement is that the sensor should be more than just a sensor—it has to be an integrated self-contained sensing module. The sensing module should be a miniature instrument that detects, conditions, digitizes, processes, and outputs information. Other important requirements for MCD sensing modules include low power consumption, small size/weight, high accuracy, stability, short response time, and several others.

Figure 14.2 summarizes 10 essential requirements for an MCD sensor. To emphasize their importance, we may call them "10 commandments" of the mobile sensor design. Each and every requirement is critical and should not be ignored. If any one of the commandments is not met, such a sensor may not be fully suitable for a mobile application.

1. Built-in signal conditioner and DSP (intelligent sensor)
2. Built-in communication circuit (I^2C, NFC, Bluetooth, etc.)
3. Integrated supporting components (optics, thermostat, blower, etc.)
4. High selectivity of the sensed signal
5. Fast speed of response
6. Miniature size
7. Low power consumption
8. High stability in changing environment
9. Time stability: no periodic recalibration
10. Low cost at sufficiently high volumes

FIGURE 14.2
10 commandments of mobile sensor design.

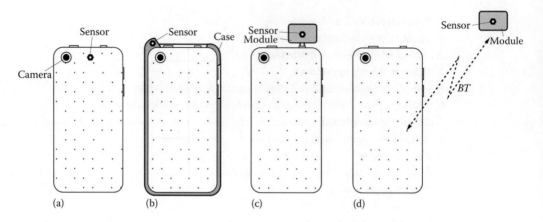

FIGURE 14.3
Options for positioning additional sensing modules with respect to the MCD. Sensor embedded into MCD: (a) sensor embedded into a protective case, (b) plug-in sensor, (c) remote sensor, and (d) Bluetooth (BT).

There are four possible ways of coupling a sensing module to an MCD. Figure 14.3 shows that the module can be embedded directly into the MCD housing (a); incorporated into a protective case (jacket) that envelops an MCD (b); also, it can be configured as an external sensing module plugged into one of the communication ports of an MCD (c); and finally the module can be totally external to and MCD (d) and communicate with it wirelessly. All these possible options are workable from the engineering standpoint, but from a convenience and practicality perspective, it appears that option B is the most attractive for coupling with a generic MCD that is used by a consumer. Positioning sensors inside a protective case allows hiding them ergonomically and inconspicuously and making the sensors instantly available whenever needed without additional actions by an operator. The "sensing case" can be easily upgraded, replaced, and adapted with no effect on the MCD. In fact, a protective sensing "smart" case is the most efficient way of augmenting a smartphone or other MCD with additional sensing capabilities. It communicates with the MCD either by wires or preferably wirelessly, for example, through NFC or Bluetooth.

14.3 Integration of Sensors

Integration of a specific detector into a sensing module is based on the reality that a sensing element rarely operates by itself—it needs numerous supporting components, such as voltage or current references, signal conditioners, heaters, multiplexers, gas blowers, lenses, and processors. To illustrate the point, Figure 14.4 shows a module with an integrated noncontact infrared (IR) body-core thermometer [1] coupled to an MCD. This IR thermometer can be directly incorporated into MCD housing option (a), or be part of a smart jacket, option (b).

The IR sensing module operates in concert with the MCD's internal digital imaging camera that functions as a viewfinder for the IR lens. In more advanced designs, the MCD camera can be much more than just a viewfinder—it can be part of an image recognition system for identifying a correct spot on the subject/object surface for temperature measurement and actuating such a measurement when the right spot is detected within the field of view of the IR lens.

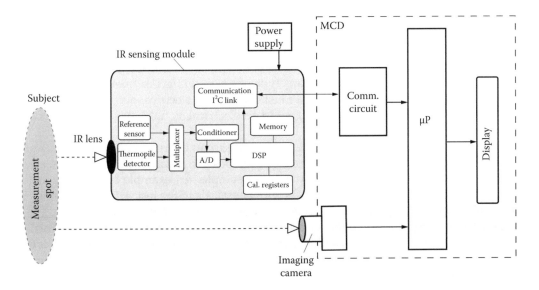

FIGURE 14.4
Block diagram of a noncontact infrared thermometer module for MCD (μP is a microprocessor).

Both the imaging and IR lenses are aimed at the subject's preferred measurement location (the spot). For a human patient, it is a spot on a right or left temple. Thermal radiation that is naturally emanated from the skin surface is focused by the narrow-angle IR lens on the surface of one or more of the thermal radiation detectors, such as a thermopile or microbolometer. The detecting element converts the radiation into minute electrical voltages that, along with the reference temperature sensor's output, are multiplexed and fed into the signal conditioner. The conditioner filters and brings the measured signals to the levels suitable for conversion to a digital format by an ADC, typically of a 15-bit resolution. The digitalized signals are processed by the DSP to compute first the skin surface temperature and, subsequently—the body inner core temperature. The processing algorithms are stored in the module memory. To enhance accuracy, certain calibrating constants are stored in the calibration registers. And finally, the computed body temperature passes to the communication circuit that converts the information to a format for the most economical and fast coupling to the MCD's internal processor. As it can be appreciated, the described sensing module is the entire noncontact IR thermometer that requires no or very little external components for interfacing with the MCD. After receiving temperature data, the MCD makes use of them; for example, the patient body temperature can be displayed, stored, plotted, sent to a remote site, or used as part of a more elaborate medical diagnostic, such a determining the metabolic activity of the patient. All sensors intended for an MCD should follow a similar approach—a complete integration into a single small housing.

14.4 Chemical and Bacteriological Sensing

While many supporting hardware components, like ADCs, DSPs, communication circuits, and software, are more or less nonspecific for a sensed signal, a sensing element is the key component for detection of a specific stimulus (analyte). Identifying and quantifying a chemical composition and detecting presence of germs or viruses are the most difficult areas of sensing. The key characteristic of such a chemical/biological sensing element is *selectivity*—an ability to reliably distinguish one type of a molecule, cell, or virus from another. Since a sensor is a converter of any type of a stimulus (action, material, energy, etc.) to electrical signal [2], a sensing element must generate an electrical output. Conversion of a chemically generated nonelectrical signal to an electrical signal is one of the most challenging engineering issues.

A material sample for detecting by sensing module can be in two phases: gaseous and liquid. The latter frequently requires use of a disposable component that carries a test reactive material for interacting with the liquid sample. An example is a blood glucose monitor, where the blood sample interacts with chemicals in the test strip and modifies its electrical or optical properties that can be measured by an electronic circuit.

The gaseous phase requires use of the olfactory (smell) sensors for detecting various gases, malodors, alcohol vapors, and, hopefully, certain components in the exhalation that may manifest onset of diseases, such as cancer. This can be done by trained dogs [3], so why not by a man-made sensor? A sensor for gases is called an *e-nose* (electronic nose).

A promising approach to a selective bacterial detection employs the affinity-based "lock-and-key" elements with the bacteria phages [4]. Phages are relatively inexpensive and quite environmentally stable and can be genetically modified. Phages can be used in amperometric (current-generating) circuits to produce electrical output when a matching bacterium is detected. At the time of this writing, the phage sensors require bacteria immobilization, which cannot be achieved in a portable mobile device.

To detect gaseous molecules several types of e-noses can be employed, among which the most popular are chemoresistors and chemical field-effect transistors (ChemFETs). A chemically selecting element in both can employ either oxides of various metals or chemical polymers (CPs). Figure 14.5 shows the operating principles of a chemoresistor and a ChemFET. In both elements, the sensing layer exhibits a conductivity property that is being modulated by the chemical molecules absorbed from the passing gas.

As the CP interacts with gaseous molecules, it can act as either an electron donor or an electron acceptor. If a p-type CP donates electrons to the gas its hole conductivity increases. Conversely, when the same CP acts as an electron acceptor its conductivity decreases. Disadvantages of a response originating in the bulk of the CP are the relatively long time constant (tens of seconds to minutes), often accompanied by hysteresis. These effects are caused by slow penetration of gases into the CP. This can be to some degree improved by

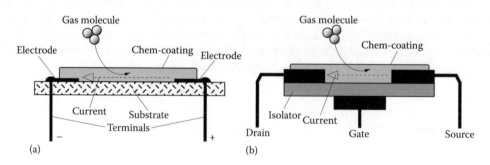

FIGURE 14.5
Chemoresistor (a) and ChemFET (b) rely on modulation of electrical conductivity of the sensing layer.

use of miniature blowers and MEMS structures and further improved during the signal processing as illustrated in the following.

Another type of CP works in a different principle—swilling of the chemically selective coating while absorbing specific analyte vapors. These sensors usually do not require heating to a high temperature; however, since the response is temperature dependent, a constant temperature of the CP should be maintained. This may be achieved by warming up the sensing element above ambient temperature (40°C, e.g.) while during measurement maintaining that temperature at a stable level. In other words, the CP should be augmented with a thermostat. In the manufacturing process, the CP film is impregnated with microscopic conductive particles [5] or carbon black powder (Figure 14.6a). Initially, a reference current is established by the monitoring circuit in the absence of the odor of interest. When the specific odor is present in the air (Figure 14.6b) and absorbed by the coating, the bulk volume of the CP coating swells somewhat, causing an increase in the average distance between the conductive particles, which, in turn, leads to an increase in electrical resistance between the terminals. Subsequently, resistance is converted to an electrical signal for processing.

Because swelling of the polymer begins immediately after exposure to the vapor, the resistance signals can be read in real time or near real time. Currently, with sensing films having thickness on the order of 1 μm, the swelling (and, therefore, resistance) response times to equilibrium film swelling values range from 0.1 to 100 s, depending on the vapor and the polymer through which the vapor must diffuse. More rapid responses to equilibrium could be simply obtained through reduction in the film thickness. Since the diffusion time is proportional to the square of the film thickness, decreasing the film thickness to the range of 0.1 μm should provide a practically quick response. At small swellings, the film returns fully to its initial unswollen state after the vapor source is removed, and the film resistance on each array element returns back to its original value. The sensitivity of the conducting polymer composite–based electronic nose compares highly favorably to other vapor-detection systems.

Let us examine two examples of the chemoresistors that are suitable for MCD applications. The first is the alcohol detector. There are several methods of detecting blood alcohol, wherein a direct blood test is the most accurate, albeit less convenient and slow. Screening for blood alcohol is used to determine whether an individual's blood alcohol content (BAC) is below or above a certain threshold value. In most practical cases, instead of sampling blood, screening is done by a breathalyzer to establish the breath alcohol content (BrAC). Conversion factors have been established to convert BrAC values into BAC values. The

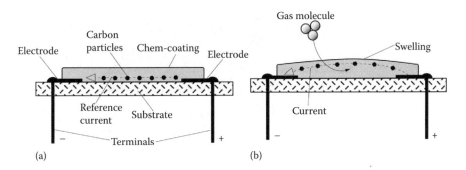

(a) (b)

FIGURE 14.6
Swell-type chemoresistor: (a) reference state and (b) detecting state.

most commonly accepted conversion factor is 2100 [6]. The alcohol-sniffing e-nose can be integrated into an MCD with not much difficulty. To detect alcohol in breath, sample gas (breath) is injected into the sensing module of the breathalizer. Of several possible ways of detecting breath alcohol, three possibilities are practical for use with the MCDs:

1. *Fuel-cell* sensor devices are based on electrochemical reactions in which alcohol in the gas phase is oxidized on a catalytic electrode surface to generate a quantitative electrical response.
2. *Semiconductor* sensing elements utilize small heated (300°C) beads of a transition metal oxide, across which a voltage is applied to produce a small standing current. The current magnitude depends on the conductivity of the surface of the bead. Since the conductivity is affected by the amount of alcohol molecules adsorbed, it can be utilized as a measure for the alcohol concentration in the gas sample.
3. *Infrared* absorption devices for breath sampling operate on the principle of infrared light being absorbed by alcohol molecules. The amount of light absorbed by the gas sample flowing though the sample cell can be taken as a measure of the alcohol content.

Another example of a chemoresistor for employing with the MCD's e-nose is a hydrogen sulfide (H_2S) detector. Hydrogen sulfide is a toxic gas which may be present in many environments and has a distinct smell of a rotten egg or decaying waste. In particular, it is responsible for a mouth malodor (halitosis). A level of H_2S gas at or above 100 parts per million (ppm) is immediately dangerous to life and health. The H_2S sensors require high sensitivity to fairly low levels of the gas and must also be able to discriminate H_2S from other gases that may be present and not give spurious readings affected by such other gases.

Just as in many breathalizers, to operate, the H_2S sensor's surface has to be heated well above ambient temperature to about 300°C. This puts a significant strain on the sensing module power supply and also prolongs the response time of the sensor. To reduce both, a modern MEMS technology may be employed. Figure 14.7 shows the concept of the MEMS H_2S sensor, where a silicon structure is formed with a substrate supporting a thin membrane having thickness on the order of 1 µm. Such a membrane has low thermal capacity and, thus, can be warmed up to high temperature in a short time by a relatively low power.

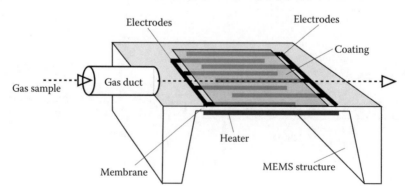

FIGURE 14.7
Concept of the H_2S MEMS variable resistance sensor.

Two interdigitized (alternating) electrodes are formed on the membrane's upper surface. A selective coating is deposited by a sputtering and oxidizing technique on the top of the electrodes. The coating has a finite resistance that varies in relation with concentration of the H_2S molecules in the gas sample. To form a sensing film, firstly a layer of a molybdenum sulfide (MoS_2) having thickness no greater than 1000 Å is sputtered over the silicon membrane and the electrodes and subsequently a layer of tungsten oxide (WO_x) of the same thickness is sputtered on. The resulting double coating is then heated in air for several hours at a temperature of around 500°C. This has both a sintering and an oxidizing effect and generates a complex combination of metal oxides [7]. This coating combination does not consist merely of separate oxides of tungsten and molybdenum, but instead an ordered structure is formed which is an inseparable combination of the oxides and is, in effect, a type of crystal structure having both types of oxide contained within the same crystal lattice. The bottom of the silicon membrane is given a heater layer that during operation brings up the membrane temperature close to 300°C. When in use, after the heater temperature is settled at a constant level, a sample of the outside gas is drawn over the membrane either by blowing into the gas duct or sucking up by a miniature blower that augments the sensing module. The gas reacts with the coating, whose electrical resistance changes accordingly to a concentration of the H_2S molecules. The resistance can be easily measured by one of the conventional methods and related to the gas concentration. This sensor responds in a few seconds and, what is also very important, has a fast clearing time, that is a quick readiness for the next measurement.

For the efficient use by an MCD, olfactory sensors of different types should be combined into a multisensor array where each element is tailored for detecting a particular odorant. Thus, the array will cover a broad spectrum of smells. Since selectivity of each individual sensing cell in the array is far from being ideal, there is always a noticeable cross talk between them. That is, each particular odorant would cause responses in several cells, albeit the levels of responses will be different. Each odorant would develop a distinct pattern in the array that can be analyzed and processed. This to some degree is analogous to noses of animals, where odor identification is a result of the pattern recognition by neural networks. Even without resorting to a neural network, a redundant olfactory array can do a pretty good job. For example, to detect freshness of fruits an array of 32 polymer sensors were employed [8]. Odors from pears of different state of ripeness were detected and reliably classified into three classes dependent on their physiological states through distinctive odor pattern formed by 32 outputs of the e-nose array.

Since the e-nose sensing cells in an array are relatively slow to respond, produce intrinsic noise, and have relatively low selectivity, the bionic methods of signal processing become more popular. They employ the adaptive and learning (trainable) neural-network software in the DSP and can yield quite impressive results by responding to dynamics of the changing outputs of the sensing array [9]. This neural processing has three advantages: faster speed response, improved signal-to-noise ratio, and better selectivity. Figure 14.8 shows an array of multiple CP sensors that are exposed to an odorant [10]. The dynamical approach utilizes time transients of the sensor responses without waiting for their settling at constant levels. Each sensor in the array is predominantly sensitive to a specific odorant and coupled to one or more inputs of the neural network and responds to the rates of the signal changes and their magnitudes. Noisy and slow sensor responses are refined by the neural system and formed as a multiple coupling of the excitatory (white circles) and inhibitory (black circles) dynamical units. Such processing improves accuracy of detection and speeds up the analyte recognition.

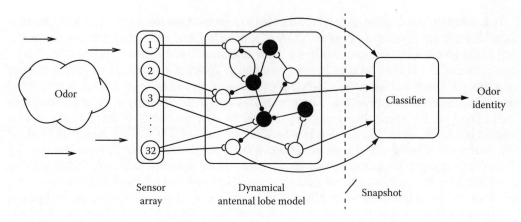

FIGURE 14.8
Dynamical processing of the e-nose signals by a neural network. (From Muezzinoglu, M. K. et al., *Neural Computation*, Vol. 21, pp. 1018–1037, 2009. With permission.)

14.5 Blood Pressure Monitoring

Classical methods of arterial pressure monitoring are based on detecting mechanical vibrations of the arterial walls when an external balancing pressure is in the range between the systolic and diastolic blood pressures. The methods are known as Korotkoff (detects sounds) and oscillometric (detects low-frequency oscillations of the arterial walls). Since they require application of an external pressure to the subcutaneous artery, an encompassing cuff and air pump have to be employed—a hardly convenient arrangement for use with an MCD. Any such external attachment is bulky, inconvenient, and thus not practical for carry-on portable devices. Alternative methods do not involve application of external pressures but rather rely on various physiological manifestations of the mean arterial pressure (MAP). Thus, they are called *indirect methods* because they do not measure pressure but rather infer it from other factors. Since all such manifestations are unique for a particular patient, these indirect methods need periodic recalibration and adaptation to the particular patient.

To understand how the arterial pressure can be monitored by indirect methods, blood flow rate and vascular resistance should be considered. Blood flow is defined as the quantity of blood passing a given point in the circulation in a given period and is normally expressed in milliliters/minute. The overall blood flow in the total circulation of an adult is about 5000 mL/min. Since the vascular system obeys Darcy's law, blood flow v can be expressed through the blood pressure differential ΔP and the vascular resistance Z across the system (similar to Ohm's law in electricity):

$$v = \frac{\Delta P}{Z}. \tag{14.1}$$

Therefore, the vascular resistance and flow define the blood pressure. This suggests two methods of inferring the arterial pressure by estimating value of a vascular resistance. Since the heart pumps blood in a pulsatile manner, the arterial pressure is not constant but with each heart stroke varies between the peak (systolic) and valley (diastolic) pressures.

The pressure wave propagates through the vascular system with the rate dependent on elasticity and size of the system that define the vascular resistance Z. The farther from the heart, the longer it takes for the pressure wave to propagate there. Thus, several indirect methods are possible of which two are the most promising:

1. *Pulse transit time* (PTT) method is based on measuring time for the arterial pulse pressure wave to travel from the aortic valve to a peripheral site (e.g., a tip of the finger). It is usually measured as time from the R wave on the electrocardiogram to a photo-plethysmographic (PPG) signal [11–13] representing the pressure wave. PTT is inversely proportional to blood pressure. Figure 14.9a shows the PTT obtained from the ECG and PPG. The longer the PTT, the smaller the Z, and the MAP smaller accordingly (considering v being more or less constant). Figure 14.10 shows that relationship. Note that the curve is patient dependent and initially

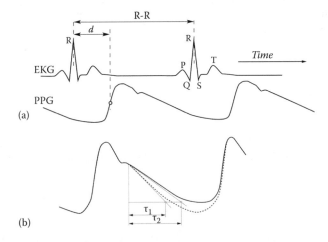

FIGURE 14.9
Analysis of the arterial pressure wave: (a) pulse transit time and (b) pulse relaxation time.

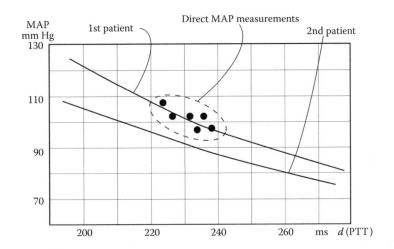

FIGURE 14.10
Experimental determination of the MAP–PTT relationship.

should be individually determined (best-fit line) by comparing the PTT readings with data as measured by a conventional arterial pressure monitor.

2. *Pulse relaxation time* (PRT) is based on the relaxation time constant of a plethysmographic signal—as measured by either a rheographic or an optical (PPG) monitoring. The relaxation is measured as a time constant τ of the near-exponential slope between the systolic and diastolic values of the pressure wave. The PRT is nearly proportional to Z; that is, the shorter the relaxation-time constant, the lower the vascular resistance Z and, subsequently, the lower MAP. Figure 14.9b shows the relaxation-time constants obtained from the PPG signal ($\tau_1 < \tau_2$). As indicated previously, this also should be individually periodically recalibrated for each particular patient by comparing with the arterial pressure obtained by a traditional method (with a pressure cuff).

The PTT method may be realized in a sensing jacket for a smartphone [14] that has embedded two dry ECG electrodes and a PPG detector (Figure 14.11). To take a measurement, the user holds the smartphone in one hand, making a palm contact the side ECG electrode while touching the circular ECG electrode with a finger of the other hand. That finger also covers the PPG detector. This coupling with the patient obtains the ECG wave between two hands. The pressure wave is detected at the finger as a PPT signal. Since ECG signals propagate instantly (at the speed of light), while the pulse pressure wave moves from the aortic valve to the finger much more slowly, the time delay d from the R wave of ECG to the fastest point of the PPT (Figure 14.9a) can be measured by the signal processor, related to Z, interpreted as MAP and communicated to the smartphone. The PPG detector is a well-known device [15]; thus, we do not need to describe it here.

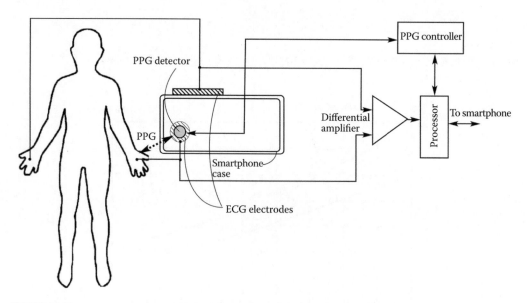

FIGURE 14.11
Block diagram of a blood pressure monitor utilizing a time delay of the peripheral arterial pressure pulse from the ECG R wave as a measure of the mean arterial pressure.

14.6 Energy Harvesting

As Figure 14.3 implies, options (a) and (c) may derive operating power for the sensing module directly from the MCD internal battery. However, options (b) and (d) must rely on other energy sources. The manually charging or recharging of batteries is always a nuisance and should be avoided whenever possible. Thus, it is beneficial to harvest energy for the sensing module automatically from sources that exist independently of the module and whose energy can be drawn and used by the sensing module inconspicuously and without any action by the user. Energy can be derived from electromagnetic fields, optically from a photovoltaic cells, from thermal sources using thermoelectric elements, and kinetically. Kinetic harvesting takes energy from the operator's natural movements by using a piezoelectric converter that is actuated by a moving spring-loaded mass. For example, walking produces somewhere in the region of 100–200 µW and intentional shaking of an MCD creates more than 3 mW.

If a sensing module is plugged into a phone jack of an MCD option (c), some useful amount of energy can be harvested from the speaker output when generating sound in the frequency range of up to 24 kHz—typically at 5 kHz. The efficiency of this energy harvesting may be achieved well over 90%, with the output voltage near 4 V and total available power around 15 mW [16]. This means that at a 4 V output, available current is up to 3.5 mA—quite sufficient for supplying many low-power sensing devices and supporting circuits.

If no electrical connection to the MCD is available, wireless power harvesting is the only available option. An attractive way is to use an NFC link that can harvest a small amount of energy for a peripheral device. Typically, NFC can supply up to 1 mA at 3 V, which may be sufficient for several low-power sensors. Power also can be derived from various external RF sources, such as wireless local area network (LAN) routers operating at a 5 GHz frequency sub-band, but the most efficient source is the emanated RF field from the MCD itself because of its very close proximity to the sensing module. Since by its very nature, MCD transmits electromagnetic radiation, part of that radiation may be captured by the sensing module, rectified, and utilized [17]. Peak RF power transmitted by an MCD is typically in the 100 mW range; thus, capturing a small percentage of that power, say

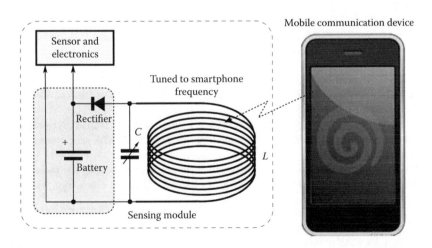

FIGURE 14.12
Simplified block diagram of harvesting power from RF transmitted by MCD.

2%–3%, can be sufficient for trickle charging a built-in secondary-type miniature battery (e.g., lithium ion). Depending on the type of the MCD, the RF frequency may be 800 MHz, 850 MHz, 1.9 GHz, and several others. Figure 14.12 shows a simplified block diagram of that type of the energy harvester. The resonant tank consisting of a coil having inductance L and capacitor C is tuned to the frequency that is transmitted by the MCD. Obviously, the MCD does not transmit all the time but it does so mostly during conversations or data transmission. This periodic RF signal is picked up by the tuning tank, rectified, and used to charge the battery or miniature supercapacitor. Thus, whenever the sensor and its supporting electronics need power, the battery can supply the required current. Every time the MCD transmits, the spent energy is replenished in the module battery.

References

1. Fraden, J. et al. Wireless communication device with integrated electromagnetic radiation sensors, U.S. Patent No. 8275413, 2012.
2. Fraden, J., *Handbook of Modern Sensors*. Fifth Ed. Springer-Verlag, 2015.
3. Willis, C. M. et al. Olfactory detection of human bladder cancer by dogs: Proof of principle study. *BMJ*, Vol. 329, September 25, 2004.
4. Van Dorst, B. et al. Recent advances in recognition elements of food and environmental biosensors: A review. *Biosensors and Bioelectronics*, Vol. 26, pp. 1178–1194, 2010.
5. Burl, M. C. et al. Assessing the ability to predict human percepts of odor quality from the detector responses of a conducting polymer composite-based electronic nose. *Sensors and Actuators B*, Vol. 72, pp. 149–159, 2001.
6. Jones, A. W., Precision, accuracy and relevance of breath alcohol measurements. *Modern Problems of Pharmacopsychiatry*, Vol. 11, pp. 65–78, 1976.
7. Jones, E. et al. Hydrogen sulfide sensor, U.S. Patent No. 4822465, 1989.
8. Oshita, S. et al. Discrimination of odors emanating from "La France" pear by semi-conducting polymer sensors. *Computers and Electronics in Agriculture*, Vol. 26, Issue 2, pp. 209–216, April 2000.
9. Rabinovich, M. et al. Transient dynamics for neural processing. *Science*, Vol. 321, pp. 48–50, July 4, 2008.
10. Muezzinoglu, M. K. et al. Chemosensor-driven artificial antennal lobe transient dynamics enable fast recognition and working memory. *Neural Computation*, Vol. 21, pp. 1018–1037, 2009.
11. Fraden, J., Smartphone case with a blood pressure monitor, U.S. Patent Application No. 61930473, 2014.
12. McCarthy, B. M. et al. An investigation of pulse transit time as a non-invasive blood pressure measurement method. *Sensors & Their Applications XVI. Journal of Physics: Conference Series*, Vol. 307, 012060, 2011.
13. Douniama, C. et al. Blood pressure estimation based on pulse transit time and compensation of vertical position. *Third Russian-Bavarian Conference on Bio-Medical Engineering*. Erlangen, Germany, 2007.
14. Gesche, H. et al. Continuous blood pressure measurement by using the pulse transit time: Comparison to a cuff-based method. *European Journal of Applied Physiology*, Vol. 112, Issue 1, pp. 309–315, 2011.
15. Allen, J., Photoplethysmography and its application in clinical physiological measurement. *Physiological Measurement*, Vol. 28, No. 3, 2007.
16. Kuo, Y.-S., Hijacking power and bandwidth from the mobile phone's audio interface. *ISLPED '10 Design Contest*, Austin, Texas, 2013.
17. Fraden, J., Protective case with sensing functions, U.S. Patent Application No. 14095016, 2013.

15

Development of Disposable Adhesive Wearable Human-Monitoring System

Alex Chun Kit Chan, Kohei Higuchi, and Kazusuke Maenaka

CONTENTS

15.1 Introduction

Over the past 60 years, life expectancy at birth has grown significantly in the world. According to the United Nations (UN), Department of Economic and Social Affairs [1], at the beginning of the 1950s, the average life expectancy of a newborn baby was 46.9 years. By 2010, average life expectancy had risen to 68.7 years, which has increased by more than 20 years. Moreover, life expectancy at birth was projected to increase to 75.9 years in 2045–2050 worldwide.

As shown in Figure 15.1, people are not only living much longer but also tend to have few children in the 21st century. From the beginning of the 1950s, a woman had 4.97 children, but this decreased to 2.5 children in 2010. In the past 60 years, the total fertility rate declined almost 50% globally. Moreover, in 2045–2050, it is expected to drop to 2.24 children per woman, which is close to the replacement level of 2.1 children per woman.

As a result of both decreasing fertility rates and successful improvement in life expectancy, the number of people aged 65 or older is growing faster than any other age group. As shown in Figure 15.2, which illustrates the population proportion of the world between 1950 and 2050, only 5.1% of the world population was aged 65 years or older in 1950s. However, this proportion had risen to 7.7% in 2010 and is projected to increase to 15.6% by 2050.

Undoubtedly, the increase of aging population is the remarkable achievement of medicine advancements and public health policies over the last few decades. However, it also becomes a challenge for the society to maintain their health, maximize their quality of life,

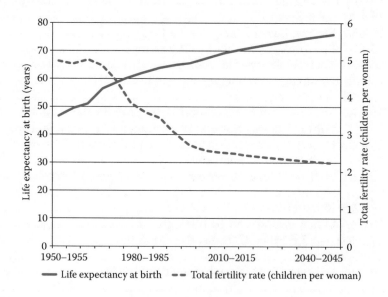

FIGURE 15.1
Life expectancy at birth and total fertility rate of the world in 1950–2050. (From United Nations, Department of Economic and Social Affairs, World Population Prospects: The 2012 Revision. Available at http://esa.un.org /unpd/wpp/index.htm, accessed on August 15, 2014.)

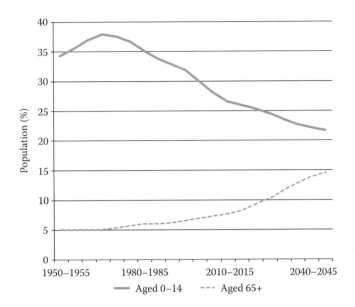

FIGURE 15.2
Population proportion of the world in 1950–2050. (From United Nations, Department of Economic and Social Affairs, World Population Prospects: The 2012 Revision. Available at http://esa.un.org/unpd/wpp/index.htm, accessed on August 15, 2014.)

and provide prolonged medical care for them. As shown in Figure 15.3, the proportion of people aged 65 or older in Europe had already climbed to more than 15% in 2010 and is projected to extend to more than 25% by 2050. Moreover, in most of the developed countries, the aging population is projected to increase to more than 20% too. The challenge is specifically more serious in Japan, which is famous for having the highest life expectancy in the world; the proportion of elderly people has already reached 23% in 2010 and is projected to grow up to 36.5% by 2050.

Besides the aging population, most of the people are having unhealthy lifestyles such as smoking, unhealthy diets, lack of exercise, and overwork in modern society. According to WHO Global Status Report 2010 [2], unhealthy lifestyles are the primarily cause of noncommunicable diseases (NCDs), comprising mainly cardiovascular diseases, cancers, diabetes, and chronic lung diseases. According to the report, NCDs are the largest global killers today, causing more deaths than all other causes combined. During 2008, 63% of deaths occurred in the world due to NCDs. Moreover, about one quarter of global NCD-related deaths take place before the age of 60. Thus, NCDs are another key challenge for health systems in the 21st century.

To overcome the low fertility rate, aging population, and NCD difficulties of Japan, Japan Science and Technology Agency (JST) supports the Exploratory Research for Advanced Technology (ERATO) project Maenaka Human-Sensing Fusion Project (Maenaka Project) with the collaboration of the Electronic Circuits Research Group of the University of Hyogo in 2008 [3]. The purpose of the project is focused not only on solving the issues of the welfare and healthcare systems but also on enhancing public safety and security. Professor

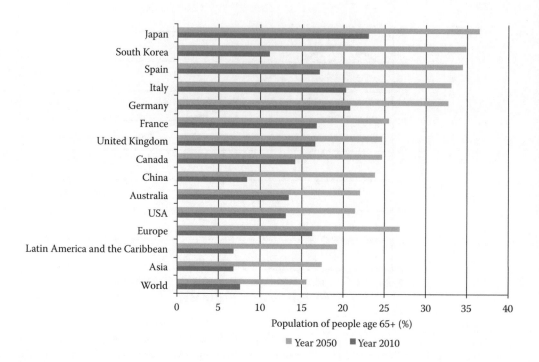

FIGURE 15.3
Proportion of people aged 65 or older in 2010 and 2050. (From United Nations, Department of Economic and Social Affairs, World Population Prospects: The 2012 Revision. Available at http://esa.un.org/unpd/wpp/index .htm, accessed on August 15, 2014.)

Kazusuke Maenaka, from the Department of Electrical Engineering and Computer Sciences of the University of Hyogo, lead the project team in integrating wireless communication unit, power-generation unit, MEMS technology, sensors, and control circuits for the development of an adhesive-type human activity-monitoring system [4–6] to realize the project goal supporting safety and healthy life for humans.

15.2 Human Activity-Monitoring System

With the support of a human activity-monitoring system, we can monitor and understand the health condition of elderly people or individuals in an emergency evacuation after a disaster. If the system detects a critical situation, it issues an appropriate alarm to their family, doctor, or hospital for providing additional healthcare attention to ensure their safety or even prevent accidents as well. Moreover, the system quantifies and performs objective analysis on the physical condition and mental stress of individuals to prevent illness or offer medical treatment in advance by detecting the disease's symptoms at an earlier stage. In addition, it supports health management and offers suitable lifestyle guidance and health improvement recommendations according to the analysis result. Thus, people can not only reduce or even eliminate unnecessary medical expenses but also reduce the burden on the public health system. Furthermore, the system could help officials or doctors visualize the population health, predict or prevent the spread of the infection of communicable disease, and suggest new public health policies by analyzing and observing the collected health data from a large number of people in a wide region.

The system is not only for elderly people but also can be used by children, teenagers, or healthy individuals to help themselves maintain their fitness to prevent NCDs and prevent mental illness from stress or overwork. The system can also be applied to specific areas such as kindergartens or nursing homes to prevent the spread of influenza, hazardous working environments to support health management and risk prediction of workers, gyms or sports centers to help training under optimal physical condition, and public transportation to understand the physical condition and mental situation of drivers to prevent the occurrence of serious accidents.

15.3 System Concept

The system concept of the adhesive-type human activity-monitoring system is shown in Figure 15.4. Numerous wireless base stations that have Internet connectivity are set up around living areas such as living rooms and offices; the adhesive-type sensing device is implemented to work in a short-distance range (5 to 10 m) wirelessly and, therefore, could greatly reduce the total power consumption of the device. When the sensing device is close to the base station, the sensed data will be transferred to the data server or cloud data center automatically through the base station and Internet connection. If the device cannot connect to the base station, the sensed data will be stored on the device memory temporarily and will be transferred to the data server once it resumes connection to the base station.

The data center should process and resolve the received data in real time to detect dangerous situations of the owners and deliver a proper alarm to their family, doctor, or hospital. Besides dangerous situations, the data center should also provide suitable feedback to the user; it should not only assist them in maintaining or improving their health condition and lifestyle but also advise them not to travel to hazardous environments excessively.

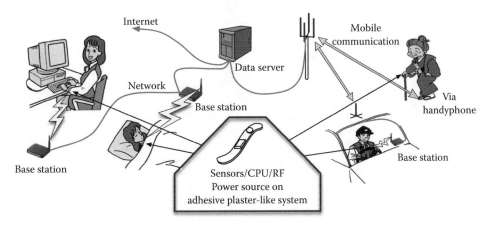

FIGURE 15.4
System concept of adhesive-type human activity-monitoring system. (From K. Maenaka et al., *IEEJ Transactions on Sensors and Micromachines*, Vol. 132, No. 12, pp. 443–450, 2012. With permission.)

Recently, most of the mobile units, such as smartphone and tablet, have high processing power with wireless and Internet connectivity; thus, the sensing device could directly communicate to the mobile unit so that it can perform real-time monitoring of the user's activity to manage their health and lifestyle and detect illness at earlier stages.

15.4 Concept of Adhesive-Type Wearable Sensing Device

To monitor human activities and health conditions continuously, the Maenaka Project proposes to develop an easy-to-use adhesive-/band-type wearable sensing device, ideally as low cost as the usual band aid that is disposable and which is integrated with multiple sensors, a control unit (complementary metal–oxide–semiconductor [CMOS]–large-scale integration [LSI]), a wireless communication unit, and an energy-harvesting power generation unit for continuous operation.

Figure 15.5 shows the conceptual diagram of the adhesive-/band-type wearable sensing device. The device should monitor multiple vital signals (electrocardiography, heart rate, body or skin temperature, blood pressure, etc.), body movement, and surrounding environmental information such as atmospheric temperature, pressure, and humidity. It is because, for example, even when just walking, our body responds differently in walking in a nice sunny day and during hot, high-humidity weather.

Moreover, the device should consume extremely low power and be small sized, lightweight, flexible, and fit enough for wearing on the human body (skin) without unpleasant

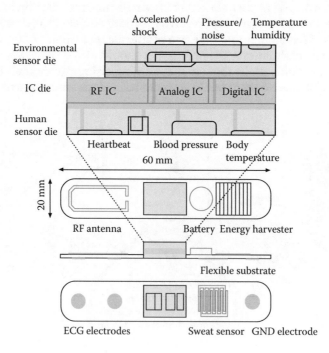

FIGURE 15.5
Conceptual diagram of adhesive-/band-type wearable sensing device.

pulling or unnecessary sensations. The user should feel comfortable or can even forget that they are wearing it on the body. And while the device is monitoring the user's activity, it should not interfere with their daily life such as sleeping, exercising, or even taking a bath [8–10]. Additionally, it should be able to operate for a long period of time and transmit the collected sensed data to the database/data server through the base station by using wireless technology. Furthermore, the device should avoid manual maintenance, including initial check, tuning, configuration, reprogramming, and deactivation. All the necessary configurations could be done via a wireless link automatically. The only maintenance needed may be installation, adding of the plug-in modules, or other operations without using specialized tools.

Unfortunately, using conventional electronics and MEMS sensor integrated circuits (ICs) combined with printed circuit board (PCB) technology falls short of the desired performance. Not only are the cost of commercially available ICs and the fabrication fee too expensive to produce these ICs as a disposable device, but also the ICs are too bulky, the interconnections are too complex, and power dissipation is one to two orders of magnitude too large for the energy-harvesting system. Therefore, the Maenaka Project worked to design a custom ASIC [11] with CMOS-LSI technology to reduce the overall device power consumption, which is lower than the threshold of energy generation, and this will make power harvesting feasible [12]. Furthermore, the custom ASIC is integrated using system-on-chip (SoC) and system-in-package (SiP) to bring down the entire device size and manufacturing cost.

15.4.1 Hybrid Implementation Model

The top view and side view of the hybrid implementation model of the adhesive-type wearable sensing device are shown in Figures 15.6 and 15.7, respectively. The reduction in power consumption is the primary goal of this custom ASIC; hence, the power dissipation of system interfaces should be reduced systematically, which requires a design with the sensors intimately integrated with the circuit chip. Therefore, the MEMS sensor chips are directly connected with the circuit chip by using the anisotropic conducting film (ACF) flip-flop bonding technology, which is currently the cheapest industrial method of creating 2.5 µm contact pitch in flip-chip assembly. Since ACF is the only flip-chip technology that does not use a mask and it can also be eliminated under a fill process, it cuts down the manufacturing cost significantly.

In addition, reducing the CMOS chip area is another option for cost reduction. This goal can be achieved by proper circuit design with the minimal number of the circuit blocks and allowing share data channels/internal buses between different blocks, which also share area spent for interconnects. For the SoC the most area-efficient, free architecture is the Wishbone bus. It allows eliminating about 8000 transistors of glue logic and 2 m of interconnection, reducing chip area to 0.3 mm^2 and, thus, reducing system price to 0.5% approximately. Additionally, to reduce the number of components and processing steps during chip assembly, the simplest die-attach technique is preferable.

Application-specific components were planned to be implemented as expansion modules SoC. Serial interface is obligatory for the expansion modules to reduce the amount of wiring. Since one-wire interface requires only one contact point per expansion module, it makes it superior to all other digital and analog interfaces when reducing number of connections is highly desired. But backward compatibility to the low-power I²C interface is also planned to accommodate a larger number of legacy modules. So each expansion module only has three signal wires and one ground wire, allowing packages with reliable large-area contacts.

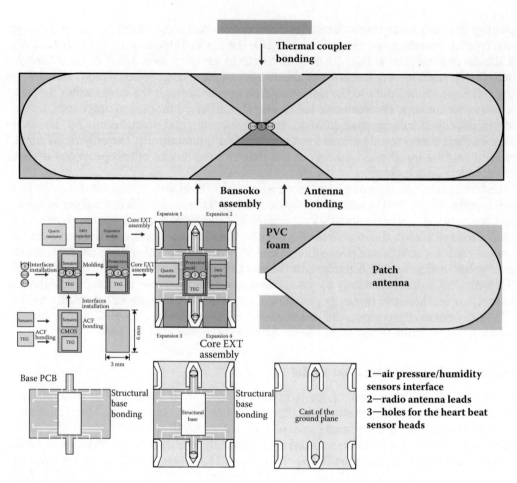

FIGURE 15.6
Top view of the hybrid implementation model.

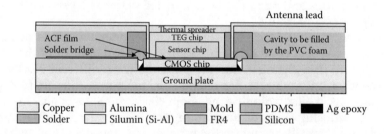

FIGURE 15.7
Side view (central section) of the hybrid implementation model.

Most of the miniaturized SoC and SiP suffer from severe electromagnetic compatibility (EMC) problems, resulting in the usage of the multiple power supplies. But each power supply requires its own decoupling capacitor. Because these capacitors do not permit on-chip implementation, each additional power supply adds one discrete capacitor with the associated two contacts to the system. To reduce the EMC problem extent, low-impedance and low-inductance ground path should be implemented.

15.4.2 CMOS System-on-Chip

Figure 15.8 shows a block diagram of the CMOS SoC; it can be roughly separated into four main blocks:

- Sensor interface and ADC
- Digital core and microcontroller unit (MCU)
- Interconnecting digital configuration bus-on-chip (Wishbone)
- RF interface (1.2 GHz radio transceiver)

15.4.2.1 Sensor Interface and ADC

The eight-channel analog multiplexer, current–voltage (C/V) converter, low-voltage band-gap reference, and the successive approximation register (SAR) ADC are designed and implemented in the sensor interface front-end. To facilitate low-voltage operation of the analog multiplexer (0.9–1.2 V range is targeted), a unique design of the fully digital gate-voltage booster is implemented. The analog multiplexer can work in the 1 V systems using low-leakage, 0.7 V threshold voltage transistors. The combination of the extremely low leakage power and low active power of the multiplexer makes it extremely suitable for the energy-harvesting system. According to measured data, the designed analog multiplexer with the capacitive load equivalent to the load of the C/V converter and the ADC (10 pF) can pass signals with frequency of up to 1 kHz with minimal distortion. Further increase in the signal frequency decreases the signal-to-noise-plus-distortion ratio (SNDR) with a slope of approximately 10 dB/decade due the influence of the analog multiplexer.

The main target of the SAR ADC design is to reduce power consumption; thus, aggressive voltage scaling is performed and it works successfully at low power operation

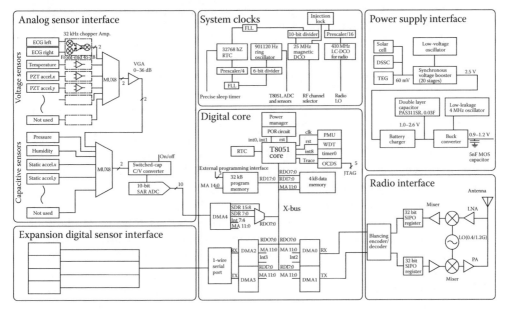

FIGURE 15.8
Block diagram of CMOS system-on-chip.

(0.1 pJ/conversion step). However, a few unresolved problems still remain, such as mismatches of the bit 4 of the switching capacitors array are too large (around 1 least significant bit [LSB]), the 1.2 V power supply voltage is higher than required (0.9 V), and the yield is too low (27% in the group of 15 chips). Therefore, design optimization is still necessary.

To operate a precision block like the ADC, voltage reference of equal precision is necessary. But classical bandgap reference is not suitable for the low-voltage SoC because of the required at least 1.6 V high power-supply voltage. Thus, the variation of the current-mode, low-voltage bandgap reference is designed and implemented. The weak point of the current-mode bandgap is the sensitivity of the output voltage to the process variation in the p-type metal–oxide–semiconductor (PMOS) parameters. The output may be still very stable around the operating point, but the initial position of the operating point is poorly known. So a digital trimming circuit is implemented in the PMOS current source to cancel the process variation.

15.4.2.2 Digital Core and MCU

Taking into account the very small computing power requirements, MEMS logic–based or reversible logic–based (e.g., split-level charge recovery logic [SCRL]) processors would be a lower-power solution. Unfortunately, the design of these processors is itself a tremendous task larger than all other components together, so it would be better to wait until reversible-logic technology becomes more mature. Currently, the T8051 core, 4 kB data memory, and power manager circuits are designed and implemented.

15.4.2.2.1 Processor Core—T8051

For the processor core, the T8051 IP core was selected. The main reason behind selection was the minimization of the gate count and, hence, static power dissipation. The T8051 is compatible on the instruction level with the legacy 8051C, so standard compilers can be used. The T8051 core was implemented with the specially designed, custom digital cell library, tailored for the low leakage power and small cell area. The very low static and dynamic power of the designed library was achieved primarily by the usage of the following techniques:

- Preference is given for the gates using stacked PMOS transistors (helps to decrease leakage power): Because PMOS can conduct less current compared to the same-size n-type metal–oxide–semiconductor (NMOS), and the maximal operating frequency is reduced. Furthermore, all gates become severely skewed, so additional buffers are introduced in the design to fix hold violations and reduce the energy spent on the glitches. This technique is useful only at low operating frequencies, because glitch-induced power dissipation is proportional to the operating frequency.

- Restriction of the usage of the inverters in the composite gates: Among digital gates, inverters not only have a disproportionally large leakage power but also a disproportionally large maximal fan-out. For the nets where large fan-out is required, separate inverters will be implemented. But because these nets are relatively rare, reduction of the overdrive of the other nets results in the decrease in the leakage power.

- Strict minimization of the gate area and diffusion area: It means usage of the minimal-size transistors for NMOS and PMOS part of the gate, merging of the power-supply terminals, usage of dummy transistors to isolate otherwise merged unipolar nodes, and unification of the gate sizes to simplify pairwise merging of the n-well and substrate contacts.

- All flip-flops having a clock internally gated (reducing clock-tree capacitance and, therefore, dynamic power): With the 630 flip-flop in the T8051 design and the typical activity of the flip-flops being 12 times the activity of the rest of the combinatorial logic, it translates to 0.012 nW/Hz reduction in the dynamic power (40% reduction if other techniques are implemented). The problem is the differential clock tree cannot be implemented any longer, so all gates are suffering a relatively high level of clock feedthrough, limiting maximal operating frequency of the system.

- Extensive usage of the upper metal levels in the cell-level routing as long as it helps to reduce total capacitance: This technique is contradictory, because it can cause routing congestion and layout expansion, ultimately resulting in the increase in the wire load and parasitic capacitance. In the currently synthesized T8051, to prevent routing congestion threat, gate density should be reduced. In the six-metal process, density falls from 70% in the standard library to the 52% in the designed library to maintain the same amount of the routing resources. But because developed gates are very small (average 61% of the standard cells), total system area actually reduces by 18%. Overall reduction in interconnects capacitance is, thus, about 2.5% compared to that in the standard library.

- Design for an aggressive voltage scaling: No cells were used with more than four transistors in series between power-supply rail and ground. This approach allows operation at 1 V power supply while avoiding triode mode of the transistors in the logic gates. Standard libraries need to oversize some gates to operate at 1 V. But the developed library is able to use only unit-size gates, reducing both static and dynamic powers.

15.4.2.2.2 4 kB Data Memory

The static random-access memory (SRAM) block is based in the eight-transistor (8T) low-voltage SRAM cell. SRAM elementary cells are assembled into the 8-bit blocks (bytes). Each byte block additionally contains local address decoder, address column line, and address row line. The total transistor count per byte is 80, resulting in 10 transistors/bit efficiency. Although this results in increased transistor area, systematic merging of the transistor wells and terminals allows average array density of 15.54 μm^2/bit, comparable to the classically built six-transistor (6T)–based high-voltage SRAM cell. This effective ten-transistor (10T) cell solution had a power advantage over classical 6T solution, because unlike a 6T cell, which is working properly starting from 1.8 V, developed SRAM can be rapidly and reliably read or written at power-supply voltages as low as 0.9 V, reducing dynamic power dissipation four times. On the contrary, the added three PMOS and one NMOS transistors per bit increase SRAM power by roughly 66%. Expected power of the proposed SRAM cell is 5/12 of the standard 6T cell if operated at the lowest possible supply voltage. Designed SRAM retains power efficiency up to the power-supply voltage of 1.4 V. The SRAM row decoder is built using a power-efficient 1:32 decoding scheme. The column decoder is implemented as two-stage 1:8:64 decoders. The column decoder is integrated with the data bus controller, so bit-line activation happens in parallel with the column address decoding. Bit-line precharging (adding a lot of circuit complexity and area) was not used. Sense amps (adding much to the SRAM area and power dissipation) were not used. Instead, all data multiplexers were implemented using tristate inverters. The small area of the multiplexers in this solution have allowed integration of data multiplexers with the column decoders, reusing intermediate decoder nodes and reducing power dissipation of the data buses. Clock gating was extensively used in the address decoders and data decoders/multiplexers to further reduce power dissipation, thus resulting in extremely compact layout of the SRAM 4 kB block.

15.4.2.3 Wishbone

As CMOS technology scales down, process variation becomes a progressively greater challenge. For most analog circuits it means increasing the design margin, leading to inevitable increase in the chip area and power dissipation. The situation with the increased design margin is most severe for the low-voltage circuits, because the small (below 0.5 V) overdrive voltage of the CMOS transistors in the typical 1 V–powered system makes current through these transistors very sensitive to the metal–oxide–semiconductor field-effect transistor (MOSFET) threshold voltage variations. For example, experimentally detected 0.1 V reduction in the threshold voltage from the 0.53 V nominal of the NMOS may result in the 61% increase of the current over the specified value. Although in most of the circuits it is possible to compensate to some degree (50%–80%) threshold voltage variation by analog feedback loops, analog feedback loops are very costly in the energy-harvesting system, because a continuous compensation approach requires continuous current flowing through the reference transistors.

The problem of the excess power dissipation arose from the fact that the analog feedback loop is capable of compensating for time-variant variation, but most of the process-dependent variation is actually time invariant. Furthermore, such process variation may be nearly identical for all circuits in the chip, regardless of their location. Theoretically, it is possible to use process variation–dependent voltage sources fed to the chip subcircuits by a dedicated distribution system. But in the mixed-signal SoC being designed, either such a distribution system requires excess shielding together with decoupling capacitors (and, therefore, have a large chip area) or it falls victim to the EMC problems. To overcome difficulties associated with centralized process variation–sensing circuitry, threshold voltages should be encoded by the dedicated ADC and passed to the subcircuits in digital form. It is even possible to measure process variation in the chip batch by the external ADC and supply the same process variation–compensation codes to all chips in the batch.

In the current work we design all analog parts with the minimal process variation margin and correct actual process variation (especially MOSFET threshold voltage variations) by trimming the width of the respective MOSFET. That should be done between the serial and parallel trimming interfaces. The serial in-chip interface requires less wiring but more glue logic (approximately 2 wires and ~1000 gates of the glue logic per component). The parallel on-chip interface requires more wiring but less glue logic (typically ~14 wires for the 8-bit-wide bus and ~200 gates of the glue logic per component). Because wires take a very small area in LSI technology, a parallel interface for the medium-size chip is preferable. Furthermore, in systematically implementing bus architecture and standard bus transceivers, design time and reliability of the system may be significantly improved if a parallel configuration interface is systematically implemented. An additional bonus is faster configuration time for the parallel configuration bus, reducing processor load and software complexity. Although many parallel on-chip bus architectures exist, if we target free architecture capable of 8-bit operation, only the Wishbone bus standard will comply. Therefore all components of the SoC are planned to be trimmed by Wishbone interface.

15.4.2.4 RF Interface (400 MHz/1.2 GHz)

For low-power radio, the primary consideration is the receiver. It is because transmitted data are processed by a stationary radio with fewer hardware limitations and, thus, larger communication range; receiver sensitivity, not the available output power, is limiting the effective communication range. If we consider the best receiver operating frequency with a fixed sensitivity and minimal power dissipation, the power dissipation of the transceiver as a function of the frequency can be approximated by

$$P_{dd} = P_{dd0} \cdot f^{4/3}. \tag{15.1}$$

But if the radio frequency is too low, radiation resistance of the antenna of a fixed size falls and the power dissipated in the transmitter must be increased to drive the inefficient antenna. Onset of this problem happens at approximately 40% of the antenna radiation efficiency that corresponds to the size of the antenna 0.20 of the free-space wavelength. So to keep constant power received by the host at lower frequency, power dissipation of the transmitter must be increased as a reverse second power of the frequency. Radiated power of the stationary host must be also increased as the inverse square of the radio frequency, creating EMC problems and even reaching SAR limitations:

$$
\begin{aligned}
P_{dd} &= P_{dd0} \cdot f^{4/3} & f &> 0.2 \cdot c/l_{\max}, \\
P_{dd} &= P_{dd0} \cdot f^{-2/3} & f &< 0.2 \cdot c/l_{\max}.
\end{aligned}
\tag{15.2}
$$

According to Equation 15.2 and Table 15.1, systems exist at an optimal operating frequency for the low-power wireless device of a given size. Operating above that critical frequency will increase power dissipation slightly due to the increase in switching losses. Operating below critical frequency, however, will cause a dramatic increase in the transmitter power due to the inferior radiation efficiency of the electrically small antenna. If the length of an adhesive-type wearable sensing device is 6 cm, the optimal frequency is 1.0 GHz. Any deviation from this frequency will impose a heavy penalty on either power dissipation of the system or the communication range.

Although the 900–1000 MHz frequency range is the most suitable range for an adhesive-type wearable device to use, the 1215–1260 MHz band, which is close to 900–1000 MHz, is the only unlicensed frequency band available for a medical telemetry data link in Japan.

TABLE 15.1

Selection of the Optimal Radio Frequency

Frequency	Antenna Efficiency	Transceiver Power if Receiver and Transmitter Are Dissipating the Same Power (mW)	Transceiver Power to Keep 10 m Communication Range (mW)	Communication Range, Indoor (m)
315 MHz	0.4%	0.37	13.0	1.2
433 MHz	1.4%	0.57	5.58	2.3
0.9 GHz	26%	1.50	1.50	10
1.4 GHz	100%	2.70	1.70	19.6
1.8 GHz	100%	3.78	2.38	19.6
2.4 GHz	100%	5.55	3.83	19.6

An additional bonus for using the 1215–1260 MHz frequency band is that another unlicensed frequency band, 402–406 MHz, can optionally support switching from the two-step conversion to the one-step conversion in the transceiver.

After considering receiver sensitivity and operating frequency, the physical (PHY) level of the radio operation, namely, the modulation scheme and the encoded data rate, needs to be decided. For low-power wireless devices, only frequency-shift keying (FSK) and binary phase-shift keying (BPSK) are practical solutions. Selection will depend on the required spurious emissions mask and adjacent channel power ratio (ACPR). Generally BPSK allows better spectrum usage in the RF domain and FSK delivers better performance in the baseband domain. If the available RF bandwidth is not an issue, FSK is strongly preferable because it saves transceiver power by selecting a less demanding local oscillator (LO) and lower bandwidth baseband circuitry. RF receiver bandwidth is inevitably larger than the limit specified by the emission mask because of the unavailability of the high-Q resonators on the CMOS chip. So RF bandwidth does not represent a limiting factor to the power dissipation of the CMOS transceiver. We also must especially comment on a myth about on/off keying (OOK) modulation, namely, that not transmitting power during logical 0 allows a twofold decrease in the transmitter power. In fact, if we look at the constellation diagram of the OOK modulation, we can see that the SNR of the OOK is 6 dB worse than BPSK or FSK modulation, so a twofold decrease in the transmitted power due to zero power at logical 0 is negated by the fourfold increase in the power at the logical 1 while maintaining the same bit error rate (BER). Regarding FSK, one should also keep in mind that low power dissipation of the FSK symbols could be achieved if the noises of in-phase (I) and quadrature (Q) channels are strongly correlated. Usually this requires that the total noise factor of the receiver should be dominated by the propagation channel noise and low-noise amplifier (LNA) noise, and not by the quadrature mixers and the baseband noise. Typically this requirement is easily met if gains of the LNA and RF–to–intermediate frequency (IF) mixers are high enough.

After determining the optimal frequency band (404/1238 MHz) and optimal modulation method (FSK), the MAC data communication protocol needs to be selected. Since the project targets a low-power sensor network, therefore "one-net" [13], which is specially designed for low-cost, battery-powered sensor networks using open-source standard, should be an acceptable selection.

15.4.3 Energy Harvesting

Because of the small-size and lightweight requirement of the adhesive-type wearable sensing device, a bulky power source is particularly undesirable, which directly limits the energy storage capability of the device. Although many different light and small, thin-film rechargeable lithium batteries have been available on the market in recent years, the current capacity is comparatively limited. On the other hand, the power source requires maintaining a long period of operation without additional maintenance such as recharging, which an appropriate proportional power source expects. Essentially, the most effective way to balance the above antithetical demands is to reduce total power consumption of the device; however, it will finally reach the minimum limit. Hence, the project focuses on energy-harvesting technology for retrieving energy from the surrounding environment, ideally as the primary energy source for the wearable sensing device [14].

Energy harvesting (also known as power harvesting or energy scavenging) is a process of capturing energy that is scattered in the surrounding environment, such as light, heat (temperature difference), wind, electromagnetic waves, and vibration, then converting and storing it as electrical energy. With energy-harvesting technology, it is achievable not only to construct batteryless or semipermanent systems such as a large-scale maintenance-free sensor network, but also to significantly reduce environmental impact due to the elimination of hazardous waste from batteries.

A conceptual diagram of energy harvesting from the human body is shown in Figure 15.9. Currently, the energy retrieved from humans is tiny, lower than 1 μW. However, the power consumption of microcomputers, wireless modules, and sensors is also becoming lower and lower nowadays; with the rapid advancement of science and technology, we believe that energy-harvesting technology is more promising for wearable sensing devices. Therefore, the Maenaka Project investigated the feasibility of solar energy [15], vibration power generation using electret and micromagnet [16–18], and thermal energy harvesting.

15.4.3.1 Solar Energy

Solar energy—the sun's energy (sunlight) can be captured and directly converted into current electricity by using solar cells or solar photovoltaic (PV) arrays. This phenomenon, called the photovoltaic effect, was discovered in 1839 by French physicist Edmond Becquerel. Today, solar electricity is a steadily growing energy technology and solar cells have found markets in variety of applications ranging from consumer electronics and small-scale distributed power systems to centralized megawatt-scale power plants. Direct utilization of solar radiation to produce electricity is close to an ideal way of utilizing nature's renewable energy flow. With photovoltaic cells, power can be produced near the end user of the electricity, thus avoiding transmission losses and costs, and the solar cell operates without noise, toxicity, and greenhouse gas emissions.

Although solar energy seems to be a practical solution for powering wearable devices, the harvestable power highly depends on the surface area of solar cell and brightness of the light source [19]. Since the wearable sensing device should be small in size, this limits the surface area of the solar cell. Moreover, if the device is used for electrocardiography recording, it is supposed to be worn on the chest; thus, it is very unlikely for the solar cell to obtain enough light to generate electric energy. Therefore, the solar energy may not be considered solely as a power source for the wearable sensing device due to the limited area of the solar cell and its high dependence on the device wearing location.

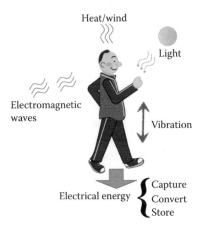

FIGURE 15.9
Conceptual diagram of energy harvesting from the human body.

15.4.3.2 Electret-Based Energy Harvesting

The electret-based energy harvester is one kind of vibration-type energy generator, which is a device that converts kinetic energy due to movement or vibration of a mass into electrical energy. An electret is a dielectric material that has a quasipermanent electric charge and the principle of an electret-based energy harvester is based on electrostatic induction. As shown in Figure 15.10, the electret energy harvester is constructed by joining a moving counterelectrode, a base electrode, and a charged electret film. The charged (negative) electret film induces an opposite (positive) electrical charge Q that is defined by the capacitance C on the counterelectrode by electrostatic induction. When the counterelectrode is moved (kinetic energy), the capacitance C is changed to $C - \Delta C$ and the amount of electrical charge Q induced on the counter electrode becomes $+(Q - \Delta Q)$. If a load R is connected to the counterelectrode, an electric current (electrical energy), which is the rate of change with time, can be retrieved. The generated current depends on the surface charge density of the electret, the rate of change of the capacitance, and the gap between the counterelectrode and electret. Moreover, in order to maximize energy harvested, etched electrets with mechanically fine pattern and alternately electric-charged area should be used on the counterelectrode and base electrode [20].

To evaluate the efficiency of the harvester, a $13 \times 12 \times 1.5$ mm^3 electret-based harvester with 57 Hz resonant frequency in the harvesting direction was fabricated. The structure of the electret-based energy harvester [21] is shown in Figure 15.11. The harvester is composed of three layers; the top layer is a silicon counterelectrode with multiple slits, the second layer is movable with a fine patterned and charged electret structure, and the bottom layer is a cover glass layer. The electret material is a 3 μm polymer film made of Cytop (CTL-809M, Asahi Glass Co., Ltd., Japan). The Cytop does not simply act as suitable electret material but also acts as an adhesive material for interlayer bonding as well [22].

When using 20 Hz external excitation frequency and a 200 μm peak-to-peak (p-p) moving distance with optimized 32 MΩ load resistance, the maximum harvested power is 1.8 nW$_{rms}$ [23]. Although, the harvested power is not much, it is expected as to be significantly enhanced by improving the manufacturing process.

15.4.3.3 Electromagnetic Energy Harvesting

The electromagnetic energy harvester is another type of vibration-type energy generator and usually has a longer lifetime than the electret-based energy harvester, which suffers from electrical charge leakage. It also has high output current because of lower output impedance characteristics. The electromagnetic energy harvester is constructed

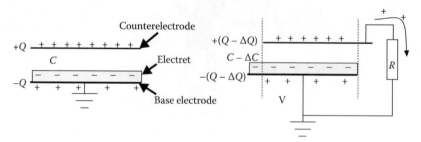

FIGURE 15.10
Principle of electret-based energy harvesting.

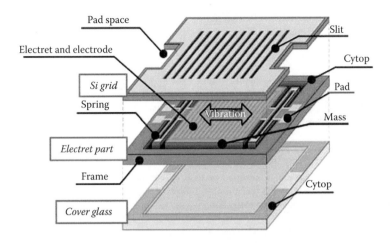

FIGURE 15.11
Schematic diagram of electret-based energy harvester. (From K. Fujii et al., "Electret based energy harvester using a shared Si electrode," Solid-State Sensors, Actuators and Microsystems Conference [TRANSDUCERS], 2011 16th International, Beijing, China, June 5–9, 2011, pp. 2634–2637, 2011. With permission.)

by combining a bidirectional movable micromagnetic array layer and serially connected microcoils layer as shown in Figure 15.12.

The working principle is based on Faraday's law of induction. When the micromagnetic array layer is moved (kinetic energy), the change in the magnetic flux induces electric current (electrical energy) on the microcoils. Therefore, the harvestable energy is directly proportional to the rate of change in magnetic flux. In other words, it depends on the mass of the micromagnetic array layer and the moving distance and frequency as well.

In order to inspect the harvested energy, an electromagnetic energy harvester was fabricated by using a sputtering process (micromagnetic array layer) and single-crystal silicon wafer with a MEMS batch fabrication process (microcoils layer) and mounted on a PCB for testing, which is shown in Figure 15.13. The micromagnetic array layer is constituted by a movable silicon mass with $10 \times 10 \times 0.5$ mm^3 dimensions combined with 24 pieces of micro-neodymium-magnet (NdFeB) which has a size of $400 \times 4000 \times 20$ μm^3 and all the NdFeBs are arranged in 400 μm lines and spaces. For the microcoils layer, 46 pieces of five-turn coils are used and connected in series to increase the electromotive voltage. The gap between the mass and microcoils layer is 30 μm, which uses polymer material to bond together. The magnetic field of the NdFeB is 2.7 T and the magnetic flux density is 5.7 mT p-p, which is measured 300 μm on top of micromagnetic layer.

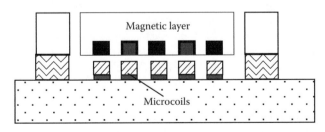

FIGURE 15.12
Schematic of electromagnetic energy harvesting. (From Y. Jiang et al., "Fabrication and evaluation of NdFeB microstructures for electromagnetic energy harvesting devices," *Proc. PowerMEMS 2009*, Washington, DC, December 1–4, 2009, pp. 582–585, 2009. With permission.)

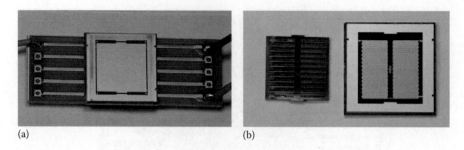

FIGURE 15.13
(a) An electromagnetic energy harvester mounted on PCB; (b) exploded view of micromagnetic and microcoils layers. (From T. Fujita, "Energy harvesters for human-monitoring applications," *IEICE Transactions on Electronics*, Vol. E96-C, No. 6, pp. 766–773, 2013. With permission.)

When the harvester was vibrated (moving) at 30 μm peak-to-peak amplitude with a 512 Ω optimum load impedance, the resonance frequency was 73 Hz, the mechanical Q factor was about 8 and the harvested power was slightly higher than 100 pW. As the vibration amplitude increased to 48 μm peak to peak, 760 pW power was obtained and became saturated as the amplitude was further increased [23].

Although the harvested power of the electromagnetic harvester is very low, it is expected to have great improvement by achieving the following:

- Shrinking down further the size of magnet
- Reducing further the spacing between the magnets
- Decreasing the gap between two layers
- Increasing the film thickness of the magnets

15.4.3.4 Thermal Energy Harvesting

The principle of thermal energy harvesting is based on the Seebeck effect, which converts a thermal gradient directly into electricity. A thermoelectric harvester is composed of serially connected p-doped and n-doped semiconductors (also called Seebeck elements). When a temperature difference is applied on both semiconductors, electric voltage is generated. The thermopower conversion efficiency not only depends on the semiconductor's characteristic but also relies on the thermal structure that keeps the temperature difference between both sides of semiconductors. Therefore, a flexible thermal structure combined with three types of flexible sheets [26] was designed to be used on wearable sensing device to harvest thermal energy between human body temperature and ambient temperature.

A conceptual diagram of flexible thermal structure is shown in Figure 15.14. The bottom layer is a thermal conductivity– and resistivity-flexible sheet directly attached on human skin; hence, it should be compatible with the human body. Besides, the thermal conductivity should be higher than 2.0 W/mK to ensure that the body temperature (heat) transfers to the Seebeck element efficiently. Moreover, when considering the possibility to embed the ECG electrodes inside the sheet, the resistivity should be higher than 20 MΩ·m. To satisfy the requirements, polydimethylsiloxane (PDMS), well known as an excellent biocompatible material, is used and mixed with graphite and alumina particles so that it has a 2.3 W/mK thermal conductivity and a resistivity of more than 50 MΩ·m.

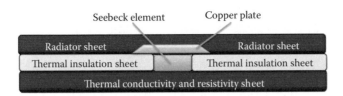

FIGURE 15.14
Conceptual diagram of flexible thermal structure with thermoelectric harvester.

The middle layer is a thermal insulation sheet used to enclose the Seebeck element to block the heat loss to the surrounding environment. It is also based on PDMS but is filled with borosilicate glass hollow particles to minimize the thermal conductivity to 0.06 W/mK. The top layer is a radiator sheet of mixed PDMS with copper particles. It is used to enhance heat dissipation from the Seebeck element to the surrounding environment; hence, it has 2.78 W/mK thermal conductivity, higher than those of other two sheets. In addition, a thin copper plate is mounted between the radiator sheet and top of the Seebeck element to further increase the heat dissipation.

A 50 × 100 mm² flexible thermal structure prototype embedded with a 3.3 × 4.2 mm² thin film thermogenerator chip was fabricated, to evaluate the thermopower. The prototype is shown in Figure 15.15. In the evaluation, a hot plate is set at 35°C to simulate the body skin temperature and the ambient temperature is set at 25°C. For 10°C temperature difference, the harvested power is 390 nW, about 1/25 of the simulation result. It may be due to heat dissipation efficiency that is lower than expected; therefore, there is a need to improve the heat dissipation and surface area of the radiator sheet.

15.4.4 MEMS Technology and Single-Chip Multisensor Integration

15.4.4.1 Sensor Integration

The most important factors for the disposable adhesive-type wearable human-monitoring system are power source, small size, lightness of weight, and low cost. In order to achieve these requirements, multisensor integration in one chip is essential. We studied a multisensor integration chip more than 10 years ago, where a 3D acceleration sensor, humidity sensor, atmospheric pressure sensor, and temperature sensor were integrated with sensor interface circuitry including power-down functionality in a 10 × 15 mm² chip [27]. The device was designed by using bulk micromachining for the sensors and bipolar technology for the circuitry. The fabricated device was advanced at that time, but it required a

FIGURE 15.15
Flexible thermal structure prototype embedded with thin film thermogenerator chip.

relatively high supply voltage (±5 V) and large operating power (several megawatts), and compared to current MEMS devices it was relatively large sized. Therefore, in the Maenaka Project, we focused on the following points for sensor integration:

1. In order to fabricate using a batch process, in which all sensors are realized simultaneously by a single fabrication process, we investigated an advanced starting wafer that is compatible for the following high-temperature process such as pn junction formation.

2. For miniaturization, the use of bulk micromachining (deep etching through the wafer) is avoided if possible.

3. For reduction in the scale of the interface circuit, the output forms of the sensors are the same and one interface circuit can be used for many kinds of sensors by multiplexing the outputs of the sensors.

4. In the current technology, since the feature sizes of LSIs and MEMS sensors are quite different, these parts are not fabricated using the same wafer.

5. The system should not be monolithic but a hybrid system.

6. Peripheral circuits should be extremely low power, and CMOS 0.15 μm technology and around 1 V operation are assumed.

15.4.4.2 Starting Wafer: Silicon-on-Honeycomb Insulator Wafer with Silicon-on-Nothing Machining for Pressure Sensor

We suggested special silicon-on-insulator (SOI) wafer named silicon-on-honeycomb insulator (SOHI) [28] with silicon-on-nothing (SON) area, which is the vacuum chamber area for the

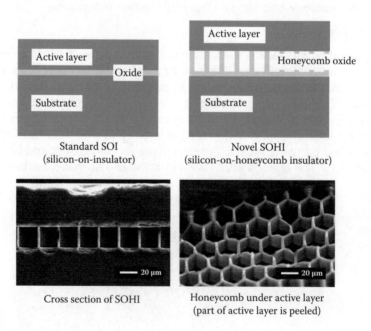

FIGURE 15.16
Image of SOHI wafer structure.

pressure sensor. Figure 15.16 shows the SOHI wafer structure, in which the active layer (thin device layer) is separated from the substrate layer by the tall honeycomb SiO_2 insulator. In the SOHI wafer, parasitic capacitance between the active and substrate layers is reduced, dramatically resulting in high SNR and low-power device for capacitance output sensors. Moreover, it has thermal isolation useful for separation of the body temperature and ambient temperature. Furthermore, although not for human-monitoring application, the structure also has high voltage resistance between the active layer and substrate layer, which is useful for large voltage operation of electrostatic actuators.

On the other hand, SON is a unique technology that uses atomic flow of Si in a high-temperature and vacuum environment. The area with small and dense pits becomes a vacuum cavity with a thin membrane as shown in Figure 15.17. This structure can be used directly for an absolute pressure sensor when the capacitive detector electrode is formed above the membrane [29]. The SOHI with SON structure does not contain any other material except Si and SiO_2; it can be used for the starting material of MEMS sensors as well as integrated circuits.

15.4.4.3 Dynamic Acceleration Sensor: Lead Zirconate Titanate as Detection Mechanism

For low-power operation, sensors with interface circuits should operate in low power consumption as well as other electrical systems. Lead Zirconate Titanate (PZT), a piezoelectric material, is a useful material for low-power operation. When a strain is applied to PZT, it generates electric power in the form of voltage or current, which results in a simple interface circuit design. Although it is said that the fusion of PZT into MEMS technology is difficult, we obtained suitable technology for incorporating PZT into standard MEMS technology successfully. All-dry processing, which is sputter deposition of PZT and dry-etching technology, has been developed and applied to dynamic acceleration sensors. Figure 15.18 shows the example of the 3D dynamic acceleration sensor. The PZT cells are deposited and patterned on the thin Si springs, which are used to suspend the large mass. When the acceleration or vibration is applied to the sensor, the springs bend according to the direction of the acceleration, resulting in stress on the PZT cells. A suitable connection between PZT cells calculates x, y, and z accelerations and individual outputs for the components of 3D acceleration can be obtained [30].

15.4.4.4 Static Acceleration Sensor: Capacitive-Type 3D Acceleration Sensor

The PZT dynamic acceleration sensor has low power consumption merit and can be used for continuous monitoring of human motion activity. However, because of the leakage current of PZT, it cannot detect static acceleration. However, in order to obtain human body postures such as sitting, standing, and lying, the detection of static acceleration is required. Therefore, a capacitive-type acceleration sensor using special SOHI wafer [31] was developed and is shown in Figure 15.19. The basic structure and operation principle are the same as for the MEMS acceleration sensor in the commercial market. The center mass is supported by the four serpentine springs and the mass has two sets of capacitor elements that are like a comb for x and y directional accelerations. The device is attached on the CMOS capacitance-to-digital converter circuits with the z electrode on the surface of the circuits. The z electrode on the CMOS substrate detects the z acceleration as well as the x and y accelerations.

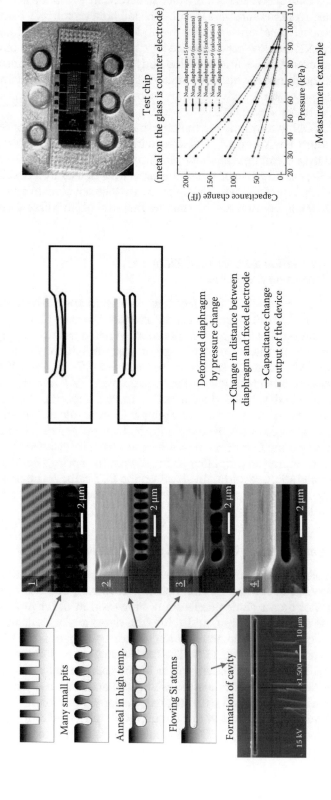

FIGURE 15.17
SON formation and application for pressure sensor.

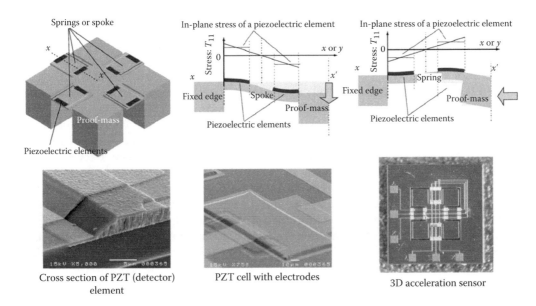

FIGURE 15.18
Structure and image of PZT 3D acceleration sensor.

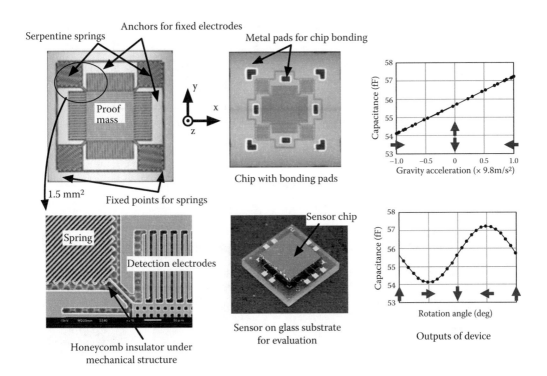

FIGURE 15.19
The image and characteristics of capacitive-type acceleration sensor.

15.4.4.5 Blood Pulse/SpO₂ Sensor

It is easy to obtain a pulse rate by ECG when the sensing device is attached on the chest or back. On the other hand, the pulse rate can also be measured by a photoreflectance change from a blood vessel in the wrist and arm. Moreover, if two different wavelengths, infrared (IR) and red, are alternatively used for photoreflectance measurement, SpO_2 can be obtained by using the difference between absorptions by oxyhemoglobin and reduced hemoglobin as shown in Figure 15.20 (pulse oximetry method). A novel reflective photo-plethysmography sensor [32] was fabricated with a ring-shaped photodetector and buried light-emitting diodes (LEDs). The ring-shaped photodetector was effective for correcting the spread of reflection light and robust for minimizing body movement artifacts.

Figure 15.21 shows the image and structure of the device. The IR and red LEDs are buried in the cavity at the center of the device and the whole device is covered by transparent resin with a photoshading wall. The example output of the pulse sensor is shown in Figure 15.22. While wearing the pulse sensor on the wrist to obtain the wrist pulse rate, if the ECG on the chest is also measured at the same time, the blood pulse delay that corresponds to

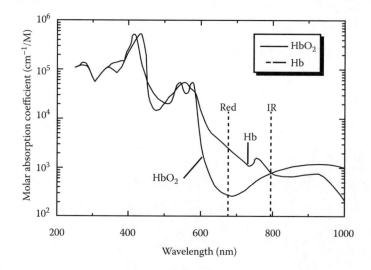

FIGURE 15.20
Absorption coefficients of oxyhemoglobin (HbO_2) and reduced hemoglobin (Hb).

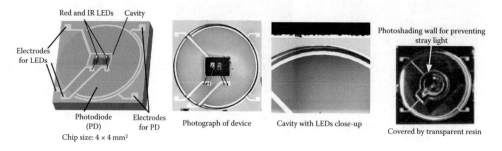

FIGURE 15.21
The image and structure of SpO_2 sensor with ring-shaped photodetector.

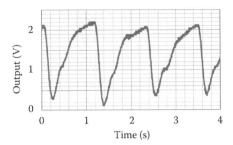

FIGURE 15.22
Output example from pulse sensor.

the pulse wave velocity (PWV) [33] between the heart and the wrist can be measured. As a result, blood pressure can be estimated by the PWV method.

15.4.4.6 Other Sensors and Technologies

The Maenaka Project introduced different kinds of novel technologies and devices for sensor integration, which included through-silicon via, glass structure, flexible sensors, resin devices, humidity sensors, gyroscopes, and so on [34–38]. The photographs and scanning electron microscope (SEM) images for these technologies and devices are shown in Figure 15.23.

15.4.5 Wireless—A 315 MHz RF Transceiver Module

The wearable sensing device consumes power for the control circuit, data sensing, data processing, and wireless communication. In most cases, wireless communication uses up most of the power, which affects the operation time of the device significantly. Therefore, it is critical to select a wireless technology that does not require high power. Moreover, it is also important that the wireless transceiver module should be small in size and lightweight to minimize the area of the device.

In 2008, the Maenaka Project attempted to select a suitable wireless technology such as Wi-Fi [39], ZigBee [40], and Bluetooth [41] to be used on wearable sensing devices. However, although different kinds of transceiver modules are available on the market, they consume too much power and require a large area for installation. Especially for peak power consumption, they may require more than 15 mA current instantaneously during wireless communication, which restricts the implementation of those technologies on the device. Since a small-sized battery such as CR2032 (coin size) can supply only less than 15 mA peak current without degrading the battery life [42], if one of those wireless modules is used, a large-size high current capability battery is required that would limit the miniaturization of the device, which is extremely undesired.

Therefore, the project turned out to realize an extremely low-power, small-sized, and lightweight transceiver module at the 315 MHz radio band [43–44]. The module is a single-chip solution that was designed and fabricated within $11 \times 11 \times 1.5$ mm^3 volume including a bipolar ASIC for radio circuit and 8-bit microcontroller (4 kB program memory and 128 bytes of RAM), and weighs only 0.3 g. For the radio circuit, the current consumptions of transmitter and receiver are 1 and 2 mA, respectively, with a 3.3 V supply voltage (not including the current of 8-bit MCU). The main characteristics of the module are shown in Table 15.2.

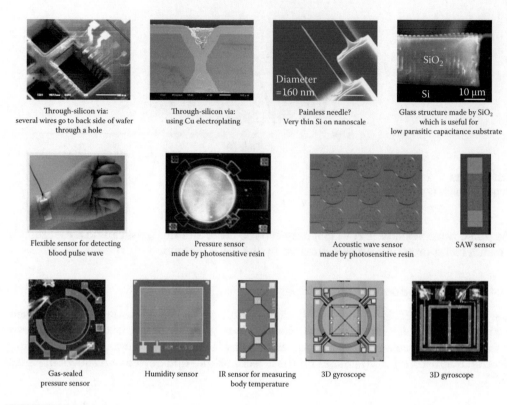

Through-silicon via: several wires go to back side of wafer through a hole

Through-silicon via: using Cu electroplating

Painless needle? Very thin Si on nanoscale

Glass structure made by SiO_2 which is useful for low parasitic capacitance substrate

Flexible sensor for detecting blood pulse wave

Pressure sensor made by photosensitive resin

Acoustic wave sensor made by photosensitive resin

SAW sensor

Gas-sealed pressure sensor

Humidity sensor

IR sensor for measuring body temperature

3D gyroscope

3D gyroscope

FIGURE 15.23

Some special technologies and sensors. (From S. Tanaka et al., "Crystallographic anisotropic etching in RIE and its application to through silicon via," The 27th Sensor Symposium, Book of Abstracts, Micromachines and Applied Systems, p. 119, 2010; J. Nakamura et al., *Microsystem Technologies*, Vol. 19, Issue 3, pp. 433–438, 2013; K. Kanda et al., *Procedia Engineering*, Vol. 25, pp. 843–846, 2011; Y. Jiang et al., *Procedia Engineering*, Vol. 5, pp. 1466–1469, 2010; X. C. Hao et al., *Sensors and Actuators A: Physical*, Vol. 205, pp. 92–102, 2013.)

TABLE 15.2

The Main Characteristics of the 315 MHz RF Transceiver Module

Item	Typical	Unit
Frequency	315	MHz
Intermediate frequency	10.7	MHz
Transmission power	−24	dBm
Receiver sensitivity	−95 (BER 1%)	dBm
Transmitter current[a]	1	mA
Receiver current[a]	2	mA
Standby current[a]	<1	μA
Size	11 × 11 × 1.5	mm³
Weight	0.3	g

[a] The current of 8-bit MCU is not included.

After that, the Maenaka Project applied the 315 MHz RF transceiver module to develop a sensor node [45–46] for performance evaluation and demonstration. The sensor node consists of four types of sensors, a three-axis accelerometer, an altimeter, a humidity sensor, and a temperature sensor. Including a coin-sized lithium battery (3.0 V, 220 mAh), the whole sensor node can be fabricated with dimensions $28 \times 20 \times 8$ mm^3 and weighs only 7 g without the antenna.

The 8-bit MCU that is included in the module is used not only to drive the RF circuit but also as an interface with the four sensors by using I^{2}C and controlling the entire sensor node operation. After information is gathered from the sensors, the MCU wraps the sensed data into a data packet and use Manchester encoding for RF transmission. When the MCU samples the three-axis accelerometer at 38 samples per second (sps) and the altimeter and humidity and temperature sensors at 2 sps and transfers the sensed data continuously, the average current consumption of the entire sensor node is 2.9 mA. For a 220 mAh battery, it can continue operating for more than 75 h. Moreover, when intermittent operation is applied and samples all the sensors at 1 sps, the average current consumption is reduced to less than 0.5 mA, for which the sensor node can continue operation for around 3 weeks.

During the data transmission test, the same 315 MHz RF transceiver module is used inside the base station to communicate with the sensor node. If the sensor node and base station both implement a half-wave dipole antenna, the transmission distance is approximately between 15 and 20 m. When using a small antenna, due to the reduction in the antenna gain, the transmission distance will reduce accordingly. For example, when using a half-wave whip antenna, the distance may reduce to few meters. For a few centimeters long without matching lead antenna, it may further reduce to less than 2 m.

For evaluating the RF characteristics when the transceiver module is worn on the human chest, a loop antenna of size 15×30 mm^2 is designed for the module. A −80 dBm RF intensity is obtained at 5 m, which is measured by a half-wave dipole antenna. Assuming the base station has −90 dBm receiver sensitivity with a half-wave dipole antenna implemented, if this loop antenna is used on wearable sensing device, the transmission distance is expected to be increased to 10 m. Although the RF performance of this loop antenna is acceptable, the size is a little too big for wearable sensing devices. Thus, the antenna structure and configuration are still being optimized.

15.4.6 The Software

An adhesive-type wearable sensing device requires a small size and lightness of weight that are comfortable enough for wearing on the human body. This design requirement limits the size of the power source, which also limits the energy storage capability; hence. optimizing power consumption is an important design goal for the device. Therefore, the device is not simply implemented with a low-power hardware design, but also the software should be using a low-power design/scheme as well. For example, the reduction in current consumption can be achieved through the following:

- Allowing intermittent operation of the MCU
- Optimizing the sampling period of sensors
- Implementing data-processing algorithms for extracting representative or critical data from the raw data to drop down the throughput such as ECG–to–heart rate algorithm and acceleration data–to–human posture algorithm
- Reduction of the wireless communication period

Besides low-power consumption design, the software should also analyze the sensed data in a comprehensive manner. Since the device measures both the vital and environmental signals, thus a variety of sensed data can be obtained simultaneously. By analyzing the data in a comprehensive manner, it is possible to discover a high-level data pattern that cannot be obtained from a single data source. In addition, as with a medical record, fingerprint, or DNA, vital signals and activity data should be considered as personal information as well; therefore, the data should be transferred under encrypted format and stored in a secured manner.

15.5 Large Module Adhesive-Type Wearable Sensing Device

While the development of the low-cost CMOS ASIC is in progress, several project members established a venture company called AffordSENS Corporation [47], to fabricate a large module prototype of an adhesive-type wearable sensing device [48–49] using commercial ICs. The device image is shown in Figure 15.24. It can be fabricated with an area of 90 × 22 mm² and its weight is lighter than 10 g. The entire device is composed of three blocks, namely, the main block, RF block, and power block. The main block includes an 8-bit MCU, an ECG circuit, a three-axis accelerometer, an altimeter, a humidity sensor, and a temperature sensor. The RF block is equipped with a Bluetooth low energy (BLE) module, while the power block is composed of a coin-sized 75 mAh rechargeable lithium-ion battery and a USB charger circuit.

For the device to be wearable on the chest and capture the ECG signal easily, a magnet connectivity structure is designed and is shown in Figure 15.25. Two magnets are installed under the power block and RF block to make contact with a disposable ECG electrode tape

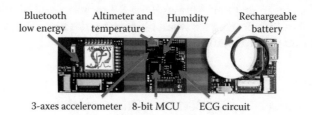

FIGURE 15.24
Image of large module prototype of adhesive-type wearable sensing device.

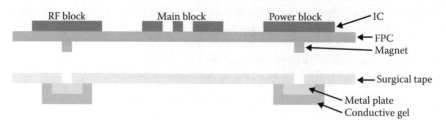

FIGURE 15.25
Magnet connectivity structure of large module prototype.

that is assembled with surgical tape, two metal plates, and conductive gels. The conductive gels are used as ECG electrodes that are placed under the surgical tape. The metal plates are set in between the conductive gels and the surgical tape, and they are used to attract the magnet for better contact. With this structure, not only can the device easily be put on/off, but it can also be reused repeatedly, while the ECG electrode tapes are disposable.

Furthermore, an iOS application was developed to display the raw sensed data on mobile devices graphically. The ECG signal was sampled at 75 Hz, the acceleration data were sampled at 25 Hz, and the other sensor data were sampled at 1 Hz. During continuous transmission of the sensed data, the device could operate for 9 h. Currently, the intermittent operation methods, ECG–to–heart rate algorithm and acceleration data–to–exercise level algorithm, are under development and will be implemented in the device firmware. By applying those algorithms, the overall data rate will be reduced significantly and the operation time is expected to increase to more than 48 h.

15.6 Conclusion

The purpose of the Maenaka Project is to achieve an adhesive-type human activity-monitoring system to overcome the global trend of fertility decline, aging population, and NCDs. The system monitors and understands not only the health condition of an individual but also their mental conditions such as stress from overwork. If a critical situation is detected, an appropriate alarm will be issued to their family, doctor, or hospital to prevent the occurrence of a serious accident and/or unnecessary events. Also, by detecting the disease symptoms at an earlier stage, medical treatment can be provided in advance that reduces not only the medical expenses but also the burden on the health system. Moreover, it supports health management and provides suitable improvement recommendation for individuals to maintain fitness.

For monitoring human activity and health condition continuously, the project works to develop a low-cost and easy-to-use adhesive-type wearable sensing device that requires extremely low power consumption and is small, lightweight, flexible, comfortable, and fit for human skin by integrating different kinds of sensors into a single chip, developing a custom low-power ASIC, researching on energy-harvesting technology, MEMS technology, low-power transceiver modules, and so on. Although the device is still under development, due to the technical enlightenments and impressive research achievements of the project, the adhesive-type device is recognized as one of the most useful devices for wearable human-monitoring system.

Acknowledgments

Maenaka Human-Sensing Fusion Project was supported by the Exploratory Research for Advanced Technology (ERATO) scheme of Japan Science and Technology Agency (JST). The authors would like to thank all the members of the project for their contribution and technical enlightenment. Moreover, special thanks to Olinver M. Vinluan, who willingly helped and has taken painstaking efforts for proofreading the manuscript.

References

1. United Nations, Department of Economic and Social Affairs, World Population Prospects: The 2012 Revision, http://esa.un.org/unpd/wpp/index.htm, accessed on August 15, 2014.
2. World Health Organization, Global Status Report on Noncommunicable Diseases 2010, http://www.who.int/nmh/publications/ncd_report2010/en/index.html, accessed on August 15, 2014.
3. Maenaka Human-Sensing Fusion Project, ERATO Japan Science and Technology Agency, http://www.jst.go.jp/erato/maenaka/e/index.html, accessed on August 15, 2014.
4. K. Maenaka, "Human sensing fusion project for safety and health society," *IEEJ Transactions on Sensors and Micromachines*, Vol. 128, pp. 419–422, 2008.
5. K. Maenaka, K. Masaki, and T. Fujita, "Application of multi-environmental sensing system in MEMS technology—Monitoring of human activity," Fourth International Conference on Networked Sensing Systems, 2007 (INSS '07), IEEE, 2007.
6. K. Maenaka, "Human activity monitoring by MEMS sensing fusion," Proc. 23rd Int. Microprocesses and Nanotechnology Conf., Kokura, Japan, 10C-2-1, November 2010.
7. K. Maenaka et al., "Low power and miniaturized 315 MHz transceiver module for bio-signal transmission [in Japanese]," *IEEJ Transactions on Sensors and Micromachines*, Vol. 132, No. 12, pp. 443–450, 2012.
8. S. Rhee, B.-H. Yang and H. H. Asada, "Artifact-resistant power-efficient design of finger-ring plethysmographic sensors," *IEEE Transactions on Biomedical Engineering*, Vol. 48, No. 7, pp. 795–805, 2001.
9. H. H. Asada and A. Reisner, "Wearable sensors for human health monitoring," *Proc. of SPIE—The International Society for Optimal Engineering*, Vol. 6174, p. 617401-1-13, 2006.
10. R. J. Oweis and A. Barhoum, "PIC microcontroller-based RF wireless ECG monitoring system," *J. Medical Engineering & Technology*, Vol. 31, No. 6, pp. 410–418, 2007.
11. H. Takao et al., "ASIC for monitoring of human motion," *Transactions of Japanese Society for Medical and Biological Engineering*, Vol. 51, M-157, 2013.
12. J. Penders et al., "Power optimization in body sensor networks: The case of an autonomous wireless EMG sensor powered by PV-cells," Engineering in Medicine and Biology Society (EMBC), 2010 Annual International Conference of the IEEE, IEEE, 2010.
13. ONE-NET Low Power Wireless Protocol project, http://one-net.sourceforge.net/, accessed on August 15, 2014.
14. P. D. Mitcheson et al., "Energy harvesting from human and machine motion for wireless electronic devices," *Proceedings of the IEEE*, Vol. 96, No. 9, pp. 1457–1486, 2008.
15. A. Shah, P. Torres, R. Tscharner, N. Wyrsch, and H. Keppner, "Photovoltaic technology: The case for thin-film solar cells," *Science*, Vol. 285, No. 5428, pp. 692–698, 1999.
16. Y. Jiang et al., "Fabrication of a vibration-driven electromagnetic energy harvester with integrated NdFeB/Ta multilayered micro-magnets," *J. Micromech. Microeng.*, Vol. 21, p. 095014, 2011.
17. Y. Jiang et al., "Fabrication of NdFeB microstructures using a silicon molding technique for NdFeB/Ta multilayered films and NdFeB magnetic powder," *J. Magnetism and Magnetic Materials*, Vol. 323, No. 21, pp. 2696–2700, 2011.
18. T. Fujita et al., "Electret based energy harvester by using silicon grid electrode," *Tech. Dig. Power MEMS*, pp. 407–410, 2010.
19. N. Oleg et al., "The availability and statistical properties of ambient light for energy-harvesting for wearable sensor nodes," SENSORCOMM 2012, The Sixth International Conference on Sensor Technologies and Applications, 2012.
20. T. Toyonaga et al., "Novel corona charging method for electret power-harvester using buried grid-electrodes," *Tech. Dig. of the 5th APCOT (2010)*, p. 290, 2010.
21. T. Fujita et al., "Selective electret charging method for energy harvesters using biased electrode," *Procedia Engineering*, Vol. 5, pp. 774–777, 2010.

22. K. W. Oh et al., "A low-temperature bonding technique using spin-on fluorocarbon polymers to assemble microsystems," *J. Micromech. Microeng.*, Vol. 12, No. 2, p. 187, 2002.
23. T. Fujita, "Energy harvesters for human-monitoring applications," *IEICE Transactions on Electronics*, Vol. E96-C, No. 6, pp. 766–773, 2013.
24. K. Fujii et al., "Electret based energy harvester using a shared Si electrode," Solid-State Sensors, Actuators and Microsystems Conference (TRANSDUCERS), 2011 16th International, Beijing, China, June 5–9, 2011, pp. 2634–2637, IEEE, 2011.
25. Y. Jiang et al., "Fabrication and evaluation of NdFeB microstructures for electromagnetic energy harvesting devices," *Proc. PowerMEMS 2009*, Washington, DC, December 1–4, 2009, pp. 582–585, 2009.
26. A. Kubota et al., "Development and evaluation of the flexible, high thermal conductivity and high resistivity substrate material [in Japanese]," *The Japan Society of Applied Physics*, 13p-F7-5, 2012
27. T. Fujita and K. Maenaka, "Integrated multi-environmental sensing-system for the intelligent data carrier," *Sensors and Actuators*, A97-98C, pp. 527–534, 2002.
28. T. Yokomatsu et al., "Novel honeycomb SOI structure with low parasitic capacitance for human-sensing accelerometer," Proc. of 2012 Fifth International Conference on Emerging Trends in Engineering and Technology, pp. 70–74, 2012.
29. X. C. Hao et al., "Application of silicon on nothing structure for developing a novel capacitive absolute pressure sensor," *IEEE Sensors J.*, Vol. 14, No. 3, pp. 808–815, 2013.
30. K. Kanda et al., "A tri-axial accelerometer with structure-based voltage operation by using Series-connected PZT elements," *Procedia Engineering*, Vol. 5, pp. 894–897, 2010.
31. Y. Miyagawa et al., "Acceleration sensor using SOI wafer with honeycomb insulator," 2012 IEEE International Conference on Systems, Man, and Cybernetics, pp. 2062–2066, 2012.
32. M. Kano et al., "Fabrication of reflectance pulse oximeter with ring-shaped photodiode," Proc. Int. Conf. IEEE Systems, Man, and Cybernetics (SMC 2013), pp. 3771–3774, 2013.
33. J. Blacher et al., "Aortic pulse wave velocity as a marker of cardiovascular risk in hypertensive patients," *Hypertension*, Vol. 33, No. 5, pp. 1111–1117, 1999.
34. S. Tanaka et al., "Crystallographic anisotropic etching in rie and its application to through silicon via," The 27th Sensor Symposium, Book of Abstracts, p. 119, 2010.
35. J. Nakamura et al., "Vertical Si nanowire with ultra-high-aspect-ratio by combined top-down processing technique," *Microsystem Technologies*, Vol. 19, Issue 3, pp. 433–438, 2013.
36. K. Kanda et al., "Silicon dioxide as mechanical structure realized by using trench-etching and thermal oxidation process," Eurosensors XXV, Athens, Greece, *Procedia Engineering*, Vol. 25, pp. 843–846, 2011.
37. Y. Jiang et al., "A PVDF-based flexible cardiorespiratory sensor with independently optimized sensitivity to heartbeat and respiration," Eurosensors XXIV, *Procedia Engineering*, Vol. 5, pp. 1466–1469, 2010.
38. X. C. Hao et al., "An analytical thermal-structural model of a gas-sealed capacitive pressure sensor with a mechanical temperature compensation structure," *Sensors and Actuators A: Physical*, Vol. 205, pp. 92–102, 2013.
39. Wi-Fi Alliance, http://www.wi-fi.org, accessed on August 15, 2014.
40. ZigBee Alliance, http://www.zigbee.org, accessed on August 15, 2014.
41. Bluetooth SIG, http://www.bluetooth.org, accessed on August 15, 2014.
42. M. Jensen, "Coin cells and peak current draw," Texas Instruments, White Paper SWRA349.
43. K. Maenaka et al., "Small sized and super low power consumption transceiver module for wireless sensor network," The 28th Sensor Symposium, 2011.
44. K. Maenaka et al., "Low power and miniaturized 315 MHz transceiver module for bio-signal transmission [in Japanese]," *IEEJ Transactions on Sensors and Micromachines*, Vol. 132, No. 12, pp. 443–450, 2012.
45. A. C. K. Chan et al., "Development of small sized, low power and long term operating wireless sensor node," Proc. 6th Asia-Pacific Conference on Transducers and Micro-Nano Technology, July 2012.

46. A. C. K. Chan et al., "Low power wireless sensor node for human centered transportation system," 2012 IEEE International Conference on Systems, Man, and Cybernetics (SMC), IEEE, 2012.
47. AffordSENS Corporation, http://www.affordsens.com, accessed on August 15, 2014.
48. A. C. K. Chan et al., "A small, wearable, stretchable electrocardiogram and physical activity monitoring system," Proceedings of the 7th International Conference on Body Area Networks, Institute for Computer Sciences, Social-Informatics and Telecommunications Engineering (ICST), 2012.
49. A. C. K. Chan et al., "Adhesive plaster-type human activity monitoring device," *Transactions of Japanese Society for Medical and Biological Engineering*, Vol. 51, M-158, 2013.

16

Drug-Delivery Systems in eMedicine and mHealth

Arni Ariani, Soegijardjo Soegijoko, and Hermawan Nagar Rasyid

CONTENTS

16.1 Introduction...333
16.2 Implantable Drug-Delivery Systems (IDDSs) 337
    16.2.1 Implantable Pump Systems..338
        16.2.1.1 Osmotic Pumps 338
        16.2.1.2 Infusion Pumps 339
    16.2.2 Micro-/Nanofabricated IDDSs...339
    16.2.3 Implantable Microfluidic Devices 340
    16.2.4 Ceramic Drug-Delivery Systems (CDDSs) 341
    16.2.5 PMMA Beads...342
    16.2.6 Discussion 343
16.3 Dermal Drug-Delivery Systems...344
    16.3.1 Microneedle Syringes...344
    16.3.2 Microneedle Patches...346
    16.3.3 Discussion 346
        16.3.3.1 Pain...346
        16.3.3.2 The Irritation of the Skin...346
        16.3.3.3 The Infection of the Skin..347
16.4 Human Body Modeling for Study of EMF Interaction...347
16.5 Power 348
16.6 Data Telemetry 349
    16.6.1 Near-Field Resonant Inductive Coupling...349
    16.6.2 Wireless Communication 350
16.7 mHealth Solutions for Drug-Delivery System...351
    16.7.1 Automatic Drug-Delivery System 351
    16.7.2 Disposable Patch Pump 351
    16.7.3 Wireless Drug Dosage Monitoring 352
16.8 Conclusion 353
References..353

16.1 Introduction

In recent years, a more significant number of people are suffering from severe life-threatening health conditions and chronic diseases worldwide, including cardiovascular diseases (hypertension, heart attack, and stroke), cancer, diabetes, melancholia, malignant lymphoma, and septicemia [1]. Some experts in the field of medical and biological sciences have a view that the primary key to controlling or even eliminating such high-risk disorders is by performing early detection and treatment of such diseases [2–3]. The administering of drugs plays a role in the treatment regimens of most of these diseases.

Currently, conventional ways of delivering drugs (i.e., oral tablets or injections) are used in treating most medical illnesses [4]. However, this method has several key disadvantages [5]:

- Some drugs are not completely absorbed from the gastrointestinal tract due to solubility issues or significant first-pass hepatic metabolism that may occur after the administration of the drug.
- The instability of drugs occurs in the gastrointestinal system (GITS).
- The efficacious concentrations of drug products in the targeted areas require systemic concentrations that can lead to toxicity or unwanted side effects.
- In order to achieve effective drug treatment, sustained and controlled manner of drug release must be maintained for periods longer than 24 h.
- Some drugs may lose their potency within a short period of time, which could lead to ineffective results.

Thus, we may conclude that these conventional methods are not effective in controlling levels of drug-delivery treatment [6–7].

Given these weaknesses, and the general dearth of progress in treatment efficacy for severe diseases, there is a growing need to develop a new approach to delivering drugs in a more targeted manner to desired tissue areas [8]. Drug-delivery methods are important because the efficacy of the drug is significantly affected by the choice of methods for delivering the drug into a human body [9]. Furthermore, a drug may not produce beneficial results and may have unwanted side effects when the wrong concentration of the drug (either lower or higher than the maximum allowable concentration) is given to a patient [9].

A drug-delivery system (DDS) refers to a group of recent technological developments in drug delivery that aim to offset the weaknesses of conventional drug-delivery methods. The key desirable feature of a DDS is the capability to control the release of the drugs at specific areas inside the body and/or at particular periods of time [10], so that patients have their physiological or pathological requirements fulfilled better than before [7,11–12].

This feature forms the basis of DDS's advantages, such as the potential for decreasing frequency of drug dosage and improving the patient compliance of medication [13], maintaining drug concentration within the therapeutic interval by reducing in vivo drug concentration fluctuation [14], and reducing some of the unwanted side effects [15–17] by locally delivering the drugs exactly at the targeted areas (either at the disease sites or in the affected cells) [18–19], as well as enhancing the bioavailability of drug products by reducing or delaying premature degradation of a drug and intensifying the uptake of a drug [19]. Drug-delivery systems has become one of the most prominent research subjects in the fields of medicine and healthcare. Twenty-five years ago, in 1990, the Food and Drug Administration (FDA) approved the first liposomal DDS amphotericin B [20]. Since then more than 10 DDSs have been commercially released for different treatments ranging from cancer to muscular degeneration (as shown in Table 16.1) [20].

The use of DDSs appears to be a valuable method in improving the efficacy of drugs by relieving pain in patients suffering from prolonged critical illnesses [19]. Their availabilities have even influenced the economics of drug development [19]. Taking a step forward to transform existing drugs into controlled release formulations may lead to the enhancement

TABLE 16.1

DDSs Available in the Market and Approved by the FDA [19]

Product Name and Category	Year of Approval	Technology	Indication
Liposomes and Micelles			
Doxil	1995	Polyethylene glycosylated (PEGylated) liposomal doxorubicin	Various types of cancer
DaunoXome	1996	Liposomal daunorubicin	Advanced human immunodeficiency virus (HIV)–associated Kaposi's sarcoma
AmBisome	1997	Liposomal amphotericin B	Fungal infections
DepoCyt	1999	Liposomal cytarabine	Lymphomatous meningitis
Visudyne	2000	Liposomal verteporfin	Age-related macular degeneration
Estrasorb	2003	Estradiol micellar nanoparticles	Moderate to severe vasomotor symptoms of menopause
DepoDur	2004	Liposomal morphine sulfate	Postoperative pain
Polymer–Drug Conjugates			
Adagen	1990	PEGylated adenosine deaminase	Adenosine deaminase deficiency causing severe combined immunodeficiency disease
Oncaspar	1994	PEGylated L-asparaginase	Acute lymphoblastic leukemia
PegIntron	2001	PEGylated interferon alfa-2b	Chronic hepatitis C
Pegasys	2002	PEGylated interferon alfa-2a	Chronic hepatitis C and B
Neulasta	2002	PEGylated granulocyte colony-stimulating factor analog	Neutropenia
Somavert	2003	PEGylated recombinant analog of the human growth hormone	Acromegaly
Macugen	2004	PEGylated anti–vascular endothelial growth factor (VEGF) aptamer	Age-related macular degeneration
Mircera	2007	PEGylated erythropoietin receptor activators	Anemia associated with chronic kidney disease
Cimzia	2008	PEGylated tumor necrosis factor alpha inhibitor	Crohn's disease
Krystexxa	2010	PEGylated urate oxidase	Gout
Omontys	2012	PEGylated peginesatide	Anemia caused by chronic kidney disease
Biodegradable Materials			
Zoladex	1989	Poly(lactic-co-glycolic acid) (PLGA)/goserelin acetate	Prostate and breast cancer
Lupron Depot	1989	PLGA/leuprolide acetate	Prostate cancer and endometriosis
Gliadel	1996	Polifeprosan 20/carmustine	High-grade and recurrent glioblastoma multiforme
Sandostatin LAR	1998	PLGA–glucose/octreotide acetate	Acromegaly
Atridox	1998	Polylactic acid (PLA)/doxycycline hyclate	Periodontal disease
Nutropin depot	1999	PLGA/recombinant human growth hormone	Growth hormone deficiency
Trelstar	2000	PLGA/triptorelin pamoate	Advanced prostate cancer

(Continued)

TABLE 16.1 (CONTINUED)

DDSs Available in the Market and Approved by the FDA [19]

Product Name and Category	Year of Approval	Technology	Indication
Arestin	2001	PLGA/minocycline	Adult periodontitis
Eligard	2002	PLGA/leuprolide acetate	Advanced prostate cancer
Risperdal Consta	2003	PLGA/risperidone	Schizophrenia and bipolar I disorder
Vivitrol	2006	PLGA/naltrexone	Alcohol dependence and opioid dependence
Somatuline	2007	PLGA/lanreotide	Acromegaly
Ozurdex	2009	PLGA/dexamethasone	Macular edema
Protein-Based DDSs			
Zevalin	2002	Anti-CD20 monoclonal antibody/ yttrium-90	Non-Hodgkin's lymphoma
Bexxar	2003	Anti-CD20 monoclonal antibody/ iodine-131	Non-Hodgkin's lymphoma
Abraxane	2005	Albumin/paclitaxel	Breast cancer
Brentuximab vedotin	2011	Anti-CD30 monoclonal antibody/ monomethyl auristatin E	Hodgkin's lymphoma and systemic anaplastic large-cell lymphoma

of drugs' performance and the extension of pharmaceutical patent life [19]. A study conducted by Verma and Garg [4] revealed that the average cost and time of the development of a new drug are significantly higher (approximately $500 million and more than 10 years) than those required to create and deliver a new DDS (ranging from $20 million to $50 million and between 3 and 4 years). Therefore, it is not surprising that the US market for advanced DDS has grown significantly by 61.3% to $121 million in 2010 (2001: $75 million) [21]. Other research conducted by Zhang et al. [22] predicted that the annual global market for polymer-based controlled release systems alone will approach $60 billion in 2010.

In a report [23], the International Telecommunications Union (ITU) predicted that there will be more than 7 billion mobile phone subscriptions in the world by the end of 2015. This report also revealed that active mobile-broadband subscription is around 78% in Europe and Americas, 42% in Asia and Pacific, and 17% in Africa. This averages to an active subscription of 46% worldwide; 87% in developed countries, and 39% in developing countries. One of the foremost reasons behind this growth was the cost reduction of mobile phone technology deployment [23].

The combination of availability and efficiency in voice and data transfer networks and rapid deployment of wireless infrastructure are the key performance drivers that accelerate the integration process of mHealth solutions in the global society [24]. *Mobile health* or *mHealth* can be defined as the utilization of wireless technology for supporting effectiveness in public health and clinical practice [25].

There are several additional benefits that can be derived from mHealth systems, including improving the accessibility of health-related data, decreasing the cost of delivering medical services, enhancing the ability of diagnosing and distinguishing the trend of diseases, providing timely access to information on public health topics, and accommodating education and training of medical staff [26].

One of the future trends in biomedical engineering will be the integration of DDS with mHealth solutions. In this chapter, we review some of the basic concepts of DDS, wireless communication in association with DDS, the permissible exposures to electromagnetic frequency (EMF) in the human body, and a number of case studies. Finally, we conclude with further ideas on expanding the implementation of the system.

16.2 Implantable Drug-Delivery Systems (IDDSs)

Nowadays, there are a number of common applications of implants, including cornea implants, heart valve implants, orthopedic implants, and a total DDS intended for supporting diagnosis and treatment of patients with various forms of disorders (for instance, pain relief for cancer) [27].

Additionally, there are a number of implanted devices that can be remotely interrogated. Some examples include a continuous blood glucose–monitoring system that uses a remotely powered implantable microsystem [28] and a heartbeat and respiration rate detection system that utilizes remote-sensor technology [29].

Furthermore, there are a number of functions related to implantable devices for restoring lost human functions such as artificial eyes, brain pacemakers for treating Parkinson's disease, cochlear implants [30], muscle stimulators, and nerve signal recording as a part of robotic prosthesis [31–32]. Finally, there is a high interest in utilizing implantable devices to administer drug delivery as required and developing technologies related to micro-drug-delivery devices [1,7].

Unlike the conventional ways of administering drugs (i.e., oral, topical, or parenteral), IDDSs are capable of maintaining the therapeutic concentration of the drugs by performing only one administration process and targeting a specific body part, which could lead to a lower systemic level of drug and the use of more effective drug choices [33]. Lyu and Untereker [34] reported the current listing of key polymers for IDDSs (as shown in Table 16.2).

TABLE 16.2

Representative Polymers Used in Approved IDDSs

Polymer	Application Examples	Type
Silicone	Catheters, lead insulation, tissue filling, adhesive, intraocular lens	Stable
Polyurethanes (ether, ester, carbonate, hydrocarbon)	Catheters, lead insulations, structural components (e.g., artifical heart), pacemaker connectors, cervical(spinal) disc replacement, sensors	Stable
Ployetheretherketone (PEEK)	Orthopedic parts	Stable
Polysulphone	Structural components	Stable
Epoxy	Structural components	Stable
Polyethylene	Joint replacement	Stable
Poly(methyl methacrylate)	Bone cement, intraocular cement	Stable
Fluoropolymers	Vascular grafts, drug delivery coating	Stable
Poly(ethylene-co-vinyl acetate)	Drug-delivery	Stable
Poly(ethylene terephthalate)	Vascular graft, artificial heart value structure	Stable
Poly(lactide) and its copolymers	Orthopedic products, drug delivery, suture	Degradable
Polyanhydride	Drug delivery	Degradable
Collagen	Tissue filling, tissue engineering scaffolds, and drug delivery	Degradable

16.2.1 Implantable Pump Systems

16.2.1.1 Osmotic Pumps

In general, osmosis is a diffusion of a solvent through a semipermeable membrane across regions from low solute concentration to a higher solute concentration such that water balance can be achieved [35–36]. As shown in Figure 16.1a, when a semipermeable membrane separates two solutions with different solute concentrations, an osmotic flow will be generated; the membrane will reject the solute but allows the passage of the solvent molecules. The region with low solute concentration corresponds to a high chemical potential, and the region with higher solute concentration corresponds to a low chemical potential [35]. Therefore, a hydrostatic pressure difference across the semipermeable membrane results; consequently, an opposite flow of solvent [35], as shown in Figure 16.1b, is obtained. This figure shows that in equilibrium, the osmotic flow equals the flow due to the hydrostatic pressure difference [35]. In other words, the difference between the osmotic pressures equals the pressure difference needed to generate the balancing flow of the two solutions [35].

Water, in many cases, is used as solvent. Various types of osmotic pump basic principles for drug delivery, in both intracorporeal and extracorporeal systems [35] are represented in Figure 16.1c through e. Fundamentally, all of the osmotic pumps depend on the solvent flow through a semipermeable membrane for their operations. Figure 16.1c shows a single-compartment system, in which the solvent inflows across the semipermeable membrane, thus moving the saturated drug (as an osmotic agent) through the outlet. Two-compartment systems are shown in Figure 16.1d (for intracorporeal system) and e (for extracorporeal system). As shown in the figures, in each two-compartment system, the compartment with the osmotic agent will enlarge due to osmotic process and, therefore, moves the fluid drug into the second compartment.

In general, an osmotic pump consists of an osmotic engine, a semipermeable membrane, a piston, and a reservoir of pharmaceuticals (drugs) [37]. During its operation, an existing osmotic gradient between an osmotic engine and moisture in the neighboring interstitial

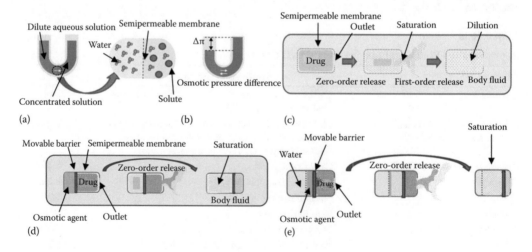

FIGURE 16.1
The basic principles of osmotic pressure: (a) osmotic flow is generated when two solvents with different solute concentrations are separated by a semipermeable membrane; (b) the osmotic pressure difference equals the pressure difference needed to generate the balancing osmotic flow. Various types of osmotic pump basic principles for drug delivery: (c) intracorporeal single-compartment system; (d) intracorporeal two-compartment system; and (e) extracorporeal two-compartment system. (Adapted from S. Herrlich et al., *Advanced Drug Delivery Reviews* 64, 1619. Copyright 2012 by Simon Herrlich. With permission.)

fluid produces a flow of water through the semipermeable membrane. Therefore, the osmotic engine expands, moves forward a piston, and finally expels the pharmaceuticals through an existing opening [37].

ALZA (currently part of Johnson & Johnson) developed the Viadur leuprolide acetate implant and Bayer marketed it until 2007 [38]. In 2000, FDA approval was granted to Viadur (that used the Duros platform) for cancer prostate treatment [38]. With a diameter of 4 mm and length of 45 mm, the Duros titanium implant delivers osmotically driven zero-order drug delivery [39–40]. A trocar is used to place the implant subdermally on the inside of upper arm [40]. Implant removal is performed after 1-year delivery duration [40]. According to a study by McNally et al [41], Viadur delivered 370 mg/mL of leuprolide in dimethyl sulfoxide continuously within 1 year at a rate of about 120 µg/day (or 0.4 µL/day) from a 150 µL drug reservoir. From both in vitro and in vivo experiments (rats and beagles), the release rate data showed zero-order delivery during 1 year [40,42]. Moreover, from monitoring of both serum testosterone and leuprolide levels, steady release rates for 12 months were obtained [40,42]. Finally, the implant was capable of suppressing testosterone concentrations in either canines or humans by producing a steady-state level of lueprolide serum [40].

16.2.1.2 Infusion Pumps

An IDDS delivers its drug hypodermically, for example, in the treatment of patients with insulin-dependent diabetes; a palm-size infusion pump is available [5]. An external wearable insulin pump is worn by the user (outside the body), while the cannula part is implanted. After the development of numerous types of external artificial beta cells (insulin pumps), there exist a number of subsequent insulin pumps, such as the ones introduced by MiniMed/Medtronic [5]. In recently developed programmable versions, the insulin pumps are integrated with continuous blood glucose monitoring. With the help of such pumps, it is possible to provide subcutaneous infusion of a minuscule dose of insulin with automatic changes in delivery rate and at specified time intervals. Moreover, the pump is also able to save previous infusion data and provide more patient comfort through short-syringe selection to break the syringe connector inside the pump. A soft cannula is inserted under the skin almost without pain; then insulin can be subcutaneously delivered for up to 3 days. A small glucose sensor is inserted subcutaneously using an automated insertion device and worn by the user, each time up to 3 days. The sensor data are sent to a transmitter attached to it; then the transmitter wirelessly sends the data to the pump to provide continuous blood glucose monitoring [43]. Panteleon et al. [44] has evaluated the use of a continuous closed-loop insulin delivery system in eight diabetic dogs. The obtained results have demonstrated that a stable glycemic control can be achieved within a specific range of gain by incorporating the combination between an automated closed-loop insulin delivery and subcutaneous insulin delivery.

16.2.2 Micro-/Nanofabricated IDDSs

A number of biomedical implants have been developed through a series of miniaturization processes [45]. Significant technological growth in MEMSs and nanoelectromechanical systems (NEMSs) in the industry has strongly influenced important results of the development. Specifically, new paradigms in the fields of biology and biomedicine have been developed through the newly created bio-MEMS-based implantable micro-drug-delivery systems [45].

Microelectromechanical systems (microsystems) can be defined as very small electromechanical devices processed using semiconductor/integrated circuit technology and built onto semiconductor chips and are measured in micrometers [46]. In general, MEMSs can be used in the miniaturization of either integrated devices or systems through combining both electrical and mechanical elements [46]. In their fabrication process, IC batch-processing techniques are implemented and their size varies from a few micrometers to millimeters [47]. Basically, MEMS applications include sensing, control, and actuating on the microscale but may produce macroscale effects as well [47].

Recently, MEMS and NEMS technology–based drug-delivery systems have become more popular, because of their interesting biomedical applications [48]. Both technologies have shown their advantages in producing various high-performance biomedical devices with small dimensions for meeting specific critical medical requirements that include location-specific drug delivery, improved bioavailability, increased therapeutic effectiveness, and reduced side effects [48]. Numerous types of medical devices have been developed using MEMS and NEMS technologies with various materials [2]. Some examples of the MEMS devices include microneedles, micropumps, microsensors, microvalves, microfluidic channels, and drug reservoirs [45].

New developments in implantable drug-delivery systems based on macrofabrication or nanofabrication technologies have existed for quite some time [49–51]. MicroCHIPS has developed minuscule microliter-sized drug reservoirs which are etched into a silicon wafer and filled with single or multiple drugs [52]. Gold or other film material covers the reservoirs and is electronically addressable on the chip [52]. When the implantation of the chip is complete, a voltage applied to the gold foil of a standard container will produce electrochemical dissolution of the film and release the drug from the container [52]. Thus, single or multiple drug delivery will be possible using controlled or pulsatile or patterned releases [53].

16.2.3 Implantable Microfluidic Devices

To obtain continuous drug delivery for a relatively long period of time from a built-in reservoir, diffusion-based drug-delivery systems are used [54–55]. The drug release time can be on the order of hours to days, depending on the diffusion coefficient of the particular drug [56]. The modification of dose initiation time in such systems is allowable, although in general the rate of delivery remains constant at a relatively slow rate [57]. To closely imitate a physiological release profile, such as insulin release, specific drug-delivery patterns are needed in some applications [57]. Some examples include a rapid dosage (bolus) delivery, a noncontinuous dosage-delivery pattern, or a pulsatile delivery pattern [57]. In vital application areas such as the brain, drugs provided by diffusion from polymer implants or bolus injections can, in general, penetrate 3 mm from the implant [58]. Therefore, instead of pure diffusion along a concentration gradient, convective transport, with an applied pressure gradient, is sometimes needed to deliver the drugs further into the tissue [59–61].

In an implantable microfluidic device, an active drug delivery from a reservoir to a target is commonly conducted with a power source [54]. Actuation can be initiated either manually [62], by applying a typical voltage [63–64], or wirelessly by making use of magnetic fields [65].

An innovative implantable drug-delivery device (with a thickness of less than 2 mm) has been developed and manufactured by Lo et al. [62] for the treatment of chronic ocular diseases (Figure 16.2). The device consists of a refillable drug reservoir, flexible cannula, check valve, and suture tabs using three layers of polydimethylsiloxane (PDMS); it is mechanically actuated by a patient's finger for dispensing a typical dose of drug from the device [62]. The study also reported that the valve closing time constants at pressures of 500 mmHg and 250 mmHg were 10.2 seconds and 14.2 seconds, respectively [62]. For replacing manual actuation, a micropump was made through water electrolysis into oxygen and hydrogen gases

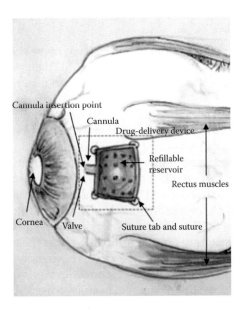

FIGURE 16.2
Application example of ocular drug-delivery system. (Adapted from R. Lo et al., *Biomedical Microdevices* 11, 962. Copyright 2009 by Ronalee Lo. With permission.)

[63–64]. Adequate pressure in the drug reservoir can be generated by the pump to accurately deliver an amount of 250 nL into the intraocular space through a Parylene cannula and a one-way check valve [62].

Pirmoradi et al. [65] has created a MEMS drug called docetaxel (DTX) in 35 days. The deformation of the magnetic PDMS membrane was caused by a magnetic field, which resulted in the release of drug to the surrounding bovine serum albinum (BSA) solution. The release rate of controlled DTX and the intensity of magnetic field are 171 ± 16.7 ng per actuation interval and 255 mT, respectively.

16.2.4 Ceramic Drug-Delivery Systems (CDDSs)

Typical implantable drug-delivery systems based on ceramics are grouped as ceramic drug-delivery systems [66]. A number of different pharmaceuticals have been delivered using these ceramics-based systems [67]. Currently, around five types of ceramics are used such as aluminium calcium phosphorous oxides, ceramic-metal hybrids, hydroxypatite, inorganic bone meal, tricalcium phosphate [67]. Two important characteristics of the CDDSs are in vivo biocompatibility (with both body fluids and tissues) and mechanical properties [5]. Both characteristics are suitable for the following therapeutic applications: bone infections, drug eluting stents (nanoporous coatings), intraocular implants for glaucoma treatment, and transurethral devices for impotence treatment [5].

These systems offered benefits like the delivery of controlled rate of drugs at targeted locations and the opportunity to protect unstable in vivo drugs [5]. Therefore, the development of IDDSs for one-time drug administration in a week to a year is likely. Various literatures comprehensively discussed both general and specific CDDSs [67–71]. Some reviews have focused on the magnetic-HAP nanoparticles [68], the microporous coating on drug-releasing vascular stents [69] and the mesoporous silica/magnetite systems [71].

16.2.5 PMMA Beads

Among two of the major complications in modern surgery are postoperative and post-traumatic infections of bone, soft tissue, and joints [72]. The invention of antibiotic-loaded PMMA beads for local delivery of antibiotics made an enormous impact on the treatment of osteomyelitis [72–73].

One of the commercial antibiotic-loaded PMMA beads that are obtainable only in certain regions of the world due to different reasons is Septopal. Unfortunately, most patients in the developing countries cannot afford to buy those beads due to relatively high prices. As a solution, the hand mixing of antibiotics into bone cement and the bead preparation are manually performed by orthopedic surgeons in the operating room [74].

Local orthopedic surgeons in Indonesia usually mixed the antibiotic (fosfomycin sodium) with the artificial resin PMMA for creating handmade antibiotic-loaded beads [74]. However, there was no scientific proof for the efficacy of these beads. As a case study, researchers have asked some orthopedic surgeons to prepare antibiotic-loaded beads based on their own version (as shown in Figure 16.3) [74]. The reported results have shown a large difference in sizes and shapes. Note that the ratio sizes of these handmade beads is quite small (varies between 2.1 and 3.8 mm) as compared to that of Septopal beads (8.6 mm). A large volume of the beads is required for efficiently releasing the antibiotics [74].

As a solution to those issues, Rasyid [74] has designed and manufactured a mould system that can be used to produce antibiotic PMMA beads when performing surgery in the operating room. The proposed system consists of a pair of PTFE plates and a set of stainless steel rings (as shown in Figure 16.4). Thirty PMMA beads can be made by using this proposed template system and are nearly spherical in shape, each with a diameter of about 8 mm. A 0.8 mm diameter stainless steel wire connects those beads to each other. The preparation of the beads will take about 30 min and orthopedic surgeons can choose to mix the bone cement with various types of antibiotics, or even without any antibiotic. The resulting low-cost antibiotic-loaded beads can be implemented in a much similar procedure as the existing commercial ones.

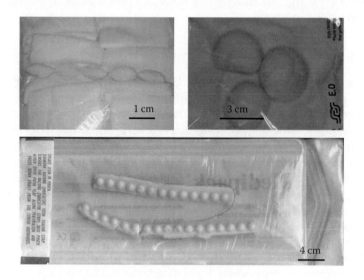

FIGURE 16.3
Handmade fosfomycin-loaded beads are prepared by local orthopedic surgeons in Indonesia. (Reprinted from H. N. Rasyid, "Low-cost antibiotic delivery system for the treatment of osteomyelitis in developing countries," PhD dissertation, University of Groningen, Netherlands, 2009, p. 19. With permission.)

(a)

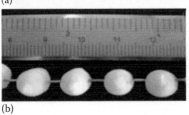
(b)

FIGURE 16.4
(a) The proposed template system; (b) a chain of beads consisting of small antibiotic PMMA beads connected via a flexible stainless steel wire. The beads are (almost) spherical, each with a diameter of 8 mm, and the distances from one bead to another vary between 3 and 4 mm. (Reprinted from H. N. Rasyid, "Low-cost antibiotic delivery system for the treatment of osteomyelitis in developing countries," PhD dissertation, University of Groningen, Netherlands, 2009, p. 87. With permission.)

16.2.6 Discussion

The improvement of safety and efficacy of drug therapy is one of the most important goals in designing s DDS and this goal can be achieved by releasing medication at a predetermined rate and/or particular site of action [5].

The advantages of using implantable devices for DDSs are the following [5]:

- The compliance of the patient: Since the implanted devices released a dose of drug at a predefined time, the patients are no longer required to remember their medication schedule. Furthermore, it also helps to reduce the required amount of drugs.

- The reduction of side effects: By sustaining drug release over prolonged periods of time at the site of action, fewer doses are needed for effective treatment, and peaks and valleys of the drug plasma levels can be avoided.

- The lowered dose of drug: If drugs are targeted to desired sites, drugs need to cross biological barriers, for instance, first-pass hepatic effects, before penetrating to the active receptor.

- The improvement of drug stability: A DDS controls the rate of drug release, preventing inappropriate rates of drug metabolism.

- Suitability over intravenous (IV) administration: Patients with chronic diseases may no longer need to stay in the hospital.

- Cases of drug allergy: In the case where adverse drug reactions occur, the removal of implanted devices can be processed immediately.

The potential disadvantages of using implantable devices for DDSs are the following [5]:

- Minor surgery needed for the insertion and/or removal of some implantable devices: This need contributes to lower user acceptance of the IDDS and increases demands for less invasive alternatives.

- The cost-effective ratio: The IDDS may not be covered by insurance and could reduce the acceptance of the product.

- Training requirement: Proper implantation and explantation of implants may need specific surgical skill training.
- More complex IDDS design and manufacturing (as compared to oral dosage): The requirements may need longer time and higher cost for obtaining the essential regulatory approval.
- Unwanted effects/adverse reactions: Patient discomfort and pain in the implantation and explantation process may lead to lower patient acceptance rates.

16.3 Dermal Drug-Delivery Systems

Two types of technologies employed for transdermal and intradermal delivery are microneedle syringes and microneedle patches [37]. The aim behind the development of microneedles was to enhance the permeability of the skin without any pain involved [37].

The thickness of stratum corneum in the upper epidermis is around 20 μm and this layer has become the barrier to the transport mechanism of drugs [37]. The use of microneedles will aid in the micrometer penetration of drug products due to the fact that some of the nerves are located at a few hundred micrometers below the skin [37]. Moreover, it helps to bypass the stratum corneum barrier and does not cause any pain [37].

Drug–polymer mixtures are used for the formation of biodegradable needles and this type of needle offers the capability to deliver a high payload of drugs to targeted areas but also suffers from poor performance of drug stability [50]. The drugs that are stored in the external reservoir or inside the hollow of the needle can be released through the skin by using hollow microneedles with reservoirs [37].

16.3.1 Microneedle Syringes

Deep reactive-ion etching (DRIE) technology has been used to fabricate most of silicon-based microneedles and has also been applied to drug-coated microneedles (solid or hollow bores with tapered or beveled tips) [50,75–77]. Though these coating microneedle arrays are not reservoir-based drug-delivery devices, integration with a reservoir system can be accomplished and an initial bolus dose can be released from these arrays [37]. Hollow microneedles with reservoirs have been implemented in various ranges of geometries including barbed tips [78], bubble pumps for delivery fluids [79], blunt cylinders [80], citadel structures [81–82], microfilters at the needle base [83], sawtooth structures [84–85], tapered cylinders [86], and volcano shapes [87–88].

Debiotech holds the license for the citadel side-opened needles of the MicroJet/Nanoject platform [37]. There are two advantages offered by the use of side-opening hollow microneedles: highly efficient skin penetration can be achieved using an ultrasharp tip and the tissue coring and/or channel blockage can be alleviated during the insertion process [81–82]. The microneedles are fabricated in a single chip with the size of 3×3 mm^2 and have a length of about 210 μm [81]. The microneedles can inject 100 μL of fluid directly into the skin in 2 s [81]. Debiotech produces an array of 25 microneedles, each of which is 300–1000 μm in length [37]. A gold film is used for sealing the drug reservoir and will be ruptured when applied to the skin [89]. Moreover, the microneedles can be integrated with the MEMS micropump for insulin delivery [90]. The researchers also pointed out the systematic or local delivery of macromolecular drugs (for instance: endocrine substances and vaccines) can be performed efficiently by using this type of microneedles [90].

Therapeutic drug intervention into the eye can be delivered using hollow microneedles [91]. The suprachoroidal space of pig, rabbit, and human cadaver eyes are the target areas in which sulforhodamine B as well as nanoparticle and micoparticle suspensions are released. The volume of drugs that is constantly released through the needles is 800–1000 μL. Small particles with the size of 20–100 nm can be quite easily flowed through the scleral tissue and larger particles with the size of 500–1000 nm can be transported through long needles. A system consisting of a miniature syringe and a PDMS drug reservoir has been established by Häfeli et al. [92] for releasing liquid formulations via hollow microneedles. Figure 16.5 presents the proposed method in the field of microneedle-based drug delivery.

The lengths of the sawtooth structures range between 150 and 350 μm and the width of the structure's base is about 250 μm [84]. The preliminary experiments have verified that the microneedles arrays are robust enough to survive skin penetration tests [84]. These structures have been integrated into the MicroJet Needle (from NanoPass Technologies Ltd.) and have been tested in several clinical trials, including the administration of local anesthesia, influenza vaccine, and the delivery of insulin [85].

Silicon, metal, and polymer are three main components for creating tapered cylinder hollow microneedles [86]. The silicon needles have an increased wall thickness at the base and a bore diameter of about 60 μm [86]. While for the metal needles, the thickness of the wall is constant at 10 μm and the bore of the needle is widened at the base [86]. The width and the length of the needles are 35–300 and 150–1000 μm, respectively [86]. In a study conducted by McAllister et al. [86], a syringe has been used to deliver insulin for regulating the blood glucose level in diabetic rats. The result demonstrated that a half hour microinfusion was able to maintain the blood glucose level at steady state for more than five hours. Another study has chosen to use an array of 16 microneedles, made from polyethylene terephthalate with a length of 50 μm and a tip diameter of 75 μm, for delivering insulin to rats with diabetes mellitus [93]. The study also revealed that blood glucose level

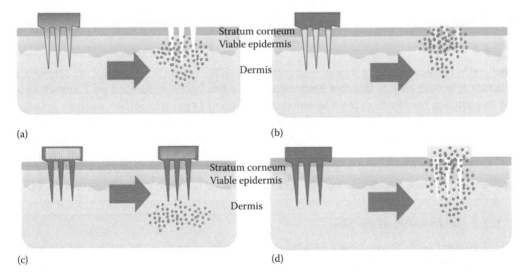

FIGURE 16.5
Methods of drug delivery using different types of microneedles: (a) coated microneedles; (b) dissolving microneedles; (c) hollow microneedles; and (d) solid microneedles. (Adapted from Y.-C. Kim et al., *Advanced Drug Delivery Reviews* 64, 1549. Copyright 2012 by Yeu-Chun Kim. With permission.)

is decreased in the first 4 hours and then stabilized for the next 4 hours. It also reported that passive diffusion reaches a peak value of 0.43 ng/mL [93].

Some researchers have injected methyl nicotinate into the body by using volcano-style needles [87–88]. There are two main components of the MEMS design: silicon microneedles array and a PDMS reservoir that permanently connected to the back of the silicon plate [88,92]. The reservoir has a dimension of 10 mm and holds an array of eight needles that have a length of 200 μm [88]. Hafeli et al. have experimented with 4 different mice to evaluate the eficiency of the microneedles for intradermal drug delivery. The device is pressed on the skin of mice for one minute with approximately a two-newton force for injecting the radioactive albumin solution [92].

16.3.2 Microneedle Patches

One of the achievements in the field of microneedle-based drug delivery is the integration of microneedle arrays into transdermal patches [37]. V-Go uses tapered cylinder hollow microneedles for continuously delivering insulin in the required amounts (at either basal rate or bolus rate) and enabling technology has been licensed from Georgia Tech [86,93–94]. The combination of the hPatch platform that is designed for injecting drugs into subcutaneous body tissues and the Micro-Trans microneedle array has been employed in this product (V-Go).

Dissolving microneedle patches for the vaccination of influenza are developed by some scientists and have been tested in mice and pigs [95–96]. Each patch consists of polyvinylpyrrolidone (PVP) microneedles with 650 μm in height and contains encapsulated 3 μg of inactivated influenza virus. The patch can be applied by a thumb and is penetrated into the skin to a depth of 200 μm. Then the drug inside the patch is deposited into the epidermis. It has been found from the study that the effectiveness of using dissolving microneedles to clear lung virus was 1000-fold more efficient than using hypodermic needles [95]. This product has facilitated the distribution of the vaccine in remote areas around the world [95].

16.3.3 Discussion

16.3.3.1 Pain

Minimally invasive interventions using microneedles have the potential to attract the patient's attention [97]. A blinded controlled study has been performed by Kaushik et al. [98] to compare the levels of pain when using different types of needles, such as an array of 400 microneedles, and a 26-gauge hypodermic needle, and a smooth surface of silicon wafer. The study revealed that patients feel less pain when pressing the microneedles, just 150 μm in length, into their skin than when injecting a drug with that particular hypodermic needle. The average pain scales with 95% confidence intervals were 0.42, 0.67, and 23.9 for the silicon wafer, the microneedles and the gauge hypodermic needle, respectively.

16.3.3.2 The Irritation of the Skin

Direct contact with these types of microneedles may cause mild, transient skin irritation and/or punctate erythema [97]. One study has reported that there is a clinical connection between the length of microneedles and the greater presence of erythema [99]. For instance, a study conducted by Bal et al. [99] recruited 18 healthy volunteers for assessing the safety of the microneedles. The researchers used solid and hollow metal microneedle arrays. The study found out that 400 μm long microneedles resulted in the induction of greater erythema and blood flow when compared with 200 μm long microneedles [99]. Another study conducted by Noh et al. [100] has reported that mild erythema will be gone within hours or may take days. In their study, in vitro experiments were performed on hairless rats and healthy participants.

16.3.3.3 The Infection of the Skin

There has been a rise in concerns that infection could be spread through residual holes in the skin made by microneedle insertion [97]. A clinical study has shown that residual holes that exist in the skin following sterile hypodermic needle insertion and withdrawal are not the sources of infection [101] and these are largely due to the innate immune defense of skin [102]. Animal and human experiments using microneedles have consistently shown that infections are unlikely to occur after the removal of microneedles [97]. It is highly recommended that the contaminated syringes and needles should be discarded into safe disposals to prevent the spread of infectious diseases in communities [101].

16.4 Human Body Modeling for Study of EMF Interaction

It is known that the human body is a good electromagnetic energy absorber due to its capability of absorbing energy generated by any transmission circuits [103]. Human tissue electrical properties need to be considered when proposing implantable telemetry modules [103]. The human body comes in heterogeneous shapes and sizes and is composed of a variety of tissues and organs [103]. All of these human body parts have specific conductivities, magnetic permeabilities, and permittivities [103]. For example, at the frequency of 150 MHz, fat has a conductivity of 0.035 S/m and a relative permittivity of 5.79, while the conductivity and the relative permittivity of blood are 1.644 S/m and 65.2, respectively [104].

Advances in wireless communication have contributed to the growth of studies related to the interaction between different types of electromagnetic fields and radiators with the human body [105–107]. Despite this trend, there are a number of barriers that are currently slowing the research process including the difficulties of defining an in vivo reference standard and restricted conditions imposed on the human body [104]. In order to overcome these barriers, some studies have been directed to assess implantable device interactions with electromagnetic fields [103]. Another indirect approach is by simulating human body responses to EMF exposure [103].

The full-segmented model, the semisegmented model, and the simple homogenous model are three types of human body basic models [103]. The homogeneous body model is frequently chosen because of its computational simplicity [108]. In this particular model, the homogeneous dielectric properties shared among surrounding and adjacent tissues and the animation software can be used to determine the shape of the body [103]. A combination of those models has been proposed to create a composite-body model with the intention to demonstrate the real condition of a human body [103].

The results of simulation modeling and the measurement of real absorption of radiofrequency energy have been compared in a study conducted by Scanlon et al. [105]. The reference system consists of radio telemetry systems at 418 MHz and 916 MHz that have been semi-implanted in the human vagina while the simulation model has involved the simulated interaction between human body and electromagnetic waves. In this study, researchers have proposed to use a 5 mm fully homogeneous model that has been divided into three different layers (skin, fat, and muscle) and three semisegmented body models that have various tissue depths from the source of radiation at excitations of 100, 200, and 300 mm, respectively. It has been demonstrated from the study that there were good agreements between practical measurements and simulations, from within 2 dB at 418 MHz to 3 dB at 916.5 MHz. Moreover, the analysis results have shown that there is not much variation in the radiation pattern and field strength at different depths of layers. Therefore the original assumption that most of the EMF energy has been absorbed by the surrounding tissues is correct.

Chirwa et al. [104] have carried out a study to investigate the use of wireless capsules to examine the human gastro-intestinal (GI) tract disorder in the intestine organ and to assess the level of EMF exposure from the device. The chosen frequency bands are varied from 150 MHz to 1.2 GHz. Compared to the proposed approaches in the study conducted by Scanlon et al. [105], the research team in this study has constructed a full-segmented body model by employing the finite-difference time-domain (FDTD) method, which is based on the collected Visible Human Project data. They considered two different performances in this study: near-field and far-field performances. Results from both of the measuring performances have shown that the maximum radiation has been detected on the body's anterior side from sources placed in the small intestine (or small bowel). Furthermore, the study found out that the frequency band between 450 and 900 MHz must be chosen to achieve effectiveness of the implanted devices. Finally, note that the overall experiment was conducted by placing the EMF sources inside intestine; therefore, there is a chance that the obtained results are not applicable to all types of implants (e.g., in the case of subcutaneous implants).

16.5 Power

To maintain the operation of implanted medical implants, the availability of efficient battery power is critical [109]. The battery must be small in size and provide a steady supply of power to the implants [110]. Some existing medical implants (not intended for DDSs) are already equipped with this kind of battery [111]. However, the battery size still needs to be reduced further [111]. Although technology for delivering reliable, long-term battery performance has been successfully deployed for particular products such as cardiac-assist devices, the fabrication of a small battery can still be quite challenging due to the need for releasing drugs from such implants [111].

One of the possible solutions for solving these issues is a wireless transcutaneous energy-transmission system for any chronically implanted medical device or in vivo medical microdevices [27]. As a result of growing use of radio-frequency inductive coupling, the use of wireless powering for implanted medical devices is increasingly feasible [33,112]. Energy can be transmitted transcutaneously through magnetic (inductive) coupling between two coils, one of which is an external transmitter coil and the other being a receiver coil within the implant [33].

Some of the application examples are atrial defibrillation [113], neurostimulation [114–115], and physiological recording [116–118]. Researchers have conducted several experiments to improve the efficiency of power transfer, either by focusing on the optimization of coil design [119] or the reduction of power consumption (as a result of a reduced effective series resistance [ESR] of the receiver coil) [120].

There is a potential possibility of further reducing or minimizing the size of IDDS devices by eliminating the battery as a power source [51]. One of the existing applications is IDSS for glaucoma treatment [33]. A pressure sensor is used to monitor the level of intra-ocular pressure (IOP) while the drug will be released locally by the system when the IOP level exceeded normal values [33]. Wireless power-transfer technology has been adopted by MicroCHIPS for numerous types of the reservoir-based system. This technology adoption has provided significant reductions in the size of the device due to the battery elimination and actually improved the feasibility of intraocular implants [121–122].

In an earlier work by Smith et al. [122], the research team has designed and implemented a new silicon-based IDDS as an ocular implant. The team also proposed the inductive link implementation for the IDDS which can be used for multiple purposes, including power-supply transfer and data-exchange process (an "address" signal is sent out to the targeted reservoir). For ensuring the safety of a person, the transmitted power must not exceed the maximum permissible RF exposure limits. The study has revealed that the communication system has transferred 500 μW of power over a short distance (300 mm) for supplying energy to the implant. The required powers for operating the receiver system and opening the cavity are ~370 μW and 90 μW, respectively.

16.6 Data Telemetry

The medical implant communication service (MICS) band and the ISM band are two frequency bands that are used for implantable medical systems [103]. The 402–405 MHz MICS band is assigned by the Federal Communications Commission (FCC) for medical implant communication services and mainly used for implantable devices in humans and animals [123]. The 6.78 MHz–245 GHz ISM band allocated by the International Telecommunication Union for industrial, scientific, and medical applications is widely applied on various types of wireless applications [124]. The two standards IEEE 802.15.1 (Bluetooth) and IEEE 802.15.4 (ZigBee) reside within the ISM band [125].

Nowadays, most of these implantable devices are facilitated with wireless communications for the transmission of data, the sending of command signals, and the transportation of power [103]. Both inductive coupling and RF communication are being used for data telemetry systems [103].

16.6.1 Near-Field Resonant Inductive Coupling

The use of batteries for implantable devices is not recommended due to limited lifetime and potential hazards associated with leakage, which can affect the surrounding tissues [126]. Therefore, there is a great need to develop and implement wireless power delivery systems. Wireless communication technique has adopted the inductive coupling principle for transmitting a series of external commands to the implant and retrieving data from the implant [103]. This technique is commonly used for particular implants that do not require a high-speed data-transmission rate, including implantable microstimulators and implantable RFID transponders [126–127]. Other researchers have adopted this technique as well, particularly for transcutaneous wireless power transfer in implantable devices [128–129]. A research by Lee et al. [126] has stated that the choice of carrier frequency will affect the system design. The benefit of choosing low carrier frequency is simpler design and less power loss, but the drawbacks of this choice is that the device needs large antenna and tuned capacitors. In contrast, the choice of high carrier frequency will lead to tremendous power loss and unwanted effects to the human tissues.

With the intention of minimizing circuitry, some designs have allocated a single carrier-frequency band for both data telemetries and power [103]. On the other hand, some models have separated data telemetries and power into different carrier-frequency bands for reducing the number of interferences and increasing the speed of data transmission [103]. A carrier frequency, which resides within either medium-frequency (MF) or high-frequency (HF) bands (including 13.56 [127], 20, and 2 MHz [126]), has been chosen for the implementation of systems for transcutaneous telemetry. In a study conducted by Lee and Lee [126], an inductive charging system has been designed by employing an E amplifier with amplitude-shift keying (ASK) modulations and a 2 MHz carrier frequency for supplying the required power to the implantable microstimulator with lower coupling efficiencies.

The disadvantages of using this inductive coupling system are a bulky system due to the replacement of antennas with inductive loop coils and a low-speed data-transmission rate [103]. There is a need to place the external coil and the internal coil in a near-field range for establishing a coupling communication session [103]. However, an accurate placement of those coils could be harder to achieve due to the human body irregularities in shapes and functioning [103].

16.6.2 Wireless Communication

The integration between several implantable medical devices and wireless RF telemetry has taken place to prevent the complications due to the use of wired implants [130–131]. The implantable antenna will transmit the data from inside the human body to the base station directly [132]. Some researchers have proposed the design and development of an implantable antenna [133–136] (not exhaustive).

The availability of wireless communication in microchip medical devices offers distinctive advantages [111]. By using wireless communication technology, the functionalities of an implantable device are significantly improved [111]. This kind of technology could be used for collecting data for further evaluation by using artificial neural network (ANN) technology [137–138], controlling the release of the drug from the reservoir, editing commands, monitoring device, observing patient responses to medication, transferring supply power to the implants, and transmitting data from the implants (for instance, status reports and/or alerts) [139].

There are three different devices involved in wireless communication: extracorporeal or interimplant communication devices [140–141] or a combination of the two. One of the case studies is the use of wireless communication for providing communication and transferring power to the ophthalmic drug-delivery implants [121].

Since there is a need for avoiding interference from other wireless signals within a specific period of time, some rules related to signal strength and frequency have been released for controlling the proliferation of wireless-enabled medical implants [142].

As mentioned previously, various medical implants have chosen to use the ISM band as a carrier frequency for data telemetry. A study conducted by Valdastri et al. [143] is focused on the design and implementation of an implantable data telemetry system that used the 2.4 GHz frequency band as a communication medium. There are three main parts in the proposed system: an implantable unit, an external host, and a user terminal. The assessment of the system is performed through a series of in vivo experiments. The results have shown that the researchers have successfully established a reliable data transfer from/to the implant in an anesthetized pig, with 13.33 μW transmission power. On the other side, the MICS band is specifically proposed for active implantable medical systems like glucose monitor [144], blood pressure monitor [145], and bionic eyes [146]. This frequency band has an available maximum bandwidth of 3 MHz and can cover an area within a 2 m radius [123]. For safety reason, the maximum power in the air is restricted to 25 μW equivalent radiated power (ERP) [147]. The calculation of ERP considered all the gain and loss factors inside the transmitter [148].

Note that the requirements for this wireless communications are an adequate power supply, reliable transmission connectivity, and biocompatible materials [109,149]. However, this proposed method suffered from poor transmission, as the transmitted signals were often unable to penetrate the biological tissues [109]. Moreover, the size of the antenna is quite big; therefore, it affects the size of the implanted devices and prevents the use in particular areas of the body, including the brain, the heart, and the spinal cord, due to safety issues [109]. Furthermore, the insulation layer for implantable antenna must be made from biocompatible material with the intention to deliver biocompatible solution and to prevent short circuit due to electrical conductivity of human tissues (for instance, see the paper of Beshchasna et al. [149]). Finally, the complexity of the human body could affect the realization of the implantable antenna [150–151].

To solve those issues, some researchers have investigated other types of wireless communication, including optical [152] and ultrasound [153]. However, the drawbacks of the alternative methods are miniaturization problems and low transmission efficiency when the invisible stream of RF signals penetrate through the human body [109].

16.7 mHealth Solutions for Drug-Delivery System

16.7.1 Automatic Drug-Delivery System

The automatic drug-delivery system proposed by Kumar and Sindhu [154] is performed based on the robotic brain's principle where each decision will be made by the machine on a case-by-case basis. A microsensor implanted inside the body can be used to automatically detect heart disease symptoms (e.g., heart attack). An idle mode will be maintained when normal conditions are detected, but if there are clinical symptoms that indicate the heart disorder, then necessary actions must be taken. The implanted capsules automatically release the drugs into either the arteries or the veins. In case that decision cannot be made, then the message will be transmitted to the smartphone from the machine and subsequently the actions are taken.

16.7.2 Disposable Patch Pump

The concept of JewelPUMP [155] is based on the previous NanoPUMP construction. A connection is established between a disposable reservoir with pumping mechanism and a reusable controller. The volume of the reservoir is 5 mL, while the amount of insulin held by the reservoir is 500 U of insulin, which will be enough for 1 week's treatment. The lifetime of the container is 2 years. The external volume of the entire pump is around 60 × 40 × 14 mm^3 and the total weight of this pump is less than 25 g.

A first sensor in the pump is used to observe the insulin temperature inside the reservoir and to automatically adjust the treatment. Moreover, it senses when overexposure to heat occurs and is designed to remind the patient about this situation (for instance, exposure to the sun for quite some time). The pump is connected via Bluetooth to an Android mobile phone for control purposes. The mobile application is intended to manage the dosage of insulin and allows a patient to send this information to the caregiver. The design is considered to be highly convenient since it comes with different colors and font types for each menu. A unique subscriber identity module (SIM) card is integrated with the pump to enable secure data exchange between the pump and the remote controller (smartphone).

16.7.3 Wireless Drug Dosage Monitoring

Figure 16.6 shows the basic concept of wireless monitoring system for drug dosage calculations. Real-time tracking process for monitoring the volume of the drugs inside transdermal and implanted drug-delivery devices can be processed through proper design and retrofit of antenna sensors [156] as shown in Figure 16.6 [157]. The examples of commercial products for transdermal drug-delivery system shown here are the MiniMed insulin pump [158] and Empi Action Patch [159]. In addition, some commercial products for IDDSs include SmartPill [160], Philips iPill [161], and MicroCHIPS [162].

Any RFID reader embedded on mobile devices (such as PDAs or smartphones) can be used as an external control and monitoring unit. This proposed approach could become one of the vital features in the sensors and potentially turned into a part of the nodes in a wireless network of a healthcare system [163].

A research team has designed and developed miniaturized helix antennas that can function as RFID wireless tag sensors when integrated with drug reservoirs [157]. The proposed design is based on the theory of electromagnetic fields and the simulation of finite-element models. An approach to encapsulating antenna sensors in a PDMS package has proved to be cost effective. In addition, the drug storage is included inside the

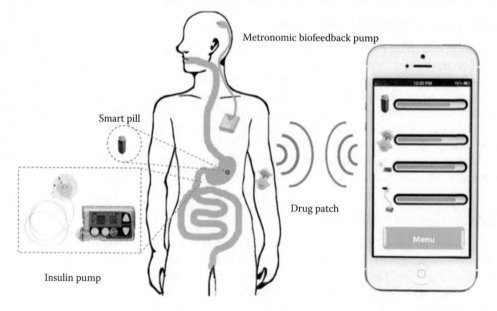

FIGURE 16.6
The proposed implementation of wireless system for monitoring the release of drugs from drug-delivery devices. (Adapted from H. Huang et al., *Translational Engineering in Health and Medicine, IEEE Journal of,* 2, pp. 1, 8. Copyright 2014 by Haiyu Huang. With permission.)

sensor. Moreover, in vitro experiments have been carried out on two prototype systems for assessing the ability of such systems to monitor drug dosage by tracing the resonant frequencies of antenna sensors that shifted from 2.4 to 2.5 GHz ISM band. Results from these experiments have shown that the sensitivities of transdermal and implanted drug-delivery devices were about 1.27 and 2.76 μL/MHz, respectively.

16.8 Conclusion

There are many issues related to conventional drug delivery systems (tablets, capsules, etc.) [5], including the patient compliance with medication, the incorrect dosage of medication, and the harmful side-effects of the drug. Furthermore, both the physicians and patients are not completely aware of the importance of sustaining optimal therapeutic blood levels. Even today, those issues are not addressed sufficiently.

Based on the presented views, there is a need for an alternative system that can be used to control the efficacy, pharmacokinetics and pharmacodynamics, and/or toxicity of drugs [9]. This system, often recognized as a DDS, is based on multidisciplinary research teams working in bioconjugate chemistry, molecular biology, polymer science, pharmaceutics, and engineering [164]. There are two elements that need to be possessed for an ideal DDS: the capabilities to target and to control the release process of drugs [165]. By having these capabilities, the high efficacy of the drugs can be preserved and the unwanted side effects can be reduced [165]. For example, in the case of cancer, the cancer drug is not only killing the cancer cells but also destroying the patient's healthy cells [166]. The temporal and/or spatial controls over the drug release are two important functions offered by controlled DDS. The release rate can be set at constant or variable values [167].

One of the high priority steps in establishing a fully smart system and its infrastructure for delivering healthcare services in the field of medicine is by combining the drug-delivery system with wireless technology [156], particularly as a part of an mHealth system.

In this chapter, we have elaborated on the recent developments in external and implanted drug-delivery systems based on various supporting technologies. A number of existing application examples of mobile health and drug-delivery systems integration have been presented. We also understand the "almost unlimited" number of continuous developments in the supporting technologies, such as microelectronics, micro- and nano-electromechanical systems, and mobile phones/smartphones as well as other telecommunications, wireless devices, and infrastructures. Therefore, in the near future, more and more new developments resulting from the integration of drug-delivery systems and their supporting technologies are expected.

References

1. V. Ranade and M. Hollinger, *Drug Delivery Systems*. CRC Press, 2004.
2. A. Nisar, N. Afzulpurkar, B. Mahaisavariya, and A. Tuantranont, "MEMS-based micropumps in drug delivery and biomedical applications," *Sensors and Actuators B: Chemical*, vol. 130, no. 2, pp. 917–942, 2008.

3. "Lifespan," https://www.lifespan.org. Accessed on May 10, 2014.
4. R. K. Verma and S. Garg, "Drug delivery technologies and future directions," *Pharmaceutical Technology On-Line*, vol. 25, no. 2, pp. 1–14, 2001.
5. L. W. Kleiner, J. C. Wright, and Y. Wang, "Evolution of implantable and insertable drug delivery systems," *Journal of Controlled Release*, vol. 181, no. 1, pp. 1–10, 2014.
6. S. Zafar Razzacki, P. K. Thwar, M. Yang, V. M. Ugaz, and M. A. Burns, "Integrated microsystems for controlled drug delivery," *Advanced Drug Delivery Reviews*, vol. 56, no. 2, pp. 185–198, 2004.
7. N.-C. Tsai and C.-Y. Sue, "Review of MEMS-based drug delivery and dosing systems," *Sensors and Actuators A: Physical*, vol. 134, no. 2, pp. 555–564, 2007.
8. J. Patil and S. Sarasija, "Pulmonary drug delivery strategies: A concise, systematic review," *Lung India*, vol. 29, no. 1, p. 44, 2012.
9. V. K. Devi, N. Jain, and K. S. Valli, "Importance of novel drug delivery systems in herbal medicines," *Pharmacognosy Reviews*, vol. 4, no. 7, p. 27, 2010.
10. H. He, X. Cao, and L. J. Lee, "Design of a novel hydrogel-based intelligent system for controlled drug release," *Journal of Controlled Release*, vol. 95, no. 3, pp. 391–402, 2004.
11. M. Bikram, A. M. Gobin, R. E. Whitmire, and J. L. West, "Temperature-sensitive hydrogels with SiO_2–Au nanoshells for controlled drug delivery," *Journal of Controlled Release*, vol. 123, no. 3, pp. 219–227, 2007.
12. B. Kundu, C. Soundrapandian, S. K. Nandi, P. Mukherjee, N. Dandapat, S. Roy, B. K. Datta, T. K. Mandal, D. Basu, and R. N. Bhattacharya, "Development of new localized drug delivery system based on ceftriaxone-sulbactam composite drug impregnated porous hydroxyapatite: A systematic approach for in vitro and in vivo animal trial," *Pharmaceutical Research*, vol. 27, no. 8, pp. 1659–1676, 2010.
13. F. Steinke, W. Andrä, R. Heide, C. Werner, and M. E. Bellemann, "Rotating magnetic macrospheres as heating mechanism for remote controlled drug release," *Journal of Magnetism and Magnetic Materials*, vol. 311, no. 1, pp. 216–218, 2007.
14. K. Koombua, R. M. Pidaparti, and G. C. Tepper, "A drug delivery system based on polymer nanotubes," in *Proceedings of the Second IEEE International Conference on Nano/Micro Engineered and Molecular Systems*, pp. 785–787, 2007.
15. X. Wang, T. Chen, Z. Yang, and W. Wang, "Study on structural optimum design of implantable drug delivery micro-system," *Simulation Modelling Practice and Theory*, vol. 15, no. 1, pp. 47–56, 2007.
16. Y. Sun, Y. Peng, Y. Chen, and A. J. Shukla, "Application of artificial neural networks in the design of controlled release drug delivery systems," *Advanced Drug Delivery Reviews*, vol. 55, no. 9, pp. 1201–1215, 2003.
17. A. Mohamad and A. Dashevsky, "pH-independent pulsatile drug delivery system based on hard gelatin capsules and coated with aqueous dispersion Aquacoat® ECD," *European Journal of Pharmaceutics and Biopharmaceutics*, vol. 64, no. 2, pp. 173–179, 2006.
18. S. Prabhu and S. Hossainy, "Modeling of degradation and drug release from a biodegradable stent coating," *Journal of Biomedical Materials Research Part A*, vol. 80, no. 3, pp. 732–741, 2007.
19. Y. Zhang, H. F. Chan, and K. W. Leong, "Advanced materials and processing for drug delivery: The past and the future," *Advanced Drug Delivery Reviews*, vol. 65, no. 1, pp. 104–120, 2013.
20. T. M. Allen and P. R. Cullis, "Drug delivery systems: Entering the mainstream," *Science*, vol. 303, no. 5665, pp. 1818–1822, 2004.
21. A. J. Almeida and E. Souto, "Solid lipid nanoparticles as a drug delivery system for peptides and proteins," *Advanced Drug Delivery Reviews*, vol. 59, no. 6, pp. 478–490, 2007.
22. L. Zhang, D. Pornpattananangkul, C.-M. Hu, and C.-M. Huang, "Development of nanoparticles for antimicrobial drug delivery," *Current Medicinal Chemistry*, vol. 17, no. 6, pp. 585–594, 2010.
23. ITU, The World in 2015: ICT Facts and Figures. International Telecommunication Union, 2015. Available at http://www.itu.int/en/ITU-D/Statistics/Pages/facts/default.aspx. Accessed on August 19, 2015.
24. C. Perera and R. Chakrabarti, "The utility of mHealth in medical imaging," *Journal of Mobile Technology in Medicine*, vol. 2, no. 3, pp. 4–6, 2013.

25. J. G. Kahn, J. S. Yang, and J. S. Kahn, "Mobile health needs and opportunities in developing countries," *Health Affairs*, vol. 29, no. 2, pp. 252–258, 2010.
26. P. Mechael, H. Batavia, N. Kaonga, S. Searle, A. Kwan, A. Goldberger, L. Fu, and J. Ossman, *Barriers and gaps affecting mHealth in low and middle income countries: Policy white paper*. The Earth Institute, Columbia University, 2010.
27. D. W. E. Dissanayake, "Modelling and analysis of wirelessly interrogated saw based micropumps for drug delivery applications," PhD dissertation, The University of Adelaide, 2010.
28. M. M. Ahmadi and G. A. Jullien, "A wireless-implantable microsystem for continuous blood glucose monitoring," *IEEE Transactions on Biomedical Circuits and Systems*, vol. 3, no. 3, pp. 169–180, 2009.
29. J. H. Choi and D. K. Kim, "A remote compact sensor for the real-time monitoring of human heartbeat and respiration rate," *IEEE Transactions on Biomedical Circuits and Systems*, vol. 3, no. 3, pp. 181–188, 2009.
30. J. Cavuoto, "Neural engineering's image problem," *IEEE Spectrum*, vol. 41, no. 4, pp. 32–37, 2004.
31. R. F. Weir, P. R. Troyk, G. A. DeMichele, D. A. Kerns, J. F. Schorsch, and H. Maas, "Implantable myoelectric sensors (IMESs) for intramuscular electromyogram recording," *IEEE Transactions on Biomedical Engineering*, vol. 56, no. 1, pp. 159–171, 2009.
32. A. Johansson, "Wireless communication with medical implants: Antennas and propagation," PhD dissertation, Lund University, 2004.
33. T. Tang, S. Smith, B. Flynn, J. Stevenson, A. Gundlach, H. Reekie, A. Murray, D. Renshaw, B. Dhillon, A. Ohtori et al., "Implementation of wireless power transfer and communications for an implantable ocular drug delivery system," *IET Nanobiotechnology*, vol. 2, no. 3, pp. 72–79, 2008.
34. S. Lyu and D. Untereker, "Degradability of polymers for implantable biomedical devices," *International Journal of Molecular Sciences*, vol. 10, no. 9, pp. 4033–4065, 2009.
35. S. Herrlich, S. Spieth, S. Messner, and R. Zengerle, "Osmotic micropumps for drug delivery," *Advanced Drug Delivery Reviews*, vol. 64, no. 14, pp. 1617–1627, 2012.
36. S. Gupta, R. P. Singh, R. Sharma, R. Kalyanwat, and P. Lokwani, "Osmotic pumps: A review," *Pharmacie Globale (IJCP) International Journal of Comprehensive Pharmacy*, vol. 2, no. 6, pp. 1–8, 2011. Available at http://www.academia.edu/1987423/OSMOTIC_PUMPS_A_REVIEW. Accessed on July 25, 2015.
37. C. L. Stevenson, J. T. Santini Jr., and R. Langer, "Reservoir-based drug delivery systems utilizing microtechnology," *Advanced Drug Delivery Reviews*, vol. 64, no. 14, pp. 1590–1602, 2012.
38. Intarcia, "Intarcia Therapeutics, Inc.," 2005, http://www.sec.gov/Archives/edgar/data/1086688 /000104746905022978/a2153238zs-1a.htm. Accessed on May 10, 2014.
39. D. L. Wise, *Handbook of Pharmaceutical Controlled Release Technology*. CRC Press, 2000.
40. J. C. Wright, S. Tao Leonard, C. L. Stevenson, J. C. Beck, G. Chen, R. M. Jao, P. A. Johnson, J. Leonard, and R. J. Skowronski, "An in vivo/in vitro comparison with a leuprolide osmotic implant for the treatment of prostate cancer," *Journal of Controlled Release*, vol. 75, no. 1, pp. 1–10, 2001.
41. E. J. McNally and J. E. Hastedt, *Protein Formulation and Delivery*. CRC Press, 2013.
42. M. J. Cukierski, P. A. Johnson, and J. C. Beck, "Chronic (60-week) toxicity study of Duros leuprolide implants in dogs," *International Journal of Toxicology*, vol. 20, no. 6, pp. 369–381, 2001.
43. "MiniMed Paradigm Revel™ insulin pump," http://www.medtronicdiabetes.com/treatment -and-products/minimed-revel-insulin-pump. Accessed on May 10, 2014.
44. A. E. Panteleon, M. Loutseiko, G. M. Steil, and K. Rebrin, "Evaluation of the effect of gain on the meal response of an automated closed-loop insulin delivery system," *Diabetes*, vol. 55, no. 7, pp. 1995–2000, 2006.
45. D. W. Dissanayake, S. Al-Sarawi, and D. Abbott, "Modelling and simulation of wirelessly and securely interrogated low-powered actuators for bio-MEMS," *Smart Materials and Structures*, vol. 20, no. 1, p. 015025, 2011.
46. A. Caballero and K. Yen, "The use of micro-electro-mechanical systems (MEMS) in the construction industry," *Proceedings of the 20th International Symposium on Automation and Robotics in Construction*, no. 1, pp. 161–165, 2003.
47. R. Sonje and S. Borde, "Micro-electromechanical systems (MEMS)," *International Journal of Modern Engineering Research*, vol. 4, no. 3, pp. 102–105, 2014.

48. M. W. Ashraf, S. Tayyaba, A. Nisar, and N. Afzulpurkar, "MEMS based system for drug delivery," in *Proceedings of the Sixth International Conference on Emerging Technologies*, pp. 82–87, 2010.
49. P. Gardner, "Microfabricated nanochannel implantable drug delivery devices: Trends, limitations and possibilities," *Expert Opinion on Drug Delivery*, vol. 3, no. 1, pp. 479–487, 2006.
50. E. E. Nuxoll and R. A. Siegel, "BioMEMS devices for drug delivery," *IEEE Engineering in Medicine and Biology Magazine*, vol. 28, no. 1, pp. 31–39, 2009.
51. E. Meng and T. Hoang, "Micro- and nano-fabricated implantable drug-delivery systems," *Therapeutic Delivery*, vol. 3, no. 12, pp. 1457–1467, 2012.
52. J. M. Maloney, "An implantable microfabricated drug delivery system," in *Proceedings of the 2003 International Mechanical Engineering Congress and Exposition*, pp. 115–116, 2003.
53. J. T. Santini, M. J. Cima, and R. Langer, "A controlled-release microchip," *Nature*, vol. 397, no. 6717, pp. 335–338, 1999.
54. N.-T. Nguyen, S. A. M. Shaegh, N. Kashaninejad, and D.-T. Phan, "Design, fabrication and characterization of drug delivery systems based on lab-on-a-chip technology," *Advanced Drug Delivery Reviews*, vol. 65, no. 11, pp. 1403–1419, 2013.
55. Y. Sultana, R. Jain, M. Aqil, and A. Ali, "Review of ocular drug delivery," *Current Drug Delivery*, vol. 3, no. 2, pp. 2017–2217, 2006.
56. Z. Wu and N.-T. Nguyen, "Convective–diffusive transport in parallel lamination micromixers," *Microfluidics and Nanofluidics*, vol. 1, no. 3, pp. 208–217, 2005.
57. A. J. Chung, D. Kim, and D. Erickson, "Electrokinetic microfluidic devices for rapid, low power drug delivery in autonomous microsystems," *Lab on a Chip*, vol. 8, no. 2, pp. 330–338, 2008.
58. C. E. Krewson and W. M. Saltzman, "Transport and elimination of recombinant human NGF during long-term delivery to the brain," *Brain Research*, vol. 727, no. 1, pp. 169–181, 1996.
59. K. Neeves, C. Lo, C. Foley, W. Saltzman, and W. Olbricht, "Fabrication and characterization of microfluidic probes for convection enhanced drug delivery," *Journal of Controlled Release*, vol. 111, no. 3, pp. 252–262, 2006.
60. C. P. Foley, N. Nishimura, K. B. Neeves, C. B. Schaffer, and W. L. Olbricht, "Flexible microfluidic devices supported by biodegradable insertion scaffolds for convection-enhanced neural drug delivery," *Biomedical Microdevices*, vol. 11, no. 4, pp. 915–924, 2009.
61. G. W. Astary, S. Kantorovich, P. R. Carney, T. H. Mareci, and M. Sarntinoranont, "Regional convection-enhanced delivery of gadolinium-labeled albumin in the rat hippocampus in vivo," *Journal of Neuroscience Methods*, vol. 187, no. 1, pp. 129–137, 2010.
62. R. Lo, P.-Y. Li, S. Saati, R. N. Agrawal, M. S. Humayun, and E. Meng, "A passive MEMS drug delivery pump for treatment of ocular diseases," *Biomedical Microdevices*, vol. 11, no. 5, pp. 959–970, 2009.
63. P.-Y. Li, J. Shih, R. Lo, S. Saati, R. Agrawal, M. S. Humayun, Y.-C. Tai, and E. Meng, "An electrochemical intraocular drug delivery device," *Sensors and Actuators A: Physical*, vol. 143, no. 1, pp. 41–48, 2008.
64. H. Gensler, R. Sheybani, P.-Y. Li, R. L. Mann, and E. Meng, "An implantable MEMS micropump system for drug delivery in small animals," *Biomedical Microdevices*, vol. 14, no. 3, pp. 483–496, 2012.
65. F. N. Pirmoradi, J. K. Jackson, H. M. Burt, and M. Chiao, "On-demand controlled release of docetaxel from a battery-less MEMS drug delivery device," *Lab on a Chip*, vol. 11, no. 16, pp. 2744–2752, 2011.
66. A. Dash and G. Cudworth II, "Therapeutic applications of implantable drug delivery systems," *Journal of Pharmacological and Toxicological Methods*, vol. 40, no. 1, pp. 1–12, 1998.
67. L. Kleiner, S. Hossainy, I. Astafieva, S. Pacetti, T. Glauser, and J. Desnoyer, "Microporous coating on medical devices," Nov. 8 2007, Patent US20070259101 A1.
68. W. Paul and C. P. Sharma, "Ceramic drug delivery: A perspective," *Journal of Biomaterials Applications*, vol. 17, no. 4, pp. 253–264, 2003.
69. H.-C. Wu, T.-W. Wang, J.-S. Sun, W.-H. Wang, and F.-H. Lin, "A novel biomagnetic nanoparticle based on hydroxyapatite," *Nanotechnology*, vol. 18, no. 16, pp. 1–9, 2007.

70. A. Aneja, J. Woodall, S. Wingerter, M. Tucci, and H. Benghuzzi, "Analysis of tobramycin release from beta tricalcium phosphate drug delivery system," *Biomedical Sciences Instrumentation*, vol. 44, no. 1, pp. 88–93, 2008.

71. K. Souza, J. Ardisson, and E. Sousa, "Study of mesoporous silica/magnetite systems in drug controlled release," *Journal of Materials Science: Materials in Medicine*, vol. 20, no. 2, pp. 507–512, 2009.

72. B. Buranapanitkit, S. Wongsiri, N. Ingviya, K. Chamniprasas, and S. Kalnauwakul, "In vitro inhibitive effect of antibiotic beads to common orthopaedic pathogens: Home-made vs. commercial beads," *The Thai Journal of Orthopaedic Surgery*, vol. 25, no. 2, pp. 48–52, 2000.

73. M. E. Kent, R. P. Rapp, and K. M. Smith, "Antibiotic beads and osteomyelitis: Here today, what's coming tomorrow?" *Orthopedics*, vol. 29, no. 7, pp. 599–603, 2006.

74. H. N. Rasyid, "Low-cost antibiotic delivery system for the treatment of osteomyelitis in developing countries," PhD dissertation, University of Groningen, Netherlands, 2009.

75. J. A. Matriano, M. Cormier, J. Johnson, W. A. Young, M. Buttery, K. Nyam, and P. E. Daddona, "Macroflux microprojection array patch technology: A new and efficient approach for intracutaneous immunization," *Pharmaceutical Research*, vol. 19, no. 1, pp. 63–70, 2002.

76. M. Staples, K. Daniel, M. J. Cima, and R. Langer, "Application of micro- and nano-electromechanical devices to drug delivery," *Pharmaceutical Research*, vol. 23, no. 5, pp. 847–863, 2006.

77. A. Arora, M. R. Prausnitz, and S. Mitragotri, "Micro-scale devices for transdermal drug delivery," *International Journal of Pharmaceutics*, vol. 364, pp. 227–236, 2008.

78. S. Chandrasekara, J. Brazzle, and A. Frazier, "Surface machined metallic microneedles," *Journal of Microelectromechanical Systems*, vol. 12, no. 1, pp. 281–288, 2003.

79. J. D. Zahn, A. P. Pisano, and D. Liepmann, "Continuous on-chip micropumping for microneedle enhanced drug delivery," *Biomedical Microdevices*, vol. 6, no. 3, pp. 183–190, 2004.

80. M. A. L. Teo, C. Shearwood, K. C. Ng, J. Lu, and S. Moochhala, "In vitro and in vivo characterization of MEMS microneedles," *Biomedical Microdevices*, vol. 7, no. 1, pp. 47–52, 2005.

81. P. Griss and G. Stemme, "Side-opened out-of-plane microneedles for microfluidic transdermal liquid transfer," *Journal of Microelectromechanical Systems*, vol. 12, no. 3, pp. 296–301, 2003.

82. N. Roxhed, T. C. Gasser, P. Griss, G. A. Holzapfel, and G. Stemme, "Penetration-enhanced ultrasharp microneedles and prediction on skin interaction for efficient transdermal drug delivery," *Journal of Microelectromechanical Systems*, vol. 16, no. 6, pp. 1429–1440, 2007.

83. J. D. Zahn, N. H. Talbot, D. Liepmann, and A. P. Pisano, "Microfabricated polysilicon microneedles for minimally invasive biomedical devices," *Biomedical Microdevices*, vol. 2, no. 4, pp. 295–303, 2000.

84. H. J. Gardeniers, R. Luttge, E. J. Berenschot, M. J. De Boer, S. Y. Yeshurun, M. Hefetz, R. van't Oever, and A. van den Berg, "Silicon micromachined hollow microneedles for trans-dermal liquid transport," *Journal of Microelectromechanical Systems*, vol. 12, no. 6, pp. 855–862, 2003.

85. P. Van Damme, F. Oosterhuis-Kafeja, M. Van der Wielen, Y. Almagor, O. Sharon, and Y. Levin, "Safety and efficacy of a novel microneedle device for dose sparing intradermal influenza vaccination in healthy adults," *Vaccine*, vol. 27, no. 3, pp. 454–459, 2009.

86. D. V. McAllister, P. M. Wang, S. P. Davis, J.-H. Park, P. J. Canatella, M. G. Allen, and M. R. Prausnitz, "Microfabricated needles for transdermal delivery of macromolecules and nanoparticles: Fabrication methods and transport studies," *Proceedings of the National Academy of Sciences*, vol. 100, no. 24, pp. 13755–13760, 2003.

87. R. K. Sivamani, B. Stoeber, G. C. Wu, H. Zhai, D. Liepmann, and H. Maibach, "Clinical microneedle injection of methyl nicotinate: Stratum corneum penetration," *Skin Research and Technology*, vol. 11, no. 2, pp. 152–156, 2005.

88. B. Stoeber and D. Liepmann, "Arrays of hollow out-of-plane microneedles for drug delivery," *Journal of Microelectromechanical Systems*, vol. 14, no. 3, pp. 472–479, 2005.

89. N. Roxhed, P. Griss, and G. Stemme, "Membrane-sealed hollow microneedles and related administration schemes for transdermal drug delivery," *Biomedical Microdevices*, vol. 10, no. 2, pp. 271–279, 2008.

90. L. Nordquist, N. Roxhed, P. Griss, and G. Stemme, "Novel microneedle patches for active insulin delivery are efficient in maintaining glycaemic control: An initial comparison with subcutaneous administration," *Pharmaceutical Research*, vol. 24, no. 7, pp. 1381–1388, 2007.

91. S. R. Patel, A. S. Lin, H. F. Edelhauser, and M. R. Prausnitz, "Suprachoroidal drug delivery to the back of the eye using hollow microneedles," *Pharmaceutical Research*, vol. 28, no. 1, pp. 166–176, 2011.

92. U. O. Häfeli, A. Mokhtari, D. Liepmann, and B. Stoeber, "In vivo evaluation of a microneedle-based miniature syringe for intradermal drug delivery," *Biomedical Microdevices*, vol. 11, no. 5, pp. 943–950, 2009.

93. S. P. Davis, W. Martanto, M. G. Allen, and M. R. Prausnitz, "Hollow metal microneedles for insulin delivery to diabetic rats," *IEEE Transactions on Biomedical Engineering*, vol. 52, no. 5, pp. 909–915, 2005.

94. "The V-Go helps control blood glucose with simple basal-bolus insulin delivery," https://www.valeritas.com/vgo. Accessed on July 10, 2014.

95. S. P. Sullivan, D. G. Koutsonanos, M. del Pilar Martin, J. W. Lee, V. Zarnitsyn, S.-O. Choi, N. Murthy, R. W. Compans, I. Skountzou, and M. R. Prausnitz, "Dissolving polymer microneedle patches for influenza vaccination," *Nature Medicine*, vol. 16, no. 8, pp. 915–920, 2010.

96. S. Kommareddy, B. C. Baudner, S. Oh, S.-y. Kwon, M. Singh, and D. T. O'Hagan, "Dissolvable microneedle patches for the delivery of cell-culture-derived influenza vaccine antigens," *Journal of Pharmaceutical Sciences*, vol. 101, no. 3, pp. 1021–1027, 2012.

97. Y.-C. Kim, J.-H. Park, and M. R. Prausnitz, "Microneedles for drug and vaccine delivery," *Advanced Drug Delivery Reviews*, vol. 64, no. 14, pp. 1547–1568, 2012.

98. S. Kaushik, A. H. Hord, D. D. Denson, D. V. McAllister, S. Smitra, M. G. Allen, and M. R. Prausnitz, "Lack of pain associated with microfabricated microneedles," *Anesthesia & Analgesia*, vol. 92, no. 2, pp. 502–504, 2001.

99. S. M. Bal, J. Caussin, S. Pavel, and J. A. Bouwstra, "In vivo assessment of safety of microneedle arrays in human skin," *European Journal of Pharmaceutical Sciences*, vol. 35, no. 3, pp. 193–202, 2008.

100. Y.-W. Noh, T.-H. Kim, J.-S. Baek, H.-H. Park, S. S. Lee, M. Han, S.-C. Shin, and C.-W. Cho, "In vitro characterization of the invasiveness of polymer microneedle against skin," *International Journal of Pharmaceutics*, vol. 397, no. 1, pp. 201–205, 2010.

101. Y. Hutin, A. Hauri, L. Chiarello, M. Catlin, B. Stilwell, T. Ghebrehiwet, and J. Garner, "Best infection control practices for intradermal, subcutaneous, and intramuscular needle injections," *Bulletin of the World Health Organization*, vol. 81, no. 7, pp. 491–500, 2003.

102. T. S. Kupper and R. C. Fuhlbrigge, "Immune surveillance in the skin: Mechanisms and clinical consequences," *Nature Reviews Immunology*, vol. 4, no. 3, pp. 211–222, 2004.

103. Q. Fang, "Body EMF absorption: A design issue for implantable medical electronics," *International Journal of Bioelectromagnetism*, vol. 12, no. 1, pp. 7–11, 2010.

104. L. C. Chirwa, P. A. Hammond, S. Roy, and D. R. Cumming, "Electromagnetic radiation from ingested sources in the human intestine between 150 MHz and 1.2 GHz," *IEEE Transactions on Biomedical Engineering*, vol. 50, no. 4, pp. 484–492, 2003.

105. W. G. Scanlon, B. Burns, and N. E. Evans, "Radiowave propagation from a tissue-implanted source at 418 MHz and 916.5 MHz," *IEEE Transactions on Biomedical Engineering*, vol. 47, no. 4, pp. 527–534, 2000.

106. W. G. Scanlon and N. E. Evans, "Numerical analysis of bodyworn UHF antenna systems," *Electronics & Communication Engineering Journal*, vol. 13, no. 2, pp. 53–64, 2001.

107. G. Iddan, G. Meron, A. Glukhovsky, and P. Swain, "Wireless capsule endoscopy," *Nature*, vol. 405, p. 417, 2000.

108. H. Yang and P. S. Hall, "On-body antennas and propagation: Recent development," *IEICE Transactions on Communications*, vol. 91, no. 6, pp. 1682–1688, 2008.

109. J. E. Ferguson and A. D. Redish, "Wireless communication with implanted medical devices using the conductive properties of the body," *Expert Review of Medical Devices*, vol. 8, no. 4, pp. 427–433, 2011.

110. J. H. Prescott, S. Lipka, S. Baldwin, N. F. Sheppard, J. M. Maloney, J. Coppeta, B. Yomtov, M. A. Staples, and J. T. Santini, "Chronic, programmed polypeptide delivery from an implanted, multireservoir microchip device," *Nature Biotechnology*, vol. 24, no. 4, 2006.

111. M. Staples, "Microchips and controlled-release drug reservoirs," *Wiley Interdisciplinary Reviews: Nanomedicine and Nanobiotechnology*, vol. 2, no. 4, pp. 400–417, 2010.

112. J. Santini and D. Ausiello, "Pre-clinical animal testing method," Oct. 29, 2007, US Patent App. 11/927,316.

113. G. Manoharan, N. Evans, B. Kidwai, D. Allen, J. Anderson, and J. Adgey, "Novel passive implantable atrial defibrillator using transcutaneous radiofrequency energy transmission successfully cardioverts atrial fibrillation," *Circulation*, vol. 108, no. 11, pp. 1382–1388, 2003.

114. J. D. Weiland, W. Liu, and M. S. Humayun, "Retinal prosthesis," *Annual Review of Biomedical Engineering*, vol. 7, no. 1, pp. 361–401, 2005.

115. K. Hungar, M. Görtz, E. Slavcheva, G. Spanier, C. Weidig, and W. Mokwa, "Production processes for a flexible retina implant," *Sensors and Actuators A: Physical*, vol. 123, pp. 172–178, 2005.

116. C.-K. Liang, J.-J. J. Chen, C.-L. Chung, C.-L. Cheng, and C.-C. Wang, "An implantable bidirectional wireless transmission system for transcutaneous biological signal recording," *Physiological Measurement*, vol. 26, no. 1, p. 83, 2005.

117. J. Csicsvari, D. A. Henze, B. Jamieson, K. D. Harris, A. Sirota, P. Barthó, K. D. Wise, and G. Buzsáki, "Massively parallel recording of unit and local field potentials with silicon-based electrodes," *Journal of Neurophysiology*, vol. 90, no. 2, pp. 1314–1323, 2003.

118. P. Valdastri, A. Menciassi, A. Arena, C. Caccamo, and P. Dario, "An implantable telemetry platform system for in vivo monitoring of physiological parameters," *IEEE Transactions on Information Technology in Biomedicine*, vol. 8, no. 3, pp. 271–278, 2004.

119. T. Leuerer and W. Mokwa, "Planar coils with magnetic layers for optimized energy transfer in telemetric systems," *Sensors and Actuators A: Physical*, vol. 116, no. 3, pp. 410–416, 2004.

120. G. A. Kendir, W. Liu, R. Bashirullah, G. Wang, M. Humayun, and J. Weiland, "An efficient inductive power link design for retinal prosthesis," in *Proceedings of the 2004 International Symposium on Circuits and Systems*, pp. 41–44, 2004.

121. S. Smith, T. Tang, J. Stevenson, B. Flynn, H. Reekie, A. Murray, A. Gundlach, D. Renshaw, B. Dhillon, A. Ohtori et al., "Miniaturised drug delivery system with wireless power transfer and communication," in *Proceedings of the IET Seminar on MEMS Sensors and Actuators*, pp. 155–162, 2006.

122. S. Smith, T. Tang, J. Terry, J. Stevenson, B. Flynn, H. Reekie, A. Murray, A. Gundlach, D. Renshaw, B. Dhillon et al., "Development of a miniaturised drug delivery system with wireless power transfer and communication," *IET Nanobiotechnology*, vol. 1, no. 5, pp. 80–86, 2007.

123. P. D. Bradley, "An ultra low power, high performance medical implant communication system (MICS) transceiver for implantable devices," in *Proceedings of the 2006 IEEE Biomedical Circuits and Systems Conference*, 2006, pp. 158–161.

124. M. Willert-Porada, Advances in Microwave and Radio Frequency Processing: Report from the Eighth International Conference on Microwave and High-Frequency Heating Held in Bayreuth, Germany, September 3–7, 2001. Springer, 2007.

125. R. Challoo, A. Oladeinde, N. Yilmazer, S. Ozcelik, and L. Challoo, "An overview and assessment of wireless technologies and co-existence of ZigBee, Bluetooth and Wi-Fi devices," *Procedia Computer Science*, vol. 12, pp. 386–391, 2012.

126. S.-Y. Lee and S.-C. Lee, "An implantable wireless bidirectional communication microstimulator for neuromuscular stimulation," *IEEE Transactions on Circuits and Systems I: Regular Papers*, vol. 52, no. 12, pp. 2526–2538, 2005.

127. H. M. Lu, C. Goldsmith, L. Cauller, and J.-B. Lee, "MEMS-based inductively coupled RFID transponder for implantable wireless sensor applications," *IEEE Transactions on Magnetics*, vol. 43, no. 6, pp. 2412–2414, 2007.

128. K. Shiba, M. Nukaya, T. Tsuji, and K. Koshiji, "Analysis of current density and specific absorption rate in biological tissue surrounding transcutaneous transformer for an artificial heart," *IEEE Transactions on Biomedical Engineering*, vol. 55, no. 1, pp. 205–213, 2008.

129. K. Shiba, T. Nagato, T. Tsuji, and K. Koshiji, "Energy transmission transformer for a wireless capsule endoscope: Analysis of specific absorption rate and current density in biological tissue," *IEEE Transactions on Biomedical Engineering*, vol. 55, no. 7, pp. 1864–1871, 2008.

130. G. E. Loeb, F. J. Richmond, and L. L. Baker, "The BION devices: Injectable interfaces with peripheral nerves and muscles," *Neurosurgical Focus*, vol. 20, no. 5, pp. 1–9, 2006.

131. M. V. Orlov, T. Szombathy, G. M. Chaudhry, and C. I. Haffajee, "Remote surveillance of implantable cardiac devices," *Pacing and Clinical Electrophysiology*, vol. 32, no. 7, pp. 928–939, 2009.

132. F. Merli and A. Skriverviky, "Design and measurement considerations for implantable antennas for telemetry applications," in *Proceedings of the Fourth European Conference on Antennas and Propagation*, pp. 1–5, 2010.

133. K. Gosalia, G. Lazzi, and M. Humayun, "Investigation of a microwave data telemetry link for a retinal prosthesis," *IEEE Transactions on Microwave Theory and Techniques*, vol. 52, no. 8, pp. 1925–1933, 2004.

134. W. Xia, K. Saito, M. Takahashi, and K. Ito, "Performances of an implanted cavity slot antenna embedded in the human arm," *IEEE Transactions on Antennas and Propagation*, vol. 57, no. 4, pp. 894–899, 2009.

135. P. M. Izdebski, H. Rajagopalan, and Y. Rahmat-Samii, "Conformal ingestible capsule antenna: A novel chandelier meandered design," *IEEE Transactions on Antennas and Propagation*, vol. 57, no. 4, pp. 900–909, 2009.

136. T. Dissanayake, K. P. Esselle, and M. R. Yuce, "Dielectric loaded impedance matching for wideband implanted antennas," *IEEE Transactions on Microwave Theory and Techniques*, vol. 57, no. 10, pp. 2480–2487, 2009.

137. F. E. Ahmed, "Artificial neural networks for diagnosis and survival prediction in colon cancer," *Molecular Cancer*, vol. 4, no. 1, p. 29, 2005.

138. S. Agatonovic-Kustrin and R. Beresford, "Basic concepts of artificial neural network (ANN) modeling and its application in pharmaceutical research," *Journal of Pharmaceutical and Biomedical Analysis*, vol. 22, no. 5, pp. 717–727, 2000.

139. K. G. Fricke, "Wireless telemetry system for implantable sensors," PhD dissertation, University of Western Ontario, 2012.

140. T. Robertson and M. J. Zdeblick, "Medical diagnostic and treatment platform using near-field wireless communication of information within a patient's body," September 1, 2006, US Patent App. 12/063,095.

141. M. Zdeblick, A. Thompson, A. Pikelny, and T. Robertson, "Pharma-informatics system," November 20, 2008, US Patent 20,080,284,599.

142. FDA, Draft Guidance for Industry and FDA Staff: Radio-frequency Wireless Technology in Medical Devices. US Department of Health and Human Services, 2007.

143. P. Valdastri, S. Rossi, A. Menciassi, V. Lionetti, F. Bernini, F. A. Recchia, and P. Dario, "An implantable ZigBee ready telemetric platform for in vivo monitoring of physiological parameters," *Sensors and Actuators A: Physical*, vol. 142, no. 1, pp. 369–378, 2008.

144. T. Basmer, D. Genschow, M. Froehlich, M. Birkholz, and F. Oder, "Energy budget of an implantable glucose measurement system," *Biomedical Technology*, vol. 57, no. 1, pp. 259–262, 2012.

145. A. Kiourti and K. S. Nikita, "Detuning issues and performance of a novel implantable antenna for telemetry applications," in *Proceedings of the Sixth European Conference on Antennas and Propagation*, pp. 746–749, 2012.

146. J. Yang, N. Tran, S. Bai, M. Fu, E. Skafidas, M. Halpern, D. Ng, and I. Mareels, "A subthreshold down converter optimized for super-low-power applications in MICS band," in *Proceedings of the 2011 IEEE Biomedical Circuits and Systems Conference*, pp. 189–192, 2011.

147. C. Furse, D. A. Christensen, and C. H. Durney, *Basic Introduction to Bioelectromagnetics*. CRC Press, 2009.

148. A. Ghasemi, A. Abedi, and F. Ghasemi, "Basic principles in radiowave propagation," in *Propagation Engineering in Wireless Communications*, pp. 23–55. Springer, 2012.

149. N. Beshchasna, B. Adolphi, S. Granovsky, J. Uhlemann, and K.-J. Wolter, "Biostability issues of flash gold surfaces," in *Proceedings of the 59th Electronic Components and Technology Conference*, pp. 1071–1079, 2009.

150. C. Gabriel, "Compilation of the dielectric properties of body tissues at RF and microwave frequencies." DTIC Document, Tech. Rep., 1996.

151. K. Ito, "Human body phantoms for evaluation of wearable and implantable antennas," in *Proceedings of the Second European Conference on Antennas and Propagation*, pp. 1–6, 2007.

152. K. Murakawa, M. Kobayashi, O. Nakamura, and S. Kawata, "A wireless near-infrared energy system for medical implants," *IEEE Engineering in Medicine and Biology Magazine*, vol. 18, no. 6, pp. 70–72, 1999.

153. D. S. Echt, M. W. Cowan, R. E. Riley, and A. F. Brisken, "Feasibility and safety of a novel technology for pacing without leads," *Heart Rhythm*, vol. 3, no. 10, pp. 1202–1206, 2006.

154. V. Kumar and S. Sindhu, "Automatic drug delivery system for the cardiac patients with the help of smart mobile phones," *International Journal of Advanced Technology & Engineering Research*, vol. 2, no. 2, pp. 154–156, 2012.

155. "JewelPUMP I," http://www.debiotech.com. Accessed on May 10, 2014.

156. H. Huang, P. Zhao, P. Chen, Y. Ren, X. Liu, M. Ferrari, Y. Hu, and D. Akinwande, "RFID tag helix antenna sensors for wireless drug dosage monitoring," *Translational Engineering in Health and Medicine, IEEE Journal of*, vol. 2, pp. 1, 8, 2014.

157. M. M. Ahmadi and G. A. Jullien, "A wireless-implantable microsystem for continuous blood glucose monitoring," *IEEE Transactions on Biomedical Circuits and Systems*, vol. 3, no. 3, pp. 169–180, 2009.

158. "MiniMed insulin pump therapy," http://www.medtronicdiabetes.com/. Accessed on May 10, 2014.

159. "Empi action patch," https://www.djoglobal.com/products/empi/empi-action-patch. Accessed on May 10, 2014.

160. "Smart pills," http://www.smartpillcorp.com/. Accessed on May 10, 2014.

161. "Philips iPill," http://www.research.philips.com/. Accessed on May 10, 2014.

162. "MicroCHIPS' drug delivery device," http://www.mchips.com/. Accessed on May 10, 2014.

163. N. Cho, J. Bae, and H.-J. Yoo, "A 10.8 mW body channel communication/MICS dual-band transceiver for a unified body sensor network controller," *IEEE Journal of Solid-State Circuits*, vol. 44, no. 12, pp. 3459–3468, 2009.

164. W. N. Charman, H.-K. Chan, B. C. Finnin, and S. A. Charman, "Drug delivery: A key factor in realising the full therapeutic potential of drugs," *Drug Development Research*, vol. 46, nos. 3–4, pp. 316–327, 1999.

165. K. K. Jain, K. K. Jain, N. Médecin, and K. K. Jain, *Drug Delivery Systems*. Springer, 2008.

166. B. Dimitrios, "Nanomedicine in cancer treatment: Drug targeting and the safety of the used materials for drug nanoencapsulation," *Biochemistry & Pharmacology: Open Access*, vol. 1, no. 5, 2012.

167. A. Keraliya Rajesh, P. Chirag, P. Pranav, K. Vipul, G. Soni Tejal, C. Patel Rajnikant, and M. Patel, "Osmotic drug delivery system as a part of modified release dosage form," *ISRN Pharmaceutics*, vol. 2012, no. 1, pp. 1–9, 2012.

17

Implantable Systems

Vincenzo Luciano, Emilio Sardini, Alessandro Dionisi,
Mauro Serpelloni, and Andrea Cadei

CONTENTS

17.1 Introduction

Nowadays, with the rapid development of bioengineering science, implantable medical devices are widely employed. These devices are implanted in the human body and they include pacemakers, smart prostheses, drug pumps, cochlear implants, implanted defibrillators, neurostimulators, bladder stimulators, nerve stimulators, diaphragm stimulators, etc. In most cases, they perform real-time control and/or real-time monitoring of several physiological parameters.

In the world, over 6 million heart patients have implanted pacemakers and about 150,000 pacemakers are surgically implanted just in the United States every year (Wang et al. 2003). The implanted defibrillators or pacemakers allow treating several heart conditions, for example, fibrillation, atrial and ventricular tachyarrhythmia, and bradycardia (Soykan 2002).

The implanted monitoring systems can communicate wirelessly with an external readout unit. By means of this architecture, the employment of transcutaneous wires is obviated; the risk of infection is avoided and measurement data can be collected with ease. The wireless communication allows the home monitoring of several physiological parameters during patient's daily activities, e.g., blood pressure, heart rate, and body temperature (Valdastri et al. 2004; Tamura et al. 2011). These data can be collected and sent to a remote unit; medical staff, through web access, will examine the measured values in order to assist the patient remotely (Halperin et al. 2008).

As shown by Troyk et al. (2007), myoelectric sensors represent another example of implantable medical devices. They are composed of several myoelectric electrodes implanted in the residual limb, for example, in patients with a prosthetic hand or arm. The electrodes detect the intramuscular myoelectric activity and these signals are elaborated to control the prosthesis.

Typically, an implantable medical device is powered by a battery, which constitutes a severe limitation. In most cases, because of the battery discharge, the implanted system must be replaced through a surgical operation. For example, although an implanted defibrillator lasts about 10 years, the battery contained in it must be substituted after about 4.7 years (Wei and Liu 2008). Hence, the patient must undergo a surgical operation that causes physical and mental pain in the patient themselves; furthermore, this represents an economic burden both for the patient and for the national health system. As stated above, the battery defines the lifetime of the entire implantable medical device. To obviate this problem, the implantable medical device should be powered through a telemetric technique or through an energy-harvesting system implanted together with the device.

In the telemetric technique, an inductive powering system composed of two coils is used; the first one is out of the body and the second one is implanted. The primary coil produces a magnetic field harvested by the secondary coil. In this way, energy is transferred within the human body. By means of this technique, the circuits of the implanted medical devices are powered wirelessly, without transcutaneous wires.

Examples of inductive power systems are reported by Morais et al. (2009), Silay et al. (2011), and Riistama et al. (2007). Morais et al. (2009) describe a smart hip prosthesis for the measurement of the joint forces and the temperature distribution in the prosthesis itself; the inductive link is performed through a coil placed in the stem of the insert. An inductive power system for cortical implant is reported by Silay et al. (2011). The device is composed of a class E power amplifier, two coils, a matching network, and a rectifier. The proposed system is embedded in a biocompatible packaging that can be placed in a cavity of the skull. Another example of an inductive power system is described by Riistama et al. (2007), who describe an implant for the measurement of electrocardiogram; the operational range is about 16 mm.

The telemetric technique requires the two coils to be close and this represents a constraint on the patient's movements. Furthermore, the operating frequency must be compatible with the tissue absorption level; in particular, the power received by the implanted device, the tissue absorption level, and the consequent tissue warming are related to the radiation frequency. Hence, the tissue energy absorption is not only a transmission problem but also a biocompatibility and safety problem. Table 17.1 shows maximum allowable exposure for human tissue as suggested by the IEEE Standard for Safety Levels (1999): as

TABLE 17.1

Maximum Permissible Values of Magnetic Field, Electric Field, and Power Density Field for Human Tissue (IEEE Standard C95.1, 1999)

Frequency Range (MHz)	Magnetic Field Strength (H) (A/m)	Electric Field Strength (E) (V/m)	Power Densities (S) of E and H Fields (mW/cm²)
0.003–0.1	163	614	(100, 1,000,000)
0.1–3.0	$16.3/f$	614	(100, $10,000/f^2$)
3–30	$16.3/f$	$1842/f$	($900/f^2$, $10,000/f^2$)
30–100	$16.3/f$	61.4	(1, $10,000/f^2$)
100–300	0.163	61.4	1.0
300–3000	–	–	$f/300$
3000–15,000	–	–	10

the radiation frequency increases, the maximum magnetic field strength and the maximum power density decrease. An example of operating frequency for data and power transfer in telemetric systems is 125 kHz (Crescini et al. 2011).

An energy-harvesting technique represents an alternative technique to power an implantable medical device. This technique is under development and it will avoid the use of a battery; its necessity for surgery for battery substitution would enable the patient to move in a free and autonomous way. The energy harvesting is a process through which energy is captured and stored from the ambient, in this case from the human body. As reported by Starner (1996), the human body is a rich reservoir of energy and the values reported are more than sufficient to supply an implantable medical device. Starner (1996) has calculated that energy of about 390 MJ is stored by an individual of 68 kg with 15% of body fat; this stored energy is converted into mechanical energy through muscles and it is partially exploited during daily activities. Starner (1996) reports the calculated available powers for some different activities; for example, the available powers are about 67, 35, and 1 W for walking, upper limb motion, and breathing, respectively.

The human body exploitable energy is classified in three different forms: thermal energy, mechanical energy, and chemical energy.

The thermal energy is harvested through a thermoelectric generator, which exploits the Seebeck effect. A thermoelectric generator is typically composed of several thermocouples and it produces a voltage proportional to the thermal gradient across the thermoelectric generator itself. Stark (2006) describes a compact thermoelectric generator composed of over 5000 thermocouples; this device has a volume of less than 1 cm^3 and is able to produce 120 μW of power at 3 V with a thermal gradient of 5 K. Hence, an implantable medical device can be powered with one or more of these generators connected in series or in parallel. For a thermoelectric generator, obtaining a high thermal gradient in the human body is difficult; the highest thermal gradient occurs near the skin surface (Yang et al. 2007).

The chemical energy is harvested by means of microbiofuel cells which are composed of a cathode, an anode, and an electrolyte. An example of an implantable glucose biofuel cell is described by von Stetten et al. (2006); the proposed device is composed of a hydrogel membrane which separates the electrodes made of activated carbon. Following a series of in vitro tests, this fuel cell is able to generate a power density and a peak power of 2 μW/cm^2 and 20 μW, respectively, for a period of 7 days.

The mechanical energy is the most available and the most easily exploitable energy source in the human body; it is usually harvested with electrostatic, electromagnetic, and piezoelectric transducers. Some transducers, especially the electromagnetic and the electrostatic ones, can be modeled with a mass–spring–damper system working in resonance conditions with the input motion.

In a generic electromagnetic harvester, the human movement produces a displacement between a coil and a permanent magnet; this displacement generates a time-variable magnetic flux and, according to Faraday's law, an induced voltage on the coil. The permanent magnet and the coil can be placed in the resonant structure in order to maximize the magnet–coil relative motion, the induced voltage, and the stored energy by the harvester. An electromagnetic transducer for the powering of a pacemaker is presented by Goto et al. (1998); the heart muscle contractions generate a relative motion between a movable permanent magnet and a fixed coil. With a heart frequency of between 0.5 and 2 Hz, the harvester is able to produce a maximum power of 200 μW and it allows powering a pacemaker without a battery. An electromagnetic harvester implanted in a hip prosthesis is reported by Morais et al. (2011). The transducer is a resonant structure, composed of two external fixed coils and a Teflon tube in which a magnet swings during the walk or whatever other activity produces a hip movement. Nasiri et al. (2011) describe a resonant electromagnetic transducer implantable in the diaphragm muscle. As the diaphragm muscle works continuously, the proposed transducer harvests energy even if the person is sleeping.

An electrostatic harvester is typically based on a variable precharged capacitor composed of a moving plate whose movement is produced by the human body activity. Examples of electrostatic transducers are reported by Tashiro et al. (2002) and Miao et al. (2004). In Tashiro et al.'s (2002) paper, the proposed transducer, composed of a honeycomb structure, exploits the heart muscle motion in order to power a cardiac pacemaker. Miao et al. (2004) describe a MEMS electrostatic transducer; it produces a power of 24 µW at an input mechanical frequency of 10 Hz.

In a piezoelectric transducer, the movement of any body part can be exploited to deform a piezoelectric material in order to produce a voltage. Some piezoelectric transducers use a resonant structure to maximize the piezoelectric material deformation. Almouhaed et al. (2010, 2011a,b), Lahuec et al. (2011), present a knee-monitoring system embedded in a total knee prosthesis (TKP). Four piezoceramics are placed in the tibial plate in order to measure the tibiofemoral forces and contemporaneously to harvest energy. During a single gait cycle, a total power of 7.2 mW is produced (1.8 mW for every piezoceramic). Potkay and Brooks (2008) describe a blood pressure sensor. The proposed device, composed of an arterial cuff in thin piezoelectric film, converts the artery contraction/expansion into electric energy.

The design of energy harvesters for implantable medical devices is usually more complicated than that for industrial applications (Mitcheson 2010). First of all, an energy harvester for an implantable medical device must have limited size. Furthermore, for mechanical transducers, the matching between the resonance frequency and the input mechanical frequency is difficult to obtain because the resonance frequency and the transducer size are, in general, inversely proportional. In particular, the human body motion frequencies are generally less than 10 Hz (Romero et al. 2009); to lower the resonance frequency, using a linear spring–mass–damper system, a reduction in stiffness or an increase in mass is required, but, as previously stated, implantable devices must have limited dimensions; hence, it is difficult to satisfy both requirements. In summary, from these considerations, when lowering the resonance frequency, the transducer size increases, making the system not suitable for the human body applications. A material with nonlinear elastic characteristics can be used to solve this problem (Ramlan et al. 2010). Another important point is that an implantable medical device must be realized with biocompatible materials; otherwise, the different elements of the implantable medical device must be placed and fully sealed in a biocompatible material packaging. Two examples of architectures for battery-less implantable medical devices are presented in this chapter. In particular, a telemetric technique and an energy-harvesting system are described.

17.2 Implantable System Architecture

Implantable systems can be defined as devices that execute autonomously their measurement functions in the human body. They are characterized by an autonomous power supply capable of measuring and transmitting data from inside the human body to a readout unit placed outside. They normally consist of a sensor module, conditioning electronics, a transmission module, and a powering system. A block diagram containing the main elements of an implantable system is shown in Figure 17.1. Common characteristics of each element are very low-power design, stand-alone configuration, minimal control, and elaboration circuits, resulting in less use of the microprocessor and power consumption, and minimal communication circuits, which require fast software and a more streamlined protocol, simple and quick to run with a low-power microprocessor.

An implantable system must be biocompatible with tissues and cells of the environment in which it works. This requires using materials that are biocompatible to embed the sensor and electronic circuits inside a component of the prosthesis that is already biocompatible. Other requirements are dimensions and frequency of the electromagnetic waves traveling through the human body. The antenna for the transmission must be small, for a telemetric system (Crescini et al. 2011) less than about 2 cm, and the transmission module must have frequencies that allow sending or receiving data through the human body.

Sensors and electronics blocks depend on the quantity to be measured. The supply block can be constituted by a battery but, as reported in Section 17.1, this can require a surgical operation to replace the battery. In the literature other power sources are proposed, such as harvesting modules and inductive links. Each of these solutions determines a specific composition of the transmission and supply modules.

The analysis of the implantable systems has led to the definition of a classification according to the type of architecture; one class is telemetric systems and the second is self-powered systems. *Telemetric systems* are defined as those that are powered inductively and interrogated wirelessly by a readout unit. *Self-powered systems* are those that have a power-harvesting module that scavenges energy for the functioning of the system from the environment. In the following subsection, the general architectures of telemetric and self-powered systems are described and discussed.

17.2.1 Telemetric Systems

The general architecture of a measurement system for the telemetric systems is shown in Figure 17.2.

The telemetric system is composed of an implanted unit and a readout unit. The sensor is a block in the implanted unit inside the human body, while the readout unit is placed outside and the communication between the two units is done with telemetric techniques. A block diagram of the implanted unit and readout unit is shown in Figure 17.2. It consists of different modules: a low-power sensor, which measures the quantity under interest; a low-power microcontroller for the analog-to-digital conversion of the data, the storage in memory, and the telemetric operation; and a transponder, which transfers the data collected to the readout unit. The two elements are connected by a wireless communication exploiting an electromagnetic field at typical frequencies of about 125 kHz (Crescini et al. 2011). The coil, connected to the transponder of the implanted module, is coupled to the

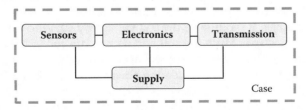

FIGURE 17.1
Block diagram of an implanted system.

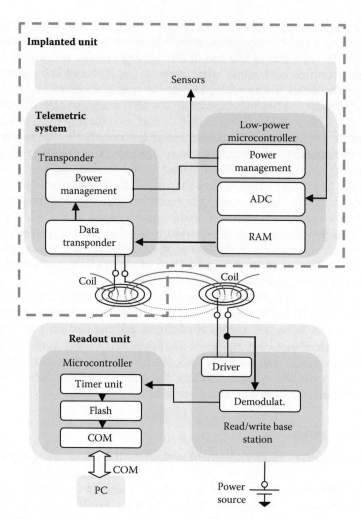

FIGURE 17.2
Block diagram of a telemetric system.

external one, receiving the power for the communication of the measurement data. Usually the readout coil is bigger than the coil of the implanted unit. This is due, on one hand, to the need to occupy a small space inside the human body and, on the other, to the difficulty of properly coupling the two windings. The readout coil, therefore, permits a greater area for the coupling of the magnetic field but at the expense of the efficiency of power transfer. To transmit data the transponder of the implanted unit modulates the magnetic field by using a damping stage. It modulates the coil voltage by varying the coil's load. A high level ("1") increases the current into the coil and damps the coil voltage. A low level ("0") decreases the current and increases the coil voltage. However, the current through the coil is never zero, so as to continuously provide the power supply. In particular the amplitude modulation is typical and the Manchester code is as well. The transponder interface can also receive data: the readout unit modulates the emitted field with short gaps, and then a gap-detection circuit in the implanted unit reveals these gaps and decodes the signal. Furthermore, as specified before, the readout unit generates the power supply, which is handled via electromagnetic field and the coil antenna of the transponder interface; then the voltage across the coil is rectified and managed by the power management module to generate a rectified voltage and current for the functioning of the electronic circuit. The power to all the internal modules of the implanted unit is supplied by the energy transmitted by the electromagnetic field generated by the read/write base station. The low-power microcontroller can include an ADC or timer units for measuring the sensor signals. The microcontroller has a volatile memory to save the data before the transmission and timer units that synchronize the data transmission. To maintain the power consumption low, the bus frequency should be low, the ADC and transmitting unit have a low-power configuration, and all the unused peripherals should be switched off.

Since the telemetric systems are wireless devices, transmitting not only the data but the power as well, the covered distance between the wireless device and the collecting data system must be short. The maximum transmitting distance depends on different factors; in Dalola et al. (2009) report a maximum distance of about 8 cm for open-field transmission. For these reasons and for energy saving, a point-to-point communication must be implemented. A point-to-point communication avoids the integration on the implanted system of circuits to manage the complexity of a network protocol and avoids complex communications such as those on multiple nodes that involve more complex software and, therefore, a longer time of execution of the same software, saving power supply and making the system compatible with the available low energy. Furthermore, the readout coil must be present and active during the measurement, conditioning, and transmitting phases; this means that the external coil, sometimes uncomfortable for the patient, must be placed close to or around the human body.

The external readout unit usually consists of a read/write base station able to supply power to the transponder driving the coil antenna and to demodulate the digital signal from the implanted unit. The readout unit is supplied by a line voltage and no low-power characteristics are required; so the microcontroller, the bus frequency, and peripherals have no functioning limits. A timer unit is used to decode the demodulated signal and the data collected are transferred to a personal computer (PC) by using a serial communication interface (SCI).

17.2.2 Power-Harvesting Systems

The general architecture of a measurement system for passive autonomous sensors is reported in Figure 17.3.

Since the possibility of substituting batteries with a harvesting system is ecologically attractive and avoids surgical operations, our analysis shows that a self-powered system equipped with a harvesting system is a viable solution for implanted systems. These self-powered systems consist of one or more sensing elements and different modules: front-end electronics, an analog-to-digital converter, an elaboration unit for managing the internal tasks, a power management block, a wireless transceiver, and storage memories. In Figure 17.3, a block diagram of a typical self-powered system is shown. The power-harvesting module, usually separated from the circuit board, collects the energy present in the environment of measurement in the form of mechanical energy, temperature difference, etc., and transforms it into electric energy.

The power-harvesting module must comply with very specific constraints, of not only space but also the compatibility of the materials and method of operation. For example, a system of power harvesting that uses the mechanical energy due to the movement of the human body cannot be excessively large, however, and must be able to operate with very

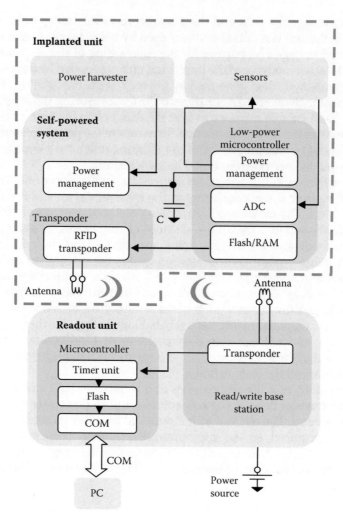

FIGURE 17.3
Block diagram of a self-powered system.

low-frequency vibrations, on the order of a few hertz. The power harvester is connected to a power management block that is very important and essential. Since the voltage and current levels of the electronic circuits exceed the possibility offered by power-harvesting systems or sometimes even by batteries, management of the power supply is required. Self-powered systems require specific levels of voltage and current obtainable by an appropriate power management block. This block commonly consists of dedicated circuits for the conditioning and/or storage of the energy harvested. First, a specific circuit can be used for matching the output electric impedance of the generator with the characteristics of the circuit load in order to have the maximum power transfer. Then, usually the power management circuit has a dedicated DC–DC converter or charge pump (CP) to provide a specific level of voltage and current at the circuit load.

The low-power microcontroller controls the sensor interface circuit, configures the front-end electronics, and converts the data coming from the sensor interface circuit and stores it in a nonvolatile memory or directly sends the data at the transponder for communication. So, in this architecture two different strategies can be implemented. The first one, with a nonvolatile memory, saves the measurement data in the implanted unit and saves the data also when the device is not powered. This means that the external readout unit is not necessary during the measuring and saving phases, increasing the possible applications and the comfort of the patient, who does not need to "wear" the readout unit constantly. All the collected data can be downloaded in a second time by using different methods. The second strategy regards the possibility of measuring when enough power is scavenged and uses it to measure and transmit the data outside. In this configuration the nonvolatile memory is not needed and the power for saving the data in the memory is not required. This means that the data, when the readout unit is not close to the implanted unit, can be lost. Specific applications can be implemented as well to measure only when a specific event happens or only when requested. This leads to power savings and avoids the loss of data. In order to reduce the energy consumption for the data transmission, some smart compression algorithms on the measurement data can be implemented as well. In fact, the system deploys strategies to reduce the power consumption; the sensor module is designed to be triggered to transmit only when required, thereby consuming less power because unnecessary transmissions are avoided.

17.3 Force Measurement inside Knee Prosthesis

In this chapter an example of implantable sensor that monitors TKP is described. It measures the forces applied on the knee prosthesis and exploits a telemetric technique for data communication and power supply (Crescini et al. 2011). Monitoring the prosthesis by an implantable sensor is very important in biomedical application. This provides several advantages such as the analysis of the wear conditions of the prosthesis caused by an incorrect use or placement, the data collection to improve future design of the prosthesis, and a better control of patient rehabilitation. Moreover, measurement devices for tibiofemoral contact stress give precise knowledge about articular movement behavior and they can be used to refine surgical instrumentation, guide postoperative physical therapy, and detect human activities that can overload the implant. Therefore, by monitoring the forces on the TKP, it is possible to ensure that the lifetime of the TKP will be greater than that available with the current prosthesis.

The TKP consists of a tibial component and a femoral component, both made up of metal alloy. These components are attached to bone by using acrylic cement and between the two components an ultrahigh-molecular-weight polyethylene (UHMWPE) insert has been embedded. The implantable sensor should be placed into the polyethylene insert, avoiding biocompatible problems.

The implantable sensor measures the forces applied on the knee insert by three magnetoresistive force transducers which consist of magnetoresistors and permanent magnets as shown in Figure 17.4. The output resistance of each sensor depends on the distance between the permanent magnet and the magnetoresistor. Two magnetoresistors are placed in the areas where the two condyles of the femoral component transmit the forces between femur and tibia, and the third magnetoresistor is placed in the central part of the insert where the forces generated have no significant effect. The third magnetoresistor works as a dummy for temperature compensation operation.

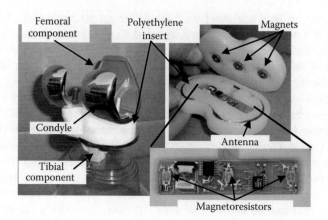

FIGURE 17.4
Example of total knee prosthesis and the implantable sensor into polyethylene insert of the TKP.

The force applied by the femoral component generates a deformation of the polyethylene insert, changing the distance between the magnet and the magnetoresistors and causing a resistance variation. The relationship between the force applied to the insert and the resistance variation of the magnetoresistors output has been experimentally evaluated:

$$F \cong 2.15 \cdot (R - R_0). \tag{17.1}$$

Equation 17.1 shows the linear relationship between the force applied and the resistance measured, where F (N) is the force applied to the polyethylene insert, R (Ω) is the resistance value associated with the insert deformation, and $R_0 = 12.5$ kΩ is the resistance value when the force applied is equal to 0 N. In this way, it has been possible to measure the forces applied to the insert by resistance measurements.

A significant issue about the implantable sensors is the relationship between output of the sensors and the temperature. In fact, it is possible that an increase in temperature of about 3°C, when the sensor is inside the human body, can be reached when a person walks for 45 min (Graichen et al. 1999). In this particular example, the magnetoresistors have a resistance drift equal to about 150 Ω/°C. For this reason, it is important to compensate for the thermal drift so as to obtain an accurate measurement equation.

The control circuit of the implantable sensor is composed of a low-power microcontroller for acquiring data with a 12-bit ADC, a 128 kB flash memory for storing the force data, and a transponder working at frequency of 125 kHz. Figure 17.4 shows the antenna implanted into the insert, which communicates the data to the readout unit by RF and, at the same time, supplies the circuit coupling with another external antenna. A damping modulator is included in the implantable sensor to transmit the data in digital mode. Furthermore, a temperature sensor has been integrated in the implantable sensor so that the microcontroller is able to measure the temperature and eventually to compensate the resistance data during the measurement activity. Outside the knee the readout unit consists of a transceiver for driving the coil antenna and for demodulating the digital signal received. The antenna is controlled by a readout unit localized around the knee as in Figure 17.5. Furthermore, the readout unit supplies the implantable system by telemetry. The readout unit is managed by a low-power microcontroller and powered by an external battery.

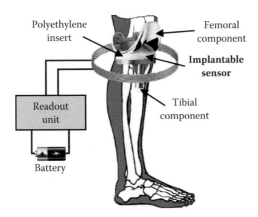

FIGURE 17.5
Prototype scheme of telemetric system for measuring the forces in TKP prosthesis.

The readout unit contains a transponder for transmitting and for receiving data by modulating the magnetic field by using a damping stage, in particular, with OOK modulation and Manchester code. The RF communication between the readout unit transceiver and implantable sensor receiver is supported by a magnetic field generated when applying to the antenna a sinusoidal voltage whose peak to peak magnitude is 80 Vpp at 125 kHz frequency. The capability to transfer data and energy through the human body is the main advantage of this solution.

The implantable sensor works based on three phases defined as stop, measure, and transmission modes. Figure 17.6 shows the three activities of the implantable sensor where the stop mode duration is 6 s, while the durations of the measure conversion and the transmit mode are about 6 and 7 ms, respectively. The communication protocol is composed of six strings of 16 bits each one. The first two are synchronization strings, and then there are three strings that contain the resistance data of three magnetoresistors and one string that provides the temperature. Finally, the data collected have been transferred to a PC by a serial interface so that the data can be analyzed by a qualified medical staff.

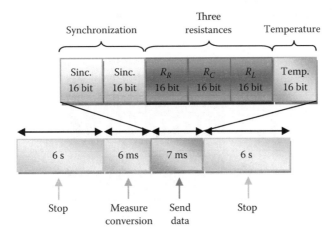

FIGURE 17.6
Functioning of implantable sensor.

17.3.1 Experimental Results

The force applied by the femoral component has been simulated using an Instron 8501 machine. This machine applied a linearly increasing force to the polyethylene insert from 0 N to 3 kN in 500 s. Figure 17.7 shows the relationship between the input orthogonal forces generated by the Instron on the TKP femoral component (input force line) and the force obtained by using Equation 17.1 considering the resistance value of the magnetoresistor output placed on the right side in the polyethylene insert (force calculated curve). The samples were acquired every 6 s when the implantable sensor was in active mode.

Possible reasons for the difference between the two trends are mainly (i) nonlinear behavior of the magnetoresistor material, (ii) a not proper temperature compensation, (iii) uncertainty in the experimental apparatus, and (iv) the hysteresis effect due to the geometry and physical characteristics of the polyethylene insert. Figure 17.8 shows the wireless transmission signals monitored when the implantable sensor is in active mode: V_{td} is the transmitted data voltage, V_{ta} is the transponder antenna voltage, V_{ra} is the signal of the differential voltage of the reader antenna, and V_{rd} is the received data.

Table 17.2 represents the power consumption during the different activities of the measurement system, when the external power from the readout unit is active. For example,

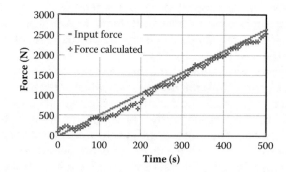

FIGURE 17.7
Comparison between the force values calculated by the resistance variation of the magnetoresistor in the right side of the insert of the implantable sensor and the input forces applied by the Instron.

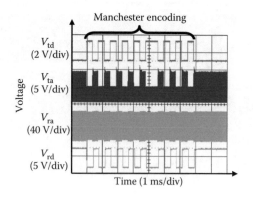

FIGURE 17.8
Wireless transmission signals of the converted data and typical implantable sensor activities.

TABLE 17.2

Power Consumption Measurements

Activity	Voltage (V)	Current (mA)	Power (mW)
Readout unit—transceiver communication	12	19.3	230
Implantable sensor—measurement and transmission	2	0.85	1.7
Implantable sensor—stop mode	2	0.16	0.3

when the implantable sensor measures and transmits the data to readout unit, the microcontroller and the transponder require a power supply of about 1.7 mW, with a current consumption of approximately 850 µA and a voltage of 2 V.

17.4 Power Harvesting in Implantable Human Total Knee Prosthesis

The fundamental requests for an implantable medical device are the capabilities of being self-powered and maintenance free. These ambitious goals excited the interest in a new and lively research field oriented toward the harvesting of energy from the human body with the aim of making the implantable (or wearable) system autonomous. Such a system, theoretically without the human supervision, should provide information about the physiological parameters concerning its application. The rapid development and reduction in size, cost, and power consumption of the wireless communications devices allow for solving the important problem of the measurement communication from the device to outside the human body. Considering the energy flow, an autonomous implantable device can be divided in two subsystems (López 2010): the power-harvesting module and the load (power-conditioning circuit [pcc], sensor, processor, and transceiver).

The first subsystem realizes the conversion of the energy from a particular domain (chemical, thermal, mechanical, etc.) to the electric domain. The second one carries out the mission of the implanted sensor, i.e., the measuring and the transmission of the data.

This section shows the attempt to make autonomous a force sensor system inserted in a TKP (shown in Section 17.3), exploiting the mechanical energy produced by the human knee joint movement.

A TKP is composed of three components: the femoral component (condyles), the tibial plate, and the tibial insert constituting the contact surface of the tibia with respect to the femur. Figure 17.9 shows a 3D computer-aided design (CAD) model of the proposed solution: a copper coil is housed in a prominence of the tibial insert and six couples of block-shaped magnets are placed into each condyle. The axes of the magnets and the coil are parallel to the tibial insert.

The energy conversion principle is based on Faraday–Neumann–Lenz's law of induction: "a time-varying magnetic induction field B, linked to a conductive path c, leads to a potential difference to the extremities of c" (Woodson and Melcher 1968). This way, when the femur moves with respect to the tibia, the magnetic induction field induces a time-varying flux and then a potential difference to the terminal of the coil.

In general, the relative motion of the condyles with respect to the tibia has six degrees of freedom and, moreover, it is very difficult to reproduce. Because of this, the complex kinematics of the electromechanical system has been reduced by a reasonable simplification:

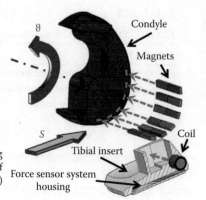

FIGURE 17.9
Cross section (in the sagittal plane) of a TKP with the energy-harvesting components. The angle ϑ and the displacement S are the degrees of freedom of the proximate relative motion of the femur (condyles) with respect to the tibial insert.

only the relative rotation ϑ and translation S (Figure 17.9), in the sagittal plane, were considered in the design of the system.

A tailored motion-control system allowed the reproduction of the gait conditions under the previous assumptions. In particular the combined motion of translation and rotation of the TKP has been reproduced with the dedicated four-bar mechanism shown in Figure 17.10. An improved design of the four-bar mechanism allows for the translation S during the rotation ϑ with the respect of the range of movement deduced by the literature (Pinskerova et al. 2000; Masouros et al. 2010).

During the gait cycle, in general, the angle ϑ is variable with the trend reported in Figure 17.11.

Due to the bigger amount of mechanical energy in the swing phase, with respect to the stance phase, in the following considerations and experimental tests, the analysis is limited only to the first one, i.e., $\vartheta_{stance} = 0$, while ϑ_{swing} is supposed linearly variable in the time (i.e., $d\vartheta_{swing}/dt$ = constant) between $\vartheta_{swing} = 0°$ and $\vartheta_{swing} = 60°$ according to the trend reported in Figure 17.11.

The discontinuous nature of the human movement and its irregularity impose the design of a power- and energy-conditioning circuit for matching the power source and the energy requested by the measurement circuit. In fact, for example, depending on the technology by which the electronic circuits are realized, the voltage supply needs to respect

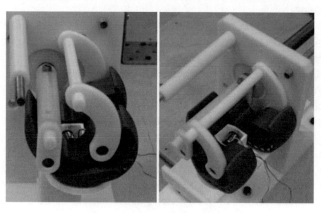

FIGURE 17.10
Prototype of the energy-harvester system implanted in the TKP.

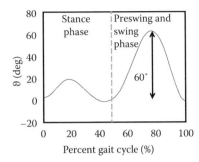

FIGURE 17.11
Knee joint angle ϑ as a function of the percent gait cycle.

a precise value or a proper range, while the voltage generated by the proposed energy harvester system is very variable and discontinuous in the time domain, due to the strict dependence on the characteristics of the knee motion, as described before. Furthermore, the energy consumption of the system is related to the time requested for measurement and also to the techniques chosen for the data processing and transmitting. Then it is necessary to establish a strategy for the functioning of the system; i.e., the autonomous force sensor has a phase in which the power-harvesting module converts the mechanical energy from the knee movement, and only when the required energy is available, a second phase in which the measurements are possible. For the proposed system, downstream of the power-harvesting module, a pcc was realized with the aim to provide the energy supply to the load with the requested characteristics of voltage and duration.

Two experimental tests were performed. The first one was conducted considering the measurement of the open circuit voltage $V_{o.c.}(t)$ induced on the coil; then in this test the energy-harvesting system is not connected to the pcc and to the load. Figure 17.12 shows the oscillating nature of this voltage due to the different couples (M1, M2, ..., M6) of magnets going near the coil (Luciano et al. 2012). In particular, the rising and falling ramps, between the broken boundary lines, delimit the complete passage, across the coil, of the generic couple Mn and the initial entrance of the following couple M(n + 1).

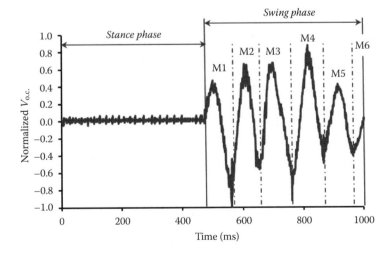

FIGURE 17.12
Normalized open-circuit voltage during a gait cycle with a step frequency of 1.0 Hz.

The second test was performed, considering the complete autonomous system, i.e., connecting the energy-harvesting system to the load using the realized pcc. From Table 17.2, the mean power consumption of the force sensor system during a single cycle of data acquisition and transmission is 1.7 mW while the time requested for the cycle is T_{cycle} = 13 ms, with a total energy consumption equal to E_L = 22.1 μJ. This load consumption was simulated using a resistive load R_L = 2.2 kΩ. The power-conditioning circuit (Figure 17.13) is composed essentially of an impedance-matching circuit, a CP, which is turned on when the input voltage is 300 mV, and a start-up capacitor. The start-up capacitor is connected, by the CP, to the energy source, when the system harvests the energy, and to the load R_L, when it executes and transmits the force measures.

The CP has a discharge start output voltage V_{CPout_1} = 2.0 V and a discharge stop voltage V_{CPout_2} = 1.4 V. The charging time T_c requested for charging the start-up capacitor is about $T_{c,zic}$ = 30.4 s if the initial voltage of the start-up capacitor is zero (zero initial condition), and $T_{c,ssc}$ = 7.6 s when it is 1.4 V (steady-state condition). Then the time requested to the patient, during the gait, for charging the start-up capacitor, is more than acceptable. The

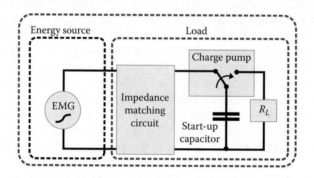

FIGURE 17.13
Block diagram of the autonomous force sensor system. EMG: electromechanical generator; R_L: resistive load equivalent to the force sensor system.

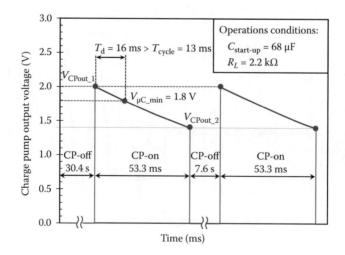

FIGURE 17.14
Operating conditions of the charge pump output voltage. CP-on: the CP connects the start-up capacitor to R_L; CP-off: the CP connects the energy source to the start-up capacitor.

capacitance $C_{\text{start-up}}$ of the start-up capacitor was deduced considering the minimum supply voltage of the processor ($V_{\mu C\_min} = 1.8$ V), the time necessary for the data acquisition ($T_{\text{cycle}} = 13$ ms), and the related energy consumption ($E_L = 22.1$ µJ). In particular, using the relation $E_L = 0.5 C_{\text{start-up}}[(V_{\text{CPout\_1}})^2 - (V_{\mu C\_min})^2]$, it is possible to deduce its minimum value ($C_{\text{start-up}} = 58$ µF). Choosing a capacitor $C = 68$ µF, a discharge time $T_d = 16$ ms is necessary to decrease the CP output voltage from $V_{\text{CPout\_1}} = 2.0$ V to $V_{\mu C\_min} = 1.8$ V. This time is greater than the time $T_{\text{cycle}} = 13$ ms necessary for a single measurement cycle (Figure 17.14).

In conclusion the energy-harvester system makes it possible to power supply a TKP implantable force sensor system, making the system autonomous.

17.5 Conclusions

Implantable medical devices are widely employed in order to monitor or to control different physiological parameters. Several implantable medical devices are powered by a battery, which constitutes a severe limitation because, in most cases, the battery defines the lifetime of the entire implantable medical device; in particular, because of the battery discharge, the implanted system must be surgically replaced. To obviate this problem, the implantable medical device should be powered through a telemetric technique or through an energy-harvesting system implanted together with the device. Several examples of implantable medical devices powered by telemetric or energy-harvesting technique are reported in the literature.

In this chapter, an example for each alternative to the battery is described; the first one is a force measurement system powered through the telemetric technique, while the second one is an energy-harvesting system which exploits the mechanical energy produced by the human knee joint movement.

The telemetric and the energy-harvesting system can theoretically operate for an indefinite time; hence, the implantable system must not be prematurely replaced. Furthermore, as the battery occupies significant space, batteryless implantable systems could be made smaller and more easily implanted. Otherwise, with a telemetric or an energy-harvesting system, energy is not always available and the measurement cannot be performed in a continuous way; the telemetric technique requires the presence of an external coil, while the energy-harvesting technique needs to store sufficient energy before powering the whole system.

The telemetric and the energy-harvesting technique, especially the second one, represent two valid alternative solutions to power an implantable medical device. The energy-harvesting technique is in course of study and development and it could enable the patient to move in a free and autonomous way.

References

Almouahed, S., M. Gouriou, C. Hamitouche, E. Stindel, and C. Roux, "Self-powered instrumented knee implant for early detection of postoperative complications," *Engineering in Medicine and Biology Society*: 5121–5124, 2010.

Almouahed, S., M. Gouriou, C. Hamitouche, E. Stindel, and C. Roux, "Design and evaluation of instrumented smart knee implant," *Biomedical Engineering* 58: 971–982, 2011a.

Almouahed, S., M. Gouriou, C. Hamitouche, E. Stindel, and C. Roux, "The use of piezoceramics as electrical energy harvesters within instrumented knee implant during walking," *Mechatronics* 16: 799–807, 2011b.

Crescini, D., E. Sardini, and M. Serpelloni, "Design and test of an autonomous sensor for force measurements in human knee implants," *Sensors and Actuators A: Physical* 166(1): 1–8, 2011.

Dalola, S., V. Ferrari, M. Guizzetti et al., "Autonomous sensor system with power harvesting for telemetric temperature measurements of pipes," *Transactions on Instrumentation and Measurement* 58(5): 1471–1478, 2009.

Goto, H., T. Sugiura, and T. Kazui, "Feasibility of the automatic generating system (AGS) for quartz watches as a leadless pacemaker power source: A preliminary report," *Engineering in Medicine and Biology Society, 1998. Proceedings of the 20th Annual International Conference of the IEEE* 1: 417–419, 1998.

Graichen, F., G. Bergmann, and A. Rohlmann, "Hip endoprosthesis for in-vivo measurement of joint force and temperature," *Journal of Biomechanics* 32: 1113–1117, 1999.

Halperin, D., T. Kohno, T. S. Heydt-Benjamin, K. Fu, and W. H. Maisel, "Security and privacy for implantable medical devices," *Pervasive Computing, IEEE* 7(1): 30–39, 2008.

IEEE Standard for Safety Levels with Respect to Human Exposure to Radio Frequency Electromagnetic Fields, 3 kHz to 300 GHz. C95.1-1999, 1999.

Lahuec, C., S. Almouahed, M. Arzel, D. Gupta et al., "A self-powered telemetry system to estimate the postoperative instability of a knee implant," *Biomedical Engineering* 58: 822–825, 2011.

López, M. T. P., Methods and circuits for the efficient management of power and energy in autonomous sensors. PhD dissertation. Departament d'Enginyeria Electronica, Universitat Politècnica de Catalunya, 2010.

Luciano, V., E. Sardini, M. Serpelloni, and G. Baronio, "Analysis of an electromechanical generator implanted in a human total knee prosthesis," *Sensors Applications Symposium IEEE*: 1–5, 2012.

Masouros, S. D., A. M. J. Bull, and A. A. Amis, "Biomechanics of the knee joint," *Orthopaedics and Trauma* 24: 84–91, 2010.

Miao, P., A. S. Holmes, E. M. Yeatman et al., "Micro-machined variable capacitors for power generation," *11th International Conference on Electrostatics*: 53–55, 2004.

Mitcheson, P. D., "Energy harvesting for human wearable and implantable bio-sensors," *Engineering in Medicine and Biology Society*: 3432–3436, 2010.

Morais, R., C. M. Frias, N. M. Silva et al., "An activation circuit for battery-powered biomedical implantable systems," *Sensors and Actuators A: Physical* 156(1): 229–236, 2009.

Morais, R., N. M. Silva, P. M. Santos et al., "Double permanent magnet vibration power generator for smart hip prosthesis," *Eurosensors XXIV, Linz, Austria, September 5–8, 2010,* 172(1): 259–268, 2011.

Nasiri, A., S. A. Zabalawi, and D. C. Jeutter, "A linear permanent magnet generator for powering implanted electronic devices," *IEEE Transactions on Power Electronic,* 26(1): 192–199, 2011.

Pinskerova, V., H. Iwaki, and M. A. Freeman, "The shapes and relative movements of the femur and tibia at the knee," *Orthopade* 29(Suppl 1): S3–S5, 2000.

Potkay, J., and K. Brooks, "An arterial cuff energy scavenger for implanted microsystems," *Bioinformatics and Biomedical Engineering*: 1580–1583, 2008.

Ramlan, R., M. J. Brennan, B. R. Mace, and I. Kovacic, "Potential benefits of a non-linear stiffness in an energy harvesting device," *Nonlinear Dynamics* 59: 545–558, 2010.

Riistama, J., J. H. S. Väisänen, H. Harjunpää et al., "Wireless and inductively powered implant for measuring electrocardiogram," *Medical and Biological Engineering and Computing* 45(12): 1163–1174, 2007.

Romero, E., R. O. Warrington, and M. R. Neuman, "Body motion for powering biomedical devices," *Engineering in Medicine and Biology Society*: 2752–2755, 2009.

Silay, K. M., C. Dehollain, and M. Declercq, "Inductive power link for a wireless cortical implant with two-body packaging," *IEEE Sensors Journal* 11(11): 2825–2833, 2011.

Soykan, O., "Power sources for implantable medical devices," *Business Briefing: Medical Device Manufacturing & Technology*: 76–79, 2002.

Stark, I., "Invited talk: Thermal energy harvesting with Thermo Life," *Wearable and Implantable Body Sensor Networks, 2006. BSN 2006. International Workshop*: 19–22, 2006.

Starner, T., "Human-powered wearable computing," *IBM Systems Journal* 35(3–4): 618–629, 1996.

Tamura, T., I. Mizukura, M. Sekine, and Y. Kimura, "Monitoring and evaluation of blood pressure changes with a home healthcare system," *IEEE Transactions on Information Technology in Biomedicine* 15(4): 602–607, 2011.

Tashiro, R., N. Kabei, K. Katayama, E. Tsuboi, and K. Tsuchiya, "Development of an electrostatic generator for a cardiac pacemaker that harnesses the ventricular wall motion," *Journal of Artificial Organs* 5(4): 0239–0245, 2002.

Troyk, P. R., G. A. DeMichele, D. A. Kerns, and R. F. Weir, "IMES: An implantable myoelectric sensor," *Engineering in Medicine and Biology Society, 2007, EMBS 2007, 29th Annual International Conference of the IEEE*: 1730–1733, 2007.

Valdastri, P., A. Menciassi, A. Arena, C. Caccamo, and P. Dario, "An implantable telemetry platform system for in vivo monitoring of physiological parameters," *IEEE Transactions on Information Technology in Biomedicine* 8(3): 271–278, 2004.

von Stetten, F., S. Kerzenmacher, A. Lorenz et al., "A one-compartment, direct glucose fuel cell for powering long-term medical implants," *MEMS 2006—19th IEEE International Conference on Micro Electro Mechanical Systems, January 22–26, 2006, Istanbul, Turkey*, 2006.

Wang, F., W. Hua, S. Zhang, D. Hu, and X. Chen, "Clinical survey of pacemakers 2000–2001," *Zhonghua Xin Lv Shi Chang Xue Za Zhi 2003* 7: 189–191, 2003.

Wei, X., and J. Liu, "Power sources and electrical recharging strategies for implantable medical devices," *Frontiers of Energy and Power Engineering in China* 2(1): 1–13, 2008.

Woodson, H. H., and J. R. Melcher, *Electromechanical Dynamics*, 3 vols. (Massachusetts Institute of Technology: MIT OpenCourseWare). http://ocw.mit.edu (accessed September 6, 2014). License: Creative Commons Attribution-NonCommercial-Share Alike, 1968.

Yang, Y., X. Wei, and J. Liu, "Suitability of a thermoelectric power generator for implantable medical electronic devices," *Journal of Physics D: Applied Physics* 40: 5790–5800, 2007.

18

Signal Processing in Implantable Neural Recording Microsystems

Sedigheh Razmpour, Mohammad Ali Shaeri,
Hossein Hosseini-Nejad, and Amir M. Sodagar

CONTENTS

18.1 Neurophysiological Background

18.1.1 Introduction

The human brain contains approximately 10^{10} to 10^{12} *neurons* [1] with 10^{14} to 10^{15} total number of neuronal connections (*synapses*) [2] in a volume of about 1500 cm^3 [3]. Understanding the physiology of the neuron, as the smallest unit of information processing and communication, is of crucial importance when studying the structure and function of the brain. Based on a simple neuronal model, a neuron integrates input signals and fires a neuronal impulse as the output in response to an adequate amount of excitation [4].

Illustrated in Figure 18.1, each neuron contains a forest of *dendrites*, a *cell body* (*soma*), and an *axon*. Dendrites are branches with short florid terminal arborization, which are capable of receiving excitatory or inhibitory signals from up to 100,000 afferents. The cell body (soma) is in charge of electrical integration of the input signals received through the dendrites. It is large in diameter (typically more than 50 μm [2]) and contains most of the cytoplasmic organelles (e.g., *nucleus*). Axon (0.2 μm to 1 mm in diameter [2]) is a long (0.1 mm to 2 m [2]) branch emanated from the base of soma and projects to the target neurons. Neuronal activities are conveyed by the axon and delivered to the proceeding neurons via *synaptic terminals*. According to the mechanism of data transmission, there are two types of synapses: chemical and electrical. *Chemical synapses* electrochemically send messages through, while *electrical synapses* (*gap junctions*) are physically connected and electrically considered as short circuits.

18.1.2 The Intracellular Neural Signal

Communications between neurons and processing of neuronal activities is realized by *action potentials* (APs), also known as *neural spikes* or simply *spikes*. Action potentials are

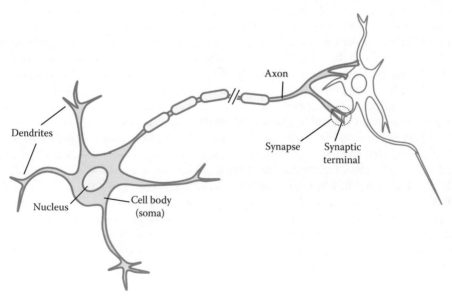

FIGURE 18.1
Anatomy of a neuron.

short electrical nerve pulses initiated in an all-or-none manner. An AP can propagate rapidly along the axon toward the target neuron(s). This section studies the electrical behavior of a neuron and how an AP is generated.

The membrane of nerve cells is made up of a lipid bilayer that does not allow the ions to permeate into or out of the cell. However, there exist ion pumps and ion channels on the membrane that let certain ions move through. In general, the role of the ion pumps and ion channels is to allow for the readjustment of the concentrations of the ions inside the cell. As a result, the concentrations of the ions inside and outside the cell will be different, and the ions consequently exhibit the tendency to diffuse from the higher-concentration medium to the lower-concentration one. The difference in the ion concentrations results in the *membrane potential* (V_m), defined as the potential inside the cell relative to the outside.

In a simplified model introduced by Hodgkin and Huxley [5] for the squid giant axon, only potassium (K^+) and sodium (Na^+) ions contribute to the potential of membrane. While the K^+ ions are more concentrated on the intracellular side, there is a rather high concentration of the Na^+ ions on the outside of the neuron.

18.1.2.1 Resting Potential

The membrane contains many *leak channels,* which are likely to be open at rest and allow the ions to leak through. In equilibrium, ionic pumps and leak channels regulate the concentration of the ions inside and outside the cell membrane. In the steady state of the membrane, net ionic flux is zero; therefore, the concentration of the ions stays fixed. This condition is called the *rest state.*

In different kinds of neurons, the *resting potential* (V_r) varies in the range of -30 to -100 mV [1] (typically -60 to -80 mV [6]).

18.1.2.2 Action Potential

At rest, when ions move only across leak channels, the membrane is *polarized* (has a negative voltage). There are also some voltage-gated ion channels in the membrane that contribute to the initiation of APs.

If the membrane is sufficiently depolarized, or in other words, if the membrane potential exceeds a certain level, known as *the threshold of firing* ($\sim$10 mV above the resting potential [2]), then the Na^+ voltage-gated channels will open, resulting in a notable increase in the inward flux of Na^+ ions. Accumulation of the Na^+ ions inside the cell charges the capacitor associated with the membrane (C_m). After depolarization, the Na^+ voltage-gated channels are inactivated and at the same time, K^+ voltage-gated channels open, leading to repolarization of the membrane. This burstlike change in the membrane potential is referred to as the action potential, or neural spike (especially the steepest part) [7]. Duration of an AP, also called the *spike time course,* is typically about 1 to 2 ms [6]. However, spikes as short as 200 µs and as long as up to 10 ms can also be found [1–2]. Peak-to-peak height of APs is in the range of 70 to 110 mV [2].

After the occurrence of an AP, the membrane needs a few milliseconds of recovery time (on the order of milliseconds), called the *refractory period,* during which a new AP cannot be initiated. The refractory period strictly limits the firing rate of a neuron to a maximum of about 1000 spikes per second [1–2]. According to van Rossum et al. [8], spike firing rates are typically in the range of 10 to 100 spikes per second.

Following the generation of an AP, it propagates along the axon until it reaches the synaptic terminals, where it is delivered to the next neurons. Because of the active mechanism of AP regeneration along the axon, the amplitude of the AP does not attenuate. Based on the electrophysiological properties of the ion channels and pumps, it is believed that the APs generated by a specific neuron all have the same wave shape independently from the amount of input impinging (all or none) [2]. Accordingly, the AP wave shape is considered as the neuron signature.

Physiological properties of the neural cell membrane are reflected in the spike wave shape as well as in the firing rate and pattern of the spike train generated by that cell [7].

18.1.3 The Extracellular Neural Signal

To study the activity of the brain (neurons), activities of large neuronal ensembles are recorded. The frequency spectrum of the useful content of an extracellularly recorded neural signal is spread from around 0 Hz to about 6 kHz [6]. Extracellular neural signals are weighted summations of activities near the recording site and are therefore called *mean*

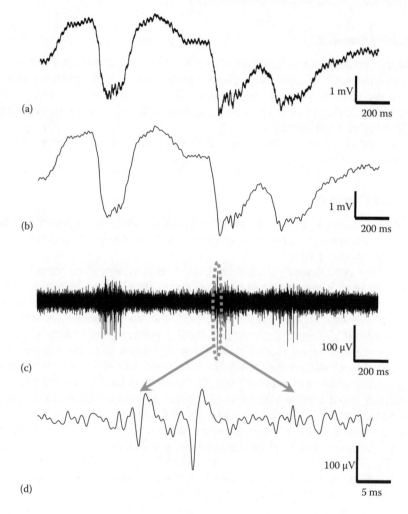

FIGURE 18.2
An extracellular neural signal: (a) the whole signal; (b) the LFP component (<150 Hz); (c) the APs (300 to 6000 Hz); and (d) close-up view of the AP component shown in (c).

extracellular field potentials (mEFPs). The frequency spectrum of an mEFP is divided into two sub-bands: the low-frequency sub-band (< 400 Hz) and the high-frequency sub-band (> 400 Hz), associated with *local field potentials* (LFPs) and action potentials, respectively. The LFP is essentially the net effect of the source and sink currents of a neural population in the vicinity of the recording site (0.25 to 3 mm) [6].

Figure 18.2 shows a raw extracellular neural signal recorded from the rat barrel cortex and its LFP and AP components.*

From the extracellular side, all of the neuronal activities (changes in membrane potentials) are current *sinks* (flow of electric current toward the cell) and *sources* (flow away from the cell), which charge or discharge the capacitor associated with the cell membrane. The extracellular potential can be recorded by implanting an electrode within the extracellular fluid.

When the recording electrode is placed near the soma, the extracellular potential recorded will be proportional to the net ionic current. Although it is believed that an extracellular action potential is proportional to the derivative of the membrane potential [7], there is still debate on the relationship between an extracellular spike and the intracellular action potential causing it [9–11]. Figure 18.3 shows an intracellular action potential and the corresponding extracellular spike.

A high-impedance microelectrode (> ~6 MΩ) with a small tip of 3 to 10 μm placed near the spiking components of a neuron (mainly the soma or the axon) may access the activity of a single neuron that can be discerned from others, the so-called *single-unit activity* (SUA). In contrast, electrodes of lower impedance have the potency to access more distant neuronal activities (140 to 300 μm away), the so called *multiunit activities* (MUAs), containing a summation of the synaptic activities and APs that are present at the recording site. To record the LFP, the impedance of the electrode needs to be even lower than that of the ones used for the recording of MUAs.

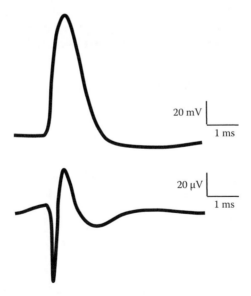

FIGURE 18.3
Top: An intracellular AP; *bottom*: the corresponding extracellular spike.

---

* Please refer to Acknowledgment.

Besides the LFP and APs, an extracellularly recorded neural signal contains nonnegligible random amplitude fluctuations, referred to as the *background noise* (BN). It is believed that different sources contribute to the background noise of a neural signal including spontaneous activities of neurons as well as weak multiunit activities of the neuronal ensembles that lie far away from the recording site [3,12].

18.2 Neural Recording Microsystems

18.2.1 Neural Recording Approaches

Nowadays, implantable microsystems developed for electrically interfacing to the brain, also referred to as *brain–machine interfaces* (BMIs), are of increasing interest to researchers in medical sciences in general, in neuroscience more specifically, and, of course, in engineering. Such microsystems are expected to have great impact on advanced research and development activities in a variety of applications, namely, curing neural disorders, diseases, and disabilities such as epilepsy, paralysis, Parkinson's disease, and blindness [13–18]. EEG is perhaps the most popular approach for neural recording, and that is because of advantages such as being noninvasive (therefore, usually risk free) and being rather convenient for the subject (e.g., patient). On the other hand, there are fully invasive approaches such intracortical neural recording using implantable microsystems specifically designed for this purpose, which are expected to provide much more precise recordings in terms of both temporal and spatial resolution [14–19]. Of the other important advantages of using implantable devices for neural recording, one can name their capability of concurrently recording neuronal activities on tens to hundreds of channels [14–19] (and even 1000+ channels [20]).

This chapter starts with the introduction of intracortical neural recording microsystems and then focuses on the neural signal processing approaches that are used in such systems in order to allow for high-density recording of neural signals from the brain.

18.2.2 Neural Recording Microsystems

Figure 18.4 shows a general block diagram of wireless implantable microsystems developed for multichannel intracortical extracellular neural recording. Such systems, in general,

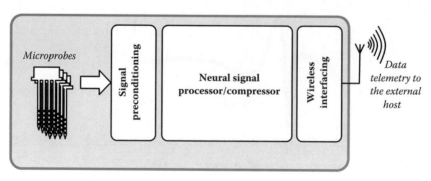

FIGURE 18.4
General block diagram of an implantable neural recording microsystem.

interface with the nervous system by using *microprobes* (also referred to as *microelectrode arrays* [MEAs]). After some analog *signal preconditioning* tasks, the recorded neural signals are delivered to a *neural signal processor/compressor* in order to make it possible to telemeter them to an external host by using a *wireless interface*.

18.2.2.1 Recording Front-End

Intracortical recording of neuronal activities is commonly performed by using penetrating microprobes. High-density neural recording from a certain region of the brain with high spatial resolution requires three-dimensional arrays of microprobes, typically made using microfabrication processes. Usually, microprobes are rigid yet tiny structures providing small conductive areas (referred to as *recording sites*) for sensing rather weak electrical/ electrochemical neuronal activities on the order of tens to hundreds of microvolts. These types of microprobes are sometimes called *passive probes* as they merely sense neural signals with no electronic circuitry on them. The single-unit neural signals intracortically recorded using penetrating microprobes generally contain all three components already introduced in the previous section (i.e., LFP, AP, and BN). Amplification of neural signals and rejection of the signal components that lie out of the frequency band of interest are considered as the first steps of the processing of the recorded neural signals. For this purpose, passive probes are usually followed by an *analog signal preconditioning module*, realizing preamplification and band-pass filtering. Typically, neural signals are amplified with a gain of 20 to 30 dB, and band-pass filtered with an upper cutoff frequency of 5 to 10 kHz. Lower cutoff frequency of the filter is set depending on whether the LFP is to be kept or rejected. For applications where recording the LFP is of interest, lower cutoff frequency is set in the sub-Hertz range [14,21]. Otherwise, the useful signal band starts from around 100 Hz [14–16] in order to remove the LFP from the neural signal being recorded. There are, however, certain types of microprobes in which active electronic circuitry is integrated for analog signal preconditioning. Figure 18.5 shows two passive probes fabricated using silicon-based microfabrication technology [22].

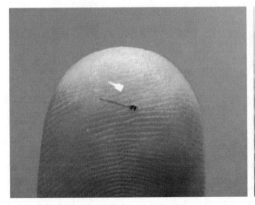

FIGURE 18.5
Passive silicon probes with four $20 \times 20 \ \mu m^2$ recording sites. (From M. Nekuyi, "Design and fabrication of penetrating passive silicon probes for neural recording," *Technical Report, Research Laboratory for Integrated Circuits and Systems (ICAS), Faculty of Electrical Engineering, K. N. Toosi University of Technology*, Tehran, Iran, Summer 2010.)

18.2.2.2 Wireless Interfacing Module

Similar to any other implantable microsystem, neural recording implants need to interface to the external world through a wireless connection. This is mainly to receive the electric energy they need for their operation as well as to exchange bidirectional data with an external host. In a neural recording microsystem, a *wireless interfacing module* is usually in charge of interfacing the system to the outside through one or more wireless channels. Telemetry of both power and data* from outside the implant (*forward telemetry*) is usually done using one wireless link. To convey data to the implant, a sinusoidal power carrier is modulated using a variety of well-known digital modulation schemes such as FSK [23], OOK [24], and PSK [25–27]. There are, however, novel ideas that are proposed for more efficient forward telemetry, combining the benefits of using sinusoidal power carriers and pulse-based digital modulation approaches [28]. The link used for forward telemetry is realized by using either inductive [14,29] or capacitive [30–31] coupling approaches, illustrated in Figure 18.6. Although telemetry of the recorded neural data to the external host (*reverse data telemetry*) is usually performed through a different wireless channel [24], there are other approaches for reverse data telemetry via the same link envisioned for forward telemetry such as *auxiliary-carrier load-shift keying* [32], resulting in more compact hardware implementation.

18.2.2.3 Neural Processing Module

Although the recorded neural signals originally include different components (e.g., LFP and AP), depending on the application a neural recording implant is designed for, certain component(s) or information of the recorded neural signals needs to be telemetered off the implant. In order to extract the useful components/information of the recorded neural signals, an analog, digital, or mixed-signal *neural processing module* (NPM) is usually envisioned on the implant [14–16,18–19]. Importance of the role of this module is more

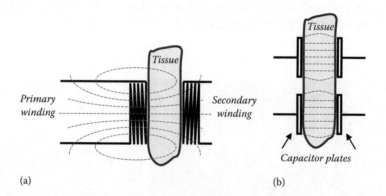

FIGURE 18.6
Illustrations of (a) an inductive link (From A. M. Sodagar et al., *IEEE Journal of Solid-State Circuits*, Vol. 44, No. 9, pp. 2591–2603, Sep. 2009; E. Ashoori et al., "Design of double layer printed spiral coils for wirelessly-powered biomedical implants," *Proceedings of the 2011 IEEE Annual International Conference of the Engineering in Medicine and Biology Society (EMBC '11)*, pp. 2882–2885, Sep. 2011.) and (b) a capacitive link (From A. M. Sodagar and P. Amiri, "Capacitive coupling for power and data telemetry to implantable biomedical microsystems," *Proceedings of the 2009 IEEE International Conference on Neural Engineering (NER '09)*, pp. 411–414, May 2009; M. Takhti et al., "Modeling of a capacitive link for data telemetry to biomedical implants," *Proceedings of the 2011 IEEE International Conference on Biomedical Circuits and Systems (BioCAS '11)*, pp. 181–184, Nov. 2011.).

---

* These are the data sent to the implant to program it for proper operation in the desired mode.

highlighted when the system is designed to record neuronal activities on multiple channels (i.e., tens to hundreds of channels) and transmit them to the outside world through a wireless link with limited bandwidth [33].

18.3 High-Density Neural Recording

When expanding the capacity of recording for an intracortical neural recording microsystem to multiple channels, the designer faces several challenges at different levels:

- At the *system level*, physical issues such as size, weight, and integration of the system become important concerns.
- At the *circuit level*, low-voltage, low-power, and low-noise operation as well as high-speed signal processing (i.e., fast enough to enable real-time signal handling) are considered among the key requirements that need to be fulfilled by circuit blocks.
- At the *signal level*, for high-density neural recording, the *function* and *architecture* of the system needs to be designed in such a way that the massively recorded neural information can be handled with acceptable quality and usually in real time.

The main scope of this chapter is focused on the techniques employed in wireless implantable neural recording microsystems in order to enable them to handle multichannel intracortical extracellular recording of neural signals (hereby referred to as *neural recording* unless otherwise stated).

18.3.1 System-Level Approaches

A variety of system-level architectural approaches have been taken in order to realize the idea of high-density neural recording using electronic hardware. Figure 18.7 illustrates four examples of system-level ideas presented to design wireless implantable multichannel neural recording microsystems:

- Site selection on the recording front-end [34], allowing for selecting a limited number of sites from among a total of so many sites potentially available to record from.
- Time-division multiplexing (TDM) [35]/frequency-division multiplexing (FDM) [36], making it possible to efficiently allocate a limited time/frequency budget for the telemetry of the so many neural channels recorded to the outside via a single wireless link, and two-step delta-sigma modulation [37], which can be an efficient way to digitize neural signals in two steps: performing the first step (delta-sigma modulation) on the implant to save area and power and moving the rest of the digitization process (decimation filtering) to the external setup where virtually unlimited power and area are available.

18.3.2 Signal-Level Approaches

It was mentioned that system-level architectural ideas help enable a neural recording implant efficiently record, process, and telemeter multiple channels of neural signals to

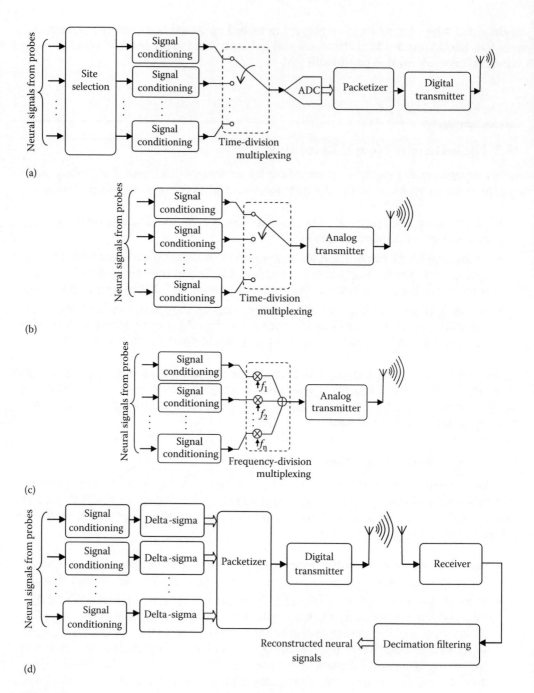

FIGURE 18.7
Examples of system-level architectural approaches to handling large number of neural channels in a neural recording implant: (a) recording-front-end site selection, (b) TDM, (c) FDM, and (d) two-step delta-sigma modulation.

the outside world. However, while in a variety of applications such as neural prostheses, recording of neuronal activities from hundreds to above 1000 channels is desirable [14–19], recording of raw neural signals (without any signal processing on the implant) using state-of-the-art size- and power-efficient neural recording microsystems is still limited to well below 100 channels [38].

From among the key restrictions in this area, one can name the limited bandwidth allocated for wireless data communication between the implant and the external world [39]. This means that in addition to efficient system-level architecture, high-density neural recording microsystems need to take advantage of signal-level techniques in order to overcome such limitations. Techniques such as extraction of useful signal components, signal compression techniques, and data reduction methods are among the successful attempts reported to efficiently utilize the limited available wireless transmission bandwidth.

The next section deals with signal-level approaches to high-density neural recording, the majority of which can be categorized into two main classes: *time-domain methods* and *transform-domain methods*.

18.4 Neural Signal Processing in the Time Domain

Action potentials are considered as the key signal components conveying the information being processed by the biological neural network. In applications such as neural prostheses, action potentials are the target components of the neural signals that are to be transmitted from the implant to the outside. Therefore, most of the neural signal processing techniques used in implantable neural recording microsystems, whether in the time domain or in the frequency domain, somehow focus on preserving the spike content of the neural signals being recorded [14–16,40–46].

In this section, the methods used in the time domain for the reduction of the amount of neural data while preserving the useful information they carry are studied. This includes methods concerning spike contents of the neural signals being recorded, such as *spike detection*, *spike extraction*, and *spike sorting*.

18.4.1 Spike Detection

It was earlier mentioned that in multichannel neural recording microsystems, concurrently reporting raw neural signals on tens to hundreds of channels does not seem to be practical when physical size, power dissipation, and wireless transmission bandwidth are concerned. Noting the fact that neural spikes are typically 1 to 2 ms wide and that their occurrence rate, also known as *spike firing rate*, barely goes beyond a few hundreds of spikes per second [8], the idea of reporting only occurrence of spikes (which is believed to be the key information to be telemetered off the implant in a wide variety of applications) will significantly save time and bit rate. Detection of the occurrence of neural spikes, which is realized by somehow discriminating between neural spikes and the background noise of a neural signal, is usually referred to as spike detection. Perhaps, the most straightforward method for spike detection is *hard thresholding*, in which the neural signal is compared with a certain threshold level. Although in a majority of cases the comparison threshold is set at a fixed value determined or programmed by the user [14–15], an appropriate threshold level can also be automatically determined or calculated for the neural signal

being processed. Figure 18.8a presents simplified functional diagrams for hard thresholding with two user-programmable fixed threshold levels [40] and also two examples for the
adaptive generation of threshold levels in digital and analog fashions [34,41]. In Olsson
and Wise's work [34] (Figure 18.8b), two threshold levels are calculated (below and above
the signal baseline or average value AVG) in the digital domain for all the signal samples
captured over a certain time window. The threshold levels are calculated as

$$THR = AVG \pm k \cdot SD, \tag{18.1}$$

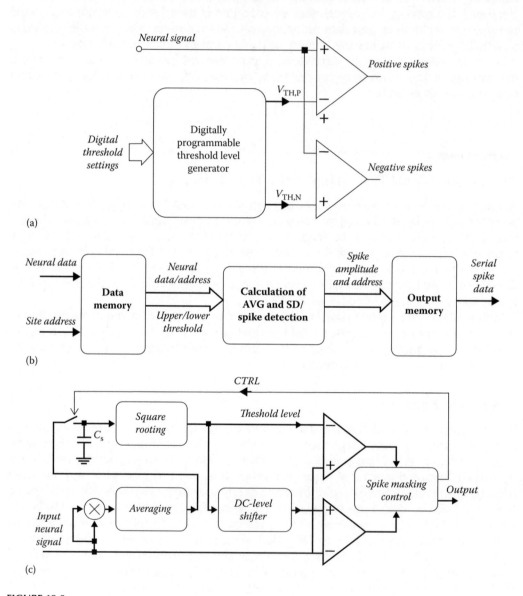

(a)

(b)

(c)

FIGURE 18.8
Examples of the generation of threshold levels for spike detection by hard thresholding: (a) digitally programmable fixed analog threshold levels, (b) calculation of adaptive threshold levels in the digital domain, and
(c) generation of adaptive threshold levels in the analog domain.

where SD is the standard deviation of the sample amplitudes and k is a constant factor (between 3 and 7) set by the user depending on the SNR of the signal. Hardware implementation of the computation of the standard deviation of the signal is not easy and straightforward. As an efficient solution in the analog domain, spike detection in Barati and Sodagar's work [41] (Figure 18.8c) is based on comparing a squared version of the input neural signal with an adaptively generated analog threshold level calculated by simply averaging a certain number of samples of the squared signal.

18.4.2 Spike Extraction

As already mentioned, reporting only the occurrence of spikes is the least expected from a high-density neural recording implant. In some works, upon the detection of a neural spike, the system telemeters the signal samples associated with the detected spike. This helps neuroscientists receive more information about the origin of the neuronal activities recorded by the system. The work by Olsson and Wise [34] detects spikes by comparing the neural signal with positive and negative threshold levels and transmits the spike samples that go beyond the threshold levels to the outside. The resulting spike wave shape reconstructed on the external setup is shown in Figure 18.9a.

In Gosselin et al.'s work [42], upon the detection of a neural spike, it is extracted from the rest of the signal by buffering enough number of samples over its entire course in order to be sent off the implant. In this approach, independently from the duration of the spike, a certain number of signal samples (16 samples prior to threshold crossing plus the

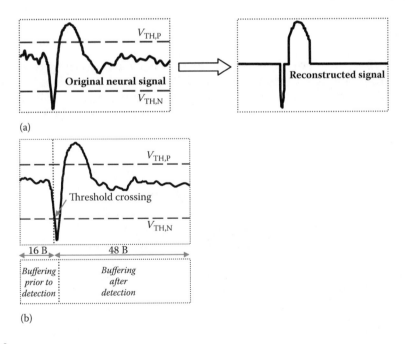

(a)

(b)

FIGURE 18.9
Spike extraction: (a) extraction of the spike samples that are beyond threshold levels (From R. H. Olsson and K. D. Wise, *IEEE Journal of Solid-State Circuits*, Vol. 40, No. 12, pp. 2796–2804, Dec. 2005.) and (b) extraction of spikes with a fixed extraction window (From B. Gosselin et al., *IEEE Transactions on Biomedical Circuits and Systems*, Vol. 3, No. 3, pp. 129–141, 2009.).

following 48 samples) are conservatively chosen in order not to miss even a small part of the longest normal spike.

Extraction of spikes in Shaeri's work [43] is performed in a more efficient way in terms of the duration of the spikes extracted. In this work, two thresholds levels, below and above the signal baseline, define a signal *baseline strip*, within which the background noise fluctuations are confined. Starting with when the neural signal goes beyond either one of the threshold levels, the extracted spike ends when it returns to within the baseline strip and stays in there for a predefined duration of time (on the order of seven cycles of the sampling clock).

18.4.3 Feature Extraction and Spike Sorting

To distinguish different spikes from each other, they do not need to be completely sent off the implant. For this purpose, certain spike features, such as the time interval between the peak and underpeak, the sharpness (slope) of the transition between the peak and underpeak, and the width of the spike, can instead be used [44,47]. Compared with the telemetry of all the spike samples, a much lower transmission bit rate will be needed to telemeter the associated features. In some works, based on the features extracted, detected spikes are first sorted (classified) on the implant, and then only a report of the class of the spikes occurred along with the corresponding occurrence time is reported to the outside [16]. This approach is referred to as spike sorting, which can significantly reduce the extent of the data telemetered. The price is, however, paid by the additional circuit complexity, chip area, and power dissipation of the on-implant spike sorting circuitry.

18.4.4 Delta Compression

As another attempt to reduce the amount of the neuronal data wirelessly transferred from a neural recording implant to the outside, the *delta compression* technique is proposed by Aziz et al. [19]. In this technique, the difference between every two consecutive samples (rather than the samples themselves) is telemetered. A simplified block diagram of the system realizing this idea is shown in Figure 18.10.

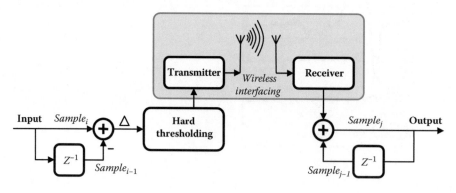

FIGURE 18.10
The delta compression technique proposed by Aziz et al. [19]. (From J. Aziz et al., *IEEE Journal of Solid-State Circuits*, Vol. 44, No. 3, pp. 995–1005, Mar. 2009.)

18.4.5 Nonlinear Quantization

ADCs are inseparable parts of high-density neural recording microsystems. Although typically linear ADCs are used to digitize neural signals, the idea of choosing a more appropriate quantization function for the specific input signal being digitized (neural signal in this specific application) can potentially result in not only more efficiently digitizing the signal but also performing some kind of data compression of reduction as well. In Judy et al.'s work [48], a signal-specific ADC with antilogarithmic quantization function is proposed for the conversion of analog neural signals to digital. In this approach, as illustrated in Figure 18.11, the nonuseful background noise with small amplitudes (around the signal baseline) is digitized with large quantization steps, while the large useful action potentials are quantized more finely. This means that, with a smaller number of *physical bits*, action potentials are still digitized with the same *effective resolution*. At the same time, the background noise is mostly rejected because of the large quantization steps around the background noise. From among the benefits of this approach, one can point to (a) reduction in the extent of the digital data carrying neural information to the outside world, (b) reduction in the background noise content of the neural signal being recorded, and (c) realization of analog-to-digital conversion and neural signal compression and denoising using a power- and area-efficient electronic hardware. In this work, two identical piecewise-linear antilogarithmic quantization functions symmetrically located around the baseline level are used to digitize neural signals.

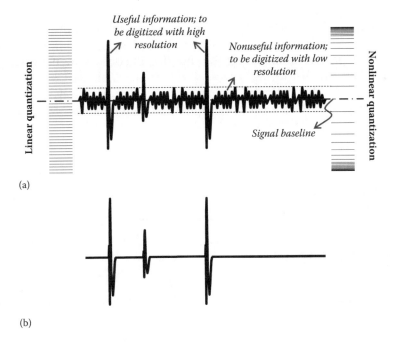

FIGURE 18.11
(a) The idea of nonlinear quantization introduced by Judy et al. [48] and (b) the resulting denoised signal.

18.5 Transform-Based Neural Signal Compression

As previously mentioned, time-domain data compression algorithms are straightforward approaches to compressing neural data, usually benefitting from simple hardware implementations. In high-density neural recording systems, as the number of recording channels increases, more efficient compression techniques (i.e., with higher compression rate and at the same time lower error) are required. Therefore, the transform-domain compression approaches that exhibit the potential to compress signals without sacrificing the signal integrity become suitable candidates for this purpose. More specifically speaking, from the signal processing standpoint, transformations with *energy preservation* and *energy compaction* properties are proper candidates for signal compression in neural recording implants. The former property guarantees that the content of the neural signal will not be lost during the compression process, and the latter ensures that the employment of that transform can potentially lead to data compression [49–50].

In addition to the signal processing properties discussed above, the transforms used for data compression in neural recording implants need to comply with the general circuit- and system-level constraints associated with implantable microsystems. Aiming at implantable applications, low computational complexity, and efficient hardware implementation (leading to small silicon area and low power consumption) are among the key requirements that the circuit implementation of any neural data compression technique needs to fulfill.

Figure 18.12 illustrates the mapping of a neural signal from the time domain to the target transformation space, where a certain number of *coefficients* correspond to the signal samples in an equivalent *time window*. The energy compaction property of the transformation used for signal compression is shown by the fact that the coefficients corresponding to the useful content of a neural signal (i.e., action potentials in this case) are of higher energy in the transform space. Compression of the neural signal being recorded is made possible by keeping the coefficients of higher significance (i.e., the ones of higher energy, carrying useful information) and discarding the less significant coefficients. Useful and nonuseful coefficients are usually discriminated by hard thresholding. This way, while considerably reducing the extent of the data sent off the neural recording microsystem, useful information of the recorded neural signal is preserved.

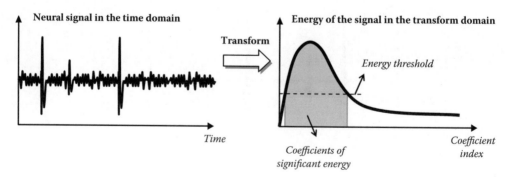

FIGURE 18.12
Illustration of the energy compaction property in the transform domain.

Figure 18.13 shows the general steps taken in order to take advantage of the energy compaction property of the transformations used for neural signal compression. Neural signals are first preconditioned in the analog domain and converted to digital. The digitized signals are then taken to the transform space and the resulting coefficients are hard-thresholded in order to achieve data reduction/compression. The compression process finally completes by using an efficient encoding technique. In this context, encoding is usually performed to reduce the amount of the data associated with the nonspike parts of the signal being telemetered. Sometimes, algorithms such as *run-length encoding* (RLE) are used to encode the number of zeros representing the denoised parts of the signal in the transform domain (i.e., the parts below the spike detection threshold level) [51–53].

18.5.1 Performance Measures

To quantify the merit of a data compression technique, it is essential to introduce figures for measuring the compression performance from different aspects. In implantable applications, performance of a data compression technique needs to be evaluated from both *signal processing* and *hardware implementation* standpoints.

The extent of the compression achieved is usually expressed by the *compression ratio* (CR) defined as

$$CR = \frac{N_B}{N_A},$$ (18.2)

where N_B and N_A are the amounts of neural data before and after the application of the data compression technique, respectively [52].

Usually, compression of a signal is achieved at the price of some extent of error between the original signal and its reconstructed form. The normalized *root-mean-square* (RMS) of *error* is a measure of the difference between the original signal and its reconstructed form (after compression), defined as

$$RMS = \frac{\sqrt{\frac{1}{N}\sum_{i=0}^{N-1}(\hat{x}_i - x_i)^2}}{x_{i(p-p)}},$$ (18.3)

where x_i is the original signal, $\hat{x}_i$ is the reconstructed signal, N is the number of signal samples taken, and $x_{i(p-p)}$ is the peak-to-peak value of the original signal [52,54]. Normalized RMS of error is used as a measure of the ability of the compression approach in preserving the shape of the neural signal (more specifically, action potentials) during the compression process.

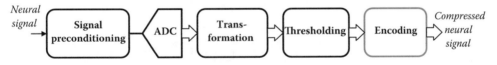

FIGURE 18.13
General block diagram of transform-domain compression approaches.

Computational complexity of a neural signal compression technique determines its merit when it is implemented using electronic hardware. More specifically speaking, a compression technique is considered suitable for use in neural recording implants when, besides its efficacy in the aforementioned signal processing aspects, its hardware implementation consumes low power, occupies small chip area, and operates in the real time.

The *discrete wavelet transform* (DWT) [51], the *Walsh–Hadamard transform* (WHT) [52], and the *discrete cosine transform* (DCT) [53] are among the most successful transformations employed for the compression of neural signals in implantable intracortical neural recording microsystems.

18.5.2 The Discrete Wavelet Transform

The *wavelet transform* in its different forms (continuous, discrete, and stationary) is a well-known transform for neural data handling. In Gosselin et al.'s work [55], the *continuous wavelet transform* (CWT) was proposed for the detection of action potentials. In Kim and Kim's [56], Quian Quiroga et al.'s [57], Obeid and Wolf's [58], Gibson et al.'s [59], and Roy and Sawan's [60] works, the DWT was proposed for high-performance spike detection and feature extraction. The DWT with Duabechies-4 basis function was introduced by Dumortier et al. [61] for neural data compression. As a different method, the DWT with Symmlet-4 basis function was used by Oweiss et al. [62] to compress neural data without any attempt to directly detect action potentials. Performance of a DWT-based compression approach is determined by the basis function used. It is believed by Oweiss et al. [62] that Symmlet-4 is advantageous over other wavelet basis functions because of having more similarity to the general wave shape of neural spikes and consequently approximating action potentials with fewer coefficients [62]. In addition to the capabilities it offers for neural signal compression, the DWT needs to be efficiently implemented using electronic hardware, too. Various schemes are available for the hardware implementation of the DWT, including the convolution-based scheme [63], lifting [64], and B-spline [65]. The convolution-based scheme implements the DWT by using a two-channel filter bank. Figure 18.14 shows the structure of this filter bank for the decomposition of the input signal into low-pass and high-pass components [63].

The lifting scheme, shown in Figure 18.15, is used to reduce computational complexity (i.e., the number of multiplications and additions) of the transform by somehow combining hardware used to realize each one of the low-pass and high-pass filters [64].

In the B-spline realization, computational complexity is even further reduced compared with the other two schemes. Realization of the DWT based on the B-spline approach uses fewer multiplications than the lifting approach, but more additions will be needed instead. A comparison between the three DWT hardware implementations discussed above is presented in Table 18.1 [62].

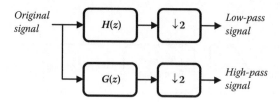

FIGURE 18.14
A two-channel decomposition filter bank.

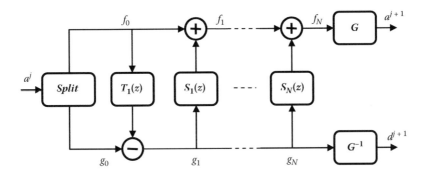

FIGURE 18.15
Lifting structure for single-level DWT implementation.

TABLE 18.1

Comparison of Arithmetic Operations for Three DWT Implementation Schemes

	Lifting	B-Spline	Convolution
Multiplications	8	6	14
Additions	8	18	14

Using the lifting structure, an efficient implementation for Symmlet-4-based DWT is presented by Oweiss et al. [62], based on which a 32-channel neural data compression system was designed by Kamboh et al. [66].

Neural data compression using the DWT with the *Haar* basis function [54] is of much less computational and hardware complexity compared with the other varieties of the DWT discussed above. This is because it is realized using merely additions and subtractions with no need for any multiplications, as illustrated in Figure 18.16. Although significantly simpler in hardware, the *discrete Haar-based wavelet transform* (DHWT) exhibits comparable signal processing performance (to the other DWT varieties) in terms of compression ratio and normalized RMS of error. In this work (i.e., that by Shaeri et al. [54]), wavelet coefficients (referred to as the *approximate* and *detail* coefficients) are first properly truncated in order to achieve some extent of data reduction, resulting in a compression rate of ~1.78. The truncated coefficients are then compared with a user-defined threshold level to detect neural spikes. Based on this approach, a neural compressor/processor is developed, in which neuronal activities on 64 channels are concurrently detected. Truncated coefficients associated with the detected action potentials are finally framed to be sent to the outside world through wireless connection.

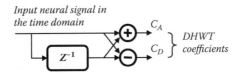

FIGURE 18.16
Single-level implementation of the Haar-based DWT.

18.5.3 The Walsh–Hadamard Transform

The *Hadamard transform* (HT) is an orthogonal transform. The nth order of the HT is defined as

$$Y = H_n \cdot X, \tag{18.4}$$

in which X is the input vector containing 2^n samples of the signal in the time domain, Y is the output vector consisting of 2^n corresponding coefficients in the transform space, and H_n is the $2^n \times 2^n$ HT matrix. The nth order of the HT matrix, H_n, is written as

$$H_n = \frac{1}{\sqrt{2^n}} \begin{pmatrix} H_{n-1} & H_{n-1} \\ H_{n-1} & -H_{n-1} \end{pmatrix}, \tag{18.5}$$

in which, $H_0 = 1$ and $n = 1, 2, \ldots$ [67]. The WHT is a variety of the HT, in which components of the matrix transform are defined based on the *Walsh functions* [67].

Since the WHT is a symmetric and orthogonal transform, the same transform matrix can be used for the inverse transform (Equations 18.4 and 18.5). In addition to orthogonality, energy preservation, and energy compaction [68], the WHT is of interest for signal compression in implantable neural recording microsystems because of the fact that it is implemented using only adders and subtractors. In Hosseini-Nejad et al.'s work [52], a 128-channel neural signal compressor was designed based on the WHT, exhibiting superior circuit-level performance compared with the other approaches already reported in the literature. Figure 18.17 shows a simplified schematic diagram for hardware implementation of the WHT in Hosseini-Nejad et al.'s work [52].

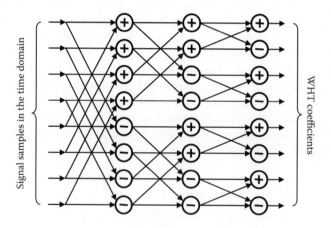

FIGURE 18.17
Block diagram of the *transform core* used in the WHT-based compression approach proposed by Hosseini-Nejad et al. [52]. (From H. Hosseini-Nejad et al., *IEEE Transactions on Biomedical Circuits and Systems*, Vol. 8, No. 1, pp. 129–137, Feb. 2014.)

18.5.4 The Discrete Cosine Transform

The DCT is a unitary transform with sinusoidal basis functions [69]. The N-point DCT is defined as

$$Y(k) = \sqrt{\frac{2}{N}} a_k \sum_{n=0}^{N-1} x(n) \cdot \cos\left[\frac{(2n+1) \cdot k\pi}{2N}\right] \quad k = 0, 1, \ldots, N-1, \tag{18.6}$$

and the inverse discrete cosine transform (IDCT) is defined as

$$x(n) = \sqrt{\frac{2}{N}} \sum_{k=0}^{N-1} a_k Y(k) \cdot \cos\left[\frac{(2n+1) \cdot k\pi}{2N}\right], \quad n = 0, 1, \ldots, N-1, \tag{18.7}$$

where $x(n)$ is the input signal sample in the time domain, n is the sample number, $Y(k)$ is the transformed signal coefficient in the DCT domain, k is the coefficient number, $a_0 = 1/\sqrt{2}$, and $a_k = 1$ for $k > 0$. The transformation relationship in Equation 18.6 means that the DCT is not so easy to implement by using standard digital hardware. To overcome this problem, a few fast algorithms have been proposed to reduce the computation complexity of the DCT by reducing the number of multiplications and additions [70–72]. In Hosseini-Nejad et al.'s [73] and Hosseini-Nejad et al.'s [53] works, DCT-based neural signal compressors are designed for the compression of 32 and 128 neural channels, respectively. Simplified block diagram of the multiplierless DCT-based compression core used in Hosseini-Nejad et al.'s work [53] is shown in Figure 18.18. In this figure, the coefficients k_1 to k_8 are fractional numbers in the form of $k/2^m$, realized by additions and logical shifts. As a sample of the neural signals processed using the transformations studied in this chapter, Figure 18.19 presents an intracortically recorded single-unit neural signal before and after the application of the DCT-based compression technique.

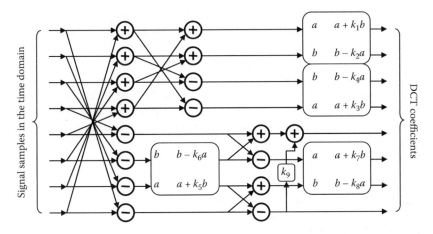

FIGURE 18.18
Block diagram of the multiplier-less DCT-based compression core used by Hosseini-Nejad et al. [52]. (From H. Hosseini-Nejad et al., *IEEE Transactions on Biomedical Circuits and Systems*, Vol. 8, No. 1, pp. 129–137, Feb. 2014.)

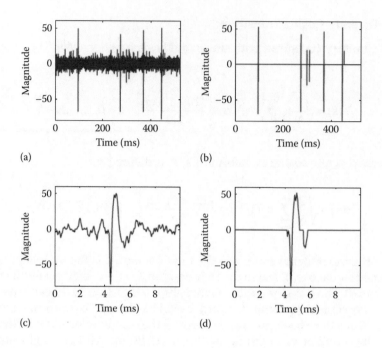

FIGURE 18.19
DCT-based neural signal compression: (a) a sample neural signal, (b) the resulting neural signal reconstructed on the receiver side, (c) close-up view of an action potential from (a), and (d) reconstructed waveform for the action potential of (c).

18.6 Data Framing

To benefit from the data reduction achieved through the compression of neural signals, efficient data framing is of crucial importance. In its most straightforward form, a packet conveys data/status of all the channels at once, independently from the activity of the channel. This is perhaps the most inefficient way of utilizing the wireless channel bandwidth for reporting neuronal activities to the outside world [15]. To be more efficient, data packets report only the sample/status associated with the activities of the channels [34,40,42]. This way, assuming that usually neural spikes do not so frequently occur (with a typical firing rate of up to 100 spikes/s [8]), considerable saving in the outgoing bit rate can be achieved. However, to successfully reconstruct neural signals/activities on the external side of the system, the samples/status need to be accompanied by timing and/or address of the activities.

18.7 Concluding Remarks

Microsystems developed to intracortically record neuronal activities of the brain are nowadays of interest in brain–machine interfacing for neuroscientific studies, for neuro-prosthetic applications, and for other applications such as advanced research in cognitive sciences. To efficiently interpret the recorded information, a large number of neural signals need to be recorded concurrently and telemetered in the real time to the outside world via a single wireless channel. This requires data compression/reduction techniques that are efficient enough from both signal processing and hardware implementation standpoints to comply with basic requirements for the design and implementation of wireless implantable microsystems.

Data compression/reduction techniques developed to handle large amount of intracortical neural signals, mostly aiming at the preservation of action potentials, were discussed. It should be emphasized that for the techniques employed for this purpose, in addition to efficacy of in signal compression, real-time operation and efficient hardware implementation are among the key aspects that are of crucial importance in implantable neural recording microsystems.

Acknowledgments

The authors would like to thank Professor E. Arabzadeh of Eccles Institute of Neuroscience, Australian National University, Canberra, Australia, and M. M. Sabri of the School of Cognitive Sciences, Institute for Research in Fundamental Sciences (IPM), Tehran, Iran, for providing the original neural data presented in Figure 18.2.

References

1. J. G. Nicholls et al., *From Neuron to Brain*, Sinauer Associates, 2012.
2. E. R. Kandel et al., *Principles of Neural Science*, Fifth Edition, McGraw-Hill, 2013.
3. G. Buzsaki, *Rhythms of the Brain*, Oxford University Press, 2006.
4. C. Koch, *Biophysics of Computation: Information Processing in Single Neurons*, Oxford University Press, 1999.
5. A. L. Hodgkin and A. F. Huxley, "A quantitative description of membrane current and its application to conduction and excitation in nerve," *Journal of Physiology*, Vol. 117, No. 4, pp. 500–544, Aug. 1952.
6. R. Q. Quiroga and S. Panzeri, *Principles of Neural Coding*, CRC Press, May 2013.
7. B. P. Bean, "The action potential in mammalian central neurons," *Nature Reviews Neuroscience*, Vol. 8, No. 6, pp. 451–465, Jun. 2007.
8. M. C. W. van Rossum et al., "Fast propagation of firing rates through layered networks of noisy neurons," *Journal of Neuroscience*, Vol. 22, No. 5, pp. 1956–1966, Mar. 2002.
9. C. Gold et al., "On the origin of the extracellular action potential waveform: A modeling study," *Journal of Neurophysiology*, Vol. 95, No. 5, pp. 3113–3128, May 2006.
10. D. Henze et al., "Intracellular features predicted by extracellular recordings in the hippocampus in vivo," *Journal of Neurophysiology*, Vol. 84, No. 1, pp. 390–400, Jul. 2000.
11. I. Cohen and R. Miles, "Contributions of intrinsic and synaptic activities to the generation of neuronal discharges in vitro hippocampus," *Journal of Physiology*, Vol. 524, No. 2, pp. 485–502, Apr. 2000.
12. A. Faisal et al., "Noise in the nervous system," *Nature Reviews Neuroscience*, Vol. 9, No. 4, pp. 292–303, Apr. 2008.

13. J. Coulombe et al., "A highly flexible system for microstimulation of the visual cortex: Design and implementation," *IEEE Transactions on Biomedical Circuits and Systems*, Vol. 1, No. 4, pp. 258–269, Dec. 2007.
14. A. M. Sodagar et al., "An implantable 64-channel wireless microsystem for single-unit neural recording," *IEEE Journal of Solid-State Circuits*, Vol. 44, No. 9, pp. 2591–2603, Sep. 2009.
15. R. R. Harrison et al., "A low-power integrated circuit for a wireless 100-electrode neural recording system," *IEEE Journal of Solid-State Circuits*, Vol. 42, No. 1, Jan. 2007.
16. M. S. Chae et al., "A 128-channel 6-mW wireless neural recording IC with spike feature extraction and UWB transmitter," *IEEE Transactions on Neural Systems and Rehabilitation Engineering*, Vol. 17, No. 4, pp. 312–321, Aug. 2009.
17. K. D. Wise et al., "Microelectrodes, microelectronics, and implantable neural microsystems," *Proceedings of the IEEE*, Vol. 96, No. 7, pp. 1184–1202, July 2008.
18. F. Shahrokhi et al., "The 128-channel fully differential digital integrated neural recording and stimulation interface," *IEEE Transactions on Biomedical Circuits and Systems*, Vol. 4, No. 3, pp. 149–161, Jun. 2010.
19. J. Aziz et al., "256-channel neural recording and delta compression microsystem with 3D electrodes," *IEEE Journal of Solid-State Circuits*, Vol. 44, No. 3, pp. 995–1005, Mar. 2009.
20. A. Bagheri et al., "1024-channel-scalable wireless neuromonitoring and neurostimulation rodent headset with nanotextured flexible microelectrodes," *Proceedings of the 2012 IEEE International Conference on Biomedical Circuits and Systems (BioCAS '12)*, pp. 184–187, Nov. 2012.
21. S. Raghunathan et al., "The design and hardware implementation of a low-power real-time seizure detection algorithm," *Journal of Neural Engineering*, Vol. 6, No. 5, p. 056005, 2009.
22. M. Nekuyi, "Design and fabrication of penetrating passive silicon probes for neural recording," *Technical Report, Research Laboratory for Integrated Circuits and Systems (ICAS), Faculty of Electrical Engineering, K. N. Toosi University of Technology*, Tehran, Iran, Summer 2010.
23. M. Ghovanloo and K. Najafi, "A wideband frequency-shift keying wireless link for inductively powered biomedical implants," *IEEE Transactions on Circuits and Systems—I: Regular Papers*, Vol. 51, No. 12, pp. 2374–2383, Dec. 2004.
24. A. M. Sodagar et al., "A wireless implantable microsystem for multichannel neural recording," *IEEE Transactions on Microwave Theory and Techniques*, Vol. 57, No. 10, pp. 2565–2573, Oct. 2009.
25. F. Asgarian and A. M. Sodagar, "A high-data-rate low-power BPSK demodulator and clock recovery circuit for implantable biomedical devices," *Proceedings of the 2009 IEEE International Conference on Neural Engineering (NER '09)*, pp. 407–410, May 2009.
26. F. Asgarian and A. M. Sodagar, "A low-power non-coherent BPSK demodulator and clock recovery circuit for high-data-rate biomedical applications," *Proceedings of the 2009 IEEE Annual International Conference of the Engineering in Medicine and Biology Society (EMBC '09)*, pp. 4840–4843, Sep. 2009.
27. F. Asgarian and A. M. Sodagar, "A carrier-frequency-independent BPSK demodulator with 100% data-rate-to-carrier-frequency ratio," *Proceedings of the 2010 IEEE International Conference on Biomedical Circuits and Systems (BioCAS '10)*, pp. 29–32, Nov. 2010.
28. R. Erfani and A. M. Sodagar, "Amplitude-engraving modulation (AEM) scheme for simultaneous power and high-rate data telemetry to biomedical implants," *Proceedings of the 2013 IEEE International Conference on Biomedical Circuits and Systems (BioCAS '13)*, pp. 290–293, Oct. 2013.
29. E. Ashoori et al., "Design of double layer printed spiral coils for wirelessly-powered biomedical implants," *Proceedings of the 2011 IEEE Annual International Conference of the Engineering in Medicine and Biology Society (EMBC '11)*, pp. 2882–2885, Sep. 2011.
30. A. M. Sodagar and P. Amiri, "Capacitive coupling for power and data telemetry to implantable biomedical microsystems," *Proceedings of the 2009 IEEE International Conference on Neural Engineering (NER '09)*, pp. 411–414, May 2009.
31. M. Takhti et al., "Modeling of a capacitive link for data telemetry to biomedical implants," *Proceedings of the 2011 IEEE International Conference on Biomedical Circuits and Systems (BioCAS '11)*, pp. 181–184, Nov. 2011.

32. M. Karimi et al., "Auxiliary-carrier load-shift keying for reverse data telemetry from biomedical implants," *Proceedings of the 2012 IEEE International Conference on Biomedical Circuits and Systems (BioCAS '12)*, pp. 220–223, Nov. 2012.

33. F. Asgarian and A. M. Sodagar, "Wireless Telemetry for Implantable Biomedical Microsystems," a chapter in *Biomedical Engineering, Trends in Electronics, Communications and Software*, Editor: A. N. Laskovski, InTech Press, pp. 21–44, 2011.

34. R. H. Olsson and K. D. Wise, "A three-dimensional neural recording microsystem with implantable data compression circuitry," *IEEE Journal of Solid-State Circuits*, Vol. 40, No. 12, pp. 2796–2804, Dec. 2005.

35. P. Mohseni et al., "Wireless multi-channel biopotential recording using and integrated FM telemetry circuit," *IEEE Transactions on Neural Systems and Rehabilitation Engineering*, Vol. 13, No. 3, pp. 263–271, Jun. 2005.

36. A. Rajabi-Tavakkol et al., "New architecture for wireless implantable neural recording microsystems based on frequency-division multiplexing," *Proceedings of the 2010 Annual International Conference of the Engineering in Medicine and Biology Society (EMBC '10)*, pp. 6449–6452, Sep. 2010.

37. A. Zabihian and A. M. Sodagar, "A new architecture for multi-channel neural recording microsystems based on delta-sigma modulation," *Proceedings of the 2009 IEEE International Conference on Biomedical Circuits and Systems (BioCAS '09)*, pp. 81–84, Nov. 2009.

38. S. B. Lee et al., "An inductively powered scalable 32-channel wireless neural recording system-on-a-chip for neuroscience applications," *IEEE Transactions on Biomedical Circuits and Systems*, Vol. 4, No. 6, pp. 360–371, Dec. 2010.

39. United States Frequency Allocation Chart, Available online at: http://www.fcc.gov/encyclopedia/radio-spectrum-allocation, accessed June 2014.

40. A. M. Sodagar et al., "A fully integrated mixed-signal neural processor for implantable multichannel cortical recording," *IEEE Transactions on Biomedical Engineering*, Vol. 54, No. 6, pp. 1075–1088, Jun. 2007.

41. S. Barati and A. M. Sodagar, "Discrete-time automatic spike detection circuit for neural recording implants," *Electronics Letters*, Vol. 47, No. 5, pp. 306–307, Mar. 2011.

42. B. Gosselin et al., "A mixed-signal multichip neural recording interface with bandwidth reduction," *IEEE Transactions on Biomedical Circuits and Systems*, Vol. 3, No. 3, pp. 129–141, 2009.

43. M. A. Shaeri, "Data Compression Technique for Neural Recording Applications," *M.Sc. Thesis, Faculty of Electrical Engineering, K. N. Toosi University of Technology*, Tehran, Iran, 2011.

44. A. M. Kamboh et al., "Computationally efficient neural feature extraction for spike sorting in implantable high-density recording systems," *IEEE Transactions on Neural Systems and Rehabilitation Engineering*, Vol. 21, No. 1, pp. 1–9, Jan. 2013.

45. J. H. Choi et al., "A new action potential detector using the MTEO and its effects on spike sorting systems at low signal-to-noise ratios," *IEEE Transactions on Biomedical Engineering*, Vol. 53, No. 4, pp. 738–746, Apr. 2006.

46. B. Gosselin and M. Sawan, "An ultralow-power CMOS automatic action potential detector," *IEEE Transactions on Neural Systems and Rehabilitation Engineering*, Vol. 17, No. 4, pp. 346–353, Aug. 2009.

47. S. Razmpour et al., "Reconfigurable biological signal co-processor for feature extraction dedicated to implantable biomedical microsystems," *Proceedings of the 2013 IEEE International Symposium on Circuits and Systems (ISCAS '13)*, pp. 861–864, May 2013.

48. M. Judy et al., "Nonlinear signal-specific ADC for efficient neural recording in brainmachine interfaces," Accepted for publication in *IEEE Transactions on Biomedical Circuits and Systems (TBioCAS)*, 2013.

49. M. Petrou and C. Petrou, *Image Processing: The Fundamentals*, John Wiley & Sons, New York, 2010.

50. R. Wang, *Introduction to Orthogonal Transforms with Applications in Data Processing and Analysis*, Cambridge University Press, New York, 2010.

51. A. M. Kamboh et al., "Area-power efficient VLSI implementation of multichannel DWT for data compression in implantable neuroprosthetics," *IEEE Transactions on Biomedical Circuits and Systems*, Vol. 1, No. 2, pp. 128–135, Jun. 2007.

52. H. Hosseini-Nejad et al., "Data compression in brain-machine/computer interfaces based on the Walsh-Hadamard transform," *IEEE Transactions on Biomedical Circuits and Systems*, Vol. 8, No. 1, pp. 129–137, Feb. 2014.

53. H. Hosseini-Nejad et al., "A 128-channel discrete cosine transform-based neural signal processor for implantable neural recording microsystems," *International Journal of Circuit Theory and Applications*, doi: 10.1002/cta.1955, 2013.

54. M. A. Shaeri et al., "A 64-channel neural signal processor/compressor based on Haar wavelet transform," *Proceedings of the 2008 Annual International Conference Engineering in Medicine and Biology Society (EMBC '08)*, Boston, Massachusetts, Sep. 2011.

55. B. Gosselin et al., "Low-power implantable microsystem intended to multichannel cortical recording," *Proceedings of the 2004 IEEE International Symposium on Circuits and Systems (ISCAS '04)*, pp. 5–8, 2004.

56. K. H. Kim and S. J. Kim, "A wavelet-based method for action potential detection from extracellular neural signal recording with low signal-to noise ratio," *IEEE Transactions on Biomedical Engineering*, Vol. 50, No. 8, pp. 999–1011, Aug. 2003.

57. R. Quian Quiroga et al., "Unsupervised spike detection and sorting with wavelets and upper paramagnetic clustering," *Neural Computation*, Vol. 16, No. 8, pp. 1661–1687, Aug. 2004.

58. I. Obeid and P. D. Wolf, "Evaluation of spike-detection algorithms for a brain-machine interface application," *IEEE Transactions on Biomedical Engineering*, Vol. 51, No. 6, pp. 905–911, Jun. 2004.

59. S. Gibson et al., "Comparison of spike-sorting algorithms for future hardware implementation," *Proceedings of the 2008 Annual International Conference of the Engineering in Medicine and Biology Society (EMBC '08)*, pp. 5015–5020, Vancouver, Canada, Aug. 2008.

60. J. F. Roy and M. Sawan, "A fully reconfigurable controller dedicated to implantable recording devices," *Proceedings of the Third IEEE International New Circuits and Systems Conference (NEWCAS '05)*, pp. 303–306, 2005.

61. C. Dumortier et al., "Wavelet transforms dedicated to compress recorded ENGs from multichannel implants: Comparative architectural study," *Proceedings of the 2006 IEEE Conference on International Symposium on Circuits and Systems (ISCAS '06)*, pp. 2129–2132, 2006.

62. K. G. Oweiss et al., "A scalable wavelet transform VLSI architecture for real-time neural signal processing in multichannel cortical implants," *IEEE Transactions on Circuits & System—I: Regular Papers*, Vol. 54, No. 6, pp. 1266–1278, June 2007.

63. S. G. Mallat, "A theory for multi-resolution signal decomposition: The wavelet representation," *IEEE Transactions on Pattern Analysis and Machine Intelligence*, Vol. 11, No. 7, pp. 674–693, 1989.

64. W. Jiang and A. Ortega, "Lifting factorization-based discrete wavelet transform architecture design," *IEEE Transactions on Circuits and Systems for Video Technology*, Vol. 11, No. 5, pp. 651–657, May 2001.

65. C.-T. Huang et al., "VLSI architecture for forward discrete wavelet transform based on B-spline factorization," *Journal of VLSI Signal Processing*, Vol. 40, No. 3, pp. 343–353, Jul. 2005.

66. A. M. Kamboh et al., "Resource constrained VLSI architecture for implantable neural data compression systems," *Proceedings of the 2009 IEEE International Symposium on Circuits and Systems*, pp. 1481–1484, 2009.

67. A. K. Jain, *Fundamentals of Digital Image Processing*, Prentice-Hall International Inc., 1989.

68. H. Kitajima, "Energy packing efficiency of the Hadamard transform," *IEEE Transactions on Communications*, Vol. 24, No. 11, pp. 1256–1258, Nov. 1976.

69. S. A. Broughton and K. Bryan, "The Discrete Cosine Transform," a chapter in *Discrete Fourier Analysis and Wavelets: Applications to Signal and Image Processing*, John Wiley & Sons, Inc., 2008.

70. C. Loeffler et al., "Practical fast 1-D DCT algorithms with 11 multiplications," *Proceedings of the 1989 International Conference on Acoustics, Speech, and Signal Processing, (ICASSP-89)*, pp. 988–991, 1989.

71. K. Z. Bukhari et al., "DCT and IDCT implementations on different FPGA technologies," *Proceedings of the 13th Annual Workshop on Circuits, Systems and Signal Processing (ProRISC 2002)*, pp. 232–235, 2002.
72. T. D. Tran, "The bin DCT: Fast multiplier-less approximation of the DCT," *IEEE Signal Processing Letters*, Vol. 7, No. 6, pp. 141–144, Jun. 2000.
73. H. Hosseini Nejad et al., "Data compression based on discrete cosine transform for implantable neural recording microsystems," *Proceedings of the 2012 IEEE International Conference on Circuits and Systems (ICCAS '12)*, pp. 209–213, 2012.

19

Electronic Health Signal Processing

Mohit Kumar, Norbert Stoll, Kerstin Thurow, and Regina Stoll

CONTENTS

19.1 Introduction

Modern medical technology results in a large volume of data per patient, making the manual processing of the data by the physician a challenging and time-consuming task. An increasing interest in information technology-aided medical decision-support systems is therefore a logical and natural research trend. Computer-aided medical diagnosis found clinical applications as early as the 1970s (de Dombal et al. 1972). The research area has exponentially grown during the last two decades from the interdisciplinary work of scientists. A review of the state-of-art technology indicates that a vast amount of the research has been dedicated; thus, computer-aided decision-support systems are becoming an inherent part of clinical medicine (Tourassi 1999; Greens 2007; Pearson et al. 2009; Stivaros et al. 2010; Belle et al. 2013). The application areas of computer-aided decision-support systems include radiology, intensive care unit, cardiovascular medicine, dental medicine, cancer, and pediatric medicine (Belle et al. 2013).

The analysis of biomedical signals plays a crucial role in medical diagnosis especially in cardiology and neurology. Therefore, an automatic analysis of signals (e.g., ECG and EEG) could functionally be included in a medical decision-support system to solve more complex and sophisticated problems (Hudson et al. 2004). The aim of automated signal data analysis is to increase the accuracy of the diagnosis while avoiding the reader's subjective point of view in data interpretation. Furthermore, the presence of noise in the signal and a large volume of the signal data might cause the errors in the manual interpretation of the data. There is no surprise in observing that signal and image-processing algorithms are integrated with the clinical decision-support systems in the general process of care (Chiarugi et al. 2008).

The computational methods and techniques in computer-aided decision support systems include rule-based systems (e.g., fuzzy), neural computations, and machine learning algorithms. The computational tools are applied to address special problems in medical data analysis (Perner 2006). The interpretation of medical data, due to the underlying uncertainties, is a challenging job. The uncertainties may arise because of a difference in the subjective behavior of individuals. Among the different computation intelligent techniques, fuzzy systems based on fuzzy set theory (Zadeh 1973, 1983) are considered a better alternative for dealing with uncertainties. The integration of fuzzy methodologies in data interpretation offers these advantages:

1. The fuzzy systems are potentially capable of handling vagueness, ambiguity, and uncertainties inherently present in multiparameter interpretation of given medical data. The uncertainty is handled by capturing the knowledge by using fuzzy sets. The data are processed through the information stored in the rule base by some inference mechanism. The approximate reasoning methods offer a robustness and flexibility to the inference procedure.

2. The interpretation system is based on simple linguistic if–then rules that can be easily understood by humans.

This has motivated many researchers to apply fuzzy techniques in medicine (Adlassnig et al. 1985; Adlassnig 1986; Wong et al. 1990; Roy and Biswas 1992; Binaghi et al. 1993; Belmonte et al. 1994; Brai et al. 1994; Fathitorbaghan and Meyer 1994; Watanabe et al. 1994; Steimann 1996; Daniels et al. 1997; Bellazzi et al. 1998, 2001; Garibaldi and Ifeachor 1999; Kuncheva and Steimann 1999).

Biomedical signals typically possess a certain degree of randomness that cannot be explained in correlation to the physiological conditions. The different analysis methods extract the features of the physiological signals by using some mathematical theories (e.g., time/frequency transformations and nonlinear analysis) such that the extracted features provide some information about the state of the patient. However, the correlations between signal features and the patients' states might be complex and uncertain due to a difference in the individual nature of different patients. Furthermore, there might be certain factors related to an individual (e.g., body conditions) affecting the correlations between signal features and the patient's states. The fuzzy filtering algorithms can be used to deal with the uncertainties associated with the analysis of physiological signals. The idea is to reduce the effect of random variations on the analysis of signal characteristics via filtering out randomness or uncertainty from the signal by using a nonlinear fuzzy filter (Kumar et al. 2007a,b). Our research group has been working on developing new biomedical signal analysis methods with their practical mobile telemedical applications in eHealth. The research results have been published by Kumar et al. (2007b, 2008, 2010c, 2012a). The approach of Kumar et al. (2007b, 2008) is classical in the sense that first the features of signal were extracted, and then fuzzy techniques were used to render robustness against uncertainties due to individual variations. Although Kumar et al. (2007b, 2008) extended the state-of-art technology by providing theoretical contributions to physiological state modeling, their telemedical applications are limited due to computational complexities and lack of an online individual adaptation mechanism. To avoid these limitations, a novel signal analysis method was suggested by Kumar et al. (2010c) that offers the following advantages:

1. There is no need for modeling the complex and uncertain relationships between signal features and the patient's states.

2. A high diagnostic efficiency is achieved as a result of mathematically formulating the problem in a sensible way by using recently developed stochastic fuzzy modeling methods (Kumar et al. 2010a).

The mathematical analysis of physiological signals in a deterministic or stochastic framework remains as an active area of research (Task Force of the European Society of Cardiology and the North American Society of Pacing and Electrophysiology 1996; Rezek and Roberts 1998; Gautama et al. 2004; Hornero et al. 2005; Ancona et al. 2006; Barbieri and Brown 2006; Ferrario et al. 2006; Hornero et al. 2006; Zhong et al. 2007; McNames and Aboy 2008; Bollt et al. 2009). The stochastic fuzzy modeling and analysis techniques have much to offer in this area (Kumar et al. 2010c, 2012a). The approach takes simultaneously the advantages of Bayesian analysis theory and fuzzy theory to formulate the patients' state prediction problem mathematically in a sensible way. It is possible to design new signal feature extraction methods with high diagnostic efficiency; thus, it is possible to build an expert system for making accurate predictions regarding the state of individuals. This chapter outlines a general framework for predicting the state of an individual based on the stochastic fuzzy analysis of biomedical signals. The framework was originally introduced by Kumar et al. (2012b) and was applied to a mobile telemedical application related to stress monitoring by Kumar et al. (2012a). The approach due to an individual-specific analysis can be potentially applied to clinical applications related to individual monitoring.

19.2 Background

This section presents the necessary mathematical background from Kumar et al. (2012a,b) required for designing an intelligent signal analysis algorithm.

19.2.1 A Takagi–Sugeno Fuzzy Filter

Consider a Takagi-Sugeno fuzzy model ($F_s: X \rightarrow Y$) that maps n-dimensional real input space ($X = X_1 \times X_2 \times \ldots \times X_n$) to one-dimensional real line. A rule of the model is represented as

$$\text{If } x_1 \text{ is } A_1 \text{ and } x_n \text{ is } A_n,$$

$$\text{then } y_f = s_0 + s_1 x_1 + \ldots + s_n x_n.$$

Here $(x_1, \ldots, x_n)$ are the model input variables, y_f is the filtered output variable, $(A_1, \ldots, A_n)$ are the linguistic terms which are represented by fuzzy sets, and $(s_0, s_1, \ldots, s_n)$ are real scalars. Given a universe of discourse X_j, a fuzzy subset A_j of X_j is characterized by a mapping:

$$\mu_{Aj} \colon X_j \to [0, 1],$$

where for $x_j \in X_j$, $\mu_{Aj}(x_j)$ can be interpreted as the degree or grade to which x_j belongs to A_j. This mapping is called as membership function of the fuzzy set. Let us define for jth input, P_j nonempty fuzzy subsets of X_j (represented by $A_{1j}, A_{2j}, \ldots, A_{Pjj}$). Let the ith rule of the rule base be represented as

$$R_i \colon \text{If } x_1 \text{ is } A_{i1} \text{ and } \ldots \text{ and } x_n \text{ is } A_{in},$$
$$\text{then } y_f = s_{i0} + s_{i1}x_1 + \ldots + s_{in}x_n,$$

where $A_{i1} \in \{A_{11}, \ldots, A_{P11}\}$, $A_{i2} \in \{A_{12}, \ldots, A_{P22}\}$, and so on. Now, the different choices of A_{i1}, $A_{i2}, \ldots, A_{in}$ lead to the $K = \prod_{j=1}^{n} P_j$ number of fuzzy rules. For a given input vector $x = [x_1 \ldots x_n]^T \in R^n$, the *degree of fulfillment* of the ith rule, by modeling the logic operator "and" using product, is given by

$$g_i(x) = \prod_{j=1}^{n} \mu A_{ij}(x_j).$$

The output of the fuzzy model to input vector x is computed by taking the weighted average of the output provided by each rule:

$$y_f = \frac{\sum_{i=1}^{K} (s_{i0} + s_{i1}x_1 + \ldots + s_{in}x_n) g_i(x)}{\sum_{i=1}^{K} g_i(x)}$$

$$= \frac{\sum_{i=1}^{K} (s_{i0} + s_{i1}x_1 + \ldots + s_{in}x_n) \prod_{j=1}^{n} \mu A_{ij}(x_j)}{\sum_{i=1}^{K} \prod_{j=1}^{n} \mu A_{ij}(x_j)}. \tag{19.1}$$

Let us define a real vector θ such that the membership functions of any type (e.g., trapezoidal and triangular) can be constructed from the elements of vector θ. To illustrate the construction of membership functions based on vector (θ), consider the following example.

19.2.1.1 Triangular Membership Functions

Let

$$\theta = \left(t_1^0, t_1^1, \ldots, t_1^{P_1-2}, t_1^{P_1-1}, \ldots t_n^0, t_n^1, \ldots t_n^{P_n-2}, t_n^{P_n-1} \right)$$

such that for ith input, $t_i^0 < t_i^1 < ... < t_i^{P_i-2} < t_i^{P_i-1}$ holds for all $i = 1, ..., n$. Now, P_i triangular membership functions for ith input ($\mu_{A1i}, \mu_{A2i}, ..., \mu_{APi}$) can be defined as

$$\mu_{A_{1i}}(x_i,\theta) = \max\left(0, \min\left(1, \frac{t_i^1 - x_i}{t_i^1 - t_i^0}\right)\right),$$

$$\mu_{A_{ji}}(x_i,\theta) = \max\left(0, \min\left(\frac{x_i - t_i^{j-2}}{t_i^{j-1} - t_i^{j-2}}, \frac{t_i^j - x_i}{t_i^j - t_i^{j-1}}\right)\right),$$

$$j = 2, ..., P_i - 1,$$

$$\mu_{A_{p_i i}}(x_i,\theta) = \max\left(0, \min\left(\frac{x_i - t_i^{P_i-2}}{t_i^{P_i-1} - t_i^{P_i-2}}, 1\right)\right).$$

For any choice of membership functions (which can be constructed from a vector θ), Equation 19.1 can be rewritten as function of θ:

$$y_f = \sum_{i=1}^{K}(s_{i0} + s_{i1}x_i + ... + s_{in}x_n)\tilde{G}_i(x,\theta),$$

$$\tilde{G}_i(x,\theta) = \frac{\prod_{j=1}^{n}\mu A_{ij}(x_j,\theta)}{\sum_{i=1}^{K}\prod_{j=1}^{n}\mu A_{ij}(x_j,\theta)}.$$

Let us introduce the following notation:

$$\alpha = \begin{bmatrix} s_{10} \\ s_{11} \\ \vdots \\ s_{1n} \\ \vdots \\ s_{K0} \\ s_{K1} \\ \vdots \\ s_{Kn} \end{bmatrix}, \quad G(x,\theta) = \begin{bmatrix} \tilde{G}_1(x,\theta) \\ x\tilde{G}_1(x,\theta) \\ \vdots \\ \tilde{G}_K(x,\theta) \\ x\tilde{G}_K(x,\theta) \end{bmatrix}.$$

Now, we have

$$y_f = G^T(x, \theta)\alpha.$$

19.2.2 Stochastic Fuzzy Modeling of Biomedical Signals

Given a short time recording of a signal $\{y(j)\}$, $j = 1, 2, \ldots$, we model the biomedical data series through a stochastic Takagi–Sugeno type fuzzy model. That is,

$$y(j) = G^T([y(j - n) \ldots y(j - 1)]^T, \theta)\alpha + v(j),$$

where $v(j)$ accounts for any uncertainty arising due to data noise and modeling errors. Defining the input vector as

$$x(j) = [y(j - n) \ldots y(j - 1)]^T \in R^n,$$

we have

$$y(j) = G^T(x(j), \theta)\alpha + v(j).$$

19.2.3 A Stochastic Mixture of Signal Data Fuzzy Models

We want to fit the input–output data pair (x, y) through a stochastic mixture of S different Takagi–Sugeno–type fuzzy models. Let $s \in \{1, 2, \ldots, S\}$ be a discrete random variable whose value represents which of the C models had generated the observed data. That is,

$$\text{If } s = 1, \text{ then } y = G^T(x, \theta^1)\alpha^1 + v^1;$$

$$\text{if } s = S, \text{ then } y = G^T(x, \theta^S)\alpha^S + v^S.$$

Here, $G^T(x, \theta^i)\alpha^i$ is the output of the ith fuzzy model [characterized by parameters (α^i, θ^i)], and v^i is the additive uncertainty associated with the modeling of data pair (x, y). We assume that $(\theta^1, \ldots, \theta^S)$ are deterministic while all other variables are random.

Let $\pi = [\pi_1 \ldots \pi_S]^T \in R^K$, with $0 \leq \pi_i \leq 1$ with $\sum_{i=1}^{S} \pi_i = 1$, be a vector defining the discrete distribution of s such that

$$p(s = 1|\pi) = \pi_1, \ldots, p(s = S|\pi) = \pi_S.$$

A symmetric Dirichlet prior with strength $c > 0$ is chosen for π,

$$p(\pi \,|\, ct) = \text{Dir}(\pi \,|\, ct), \quad \text{where } t = \left[\frac{1}{S} \cdots \frac{1}{S}\right]^T \in R^S,$$

$$= \frac{\Gamma(c)}{\left[\Gamma\left(\dfrac{c}{S}\right)\right]^S} \pi_1^{(c/S)-1} \ldots \pi_S^{(c/S)-1},$$

where $\Gamma(\cdot)$ is the gamma function. The uncertainty v^i is assumed to be Gaussian distributed with zero mean and a variance of $1/\varphi^i$, where φ^i is further assumed to be gamma distributed. That is,

$$p(v^i|\varphi^i) = N(0, (\varphi^i)^{-1}),$$

$$p\left(\varphi^i \middle| a_\varphi^i, b_\varphi^i\right) = Ga\left(\varphi^i \middle| a_\varphi^i, b_\varphi^i\right),$$

where the expressions for Gaussian and the gamma distribution are given as

$$N(x|\mu, \Lambda) = \frac{1}{\sqrt{(2\,pi)^n |\Lambda|}} \exp\left[-\frac{1}{2}(x-\mu)^T \Lambda^{-1}(x-\mu)\right],$$

$$Ga(z|a,b) = \frac{b^a}{\Gamma(a)} z^{a-1} \exp(-bz),$$

where $x \in R^n$, $\mu \in R^n$, $\Lambda \in R^{n \times n} > 0$, and (z, a, b) are positive scalars. The prior placed on the consequents of ith fuzzy model (i.e., α^i) is as follows:

$$p\left(\alpha^i | \mu_\alpha^i, \Lambda_\alpha^i\right) = N\left(\alpha^i | \mu_\alpha^i, \left(\Lambda_\alpha^i\right)^{-1}\right).$$

The conditional probability of y is given as

$$p(y|s = i, G(x, \theta^i), \alpha^i, \varphi^i) = N(G^T(x, \theta^i)\, \alpha^i, (\varphi^i)^{-1}).$$

For simplicity, define

$$\varphi = \{\varphi^1, \dots, \varphi^S\}, \alpha = \{\alpha^1, \dots, \alpha^S\},$$

$$G_x = \{G(x, \theta^1), \dots, G(x, \theta^S)\}.$$

The marginal probability can be expressed as

$$p(y|G_x) = \int \partial \pi \partial \varphi \partial \alpha\, p(y, \pi, \varphi, \alpha|G_x)$$

$$= \int \partial \Omega\, p(y, \Omega|G_x),$$

where $\Omega = \{\pi, \varphi, \alpha\}$ has been defined for simplicity.

Let $q(\Omega)$ be an arbitrary distribution. The log marginal probability of y can be written as

$$\log\left[p(y|G_x)\right] = \int \partial \Omega q(\Omega) \log\left[\frac{p(y, \Omega|G_x)}{q(\Omega)}\right]$$

$$+ \int \partial \Omega q(\Omega) \log\left[\frac{q(\Omega)}{p(\Omega|y, G_x)}\right].$$

Let $q(s)$ be an arbitrary probability mass function with $\sum_{i=1}^{S} q(s=i) = 1$. Consider

$$\log\left[\frac{p(y,\Omega\,|\,G_x)}{q(\Omega)}\right] = \log\left[\frac{p(\Omega\,|\,G_x)p(y\,|\,\Omega,G_x)}{q(\Omega)}\right]$$

$$= \log\left[\frac{p(\Omega)}{q(\Omega)}\right] + \log[p(y\,|\,\Omega,G_x)]$$

$$= \log\left[\frac{p(\Omega)}{q(\Omega)}\right] + \sum_{i=1}^{S} q(s=i)\log\left[\frac{p(y,s=i\,|\,\Omega,G_x)}{q(s=i)}\right]$$

$$+ \sum_{i=1}^{S} q(s=i)\log\left[\frac{q(s=i)}{p(s=i\,|\,y,\Omega,G_x)}\right].$$

Define

$$\mathcal{F}(q(\Omega),q(s)) = \int \partial\Omega q(\Omega)\log\left[\frac{p(\Omega)}{q(\Omega)}\right]$$

$$+ \int \partial\Omega q(\Omega)\sum_{i=1}^{S} q(s=i)\log\left[\frac{p(y,s=i\,|\,\Omega,G_x)}{q(s=i)}\right]$$

to rewrite the logarithmic marginal probability as

$$\log(p(y\,|\,G_x)) = \mathcal{F}(q(\Omega),q(s))$$

$$+ \underbrace{\int \partial\Omega q(\Omega)\sum_{i=1}^{S} q(s=i)\log\left[\frac{q(s=i)}{p(s=i\,|\,y,\Omega,G_x)}\right]}_{\text{mismatch between } q(s) \text{ and posterior of } s}$$

$$+ \underbrace{\int \partial\Omega q(\Omega)\log\left[\frac{q(\Omega)}{p(\Omega\,|\,y,G_x)}\right]}_{\text{mismatch between } q(\Omega) \text{ and posterior of } \Omega}$$

The above expression shows that $\log[p(y\,|\,G_x)]$ is equal to the sum of $\mathcal{F}$ (referred to as negative free energy) and a measure of the mismatch between arbitrary distributions and true posteriors (where the mismatch between arbitrary and true posterior has been assessed in term of Kullback–Leibler divergence). The arbitrary distributions can be made to approximate the true posteriors via minimizing their mismatch, i.e., via maximizing $\mathcal{F}$ over $(q(\Omega), q(s))$. Consider

$$\log[p(y,s=i\,|\,\Omega,G_x)] = \log[p(s=i\,|\,\pi)] - \frac{\log(2pi)}{2}$$

$$+ \frac{\log(\varphi^i)}{2} - \frac{\varphi^i}{2}\left|y - G^T(x,\theta^i)\alpha^i\right|^2$$

and evaluate $\mathcal{F}$ as

$$\mathcal{F}(q(\Omega), q(s)) = \int \partial \Omega q(\Omega) \log \left[\frac{p(\Omega)}{q(\Omega)} \right]$$

$$+ \sum_{i=1}^{S} q(s=i) \int \partial \pi q(\pi) \log \left[\frac{p(s=i \mid \pi)}{q(s=i)} \right] - \frac{\log(2pi)}{2}$$

$$+ \frac{1}{2} \sum_{i=1}^{S} q(s=i) \int \partial \varphi^i q(\varphi^i) \log(\varphi^i)$$

$$- \frac{1}{2} \sum_{i=1}^{S} q(s=i) \int \partial \varphi^i q(\varphi^i) \partial \alpha^i q(\alpha^i) \varphi^i \left| y - G^T (x, \theta^i) \alpha^i \right|^2.$$

19.3 Variational Bayesian Inference of Stochastic Mixture of Signal Data Models

Variational Bayes method obtains the approximate posterior distributions via maximizing the lower bound on logarithmic marginal probability. We optimize the lower bound $\mathcal{F}$ with respect to each distribution in turn. The analytical derivations require the following commonly used independence assumption (referred to as mean-field approximation) on approximated posterior distributions:

$$q(\pi, \alpha, \varphi) = q(\pi)q(\alpha)q(\varphi).$$

It was shown by Kumar et al. (2012a,b) that under mean-field assumption, $\mathcal{F}$ could be analytically optimized with respect to each variational distribution as described in the following.

19.3.1 Optimization with respect to $q(\pi)$

$\mathcal{F}$ can be expressed as

$$\mathcal{F} = \int \partial \pi q(\pi) \log \left[\frac{p(\pi)}{q(\pi)} \right]$$

$$+ \sum_{i=1}^{S} q(s=i) \int \partial \pi q(\pi) \log [p(s=i \mid \pi)] + \text{cons}\{\pi\},$$

where cons$\{\pi\}$ represents all π-independent terms. That is,

$$\mathcal{F} = \int \partial \pi q(\pi) \left[\log \left(\prod_{i=1}^{K} \pi_i^{(c/K)+q(s_x=i)-1} \right) \right] - \int \partial \pi q(\pi) \log [q(\pi)] + \text{cons}\{\pi\}.$$

It follows immediately that the optimal distribution that maximizes $\mathcal{F}$ over $q(\pi)$ is given as

$$q*(\pi) = \text{Dir}\,(\pi \mid \hat{c}\hat{t}), \qquad \hat{t} = [\hat{t}_1 \ldots \hat{t}_K], \qquad \sum_{i=1}^{K} \hat{t}_i = 1,$$

such that

$$\hat{c}\hat{t}_i = \frac{c}{S} + q(s = i).$$

This implies that

$$\hat{c} = c + 1, \qquad \hat{t}_i = \frac{1}{c+1}\left[\frac{c}{S} + q(s = i)\right].$$

19.3.2 Optimization with respect to $q(\alpha^i)$

$\mathcal{F}$ is expressed in terms of α^i-dependent terms as follows:

$$\begin{aligned}
\mathcal{F} = &\int \partial\alpha^i q(\alpha^i) \log\left[\frac{p(\alpha^i)}{q(\alpha^i)}\right] \\
&- \frac{q(s = i)}{2}\left\langle \varphi^i \right\rangle_{q(\varphi^i)} \int \partial\alpha^i q(\alpha^i)\left|y - G^T(x, \theta^i)\alpha^i\right|^2 \\
&+ \text{cons}\,\{\alpha^i\}
\end{aligned}$$

where $\langle f(x)\rangle_{g(x)} = \int \partial x g(x) f(x)$. That is,

$$\begin{aligned}
\mathcal{F} = &-\frac{1}{2}\int \partial\alpha^i q(\alpha^i)\left[\left(\alpha^i - \mu_\alpha^i\right)^T \Lambda_\alpha^i \left(\alpha^i - \mu_\alpha^i\right)\right] \\
&- \frac{q(s = i)}{2}\left\langle \varphi^i \right\rangle_{q(\varphi^i)} \int \partial\alpha^i q(\alpha^i)\left|y - G^T(x, \theta^i)\alpha^i\right|^2 \\
&- \int \partial\alpha^i q(\alpha^i)\log[q(\alpha^i)] + \text{cons}\,\{\alpha^i\}
\end{aligned}$$

This implies that the optimal distribution for α^i is given as

$$q*(\alpha^i) = N(\alpha^i \mid \hat{\mu}_\alpha^i,\ (\hat{\Lambda}_\alpha^i) - 1)$$

such that

$$\hat{\Lambda}_\alpha^i = \Lambda_\alpha^i + q(s=i)\langle \varphi^i \rangle_{q(\varphi^i)} G(x,\theta^i)G^T(x,\theta^i),$$

$$\hat{\mu}_\alpha^i = \left(\hat{\Lambda}_\alpha^i\right)^{-1}\left\{\Lambda_\alpha^i\mu_\alpha^i + q(s=i)\langle \varphi^i \rangle_{q(\varphi^i)} G(x,\theta^i)y\right\}.$$

19.3.3 Optimization with respect to $q(\varphi^i)$

All φ^i-dependent terms of $\mathcal{F}$ are collected as follows:

$$\mathcal{F} = \int \partial\varphi^i q(\varphi^i)\log\left[\frac{p(\varphi^i)}{q(\varphi^i)}\right] + \frac{q(s=i)}{2}\int \partial\varphi^i q(\varphi^i)\log(\varphi^i)$$
$$- \frac{q(s=i)}{2}\left\langle |y-G^T(x,\theta^i)\alpha^i|^2\right\rangle_{q(\alpha^i)}\int \partial\varphi^i q(\varphi^i)\varphi^i$$
$$+ \mathrm{cons}\{\varphi^i\}.$$

That is,

$$\mathcal{F} = \int \partial\varphi^i q(\varphi^i)\left\{\left[a_\varphi^i + \frac{q(s=i)}{2} - 1\right]\log(\varphi^i)\right.$$
$$\left. - \left[b_\varphi^i + \frac{q(s=i)}{2}\left\langle |y-G^T(x,\theta^i)\alpha^i|^2\right\rangle_{q(\alpha^i)}\right]\varphi^i\right\}$$
$$- \int \partial\varphi^i q(\varphi^i)\log\left[q(\varphi^i)\right] + \mathrm{cons}\{\varphi^i\}.$$

The optimal expression for $q(\varphi^i)$ follows as

$$q*(\varphi^i) = Ga\left(\varphi^i \mid \hat{a}_\varphi^i, \hat{b}_\varphi^i\right)$$

such that

$$\hat{a}_\varphi^i = a_\varphi^i + \frac{q(s=i)}{2},$$

$$\hat{b}_\varphi^i = b_\varphi^i + \frac{q(s=i)}{2}\left\langle |y-G^T(x,\theta^i)\alpha^i|^2\right\rangle_{q(\alpha^i)}.$$

19.3.4 Optimization with respect to $q(s)$

$\mathcal{F}$ can be expressed as follows:

$$
\begin{aligned}
\mathcal{F} = \sum_{i=1}^{S} q(s=i) \int \partial \pi q(\pi) \log\left[\frac{p(s=i\,|\,\pi)}{q(s=i)} \right] \\
+ \frac{1}{2} \sum_{i=1}^{S} q(s=i) \left\langle \log(\varphi^i) \right\rangle_{q(\varphi^i)} \\
- \frac{1}{2} \sum_{i=1}^{S} q(s=i) \left\langle (\varphi^i) \right\rangle_{q(\varphi^i)} \left\langle \left| y - G^T(x,\theta^i)\alpha^i \right|^2 \right\rangle_{q(\alpha^i)} \\
+ \mathrm{cons}\{s\}.
\end{aligned}
$$

As per a result of Dirichlet distribution,

$$
\int \partial \pi q(\pi) \log[p(s=i\,|\,\pi)] = \Psi(\hat{c}\hat{t}_i) - \Psi(\hat{c}).
$$

$\mathcal{F}$ will be maximized over $q(s)$ if

$$
\begin{aligned}
q*(s=i) \propto \exp\Bigg[\Psi(\hat{c}\hat{t}_i) - \Psi(\hat{c}) + \frac{1}{2} \left\langle \log(\varphi^i) \right\rangle_{q(\phi^i)} \\
- \frac{1}{2} \left\langle \varphi^i \right\rangle_{q(\varphi^i)} \left\langle \left| y - G^T(x,\theta^i)\alpha^i \right|^2 \right\rangle_{q(\alpha^i)} \Bigg].
\end{aligned}
$$

That is,

$$
\begin{aligned}
q*(s=i) = \frac{1}{\mathcal{Z}} \exp\Bigg[\Psi(\hat{c}\hat{t}_i) - \Psi(\hat{c}) + \frac{1}{2} \left\langle \log(\varphi^i) \right\rangle_{q(\varphi^i)} \\
- \frac{1}{2} \left\langle \varphi^i \right\rangle_{q(\varphi^i)} \left\langle \left| y - G^T(x,\theta^i)\alpha^i \right|^2 \right\rangle_{q(\alpha^i)} \Bigg],
\end{aligned}
$$

where $\mathcal{Z}$ is the normalization constant such that $\sum_{i=1}^{S} q*(s=i) = 1$.

19.3.5 Summary

Please note that

$$
\left\langle \log(\varphi^i) \right\rangle_{q(\varphi^i)} = \Psi\left(\hat{a}_\varphi^i \right) - \log\left(\hat{b}_\varphi^i \right),
$$

$$\left\langle (\varphi^i) \right\rangle_{q(\varphi^i)} = \frac{\hat{a}^i_\varphi}{\hat{b}^i_\varphi},$$

$$\left\langle \left| y - G^T(x,\theta^i)\alpha^i \right|^2 \right\rangle_{q(\alpha^i)} = \left| y - G^T(x,\theta^i)\hat{\mu}^i_\alpha \right|^2 + Tr\left[\left(\hat{\Lambda}^i_\alpha \right)^{-1} G(x,\theta^i)G^T(x,\theta^i) \right].$$

The results of Bayesian inference of fuzzy models mixture can be summarized as a set of equations for estimating the hyperparameters of posterior distributions:

$$\hat{c} = c + 1,$$

$$\hat{t}_i = \frac{1}{c+1}\left[\frac{c}{S} + q*(s=i) \right],$$

$$\hat{\Lambda}^i_\alpha = \Lambda^i_\alpha + q*(s=i)\frac{\hat{a}^i_\varphi}{\hat{b}^i_\varphi} G(x,\theta^i)G^T(x,\theta^i),$$

$$\hat{\mu}^i_\alpha = \left(\hat{\Lambda}^i_\alpha \right)^{-1}\left[\Lambda^i_\alpha\mu^i_\alpha + q*(s=i)\frac{\hat{a}^i_\varphi}{\hat{b}^i_\varphi} G(x,\theta^i)y \right],$$

$$\hat{a}^i_\varphi = a^i_\varphi + \frac{q*(s=i)}{2},$$

$$\hat{b}^i_\varphi = b^i_\varphi + \frac{q*(s=i)}{2}\left| y - G^T(x,\theta^i)\hat{\mu}^i_\alpha \right|^2$$
$$+ \frac{q*(s=i)}{2}Tr\left[\left(\hat{\Lambda}^i_\alpha \right)^{-1} G(x,\theta^i)G^T(x,\theta^i) \right],$$

$$q*(s=i) = \frac{1}{\mathcal{Z}}\exp\left\{ \Psi(\hat{c}\hat{t}_i) - \Psi(\hat{c}) + \frac{1}{2}\left[\Psi\left(\hat{a}^i_\varphi \right) - \log\left(\hat{b}^i_\varphi \right) \right] \right.$$
$$- \frac{1}{2}\frac{\hat{a}^i_\varphi}{\hat{b}^i_\varphi}\left| y - G^T(x,\theta^i)\hat{\mu}^i_\alpha \right|^2$$
$$\left. - \frac{1}{2}\frac{\hat{a}^i_\varphi}{\hat{b}^i_\varphi}Tr\left[\left(\hat{\Lambda}^i_\alpha \right)^{-1} G(x,\theta^i)G^T(x,\theta^i) \right] \right\}.$$

Algorithm 1 finally presents the variational Bayesian inference of mixture of signal data models. Given N pairs of input–output physiological data $\{x(k),y(k)\}_{k=1}^N$, a stochastic fuzzy model of the physiological signal can be developed via using algorithm 1 where $S = 1$ (i.e., the identification of a single fuzzy model is of concern) and posteriors estimated (using Algorithm 1) with kth data pair $(x(k), y(k))$ become prior for $(k + 1)$th data pair. Such a method is formally stated as Algorithm 2.

ALGORITHM 1: VARIATIONAL BAYESIAN INFERENCE OF MIXTURE OF SIGNAL DATA MODELS (KUMAR ET AL. [2012B])

Require: input–output data pair (x, y); vectors $\left\{ G(x, \theta^i)_{i=1}^S \right\}$; and priors c, $\left\{ \left(u_\alpha^i, \Lambda_\alpha^i \right)_{i=1}^K \right\}$, and $\left\{ \left(a_\varphi^i, b_\varphi^i \right)_{i=1}^K \right\}$

1. Set iteration count $k = 0$, choose tolerance limit (say, equal to 10^{-8}), and make initial guesses about the posterior parameters:

$$q*(s = i)|_0 = \frac{1}{S}, \quad \hat{t}_i|_0 = \frac{1}{S}, \quad \hat{\Lambda}_\alpha^i|_0 = \Lambda_\alpha^i, \quad \hat{\mu}_\alpha^i|_0 = \mu_\alpha^i,$$

$$\hat{a}_\varphi^i|_0 = a_\varphi^i, \quad \hat{b}_\varphi^i|_0 = b_\varphi^i.$$

2. **While** $(k < k_{max})$ **do.**
3. **While** $(k = 0)$ or $(\text{abs}(y_f|_{k+1} - y_f|_k) > 1e - 8)$ **do.**
4. Update the hyperparameters of posterior distributions as per the equations provided in Subsection 19.3.5. Once a hyperparameter has been updated, its updated value should be used in subsequent equations. Let $\hat{c}|_{k+1}$, $\hat{t}_i|_{k+1}$, $\hat{\Lambda}_\alpha^i|_{k+1}$, $\hat{\mu}_\alpha^i|_{k+1}$, $\hat{a}_\varphi^i|_{k+1}$, $\hat{b}_\varphi^i|_{k+1}$, and $q*(s = i)|_{k+1}$ denote the updated values of hyperparameters.
5. $y_f|_{k+1} = \sum_{i=1}^K \left(G^T(x, \theta^i) \hat{\mu}_\alpha^i|_{k+1} \right) q*(s = i)|_{k+1}.$
6. $k \leftarrow k + 1.$
7. **End while.**
8. **End while.**
9. **Return** updated values of hyperparameters of posterior distributions.

19.4 A Case Study

A case study by Kumar et al. (2012b) is considered to demonstrate the application of the framework of intelligent signal analysis for medical decision support systems. It is required to design an expert system for analyzing the heartbeat intervals of a subject to predict the corresponding state of the subject. The application example involves the study of a subject in tilt-table experiment. The participant lays supine on the tilting table for 10 min followed by a 10 min head-up tilt at 70°. The R-R intervals (i.e., the time in milliseconds between consecutive R waves of an electrocardiogram) were recorded during the experiment. For each of the two states (i.e., rest and tilt), the initial and final half-minute-long data of R-R intervals were ignored and rest 9 min long data were considered. The method was applied to predict the state of the subject as follows.

ALGORITHM 2: STOCHASTIC FUZZY MODELING OF SIGNAL DATA (KUMAR ET AL. [2012b])

Require: N pairs of input–output data pair $\{x(k), y(k)\}_{k=1}^{N}$ and vectors $\{G(x(k), \theta^1)\}_{k=1}^{N}$

1. Choose the number of iterations N_{itr}.
2. Set iteration count $ic = 0$ and data index $k = 1$, and choose the initial guesses as follows:

$$\hat{\Lambda}_\alpha^1|_0 = I, \quad \hat{\mu}_\alpha^1|_0 = 0, \quad \hat{a}_\varphi^1|_0 = \hat{b}_\varphi^1|_0 = 10^{-3}.$$

3. **While** $(ic < N_{itr})$ **do.**
4. **While** $(k \le N)$ **do.**
5. Run Algorithm 1 on $(x(k), y(k))$ with $G(x(k), \theta^1)$ and priors $c = 1$, $\left(\hat{\mu}_\alpha^1|_{k-1}, \hat{\Lambda}_\alpha^1|_{k-1}\right)$, and $\left(\hat{a}_\varphi^1|_{k-1}, \hat{b}_\varphi^1|_{k-1}\right)$. Denote the hyperparameters returned by Algorithm 1 as $\hat{\Lambda}_\alpha^1|_k \hat{\mu}_\alpha^1|_k, \hat{a}_\varphi^1|_k, \hat{b}_\varphi^1|_k$
6. $k \leftarrow k + 1$.
7. **End while.**
8. Reset data index $k = 1$, and the initial guesses as follows:

$$\hat{\Lambda}_\alpha^1|_0 = \hat{\Lambda}_\alpha^1|_N, \quad \hat{\mu}_\alpha^1|_0 = \hat{\mu}_\alpha^1|_N, \quad \hat{a}_\varphi^1|_0 = \hat{a}_\varphi^1|_N, \quad \hat{b}_\varphi^1|_0 = \hat{b}_\varphi^1|_N.$$

9. $ic \leftarrow ic + 1$.
10. **End while.**
11. **Return** $\hat{\Lambda}_\alpha^1|_N, \hat{\mu}_\alpha^1|_N, \hat{a}_\varphi^1|_N$, and $\hat{b}_\varphi^1|_N$.

1. A fuzzy model of Takagi–Sugeno type (described in Subsection 19.2.1), defining two triangular membership functions on each of its two inputs, was considered. The model was meant to approximate the functional relationship between RR_j (i.e., jth R-R interval) and (RR_{j-1}, RR_{j-2}) (i.e., two R-R intervals previous to jth one). For each of the two states, first 3 min long data segment was used to identify the fuzzy model via running Algorithm 2 with $N_{itr} = 1$. Let $\left(\Lambda_\alpha^{rest}, \mu_\alpha^{rest}, a_\varphi^{rest}, b_\varphi^{rest}\right)$ and $\left(\Lambda_\alpha^{tilt}, \mu_\alpha^{tilt}, a_\varphi^{tilt}, b_\varphi^{tilt}\right)$ denote the model parameters (returned by Algorithm 2) corresponding to the states of rest and tilt, respectively.

2. The aim was to analyze 6 min long heartbeat intervals data of each state by using Algorithm 1 and to predict the corresponding state of the subject from the data. Algorithm 1 outputs $q(s = i)$, i.e., posterior probability that current input–output data pair has been generated by the ith state model. This is how the more probable state of the subject can be predicted from the analysis of physiological data. It makes sense to calculate the average posterior probabilities over a finite number of data points for assessing the state of the subject during the time to which analyzed data correspond. In the current study, the posterior probabilities were averaged over 72 consecutive heartbeats (i.e., data points) to calculate the average probabilities of states of rest and tilt.

Figure 19.1 displays the physiological data and the predicted state calculated via Algorithm 1. The time point at which the state of the subject changes from rest to tilt was marked by a vertical line in Figure 19.1. The lower graphic in this figure clearly shows the higher probabilities of rest state than of tilt state before the switching point and vice versa after the switching point. This clearly demonstrates the application of the signal

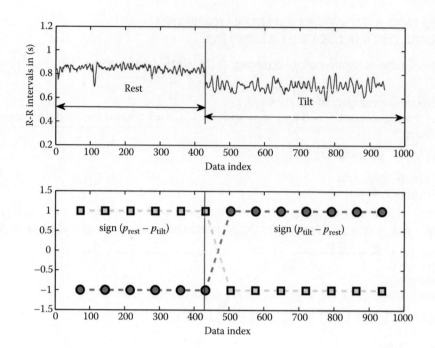

FIGURE 19.1
Prediction of the state of the subject based on the analysis of heartbeat intervals data.

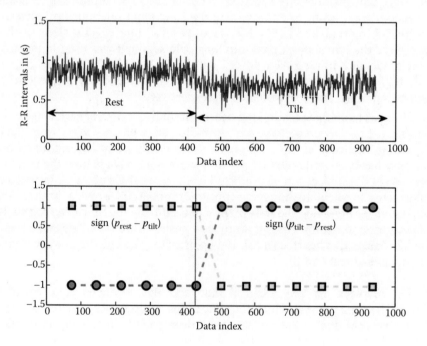

FIGURE 19.2
Prediction of the state of the subject based on the analysis of noisy heartbeat intervals data.

algorithm in predicting the state of an individual. To demonstrate the robustness of the method against signal noise, the experimentally measured heartbeat intervals were contaminated with a random Gaussian noise of mean zero and with a variance of 0.01. The prediction results in this case are shown in Figure 19.2. An accurate prediction of the state of the object even in the presence of signal noise verifies the robustness of the signal analysis method.

19.5 Concluding Remarks

The mathematical analysis of biomedical signals is meant for extracting the signal features relevant for functional state assessment. The stochastic fuzzy modeling and analysis techniques have much to offer in this area. The approach takes simultaneously the advantages of Bayesian analysis theory and fuzzy theory to formulate the patients' state prediction problem mathematically in a sensible way. It is possible to design new signal feature extraction methods with high diagnostic efficiency; thus, it is possible to build an expert system for making accurate predictions regarding the state of individuals.

References

Adlassnig, K. P. 1986. Fuzzy set theory in medical diagnosis. *IEEE Transactions on Systems, Man, and Cybernetics* 16(2):260–265.

Adlassnig, K. P., G. Kolarz, W. Sheithauer, H. Effenberger, and G. Grabner. 1985. CADIAG: Approaches to computer-assisted medical diagnosis. *Computers in Biology and Medicine* 15(5):315–335.

Ancona, N., L. Angelini, M. De Tommaso, D. Marinazzo, L. Nitti, M. Pellicoro, and S. Stramaglia. 2006. Measuring randomness by leave-one-out prediction error: Analysis of EEG after painful stimulation. *Physica A: Statistical Mechanics and its Applications* 365(2):491–498.

Barbieri, R., and E. N. Brown. 2006. Analysis of heartbeat dynamics by point process adaptive filtering. *IEEE Transactions on Biomedical Engineering* 53(1):4–12.

Bellazzi, R., R. Guglielmann, and L. Ironi. 2001. Learning from biomedical time series through the integration of qualitative models and fuzzy systems. *Artificial Intelligence in Medicine* 21:215–220.

Bellazzi, R., L. Ironi, R. Guglielmann, and M. Stefanelli. 1998. Qualitative models and fuzzy systems: An integrated approach for learning from data. *Artificial Intelligence in Medicine* 14:5–28.

Belle, A., M. A. Kon, and K. Najarian. 2013. Biomedical informatics for computer-aided decision support systems: A survey. *The Scientific World Journal* 2013.

Belmonte, M., C. Sierra, and R. L. de Mántaras. 1994. RENOIR: An expert system using fuzzy logic for rheumatology diagnosis. *International Journal of Intelligent Systems* 9(11):985–1000.

Binaghi, E., O. De Giorgi, G. Maggi, T. Motta, and A. Rampini. 1993. Computer assisted diagnosis of postmenopausal osteoporosis using a fuzzy expert system shell. *Computers and Biomedical Research* 26(6):498–516.

Bollt, E. M., J. D. Skufca, and S. J. McGregor. 2009. Control entropy: A complexity measure for nonstationary signals. *Mathematical Biosciences and Engineering* 6(1):1–25.

Brai, A., J.-F. Vibert, and R. Koutlidis. 1994. An expert system for the analysis and interpretation of evoked potentials based on fuzzy classification: Application to brainstem auditory evoked potentials. *Computers and Biomedical Research* 27(5):351–366.

Chiarugi, F., S. Colantonio, D. Emmanouilidou, D. Moroni, and O. Salvetti. 2008. Biomedical signal and image processing for decision support in heart failure. In *Advances in Mass Data Analysis of Images and Signals in Medicine, Biotechnology, Chemistry and Food Industry*, ed. P. Perner and O. Salvetti, vol. 5108 of *Lecture Notes in Computer Science*, 38–51. Springer, Berlin, Heidelberg.

Daniels, J. E., R. M. Cayton, M. J. Chappel, and T. Tjahjadi. 1997. CADOSA: A fuzzy expert system for differential diagnosis of obstructive sleep apnoea and related conditions. *Expert Systems with Applications* 12(2):163–177.

de Dombal, F. T., D. J. Leaper, J. R. Staniland, A. P. McCann, and J. C. Horrocks. 1972. Computer-aided diagnosis of acute abdominal pain. *BMJ* 2(5804):9–13.

Fathitorbaghan, M., and D. Meyer. 1994. MEDUSA: A fuzzy expert system for medical diagnosis of acute abdominal pain. *Methods of Information in Medicine* 33(5):522–529.

Ferrario, M., M. G. Signorini, G. Magenes, and S. Cerutti. 2006. Comparison of entropy-based regularity estimators: Application to the fetal heart rate signal for the identification of fetal distress. *IEEE Transactions on Biomedical Engineering* 53(1):119–125.

Garibaldi, J. M., and E. C. Ifeachor. 1999. Application of simulated annealing fuzzy model tuning to umbilical cord acid-base interpretation. *IEEE Transactions on Fuzzy Systems* 7(1):72–84.

Gautama, T., D. P. Mandic, and M. M. Van Hulle. 2004. A novel method for determining the nature of time series. *IEEE Transactions on Biomedical Engineering* 51(5):728–736.

Greens, R. A. 2007. *Clinical Decision Support: The Road Ahead.* Academic Press, Waltham.

Hornero, R., D. Abásolo, N. Jimeno, C. I. Sánchez, J. Poza, and M. Aboy. 2006. Variability, regularity, and complexity of time series generated by schizophrenic patients and control subjects. *IEEE Transactions on Biomedical Engineering* 53(2):210–218.

Hornero, R., M. Aboy, D. Abásolo, J. McNames, and B. Goldstein. 2005. Interpretation of approximate entropy: Analysis of intracranial pressure approximate entropy during acute intracranial hypertension. *IEEE Transactions on Biomedical Engineering* 52(10):1671–1680.

Hudson, D. L., M. E. Cohen, W. Meecham, and M. Kramer. 2004. Inclusion of signal analysis in a hybrid medical decision support system. *Methods of Information in Medicine* 43(1):79–82.

Kumar, M., D. Arndt, S. Kreuzfeld, K. Thurow, N. Stoll, and R. Stoll. 2008. Fuzzy techniques for subjective workload score modelling under uncertainties. *IEEE Transactions on Systems, Man, and Cybernetics—Part B: Cybernetics* 38(6):1449–1464.

Kumar, M., S. Neubert, S. Behrendt, A. Rieger, M. Weippert, N. Stoll, K. Thurow, and R. Stoll. 2012a. Stress monitoring based on stochastic fuzzy analysis of heartbeat intervals. *IEEE Transactions on Fuzzy Systems* 20(4):746–759.

Kumar, M., N. Stoll, D. Kaber, K. Thurow, and R. Stoll. 2007a. Fuzzy filtering for an intelligent interpretation of medical data. In *Proceedings of IEEE International Conference on Automation Science and Engineering (CASE 2007)*, 225–230. Scottsdale.

Kumar, M., N. Stoll, and R. Stoll. 2010a. Variational Bayes for a mixed stochastic/deterministic fuzzy filter. *IEEE Transactions on Fuzzy Systems* 18(4):787–801.

Kumar, M., N. Stoll, K. Thurow, and R. Stoll. 2012b. Physiological signals to individual assessment for application in wireless health systems. In *Proceedings of the Ninth International Multi-Conference on Systems, Signals and Devices (SSD)*, 1–6.

Kumar, M., M. Weippert, D. Arndt, S. Kreuzfeld, K. Thurow, N. Stoll, and R. Stoll. 2010b. Fuzzy filtering for physiological signal analysis. *IEEE Transactions on Fuzzy Systems* 18(1):208–216.

Kumar, M., M. Weippert, N. Stoll, and R. Stoll. 2010c. A mixture of fuzzy filters applied to the analysis of heartbeat intervals. *Fuzzy Optimization and Decision Making* 9(4):383–412.

Kumar, M., M. Weippert, R. Vilbrandt, S. Kreuzfeld, and R. Stoll. 2007b. Fuzzy evaluation of heart rate signals for mental stress assessment. *IEEE Transactions on Fuzzy Systems* 15(5):791–808.

Kuncheva, L., and F. Steimann. 1999. Fuzzy diagnosis. *Artificial Intelligence in Medicine* 16(2):121–128.

McNames, J., and M. Aboy. 2008. Statistical modeling of cardiovascular signals and parameter estimation based on the extended Kalman filter. *IEEE Transactions on Biomedical Engineering* 55(1):119–129.

Pearson, S.-A., A. Moxey, J. Robertson, I. Hains, M. Williamson, J. Reeve, and D. Newby. 2009. Do computerised clinical decision support systems for prescribing change practice? A systematic review of the literature (1990–2007). *BMC Health Services Research* 9(1):154.

Perner, P. 2006. Intelligent data analysis in medicine—Recent advances. *Artificial Intelligence in Medicine* 37(1):1–5.

Rezek, I. A., and S. J. Roberts. 1998. Stochastic complexity measures for physiological signal analysis. *IEEE Transactions on Biomedical Engineering* 45(9):1186–1191.

Roy, M. K., and R. Biswas. 1992. I-v fuzzy relations and Sanchez's approach for medical diagnosis. *Fuzzy Sets and Systems* 42:35–38.

Steimann, F. 1996. The interpretation of time-varying data with DIAMON-1. *Artificial Intelligence in Medicine* 8(4):343–357.

Stivaros, S. M., A. Gledson, G. Nenadic, X.-J. Zeng, J. Keane, and A. Jackson. 2010. Decision support systems for clinical radiological practice towards the next generation. *British Journal of Radiology* 83(995):904–914.

Task Force of the European Society of Cardiology and the North American Society of Pacing and Electrophysiology. 1996. Heart rate variability: Standards of measurement, physiological interpretation, and clinical use. *European Heart Journal* 17:354–381.

Tourassi, G. D. 1999. Journey toward computer-aided diagnosis: Role of image texture analysis. *Radiology* 213(2):317–320.

Watanabe, H., W. J. Yakowenko, Y. M. Kim, J. Anbe, and T. Tobi. 1994. Application of a fuzzy discrimination analysis for diagnosis of valvular heart disease. *IEEE Transactions on Fuzzy Systems* 2(4):267–276.

Wong, W. S. F., K. S. Leung, and Y. T. So. 1990. The recent development and evaluation of a medical expert system (ABVAB). *International Journal of BioMedical Computing* 25(2–3):223–229.

Zadeh, L. A. 1973. Outline of a new approach to the analysis of complex systems and decision processes. *IEEE Transactions on Systems, Man, and Cybernetics* 3:28–44.

Zadeh, L. A. 1983. The role of fuzzy logic in the management of uncertainty in expert systems. *Fuzzy Sets Systems* 11:199–227.

Zhong, Y., K. M. Jan, K. H. Ju, and K. H. Chon. 2007. Representation of time varying nonlinear systems with time-varying principal dynamic modes. *IEEE Transactions on Biomedical Engineering* 54(11):1983–1992.

Section IV

Implementation of eMedicine and Telemedicine

20

Telecardiology

Rajarshi Gupta

CONTENTS

20.1 Introduction

Biotelemetry is a common technique which involves collection of physiological data at one place and their transmission by using suitable communication media to another place for recording, interpretation, and analysis [1–2]. In clinical practice, however, the term *telemedicine* [3] is popular. Telemedicine is the integrated technology platform where a remote patient can be monitored through a communication link and examined by a physician. The American Telemedicine Association (ATA) [4] is one of the leading professional organizations promoting the use of remote diagnostics to improve quality, equity, and affordability of healthcare. ATA defines *telemedicine* as, "Telemedicine is the use of medical information exchanged from one site to another via electronic communications to improve patients' health status." *Telehealth*, a term closely associated with telemedicine, is often used to encompass a broader definition of remote healthcare that does not always involve clinical services. Videoconferencing, transmission of still images, eHealth including patient portals, remote monitoring of vital signs, continuing medical education, and nursing call centers are all considered part of telemedicine and telehealth. Use of information and communication technology (ICT) in healthcare service is a well-established practice in advanced nations for delivering a quality healthcare service to the common people. Telecardiology is a specialized application of telemedicine applied to the monitoring of cardiac functions of a remote patient. Apart from regular clinical usage, telecardiology is successfully applied in periodic health monitoring of elderly citizens and pre- and posthospital checkups of cardiac patients as a part of improvement of quality of life (QoL) service. Through many studies, it has been statistically established that telecardiology can significantly increase the survival rate of an emergency patient brought to a hospital by mobilizing the resources and therapeutic actions specific to the individual's requirements. The scope of this chapter, however, is to cover the technological aspects of telecardiology systems in terms of medical signal acquisition from the patients, their preprocessing, transmission of the signals, presentation to the medical professional for evaluation (sometimes supported by computerized analysis), and feedback. The layout of this chapter is as follows: Section 20.2 describes the overall functioning of a telecardiology system by using a block diagram. Sections 20.3 and 20.4 deal with digital acquisition of medical signals from patients and their preprocessing and compression techniques. ICT plays a vital role in transmission of medical data from the patient end to the expert end. The different ICTs for telecardiology practice are discussed in Section 20.5. Section 20.6 deals with storage of medical data and their access by experts for analysis. Two peripheral issues, namely, data security in telecardiology and computerized processing of medical data for assisted diagnosis, are touched upon in Sections 20.7 and 20.8, respectively. Section 20.9 provides a snapshot of trends in telecardiology systems in terms of faster and secure data communication, portability, and use of advanced gadgets.

20.2 Components of a Telecardiology System and Functioning

A typical telecardiology system consists of the following functional modules: (i) acquisition and preprocessing module, (ii) communication systems module, and (iii) presentation module. A modular block diagram of a representative telecardiology system is shown in Figure 20.1. The signal-acquisition module is attached to the patient's body and collects relevant medical signals (ECG, PPG, blood pressure [BP], etc.) in digital format for off-line

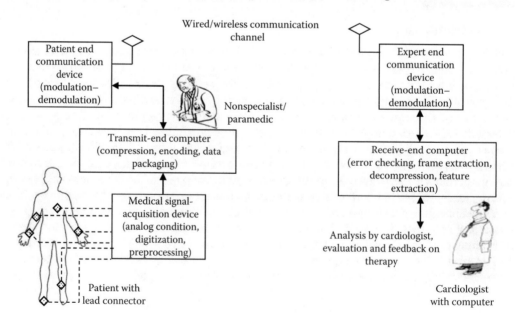

FIGURE 20.1
Block schematic of a telecardiology system.

or real-time transmission. In some systems, CT scan images of the heart and the patient and few additional clinical parameters, like body temperature, are also collected. The acquisition module contains a high-end processor for multiplexed collection of the medical data and suitable buffering. During collection, a preprocessing operation is carried out to remove unwanted noise and redundant information from the data. The same processor executes real-time or offline compression of the acquired medical signals (and images) and source encoding for the transceiver. A pair of communication module connects the patient-end module to the remote-end (expert-end) computer by using a media and performs modulation–demodulation of the data for an error-free communication.

The choice of communication module largely depends on the distance between the patient and expert and other factors like mobility of the patient, volume of medical data to be transmitted, and real-time or offline transfer. The signal extraction (decompression) is carried out at the expert-end computer before being presented to the cardiologist for clinical evaluation. Sometimes, the extracted signals are digitally processed to reveal elementary pathological information and supplied to the cardiologist as additional aid to support diagnosis. The diagnosis report is sent back to the patient end as either text message or voice data (in real-time interactive mode).

20.3 Cardiovascular Signal Acquisition

Cardiac signal acquisition from the patient body is the most critical part of the entire telecardiology system since it determines the selectivity of the acquisition device towards the medical signal and accuracy in processing and diagnosis in subsequent stages. The prime cardiovascular signal for clinical assessment is the 12-lead ECG. Very often, a cardiologist would require other ECG-related signals [5] like PPG, blood pressure, and respiration signals to correlate clinical features and symptoms. Simultaneous acquisition of all these signals puts a high data burden on preprocessing and subsequent compression stages at the transmit-end processor. An advantageous way is to derive some of these surrogates from the ECG and PPG at the receiving-end system and present these to the expert. An important criterion that determines the robustness and accuracy of the acquisition hardware is the ability to reject artifacts in individual medical signals by their filtering and amplification.

Biomedical signals are contaminated with noise and artifacts (often the two words are used interchangeably). For ECG, the artifacts are power-line interferences (PLIs), electrode pop or contact noise, baseline wander, motion artifacts, electrosurgical noise, medical amplifier noise, and quantization noise [6]. For faithful collection of medical signals the basic desirable characteristics of a biopotential acquisition circuit are

- The pathological information should be undisturbed by the amplifier and other electronic components' characteristics;
- Protection of the patient from accidental surge and power-line current and protection of acquisition circuit (amplifier) from high input voltages (defibrillation applications); and
- Best possible separation of acquired signal and unwanted artifacts.
- To facilitate sampling at 500 Hz to 1 kHz by the processor in the next stage.

A schematic block diagram of a multichannel digital biomedical signal-acquisition system is shown in Figure 20.2. This can be divided into two major sub-blocks, namely, analog conditioning (also called analog front end) and digital domain processing. The objective of the analog condition is to convert the raw medical signal into bipolar (±10 V) or unipolar (0–5 V) signal. The conditioned output of each signal is fed to the multichannel ADC followed by a high-end processor.

The analog conditioning part consists of patient isolation, medical amplifiers, and filters for using the clinical bandwidth of the respective signals and discarding unwanted noise. In the first stage, isolation amplifiers protect the patients from accidental currents passing through their body from the acquisition-side power-line. International Electrotechnical Commission (IEC) 60601 standard [7] limits a patient auxiliary current to 100 µA at 0.1 Hz. Patient safety can be achieved in two ways. The first, electrically separating the input stage of the isolation amplifiers from the output stage. The input stage has a separate floating power supply and a ground that are connected to the output side of the amplifier by a high resistance (1000 MΩ or more) and low parallel capacitance (1 pF). The input connectors of the signal are isolated from the output stage by a similar arrangement. Optical, inductive, or capacitive are the coupling variants used in commercial isolation amplifiers. The second method involves protection from accidental surge or transient that may occur from the biomedical equipment by using current-limiting devices.

The medical amplifier is a low-offset, high-CMRR instrumentation amplifier (INA) to provide a total gain of 2000 for ECG acquisition in the range of 0–5 V. The total amplification may be distributed in a preamplifier and a final amplifier stage, with a filter in between. A practical circuit for single-channel ECG acquisition is shown in Figure 20.3. The first stage consists of a DC coupled INA using the AD620, a widely popular amplifier

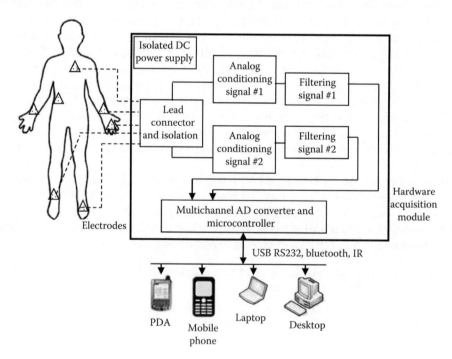

FIGURE 20.2
Block schematic of a digital biomedical signal-acquisition system.

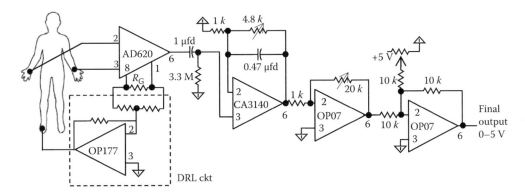

FIGURE 20.3
A practical ECG amplifier circuit.

used for medical applications with the advantages of having a high CMRR and a low input offset and using a single resistance to get a wide range of gains. The first-stage gain is set low (nominally ≈ 10–20) to avoid saturation of the output due to electrode offset potential, which may be around 300 mV. The gain setting can be done by the following formula:

$$R_G = \frac{49.4\,k}{G-1}. \tag{20.1}$$

A fraction of the output from within the AD620 is connected to the right leg of the subject with a negative feedback. This is called a driven-right-leg (DRL) circuit and is used to minimize the power-line interference. The output of the first stage is AC coupled to the second stage and high-pass filtered to block the DC and frequencies of up to 0.05 Hz by using a 1 μF and 3.3 MΩ combination. The second stage uses a CA3140 to implement a low-pass filter (cutoff at 70 Hz) and a small gain of 5. The final amplifier with a gain of 20 may be designed using an OP07 to implement an inverting amplifier followed by an adder (DC shift of +2.5 V) to obtain a unipolar output of 0–5 V. Although the clinical bandwidth of the ECG is 0–130 Hz, the proposed amplifier used 0.05–70 Hz to minimize muscle noise, which creeps in at higher than 100 Hz. Some design aspects of biomedical amplifiers are available in Clifford et al.'s [8], Bronzino's [9], and Webster's [10] books.

A photoplethysmogram (PPG) is the second important medical signal often dealt with in telecardiology applications. In recent years it has drawn significant interest from biomedical science researchers due to low cost, easier (than ECG) acquisition, and scope of generating more information (like oxygen saturation [SpO_2], blood pressure, cardiac output, and autonomic function of the heart) from the signal [11]. The PPG signal is normally acquired from fingertips and earlobe locations. However, other body sites and multiple-finger acquisition are also reported [12]. The sensor operates in either of two modes, namely, transmittance and reflectance [13]. A pair of matched LEDs and a sensing photodiode operating in the red and/or near-infrared wavelength (0.64 to 0.9 μm) are involved in the sensing process. In the transmittance mode, the intensity of the light passing through the tissue is attenuated by the amount of blood present in the tissue. This varies with the arterial pulse and is used as a measure to indicate the pulse rate. In the reflectance mode, the light from the LED enters the finger tissue and scattered by both the moving red blood cells and the nonmoving tissue, and a part of this backscattered light is detected by the photodiode. The sensor package is

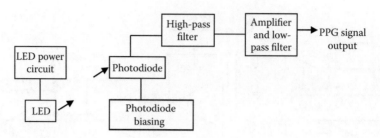

FIGURE 20.4
Block diagram of analog signal conditioning of photoplethysmogram.

carefully designed to filter interfering (visible) light. The raw PPG signal is a low-frequency wave (typically 1 Hz) superimposed on a semi-DC component that varies with the respiration and thermoregulation factors. The output signal is suitably filtered (0.05 Hz high pass to remove baseline wander and low pass to remove power-line interference) and amplified to get a clean signal. A block diagram of an analog conditioning circuit for PPG signal is shown in Figure 20.4.

Among the other nonelectrical medical signals optionally acquired in telecardiology applications are respiratory signals and BP, which provide additional information to the cardiologist regarding pulmonary and circulatory functions [14–15]. Continuous pulmonary gas-flow measurements require the sensors to be placed in the nasal cavities. Thermal convection flowmeters and transit-time ultrasonic flowmeters are used. For continuous BP measurements, intravascular or extravascular techniques are used.

The second sub-block in Figure 20.2 performs the digital processing of the signals. A high-end microcontroller unit performs sampling of the analog-conditioned units, quantization (typically 10–12 bits), and compression to transfer the encoded data to the local data-collection unit, which can be a PDA, mobile phone, or desktop or laptop computer. For short-range telecardiology applications, the signal-acquisition module integrates the analog front end, processor module, and transceiver interface in a portable stand-alone unit.

20.4 Signal Compression in Telecardiology

In the context of telecardiology, medical signal compression [16] assumes great importance. An example would reveal its necessity. A cardiac patient monitored with 12-lead ECG at 10-bit resolution and 360 Hz sampling will generate nearly 3 GB data in 24 h. For multiparameter (ECG, PPG, respiration, etc.) monitoring, which is often the case in a clinical setup, this data volume will be nearly 15 GB. For this example, data generated from the patient in 1 min = 83333.33 kbits. Assuming a moderate link speed of 200 kbps, the time to transfer this data will be 83333.33/12000 min = 6.94 min, which means six times the time of acquisition. Hence, compression of biomedical data can provide twofold benefits, namely, lesser time of transmission in health-monitoring applications and enhanced utilization of the communication link.

Biomedical data compression is a well-explored area over the last five decades and many algorithms have been reported in literature, with varied degrees of accuracy of the retrieved signal. This section describes the different compression techniques in the context of telecardiology applications. The important parameters that influence the design of software and hardware for such compression application are

- Quantization level of acquisition and real-time or offline mode of compression, which has a direct relation to the buffer requirement/memory engagement at the transmit-end module;
- Data overhead in the transmitted packet, which has a direct relation to the latency in real-time monitoring application; and
- Computational complexity of the compression algorithm, which has a direct relation to the retrieval/unfolding time before signal extraction at the receive-end processor.

The lossy compression techniques can be classified into run-length encoding (RLE), pattern substitution, inter- or intrabeat differencing, and diatomic coding. Some of these techniques are used as a part of multistage (or multipass) lossy compression techniques. The lossless compression techniques cannot achieve high compression and, thus, are unsuitable for telecardiology applications. Instead, lossy compression techniques, which are irreversible in the sense that the original signal cannot be perfectly reconstructed, are more popular in telecardiology applications. In context of biomedical signals, the compression techniques are generally classified under three broad categories, namely, direct-compression, parameter-extraction, and transform-domain techniques. The direct-compression techniques utilize the statistical correlation among a group of neighboring samples in a block of data. These techniques are computationally simpler and easy to implement using low-end processors. For parameter-extraction techniques, clinical signatures of ECG data points like peak locations, zero crossing points, and slope values are computed and stored. Transform-domain compression techniques employ a linear transform to map the signal representation in a different domain and encode the expansion coefficients. These two methods require large memory for buffering the data and demand high computational complexity. Before going to the brief review of compression techniques, let us have a quick look at the compression performance indices that have been used in literature. Compression ratio (CR) provides a direct measure of the compressor's performance:

$$CR = \frac{\text{input raw data size}}{\text{compressed data size}}. \tag{20.2}$$

The receiving-side decompression performance is indicated by a number of indices. Reconstruction quality in terms of sample-to-sample error is indicated by percentage root-mean-square difference (PRD) and percentage root-mean-square difference normalized (PRDN), defined as

$$PRD = 100 \times \sqrt{\frac{\sum (x[k] - x_r[k])^2}{\sum x^2[k]}},$$

$$PRDN = 100 \times \sqrt{\frac{\sum (x[k] - x_r[k])^2}{\sum (x[k] - \bar{x})^2}}. \tag{20.3}$$

A combined indicator of compression and reconstruction, named quality score (QS), defined as the ratio of CR and PRD, is used. The error estimates are expressed in terms of mean square error (MSE), normalized mean square error (NMSE), root-mean-square error (RMSE), normalized root-mean-square error (NRMSE), maximum absolute error (e_{max}), and SNR, defined, respectively, as

$$\text{MSE} = \frac{1}{N} \sum (x[n] - x_r[n])^2,$$

$$\text{NMSE} = \frac{\sum (x[n] - x_r[n])^2}{\sum (x[n])^2},$$

$$\text{RMSE} = \sqrt{\frac{\sum (x[n] - x_r[n])^2}{N}}, \qquad (20.4)$$

$$\text{NRMSE} = \frac{\sqrt{\sum (x[n] - x_r[n])^2}}{\sqrt{\sum (x[n])^2}},$$

$$e_{max} = \max(x[n] - x_r[n]),$$

$$\text{SNR} = 10 \log \frac{\sum (x[n] - \bar{x})^2}{\sum (x[n] - x_r[n])^2},$$

where for a length of data set N, $x[n]$ = raw sample; $x_r[n]$ = reconstructed sample; $\bar{x}$ = mean of raw sample data set; and the 'max' operator extracts the maximum value.

A wide review of ECG compression techniques is available in Jalaleddine et al.'s paper [17]. Historically, the earlier compression techniques employed the principle of "tolerance comparison." Amplitude zonal time epoch coding (AZTEC), scan-along polynomial approximation (SAPA), turning point (TP), and coordinate time reduction encoding system (CORTES) are some of the tolerance comparison methods used in ECG compression. The fan algorithm [18,19] was developed for ECG transmission. The technique uses a "coverage angle" between two samples to include all intermediate points by using a "minimum-slope principle." These intermediate points are considered redundant within this coverage angle. The reconstruction is done by interpolation and expansion of the line into discrete points. Delta encoding [20–21] is the general technique which adopts encoding of first difference of the ECG samples. An exhaustive discussion on direct-compression techniques is available in Kumar et al.'s [22] and Kulkarni et al.'s [23] papers.

The delta coding employs first difference (FD) of the ECG data set, given as

$$\Delta(i) = x(i + 1) - x(i). \qquad (20.5)$$

A typical lead II ECG plot with corresponding normalized first difference array plot is shown in Figure 20.5. This gives a very good measure of the QRS regions which contain high fluctuations in the signal. This measure has been used by many researchers for telecardiology applications. Furht and Perez [24] describe an adaptive thresholding scheme which adjusts the threshold values based on mean, standard deviation, and third

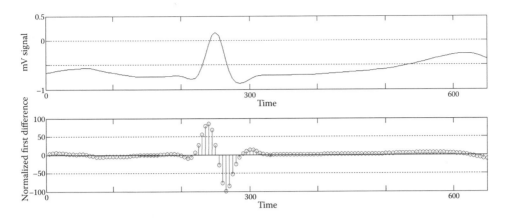

FIGURE 20.5
ECG plot with normalized first difference.

moment of the window. For reconstruction, a smooth-curve approximation is applied. A CR of 5.79 and PRD of 5.56 were obtained with Massachusetts Institute of Technology (MIT)–Beth Israel Hospital (BIH) arrhythmia data with 8-bit quantization and 500 Hz sampling. A PC-based off-line compression using delta encoding for Global System for Mobile Communication (GSM)–based telecardiology service is described by Mitra et al. [25]. The compression stages used normalization of FD, sign encoding, pairing of successive FDs in P and T regions, and finally RLE in equipotential regions. The authors report a CR of 43.54 and PRD of 1.73 using noisy signal from MIT-BIH arrhythmia database. Using data at 8-bit quantization, the same technique yields a CR of 4.68 and PRD of 0.739. A fixed thresholding enhances the CR in a few cases, at the cost of poor PRD [26]. An application of variable-length encoding is described by Hamilton and Tompkins [27] for real-time ECG compression using FD encoding. The coder performs a continuous beat alignment over a length of the signal and an average beat estimate is subtracted from the following incoming beats. The residual signal is encoded using Huffman coding with a maximum bit length of 8. The final coded bitstream contained averaged beat and its type, followed by residual encoded beats and their types and difference between present R-R and previous R-R interval. Figure 20.6 shows a plot of typical beat alignment using lead II ECG data and the corresponding average beat estimate with residual signal, which is significant only in QRS regions. The authors achieved a time reduction of 80% for Holter data storage using a PC-based system. The method is unsuitable for real-time ECG transmission and requires a large memory buffer.

Wavelet transform (WT) or sub-band coding (SBC) is perhaps the most exploited time-frequency signal analysis tool in the last two decades for biosignal and image compression applications. The basic building block of a WT analysis tool [28] is a combination of high-pass and low-pass digital filter bank that decomposes the signal up to the desired level by using a mother wavelet (most common being Haar and Coiflet).

Figure 20.7 shows the two-level decomposition structure of an SBC structure. At each level, a down-sampling factor of 2 is applied to keep the total number of samples in the derived data set the same as the original data set. The derived coefficients contain different frequency band information of the signal. For data compression, many coefficients in the transformed domain become insignificant or redundant. A suitable "thresholding" mechanism to set these coefficients to zero value will not hamper the clinical information

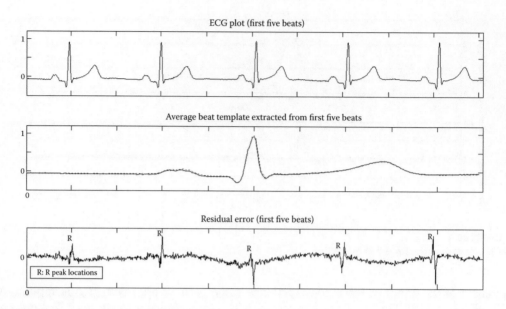

FIGURE 20.6
Beat alignment and average template generation for residual encoding.

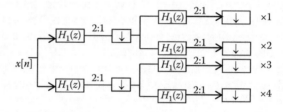

FIGURE 20.7
Two-level wavelet decomposition structure.

from the reconstructed signal. The significant coefficients are actually coded using a suitable coding method like differential pulse-code modulation (DPCM), an entropy coding for exploiting the signal property at a particular level of decomposition. Kim et al. [29] report a wavelet-based technique which provides low delay for continuous ECG transmission by using flexible bit allocation strategy at different zones of the ECG waveform. The ECG beats are segmented into "complex" and "plain" blocks, based on standard deviation over a fixed block length. Accordingly, more bits are allocated to complex blocks, which contain more information, and fewer bits are allocated to plain blocks, which contain less information. Accordingly the frame size is adjusted. For each block, biorthogonal-spline wavelet 4.4 (bior4.4) with five-layer decomposition level is applied. The header of a frame is also encoded to include block type, size, mean amplitude, size of encoded significance map, etc., for proper decoding. The coded bitstream contains significant wavelet coefficients and RLE stream of significant coefficients. In another wavelet-based approach [30–31], the multilead ECG signal is fragmented into beats and a dynamic beat template is generated. The subtracted beat is coded by Coiflet expansion and thresholded, depending upon fixed or variable bit rate of transmission. Again, the transmission mode is determined by predefined PRD outcome. The transmitted packet is formed with the fields like template (new or old), leads, number of coefficients, and bit allocations for adaptive pulse-code modulator (APCM). A wavelet-based compression using uniform scalar zero zone quantizer (USZZQ) technique is described by Manikandan and Dandapat [32]. Biorthogonal 9/7 filter set is used for up to fifth level of decomposition. Energy packing efficiency (EPE) has been used as the principal parameter for selecting and thresholding of wavelet coefficients, followed by quantization to retain the significant coefficients only. Finally, Huffman coding on coefficient differences is utilized for compression. In a similar approach the authors use bior4.4 up to fifth level of decomposition, followed by thresholding and selection of coefficients based on EPE. In Rajoub's work [33] set partitioning in hierarchical trees (SPIHT) algorithm is implemented after wavelet decomposition using the biorthogonal 9/7 wavelet. The inherent similarities between sub-bands after the decomposition are computed and encoded as most significant coefficients in priority for transmission. The discrete cosine transform (DCT) is among the other transform-domain coding techniques used in telecardiology. Allen and Belina [34] describe an approach using variable bit quantization to preserve the high-order (or significant) as well as low-order (insignificant) DCT coefficients corresponding to the power spectrum of ECG. The first 20% coefficients, which retain the majority of the signal power, are coded with full accuracy. The authors report a CR value of 2.8 with 4-bit quantization level. A real-time application for ECG transmission using Huffman coding of DCT coefficients is described by Lee et al. [35]. The ECG is down sampled and R peaks are detected using the downslope trace waveform (DSTW) algorithm. The DCT is applied on first-order backward difference 1-byte data. To compensate for the information loss during quantization, a data-cascading approach is used to add up the residuals and include them in the data trail. Finally, Huffman coding is applied on the entire frame. The algorithm was tested with different window sizes and achieves an average CR of 21.30 and PRD of 1.75. The property of the DCT for energy compaction efficiency in ECG compression is reported in a mobile phone platform [36]. The DCT coefficients corresponding to total energy in a fixed window block are sorted and below a fixed threshold are eliminated, followed by RLE. An optimal quantization scheme for DCT coefficients is proposed in Batista et al.'s paper [37], where a quantization vector associated with a threshold vector is used to minimize the estimated entropy of the quantization coefficients for a given distortion.

The distortion measure in the reconstructed data is important for its clinical acceptability and interpretation by a cardiologist. The popular parameters PRD, PRDN, and MSE provide only a first-hand statistical measure of the reconstruction quality of the ECG. Many compression methodologies devised in recent times were targeted to allow a maximum PRD so that the reconstructed signal is not significantly distorted. From the viewpoint of diagnosis, these parameters are not relevant [38]. The distortions in the reconstructed ECG were defined by weighted diagnostic distortion (WDD) measure, which provides a set of parameters defining deviations in the wave signatures. The parameters were evaluated using mean opinion score (MOS) with cardiologists, which are of semiblind and blind types. The weighted MOS error was taken as the quality measure. Manikandan and Dandapat [39] statistically show that different PRD values (PRD, PRDN, etc.) and statistical error indices cannot adequately represent the local distortion in a particular ECG beat and are not correlated with diagnostic features of the ECG in compression application. Instead, they proposed a diagnostic distortion (wavelet energy–based diagnostic distortion [WEDD]) measure based on the energy content in the wavelet-transformed coefficients, where the error between the coefficients of the original and reconstructed ECG were computed to get wavelet PRD (WPRD) in different bands. Based on the energy concentration of different ECG wave components in different sub-bands, a dynamic weight was assigned to get WEDD, which showed a better measure of local distortions.

Up to now, only a few publications are available on PPG signal compression. In Reddy et al.'s paper [40], a Fourier transform (FT)–based technique is described. The PPG signal was segmented on beats by detecting the peaks and FT was applied on a cycle-to-cycle basis. The significant number of coefficients is determined for different NRMSE values. The remaining coefficients are discarded, thus achieving a compression. However, the error in determining the period of the signal affects the percentage reconstruction error. The PPG is a very low-frequency signal (around 1 Hz); the delta modulation is a commonly used [41] technique. Use of Huffman coding for PPG compression is illustrated by Gupta [42].

20.5 Role of Information and Communication Technology in Telecardiology

From the perspective of telecardiology, there are three broad categories of communication needs, namely, (a) patient monitoring by using short-range communication for clinical setups; (b) use of Internet and mobile communication technology for day-to-day patient consultation, telemedicine, and pre- and posthospital checkup of patients; and (c) emergency healthcare using satellite communication, public telephone, and GSM networks. Factors like mobility of the patient or the consulting cardiologist and continuous or intermittent transmission govern the choice of communication link, protocols, and gadgets for communication. This section will briefly describe the various modes of communication, protocols, and framework of telecardiology applications in the light of the three categories mentioned above.

Personal healthcare system using short-range communication is practiced in monitoring of noncritical (elderly) patients. These systems have become more popular with the advent of low-power, wireless interface–compatible, and high-speed processors for short-range monitoring in a clinical or home environment. Three wireless technologies, namely, ZigBee (IEEE 802.15.4), Wi-Fi (wireless LAN–IEEE 802.11), and Bluetooth (IEEE 802.15.1), dominate the choice of communication network backbone for various short-range applications. Among these, ZigBee technology [43,44] has shown great promise in its acceptance in healthcare applications due to its useful features like reliability, low latency (typically 15 ms), high security (use of Advanced Encryption Standard [AES]), low power, interoperability (vendor independence), and inexpensive implementation. Figure 20.8 represents the ZigBee networking protocol layers. The two bottom layers, i.e., physical (PHY) and medium access control (MAC), are defined by IEEE 802.15.4 standard, initially released in 2003. The ZigBee standard defines the topmost layers, i.e., the application (APL) and network (NWK) layers, and adopts MAC and PHY layers as part of the ZigBee networking protocol. Hence, all ZigBee devices conform to IEEE 802.15.4 standard. The capability of customizing the upper part of the ZigBee stack layer is perhaps the greatest advantage of developing demand-specific applications using ZigBee networking. Among the three frequency bands offered by ZigBee, the 2.4 GHz band is used worldwide by RF device manufacturers and application engineers for developing short-range applications.

A special form of wireless sensor network, the Body Area Network (BAN), offers an important application in healthcare systems. Yuce [45] presents an implementation of three-layer wireless BAN for collecting medical data from multiple patients and presenting these

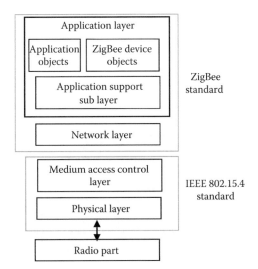

FIGURE 20.8
ZigBee wireless networking protocol layer.

to a centralized station for monitoring. A set of wearable sensors (called nodes) collect the medical signals (mainly ECG, SpO_2, and respiration signal) from the different parts of the patient's body and transfer these to a central coordinator unit (CCU). The CCU repackets and routes the data to a secondary network to the medical expert. In each node, the sensor, signal conditioning and the RF module are integrated in a very lightweight miniature circuit. There can be two different network configurations based on how the patient data are finally transferred to the expert-end computer. In the first configuration (Figure 20.9a), one CCU is dedicated to a single patient, intended for home care or individual patient in medical center. The first layer of the network exists between the sensor nodes and the CCU on the patient's body, called the intra-body network. The CCU delivers the data to the local PC through USB or RS-232 protocol. The sensor nodes communicate in the medical implant communication service (MICS) band with the CCU in either star or mesh configuration. The MICS band requires very low power (25 μW) and operates in the 402–405 MHz range [46,47] up to a distance of 10 m maximum. The analog front ends at nodes draw very low power, typically 40 μA when active and 1 μA in "sleep" modes. The second network exists between the local PC and remote server, which may be connected in a LAN. In another configuration (Figure 20.9b), one CCU is shared by multiple nodes from different patients. This CCU may be either directly connected to the local PC or may use other CCUs through a multihop network to transfer the information to the local PC. For this network the CCU operates in the wireless medical telemetry services (WMTS) band, at 608–614 MHz or ZigBee (IEEE 802.15.4) to transfer the data to a local computer. The second layer of the wireless network can span up to 100 m. The third level of network exists between the local and remote computer. This configuration represented is a three-tier BAN.

The operational complexity of the nodes depends on certain factors, like the number of medical signals being acquired, collection of priority or nonpriority medical data, and, access request mode from the CCU. Since the collected medical signals are of different natures, their sampling frequencies and collection priorities are also different. The CCUs collect data from the nodes by using time division multiple access (TDMA), polling, and contention-based (typically carrier sense multiple access, or CSMA) protocols. The first

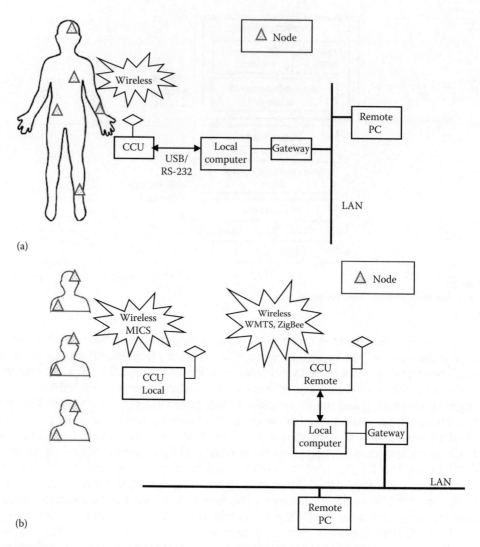

FIGURE 20.9
BAN architecture for patient-monitoring application: (a) for single/dedicated patient monitoring and (b) for multiple patient monitoring.

two types are simple and centralized and offer a fixed delay and frame structure of the packet collected from each node. Additionally, the nodes are operated in sleep mode once these finish transmission of data. The contention-based access approach increases the firmware complexity as well as power consumption in nodes. Here, prior to the transmission, the node checks the status of the communication channel and, if found free, transfers the packet. Otherwise, it waits for a certain time and reschedules the transmission at a later time. Since most of the WBAN devices are battery driven, power optimization is one of the prime criteria. The power consumption in nodes or CCUs can be analyzed from three angles, namely, sensing of medical signals, data processing, and communication. Among these, the communication consumes the maximum energy. The power consumption in the nodes will be low for polling and TDMA technique. The design aspects of BAN performance evaluation in terms of energy optimization, security, protocol, and latency in medical applications are provided by Latre et al. [48] and Khan et al. [49]. A number of ongoing research projects on implementation of WBAN for remote health monitoring indicate that the technology promises to revolutionize healthcare by exploiting low-cost, flexible, and portable systems and easy interfacing with commercial RF modules in the medical ISM band. Inexpensive commercial RF modules are readily available to develop custom-made monitoring applications. Analog Devices [50] and Texas Instruments (TI) [51] are the forerunners in popularizing ultralow-power microcontrollers, RF modules, and integrated sensor assemblies. In many reported works [52–56], CC2420 and CC2500 from TI have been used by researchers as the RF modules. These series of RF modules provide serial peripheral interface (SPI), which is supported by most of the low-power microcontrollers from Atmel (Mega series), TI (MSP series), or Microchip Corporation (PIC series).

Pre- and posthospital checkups of patients using public domain communication networks like GSM, general packet radio service (GPRS), and the Internet can reduce the necessity of the patients' visit to the healthcare center. This practice has become very popular with the expansion and awareness of Internet technology, and the handheld gadgets are empowered with easy to browse connectivity with the World Wide Web (WWW). These systems are primarily designed to handle a single patient at a time. A block schematic of typical architecture is shown in Figure 20.10. The patient's ECG is collected by a hardware acquisition module (HAM), which transfers the data to the local device for buffering, compression, and transmission. The local acquisition device may be a computer, a mobile phone, or a PDA, which accepts data from the HAM through a wired (USB, RS-232) or wireless (Bluetooth) link. The local device communicates the signal to a remote server by using the integrated services digital network (ISDN) telephone line, transmission control protocol (TCP)/internet protocol (IP), GPRS, or GSM communication for expert analysis and feedback. These applications are categorized under noncritical and semicritical conditions of the patient. In noncritical applications, the transfer of patients' data is intermittent and communication with the expert may be prescheduled. A real-time videoconference link with the expert-end physician is often established. The cardiologist can converse with the patient and suggest some therapies. The computational burden on the local-end device microcontroller is limited to

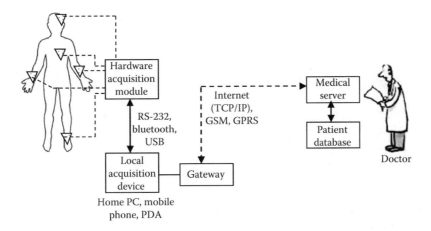

FIGURE 20.10
Block schematic of a remote health-monitoring system using public communication network.

patient data collection, compression, and bidirectional transfer (including the video). A lot of implementation [57–60] can be found in the literature for such noncritical applications. For the semicritical condition of the patients, however, the acquisition is continuous and the local signal-collecting device analyzes the patient's data to check whether any predefined alarm condition has occurred. On such occurrences, alarm messages are generated suitably to alert medical personnel, even relatives of the patient. A short strip of vital parameters (like pulse rate, heart rate [HR] variability, and five to six cycles of ECG) are transmitted to the physician to assess the criticality. Salvador et al. [61] describe a GSM- and Internet-based system for monitoring posthospital checkup of arrhythmia patients by using a mobile phone which used GSM communication to connect with a remote server. A GSM-based 24 h continuous patient-monitoring service is reported by von Wagner et al. [62]. The local acquisition device acquires and processes the single-lead ECG by using fuzzy conditional statements (FCSs) and then transfers the analyzed value to a central base station. An automatic classification of ECG using a remote processing server is described by Bousseljot et al. [63]. The patient ECG is compared with 24,000 prestored and clinically validated ECG data to classify it into one of the four abnormal/normal classes. A MATLAB web server toolbox–based application is described where a client PC can be used to compress and transfer the 12-lead ECG data by using a secured gateway for remote-end processing. The processed results can be directly accessed in the graphical user interface (GUI) on the client PC [64]. The handheld devices like PDA or mobile phones have been used in some applications [65–66] to process the data locally for abnormal beats and transfer the information through GPRS to a remote server for necessary action.

Mobile telecardiology refers to a situation where the patient is in mobile condition, namely, in an ambulance being transferred to a hospital or in a high-altitude aircraft or mid-ocean ship. A dedicated high-bandwidth satellite link is established to transfer the patient's ECG and a few vital signs along with a video link to facilitate a real-time consultation between an emergency medical technician or paramedic and the base station cardiologist. The main design challenges of the successful implementation are wireless channel capacity to allow multiplexed transmission of medical data and video signals, reliability, and latency in transmission. For such applications, the standard communications protocol for computer-assisted electrocardiography (abbreviated as SCP-ECG protocol) [67] was adopted by Association for the Advancement in Medical Instrumentation (AAMI) in the year 1997. The SCP-ECG protocol defines the content and structure of the ECG data packets transferred between two long-distance communication devices. Murakami et al. [68] report a successful implementation of mobile satellite communication for transfer of color image, audio, three-channel ECG, and blood pressure. A moving vehicle (mobile station) carrying a critical patient is set up with a PC-based real-time acquisition system to transfer the data in a 10–100 kbps bandwidth link to a base station (hospital). To ensure the data reliability, automatic repeat request (ARR) was implemented in the error-control frame, which also increased the transmission delay. An ambulance equipped with simultaneous recording of 12-lead ECG and transmission using two different speeds, 2400 and 9600 bps, and conforming to the SCP-ECG protocol is reported by Zywietz et al. [69]. Delay of transmission was assessed based on mobile and static condition of the ambulance car. In another application [70], a portable medical data-collection unit installed inside an ambulance collects the medical signals (ECG, SpO_2, HR, BP, and temperature) and still images of the patient and transfers them through a GSM-based system for real-time consultation over the wireless link. The base station PC provides a GUI-based front end for the cardiologist to interact with the ambulance-end paramedic along with the display of vital parameters and ECG lead plot. The received data are also stored in database files for future use.

20.6 Electronic Health Records, Medical Information System, and Interface with Medical Professionals

Efficient record keeping of patients' medical data is an essential component in telecardiology applications. With the introduction of IT-enabled tools for healthcare information systems, the trend is toward achieving an electronic database system for managing, securing, sharing, and administering patient records. Patient records refer to medical data, personal information, past ailments and corresponding medical data, treatments undergone, and additional information like allergy toward specific drugs and chemicals. Electronic health records (EHRs) can help to achieve this with the following main benefits:

- Better access to patient records: This saves significant time for the physician or other authorized user.

- Improved documentation: Paper-based records are prone to degrade with time and various environmental factors.

- Quality and legibility: Annotations of patient records by the attending specialist doctor helps in the future diagnosis. EHR tools enable an expert to visualize the medical data in different formats and resolutions to have an in-depth study of the patient.

- Savings in administrative and treatment costs: Although the initial setup cost is high, EHR can significantly reduce the administrative costs by efficient file exchange between different sections and communication to external worlds (government agencies and insurance). For patient parties, the burden of treatment cost is also lowered.

- Confidentiality and security: Patients' records are to be safeguarded for their confidentiality and security as per the prevailing information security laws. This imposes assignment of different access levels to the potential users.

- Medical and academic research and development (R&D): The biomedical engineering research community is greatly benefited by the EHRs by using these in their research. The medical science community can benefit from past treatment procedures and patients' responses in certain diseases. All over the world, different medical signal databases are being developed to facilitate researchers' use of these as benchmark databases for testing and validating their experimentations. Some of these databases (like PhysioNet) are open access.

All these lead to an efficient functioning of the entire healthcare system. The conceptual block diagram of a healthcare information system network is shown in Figure 20.11. All patient records are stored at a central server, named electronic patient record (EPR) database. In the context of telecardiology applications, the received medical data are stored in a different format in the database files, normally in compressed form. The cardiologist can access the data from a console terminal from the hospital. The real-time monitoring application necessitates continuous streaming of medical data to the terminal of the cardiologist and background saving of data in the medical server. For off-line application, the cardiologist can access the patient record in his PDA or mobile phone enabled with secured gateway-access connectivity to the medical server. Other users can be administrative sections, etc. In addition to that, message alerts and event monitoring can be configured in the information

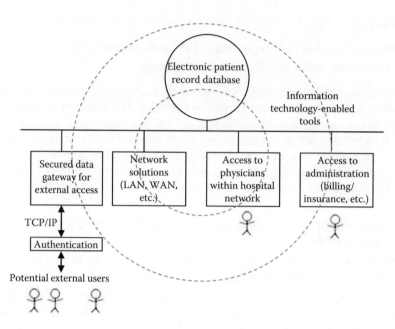

FIGURE 20.11
Healthcare information system network.

technology (IT) software to trigger a message to the cardiologist in critical conditions of the patient. An Internet-based ECG medical information system is described by James et al. [71]. The two main components are a database server and a patient server. The patient server directly acquires ECG from a remote patient module with a built-in network interface card (NIC) using TCP/IP connection. The patient server operates in two modes, namely, transmission (for collecting medical data) and configuration (operation settings, patent information entry, etc.). The acquired medical data are transferred to the central database server which can be accessed by standard web-based front end in real time by technicians, cardiologists, and administrators, each with specific functional attributes. The technician's access is restricted to checking and validating the records, whereas a cardiologist can access present and past records and annotate them. The user interface provides three types of functionalities, namely by (1) searching followed by viewing and annotating and administration functions such as addition of new patients or deletion of existing data, (2) controlling functions, and (3) setup of patient server data security levels. In another web-based telecardiology application, an ECG and echocardiograph data transmission and archiving system is described [72]. The information management system uses a common mailbox to access medical and other patient records by potential users, namely, administrator and cardiologist.

The device-level interface with the medical experts should provide mobile and hassle-free access to the medical data for diagnosis/analysis. Over the last decade, use of handheld gadgets for delivering various services has been significantly improved due to fast access to communication channels, powerful software, and easy operability. In this regard, mobile phones and PDAs have made a very strong presence in the healthcare sector. A recent study [73] revealed that more than 85% of the physicians use PDAs for different purposes, which include drug reference websites, transfer prescriptions, and access to medical data of patients. The study also indicates that many physicians (predominantly cardiologists and pediatricians) closely monitor their patients in ICUs, especially those who are critical, while they are away from the hospital. While accessing patients' medical data, a physician has to pass through an authentication process for privacy and security issues. The PDAs from IBM, Handspring, Symbol, and Psion are the most popularly used in the medical fraternity. The mobile phone is the other popular handheld device which gained popularity due to its direct connectivity, in various formats, with the outside world. Accessibility through GSM, GPRS, and the Internet has provided an opportunity for the cardiologists to access the patient data through the public network.

20.7 Data Security and Privacy Issues in Telecardiology

Under the U.S. Health Insurance Portability and Accountability Act (HIPAA) introduced in 1996, confidentiality of the patients' identity and medical data is of the utmost importance in healthcare. In the telecardiology application, patients' data are transferred through different wireless and public domain networks and finally get archived in the consulting physician–end server. Thus, security levels are required for the wireless transfer mechanism as well as the data files in the EPR data server. Various encryption methods are used for transmission of the compressed ECG packets. AES is one of the latest security protocols used in many wireless networks (WLAN, BAN, ZigBee, etc.). Some other encryption standards are Wi-Fi protected access (WPA), ad hoc on-demand distance vector (AODV), etc. For the 3G mobile platform, the subscriber identity module (SIM) and removable user identity module (RUIM) are the current authentication procedures.

Some literature is available on additional security level using joint encoding and encryption for secured ECG transmission over wireless media [74–78]. In Sufi and Khalil's work [74], the ECG samples are encoded using delta modulation to convert the encoded elements into 8-bit American Standard Code for Information Interchange (ASCII) characters. A permutation cipher is used to further encrypt the ASCII characters. In Sufi and Khalil's work [75], the detected principal ECG features (P, QRS, and T waves) are extracted and mixed with the corresponding noisy templates. The "modified" waves are sent instead of the original wave. In the receiving side, the physician with the template keys are able to extract (decrypt) only the ECG. Sufi et al. [76] describe a wavelet decomposition and selective (significant) coefficient encryption using a symmetric algorithm.

In telecardiology systems, the patients' records are stored in a central server, which is vulnerable to different kinds of attacks. Encryption and time stamping is performed while storing the patients' records to the central database. Without the decryption key, it is not possible to access the data. To provide confidentiality without hampering the usability to the authorized user, certain security policies are enforced, some of which are as follows:

- Access control: The data users are patients themselves (or an authenticated relative), treating cardiologist, refereeing physician(s), and administration only. For each of these users, there is a predefined level of access and modification rights defined by rules for opening and accessing the patients' records.

- Consent and notification: The patients should provide an informed consent to the users who are given the access right to his/her medical record.

- Attribution and audit: Each and every read, delete, and write operation on a patient's record is attributed in a separate database to audit the access level from a potential user.

20.8 Medical Signal Analysis in Telecardiology Systems

Analysis of medical data for pathological feature extraction aims to support clinical diagnosis by experts. In telecardiology applications, the received data packets at the expert-end are tested with error-check mechanisms for their reliability, followed by the decompression program. The extracted medical data are then stored in database files and a graphical time plane presentation is generated on the receiving-end computer, mobile phone, PDA, etc., with the cardiologist to facilitate visual analysis. In addition, prime clinical signatures are extracted from the reconstructed data. In the context of an ECG signal, the clinical signatures of interest are R-R interval (heart rate); P-wave duration and height; T-wave duration and height; QRS width; PR, ST, and QT segments; QT intervals; wave morphologies such as positivity (or negativity) of T wave; ST elevation; etc. For PPG signals the positions of peaks and dicrotic notch may be of interest. A set of useful clinical signatures can be obtained from PPG analysis, like pulse transit time, arterial blood pressure, SpO_2 level, and heart rate.

Medical data analysis using digital computers is quite a mature subject and a plethora of algorithms are already available in the literature. The speed of a today's mobile phone processors is high enough to get a real-time computation of all the mentioned clinical signatures from a set of multilead ECG and PPG within a few seconds. Figure 20.12 represents the typical stages of ECG feature extraction using time-domain analysis. The first stage (preprocessing) discards the artifacts and enhances the zone of interest. R peak (or the QRS complex) is the most prominent and highlighting segment in an ECG beat. Hence, many of the ECG feature extraction algorithms start with detection of R peaks and subsequent exploration of other wave features in their neighborhood by using "window search."

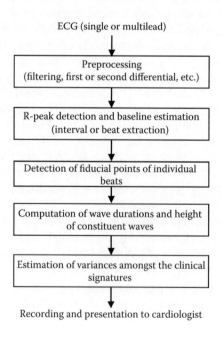

FIGURE 20.12
ECG feature extraction stages.

Hence, accurate detection of R peak is of the utmost importance in ECG feature extraction. The performance indices of R-peak detection by an algorithm are described by parameters like sensitivity, predictivity, and specificity [79], where the parameters refer to the location of R peaks, their correct detection (true positive) or misdetection (false negative), etc. A comprehensive review of software-based QRS detection methods is available in Kohler et al.'s paper [80]. QRS detection software can be broadly classified to adopt one of the following categories:

- Digital filtering and differentiation [81–83]
- Template matching [84–86]
- Wavelet and nonlinear transform [87–89]
- Neural network and genetic algorithm [90–92]

The objective of the feature extraction algorithms is to delineate the onset and offset points of the individual waves from an ECG waveform for computation of clinical signatures. Feature extraction methods from medical signals are broadly classified under the three following categories:

- Time-domain methods [93–94]
- Transform-domain analysis [95–96]
- Intelligent methods based on principal component analysis (PCA), neural network, and support vector machine [97–98]

The time-domain approaches are computationally simpler and require the least buffering of the processing device. The transform-domain approaches aim to reveal the ECG peaks, in the transform-domain coefficients, and exploit the nonstationarity nature of the signal. Intelligence-based approaches are computationally the most expensive but provide high accuracy in feature extraction tasks.

In the context of emergency telecardiology applications, note that a significant time is consumed in the error-check mechanism followed by decompression of the received data, before the cardiologist can access the medical signal. Hence, efficiency in cardiovascular diagnosis is more related to faster extraction of clinical signatures from medical signals. Recent work on detecting abnormalities from the compressed ECG has opened up a new area in telemonitoring applications. Detecting abnormalities from the compressed ECG has the following important advantages:

- Less memory required to handle the entire data buffer
- Minimum input–output (IO) read operations by the processor
- Less internal resource allocation for the processing device

A few publications on analysis of compressed ECG are available in the literature [99–101]. In general there are two approaches to detecting abnormalities from compressed ECG, namely, direct approach and intelligent technique. In direct approaches, the character patterns of a normal person's ECG are ascertained. The deviation from this pattern due to a pre-assessed abnormality generates a new pattern. The occurrence of this new pattern, its recurrence frequency (instant detection), is one way to detect abnormality under the direct method. However, the success of this method greatly depends on training of the algorithm with the patient's ECG and the detection is highly person specific. Also, detecting more than one abnormality requires more intensive training and application of classification techniques on the processor. Hence, these techniques are generally applied on the mobile phone of patient and only the outcome is reported to the healthcare unit attendant. The key idea of the intelligent agent–based approach is to extract "attributes" from the compressed ECG character blocks, followed by a minimization [96] of these attributes which correspond to certain predefined diseases. PCA, neural networks, and clustering techniques are used for feature optimization, training, and zone of interest section.

In the multiparameter-sensing scenario, the real challenge is to collate the extracted features from the individual medical signals to arrive at a semantic level of information. In prevailing practices, a cardiologist makes the diagnostic decision based on clinical features from individual cardiac and hemodynamic medical signals. Data fusion techniques from heterogeneous medical sensors can be useful in reducing the number of pathological signatures [102–103] and provide a guide to the treating physician. Kannathal et al. [104] describe an implementation of multisensor information fusion using four ICU health parameters, namely, ECG, BP, SpO_2, and respiration signal. A competitive integration approach followed by a fuzzy logic–based decision rule generation is described. In the first stage, the competitive integration provides a refined data on a particular measurement as well as reliability of sensor data (sensor discrepancy). A combined patient deterioration index is computed from the fuzzy logic decision functions. However, the computational complexity in multisensor systems is high and needs a high-end processor for centralized computation.

20.9 Trends in Telecardiology Systems

Considering the current trend in the increased number of cardiovascular diseases all over the world, telecardiology practice will continue to be one of the major modes of healthcare-delivery services in the future. The major technological revolutions that are expected to make the most impact on telecardiology systems are the following:

- Development of smaller, low-power, smart, wearable sensors: These will allow normal activity of a cardiac patient with minimum discomfort. The effort is toward achieving an integrated circuit containing the sensors, electronics, and transmitter part within a small area that can be fixed on the vest of a patient. The sensors will pick up ECG, BP, blood oxygen content, and temperature from the patient. Different energy-scavenging techniques are already in use to power this tiny module from the body temperature or ambient sources.

- Secured wireless communication: Parameters like power, transmission range, and bandwidth are the most critical factors in the choice of communication media for healthcare applications. For personal healthcare, Wi-Fi (WLAN), Bluetooth, and ZigBee are already implemented for short-range monitoring, each having its own security and authentication mechanism. An important factor will be connectivity between these short ranges with public communication networks like GSM, CDMA and GPRS, and the Internet for carrying medical signals. Hence, security and confidentiality issues regarding patient data will be more important, since in today's context the physical boundaries between private and public networks no longer exist. Considering the performance and prices offered by commercial wireless devices, it has not always been possible to achieve an absolutely secured environment in wireless healthcare applications. A lot of research is currently being undertaken by network developers and healthcare policy makers to develop newer security standards in portable devices.

- Computing in medical telematics: Today's healthcare systems employ complex computing tasks in each node through which the medical information passes, from the patient's body to the medical server. In each device, the information content is modified to cater to the source and destination requirements in terms of timely delivery, authentication, compression, and feature extraction, in addition to its own functional maintenance (power optimization, etc.). To reduce the burden of communication and computing, a new paradigm of computing, namely, "context-aware sensing," is growing up, where the medical signal-collecting device computes the data for abnormality detection (case specific and limited) and the abnormality occurrence is transmitted. This relieves the medical server from the computing burden of feature extraction.

  Computing plays an important role in automatic decision-support systems for assisted diagnosis. Complex algorithms based on artificial intelligence (AI) techniques have already been developed to produce results equally consistent with cardiologists' diagnoses. The clinical acceptability of these algorithms is gradually gaining popularity. This has enhanced the large-scale usability of the algorithms.

- Interactive platforms: Extending the concept of anytime, anywhere data access has prompted the use of handheld devices for fast and secured access of medical data. Mobile phones and PDAs are the most popular of handheld devices being adopted by physicians for patients' data access. Research is going on to develop more powerful, user-friendly software platform in mobile devices and data visualization tools are getting more user friendly for interactive and flexible ways to access data.

In this chapter, the main technological areas related to the telecardiology applications are described in brief. In today's context the main challenges are interoperability between the systems and secured protocols for efficient data exchange. The origin of remote healthcare initially started with challenges which were more technological in nature. Some of the technologies are nearing their saturation level of development. Now the challenges are shifting more toward human engineering and societal factors. From a socioeconomic point of view, the challenge is cost of service delivery and affordability. In developing nations the telecardiology service is still confined to only a very low fraction of the total population. Awareness building among the parties (patients, doctors, and healthcare policy makers) may provide the benefit of this technology for fullest utilization toward mankind in the future.

Acknowledgments

The author would like to acknowledge Dr. Samarjit Sengupta, Dr. Madhuchhanda Mitra, and Dr. Jitendranath Bera, faculty members of Department of Applied Physics, University of Calcutta, India.

References

1. N.F. Guler and E.D. Ubeyli, "Theory and applications of biotelemetry," *Jour. of Medical Systems*, Vol. 26, No. 2, pp. 159–178, April 2002.
2. R.S. Mackay, *Bio-medical telemetry: Sensing and transmitting biological information from animals and man*, IEEE Press, Piscataway, New Jersey, 1993.
3. B. Fong, A.C.M. Fong, and C.K. Li, *Telemedicine technologies*, John Wiley & Sons, West Sussex, U.K., 2011.
4. http://www.americantelemed.org/, accessed on August 16, 2014.
5. P. Sleight and B. Casadei, "Relationship between heart rate, respiration and blood pressure variabilities," *Heart rate variability*, Futura Publishing, Mount Kisco, New York, pp. 311–327, 1995.
6. R. Gupta, M. Mitra, and J.N. Bera, *ECG acquisition and automated signal processing*, Springer India, New Delhi, India, pp. 51–60, 2014.
7. AAMI Standards and Recommended Practices, Medical electrical equipment—Part 1: General requirements of basic safety and essential performance. Available at http://my.aami.org, accessed on August 16, 2014.
8. G.D. Clifford, F. Azuaje, and P.E. McSharry (Eds.), *Advanced methods and tools for ECG data analysis*, Artech House, Inc., Norwood, Massachusetts, pp. 41–50, 2006.
9. J.D. Bronzino (Ed.), *The biomedical engineering handbook*, Second Ed., CRC Press LLC, Boca Raton, Florida, 2000.
10. J.G. Webster (Ed.), *Medical Instrumentation application and design*, Fourth Ed., John Wiley & Sons, Hoboken, New Jersey 2010.
11. J. Allen, "Plethysmography and its application in clinical physiological measurement," *Physiol. Meas.*, Vol. 28, No. 3, pp. R1–R39, 2007.
12. E. Tur, M. Tur, H.I. Maibach, and R.H. Guy, "Basal perfusion of the cutaneous microcirculation: Measurements as a function of anatomic position," *Jour. of Investigative Dermatology*, Vol. 81, No. 5, pp. 442–446, 1983.
13. J.G. Webster (Ed.), *Design of pulse oximeters*, IOP Publishing, Bristol, U.K., 1997.
14. S. Rithalia, M. Sun, and R. Jones, "Blood pressure measurement," *Measurement, instrumentation, and sensors handbook: Electromagnetic, optical, radiation, chemical, and biomedical measurement*, J. Webster and H. Eren (Eds.), Second Ed., CRC Press, Boca Raton, Florida, pp. 65.1–65.24, 2014.
15. T. Towaya, T. Tamura, and P.A. Oberg, *Biomedical sensors and instruments*, Second Ed., CRC Press, Boca Raton, Florida, 2011.
16. B. Singh, A. Kaur, and J. Singh. "A review of ECG data compression techniques," *International Jour. of Comp. Appl.*, Vol. 116, No. 11, pp. 39–44, 2015.
17. S.M.S. Jalaleddine, C.G. Hutchens, R.D. Strattan, and W.A. Coberly, "ECG data compression techniques—A unified approach," *IEEE Trans. Biomed. Eng.*, Vol. 37, No. 4, pp. 329–343, April 1990.
18. L.W. Gardenhire, "ECG compression for biomedical telemetry," *Biomedical telemetry*, C. A. Caceres (Ed.), Academic Press, New York, Chapter 11, 1965.
19. L.W. Gardenhire, "Redundancy reduction—The key to adaptive telemetry," *Proc. 1964 National Telemetry Conf.*, pp. 1–16, 1964.
20. H. K. Wolf, J. Sherwood, and P. M. Rautaharju, "Digital transmission of electrocardiograms—A new approach," *Proc. Fourth Canadian Medicine and Biology Conf.*, pp. 39a–39b, 1972.
21. D. Stewart, G. E. Dower, and O. Suranyi, "An ECG compression code," *Jour. of Electrocardiology*, Vol. 6, No. 2, pp. 175–176, 1973.
22. V. Kumar, S.C. Saxena, and V.K. Giri, "Direct data compression of ECG signal for telemedicine," *International Jour. of Systems Science*, Vol. 37, No. 1, pp. 45–63, January 2006.
23. P.K. Kulkarni, V. Kumar, and H.K. Verma, "Direct data compression techniques for ECG signals: Effect of sampling frequency on performance," *International Jour. of Systems Science*, Vol. 28, No. 3, pp. 217–228, 1997.

24. B. Furht and A. Perez, "An adaptive real-time ECG compression algorithm with variable threshold," *IEEE Trans. Biomed. Eng.*, Vol. 35, No. 6, pp. 489–494, June 1988.

25. M. Mitra, J.N. Bera, and R. Gupta, "Electrocardiogram compression technique for global system of mobile-based offline telecardiology application for rural clinics in India," *IET Science, Measurement & Technology*, Vol. 6, No. 6, pp. 412–419, November 2012.

26. R. Gupta and M. Mitra, "An ECG compression technique for telecardiology application," *Annual IEEE India Conference (INDICON 2011)*, Hyderabad, India, pp. 1–4, December 2011.

27. P.S. Hamilton and W.J. Tompkins, "Compression of the ambulatory ECG by average beat subtraction and residual differencing," *IEEE Trans. Biomed. Eng.*, Vol. 38, No. 3, pp. 253–259, March 1991.

28. G. Strang and T. Nguyen, *Wavelets and filter banks*, Wellesley-Cambridge Press, Wellesley, Massachusetts, 1996.

29. B.S. Kim, S.K. Yoo, and M.H. Lee, "Wavelet-based low-delay ECG data compression algorithm for continuous ECG transmission," *IEEE Trans. Inf. Tech. in Biomed.*, Vol. 10, No. 1, pp. 77–83, January 2006.

30. A. Alesanco, S. Olmos, R.S.H. Istepanian, and J. Garcia, "Enhanced real-time ECG coder for packetized telecardiology applications," *IEEE Trans. Inf. Tech in Biomed.*, Vol. 10, No. 2, pp. 229–236, April 2006.

31. A. Alesanco, J. Garcia, S. Olmos, and R.S.H. Istepanian, "Resilient ECG wavelet coding for wireless real-time telecardiology applications," *M-Health emerging mobile health systems*, R.H. Istepanian, S. Laxminarayan, and C.S. Pattichis (Eds.), Springer, New York, New York, pp. 293–312, 2006.

32. M.S. Manikandan and S. Dandapat, "Wavelet threshold based ECG compression using USZZQ and Huffman coding of DSM," *Biomed. Sig. Proc. Control*, Vol. 1, No. 4, pp. 261–270, October 2006.

33. B.A. Rajoub, "An efficient coding algorithm for the compression of ECG signals using the wavelet transform," *IEEE Trans. Biomed. Eng.*, Vol. 49, No. 4, pp. 355–362, April 2002.

34. V.A. Allen and J. Belina, "ECG data compression using the discrete cosine transform (DCT)," *Proc. of Computers in Cardiology*, Durham, North Carolina, pp. 687–690, October 1992.

35. S. Lee, J. Kim, and J.H. Lee, "A real-time ECG data compression and transmission algorithm for an e-health device," *IEEE Trans. Biomed. Eng.*, Vol. 58, No. 9, pp. 2448–2455, September 2011.

36. I.M. Rezazadeh, S. Parvaresh, M.E.M.E. Zargar, and J. Proulx, "ECG data compression for mobile phone tele-cardiology applications using .NET framework," First Middle East Conference on Biomedical Engineering (MECBME), Sharjah, United Arab Emirates, pp. 204–207, February 2011.

37. L.V. Batista, E.U. Melcher, and L.C. Carvalho, "Compression of ECG signals by optimized quantization of the discrete cosine transform coefficients," *Medical Eng. Phy.*, Vol. 23, No. 2, pp. 127–134, March 2001.

38. Y. Zigel, A. Cohen, and A. Katz, "The weighted diagnostic distortion (WDD) measure for ECG signal compression," *IEEE Trans. Biomed. Eng.*, Vol. 47, No. 11, pp. 1422–1430, November 2000.

39. S. Manikandan and S. Dandapat, "Wavelet energy based diagnostic distortion measure for ECG," *Biomed. Signal. Proc. Control*, Vol. 2, No. 2, pp. 80–96, April 2007.

40. K.A. Reddy, B. George, and V.J. Kumar, "Use of Fourier series analysis for motion artifact reduction and data compression of photoplethysmographic signals," *IEEE Tran. in Inst. Meas.*, Vol. 58, No. 5, pp. 1706–1711, May 2009.

41. K.S. Chong, K.B. Gan, E. Zahedi, and M.A.M. Ali, "Data compression technique for high resolution wireless photoplethysmograph recording system," *2013 IEEE International Conference on Space Science and Communication (IconSpace)*, Melaka, Malaysia, pp. 345–349, July 2013.

42. R. Gupta, "Lossless compression technique for real time photoplethysmographic measurements," *IEEE Tran. Inst. Meas.* (accepted for publication), doi: 10.1109/TIM.2014.2362837.

43. H. Labiod, H. Afifi, and C. De. Santis, *Wi-Fi, Bluetooth, ZigBee and WiMAX*, Springer, Dordrecht, Netherlands, 2007.

44. S. Farahani, *ZigBee wireless networks and transceivers*, Newnes, Elsevier, Burlington, Massachusetts, 2008.
45. M.R. Yuce, "Implementation of wireless body area networks for healthcare," *Sensors and Actuators A: Physical*, Vol. 162, No. 1, pp. 116–129, July 2010.
46. FCC Rules and Regulations, "MICS Band Plan," Table of Frequency Allocations, Part 95, January 2003.
47. S. Hanna, "Regulations and standards for wireless medical applications," Third International Symposium on Medical Information & Communication Technology, Montreal, Canada, pp. 1–5, February 2009.
48. B. Latre, B. Braem, I. Moerman, C. Blondia, and P. Demeester, "A survey on wireless body area networks," *Wireless Networks*, Vol. 17, No. 1, pp. 1–18, January 2011.
49. J.Y. Khan, M.R. Yuce, G. Bugler, and B. Harding, "Wireless body area networks (WBAN) design techniques and performance evaluation," *Jour. of Medical Syst.*, Vol. 36, No. 3, pp. 1441–1457, June 2012.
50. Analog Devices: http://www.analog.com, accessed on August 16, 2014.
51. Texas Instruments: http://www.ti.com, accessed on August 16, 2014.
52. A. Milenkovic, C. Otto, and E. Jovanov, "Wireless sensor networks for personal health monitoring: Issues and an implementation," *Computer Comm.*, Vol. 29, Nos. 13–14, pp. 2521–2533, August 2006.
53. Y.C. Du, Y.Y. Lee, Y.Y. Lu, C.H. Lin, M.J. Wu, C.L. Chen, and T. Chen, "Development of a telecare system based on ZigBee mesh network for monitoring blood pressure of patients with hemodialysis in health care centers," *Jour. of Med. Syst.*, Vol. 35, No. 5, pp. 877–883, October 2011.
54. S.K. Chen, T. Kao, C.T. Chan, C.N. Huang, C.Y. Chiang, C.Y. Lai, T.H. Tung, and P.C. Wang, "A reliable transmission protocol for ZigBee based wireless patient monitoring," *IEEE Trans. on Inf. Tech. in Biomed.*, Vol. 16, No. 1, pp. 6–16, January 2012.
55. B. Gyselinckx, J. Penders, and R. Vullers, "Potential and challenges of body area networks for cardiac monitoring," *Jour. of Electrocardiology*, Vol. 40, No. 6, Suppl. 1, pp. S165–S168, November–December 2007.
56. L. Boquete, J.M.R. Ascariz, J. Cantos, R. Barea, J.M. Miguel, S. Ortega, and N. Peixoto, "A portable wireless biometric multi-channel system," *Measurement*, Vol. 45, No. 6, pp. 1587–1598, July 2012.
57. Y.H. Nam, Z. Halm, Y.J. Chee, and K.S. Park, "Development of remote diagnosis system integrating digital telemetry for medicine," *Proc. 20th IEEE International Conference of the EMBS*, Vol. 3, pp. 1170–1173, November 1988.
58. A. Hernandez, F. Mora, G. Villegas, G. Passariello, and G. Carrault, "Real-time ECG transmission via Internet for non-clinical application," *IEEE Trans. Inf. Tech. in Biomed.*, Vol. 5, No. 3, pp. 253–257, September 2001.
59. Y. Jasemian, E. Toft, and L. Arendt-Nielsen, "Real-time remote monitoring cardiac patients at distance," Second Open ECG Workshop, Berlin, Germany, pp. 48–50, April 2004.
60. S. Bonho, D. Kolm, J.F.R. Baggio, and R. Moraes, "Microprocessor-based system to ECG monitoring through Internet," *World Congress on Medical Physics and Biomedical Engineering 2006*, IFMBE Proceedings, Vol. 14, pp. 4008–4011, 2007.
61. C.H. Salvador, M.P. Carrasco, M.A.G. de-Mingo, A.M. Carrero, J.M. Montes, L.S. Martín, M.A. Cavero, I.F. Lozano, and J.L. Monteagudo, "Airmed-cardio: A GSM and Internet services-based system for out-of-hospital follow-up of cardiac patients," *IEEE Trans. Inf. Tech. in Biomed.*, Vol. 9, No. 1, pp. 73–85, March 2005.
62. R.G. von Wagner, S. Schubert, L.F. Ngambia, C. Morgenstern, and A. Bolz, "Concept for an event-triggered electrocardiographic telemetry-system using GSM for supervision of cardiac patients," *IEEE/AFCEA Conference on Information Systems for Enhanced Public Safety and Security (EUROCOMM 2000)*, pp. 374–377, May 2000.
63. R. Bousseljot, U. Grieger, D. Kreiseler, and L. Schmitz, "Internet-based ECG-evaluation and follow-up," Second Open ECG Workshop, Berlin, Germany, pp. 53–54, April 2004.

64. J. García, I. Martínez, L. Sörnmo, S. Olmos, A. Mur, and P. Laguna, "Remote processing server for ECG-based clinical diagnosis support," *IEEE Trans. Inf. Tech. in Biomed.*, Vol. 6, No. 4, pp. 277–284, December 2002.

65. J. Rodríguez, L. Dranca, A. Goñi, and A. Illarramendi, "A wireless application that monitors ECG signals on-line: Architecture and performance," *Enterprise Information Systems*, Vol. VI, pp. 267–274, 2006.

66. C. Wen, M.F. Yeh, K.C. Chang, and R.G. Lee, "Real-time ECG telemonitoring system design with mobile phone platform," *Measurement*, Vol. 41, No. 4, pp. 463–470, May 2008.

67. AAMI SCP-ECG Protocol: http://www.san-ei.com/officemedic/AAMI.SCPECG.v1.2.pdf, accessed on August 16, 2014.

68. H. Murakami, K. Shimuju, K. Yamamoto, T. Mikami, N. Hishimaya, and K. Kondo, "Telemedicine using mobile satellite communication," *IEEE Trans. Biomed. Eng.*, Vol. 41, No. 5, pp. 488–497, May 1994.

69. C. Zywietz, V. Mertins, D. Assanelli, and C. Malossi, "Digital ECG transmission from ambulance cars with application of the European Standard Communications Protocol SCP-ECG," *Proc. of Computers in Cardiology*, Hannover, Germany, pp. 341–344, September 1994.

70. S. Pavlopoulos, E. Kyriacou, A. Berler, S. Dembeyiotis, and D. Koutsouris, "A novel emergency telemedicine system based on wireless communication technology—AMBULANCE," *IEEE Trans. Inf. Tech. in Biomed.*, Vol. 2, No. 4, pp. 261–267, December 1998.

71. D.A. James, D. Rowlands, R. Mahnovetski, J. Channells, and T. Cutmore, "Internet based ECG medical information system," *Australasian Phy. & Eng. Sc. Med.*, Vol. 26, No. 1, pp. 25–29, November 2003.

72. C. Costa and J.L. Oliveira, "Telecardiology through ubiquitous Internet services," *Int. Jour. of Medical Informatics*, Vol. 81, No. 9, pp. 612–621, September 2012.

73. K. Beaver (Ed.), *Healthcare information systems*, Second Ed., *Best Practices Series*, Auerbach, CRC Press, Boca Raton, Florida, 2003.

74. F. Sufi and I. Khalil, "Enforcing secured ECG transmission for real time telemonitoring: A joint encoding, compression, encryption mechanism," *Security and Comm. Networks*, Vol. 1, No. 5, pp. 389–405, 2008.

75. F. Sufi and I. Khalil, "A new feature detection mechanism and its application in secured ECG transmission with noise masking," *Jour. of Medical Systems*, Vol. 33, No. 2, pp. 121–132, April 2009.

76. F. Sufi, S. Mahmoud, and I. Khalil, "A new ECG obfuscation method: A joint feature extraction and corruption approach," *Proc. International Conference on Technology and Applications in Biomedicine, 2008 (ITAB 2008)*, Shenzhen, China, pp. 334–337, May 2008.

77. F. Sufi, S. Mahmoud, and I. Khalil, "A novel wavelet packet based anti spoofing technique to secure ECG data," *International Jour. of Biometrics*, Vol. 1, No. 2, pp. 191–208, 2008.

78. F. Sufi, S. Mahmoud, and I. Khalil, "A wavelet based secured ECG distribution technique for patient centric approach," *Fifth International Summer School and Symposium on Medical Devices and Biosensors*, ISSS-MDBS 2008, Hong Kong, China, pp. 301–304, June 2008.

79. R.M. Rangayyan, *Biomedical signal analysis: A case study approach*, Wiley, Singapore, 2002.

80. B.U. Kohler, C. Hennig, and R. Orglmeister, "Principles of software QRS detection," *IEEE Eng. in Med. Biology*, Vol. 21, No. 1, pp. 42–57, January–February 2002.

81. M. Okada, "A digital filter for the QRS complex detection," *IEEE Tran. on Biomed. Eng.*, Vol. BME-26, No. 12, pp. 700–703, December 1979.

82. A. Ligtenberg and M. Kunt, "A robust-digital QRS detection algorithm for arrhythmia monitoring," *Computers Biomed. Research*, Vol. 16, No. 3, pp. 273–286, June 1983.

83. J. Pan and W. J. Tompkins, "A real time QRS detection algorithm," *IEEE Tran. Biomed. Eng.*, Vol. BME-32, No. 3, pp. 230–236, March 1985.

84. P.E. Trahanias, "An approach to QRS detection using mathematical morphology," *IEEE Trans. on Biomed. Eng.*, Vol. 40, No. 2, pp. 201–205, February 1993.

85. A. Ghaffari, H. Golbayani, and M. Ghasemi, "A new mathematical based QRS detector using continuous wavelet transform," *Computers Elect. Eng.*, Vol. 34, No. 2, pp. 81–91, March 2008.

86. Md. S. Islam and N. Alajlan, "A morphology alignment method for resampled heartbeat signals," *Biomedical Signal Proc. Control*, Vol. 8, No. 3, pp. 315–324, May 2013.

87. S. Suppappola and Y. Sun, "Nonlinear transforms of ECG signals for digital QRS detection: A quantitative analysis," *IEEE Trans. on Biomed. Eng.*, Vol. 41, No. 4, pp. 397–400, April 1994.

88. D.S. Benitez, P.A. Gaydecki, A. Zaidi, and A.P. Fitzpatrick, "A new QRS detection algorithm based on the Hilbert transform," *Proc. Computers in Cardiology 2000*, Cambridge, Massachusetts, pp. 379–382, September 2000.

89. S. Kadambe, R. Murray, and G. B. Bartels, "Wavelet transform-based QRS complex detector," *IEEE Trans. on Biomed. Eng.*, Vol. 46, No. 7, pp. 838–848, July 1999.

90. K. Zhu, P.D. Noakes, and A.D.P. Green, "ECG monitoring with artificial neural networks," *Proc. Second International Conference on Artificial Neural Networks*, Colchester, U.K., pp. 205–209, November 1991.

91. Q. Xue, Y.H. Hu, and W.J. Tompkins, "Neural-network-based adaptive matched filtering for QRS detection," *IEEE Trans. Biomed. Eng.*, Vol. 39, No. 4, pp. 317–329, April 1992.

92. R. Poli, S. Cagnoni, and G. Valli, "Genetic design of optimum linear and nonlinear QRS detectors," *IEEE Tran. on Biomed. Eng.*, Vol. 42, No. 11, pp. 1137–1141, November 1995.

93. E.B. Mazomenos, T. Chen, A. Acharyya, A. Bhattacharya, J. Rosengarten, and K. Maharatna, "A time-domain morphology and gradient based algorithm for ECG feature extraction," *Proc. IEEE International Conference on Industrial Technology (ICIT)*, Athens, Greece, pp. 117–122, March 2012.

94. H.K. Chatterjee, R. Gupta, and M. Mitra, "A statistical approach for determination of time plane features from digitized ECG," *Computers in Biology Med.*, Vol. 41, No. 5, pp. 278–284, May 2011.

95. C. Li, C. Zheng, and C. Tai, "Detection of ECG characteristic points using wavelet transform," *IEEE Trans. Biomed. Eng.*, Vol. 42, No. 1, pp. 21–28, January 1995.

96. D. Benitez, P.A. Gaydecki, A. Zaidi, and A.P. Fitzpatrick, "The use of the Hilbert transform in ECG signal analysis," *Computers in Biology and Med.*, Vol. 31, pp. 399–406, September 2001.

97. R.J. Martis, C. Chakraborty, and A.K. Roy, "An integrated ECG feature extraction scheme using PCA and wavelet transform," *Proc. Annual IEEE India Conference (INDICON)*, Gujarat, India, pp. 1–4, December 2009.

98. R. Ghongade and A.A. Ghatol, "Performance analysis of feature extraction schemes for artificial neural network based ECG classification," *Proc. International Conference on Computational Intelligence and Multimedia Applications*, Tamil Nadu, India, Vol. 2, pp. 486–490, December 2007.

99. F. Sufi and I. Khalil, "Diagnosis of cardiovascular abnormalities from compressed ECG: A data mining based approach," *IEEE Trans. Inf. Tech. in Biomed.*, Vol. 15, No. 1, pp. 33–39, January 2011.

100. F. Sufi, Q. Fang, I. Khalil, and S. Mahmoud, "Novel methods of faster cardiovascular diagnosis in wireless telecardiology," *IEEE Jour. on Selected Areas in Comm.*, Vol. 27, No. 4, pp. 537–552, May 2009.

101. A. Abida, I. Khalil, and F. Sufi, "Cardiac abnormalities detection from compressed ECG in wireless telemonitoring using principal components analysis (PCA)," *Proc. International Conference on Intelligent Sensors, Sensor Networks and Information Processing 2009 (ISSNIP 2009)*, Melbourne, Australia, pp. 207–212, December 2009.

102. W.T. Sung and K.Y. Chang, "Evidence-based multi-sensor information fusion for remote healthcare systems," *Sensors and Actuators A: Physical*, Vol. 204, pp. 1–19, December 2013.

103. A.I. Hernández, G. Carrault, F. Mora, L. Thoraval, G. Passariello, and J.M. Schleich, "Multisensor fusion for atrial and ventricular activity detection in coronary care monitoring," *IEEE Trans. Biomed. Eng.*, Vol. 46, No. 10, pp. 1186–1190, October 1999.

104. N. Kannathal, U.R. Acharya, E.Y.K. Ng, S.M. Krishnan, L.C. Min, and S. Lxminarayan, "Cardiac health diagnosis using data fusion of cardiovascular and heamodynamic signals," *Computer Methods Prog. Biomed.*, Vol. 82, No. 2, pp. 87–96, May 2006.

Further Reading

Bronzino J.D. (Ed.), *The biomedical engineering handbook*, Second Ed., CRC Press LLC, Boca Raton, Florida, 2000.

Criste B.L., *Introduction to biomedical instrumentation: The technology of patient care*, Cambridge University Press, Cambridge, U.K., 2009.

Sörnmo L. and P. Laguna, *Bioelectrical signal processing in cardiac and neurological applications*, Elsevier Academic Press, Waltham, Massachusetts, 2005.

Tze D., H. Lai, R. Begg, and M. Palaniswami, *Healthcare sensor networks: Challenges towards practical implementation*, CRC Press, Boca Raton, Florida, 2012.

Webster J.G. (Ed.), *Medical instrumentation: Application and design*, Fourth Ed., John Wiley & Sons, Hoboken, New Jersey, 2010.

Xiao Y. and H. Chen (Eds.), *Mobile telemedicine: A computing and networking perspective*, CRC Press, Boca Raton, Florida, 2008.

Yang G.Z. (Ed.), *Body sensor networks*, Springer Verlag, London, 2006.

21

Telecardiology Tools and Devices

Axel Müller, Jörg Otto Schwab, Christian Zugck,
Johannes Schweizer, and Thomas M. Helms

CONTENTS

21.1 Introduction

Heart and circulatory diseases are the leading cause of death in industrialized countries. Due to the demographic development of increasing numbers of elderly and polymorbid patients, heart and circulatory diseases are becoming more important. Major advances in cardiologic diagnosis and treatment have been achieved (e.g., pacemaker [PM] and implantable cardioverter/defibrillator [ICD] devices, coronary angiography, and drugs). Thus, the treatment of coronary heart disease by catheter interventions and bypass surgery has significantly improved over the past decades. However, this also raises the costs for the healthcare system. New diagnostic and therapeutic methods require a higher specialization of the medical staff. Thus, a challenge is posed by the demand for ensuring countrywide coverage of highly specialized cardiac care. Considering this as well as the enormous developments in information and communication technology in the last two decades, telemedicine poses additional possibilities in the diagnosis and treatment of patients with heart and circulatory diseases, especially in the area of outpatient care. The use and application of telemedicine can also be extended to the prevention and rehabilitation of cardiac patients.

This chapter will introduce telemedical concepts on the diagnosis of cardiac arrhythmias and monitoring of patients with cardiovascular implantable electronic devices (CIEDs) and chronic heart failure as well as on the monitoring of patients with arterial hypertension. For these indications, various telemedicine basic concepts exist and a range of experiences have been made. In addition, telemedical concepts on coagulation management and monitoring of patients with valvular heart disease, prosthetic heart valves, or heart transplants have been developed.

21.2 Telemedical ECG Monitoring

The ECG is of particular significance in cardiology. First of all, it is an important tool in the diagnosis of cardiac arrhythmias. Secondly, it has a significant place in the detection of acute coronary syndromes, in particular the ST segment elevation in myocardial infarction. The most important indication in telemonitoring of patients with cardiac arrhythmias is the detection of symptomatic (sinus tachycardia, atrioventricular [AV] reentrant tachycardia, and premature ventricular contractions [PVCs]), asymptomatic, or slightly symptomatic arrhythmias (e.g., paroxysmal atrial fibrillation) and the clarification of some uncharacteristic symptoms such as palpitations, tachycardia, dizziness, transient weakness, nausea, or transient hot flashes that suggest a correlation with the occurrence of cardiac arrhythmias. In clinical practice, the detection of atrial fibrillation is crucial. Approximately 1.5%–2% of the population suffers from atrial fibrillation. Atrial fibrillation is associated with a fivefold increased risk of stroke and a threefold increased risk of the development of heart failure [1].

Using the tele-ECG, the sensitivity could be improved in the diagnosis of atrial fibrillation in emergency patients [2]. Further indications include the monitoring of patients after therapeutic measures, such as cardioversion, catheter ablation, or medications. Patients can be supported by telemedicine in self-management (e.g., pill-in-the-pocket concept or anticoagulation for patients with atrial fibrillation). Thus, unnecessary emergency consultations or hospitalizations can be avoided [3–5].

Also, monitoring of patients in clinical trials (e.g., in patients with atrial fibrillation) by means of periodic or event-triggered transmission of tele-ECGs is possible [6–7].

The telephone transmission of an ECG dates back to the early days of ECG diagnostics. The pioneer of ECG diagnosis, Willem Einthoven, already submitted an ECG from the hospital to his laboratory 1.5 km away [8].

The goal of telemedical ECG monitoring is to record an ECG at any place and time and the possibility of transmitting it to the supervising physician immediately. This requires that the patient is able to record an ECG himself or herself or that it is recorded and transmitted automatically. The transmission of tele-ECGs to the hospital, the general practitioner (GP), the cardiologist, or a telemedical service center where it is evaluated occurs via fax or via the Internet (e-mail).

The tele-ECG can be derived by direct skin contact of metal electrodes or adhesive electrodes. Recordings of 1-lead, 2-lead, 3-lead, and modified 12-lead tele-ECG are possible. In the tele-ECG, measurements (RR intervals, PQ interval, QRS width, and QT interval) are possible. The tele-ECGs can be transmitted and stored asynchronously or timely (synchronous). The tele-ECG can be transmitted continuously or event triggered. The routes of transmission are analog (frequency modulated), via infrared interfaces, via Bluetooth, or directly via mobile phone. For the transfer, fixed and mobile networks can be used. Recent studies also test NFC for transferring data to a smartphone [9].

Different manufacturers offer various devices. The devices can be classified into the following groups of devices:

- ECG event recorders (1- or 12-lead ECG)
- External loop recorders (1-, 2-, or 3-lead ECG)
- Implantable loop recorders (e.g., Reveal with CareLink system, Medtronic; Housecall Plus, St. Jude Medical Company; and BioMonitor with Home Monitoring system, Biotronik)

For recording, storage, transmission, and evaluation of tele-ECGs, different concepts exist as presented in Figure 21.1.

When using ECG event recorders, tele-ECGs are recorded by the patients themselves. The prerequisite is that the patient is symptomatic and can operate the ECG-monitoring system himself or herself. To obtain a symptom–ECG correlation, the clinical symptoms (e.g., palpitations, rapid heartbeat, dizziness, temporary weakness, nausea, or temporary hot flashes) must last long enough to record a tele-ECG. The advantages of ECG-monitoring cards with one-channel recording are the compact design (check card) and ease of use. This allows them to be worn by the patient continuously. The recording of the ECG via metal electrodes is straightforward (Figure 21.2). It can be transmitted to a landline phone or a mobile phone. In addition to the acoustic signal transmission (analog or digital), transmission via infrared or Bluetooth is possible.

In the diagnosis of cardiac arrhythmia one- to three-lead ECG recordings are sufficient (Figure 21.3).

The use of the ECG-monitoring card compared to the 24 h Holter monitoring was more effective in patients with palpitations. The diagnosis was made faster and it was, therefore, more cost effective [10]. Under certain conditions ST segment changes can be evaluated with the modified 12-channel tele-ECG. Schwaab et al. [11] were able to demonstrate a high correlation of ST segment changes when comparing 12-lead ECGs to self-recorded 12-lead tele-ECGs.

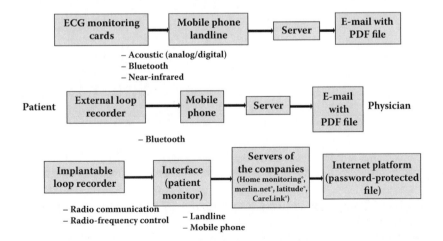

FIGURE 21.1
Concept of recording, storing, transmitting, and evaluating tele-ECGs.

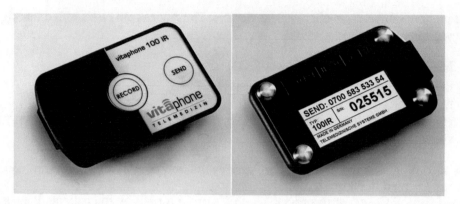

FIGURE 21.2
Front side and backside of an ECG-monitoring card (100 IR; company vitaphone, Germany) for the recording of a one-lead tele-ECGs.

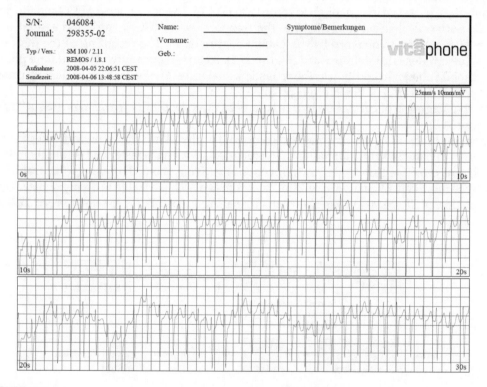

FIGURE 21.3
With an ECG-monitoring card IR 100 (company vitaphone, Germany) recorded one-lead ECG—evidence of a supraventricular tachycardia with a rate of 210 heartbeats/min in a patient with intermittent palpitations.

The analysis of ST segment changes (in particular ST segment elevation) is used in the workup of angina-like complaints or in the diagnosis of an acute coronary syndrome. Using transtelephonic transmission of a modified 12-lead tele-ECG ST segment elevation, myocardial infarction be diagnosed safely and quickly [12]. This method is used to ensure early intervention for patients with serious coronary problems. In addition, telemedical monitoring with the ability to transfer a modified 12-lead ECG can be used in the follow-up care after an acute coronary syndrome. The acceptance of this method by patients was shown to be high in a study of Katalinic et al. [13]. However, most patients have waited too long after the onset of symptoms (angina pectoris) until they transmitted an ECG. Only 23% of telemedically monitored patients contacted the service center within the first hour after the onset of symptom [13]. This shows that in addition to the use of technical equipment, the education and training of patients is critical to the success of telemedical monitoring concepts.

External loop recorders can record one-, two-, or three-lead tele-ECGs. The transfer can occur analogously via telephone (audible) or via Bluetooth. Many devices have integrated algorithms for automatic ECG analysis (e.g., bradycardia, tachycardia, asystole, atrial fibrillation, and ventricular tachycardia, in addition to the possibility of manual activation by the patient). This allows for the automated recording and transmission of events in asymptomatic patients. An additional advantage of loop recorders is in the recording of the ECG sequences before the occurrence of an event. Through the tele-ECG transmission, the period between the beginning of the recording and the provision of the diagnosis could be significantly shortened. In a study by Leshem-Rubinow et al. [14], with 604 patients, the mean time between the tele-ECG transmissions to the definitive ECG assessment took 16 h. Through the use of telemedicine, the provision of the diagnosis in patients with palpitations, syncope, or chest pain was made much faster and targeted treatment began sooner [14].

Another difficulty posed is the clarification of presyncope or syncope symptoms. Since these symptoms represent a variety of mechanisms and clinical pictures, in addition to the 12-lead standard ECG and a 24 h Holter ECG, further diagnostic measures are indicated [15].

ECG event recorders are probably less suitable here, because the patient must be able to activate the system himself or herself. An alternative could be implantable loop recorders.

The indications are analogous to the external loop recorders. Implantable loop recorders are particularly suitable for repeated syncope with secondary injuries. Even though newer implantable systems have a compact design, the disadvantage of these systems, the need for surgical implantation, remains. A problem of the external or internal loop recorder is the specificity and sensitivity of automatically transmitted episodes. The automatic detection rates of atrial fibrillation of external loop recorders reached a sensitivity of 100% and a specificity of 50%. This leads to a considerable number of false-positive detections [16].

The problem of the high percentage of false-positive findings can be found in implantable loop recorders as well. Here the potential problem is that due to the limited storage capacity of the recorder, relevant events may be overwritten and, therefore, cannot be evaluated. Through telemonitoring, the events are transferred immediately or in the intervals to free up storage capacity relatively quickly. Furukawa et al. [17] were able to achieve an average period of 28 days between the implantation of loop recorders in patients with syncope or palpitations and the first relevant event detected through the use of telemonitoring in combination with an implantable device. In clinical practice, the implantable recorders are routinely read every 3 months. On the other hand, reprogramming of the device (e.g., due to small amplitudes) was possible early on. This allowed for the reduction of the recording of false positive events [17].

In addition, a "filter function" by a service center is possible by the evaluation of automatically transmitted tele-ECG episodes at the service center. This can reduce the number of false positives as well [18]. A combination of event recorder with emergency options (emergency call button and GP location) was developed with the Heart Handy concept. This emergency management through bidirectional communication between physician and patient is possible. In emergency situation, this can reduce the time factor [19].

Today, many ECG-monitoring systems are used in conjunction with smartphones. The data are transmitted via Bluetooth. The GPS function is integrated in the smartphone. Through special apps (e.g., with emergency dispatch alarming), emergency management will be possible in the future.

Other new concepts combine ECG recording and transmission by integrating a mobile phone card in the system. This makes the application even simpler for the patient to use. Despite the positive experience with the use of tele-ECG, especially in the diagnosis of cardiac arrhythmias, one must be made aware of its limitations. One problem is the increased number of artifact recordings. This causes a large number of recorded ECGs to be transmitted, in particular, automatically recorded events. ECG event recorders and loop recorders are not suitable for emergency patients. This means that tele-ECG monitoring cannot be used on patients with vital risks. Another problem will be the separation of medical products from products for the health and fitness sector. Various manufacturers will integrate monitoring functions (e.g., ECG and pulse recording) into smartphones, and software (apps) is offered in this sector. A clear distinction in terms of indication and certification as a medical device will become necessary here.

In summary, from today's medical perspective the following requirements can now be defined for equipment for telemedical ECG monitoring:

- Easy and safe handling (creating, recording, and transmission), especially for older people
- Compact design of the devices (light weight and small size)
- Avoiding of adhesive electrodes for long-term use; in the future, possible integration of the electrodes into the clothing
- Transmission of data via smartphones or integration of the transmission unit in the system
- Low power consumption for longer battery life
- High sensitivity for the automatic detection of arrhythmias (e.g., bradycardia, tachycardia, asystole, atrial fibrillation, and ventricular tachycardia) with a reasonable specificity (few false positives)
- Secure encryption of data in transit
- Approval as a medical device

21.3 Telemonitoring in Patients with CIEDs

21.3.1 Introduction

For over 50 years, pacemakers have been used for treatment of bradycardic arrhythmias. Since that time, tremendous advances have been made in technology development (such as dual-chamber pacemakers, rate adaptation, and Holter functions). In 1980, the first ICD was implanted for secondary prevention of sudden cardiac death. Meanwhile, technological developments have led to reduced aggregate sizes and the capability of uncomplicated transvenous implantation. According to the results of the Multicenter Automatic Defibrillator Implantation Trial (MADIT) II study that was published in 2002, the indication for ICD implantation was extended to include primary prevention of sudden cardiac death in patients with coronary heart disease [20]. ICD implantation was able to improve survival rates in these patients. In the '90s of the last century, cardiac resynchronization therapy (CRT) was introduced for treatment of chronic heart failure. Using CRT, improvement of heart failure-related symptoms was achieved in patients with broad chamber complexes and poor left ventricular pump function [21].

There are currently established indications defined in the guidelines for treatment with antibradycardia pacemakers, ICDs, and CRTs (CIEDs) [22–24].

In recent years, there has been a steady rise in the number of implantations of pacemakers, ICDs, and CRT systems. Since 2001, a sharp increase in the numbers of ICD and CRT implantations has been observed in the United States. Treatment with CIEDs is increasingly performed in elderly and multimorbid patients [25].

Despite improvements in surgical and instrumentation techniques, complications must be anticipated. Thus, in a study that included 440 ICD patients, Alter et al. [26] were able to determine a complication rate during a 46 (plus/minus 36) month follow-up period of 31%. This was related to perioperative complications (10%), inappropriate shock deliveries (12%), ICD lead-related complications (12%), and aggregate-related complications (6%) [26]. In an analysis by the U.S. FDA, the annual rates of ICDs that were explanted due to technical defects were 1.4 to 9.0 per 1000 pacemaker implantations and 7.9 to 38.6 per 1000 ICD implantations. While the number of pacemaker malfunctions in the period from 1990 to 2002 decreased, the rate for ICDs increased. The rate of device-associated malfunction was significantly higher for ICD systems [27].

In this context, the problems experienced in recent years with Sprint Fidelis (from Medtronic) and Riata (from St. Jude Medical) ICD leads will come to mind [28].

From the problems and developments presented, the requirement arises for safe and effective monitoring of patients with CIEDs. Here, telemonitoring can make a significant contribution.

With regard to the method of telemonitoring, a distinction is made between remote follow-up and remote monitoring [29–31].

With remote follow-up, data (such as data on aggregate status, impedances, or mode switch episodes) are transferred at defined time points from the implant to the doctor. The initialization of the data transfer may be provided automatically or by the patients themselves.

Remote monitoring allows regular (for example, daily) automatic transmission of predetermined data from the implant to the doctor, independent of any action by the patient. Data transmitted are not only data relating to system health (for example, battery status, impedances, and thresholds) but also the clinical data (such as heart rate, atrial fibrillation burden, patient activity, and treatment administrations). However, remote monitoring does not mean continuous monitoring of patients.

21.3.2 Telemonitoring Methods in Patients with CIEDs

Initial studies on transtelephonic querying of bradycardia pacemakers were already performed in the mid-'70s and '80s of the last century [32]. The CareLink 2090 system (Medtronic) and the Housecall system (St. Jude Medical) for the first time made systems available that could remotely query pacemakers or ICDs in the sense of remote follow-up. With the CareLink system, a connection between a programmable device and a computer at a center was established by a telephone line. This made it possible to overcome the physical distance between two different examiners. Consultation without active intervention in the programming was thus possible. The company St. Jude Medical developed the Housecall system for data transfer from the ICD to the doctor. The new feature was that patients themselves transmitted the data from home via a special access device. Determining thresholds and ICD programming is not possible.

In the 1990s, the company Biotronik began developing its Home Monitoring technology. The first pacemakers with Home Monitoring were implanted in 2000. Data transfer is automatic, that is, without the assistance of the patient, by the CIED to a transmission device (CardioMessenger). The CardioMessenger, in turn, transfers the data via mobile communication to a technical service center. There, the data are processed and transmitted to the doctor by SMS text messaging or by fax. In addition, the doctor can view patient data via a password-protected Internet platform. In the meantime, all CIED manufacturers (BIOTRONIK, Medtronic, St. Jude Medical, Boston Scientific, and the Sorin Group) have introduced systems for telemedical monitoring of patients. The systems differ in terms of data transmission from the implant to an interface (patient monitor).

An Internet-based presentation to the doctor is now standard. In addition, manufacturers have implemented additional features, especially for the monitoring of patients with ICDs and CRT systems (Table 21.1).

All systems are based on the same technical principle with the following individual components: the implant (with active or passive data transfer), the patient monitor (the interface for receiving data from the device and forwarding to mobile or landline phone), data transmission (a fixed network or mobile), the manufacturer's data server unit (the technical service center), and the Internet platform for data presentation to the physician (Figure 21.4).

The Home Monitoring system is the system that has been used the longest in this form in clinical practice. The devices are equipped with an antenna in the header.

Data transfer from the implant to the CardioMessenger is provided via ultralow-power active medical implants (ULP-AMIs) operating in the 402–405 MHz band, which is worldwide standardized; its terms of use are laid down in relevant standards. The devices' reduction of battery life by the energy consumption required for data transmission is minimal (about 1%–2% of battery capacity over its entire lifetime). The implant sends data at a time specified by the doctor (usually at night when the patient is asleep) to the CardioMessenger on a daily basis. From there, the data are automatically sent to the technical service center by mobile communication. The doctor can view the data on a password-protected Internet platform. By using the different colors corresponding to traffic light, the importance of the information received is visualized (Figure 21.5).

The doctor can individually define certain events that he or she would like, in addition, to be informed of by SMS text messaging, e-mail, or fax (event reports).

A major advantage of the Home Monitoring system is that the data transfer is done completely automatically, without the need for the patient to actively participate. Furthermore, the system has the option of defining patient-specific data transmission and the transmission of IEGM and other diagnostic data (Heart Failure Monitor) with ICD and CRT systems.

A disadvantage appears to be that only devices with built-in transmitter units can be telemedically monitored with this system.

The CareLink system from Medtronic was a further development of the Remote View system. With the help of a patient monitor (the CareLink Monitor), the patient can retrieve data from the device (pacemakers, ICDs, or CRT systems). Here, the implants do not require special transmitter or receiver units, so that patients with older devices can be monitored as well. From the CareLink Monitor, the data are automatically transmitted by landline telephone to the data server and can be retrieved by the doctor via an Internet platform.

TABLE 21.1

Overview of Various Systems for Telemonitoring of Patients with CIEDs

System	Home Monitoring	CareLink	Merlin.net	Latitude	SmartView
Manufacturer	Biotronik	Medtronic	St. Jude Medical	Boston Scientific	Sorin Group
Implementation of the system	2001 (Germany)	2002 (United States)	2007 (United States)	2005 (United States)	2013 (United States)
Devices	PM, ICD, and CRT	PM, ICD, and CRT (backward compatible)	Specific ICD and CRT	Specific ICD and CRT	Specific ICD and CRT
Interface	CardioMessenger	CareLink-monitor	Merlin@home	Latitude Communicator	SmartView monitor
Data transmission from device to interface	Automatic	Manual and automatic	Manual and automatic	Manual and automatic	Manual and automatic
Data transfer	Daily or in case of events	Fixed dates or in case of events	Fixed dates or in case of events	Fixed dates or in case of events	Fixed dates or in case of events
Transfer to data server	GSM and GPRS	Telephone line	Telephone line and GSM	Telephone line and GSM	Telephone line and GSM
Data presentation to the physician	Internet and alerts transmission by SMS, e-mail, and fax	Internet and alerts transmission by SMS and fax	Internet and alerts transmission by SMS, e-mail, fax, and smartphone	Internet and alerts transmission by fax	Internet and alerts transmission by SMS, e-mail, and fax
Integration into the EHR	Possible	Possible	Possible	Possible	Possible
Availability	At least worldwide	At least worldwide (analog line/connection)	At least worldwide	At least worldwide	At least worldwide
Specifics	Heart Failure Monitor IEGM-Online transmission	OptiLink system (intrathoracic impedance measurement), intracardiac electrocardiogram (IEGM), and Cardiac Compass	Holistic data management system and line transmission	Integration of external sensors (weight scale and blood pressure monitor) possible	Automatic optimization of CRT

Source: Müller, A. et al., *Kardiologe*, 7, 181–193, 2013.

This allows transfer of information on system and diagnostic data (Cardiac Compass) or on events and queried IEGMs.

Another option for telemedical monitoring is the OptiLink system. The OptiLink system consists of the two components OptiVol and CareLink. OptiVol was developed for monitoring and early alerting of cardiac decompensation risks in patients with CRT systems. The system uses intrathoracic impedance measurements to evaluate the fluid content of the lungs. This makes possible earlier detection of cardiac decompensation in this high-risk group of patients [35].

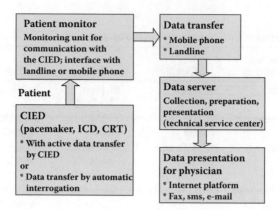

FIGURE 21.4
Schematic representation of a telemedical monitoring system for patients with CIEDs. (Adapted from Müller, A. et al., Remote monitoring in patients with pacemakers and implantable cardioverter-defibrillators: New perspectives for complex therapy management, in *Modern Pacemakers—Present and Future*, M. Kumar, Ed., InTech–Open Access Publisher, pp. 147–166, 2011.)

The Housecall Plus system was developed as a remote-monitoring system and follow-up system for query of ICDs and CRT-ICDs in the home. First, a telephone contact with the patient is made. Then the patient retrieves the data from the ICD via a patient access device (Housecall Plus transmitter) and transmits them over the phone to the doctor. This allows a timely control of the ICD with query of IEGMs and analysis of the current intracardiac ECG.

The Housecall Plus system from St. Jude Medical was developed further into the telemedical patient data management system Merlin.net. The core of the system is the monitor Merlin@home, which stays with the patient. The daily transfer of data from the devices is enabled by RF telemetry to the Merlin@home transmitter and from there via telephone line to the Internet-connected Merlin.net server. All RF telemetry-capable devices are compatible with Merlin@home. The doctor can view the transmitted data on an appropriately protected Internet platform. New in this system is the possibility of integrating the transmitted data into an electronic patient record, which allows comprehensive patient management. The data obtained during outpatient follow-up care can also be transferred to this system. In addition, there is a hotline that supports both the patient and the doctor in the case of technical questions. Through an automated system, the patient can receive a reminder of an impending transfer, information on a missed telemedical follow-up visit, confirmation of a successful data transmission, and notification of necessary contact with the clinic. In cases of defined events, the doctor is notified by e-mail, SMS text messaging, or fax.

Experience with Boston Scientific's Latitude system is mainly from the United States. The concept integrates remote interrogation of ICD and CRT systems (remote follow-up), telemedical monitoring, and heart failure management. In addition, patient-initiated queries are possible. The Latitude communicator is used as patient monitor. Data from external devices, such as from a weight scale or a blood pressure monitor, can be transferred to the Latitude communicator via wireless Bluetooth. The data are forwarded to the data server by landline telephone. The data are provided to the doctor on an Internet platform. Several doctors can have access to the patient data.

An advantage of the Latitude system is the ability to integrate external devices such as a weight scale and a blood pressure monitor.

FIGURE 21.5
Password-protected Internet platform of the Home Monitoring system (by the Biotronik company in Germany) and representation of the patient and the alarms in the different colors corresponding to traffic lights.

The SmartView system from the company Sorin Group has only recently been introduced to clinical practice. It allows telemedical monitoring of the company's latest generation of ICDs and CRT systems.

21.3.3 Clinical Data for Telemedical Monitoring of Patients with CIEDs

Telemonitoring of patients with CIEDs can be divided into the following four groups: device management, arrhythmia management, heart failure management, and patient-based management (Figure 21.6).

The first clinical trials on telemonitoring of patients with CIEDs were conducted to address issues of technical feasibility and safety of data transmission and to address questions related to improving treatment management and patient safety. Casuistry studies have described the efficiency of a telemedical monitoring including IEGM transmission with respect to the detection of technical problems (twiddler's syndrome) or misclassified ventricular tachycardia in ICD patients [36–37].

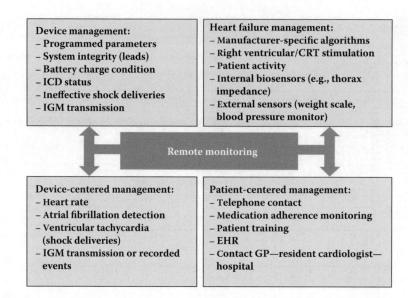

FIGURE 21.6
Various aspects of telemonitoring of patients with CIEDs.

The safety of patients with CIEDs can be improved by the timely detection of device and electrode malfunction (device management). These are rare events and in the early stages are often clinically asymptomatic [38]. In the Lumos-T Safely Reduces Routine Office Device Follow-up (TRUST) study with 1339 patients, it was shown that early detection of such failures is possible through telemonitoring. The malfunctions were detected in the TRUST trial in the telemonitoring group after 1 day on average, compared to 5 days in the control group [39]. By telemedical monitoring of ICD patients, inappropriate shocks can be identified and reduced [40]. Among patients in whom an ICD is implanted, shocks, appropriate or inappropriate, always represent a major problem, as they are associated with poor prognosis [41].

In patients with CIEDs, paroxysmal atrial fibrillation can be detected early by telemonitoring and, thus, may minimize serious complications such as ischemic stroke [42]. The detection of atrial and ventricular arrhythmias can be supported by IEGM transmitted by telemonitoring. The Reliability of IEGM Online Interpretation (RIONI) study showed that the IEGM transmitted by telemonitoring is concordant with a conventional 30 s IEGM [43].

In patients with chronic heart failure, early detection of cardiac decompensation is crucial. Clinical symptoms do not occur until relatively late. To estimate the compensation status, information on various device-specific parameters can be used telemetrically.

The following parameters have been determined as predictors for the occurrence of cardiovascular events or impending cardiac decompensation: the occurrence of atrial tachyarrhythmias, increased average heart rate, decreased heart-rate variability, reduced patient activity detected by integrated accelerometer, and the increased occurrence of ventricular extrasystolic heartbeats, and/or sustained ventricular tachycardias. In CRT systems, data on the function of the left ventricular electrode (stimulation percentage and stimulus threshold) deliver important information about the effectiveness of the electrical therapy for heart failure [44]. In two nonrandomized studies with a total of 1071 patients, Sack [45] and Whellan [46] were able to develop diagnostic algorithms for early detection of cardiac decompensation in patients with CRT systems by using the above parameters.

These algorithms predict cardiac decompensation at 4.99-fold or 7.15-fold probability and allow an early start of treatment [45–46].

Initial results of the Monitoring Resynchronization Devices and Cardiac Patients (MORE-CARE) study in patients after CRT implantation for moderate or severe chronic heart failure show that the time period between the device-detected events and clinical decision is considerably shorter with telemonitoring. In the telemonitoring group, the in-hospital visits were significantly reduced [47]. In the recently published study "Influence of Home Monitoring on the clinical management of heart failure patients with impaired left ventricular function (IN-TIME)," Home Monitoring was able to significantly reduce the number of patients with deteriorated clinical status and the total and cardiovascular mortality (by over 50%) compared to those for normal treatment of CRT patients [48].

In addition to the possibility of improved patient safety, early detection of relevant clinical events, and the personalized outpatient follow-up intervals, a reduction in cost for the health system and the patient is another feature of the use of telemonitoring.

In the Remote Follow-up for ICD Therapy in Patients Meeting MADIT II Criteria (REFORM) study, an annual outpatient follow-up of 155 patients who were provided with an ICD for primary prevention was compared to remote monitoring with the usual quarterly follow-up. Compared to the standard group (3-month follow-up), the remote monitoring group showed a slight increase in unscheduled follow-ups (0.64 versus 0.27 per patient-year, $p = 0.03$) and a significant reduction in the total number of ICD follow-ups (1.6 versus 3.85 per patient-year, $p < 0.001$). There were no significant differences in mortality, hospitalization rate, or length of hospitalizations between the two groups [49]. The results of the REFORM study show that with telemonitoring, the planned follow-up intervals can be made more personalized than in patients with CIEDs. This can become a cost saving.

The benefits of telemonitoring in clinical practice are improved patient safety (aggregate and electrode malfunction), detection of clinically relevant events (atrial and ventricular arrhythmias), early detection of cardiac decompensation in heart failure patients with ICD or CRT systems, and personalized follow-up intervals. This also makes cost savings appear potentially possible.

Telemonitoring cannot replace clinical observation of the course of structural heart disease and therapy monitoring in the affected patients. Telemonitoring is also not an emergency system.

Problems with telemonitoring in patients with CIEDs may result from the large number of transmitted data sets. In a database analysis of Lazarus [38] in 11,624 patients with pacemakers, ICDs, and CRT systems, 3,004,763 data transfers occurred. 47.6% of patients had no events [38]. In order to achieve an efficient analysis and selection of the relevant data, approaches with specially trained nurses were developed. The nurses make a preselection of clinically relevant events so that the doctor has to examine special reports only [50]. An alternative for patient monitoring is the integration of a special telemedical service center alongside (monitoring by the) implant centers and the resident cardiologist. In this way, telemonitoring opens up the opportunity for networked follow-up between the general physicians, the resident cardiologists, and the hospitals.

For all patients with implants, telemonitoring is available as an option and should be taken advantage of in routine clinical practice. Because of the state of current studies, the use of telemonitoring is recommended, in particular in patients with ICD and CRT systems. This is in context of current national and international guidelines of various professional societies [24,33,51]. In its current guideline for telemonitoring in patients with CIEDs, the European Society of Cardiology put forth a class of recommendation IIA with a level of evidence A [24].

21.4 Telemonitoring in Patients with Chronic Heart Failure

21.4.1 Goals of Telemonitoring in Patients with Chronic Heart Failure

Chronic heart failure is increasingly becoming a challenge for health systems in Western industrialized countries. For one reason, it results from the frequency of chronic cardiac and circulatory diseases (such as coronary and hypertensive heart disease), which are among the most common causes of chronic heart failure, and for another reason, from the demographic progression towards a high proportion of old and very old people who are more frequently affected by heart failure. The prognosis of chronic heart failure is dependent on the stage, and, despite treatment progress, with higher New York Heart Association (NYHA) stages it is worse than for many cancers [52–53].

The patients are threatened by sudden cardiac death and the frequent occurrence of recurrent cardiac decompensation. The professional societies have defined different treatment goals for patients with chronic heart failure. These include improving clinical symptoms and reducing morbidity and mortality. The reduction in morbidity can be measured primarily by the reduction in hospitalization rates [54–55]. Repeated hospitalizations are also a major cost factor in the treatment of patients with chronic heart failure.

The situation can be improved for patients with chronic heart failure by guideline-based treatment [56–57]. Causes of the lack of adherence to guidelines can be attributed to doctors (lack of knowledge of the guidelines) and to the patients (lack of compliance, e.g., with taking medication, and lack of knowledge of behavior or self-management). Therefore, the aim of special care programs for patients with chronic heart failure must be both the implementation of guideline-based therapies and the compensation for shortcomings.

21.4.2 Telemonitoring Method in Patients with Chronic Heart Failure

In order to improve the situation of patients with chronic heart failure, structured treatment programs were established. The aims of the programs are to implement guideline-based diagnosis and treatment, to connect primary care and specialist medical services, and to reduce costs by decreasing the number of hospital stays. To this end, different methodological approaches may be applied. Aside from the classic ambulatory care concept between the general practitioner and the cardiologist, new concepts, such as care by heart failure nurses or telemonitoring, have been established in recent years.

21.4.2.1 Support Concept with Heart Failure Nurses

The concept of care for chronic heart failure patients by specially trained nurses (heart failure nurses) consists mainly of telephonic compliance management and patient education. Heart failure nurses regularly contact the patient by phone. Home visits are also possible. In addition to reviewing the current patient status (symptoms of cardiac decompensation) and their taking of medication, the patients are educated on selected topics (causes and symptoms of heart failure, taking medication, diet, fluid intake, physical activity, etc.). Ideally, care is already initiated during the hospital stay. As part of the interdisciplinary network of Würzburg (Würzburg HeartNetCare-HF), a complex nurse-based telephone monitoring and telephone training was established. Although in a randomized study with 715 patients this approach did not lead to reduction in mortality or number of hospitalization stays, an improvement of NYHA classification and the psychometric parameters was achieved in the group with a nurse-coordinated disease management program, resulting in improvement of the quality of life in these patients [58].

21.4.2.2 Care Concepts with Telemonitoring

Telemonitoring in patients with chronic heart failure was described in the Trans-European Network–Home-Care Management System (TEN-HMS) study [59]. Selected vital parameters (body weight, ECG, heart rate, blood pressure, thoracic impedance, respiratory rate, oxygen saturation, etc.) are regularly measured by the patients at home with external sensors. Then, the data are transtelephonically transmitted by a modem to the data server. Lastly, the care team (care manager, and physician) evaluates the data, and a telephone consultation with the patient is performed where needed [59].

Figure 21.7 shows the complex telemedical care concept that was used in the TEN-HMS study.

Favorable systems proved to be those that automatically transmit vital parameters (for example, using Bluetooth via a modem) to the telemedical service center. In the service center, the data are integrated in an electronic medical record (EHR) (Figure 21.8).

With automatic information processing systems, alarms can be triggered when values fall below or above individually set thresholds. Usually, when alarms are triggered (for example, in case of exceeding weight), the service center contacts the patient by phone. With the help of a structured telephone interview, the patient's current condition can be assessed and appropriate measures can be initiated (such as presentation to the family doctor or cardiologist). Aside from using external sensors (ECG, scales, and blood pressure monitor), telemonitoring can also be conducted in patients with CIEDs in the form of device-based remote monitoring (Figure 21.8).

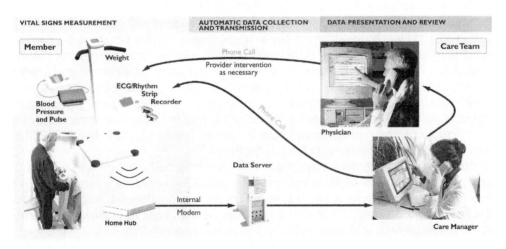

FIGURE 21.7
Complex telemedical care concept in the TEN-HMS study with devices for detection of various vital parameters in patients, data collection and transmission, and evaluation. (From Cleland, J.G.F. et al., *J Am Coll Cardiol*, 45, 1654–1664, 2005. With permission.)

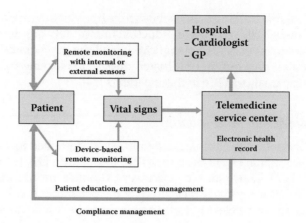

FIGURE 21.8
Schematic representation of telemonitoring components in patients with chronic heart failure.

In addition to monitoring vital signs, the data from the telemedical service center are also made available to the attending physicians and specialists, as well as to the clinics. This way, the support networking approach connects the individual participants. The telemedical service center also enables compliance management (such as checking intake of medication) and training programs for the patients (Figure 21.8). In some cases, emergency care management is provided with available appropriate equipment at the service center (staffed 24 h/365 days).

21.4.3 Current Study Status

Cleland et al. [59] comprehensively developed the concept of telemonitoring in patients with chronic heart failure in the 2005 TEN-HMS study.

The study was designed as a three-arm prospective randomized multicenter study. The inclusion criteria were hospitalization due to worsening of chronic heart failure, persistent symptoms of heart failure, an ejection fraction <40%, and treatment requiring diuretics. A total of 426 patients were included in the study. After randomization, 85 patients were assigned to the standard-care group, 173 patients to the nurse telephone support group, and 168 patients to the home telemonitoring group. Patients in the nurse telephone support group were contacted once a month by phone by a specially trained nurse regarding their clinical symptoms and current medication. In case of problems, the general practitioner was consulted. The patients could also call the nurse [59].

In contrast, a comprehensive telemonitoring was established for the home telemonitoring group that included weight, blood pressure, and ECG monitoring (Figure 21.7). Vital signs were transmitted twice a day to a service center. When values fell below or exceeded defined thresholds, an automatic alarm was triggered. In case of problems, immediate intervention or contacting the general practitioner was possible [59].

Patients in the nurse telephone support and in the home telemonitoring groups had lower heart failure-related mortalities and fewer hospitalizations than patients receiving standard treatment. There was no significant difference between the nurse telephone support and the home monitoring groups, but the hospital stays were 6 days shorter in the home monitoring group, so the patient could be discharged faster to outpatient care [59].

Similar results were obtained in other studies, such as the Home-HF and the Home or Hospital in Heart Failure (HHH) study [60–61]. The use of telemonitoring mainly reduced the frequency and duration of hospitalizations and emergency room visits for cardiac decompensation [60].

In two large meta-analyses with 6258 and 8323 patients, reduced mortality and rehospitalization rates were obtained by telemonitoring compared to the standard treatment in patients with chronic heart failure [62–63].

The included studies primarily monitored patients' clinical symptoms and body weight [62]. Today, both of these meta-analyses are assessed rather critically in terms of study selection (combination of randomized trials and cohort studies, patient selection, and definition of *usual care*) [64].

Because of the positive results of the meta-analyses, the results of the 2010 and 2011 published Telemonitoring to Improve Heart Failure Outcomes (Tele-HF) study and the Telemedical Interventional Monitoring in Heart Failure (TIM-HF) study were all the more surprising [65–66].

In the Tele-HF trial, 1653 patients were randomized into two groups after hospital treatment for heart failure (the telemonitoring group with 826 patients and standard treatment group with 827 patients). The primary end point of the study was defined as readmission to the hospital for any (cause) or death within a period of 180 days. For telemonitoring, a commercial system (Tel-Assurance, Pharos Innovations, United States) was used. With the help of an automatic speech-recognition system, questions regarding the health status and symptoms of heart failure were asked daily and regarding depression once a month. The data were evaluated daily on weekdays by a coordinator and, where needed, discussed with a doctor. There was no personal telephone contact with a nurse or a doctor. The primary end point (hospitalization or death from any cause within 180 days) was achieved for 52.3% of patients in the telemonitoring group and for 51.5% of patients in the standard treatment group ($p = 0.75$). There was also no significant difference between the two groups in terms of mortality [65].

In the TIM-HF study, 710 patients with heart failure (NYHA class II or III) and with an ejection fraction below or equal to 35% were randomized into two groups. The first group received standard treatment. The second group was monitored by complex remote telemedical management. Thereby body weight, ECG, and blood pressure were transmitted daily to a telemedical service center through a Bluetooth interface by using a PDA system. The telemedical service center was continuously staffed and monitored by a physician. An emergency system was integrated into the study design. The complex telemonitoring is illustrated in Figure 21.9 [66].

The primary end point in the TIM-HF trial was death from any cause. The secondary end point was a combined end point of cardiovascular death and hospitalization for heart failure. The average follow-up was 26 months (minimum 12 months). No significant difference between the two groups was achieved in terms of both the primary and the secondary end point. Both groups received very good drug treatment for heart failure (angiotensin-converting enzyme [ACE] inhibitors: 96.6% or 94.1%; beta blockers: 92.1% or 93.0%; and aldosterone antagonists: 65.3% or 63.2%) [66].

The results of these two recent large randomized prospective studies on telemonitoring of patients with chronic heart failure seem surprising against the background of the meta-analyses and require commentary. The main problem with the Tele-HF study is the method of telemonitoring. The system using an interactive voice-recognition program without personal telephonic interaction and without transmission of vital signs is inferior to the support approach involving heart failure nurses and does not meet the criteria for

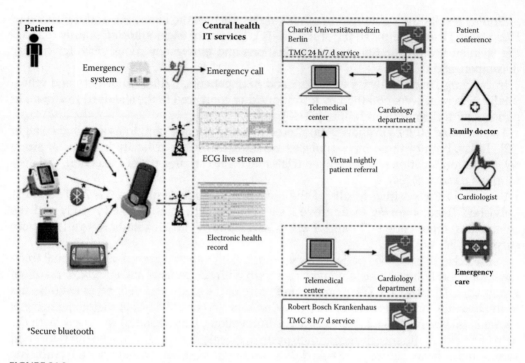

FIGURE 21.9
Telemonitoring approach in the TIM-HF study. (From Koehler, F. et al., *Circulation*, 123, 1873–1880, 2011. With permission.)

telemonitoring in accordance with the TEN-HMS study. In the telemonitoring group in the Tele-HF study, only 85.6% of the enrolled patients used the system only once and 14% of patients in the intervention arm not at all. At the end of the study only 55.3% of patients used the system three times a week [65].

The main problem with the TIM-HF study is primarily with already optimal medical treatment of the enrolled patients, of whom additionally more than 40% were supplied with an ICD. This results in a relatively low mortality rate. However, this does not reflect the reality of care. In addition, subgroup analyses are missing regarding patients who benefit less or more from telemonitoring. Subgroup analyses show that patients with cardiac decompensation or ICD implantation prior to randomization benefit from telemonitoring. In contrast, patients with depression have no benefit [67].

At present it is not yet clear which groups of patients benefit from telemonitoring. It is certain that not all patients benefit from a complex care program. This aspect needs to be worked out in detail in further studies. Thereby, targeted groups of patients (for example, grouped according to NYHA class and comorbidities) who have potential benefit must be identified. Other questions pertain to the duration and intensity of telemonitoring in patients with chronic heart failure. The current study situation cannot take a position on this. Telemonitoring in patients with chronic heart failure has taken quite an eventful path in recent years. From the results of studies and meta-analyses, questions and problems arose that have contributed to the acceptance of telemonitoring by physicians and payers being nonuniform. The recommendations of the European Society of Cardiology currently do not provide recommendations for the use of telemonitoring in patients with chronic heart failure [55].

In Germany, telemonitoring in patients with chronic heart failure is already a reality of care. Most telemedical care programs represent so-called island solutions. However, the final result in terms of medical and health economic benefits remains to be seen [68].

Despite initial additional expenses, telemonitoring appears to allow medium- and long-term cost reduction in the care of patients with chronic heart failure [69–70].

Alternative sensors represent another approach for future solutions. So far, external sensors (body weight scales, ECG, and blood pressure monitor) were mainly used in patients without CIEDs. As part of the Multisensor Monitoring in the Congestive Heart Failure (MUSIC) study, a noninvasive monitoring system was used. This allows monitoring of various vital signs (heart rate, respiratory rate, body impedance, position, and activity). By linking the data, an algorithm that predicts cardiac decompensation could be determined. The goal is to improve the specificity and sensitivity of monitoring systems [71–72].

Alternatively, implantable internal sensors are conceivable. Due to developments, in particular in nanotechnology, these can be made very small in the future and implanted with minimal invasiveness. First results were obtained with a right ventricular pressure sensor and a pulmonary artery pressure sensor [73]. For hemodynamic monitoring, a sensor was implanted into the pulmonary artery. The sensor wirelessly transmits the data to a receiving device. The CardioMEMS Heart Sensor Allows Monitoring of Pressure to Improve Outcomes in NYHA Class III Heart Failure Patients (CHAMPION) study demonstrated in NYHA stage III patients that hemodynamic monitoring significantly reduces hospitalization compared to the usual care [74]. In addition, by linking various vital parameters, the sensitivity and specificity of the detection of cardiac decompensation was improved. In patients with CIEDs, this was successfully tested in an approach that determined a risk factor for cardiac decompensation from various vital parameters (intrathoracic impedance measurement, physical activity, nocturnal heart rate, heart rate variability, atrial fibrillation burden, percentage of biventricular stimulation, and occurrence of ventricular tachycardia) [75].

21.5 Telemonitoring in Patients with Arterial Hypertension

Arterial hypertension is an independent risk factor for major cardiovascular events (stroke, myocardial infarction, sudden cardiac death, and peripheral arterial occlusive disease) and for terminal renal failure [76]. Many patients with elevated blood pressure have additional cardiovascular risk factors. The control of blood pressure is more difficult in high-risk patients compared to that in low-risk patients. For diagnosis or monitoring of patients with arterial hypertension, in the outpatient setting two options are available. Ambulatory blood pressure monitoring (ABPM) regularly measures blood pressure over a period of 24 h (for example, during the day every 15 min and during the night every 30 min, or every 20 min during the day and night). During this time, the patient receives a portable blood pressure monitor with a cuff. The measurements are automatic and are stored in the device. The evaluation of the data is performed by the doctor on the computer following the examination and transmission of data. The other method is home blood pressure monitoring (HBPM), which is usually performed as self-management by the patient. With this system, the patient can monitor blood pressure in a familiar environment at different times of the day and for several days. The data are recorded and documented by the patient himself. However, with this method, blood (pressure) values cannot be collected during everyday activities or at night [76]. Blood pressure measurement in the familiar environment, in contrast to blood pressure measurement by the doctor, excludes the "white-coat effect," which shows increased values during office blood pressure measurement [77].

Home blood pressure telemonitoring (HBPT) offers the possibility of transmitting measured data over the phone, a modem, or the Internet by the patients at home to the doctor or a service center [78]. In previous studies, the blood pressure data were transmitted over the telephone by the patient. Newer systems transmit data independently from the patient directly via interfaces (e.g., Bluetooth) and a modem to a secure web-based system. This provides the physician with direct data access. In addition, automatically preset alarms can be generated. The alarms can also be sent together with appropriate instructions directly to the patient. In more recent studies blood pressure monitors (blood pressure meters with computer) and mobile phones have been used for wireless data transmission [77]. New systems that are still in the experimental stage on animals have taken advantage of implants with wireless data transmission. In one system, a miniaturized pressure sensor was implanted into the femoral artery of sheep. Through a transponder unit, the data are transmitted from the sensor and at the same time energy is transmitted to the sensor system. The data transmission is performed telemedically from the readout station to an evaluation unit [79]. The advantage of these systems is the ability to continuously measure blood pressure, even during a physical activity and at night. The disadvantage is the invasiveness of the method. The extent to which implantable systems will prevail in the future depends primarily on the handling and safety of the sensors.

Recent studies have demonstrated benefits of telemonitoring in patients with arterial hypertension [80–82]. However, the design was not uniform in the studies. The transmission of blood pressure values only appears to not add benefits. Important is an intervention program as part of telemonitoring [80–82]. In the future, intelligent systems must be developed that automatically provide the patient with appropriate recommendations (blood pressure measurement, physician consultation, and so on). Currently, the costs of telemonitoring are still higher than of conventional treatment [78].

21.6 Conclusions

Telemonitoring in cardiology today means combining complex technical processes with intelligent evaluation and intervention algorithms for remote monitoring of patients. Telemonitoring is thus more than the transfer of vital sign data. The challenge is that of complex patient management.

Telemedical ECG monitoring is a diagnostic tool for diagnosis and monitoring after interventions. The advantage is the capability of recording and transmitting an ECG at any time and at any location.

Remote monitoring in patients with CIEDs is already assessed as useful by various cardiology societies. The technology is available and is used in practice. Innovation barriers to a widespread use are structural problems (ambulance-hospital) and reimbursement. However, in the near future the technology is likely to continue to be successful.

Monitoring of patients with chronic heart failure is a major problem because of the high number of affected patients and the resulting hospitalizations and healthcare costs. Initially, a positive effect through complex telemonitoring was seen. However, newer studies did not show benefit in comparison to the usual care. The challenge in the future is likely to be the differential selection of patients who will benefit from this methodology. Secondly, in order to identify early-stage cardiac decompensation, new external or implantable sensors must be developed with corresponding specificity and sensitivity. New randomized trials are needed here.

For the treatment of patients with arterial hypertension, telemedical management offers the opportunity to improve blood pressure control.

However, telemedicine in cardiology today should not be seen primarily as a tool to reduce healthcare costs. The improvement of a patient's condition and the reduction of morbidity and mortality can secondarily contribute to higher cost effectiveness. In the future, telemonitoring must differentiate itself from fitness and lifestyle products that are already available as apps or in smartphones. Telemedicine makes a contribution to the diagnosis, treatment, rehabilitation, and prophylaxis of heart and circulatory diseases.

References

1. Camm, A.J. et al., 2012 focused update of the ESC Guidelines for the management of atrial fibrillation: An update of the 2010 ESC Guidelines for the management of atrial fibrillation, *Eur Heart J*, 33, 2719–2747, 2012.
2. Brunetti, N.D. et al., Atrial fibrillation with symptoms other than palpitations: Incremental diagnostic sensitivity with at-home tele-cardiology assessment for emergency medical service, *Eur J Prev Cardiol*, 19, 306–313, 2012.
3. Senatore, G. et al., Role of transtelephonic electrocardiographic monitoring in detecting short-term arrhythmia recurrences after radiofrequency ablation in patients with atrial fibrillation, *J Am Coll Cardiol*, 45, 873–876, 2005.
4. Liu, J. et al., The value of transtelephonic electrocardiogram monitoring system during the "Blanking Period" after ablation of atrial fibrillation, *J Electrocardiol*, 43, 667–672, 2010.
5. Shacham, J. et al., Telemedicine for diagnosing and managing paroxysmal atrial fibrillation in outpatients: The phone in the pocket, *Int J Cardiol*, 157, 91–95, 2012.
6. Patten, M. et al., Event-recorder monitoring in the diagnosis of atrial fibrillation in symptomatic patients: Subanalysis of the SOPAT trial, *J Cardiovasc Electrophysiol*, 17, 1216–1220, 2006.
7. Fetsch, T. et al., Prevention of atrial fibrillation after cardioversion: Results of the PAFAC trial, *Eur Heart J*, 25, 1385–1394, 2004.
8. Einthoven, W., Le télécardiogramme, *Arch Intern Physiol*, 4, 132–164, 1906.
9. Morak, J., Kollmann, A., and Schreier, G., Feasibility and usabilility of a home monitoring concept based on mobile phones and near field communication (NFC) technology, in *Medinfo 2007*, K. Kuhn, Ed., IOS Press, pp. 112–116, 2007.
10. Scalvini, S. et al., Cardiac event recording yields more diagnoses than 24-hour Holter monitoring in patients with palpitations, *J Telemed Telecare*, 11(Suppl 1), 14–16, 2005.
11. Schwaab, B. et al., Validation of 12-lead tele-electrocardiogram transmission in the real-life scenario of acute coronary syndrome, *J Telemed Telecare*, 12, 315–318, 2006.
12. Mischke, K. et al., Telephonic transmission of 12-lead electrocardiograms during acute myocardial infarction, *J Telemed Telecare*, 11, 185–190, 2005.
13. Katalinic, A. et al., The TeleGuard trial of additional telemedicine care in CAD patients: Utilization of the system, *J Telemed Telecare*, 14, 17–21, 2008.
14. Leshem-Rubinow, E. et al., New real-time loop recorder diagnosis of symptomatic arrhythmia via telemedicine, *Clin Cardiol*, 34, 420–425, 2011.
15. Brignole, M. et al., Indications for the use of diagnostic implantable and external ECG loop recorders, *Europace*, 11, 671–687, 2009.

16. Müller, A. et al., Reliability of an external loop recorder for automatic recognition and transtelephonic transmission of atrial fibrillation, *J Telemed Telecare*, 5, 391–396, 2009.

17. Furukawa, T. et al., Effectiveness of remote monitoring in the management of syncope and palpitations, *Europace*, 13, 431–437, 2011.

18. Arrocha, A. et al., Remote electrocardiographic monitoring with a wireless implantable loop recorder: Minimizing the data review burden, *Pacing Clin Electrophysiol*, 33, 1347–1352, 2010.

19. Sack, S. et al., Das Herz Handy—Ein neues telemedizinisches Servicekonzept für Herzpatienten, *Herzschr Elektrophys*, 16, 165–175, 2005.

20. Moss, A.J. et al., Prophylactic implantation of a defibrillator in patients with myocardial infarction and reduced ejection fraction, *N Engl J Med*, 346, 877–883, 2002.

21. Bristow, M.R. et al., Cardiac-resynchronization therapy with or without an implantable defibrillator in advanced chronic heart failure, *N Engl J Med*, 350, 2140–2150, 2004.

22. Epstein, A.E. et al., ACC/AHA/HRS 2008 Guidelines for device-based therapy of cardiac rhythm abnormalities: A report of the American College of Cardiology/American Heart Association Task Force on Practice Guidelines (writing committee to revise the ACC/AHA/NASPE 2002 guideline update for implantation of cardiac pacemakers and antiarrhythmia devices) developed in collaboration with the American Association for Thoracic Surgery and Society of Thoracic Surgeons, *J Am Coll Cardiol*, 51, e1–e62, 2008.

23. Tracy, C.M. et al., 2012 ACCF/AHA/HRS focused update of the 2008 guidelines for device-based therapy of cardiac rhythm abnormalities: A report of the American College of Cardiology Foundation/American Heart Association Task Force on Practice Guidelines, *J Am Coll Cardiol*, 60, 1297–1313, 2012.

24. Brignole, M. et al., 2013 ESC Guidelines on cardiac pacing and cardiac resynchronization therapy, *Europace*, 15, 1070–1118, 2013.

25. Zhan, C. et al., Cardiac device implantation in the United States from 1997 through 2004: A population-based analysis, *J Gen Intern Med*, 23(Suppl 1), 13–19, 2007.

26. Alter, P. et al., Complications of implantable cardioverter defibrillator therapy in 440 consecutive patients, *PACE*, 28, 926–932, 2005.

27. Maisel, W.H. et al., Pacemaker and ICD generator malfunctions: Analysis of Food and Drug Administration annual reports, *JAMA*, 295, 1901–1906, 2006.

28. Liu, J. et al., Class I recall of defibrillator leads: A comparison of the Sprint Fidelis and Riata families, *Heart Rhythm*, 9(8), 1251–1255, 2012.

29. Dubner, S. et al., ISHNE/EHRA expert consensus on remote monitoring of cardiovascular implantable electronic devices (CIEDs), *Europace*, 14, 278–293, 2012.

30. Wilkoff, B.L. et al., HRS/EHRA expert consensus on the monitoring of cardiovascular implantable electronic devices (CIEDs): Description of techniques, indications, personnel, frequency and ethical considerations, *Europace*, 10, 707–725, 2008.

31. Heidbuchel, H., Telemonitoring of implantable cardiac devices: Hurdles towards personalised medicine, *Heart*, 97, 931–939, 2011.

32. Dreifus, L.S. et al., Transtelephonic monitoring of 25,919 implanted pacemakers, *PACE*, 9, 371–378, 1986.

33. Müller, A. et al., Empfehlungen zum Telemonitoring von Patienten mit implantierten Herzschrittmachern, Defibrillatoren und kardialen Resynchronisationssystemen, *Kardiologe*, 7, 181–193, 2013.

34. Müller, A. et al., Remote monitoring in patients with pacemakers and implantable cardioverter-defibrillators: New perspectives for complex therapy management, in *Modern Pacemakers—Present and Future*, M. Kumar, Ed., InTech–Open Access Publisher, pp. 147–166, 2011.

35. Maines, M. et al., Usefulness of intrathoracic fluids accumulation monitoring with an implantable biventricular defibrillator in reducing hospitalizations in patients with heart failure: A case-control study, *J Interv Card Electrophysiol*, 19, 201–207, 2007.

36. Scholten, M.F. et al., Twiddler's syndrome detected by home monitoring device, *PACE*, 27, 1151–1152. 2004.

37. Ritter, O., and Bauer, W.R., Use of "IEGM Online" in ICD patients—Early detection of inappropriate classified ventricular tachycardia via Home Monitoring, *Clin Res Cardiol*, 95, 368–372, 2006.

38. Lazarus, A., Remote, wireless, ambulatory monitoring of implantable pacemakers, cardioverter defibrillators, and cardiac resynchronization therapy systems: Analysis of a worldwide database, *PACE*, 30, S2–S12, 2007.

39. Varma, N. et al., Automatic remote monitoring of cardioverter-defibrillator lead and generator performance: The Lumos-T Safely Reduces Routine Office Device Follow-up (TRUST) trial, *Circ Arrhythm Electrophysiol*, 3, 428–436, 2010.

40. Res, J.C.J., Theuns, D.A.M.J., and Jordaens, L., The role of remote monitoring in the reduction of inappropriate implantable cardioverter defibrillator therapies, *Clin Res Cardiol*, 95(Suppl 3), III/17–III/21, 2006.

41. Sweeney, M.O. et al., Differences in effects of electrical therapy type for ventricular arrhythmias on mortality in implantable cardioverter-defibrillator patients, *Heart Rhythm*, 7(3), 353–360, 2010.

42. Ricci, R.P., Disease management: Atrial fibrillation and Home Monitoring, *Europace*, 15, i35–i39, 2013.

43. Perings, C. et al., Remote monitoring of implantable-cardioverter defibrillators: Results from the Reliability of IEGM Online Interpretation (RIONI) study, *Europace*, 13, 221–229, 2011.

44. Koplan, B.A. et al., Heart failure decompensation and all-cause mortality in relation to biventricular pacing in patients with heart failure: Is a goal of 100% biventricular pacing necessary?, *J Am Coll Cardiol*, 53, 355–360, 2009.

45. Sack, S. et al., Potential value of automated daily screening of cardiac resynchronization therapy defibrillator diagnostics for prediction of major cardiovascular events: Results from Home-CARE (Home Monitoring in Cardiac Resynchronization Therapy) study, *Eur J Heart Fail*, 13, 1019–1027, 2011.

46. Whellan, D.J. et al., Combined heart failure device diagnostics identify patients at higher risk of subsequent heart failure hospitalizations, *J Am Coll Cardiol*, 55, 1803–1810, 2010.

47. Boriani, G. et al., The MOnitoring Resynchronization dEvices and CARdiac patiEnts (MORE-CARE) randomized controlled trial: Phase 1 results on dynamics of early intervention with remote monitoring, *J Med Internet Res*, 15(8), e167, 2013.

48. Hindricks, G., Influence of Home Monitoring on the clinical management of heart failure patients with impaired left ventricular function, ESC Congress, Amsterdam, September 1, 2013.

49. Hindricks, G. et al., Quarterly vs. yearly clinical follow-up of remotely monitored recipients of prophylactic implantable cardioverter-defibrillators: Results of the REFORM trial, *Eur Heart J*, 35(2), 98–105, 2014.

50. Ricci, R.P., Morichelli, L., and Santini, M., Home monitoring remote control of pacemaker and implantable cardioverter defibrillator patients in clinical practice: Impact on medical management and health-care resource utilization, *Europace*, 10, 164–170, 2008.

51. Yee, R. et al., Canadian Cardiovascular Society/Canadian Heart Rhythm Society position statement on the use of remote monitoring for cardiovascular implantable electronic device follow-up, *Canadian J Cardiol*, 29, 644–651, 2013.

52. Barker, W.H., Mullooly, J.P., and Getchell, W., Changing incidence and survival for heart failure in a well-defined older population, 1970–1974 and 1990–1994, *Circulation*, 113, 799–805, 2006.

53. McMurray, J.J., and Stewart, S., Epidemiology, aetiology and prognosis of heart failure, *Heart*, 83, 596–602, 2000.

54. Yancy, C.W. et al., 2013 ACCF/AHA guideline for the management of heart failure: A report of the American College of Cardiology Foundation/American Heart Association Task Force on Practice Guidelines, *J Am Coll Cardiol*, 62(16), 147–239, 2013.

55. McMurray, J.J. et al., ESC guidelines for the diagnosis and treatment of acute and chronic heart failure 2012: The Task Force for the Diagnosis and Treatment of Acute and Chronic Heart Failure 2012 of the European Society of Cardiology, *Eur J Heart Fail*, 14(8), 803–869, 2012.

56. Störk, S. et al., Pharmacotherapy according to treatment guidelines is associated with lower mortality in a community-based sample of patients with chronic heart failure—A prospective cohort study, *Eur J Heart Fail*, 10, 1236–1245, 2008.
57. Zugck, C. et al., Implementation of pharmacotherapy guidelines in heart failure: Experience from the German Competence Network Heart Failure, *Clin Res Cardiol*, 101(4), 263–272, 2012.
58. Angermann, C.E. et al., Competence Network Heart Failure; Mode of action and effects of standardized collaborative disease management on mortality and morbidity in patients with systolic heart failure: The Interdisciplinary Network for Heart Failure (INH) study, *Circ Heart Fail*, 5(1), 25–35, 2012.
59. Cleland, J.G.F. et al., Noninvasive home telemonitoring for patients with heart failure at high risk of recurrent admission and death, *J Am Coll Cardiol*, 45, 1654–1664, 2005.
60. Dar, O. J. et al., A randomized trial of home monitoring in a typical elderly heart failure population in North West London: Results of the Home-HF study, *Eur J Heart Fail*, 11, 319–325, 2009.
61. Mortara, A. et al., Home telemonitoring in heart failure patients: The HHH study (Home or Hospital in Heart Failure), *Eur J Heart Fail*, 11, 312–318, 2009.
62. Klersy, C. et al., A meta-analysis of remote monitoring of heart failure patients, *J Am Coll Cardiol*, 54, 1683–1694, 2009.
63. Inglis, S.C. et al., Which components of heart failure programmes are effective? A systematic review and meta-analysis of the outcomes of structured telephone support or telemonitoring as the primary component of chronic heart failure management in 8323 patients: Abridged Cochrane Review, *Eur J Heart Fail*, 13(9), 1028–1240, 2011.
64. Gurné, O. et al., A critical review on telemonitoring in heart failure, *Acta Cardiol*, 67(4), 439–444, 2012.
65. Chaudhry, S.I. et al., Telemonitoring in patients with heart failure, *N Engl J Med*, 363, 2301–2309, 2010.
66. Koehler, F. et al., Impact of remote telemedical management on mortality and hospitalizations in ambulatory patients with chronic heart failure, *Circulation*, 123, 1873–1880, 2011.
67. Koehler, F. et al., Telemedicine in heart failure: Pre-specified and exploratory subgroup analyses from the TIM-HF trial, *Int J Cardiol*, 161(3), 143–150, 2012.
68. Müller, A. et al., Telemedical support in patients with chronic heart failure: Experience from different projects in Germany, *Int J Telemed Appl*, pii: 181806. Epub August 12, 2010.
69. Seto, E., Cost comparison between telemonitoring and usual care of heart failure: A systematic review, *Telemed J E Health*, 14, 679–683, 2008.
70. Sohn, S. et al., Costs and benefits of personalized health care for patients with chronic heart failure in the care and education program "Telemedicine for the Heart," *Telemed J E Health*, 18(3), 198–204, 2012.
71. Anand, I.S. et al., Design of the Multi-Sensor Monitoring in Congestive Heart Failure (MUSIC) study: Prospective trial to assess the utility of continuous wireless physiologic monitoring in heart failure, *J Card Fail*, 17(1), 11–16, 2011.
72. Anand, I.S. et al., Design and performance of a multisensor heart failure monitoring algorithm: Results from the multisensor monitoring in congestive heart failure (MUSIC) study, *J Card Fail*, 18(4), 289–295, 2012.
73. Abraham, W.T., Disease management: Remote monitoring in heart failure patients with implantable defibrillators, resynchronization devices, and haemodynamic monitors, *Europace*, 15(Suppl 1), i40–i46, 2013.
74. Abraham, W.T. et al., CHAMPION Trial Study Group Wireless pulmonary artery haemodynamic monitoring in chronic heart failure: A randomised controlled trial, *Lancet*, 377(9766), 658–666, 2011.
75. Cowie, M.R. et al., Development and validation of an integrated diagnostic algorithm derived from parameters monitored in implantable devices for identifying patients at risk for heart failure hospitalization in an ambulatory setting, *Eur Heart J*, 34, 2472–2480, 2013.
76. Mancia, G. et al., 2013 ESH/ESC Guidelines for the management of arterial hypertension, *Eur Heart J*, 34, 2159–2219, 2013.

77. AbuDagga, A., Resnick, H.E., and Alwan, M., Impact of blood pressure telemonitoring on hypertension outcomes: A literature review, *Telemed J E Health*, 16(7), 830–838, 2010.
78. Omboni, S., Gazzola, T., and Carabelli, G., Clinical usefulness and cost effectiveness of home blood pressure telemonitoring: Meta-analysis of randomized controlled studies, *J Hypertens*, 31(3), 455–567, 2013.
79. Cleven, N.J. et al., A novel fully implantable wireless sensor system for monitoring hypertension patients, *IEEE Trans Biomed Eng*, 59(11), 3124–3130, 2012.
80. Logan, A.G. et al., Effect of home blood pressure telemonitoring with self-care support on uncontrolled systolic hypertension in diabetics, *Hypertension*, 60(1), 51–57, 2012.
81. Bove, A.A., Homko C.J., and Santamore, W.P., Managing hypertension in urban underserved subjects using telemedicine—A clinical trial, *Heart J*, 165, 615–621, 2013.
82. Margolis, K.L. et al., Effect of home blood pressure telemonitoring and pharmacist management on blood pressure control: A cluster randomized clinical trial, *JAMA*, 310(1), 46–56, 2013.

22

Teleradiology

Liam Caffery

CONTENTS

22.1 Introduction

Although teleradiology has matured over the last decade and is now a routine part of hospital and private radiology services, it continues to evolve. This chapter investigates contemporary themes in teleradiology and aims to provide an up-to-date synthesis and critical analysis of these topics. This investigation has established reasons why teleradiology has reshaped the practice of radiology. To enhance understanding, contextual information on clinical teleradiology, the digital imaging and communication in medicine (DICOM) standard, digital image fundamentals, and monitor characteristics is also presented.

22.2 Background

Teleradiology is an umbrella term for the electronic transfer of radiographic images from one geographic location to another. Once transferred, images can be reviewed for primary reporting and interpretation by a radiologist or the clinical review of images by the patient's treating doctor. This is an important distinction as the regulatory oversight of teleradiology is nearly always specific to primary diagnosis.

22.2.1 Definitions

Teleradiology services used for primary diagnosis are classified as either intramural or extramural [1]. Intramural or intraorganizational teleradiology is where the teleradiologist works for the organization that is directly responsible for the patient's care, even if the reporting occurs in a different location. Extramural or outsourced teleradiology is where the radiologists are employed by a company distinct to the organization caring for the patient. This designation has implications for clinical practice—for example, it may influence whether the teleradiologist has access only to the patient's images or alternatively to the images, the hospital's radiology information system (RIS), the patient's electronic health records (EHR), to prior imaging (stored on the hospital picture archiving and communication system [PACS]), and other diagnostics results. The designation may also dictate whether both the on-site radiologist and the teleradiologist adhere to the same policy and procedure, e.g., the mandatory peer review of reports.

22.2.2 Clinical Teleradiology

There are different scenarios where teleradiology is used for primary diagnosis, including

- Provision of interpretation services to underserved regions;
- At-home review by an on-call radiologist;
- Subspecialist reporting;
- Consolidation of interpretation service for multiple image-acquisition sites;
- Interpretation-only services for on-call coverage; and
- Interpretation-only services for routine in-hours reporting.

As with most specialists, radiologists tend to live in metropolitan areas, leaving regional and rural populations underserved. This problem has been exacerbated by a shortage of qualified radiologists in some countries [2]. Routine radiological reporting of images acquired in rural practices can be performed by teleradiology [3–5]. This may be as an intramural or extramural service.

The at-home review of images is the second most frequent form of teleradiology use according to a recent survey of European radiologists [6]. The survey revealed 44% of radiologists use teleradiology for this purpose. Accessing images at home is quite easily achieved. However, communicating the diagnosis to the referring doctor is not as straightforward. In the majority of cases, surveyed radiologists communicated reports to the referring doctor via telephone. Other means of communicating the diagnosis included entering the report directly into the hospital's RIS or by e-mailing the referring doctor.

Radiology is becoming increasingly subspecialized. In some jurisdictions 80% of radiology trainees are undertaking fellowship training [7]. Researchers have observed improved diagnostic accuracy by subspecialists when compared to generalist reporting [8–10]. Conversely, subspecialization can also create voids of knowledge [11]. Teleradiology can be used successfully to access subspecialists in external organizations [9]. However, second opinions from a fellowship-trained colleague are more widespread.

Teleradiology means that the radiologist no longer needs to be located at the site of image acquisition. Therefore, radiologists can perform reporting in a hub-and-spoke model where a hub reporting site services multiple image-acquisition departments. This model of teleradiology has been used extensively by extramural service providers but can also be used for intramural services. For example, radiologists at Kaiser Permanente provide an after-hours radiology service using a hub-and-spoke teleradiology model [12]. Two radiologists perform reporting for 11 image-acquisition sites. They can be physically located at any of the 11 facilities. Their average report turnaround time is 19 min. This compares favorably to their previous service model, where 11 on-call radiologists (one per site) provided the after-hours service by returning to the hospital for emergency reads. The teleradiology model reduces the amount of on-calls a radiologist has to perform.

The extramural provision of after-hours reporting is a widespread use of teleradiology— more so in the United States than in Europe, Canada, or Australia. This type of service— known frequently as "nighthawking"—is estimated to be used by 50% of all U.S. hospitals to provide out-of-hours reporting [13]. The use of extramural, off-site radiologists can be implemented to take advantage of different time zones. For example, radiologists working during the day in Sydney, Australia, can use teleradiology to perform emergency reporting for a hospital located in Uppsala, Sweden [14]. Similarly, Indian radiologists can cover out-of-hours reporting for the United States [15]. The use of extramural, off-site reporting avoids the need for a hospital to have one of their on-site radiologists provide an on-call service. Hence, the use of a nighthawking seemed to be a good idea to many on-site radiologists. However, there is now concern that nighthawking services will encroach on daytime reporting and eventually displace local radiology. Hospital administrators who have experienced the cost and turnaround-time benefits of nighthawking are increasingly investigating if daytime services can be offered by teleradiology groups [16]. Despite the potential, the routine in-hours reporting by teleradiology is not as widespread as nighthawking but its use is expanding in certain scenarios. For example, teleradiology can be used to cover for temporary capacity problems such as holidays and staff shortages [6] or the lack of a particular skill set, e.g., magnetic resonance imaging (MRI) credentials within existing on-site radiologists [17]. Due to higher rates of subspecialization, others have found it harder to find an on-site radiologist who will report plain films. Selective outsourcing of some services—for example, plain-film reporting—is another scenario that can be undertaken using off-site teleradiology reporting [17].

Teleradiology is also used for secondary review by a patient's treating doctor (as opposed to primary diagnosis). The transmission of images (and radiologist's report) when a patient's care is transferred from one facility to another can avoid the need to reimage a patient, thus saving both radiation exposure and cost [18].

Teleradiology also facilitates specialist clinical review such as when a neurosurgeon at a tertiary facility reviews CT scans acquired at regional site to advise on whether a patient should be transferred. Ashkenazi et al. [19] reported that when teleradiology was used, around 40% of transfers to a tertiary facility were avoided. The use of teleradiology also significantly reduced adverse events—such as hypoxia—when a patient was actually transferred [20].

Teleradiology enables telemedicine consultations. For example, there has been a rapid expansion of telestroke networks aimed at the acute treatment of stroke [21]. Teleradiology is needed for neurologists to assess the presence or absence of radiological contraindications to thrombolysis [22]. Similarly, remote fracture clinics by an orthopedic surgeon are reliant on teleradiology [23]. Remote fracture clinics have an orthopedic surgeon consulting with a patient via videoconferencing. The aim of these clinics is to avoid patient travel for what is often a very short consultation. To accurately assess a healing fracture, a recent radiograph is necessary. A patient will have their radiograph performed at the remote site; the images will be transmitted by teleradiology to the orthopedic surgeon. The surgeon will assess the radiograph and consult with the patient.

22.2.3 Technology Considerations

Teleradiology is underpinned by the use of standards. Some of these standards are imaging specific standards—for example, DICOM; others are broader health standards—for example, HL7; while others, still, are common information technology standards such as TCP and IP, used in image transmission, and virtual private networks (VPNs), used in encryption of patient information during transmission. The widespread adoption of these standards coupled with the availability of high-bandwidth Internet connections means that the problem of securely transferring radiographic images and radiological reports in a timely manner from one location to another has largely been solved.

Image transfer is achieved predominantly using DICOM [6]. However, DICOM is not a single means of implementing image transfer. The various ways of implementing DICOM-based image transfer include

- Dedicated teleradiology equipment;
- A remotely connected modality storing images on a networked PACS;
- PACS-to-PACS image transfer;
- A remote review station retrieving images from a networked image store;
- A web client viewing images from a distant image store; and
- The transmission of images as an e-mail attachment.

Only 16% of teleradiology services use dedicated teleradiology equipment [6]. Instead, existing PACS infrastructure is often leveraged to provide teleradiology [24]. Many radiology departments are PACS based with DICOM-compliant modalities sending images to a central image archive and the images viewed by radiologists and other clinicians on review stations. Images transmitted from remote sites can be archived and reviewed on the same equipment as locally acquired images. The use of existing PACS equipment reduces the cost of implementing a teleradiology service, does not require substantive changes in work practices, and does not require additional training on a separate teleradiology application. PACS-to-PACS image transfer, remote imaging modality, remote workstation, or web-based access to images are all means of using PACS to facilitate teleradiology. This architecture is common when one organization has several geographically separate facilities. In Europe, the practice of teleradiology is closely related to PACS installations [6].

Germany has adopted the DICOM e-mail standard [25] for implementing teleradiology networks [26]. Using e-mail as a transport mechanism for DICOM files allows ad hoc connections between sites to be established simply. It avoids the need for permanent, continuous network connectivity between locations. E-mail also transports ancillary information such as radiological reports and patient history. Encryption to ensure patient privacy is already addressed in standard e-mail protocols.

22.2.3.1 DICOM

DICOM is considered to be a key determinant of the success of teleradiology [27]. The widespread use of DICOM has realized a standardized image file format and transfer protocol, resulting in interoperability between equipment vendors and institutions. Today, nearly all image-acquisition modalities, PACS, and teleradiology equipment are DICOM compliant, which is testament to the success of DICOM.

What exactly is the DICOM standard? The DICOM standard is a document which defines file formats and message exchange. Message exchange is used to negotiate the transfer of a DICOM file from one DICOM-compliant device to another. The DICOM standard document is used by imaging equipment manufacturers to implement a consistent and standardized way of creating and transferring DICOM files in their software and products. In addition, DICOM defines work-flow exchange—for example, how to transfer patient demographics from one device to another. Further, it defines ways of ensuring that an image is consistently displayed, regardless of the monitor on which it is viewed. The stated goals of the DICOM standards committee *are to achieve compatibility and to improve work-flow efficiency between imaging systems and other information systems in healthcare environments worldwide.*

The first version of DICOM was released in 1985 as ACR/NEMA 1.0. It was jointly developed by the American College of Radiology (ACR) and the National Electrical Manufacturers Association (NEMA) (who represented imaging equipment vendors). Prior to the release of the standard, images were stored in proprietary formats. Transmission of images was not via data networks but achieved by copying the image file to removable media and physically transporting the removable media to another imaging device. DICOM has realized its objective of network transfer and a DICOM application is designed to work with a standard TCP/IP network.

The standard was renamed DICOM 3.0 at its third iteration. Since then no major versions changes have occurred; instead, supplements and corrections are released yearly. Hence, the version of DICOM is identified by year suffix. DICOM is also known as NEMA standard PS 3. The most recent version is DICOM 3.0 2011; it has 18 base parts (Table 22.1) (PS 3.9 and PS 3.13 have been retired).

The development and maintenance of the DICOM standards is done by the DICOM Standards Committee, which is made up of three groups: manufacturers, user groups, and general interest groups. User groups include the ACR, the European Society of Radiology, and the Society for Imaging Informatics in Medicine; whereas general interest groups include Canada Health Infoway, the National Cancer Institute, and the Web3D Consortium. The standards committee coordinate the effort of working groups (WGs) who advance categories of work. The current working groups are listed in Table 22.2. Notably, DICOM has expanded beyond radiology and now includes working groups from other imaging specialties—for example, WG-09, Ophthalmology; WG-19, Dermatologic Standards; WG-22, Dentistry; WG-24, Surgery; and WG-26, Pathology.

DICOM defines a file format which has two parts. The first is text-based metadata, which describe patient, study, equipment, and image attributes. The second is the pixel data of

TABLE 22.1

Parts of the DICOM Base Standard

Section	Section Title
PS 3.1	Introduction and overview
PS 3.2	Conformance
PS 3.3	Information object definitions
PS 3.4	Service class specifications
PS 3.5	Data structures and encoding
PS 3.6	Data dictionary
PS 3.7	Message exchange
PS 3.8	Network communication support for message exchange
PS 3.10	Media storage and file format for data interchange
PS 3.11	Media storage application profiles
PS 3.12	Media formats and physical media for data interchange
PS 3.14	Grayscale standard display function
PS 3.15	Security profiles
PS 3.16	Content mapping resource
PS 3.17	Explanatory information
PS 3.18	Web access to DICOM persistent objects (WADO)
PS 3.19	Application hosting
PS 3.20	Transformation of DICOM to and from HL7 standards

the image, which can be in any standard image file formats—for example, JPEG. The two parts are melded into a single file. The metadata are described in the DICOM standard as an image object definition (IOD). There are different IODs for each modality type—for example, DICOM defines a CT image IOD and a magnetic resonance (MR) image IOD because CT requires parameters such as spiral pitch, gantry tilt, and table position, which are irrelevant to any other modality. Similarly parameters such as MR sequence type and echo time are relevant only to magnetic resonance imaging.

DICOM network services involve transmitting a DICOM file (often called a DICOM object) from one medical imaging device to another for various purposes—for example, storing the file from a modality to a PACS, printing the image to a networked DICOM printer or querying the PACS for a particular image, and retrieving it to a workstation for viewing. These transactions occur between two DICOM-compliant devices that are connected on a data network. DICOM network services are based on client/server architecture. DICOM defines a service class for each of the network services. For the above examples, storage service class, the query/retrieve service class, and the print management class are used. For each service class there is a defined set of messages and responses. These messages and response are exchanged between two DICOM devices and are used to negotiate an association between devices, transfer the DICOM object, and close the connection between devices. The two communicating devices must agree on parameters such as the image type and format of the pixel data before an association is established.

DICOM also defines network services for the exchange of patient demographics. One such service is the DICOM modality worklist (DMWL). The DMWL achieves integration between a RIS, a PACS, and imaging modalities and avoids the need for rekeying of patient demographics. This results in both work-flow efficiency and improved accuracy of metadata.

TABLE 22.2

DICOM Standards Committee WGs

Committee	Field of Emphasis
WG-01	Cardiac and Vascular Information
WG-02	Projection Radiography and Angiography
WG-03	Nuclear Medicine
WG-04	Compression
WG-05	Exchange Media
WG-06	Base Standard
WG-07	Radiotherapy
WG-08	Structured Reporting
WG-09	Ophthalmology
WG-10	Strategic Advisory
WG-11	Display Function Standard
WG-12	Ultrasound
WG-13	Visible Light
WG-14	Security
WG-15	Digital Mammography and CAD
WG-16	Magnetic Resonance
WG-17	3D
WG-18	Clinical Trials and Education
WG-19	Dermatologic Standards
WG-20	Integration of Imaging and Information Systems
WG-21	Computed Tomography
WG-22	Dentistry
WG-23	Application Hosting
WG-24	Surgery
WG-25	Veterinary Medicine
WG-26	Pathology
WG-27	Web Technology for DICOM
WG-28	Physics
WG-29	Education, Communication, and Outreach

DICOM also defines methods for ensuring image-viewing quality. DICOM has defined a grayscale standard display function (GSDF) for facilitating a monitor displaying in a visually consistent manner. Review station software allows users to manipulate images—for example, windowing, flipping or rotating, or graphical transformations (annotations and overlays). The use of soft-copy presentation states allows these transformations to be stored and reproduced. The presentation state is stored as a separate (but linked) DICOM object to the image. Some review stations automatically display the image with the presentation states applied.

The DICOM standard now includes PS 3.18 (WADO). The inclusion of WADO has facilitated web-based PACS and teleradiology. WADO allows a clinician using a web application within a standard browser to query, retrieve, and display DICOM images that are stored on a central image archive. This represents a way of leveraging existing PACS infrastructure to provide teleradiology services.

DICOM is an active standard and continues to evolve to meet the changing needs of medical imaging. This progression combined with DICOM's near-ubiquitous adoption means it will continue to be influential in the practice of teleradiology.

22.3 The Commoditization War

The practice of radiology has drastically changed over recent years due in a large part to the secondary effects of teleradiology. The ability to transmit images from one geographic location to another location (potentially an international destination) presented commercial opportunities, which, in turn, led to the corporatization and commoditization of radiology. Radiology became commercialized at a speed that has rarely been seen [13]. The speed of change meant that the regulatory and professional bodies were unable to keep pace with potential quality and medicolegal issues that arose. Hence, teleradiology services operated for some time with little or no formal controls [28].

There is little doubt that teleradiology and the reshaping of radiology have been advantageous to some radiologists and detrimental to others, a fact that is highlighted by the title of a commentary in the *American Journal of Radiology*, "Teleradiology Coming of Age: Winners and Losers" [16]. Commoditization is perceived by some to devalue the profession of radiology, likening it to a factory production line rather than a medical specialty [29]. Further, commoditization threatens remuneration of on-site radiologists [29] and also threatens the existence of local radiology practice [13]. Battle lines were drawn between teleradiologists and their on-site counterparts. The threat to teleradiology was counted. Firstly, on-site radiologists promote the value of supplementary aspects of their profession, such as participating in multidisciplinary meetings; overseeing appropriate imaging; and training of registrars. Secondly, other professional bodies develop practice guidelines and position statements on teleradiology.

Why did teleradiology contribute to the commoditization of radiology? Teleradiology allows the

- Separation of image acquisition from image interpretation, thereby permitting radiologists to work outside of hospital or other clinical setting, which, in turn, transformed radiology into an interpretation-only service;

- Consolidation of multiple image-acquisition sites with a single image-interpretation site, thereby allowing economies of scale (to illustrate the scale, the average nighthawk radiologist holds licenses in 38 U.S. states and is on staff of more than 400 hospitals) [30]; and

- Globalization of radiology, which allows access to alternative (and often cheaper) labor markets.

Some commentators consider the introduction of PACS as the beginning of the "invisible radiologist" and a precursor to outsourced teleradiology and commoditization [31]. The digital distribution of images and reports reduced the need for personnel interaction between referring clinicians and their radiology colleagues [16]. This established a norm of referring clinician getting reports from an unknown radiologist and potentially created a precedent that increased the acceptability of outsourced teleradiology reporting.

Proponents of teleradiology argue that it improves coverage of services, improves subspecialty expertise, and improves the timeliness of interpretation [29]. All of these can result in better services and patient outcomes [2]. In addition, teleradiology affords lifestyle convenience for on-site radiologists when out-of-hours coverage is provided by a teleradiologist. Advocates argue that teleradiology not only reduces cost and interpretation time but also provides better quality of report because local radiology cannot offer the full range of subspecialty services that increasingly are being demanded of them [16]. Others argue that the ongoing uptake of teleradiology is testament to the quality of reporting [2]. Some studies show that teleradiology reports are highly accurate and the discordance between preliminary teleradiology report and the final report of an on-site radiologist is very low [32–34]. Other authors report major discrepancies in reports [35].

In a recent survey, a majority of respondents felt outsourcing would reduce the quality of radiology reports, threaten a radiologist's income, and compromise patient care [6]. There were several reasons contributing to perceived reduction in quality, including insufficient quality assurance program within the outsourcing company; the lack of contact between the teleradiologist and the referring clinician; and the lack of access the teleradiologists has to relevant patient information. The discussion between the radiologist and referring clinician has been shown to result in a change of the diagnosis in half of patients and a change of treatment in 60% of cases discussed [36]. The report quality may be compromised if the teleradiologist reports without referral information, prior imaging, pathology results, and the patient's medical record [37]. This is often due to lack of integration between teleradiology systems and other hospital information system or the inability to access these systems from outside the hospital [17]. The lack of supporting information may result in overly cautious, noncommittal reports and recommendations for further unnecessary imaging [28].

To compete with teleradiology, on-site radiologists have had to provide excellent service, develop subspecialty expertise, and develop an identity with patients and clinicians to demonstrate the added advantage of local radiology [29]. The loss of ancillary tasks when an interpretation-only teleradiology service is used has been one way of demonstrating the added value of local radiology. The Vancouver workload study highlights these ancillary roles and the time undertaken on them [38]. The authors found that a radiologist spends 87.7% of their time on clinical activity but only 36.4% on image interpretation. Other clinical activities include protocoling requisition, supervising studies, interventional procedures, consulting with clinicians, and direct patient care. They also observed that radiologists communicate with other healthcare professionals on average six times per hour. The wider responsibilities of radiologist may also include optimizing imaging protocols to protect patients from unnecessary or inappropriate investigations and unjustified radiation dose [1].

Regulatory control has now caught up with practice of teleradiology. Professional bodies have established standards, policies, and protocols for aspects of teleradiology including quality assurance, training, technical standards, accreditation, and licensure [2]. These include the Royal College of Radiologists' "Standards for the Provision of Teleradiology within the United Kingdom" [39], the European Society of Radiology's "Teleradiology in the European Union" [28], the Canadian Association of Radiologists' "Standards for Teleradiology" [40], the Royal Australian and New Zealand College of Radiologists' "Position on Teleradiology" [41], and the technical standards of the American College of Radiology, in association with the American Association of Physicists in Medicine (AAPM) and the Society for Imaging Informatics in Medicine (SIIM) [42–43]. These standards aim at providing best practice guidelines on ensuring high-quality teleradiology. Recommendations include limiting the use of teleradiology to scenarios where the radiologist has access to previous imaging and ancillary diagnostic data; ensuring that the teleradiologist is available for consultation with the referrer; mandating that the teleradiologist is responsible for the quality of images interpreted; and ensuring that teleradiologists use a quality-control mechanism such as clinical auditing. Administrative and logistical processes are also recommended—for example, ensuring that the teleradiologists include referral details as part of the report and employing report acknowledgment systems to allow the referring clinician to notify the radiologist that they have read the radiologist's report. To ensure that local radiology and teleradiology are self-regulating, it is now commonly recommended that the local radiologists are involved in decision-making processes of outsourcing.

The regulations governing teleradiology go a long way to ensure patient safety and to safeguard against reporting being compromised by the lack of information and communication, substandard equipment, and ambiguous medicolegal situations. Local radiology practice is afforded a level of protection under these guidelines given that they have a degree of control over the engagement of teleradiology.

22.4 Mobile Teleradiology

The use of handheld devices such as smartphones and tablet computers has been expanding in medical imaging applications. The suitability of these devices for teleradiology needs to be considered in the context of their use for either primary diagnosis or clinical review. Further, the appropriateness of a handheld device for teleradiology will depend on its display characteristics, the performance of the network connections, the video-processing capability, the functionality of the image-viewing software, ergonomics, and environmental factors, e.g., ambient lighting. To understand the suitability of viewing radiographic images on a mobile device, it is important to understand the fundamental properties of radiographic images and to understand the characteristics of a traditional radiologist workstation.

22.4.1 Digital Image Fundamentals

Radiographic images are made up of columns and rows of pixels (picture elements). Each pixel is one homogeneous shade of gray in a grayscale image or one discrete color in a color image. The columns and rows are also known as the image matrix. The image matrix is expressed as columns multiplied by rows, e.g., 512×512. Radiographic images fall into two main categories depending on image matrix. The first category is small-matrix images. CT, MRI, ultrasound, nuclear medicine, and angiography are modalities that produce small-matrix images. These images are less than 1024×1024 in size but often smaller—for example, 256×256 or 512×512. The second category is large-matrix images—for example, computed radiography (CR) or digital mammography. Matrix sizes for these modalities are typically 2048×2048 and 4096×4096, respectively.

22.4.2 Monitor Characteristics

Quality metrics for a monitor include physical size, resolution, pixel pitch, luminance, and luminance ratio (LR).

The physical size of a monitor is measured diagonally and often expressed in inches even in countries that use the metric system. The resolution is the number of columns and rows that a monitor displays and is also known as the display matrix or screen resolution. Screen resolution is expressed as width multiplied by height (2048×1536) or in megapixels (MP). A 2 MP display typically has a screen resolution of 1600×1200, a 3 MP display has a screen resolution of 2048×1536, and a 5 MP display has a screen resolution of 2048×2560. The ideal way to display an image is when one pixel of the image corresponds to one pixel of the display matrix. If the image matrix size is greater than the display matrix, the image is rescaled (a number of image pixels are represented by one pixel on the display). A 2 MP (or greater) monitor can display small-matrix images without rescaling; however, large-matrix images need to be rescaled unless a 5 MP monitor is used.

Pixel pitch is the physical size of an individual pixel; it can be expressed in microns (μm) (1 mm = 1000 μm) or pixels per inch (ppi). Given that there are 25.4 mm in an inch, a display with a pixel pitch of 264 ppi is the same as a pixel pitch of 96 μm (25.4 mm/inch ÷ 264 ppi × 1000). The optimum pixel pitch of a display is achieved when the pixel pitch is small enough to remove artifact, e.g., staircase effect on oblique lines. The viewing distance also affects the visualization of an artifact. Handheld devices are typically viewed closer to the eye than desktop workstations; hence, artifacts are more apparent on handheld devices.

Luminance is the intensity of light emitted from the monitor and is measured in candelas per meter squared. Generally, higher luminance correlates with a higher quality of a monitor. Luminance is important as it allows the human eye to distinguish different shades of gray in an image. This ability is lost if the display is too dull. The maximum luminance of a monitor (L_{max}) is obtained by measuring the luminance of white pixels, whereas the minimum luminance (L_{min}) is obtained by measuring the luminance of black pixels (with some compensation for ambient lighting). Luminance ratio is the ratio of L_{max} to L_{min} and must be large for good image contrast. L_{min} is an important characteristic because the ability of the human eye to perceive contrast changes in dark regions is very poor. Hence, L_{min} should not be extremely low.

The American College of Radiology in association with AAPM and SIIM has published a technical standard which identifies monitor characteristics values that should be employed for primary displays. Secondary displays used for clinical review do not need to adhere to the standards for primary displays; however, some recommendations for secondary display are also included (Table 22.3) [43].

22.4.3 Workstation Characteristics

Workstations used by radiologists for primary diagnosis are typically high-end PCs running an image-viewing-and-manipulation software. This software allows the user to search for studies based on different criteria—for example, patient identifier, patient name, date, or modality—retrieve the image from an archive, and display the retrieved image file. Image files are normally in DICOM format. The software allows the user to navigate and manipulate images. Functions such as window and level adjustment, zooming, panning, rotation, and flipping are common. Other functions used by radiologists may include linear, region of interest (ROI), and pixel value–measurement tools. The software displays not only the image but also the associated patient demographics (sourced from the header of the DICOM file) and, where appropriate, a flag to indicate the use of lossy compression

TABLE 22.3

ACR-AAPM-SIIM Technical Standards for Displays

Characteristic	Primary	Secondary
Physical size	Not specified	Not specified
Resolution	Not specified	Not specified
Pixel pitch	<210 μm	<300 μm
Maximum luminance	≥350 cd/m²	≥250 cd/m²
Minimum luminance	≥1.0 cd/m²	≥0.8 cd/m²
Luminance ratio	>250:1	Not specified

and image exposure values. Some software includes advanced postprocessing tools for 3D rendering or multiplanar reconstruction (MPR).

The workstation is connected via a network interface card (NIC) to an imaging network that allows connectivity with a PACS or other image archive, e.g., teleradiology infrastructure or an imaging modality. These archives may be attached to a LAN or a WAN or a broadband Internet connection. The workstation typically connects to a LAN at a speed of 1 Gigabit per second (Gb/s) and to a WAN at a speed governed by the type of broadband connection. Broadband speeds are not specific and are influenced by factors including the plan offered by the Internet service provider (ISP); congestion on the network; and distance from the telephone exchange or antenna. Broadband connections are asynchronous and have download speeds that are faster than upload speeds. Mobile broadband services (3G and 4G) have greater bandwidth variation with speed differing considerably between indoor and outdoor and stationary or moving devices. Typical speed ranges are listed in Table 22.4.

A workstation does differ from a PC in terms of its monitors and video card. Most radiologists prefer to have two or three display monitors. Two of the monitors are primary displays and the third is a commercial-grade monitor (Figure 22.1). The medical-grade monitor is used to view images, while the third monitor will run ancillary applications for digital dictation, voice recognition, RIS, e-mail, and web browsing. Having the two medical-grade monitors allows the radiologist to easily compare prior imaging or view multiple series within one study. The medical-grade monitors may be color or grayscale. A 3 MP monitor is often used; however, 2 and 5 MP monitors are not uncommon in radiology.

The video cards are dual head to support the multiple-monitor configurations and the video card often is equipped with additional memory, thereby allowing it to perform supplementary graphics processing. Supplementary graphics processing is necessary to maintain performance during advanced postprocessing or the stack-mode viewing of large image sets.

TABLE 22.4

Typical Broadband Internet Speeds

Broadband Connection	Download Speed	Upload Speed
Asymmetric digital subscriber line (ADSL)	256 kb/s–8 Mb/s	64 kb/s–1 Mb/s
ADSL2+	5–20 Mb/s	0.5–2.5 Mb/s
3G	5–7 Mb/s	384 kb/s–2 Mb/s
4G	20–40 Mb/s	1–10 Mb/s

FIGURE 22.1
Primary workstation with a three-monitor configuration.

FIGURE 22.2
External photometer used to calibrate a monitor.

Primary diagnosis workstations need to be calibrated to and conform to the current DICOM GSDF perceptual linearization methods [42]. The GSDF perceptual linearization method is described in part 14 of the DICOM standard [44]. The GSDF has been developed to ensure consistency of display, regardless of which monitor images are viewed on, and has been shown to increase diagnostic accuracy [45]. Calibration can be done manually with an external photometer (Figure 22.2); however, most medical-grade monitors have a built-in photometer and associated calibration software. The luminance response of the display monitor is measured with the photometer and this information is fed into the calibration software. The calibration software applies a correction to compensate for differences between the measured and the ideal characteristics. The software will also adjust the backlight level of the monitor to ensure consistent luminance. It is important to continually adjust the monitor's luminance as it will decay over time.

22.5 Teleradiology on Handheld Devices

This section explores the use of handheld devices for teleradiology. Firstly, the suitability of the hardware is examined, and secondly, current studies on the efficacy of radiological diagnosis on handheld devices are reviewed.

22.5.1 Handheld Device Characteristics

The variety of commercially available handheld devices is vast and includes tablet computers and smartphones. Popular tablets would be well known to most people and include (but are not limited to) Apple iPad, Samsung Galaxy, Microsoft Surface, Google Nexus, Acer Iconia, HTC EVO View, Asus MeMO, Sony Xperia, Huawei MediaPad, Panasonic Toughpad, and Lenovo ThinkPad. Most of these manufacturers also have a smartphone range of products. While there are a large number of manufacturers, the OSs on these devices are limited to Apple iOS, BlackBerry OS, Microsoft Windows, and Google Android. Tablet computers have less powerful processors than desktop PCs. At the time of writing, a typical PC had a processor speed of 3.0–4.7 GHz compared to a tablet's 1.3–2.2 GHz. Also, tablets have substantially less memory. RAM is normally 1 GB in a tablet and up to 16 GB in a desktop PC. The storage capacity of a tablet device ranges from 16–64 GB, whereas a PC has between 512 GB and 2 TB of storage. Tablet computers display size ranges from 7″ to 10″. The screen size of a smartphone is around 3.5″ to 5″ in size. Smartphones are considered as being just too small to adequately view radiographic images [46]. Computer programs or applications—for example, the PACS software that runs on a desktop workstation—cannot be installed on a tablet or smartphone. Miniapplications or apps are specifically developed for mobile devices.

22.5.1.1 Hardware

A comparison of display characteristics of a typical primary desktop workstation and handheld Apple devices (Table 22.5) would at first glance indicate that the only difference is the physical size of the display and the Apple devices exceed the workstation in some characteristics, e.g., luminance and pixel pitch. However, a more in-depth analysis reveals other characteristics that are important in teleradiology.

While handheld devices are comparable to primary desktop workstations in display resolution and luminance ratio, they do not conform to DICOM GSDF, with some authors reporting a deviation of 30% from the GSDF [47]. Also, the high luminance ratio of handheld devices is achieved as a result of having very low minimum luminance, with L_{min} values below those recommended in the ACR-AAPM-SIIM technical guidelines [43]. Further, the displays of handheld devices cannot be calibrated to the DICOM GSDF because firstly, the video subsystems cannot be altered by software and secondly, because the devices do not have a built-in photometer or the ports to physically attach an external photometer [46]. There are visual calibration tools for iPads based on a user's interaction with test patterns

TABLE 22.5

Display Characteristics of Primary Desktop Workstation and Apple Handheld Devices

Device	Physical Size	Display Resolution	Pixel Pitch (μm)	Luminance Ratio
Standard 3 MP primary display monitor	20.8″	2048 × 1536	210	750:1
iPad (with Retina display)[a]	9.7″	2048 × 1536	96	780:1
iPad mini (with Retina display)	7.9″	2048 × 1536	78	810:1
iPhone 5[b]	4″	1136 × 640	78	1320:1

[a] Technical specifications from the Apple website, https://www.apple.com/au/ipad/compare/, accessed March 2014.

[b] Technical specifications from the Apple website, https://www.apple.com/au/iphone/compare/, accessed March 2014.

but they have not gained widespread acceptance as has the DICOM GSDF [48]. The developers of some image-viewing apps integrate visual calibration tools into their software and block the display of medical images if the users fail to accurately perceive contrast changes at various gray levels. Despite the recognized limitation of handheld devices, the U.S. FDA has approved an app (for use on an iPad or iPhone) for primary diagnosis.* Notably, the approval is for small-matrix images only and only in situations when a primary display is not available. The iPad does not meet the required minimum diagonal screen size of 15″ required by the quality assurance guidelines and, hence, is unlikely to be approved for primary interpretation in Germany [49].

Handheld devices have limitations in processing power, storage capacity, memory, video processing, and network speed when compared to desktop workstations. The limited RAM and storage capacity of mobile devices inhibits the ability to cache (temporarily store) images when a mobile device is used to review images [50]. Caching is similar to buffering a video stream and is used to allow the smooth transition from image to image when scrolling through image sets. Handheld devices use either a mobile broadband (3G or 4G) or wireless LAN (Wi-Fi) connection for network connectivity. Wi-Fi protocols have gone through a number of iterations and the latest version 802.11ac has a theoretical bandwidth approaching that of a wired 1 Gb/s Ethernet connection; however, in real-world situations a wired connection is almost always faster due to protocol and security overhead of wireless protocols. The speed and coverage of mobile broadband networks have been identified as a limitation to using handheld devices for teleradiology [51]. Mobile broadband has improved dramatically with the advent of 4G networks but coverage is still limited. A 4G device will automatically revert to 3G when there is no available 4G connection. 3G networks are noticeably slower than ADSL2+ and can be unpredictable in signal quality. Handheld devices do not have separate graphics cards; therefore, their graphics processing power is lower than that of a desktop workstation with a dedicated graphics card.

22.5.2 Clinical Efficacy

Before a handheld device can be incorporated into patient care, performance under clinical conditions must be validated [52]. Hence, the diagnostic accuracy and utility of handheld devices for teleradiology have been evaluated in a number of published studies. None of these studies are comprehensive systemic evaluations; instead, they tend to focus on a single pathology, modality, or situation. For example, iPads have been evaluated by Panughpath et al. [53] for diagnosing intracranial hemorrhages on CT; Abboud et al. [52] for screening for tuberculosis on chest radiographs; Johnson et al. [54] for diagnosing pulmonary embolism on CT; and McNulty et al. [55] for diagnosing spinal trauma on MRI. The iPad is most often assessed in these studies because it is the most accepted alternative display to the PACS workstation [55]. However, evaluation of iPhones [56–57] and PDAs [58] has also been undertaken. Most authors recognize that handheld devices have little application or advantage in the primary reading environment [49]. Instead, they will most likely be used for out-of-hours emergency interpretation in an on-call situation. The published studies try to emulate this situation. For example, Panughpath et al. [53] use a secondary LCD display as the control for their comparative diagnostic accuracy study based on the rationale that this type of display

---

* FDA News Release, February 4, 2001, available at http://www.fda.gov/NewsEvents/Newsroom/Press Announcements/ucm242295.htm, accessed March 2014.

is the most likely to be used by an on-call radiologist. Similarly, the chosen data set of images was selected to represent common acute conditions found in emergency department imaging [49,50,59].

In the published literature there is little homogeneity in the study designs used to test the efficacy of handheld devices for diagnostic accuracy, with some authors reporting receiver operating characteristic (ROC) [51,55,60]; others, sensitivity and specificity only [56]; others, accuracy rate [53–54]; others, the number of major or minor discrepancies in radiological reports [50,59]; and others, correlation coefficient [49]. Further, both primary and secondary displays have been used as a control display; the gold-standard diagnosis has varied between studies and different reader and study assignments have been employed. For example, the Dorfman–Berbaum–Metz multireader, multicase was used by Toomey et al. [58] and McNulty et al. [55], whereas the same data set was read on both handheld and intervention displays in other studies [51,61]. In a further variation in study design, the data set was read on both handheld and intervention displays and the diagnoses were compared to a gold-standard diagnosis usually on a PACS workstation [50,56]. The lack of homogeneity prevents the consolidation of results and the extrapolation of findings to other diseases and situations.

Despite the variations in study design, when significance has been tested, all authors (with the exception of one) report no significant statistical difference in the diagnostic accuracy of a handheld device compared to that of a control read [49,51,54–55,57–58,61]. Further, high sensitivity, specificity, and accuracy have been reported when using handheld devices [46,51,54,61]. Resultantly, the conclusion of most authors is that a handheld device can be used for accurate interpretation of (some) radiographic images. However, they often advise a cautious approach to the use of handheld devices for primary interpretation by recommending that they be used to provide initial or preliminary readings only. The one opposing view resulted from a significantly smaller area under the ROC for an iPad when compared to a gamma-calibrated display for the detection of cerebral infarction on brain CT [60].

One of the difficulties of comparative diagnostic accuracy studies is accounting for interobserver agreement. Johnson et al. [54] and McLaughlin et al. [59] suggested that reporting errors are attributable to the reader's interpretation style, not the display, a view supported by the high intraobserver variability reported by Tewes et al. [49]. Pathologies such as mucosal thickening of the sinuses and a small arachnoid cyst were counted as discrepancies in McLaughlin et al.'s study [59] but the need to report these findings is not universally required.

Secondary observations of the use of handheld devices were elicited in many of the studies. The screen size of the iPhone was thought to be responsible for imprecise measurements with significant difference in linear measurement observed between an iPhone and a primary display [56]. The various software applications were criticized by a number of the authors for difficulty in scrolling and touch movements [51]; cumbersome user interface, especially when trying to compare previous imaging [50]; the lack of postprocessing tools [50,56,59]; and instability, especially in large studies. Limitations in network coverage and speed and potential effect of ambient lighting on diagnosis were also noted. The inability to access clinical systems [59] and the increased time to perform a read compared to that on a primary workstation [60] were other limitations of using handheld devices for teleradiology. The portability of handheld devices was seen as a major advantage [51].

The discussion so far has centered on using handheld devices for primary diagnosis. But handheld devices are also used for the clinical review of imaging. The portability of handheld devices can be used to allow patients confined to bed to see their imaging [62], a practice clinicians found to improve their patients' satisfaction, understanding, and overall experience.

22.6 Conclusions

This chapter has presented a synthesis and critical appraisal of contemporary teleradiology themes. The influence teleradiology has had on reshaping the practice of radiology as a whole was established. The use of handheld devices, e.g., iPads, for teleradiology have been analyzed in detail. This chapter has also investigated clinical scenarios where teleradiology is used. Further, background information on DICOM, workstation characteristics, digital image fundamentals, and display characteristics was presented.

Teleradiology, perhaps unintentionally, has reshaped the practice of radiology. The ability to transmit images from one location to another presented business opportunities for some, which led to radiology being corporatized and commoditized. Proponents of teleradiology argue that it benefits the patient by improving coverage of services, subspecialty expertise, and speed of interpretation. However, teleradiology threatens the income and existence of on-site radiology. To compete with teleradiologists, local radiologists have had to improve their visibility and communication with referring doctors. In addition, they have emphasized and participated in the value-adding, ancillary components of their profession—for example, overseeing appropriate imaging, contributing to multidisciplinary meetings, and educating trainees. Professional bodies have now developed practice standards and policies in response to concerns about the practice of teleradiology. The regulatory oversight of teleradiology, the resurgence in noninterpretation components of the radiology profession, and the competition between teleradiologist and local radiologist should all contribute to better radiology and resultantly better patient outcomes.

Mobile teleradiology holds the promise of increasing the availability of a radiologist for consultation and reducing the need for reporting from a fixed device. While handheld devices are comparable to primary desktop workstations in some qualities, they do not conform to DICOM GSDF and are unable to be calibrated. Despite this limitation the FDA has given approval for iPads and iPhones with particular software apps to be used for primary diagnosis. The diagnostic efficacy has been tested in a number of studies. The overwhelming results show that the accuracy of diagnosis on a handheld device matches that of a primary or secondary display. However, many identified limitations of using handheld devices, ranging from restricted network coverage, limited bandwidth, cumbersome user interface, and imprecise measurement accuracy. The use of handheld devices for primary diagnoses is still practiced with caution, with many authors recommending that they be used to provide initial or preliminary reports only. The portability of handheld devices is seen as a major advantage and can be used to show patients confined to bed their imaging in order to improve patient understanding and overall experience.

Abbreviations and Glossary

bandwidth: the data-carrying capacity of a network connection; also called *throughput*. It is measured in volume per unit of time. Bandwidth is often measured in bits per second (bit/s or b/s or bps). It is important to note that bandwidth uses bits (abbreviated as b) as the volume measure as opposed to bytes (abbreviated as B). A byte is equal to 8 bits. We often see bandwidth expressed in kilobits per second (kb/s or kbits/s or kbps) or megabits per second (Mb/s or Mbits/s or Mbs) or gigabits per second (Gb/s or Gbits/s or Gbps). To compare, 1 GB/s = 1000 Mb/s = 1,000,000 bps.

digital imaging and communication in medicine (DICOM): an international standard that was originally developed by ACR and NEMA. DICOM has been adopted by the International Standards Organization (ISO) and is universally adopted by modality, PACS, and teleradiology vendors. DICOM is a broad standard that defines an image file format, metadata, image transmission protocols, work-flow management protocols, and display calibration methods.

health level 7 (HL7): a health standard for the transfer of text-based information from one information system to another—for example, transferring a patient's demographics from a patient management system to the RIS, or transferring a radiological report from the RIS to an electronic medical record.

luminance: the intensity of light emitted per unit area; measured in candelas per meter squared (cd/m^2). Luminance is used as a quality metric when assessing display devices. The higher the luminance, the better. Luminance is analogous to brightness.

luminance ratio (LR): the ratio of the maximum luminance (white pixel) to the minimum luminance (black pixel). Luminance ratio is used as a quality metric when assessing display devices. Generally, the higher the LR, the better is the quality of the display. However, LR, maximum luminance, and minimum luminance should all be evaluated, as a high LR can result from a very low minimum luminance value, which is not recommended for primary displays.

picture archiving and communication system (PACS): a computer system, both the hardware and the software, for the storage, distribution, and review of digital medical images. PACS facilitates a "filmless" imaging service. Teleradiology and PACS are often integrated and will share infrastructure—for example, primary review stations and image archives.

pixel pitch: the physical size of a pixel.

radiology information system (RIS): a computerized information system used to support radiology services. The function of the RIS is patient registration, order entry, scheduling, patient tracking, billing, report entry, and report distribution. The RIS is normally integrated with PACS. It is often said that a RIS manages text-based information, whereas a PACS manages images.

receiver operating characteristic (ROC) curve: a graphical plot of true positive rate (sensitivity) versus false positive rate (1 – specificity). It is used in research to assess performance of a diagnostic test. Comparison of two diagnostic tests can be done by calculating and comparing the areas under the curve (AUCs) for the ROCs of the tests.

spatial resolution: a quality metric related to the size of the pixel in an image. A smaller pixel size is synonymous with a lager image matrix or more pixels in an image.

As the pixel size decreases, the spatial resolution is said to increase and visually the image becomes clearer and diagnostically the viewer is able to resolve smaller objects within the image.

References

1. European Society of Radiology White Paper on Teleradiology: An Update from the Teleradiology Subgroup. *Insights into Imaging.* Jan 18, 2014.
2. Adler J, Yu C, and Datta M. The Changing Face of Radiology: From Local Practice to Global Network. *The Medical Journal of Australia.* Jan 5, 2009; 190(1):20–23.
3. Char A, Kalyanpur A, Puttanna Gowda VN, Bharathi A, and Singh J. Teleradiology in an Inaccessible Area of Northern India. *Journal of Telemedicine and Telecare.* 2010; 16(3):110–113.
4. Lee JK, Renner JB, Saunders BF, Stamford PP, Bickford TR, Johnston RE et al. Effect of Real-Time Teleradiology on the Practice of the Emergency Department Physician in a Rural Setting: Initial Experience. *Academic Radiology.* Aug 1998; 5(8):533–538.
5. O'Leary B. Lending Rural Hospitals a Hand: By Networking Its PACS with Eight Remote Hospitals, Kalispell Regional Medical Center Brought Radiologists (Virtually) to Them. *Healthcare Informatics.* Apr 2006; 23(4):52–53.
6. Ranschaert ER and Binkhuysen FH. European Teleradiology Now and in the Future: Results of an Online Survey. *Insights into Imaging.* Feb 2013; 4(1):93–102.
7. Ryan J, Khandelwal A, and Fasih N. Trends in Radiology Fellowship Training: A Canadian Review 2009–2011. *Canadian Association of Radiologists Journal = Journal l'Association Canadienne des Radiologistes.* Aug 2013; 64(3):176–179.
8. Eakins C, Ellis WD, Pruthi S, Johnson DP, Hernanz-Schulman M, Yu C et al. Second Opinion Interpretations by Specialty Radiologists at a Pediatric Hospital: Rate of Disagreement and Clinical Implications. *American Journal of Roentgenology.* Oct 2012; 199(4):916–920.
9. Leung RS, Fairhurst J, Johnson K, Landes C, Moon L, Sprigg A et al. Teleradiology: A Modern Approach to Diagnosis, Training, and Research in Child Abuse? *Clinical Radiology.* Jun 2011; 66(6):546–550.
10. Zan E, Yousem DM, Carone M, and Lewin JS. Second-Opinion Consultations in Neuroradiology. *Radiology.* Apr 2010; 255(1):135–141.
11. Ng KL, Yazer J, Abdolell M, and Brown P. National Survey to Identify Subspecialties at Risk for Physician Shortages in Canadian Academic Radiology Departments. *Canadian Association of Radiologists Journal = Journal l'Association Canadienne des Radiologistes.* Dec 2010; 61(5):252–257.
12. Horn B, Chang D, Bendelstein J, and Hiatt JC. Implementation of a Teleradiology System to Improve After-Hours Radiology Services in Kaiser Permanente Southern California. *The Permanente Journal.* Spring 2006; 10(1):47–50.
13. Lewis RS, Sunshine JH, and Bhargavan M. Radiology Practices' Use of External Off-Hours Teleradiology Services in 2007 and Changes Since 2003. *American Journal of Roentgenology.* Nov 2009; 193(5):1333–1339.
14. Eklof H, Radecka E, and Liss P. Teleradiology Uppsala–Sydney for Nighttime Emergencies: Preliminary Experience. *Acta Radiologica.* Oct 2007; 48(8):851–853.
15. Kalyanpur A, Weinberg J, Neklesa V, Brink JA, and Forman HP. Emergency Radiology Coverage: Technical and Clinical Feasibility of an International Teleradiology Model. *Emergency Radiology.* Dec 2003; 10(3):115–118.
16. Boland GW. Teleradiology Coming of Age: Winners and Losers. *American Journal of Roentgenology.* May 2008; 190(5):1161–1162.
17. Davis A. Outsourced Radiology: Will Doctors Be Deskilled? *BMJ.* 2008; 337:a785.

18. Flanagan PT, Relyea-Chew A, Gross JA, and Gunn ML. Using the Internet for Image Transfer in a Regional Trauma Network: Effect on CT Repeat Rate, Cost, and Radiation Exposure. *Journal of the American College of Radiology.* Sep 2012; 9(9):648–656.

19. Ashkenazi I, Haspel J, Alfici R, Kessel B, Khashan T, and Oren M. Effect of Teleradiology upon Pattern of Transfer of Head Injured Patients from a Rural General Hospital to a Neurosurgical Referral Centre. *Emergency Medicine Journal.* Aug 2007; 24(8):550–552.

20. Poon WS and Goh KY. The Impact of Teleradiology on the Inter-Hospital Transfer of Neurosurgical Patients and Their Outcome. *Hong Kong Medical Journal = Xianggang yi xue za zhi Hong Kong Academy of Medicine.* Sep 1998; 4(3):293–295.

21. Rubin MN and Demaerschalk BM. The Use of Telemedicine in the Management of Acute Stroke. *Neurosurgical Focus.* Jan 2014; 36(1).

22. Demaerschalk BM, Bobrow BJ, Raman R, Ernstrom K, Hoxworth JM, Patel AC et al. CT Interpretation in a Telestroke Network Agreement among a Spoke Radiologist, Hub Vascular Neurologist, and Hub Neuroradiologist. *Stroke.* Nov 2012; 43(11):3095–3097.

23. McGill AF and North JB. Teleconference Fracture Clinics: A Trial for Rural Hospitals. *ANZ Journal of Surgery.* Jan–Feb 2012; 82(1–2):2–3.

24. Caffery L and Manthey K. Implementation of a Web-Based Teleradiology Management System. *Journal of Telemedicine and Telecare.* 2004; 10(Suppl 1):22–25.

25. Digital Imaging and Communication in Medicine (DICOM) Supplement 113: Email Transport. 2006. Available at http://medical.nema.org/Dicom/2011/11_14pu.pdf (Last accessed March 2014).

26. Weisser G, Engelmann U, Ruggiero S, Runa A, Schroter A, Baur S et al. Teleradiology Applications with DICOM-E-Mail. *European Radiology.* May 2007; 17(5):1331–1340.

27. Bashshur RL, Shannon G, Krupinski EA, and Grigsby J. Sustaining and Realizing the Promise of Telemedicine. *Telemedicine and E-Health.* May 2013; 19(5):339–345.

28. European Society of Radiology. Teleradiology in the European Union. 2006. Available at https://www.myesr.org/html/img/pool/ESR_2006_VII_Telerad_Summary_Web.pdf (Last accessed March 2014).

29. Borgstede JP. Radiology: Commodity or Specialty. *Radiology.* Jun 2008; 247(3):613–616.

30. Steinbrook R. The Age of Teleradiology. *The New England Journal of Medicine.* Jul 5, 2007; 357(1):5–7.

31. Bradley WG. Off-Site Teleradiology: The Pros. *Radiology.* Aug 2008; 248(2):337–341.

32. Agrawal A, Agrawal A, Pandit M, and Kalyanpur A. Systematic Survey of Discrepancy Rates in an International Teleradiology Service. *Emergency Radiology.* Jan 2011; 18(1):23–29.

33. Hohmann J, de Villiers P, Urigo C, Sarpi D, Newerla C, and Brookes J. Quality Assessment of Out Sourced After-Hours Computed Tomography Teleradiology Reports in a Central London University Hospital. *European Journal of Radiology.* Aug 2012; 81(8):e875–879.

34. Wong WS, Roubal I, Jackson DB, Paik WN, and Wong VK. Outsourced Teleradiology Imaging Services: An Analysis of Discordant Interpretation in 124,870 Cases. *Journal of the American College of Radiology.* Jun 2005; 2(6):478–484.

35. Platts-Mills TF, Hendey GW, and Ferguson B. Teleradiology Interpretations of Emergency Department Computed Tomography Scans. *The Journal of Emergency Medicine.* Feb 2010; 38(2):188–195.

36. Thrall JH. Teleradiology Part II: Limitations, Risks, and Opportunities. *Radiology.* Aug 2007; 244(2):325–328.

37. Dixon AK and FitzGerald R. Outsourcing and Teleradiology: Potential Benefits, Risks and Solutions from a UK/European Perspective. *Journal of the American College of Radiology.* Jan 2008; 5(1):12–18.

38. Dhanoa D, Dhesi TS, Burton KR, Nicolaou S, and Liang T. The Evolving Role of the Radiologist: The Vancouver Workload Utilization Evaluation Study. *Journal of the American College of Radiology.* Oct 2013; 10(10):764–769.

39. The Royal College of Radiologists. Standards for the Provision of Teleradiology within the United Kingdom. 2010. Available at https://www.rcr.ac.uk/docs/radiology/pdf/BFCR%2810%297_Stand_telerad.pdf (Last accessed March 2014).

40. Canadian Association of Radiologists. CAR Standards for Teleradiology. 2008. Available at http://www.car.ca/uploads/standards%20guidelines/Standard_Teleradiology_EN.pdf (Last accessed March 2014).

41. The Royal Australian and New Zealand College of Radiology. Position on Teleradiology. 2001. Available at http://www.ranzcr.edu.au (Last accessed March 2014).

42. Andriole KP, Ruckdeschel TG, Flynn MJ, Hangiandreou NJ, Jones AK, Krupinski E et al. ACR-AAPM-SIIM Practice Guideline for Digital Radiography. *Journal of Digital Imaging.* Feb 2013; 26(1):26–37.

43. Norweck JT, Seibert JA, Andriole KP, Clunie DA, Curran BH, Flynn MJ et al. ACR-AAPM-SIIM Technical Standard for Electronic Practice of Medical Imaging. *Journal of Digital Imaging.* Feb 2013; 26(1):38–52.

44. Digital Imaging and Communication in Medicine (DICOM) Part 14: Grayscale Standard Display Function. 2011. Available at http://medical.nema.org/Dicom/2011/11_14pu.pdf (Last accessed March 2014).

45. Krupinski EA and Roehrig H. The Influence of a Perceptually Linearized Display on Observer Performance and Visual Search. *Academic Radiology.* Jan 2000; 7(1):8–13.

46. Choudhri AF, Shih G, and Kim W. American College of Radiology. Mobile Devices. 2013. Available at http://www.acr.org/~/media/ACR/Documents/PDF/Advocacy/IT%20Reference%20Guide/IT%20Ref%20Guide%20%20Mobile%20Devices.pdf (Last accessed March 2014).

47. Yamazaki A, Liu P, Cheng WC, and Badano A. Image Quality Characteristics of Handheld Display Devices for Medical Imaging. *PLoS One.* 2013; 8(11):e79243.

48. De Paepe L, De Bock P, Vanovermeire O, and Kimpe T. Performance Evaluation of a Visual Display Calibration Algorithm for iPad. Medical Imaging 2012: Advanced PACS-Based Imaging Informatics and Therapeutic Applications. *Proceedings of SPIE 2012.* February 16, 2012; 8319.

49. Tewes S, Rodt T, Marquardt S, Evangelidou E, Wacker FK, and von Falck C. Evaluation of the Use of a Tablet Computer with a High-Resolution Display for Interpreting Emergency CT Scans. *RöFo: Fortschritte auf dem Gebiete der Rontgenstrahlen und der Nuklearmedizin.* Nov 2013; 185(11):1063–1069.

50. John S, Poh AC, Lim TC, Chan EH, and Chong IR. The iPad Tablet Computer for Mobile On-Call Radiology Diagnosis? Auditing Discrepancy in CT and MRI Reporting. *Journal of Digital Imaging.* Oct 2012; 25(5):628–634.

51. Park JB, Choi HJ, Lee JH, and Kang BS. An Assessment of the iPad 2 as a CT Teleradiology Tool Using Brain CT with Subtle Intracranial Hemorrhage under Conventional Illumination. *Journal of Digital Imaging.* Aug 2013; 26(4):683–690.

52. Abboud S, Weiss F, Siegel E, and Jeudy J. TB or Not TB: Interreader and Intrareader Variability in Screening Diagnosis on an iPad versus a Traditional Display. *Journal of American College of Radiology.* Jan 2013; 10(1):42–44.

53. Panughpath SG, Kumar S, and Kalyanpur A. Utility of Mobile Devices in the Computerized Tomography Evaluation of Intracranial Hemorrhage. *Indian Journal of Radiology and Imaging.* Jan 2013; 23(1):4–7.

54. Johnson PT, Zimmerman SL, Heath D, Eng J, Horton KM, Scott WW et al. The iPad as a Mobile Device for CT Display and Interpretation: Diagnostic Accuracy for Identification of Pulmonary Embolism. *Emergency Radiology.* Aug 2012; 19(4):323–327.

55. McNulty JP, Ryan JT, Evanoff MG, and Rainford LA. Flexible Image Evaluation: iPad versus Secondary-Class Monitors for Review of MR Spinal Emergency Cases, a Comparative Study. *Academic Radiology.* Aug 2012; 19(8):1023–1028.

56. Choudhri AF, Carr TM, III, Ho CP, Stone JR, Gay SB, and Lambert DL. Handheld Device Review of Abdominal CT for the Evaluation of Acute Appendicitis. *Journal of Digital Imaging.* Aug 2012; 25(4):492–496.

57. Modi J, Sharma P, Earl A, Simpson M, Mitchell JR, and Goyal M. iPhone-Based Teleradiology for the Diagnosis of Acute Cervico-Dorsal Spine Trauma. *The Canadian Journal of Neurological Sciences = Le Journal Canadien des Sciences Neurologiques.* Nov 2010; 37(6):849–854.

58. Toomey RJ, Ryan JT, McEntee MF, Evanoff MG, Chakraborty DP, McNulty JP et al. Diagnostic Efficacy of Handheld Devices for Emergency Radiologic Consultation. *American Journal of Roentgenology.* Feb 2010; 194(2):469–474.
59. McLaughlin P, Neill SO, Fanning N, McGarrigle AM, Connor OJ, Wyse G et al. Emergency CT Brain: Preliminary Interpretation with a Tablet Device: Image Quality and Diagnostic Performance of the Apple iPad. *Emergency Radiology.* Apr 2012; 19(2):127–133.
60. Yoshimura K, Nihashi T, Ikeda M, Ando Y, Kawai H, Kawakami K et al. Comparison of Liquid Crystal Display Monitors Calibrated with Gray-Scale Standard Display Function and with Gamma 2.2 and iPad: Observer Performance in Detection of Cerebral Infarction on Brain CT. *American Journal of Roentgenology.* Jun 2013; 200(6):1304–1309.
61. Panughpath SG, Kumar S, and Kalyanpur A. Utility of Mobile Devices in the Computerized Tomography Evaluation of Intracranial Hemorrhage. *Indian Journal of Radiology and Imaging.* Jan 2013; 23(1):4–7.
62. Furness ND, Bradford OJ, and Paterson MP. Tablets in Trauma: Using Mobile Computing Platforms to Improve Patient Understanding and Experience. *Orthopedics.* Mar 2013; 36(3): 205–208.

Further Reading

Kumar S and Krupinski E (eds). *Teleradiology.* New York: Springer; 2008.
Pianykh OS. DICOM and Teleradiology, in *Digital Imaging and Communication in Medicine (DICOM): A Practical Introduction and Survival Guide,* Second Edition. New York: Springer; 2012.

Technical Guidelines

Norweck JT, Seibert JA, Andriole KP, Clunie DA, Curran BH, Flynn MJ et al. ACR-AAPM-SIIM Technical Standard for Electronic Practice of Medical Imaging. *Journal of Digital Imaging.* 2013; 26(1):38–52.

23

Teledermatology

Soegijardjo Soegijoko, Arni Ariani, and Sugiyantini

CONTENTS

23.1   Introduction

Skin conditions can lead to either physical or psychological consequences or even death. Findings from a survey in Italy showed that patients with vitiligo have experienced worries of getting severe vitiligo (60%), anger (37%), embarrassment (34%), depression (31%), adverse effects on social life (28%), and shame (28%) [1]. Certain studies revealed that patients with psoriasis are humiliated with their skin's appearance [2], experienced lower quality of life [3], faced the possibility of unemployment, and struggled with financial issues [4–5].

Baibergenova et al. [6] examined mortality data from an 8-year period (from 2000 to 2007) in Canada released that there were 115 deaths related to pemphigoid, 84 deaths associated with pemphigus, and 44 deaths from toxic epidermal necrolysis. A 2010–2011 report by the World Health Organization estimated that there would be 33,409 deaths in the African region in 2013 due to skin conditions, accounting for 31.15% out of 106,089 deaths associated with skin conditions in the world [7].

Findings from one study revealed that the total costs of all skin cancers in England in 2002 were around £192.585 million. Indirect costs accounted for approximately £111.01 million (the income losses due to illnesses or deaths) and direct costs of £81.575 million (the healthcare costs of treating skin cancers) [8]. The economic burden of melanoma reported by the Lewin Group, Inc., in the United States was $39.3 billion in the year of 2004, including $10.2 billion lost productivity costs, $10.4 million in ambulatory care costs, $9 billion in hospital inpatient charges, $9.7 billion in prescription drug and OTC product costs [9].

A dermatologist is a physician who specializes in skin diseases and disorders [10]. The practice of dermatology includes the treatment of skin conditions (for instance, skin cancers, skin diseases that are accompanied by inflammation and blisters, skin infections, and hair and nail disorders) [11]. The United States had a ratio of 28,563 people per dermatologist in 2010 [12]. Furthermore, there were 57.8% out of 10,815 dermatologists aged less than 55 years and 42.2% out of 10,815 dermatologists aged 55 years and over [12]. The reasons behind the shortage of dermatologists are an insufficient number of dermatologists in certain geographical regions and a growing number of cosmetic dermatologists [13]. These situations have increased waiting times for those needing a dermatologist's opinion.

Teledermatology is one of the key solutions to solving the problem of inadequate access to dermatological services. It enables the delivery of dermatological services to remote and distant locations via telecommunication infrastructures [14–16]. It supports remote sessions between a patient (and the general practitioner) and a dermatologist for initial diagnosis and follow-up care [17]. It also provides dermatological education for primary care physicians [18–20]. It offers a great benefit in terms of cost savings for the patient and the healthcare system [20–24].

23.2   Structure of the Skin

In terms of size and weight, the integument or skin is a massive organ in the human body [25]. The mass and the surface area of an adult's skin are approximately 3 kg and 2 m², respectively [26]. It has a number of essential roles, including preserving a barrier against external threats (mechanical, thermal and physical injury, and harmful agents), maintaining the moisture levels in the human body, minimizing the effects of ultraviolet

(UV) radiation exposure, serving as a sensory organ, acting as thermoregulation, providing immunological surveillance, synthesizing vitamin D3 (cholecalciferol), and relating to physical appearances and social behaviors [25–26]. It can be divided into three primary layers: the epidermis, the dermis, and the hypodermis [25], as shown in Figure 23.1 [27].

The epidermis, the outer layer of the skin, has a thickness that ranges between 0.05 and 1.5 mm. It consists of four sublayers: a horny layer (stratum corneum), a granular cell layer (stratum granulosum), a prickle cell layer (stratum spinosum), and a germinativum cell layer (stratum basale) [25]. Those sublayers are composed of four types of cells: keratinocytes, melanocytes, Merkel cells, and Langerhans cells. Keratin, which is produced by keratinocytes, provides waterproof barriers and enhances the strength and flexibility of the skin [25]. Melanin, which is constructed by melanocytes, determines the color of the skin [25]. Merkel cells deal with touch reception [25]. Langerhans cells produce antigens for intensifying the immune system [26].

The dermis has a thickness that varies between 0.6 and 3 mm. It comprises three main parts: collagen fibers (proteins that contribute to the skin's strength and toughness), elastic fibers (proteins that are responsible for preserving the skin's flexibility and elasticity), and structural proteoglycans (proteins that produce a hydrated and viscous fluid) [25]. It also consists of mast cells and macrophages. It has two main sublayers: a thin papillary layer and a thicker reticular layer [25].

The hypodermis has a responsibility to retain the body's heat [28]. It is mainly composed of loose connectivity tissue and fat, with a thickness that is nearly 3 cm on the abdomen part [25]. It also comprises a capillary network, the hair follicles, nerve fibers, and sweat glands [26].

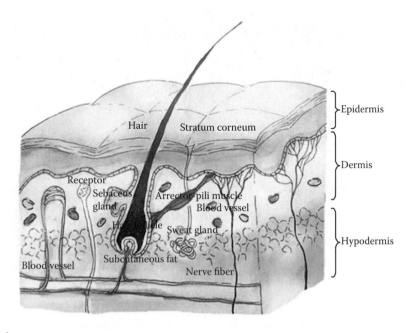

FIGURE 23.1
The cross section of skin shows the three top main layers, including the epidermis, the dermis, and the hypodermis. (Illustrator A. P. Koesoema, "Skin," 2014.)

23.3 Common Skin Diseases

23.3.1 Eczema

There are three specific stages related to the development of eczema: infancy, childhood, and adolescence/adulthood [29]. In a study conducted by Bieber [30], approximately 60% of all eczema cases occurred in the infancy period (infantile eczema). Findings from another study showed that around 45% of all eczema cases commonly occurred in children aged between 2 and 6 months and are characterized by itchy, redness, and small bumps on the cheeks, forehead, or scalp and also may spread down to the trunk [31]. The development of eczema in children 4 to 10 years of age is distinguished by raised, itchy, and scaly bumps on the face, trunk, or both of the body parts and also followed by dry and thickened skin [31]. The development of eczema in the adolescent group occurred during or after puberty and is identified by itchy, dry, scaly skin that may persist until adulthood [31]. Regardless of age onset, there are two common conditions that appeared during every single stage: pruritic erythematous papules and plaques with secondary skin peeling [31].

One of the most common eczema cases is atopic dermatitis (AD). AD is associated with a genetic predisposition to develop hypersensitivity skin reactions [32]. In this case, the symptoms of eczema can come and go and sometimes get worse by contact with allergens or irritants, emotional stress, excessive drying out of the skin, heat, infections, and sweating [32]. The appearance of AD in children and adults can be seen on various parts of the body (on the elbow and knee folds, on the wrists and ankles, and on the face and neck) [32–34] (Figure 23.2). A study conducted by Horii et al. [35] reported that in a period

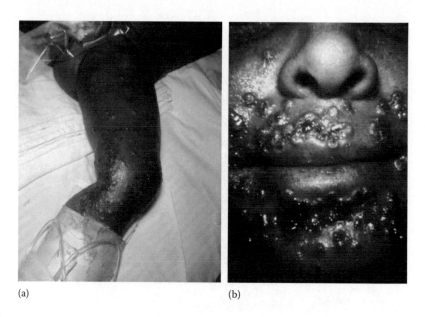

(a)   (b)

FIGURE 23.2
Some examples of eczema: (a) a child had suffered from eczema vaccinatum after receiving a smallpox vaccination (From CDC, "ID: 13325," 1969. [Online]. Available at http://phil.cdc.gov, accessed on Aug. 18, 2014); and (b) a woman got her vaccination and 8 days later was diagnosed with moderately severe eczema vaccinatum (From CDC and A. E. Kaye, "ID: 3305," 1969. [Online]. Available at http://phil.cdc.gov, accessed on March 18, 2014).

of 8 years (from 1997 to 2004), there were 7.4 million visits to U.S. physicians by children aged under 18 years for AD. Other studies also revealed that the prevalence of AD in the United States [36] and Europe [37] were around 17% and 10%–20%, respectively. The total cost of AD treatment in the United States is estimated between $364 million and $3.8 billion annually [38–39]. A small number of people may experience adverse reactions after receiving smallpox vaccination ranging from mild to rare life-threatening side effects [40]. Acute side effects of smallpox vaccine include death, eczema vaccinatum, progressive vaccinia, and postvaccinial encephalitis [40]. The finding from this study showed that eczema vaccinatum occurs in individuals previously diagnosed with skin conditions [40]. Some example cases can be seen in Figure 23.2.

23.3.2 Fungal/Yeast Infections

The three most widely known groups of fungal diseases are [41]

- Invasive life-threatening infections, particularly those caused by *Candida* and *Aspergillus* spp.;
- Superficial infections including athlete's foot, scalp ringworm, nail infections, and thrush; and
- Allergic fungal sinusitis and allergic fungal respiratory diseases.

According to a Health Protection Agency analysis of 2002 census data, there were nearly 10,000 patients with a deep fungal infection in the United Kingdom and nearly 50% of them died as a result [41]. The study also revealed that the total budget required for antifungal drugs exceeded €90 million in 2002 [41].

Candidiasis could be classified into two categories: deep and superficial [42]. Deep candidiasis appeared in the cerebrospinal compartment, gastrointestinal and urinary tracts, heart, and lungs, while superficial candidiasis can be found in the nails, oral cavity, skin, and vaginal mucous membranes [42]. A number of studies have shown that invasive candidiasis may lead to severe organ damage [43] and even death [44–46]. Some example cases can be seen in Figure 9.3. Oral candidiasis (also known as oral thrush) is one of the earliest HIV symptoms [47]. Oral candidiasis can be divided into four categories including angular cheilitis and atrophic candidiasis, hyperplastic candidiasis and pseudomembranous candidiasis [47]. According to data in 2000, the attributable increase in mortality rate due to invasive candidiasis was 14.5% (range of 12.1% to 16.9%), the average increase in length of stay was 10.1 days (range of 8.9 to 11.3 days), and the average increase in the costs of hospitalization was $39,331 (range of $33,604 to $45,602) among adult patients in the United States [44]. Some examples of fungal diseases are vulvitis, or vulvovaginal candidiasis due to *Candida albicans* in the female's perineum [48] and oral candidiasis infection found in patient with human immunodeficiency virus (HIV)/acquired immune deficiency syndrome (AIDS) [49].

23.3.3 Bacterial Infections

The three most common bacterial skin infections treated by family physicians are cellulitis, impetigo, and folliculitis [50].

Cellulitis is a skin infection that spreads into the deeper layers of the skin—the dermis and subcutaneous tissue—and is caused by several types of bacteria (*Streptococcus* or *Staphylococcus* species) [50]. According to 2011–2012 data in Australia, cellulitis is the fourth highest cause of preventable hospitalization and accounted for 45,136 out of 342,278 hospitalizations (13.19%) [51]. The cost of cellulitis treatment can vary depending on the severity of disease and the mechanism of treatment. In the United Kingdom, cellulitis is responsible for 82,113 hospital admissions, with the average length of stay around 7.2 days at the estimated cost of €133 million [52].

Impetigo is a contagious skin infection that most commonly occurs in children and is caused by causative organisms (*Staphylococcus aureus* or streptococci, alone and in combination) [53–55]. It can be categorized into two groups: bullous and nonbullous (i.e., impetigo contagiosa). *Staphylococcus aureus*, which produces exfoliative toxins, is considered to be the cause of bullous impetigo, whereas *Staphylococcus aureus* alone or in combination with *Streptococcus pyogenes* is the common cause of nonbullous impetigo [56–57]. Impetigo commonly spreads through direct skin contact with the infected person [53]. Impetigo accounted for a total of 130,095 cases in children aged between 0 and up to 4 years who attended general practitioner clinics in the United Kingdom from 1995 to 2010 [58]. An example of impetigo can be seen in Figure 23.3a [59].

Folliculitis is characterized by inflammation of the hair follicle [60] and is usually caused by *Staphylococcus aureus* or methicillin-resistant *Staphylococcus aureus* (MRSA). It can be found in certain parts of the body including scalp, face, neck, and buttocks [61]. There are two different types of folliculitis: superficial folliculitis (ostiofolliculitis) (Figure 23.3b) [62] or deep folliculitis (such as furuncle and carbuncle) [62]. A study by the Centers for Disease Control and Prevention (CDC) has reported that MRSA folliculitis is most often seen in children aged 6 to 11 years and women aged 60 years and over [63].

23.3.4 Viral Infections

Viral diseases such as human papillomavirus (HPV) and herpes simplex virus (HSV) infections could cause long-term or chronic diseases. HPV is a highly contagious virus that can

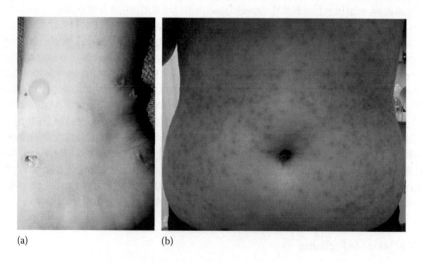

(a)   (b)

FIGURE 23.3
Some examples of bacterial infections: (a) the specialist first thought that the patient suffered from syphilis, but at the end, it was impetigo (From CDC, "ID: 5153," 1975. [Online]. Available at http://phil.cdc.gov, accessed on March 18, 2014); and (b) the related patient had suffered from hot tub folliculitis (From J. Heilman, "Hot tub folliculitis," 2013. [Online]. Available at http://commons.wikimedia.org/wiki/File:Folliculitis.JPG, accessed on March 18, 2014).

cause warts on different parts of the body [64] (for example, flat warts, genital warts, plantar warts, and oral warts) and various forms of cancers (for example, anal cancer, cervical cancer, oropharyngeal cancers, penile cancer, vaginal cancer, and vulvar cancer) [65].

Genital warts are commonly due to HPV-6 and HPV-11 [66] and spread through sexual contacts [67]. The warts are characterized by abnormal skin outgrowths (mostly small cauliflower-like bumps) in genital areas [68]. Kjær et al. [69] conducted a population-based cross-sectional study of populations in northern Europe to assess the incidence of genital warts among women aged between 18 and 45 years. The study revealed that the incidence rate was 10.6% (7351 out of 69,147 persons). A study among individuals with private insurance reported that the total estimated cost of genital wart diagnosis and treatment in the United States for the year 2000 was $140 million [70].

Herpes is a skin infection caused by either HSV-1 or HSV-2 [71]. The disease is usually transmitted from one person to another through direct skin contact and characterized by an individual blister or a cluster of blisters [72]. Furthermore, it can be passed on to the baby during pregnancy or labor or after the birth [73]. Figure 23.4 shows the case of herpes zoster (also known as shingles) [74]. The cause of shingles is the reactivation of varicella-zoster virus (VZV) [75]. The majority of people diagnosed with shingles are individuals with lower cell-mediated immune responses (such as older people) and individuals diagnosed with immunosuppressive disorders [75]. Owusu-Edusei et al. [76] conducted a study to estimate the lifetime medical costs attributable to sexually transmitted infections (STIs) in the United States. The study also revealed that the estimated cost in 2008 of treating patients with HSV-2 infection was $540.7 million.

Some examples of viral infections are incidence of venereal warts on a male patient and found in the anal–perineum area [75] and herpes zoster on some parts of the patient's body [76].

23.3.5   Parasitic Infections

A parasitic disease is an infectious disease caused by organisms called parasites. Parasites are transmitted to humans through the mouth or a break in the skin and are considered harmful to their hosts [77]. Parasitic infections are commonly found in countries with tropical climates, poor socioeconomic environments, and limited healthcare infrastructures [78].

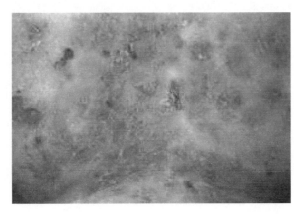

FIGURE 23.4
An example of viral infection is incidence of herpes zoster on some parts of the patient's body [74].

Scabies is one of the parasitic diseases caused by a tiny mite called *Sarcoptes scabiei* var. *hominis* [79,80] (Figure 23.5) and is transmitted through prolonged direct skin contact with the infected person [81]. A study conducted by Hay et al. [82] reported that the prevalence rates of scabies ranged from 0.4% to 31% between 2005 and 2010 in nine developing countries. A study conducted in Malaysia by Zayyid et al. [83] found that the presence of scabies among boys and girls in a welfare home was 50% (26 out of 52 persons) and 16% (11 out of 68 persons), respectively. The main contributing factors in transmitting scabies are poverty and overcrowded living environments [84]. Based on 2003 data in Canada, the total cost of controlling a scabies outbreak in a long-term care facility was CAD$200,000 [85].

23.3.6 Autoimmune Disease

Vitiligo is an autoimmune disease in which melanocytes are attacked and destroyed by the immune system, resulting in white skin patches that may gradually spread over the body [86–87] (Figure 23.6). Shajil et al. [88] assessed the incidence of vitiligo in a sample

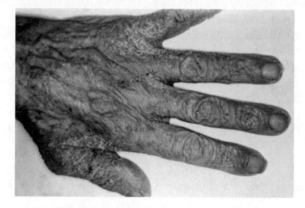

FIGURE 23.5
Scabies, which is caused by *Sarcoptes scabiei* var. *hominis*, on the hand of a patient. (From CDC, "ID: 4800," 1975. [Online]. Available at http://phil.cdc.gov.)

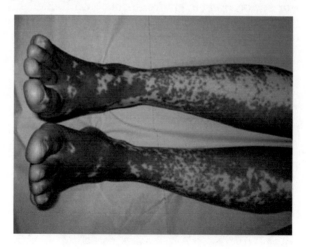

FIGURE 23.6
The incidence of vitiligo on darker skin. (From G.D. Oger, "Vitiligo," 2009. [Online]. Available at http://commons.wikimedia.org/wiki/File:Vitiligo.JPG, accessed on March 18, 2014.)

of 424 patients in Vadodara, India. The results revealed that nearly 52.36% of the patients were affected by vitiligo vulgaris and followed by focal vitiligo, at about 28.54%. Vitiligo affects both men and women equally [89], but women tend to seek specialist care [90]. A study by Augustin et al. [91] has shown that vitiligo can lead to socially detrimental outcomes [91]. Furthermore, women in certain countries have faced difficulties in finding a marital partner [92], pursuing studies, and applying for a job [93].

23.3.7 Miscellaneous Skin Diseases

Acne affects people from different ethnic groups [94] and usually starts in puberty [95]. There are four contributing factors in acne development: abnormal follicular epithelial desquamation, hyperactivity of the sebaceous glands, proliferation of *Propionibacterium acnes*, and follicular formation [96]. Furthermore, other studies in the Middle East have reported that the formation of acne can be aggravated by emotional stress, premenstrual factors in females, and certain types of food [97–98].

Nearly 62% of young Americans aged 18 and over sought treatment for acne vulgaris in 2004 [99]. Bickers et al. [9] assessed the economic burden of skin conditions in the United States in 2004 and have found that the direct and indirect costs associated with acne (cystic and vulgaris) were $2.5 billion and $619 million, respectively.

23.4 Teledermatology Models

There are three main forms of teledermatology practiced worldwide: store-and-forward (S&F), live-interactive (LI) [100–101], and hybrid models [102]. The possible implementation of teledermatology system can be seen in Figure 23.7.

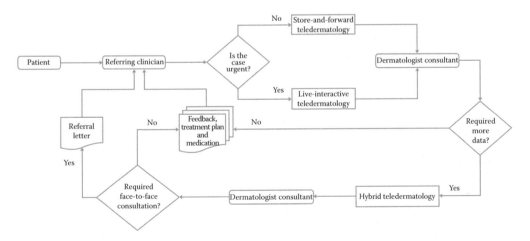

FIGURE 23.7
The proposed teledermatology system will use the existing telecommunication infrastructures in that area and off-the-shelf products (for instance, digital camera, webcam, or mobile devices). The referring clinician could use the system to store the required data while performing all the routine checkups. The choice of the teledermatology model is dependent on the patient's condition. Dermatologist consultants would give their opinions and it is possible to request face-to-face consultation for particular cases.

23.4.1 Store-and-Forward Teledermatology

Store-and-forward teledermatology can be defined as asynchronous transmission of medical records, including still digital images, patient histories, and specialist recommendations, with the intention of providing dermatology treatment and services [103].

The specialist at the referral center takes and saves a picture of the infected skin in JPG format [104], gathers medical histories, and then transmits those data through a secure and encrypted Internet connection [103]. The DICOM format applied in the process links the images with the patient's medical history [105]. The recommendation is sent out to the referral specialist after the stored data are evaluated asynchronously by the dermatologist consultant [23].

The advantage of using the S&F model is that skin lesions can clearly be seen from the taken images [104]. The disadvantage of applying this type of dermatology is a difficulty of addressing gaps in medical histories [104]. Nowadays, the S&F model has been modified by involving dermatologist and skin experts in online discussion groups with the intention to determine the care best suited for unique skin diseases [104].

23.4.2 Live-Interactive Teledermatology

Live-interactive teledermatology can be defined as synchronous interaction via video conferencing or web conferencing between a patient (or representative) and a dermatologist consultant [103,106]. The dermatologist consultant assesses the patient's skin condition, records any clinical conditions in medical history, and provides recommendations regarding the treatment of skin diseases in real time [106].

The advantage of implementing the LI model is that the missing data/gaps in the patient's medical record can be clarified directly with the patient or the referring specialist. The disadvantage of performing live communication is that the dermatologist consultant finds difficulty in providing accurate diagnosis of skin lesions when the video streaming quality is poor [105].

23.4.3 Hybrid Teledermatology

Hybrid teledermatology can be defined as a combination of still digital images and a video conferencing application for providing advanced dermatology care [100]. This model is quite ideal for both of the users (the patient with the accompaniment of the referring specialist and the dermatologist consultant) due to the finest quality of equipment that delivers images through standard protocol and high-quality video sessions [105,107].

23.5   The Implementation of Teledermatology Models (Existing Applications)

By the end of 2015, it is predicted that there will be more than 7 billion mobile phone subscriptions in the world, a penetration rate of 97% up from 738 million subscriptions in 2000. Active mobile-broadband subscription is around 78% in Europe and the Americas, 42% in Asia and the Pacific, and 17% in Africa [108]. This averages to an active subscription of 46% worldwide; 87% in developed countries, and 39% in developing countries. The key reason behind this growth was largely driven by significant cost reduction of deployment of mobile technology [108]. Nevertheless, the mobile broadband prices are twice as expensive in developing countries compared to the developed countries. The global internet usage grew 7 folds from 6.5% to 43% between 2000 and 2015. 3G internet coverage is 29% in rural areas with population of 3.4 billion worldwide, and 89% in urban areas with population of 4 billion [108]. An interesting fact is that the global average fixed broadband plan is 1, 7 times higher than the average price of a comparable mobile broadband plan [108] adding to the wider proliferation of mobile technologies.

The increase in availability and efficiency of voice and data transfer networks along with rapid wireless infrastructure deployments will likely accelerate the process of mHealth solution integration throughout the world [109]. One of the perfect candidates for the implementation of mHealth is mobile teledermatology [110]. Numerous studies have demonstrated the successful implementation of mobile teledermatology for collecting and transmitting patients' medical records [111–114].

In the last decade alone, the potential uses of mobile phones have been proven for diagnosing teledermatoscopy images [114–115]; treating patients with various skin conditions, including psoriasis [116], melanoma [113], wounds [111], or pigmented lesions; and performing follow-up treatment/care for a patient who requires systemic treatment [114]. As mobile teledermatology does not involve the use of computer technology and Internet connection via cable modem, it offers a huge potential for ministering patients in areas with limited resources.

Tran et al. [17] conducted a study to determine the feasibility of mobile dermatology in Egypt. The proposed system consisted of the latest-generation mobile phones with embedded software and wireless connection. A survey questionnaire is created in multiple-choice format and displayed on the mobile phone's screen. The on-site dermatologist (junior physician) selected the best possible answer from each question, captured an image of the skin lesion, and immediately transferred those collected data to the online database via mobile Internet connection. Diagnostic agreements between the junior physician who provided face-to-face consultation with the patient and the two senior dermatologist consultants who performed on line evaluation were 77% (23 out of 30) and 73% (22 out of 30), respectively. Even though this study involved 30 patients only, the achieved results suggest the continuation of system development and implementation in such areas with lack of access to dermatological services.

23.6 Technical Aspects

23.6.1 Acquisition of Images

The quality of images is one of the major factors in improving the diagnostic accuracy in teledermatology [117].

23.6.1.1 Digital Camera

Digital cameras can be divided into three basic categories: single-lens reflex (SLR), compact, and subcompact cameras [117].

The SLR camera allows various types of lenses and flash operations [118] and offers durability along with high-quality images [119]. Macrophotography can be performed by mounting a macrolens to the SLR camera body [118]. By choosing this type of lens, the texture characteristics of lesions are perfectly captured [118]. To solve the small space problem, a dermatologist takes images of the patient's entire body by using a wide-angle lens [118]. By utilizing this lens, the distribution and pattern of skin rashes over the body can be clearly seen and also the changes following liposuction can be further evaluated [118]. Ring flashes are ideal in dermatological photography as these enable the dermatologist to perfectly photograph either skin [119] or oral mucosal lesions [118].

Other advantages of SLR cameras are improved start-up times, faster shutter, and shooting speeds when compared with other types of digital cameras [118]. For example, the Canon EOS 5D Mark III has a shutter speed of 1/8000 s and takes 6.0 frames per second (fps) high-speed continuous shooting. Although the SLR cameras provide excellent quality of the taken images, the prices are quite expensive [118]. The prices ranged from $560 to $1400 for an SLR camera without a lens [118].

The subcompact and compact cameras have distinct advantages when compared with a SLR camera. The subcompact and compact cameras are portable, lightweight, and small in size; have a large display; and are affordable in cost [119]. However, these types of cameras often suffer from two types of delays: shutter lag and shot-to-shot delays [118]. For example, the Panasonic Lumix FT25 has a shutter speed of 1/1300 s and a shooting speed of 1.3 fps. Another disadvantage of subcompact and compact cameras is the absence of hot shoe support. Without this feature, these types of cameras tend to produce under- or overexposed photos of lesions [118].

A study has been conducted by Lassiera et al. [120] to determine the diagnostic concordance between S&F teledermatology and face-to-face consultation. This study was held between April 2008, and July 2010, and recruited a total of 120 patients. Each patient attended two different sessions: S&F teledermatology and in-person consultation. In S&F teledermatology, a nurse is responsible for capturing the skin diseases by using Canon Digital IXUS 75 and Nikon Coolpix P900 and reporting the patient's health condition. The collected data are then viewed in computer by a dermatologist consultant. In the other session, the same dermatologist who previously delivered dermatology care via store-and-forward technology now continues the examination of skin conditions by seeing the patient directly. The diagnostic concordance between the S&F teledermatology and face-to-face diagnosis was 76% for inflammatory dermatoses in children and 83% for infections and infestations in adults.

23.6.1.2 Digital Video Camera

The following recommendations of video camera are considered by ATA when conducting LI teledermatology [121]:

- The video resolution is either standard definition or high definition.
- The video camera allows users to scan, zoom, autofocus, and freeze the frame when capturing the skin lesions.
- An internal lighting source can be used to illuminate the lesions.
- A polarizing filter can be used if necessary.

There are a number of factors that affected the image quality in LI teledermatology: the bandwidth (connection speed), on-site physician experience with the system, and whether or not the freeze-frame is added for further diagnosis [121]. A bandwidth of at least 384 kbps is required to ensure that the video image quality remains within acceptable levels [121]. Freeze-frame captures are quite important in situations where the image quality is degraded due to slow connection speed [121].

A study has been conducted by Lowitt et al. [122] to measure the diagnostic concordance between LI teledermatology and in-person consultation. A total of 139 patients were recruited in this study, but only 102 patients participated in both LI teledermatology and in-person consultation. The videoconference consultations were conducted via either T1 line or 1/4 T1 line. The study reported that the diagnostic agreement between the LI teledermatology and in-person examination was quite reasonable (80%, 104 out of 130 consultations).

23.6.2 Data Display

Improved quality and diagnostic confidence can be achieved by using a high-definition monitor [123]. A LED-backlit LCD screen was chosen as one of the preferred devices for many of dermatologists for treating a patient through online communication [123]. It has a resolution that is close to the collected image's resolution and a high level of luminance and contrast and displays the images with remarkable sharpness and color depth [123]. A number of studies have revealed that image degradation quality could occur if the collected images are displayed on an inferior monitor [123].

23.6.3 Data Storage, Retrieval, and Transmission

Particular data that must be recorded on a patient's medical history are the demographic characteristic, the condition duration, the history and description of symptoms, the factor of aggravating circumstances, the distribution of the lesions, the diagnosis of suspected diseases, and the treatment of previous infections [117]. General data that must be stored on a patient's medical history are the personal background of the patient, the current use of medication, and the history of allergies and family diseases [117]. Furthermore, any information that affected the diagnostic accuracy must be provided to the referring dermatologist [117]. Then, the collected data can be transmitted and retrieved after being stored in the database [117].

An informed consent form must be provided and signed by both parties (the medical staff and the patient) before running the session [124], most importantly if the images taken include the facial region. A captured image that shows the identity of the patient is unavoidable in certain cases, but if necessary, it must be kept to a minimum [117]. All images must be stored in the JPEG (.jpg) format and have a resolution of 1280 × 960 pixels [125].

The teledermatology protocols should be reasonably informed to the related patient before the video session takes place [117]. Moreover, proper training must be provided to the qualified medical staff (either primary care physicians or nurses) for the effectiveness of the teleconsultation [126].

There are four main issues that need to be addressed when running the LI teledermatology session: minimal privacy conditions, comfort, lighting, and adequate physical space [127]. Those issues are not really crucial in the implementation of S&F teledermatology since it requires technical skills in photography only [117].

There is a new concern about security since the process of collecting, transmitting, and storing patients' medical records has shifted from being paper based to the electronic version. Breaches of medical record confidentiality could cause serious damage to the implementation of teledermatology and a distrustful feeling from the related patient. The privacy and confidentiality of medical records can be maintained by transferring "encryption keys" via a secured channel [128]. There are three major aspects offered by cryptography when securing data transmission: confidentiality, identification authenticity, and integrity of the data [129].

Furthermore, current regulatory compliance governing the confidentiality of sensitive data (including medical records) comprises the following main points: secure protocol implementation (e.g., hypertext transfer protocol secure [https]), user and password authentication, access traceability records, and intranet implementation [130]. Legislation in the United States [131] and Europe [132] authorizes the sharing of medical records across and between colleagues when considering the patient's state of care [133]. The completed and signed consent form must be collected from the patients if their medical records or photographs are used for the purposes of education and training [134].

23.7 Teledermatology Evaluation

The proper evaluation of teledermatology has involved five key points, including reliability, accuracy, outcomes, costs, and end user's satisfaction [135–136]. The detailed breakdown for each factor is discussed below.

23.7.1 Diagnostic Reliability

Diagnostic reliability can be defined as the measurement of diagnostic concurrence [103]. It can be divided into two types: intra- and interobserver reliabilities [137]. Intraobserver reliability is the ability of an examiner to obtain the same diagnosis results in two different sessions, while interobserver reliability is the ability of at least two examiners to derive similar diagnosis results from the same individual [137]. Complete agreement, which refers to the likelihood of parity in diagnostic categorization, will be evaluated for each of the reliability measurements [103].

In interobserver studies, the results varied from 53% to 91.6% for S&F teledermatology [14,138–143] and from 60% to 80% for LI teledermatology [14,144–146], as summarized in Tables 23.1 and 23.2. Based on the presented results, we can conclude that S&F teledermatology is more reliable for observing skin condition as compared to LI teledermatology.

In intraobserver studies, the agreements between an in-person dermatologist and a teledermatologist (who provides S&F consultation) range from 70% to 82% for complete agreement [139,142–143,147–148], as seen in Table 23.3.

TABLE 23.1

Interobserver Diagnostic Reliabilities of S&F Teledermatology

Researchers	Conducted in Year	Country	Patient	Data	Complete Agreement
Oakley et al. [138]	2004	New Zealand	73	109 skin lesions	53%
Ebner et al. [139]	2005	Austria	58	58 cases	71%–76%
Aguilera et al. [140]	2005	Spain	170	170 cases	72%
Silva et al. [142]	2007	Brazil	60	60 cases	86.6%–91.6%
Edison et al. [14]	2007	United States	110	110 cases	73%
Tan et al. [141]	2008	New Zealand	200	491 lesions	74%
Ruiz et al. [143]	2009	Colombia	82	171 dermatologic diseases and 1 normal skin case	80.8%–86.6%

TABLE 23.2

Interobserver Diagnostic Reliabilities of LI Teledermatology

Researchers	Conducted in Year	Country	Patient	Data	Complete Agreement
Nordal et al. [144]	1994–1995	Norway	121	112 cases	72%
Lesher et al. [145]	1997	United States	60	68 cutaneous problems	78%
Loane et al. [146]	1998	Ireland	351	427 diagnoses	67%
Edison et al. [14]	2007	United States	110	110 cases	80%

TABLE 23.3

Intraobserver Diagnostic Agreements of S&F and LI Teledermatology

Researchers	Conducted in Year	Country	Patient	Data	Complete Agreement
Pak et al. [147]	1999–2000	United States	404	404 cases	70%
Ebner et al. [139]	2005	Austria	58	58 cases	74%
Heffner et al. [148]	2006–2007	United States	135	135 cases	82%
Ruiz et al. [143]	2009	Columbia	83	171 dermatologic diseases and 1 normal skin case	80.8%–86.6%
Tan et al. [141]	2008	New Zealand	200	491 lesions	75.5%–82.2%

23.7.2 Diagnostic Accuracy

It is essential to validate the accuracy of diagnostic measurements by referring to a gold standard [103]. One study has revealed that the diagnostic accuracy of S&F teledermatology was significantly greater than a face-to-face session [149]. Based on the obtained results, the researchers believed that other opinions provided by specialist through S&F teledermatology may be beneficial for both the local dermatologists and the patients, especially in managing difficult cases [149]. Several studies have shown that the diagnostic accuracies of S&F teledermatology were comparable to those obtained by an in-person dermatologist [150–151]. Other studies have reported that the diagnostic accuracies of S&F teledermatology were significantly lower than those of direct dermatology consultations [152–153]. The diagnostic accuracy comparison between face-to-face consultation and S&F teledermatology can be seen in Table 23.4.

TABLE 23.4

Diagnostic Accuracy Rates between Face-to-Face Consultation and S&F Teledermatology

Researchers	Conducted in Year	Country	Patient	Valid Cases	Clinic Based	S&F Teledermatology
Barnard and Goldyne [150]	2000	United States	50	33	84%–88%	73%–90%
Warshaw et al. [152]	2002–2005	United States	542	542	80.26%	64.02%
Warshaw et al. [153]	2002–2005	United States	728	716	56.32%–76.10%	42.99%–59.48%
Lozzi et al. [149]	2004–2005	Italy	33	33	30.3%–42.4%	78.8%

Differences in diagnostic accuracies from one study to the other studies were caused by a number of factors [103]. First, each study used different gold standards for evaluating the diagnostic accuracies in various settings (from in-person consultation to a clinical pathologic session). Second, the distribution of skin lesions in specified populations has varied enormously from one study to another.

23.7.3   Outcomes

Intermediate results from the implementation of teledermatology model are preventable clinic visits [154], time to first intervention [155–157], and consultation durations [158–161].

A study conducted by Eminovic et al. [154] has reported that the percentages of preventable dermatology visits were around 39% for patients who had received S&F teledermatology and 18.3% for patients who had attended dermatology clinics. A study by Loane et al. [155] has revealed that the percentage of avoidable dermatology visits were approximately 44% for LI teledermatology and 30% for in-person examination.

A number of studies have revealed that the time to first intervention was significantly shorter in S&F teledermatology than in a conventional clinical visit: 2 days versus 17 days in a study that was organized by van der Akker et al. [155]; an average of 41 days versus 127 days in a study that is organized by Whited et al. [156]; and a median of 12 days versus a median of 88 days in a study that was conducted by Moreno-Ramirez et al. [157].

The duration of consultation in S&F teledermatology sessions was quite varied. The average time spent by a dermatologist for performing an S&F teledermatology session was approximately 10 min [155]. The longest time required by a dermatologist to carry out a complete procedure of S&F teledermatology with an average of 11 min [158]. Furthermore, Berghout et al.'s study [158] also revealed that the referring dermatologists spent 41% of the consultation time (around 4:43 min) collecting the patient's medical history and 28% of the time (approximately 3:12 min) filling out an electronic version of referral form. The total time (including actual consultation time, travel time, and waiting time) required for LI teledermatology consultation was quite steady and shorter than that for conventional clinical care: 51 min versus 259 min in a study by Oakley et al. [22] and 52:59 min versus to 259:18 min in a study by Loane et al. [159].

23.7.4   Cost Effectiveness

Moreno-Ramirez et al. [21] conducted a study that intended to evaluate the effectiveness of cost of S&F teledermatology in managing skin cancer cases. The study revealed that patients undergoing S&F teledermatology spent less money than those who attended clinical care (€79.78 versus €129.37 for each individual patient).

Another study conducted by Whited et al. [24] stated that the use of S&F teledermatology in a U.S. Department of Veterans Affairs (VA) healthcare setting was quite costly when compared with in-person consultation ($36.40 versus $21.40 per patient). However, remote consultation via store-and-forward technology was more cost effective when considering the cost that must be spent for travel and related to lost productive time, costing at most $0.17 per patient per day until the first intervention session.

Several studies in the past have shown that the implementation of LI teledermatology was quite expensive compared to conventional care: £180.22 versus £48.77 in a study by Loane et al. [160] and £132.10 versus £48.73 in a study by Wootton et al. [106]. The researchers argued that the costs for providing LI tedelermatology might be lower than for attending in-person consultation [106]. The reasons behind this argument were the budget for travelling and the productivity losses due short-term absenteeism [106]. In another study, it was found that LI teledermatology was cheaper than in-site consultation (hourly rates for consultation: $274 versus $346) [161].

23.7.5 End Users' Satisfaction

The assessment of users' acceptance and satisfaction with teledermatology system is essential when considering the impacts on data quality [162]. A questionnaire is usually used for measuring the users' satisfaction and may be beneficial for guiding future studies in the development of assessment strategies [162].

23.7.5.1 Patients' Satisfaction

Patients' satisfaction with S&F teledermatology has been extensively evaluated in numerous studies [126,163–165]. In general, most of the patients were satisfied with the use of store-and-forward technology in providing dermatology care.

Hsueh et al. [156] conducted a satisfaction survey among all the veterans admitted to either teledermatology care (504 people) or face-to-face (FTF) dermatology care (196 people). Of 503 veterans answering their preference of care, 332 (66%) preferred teledermatology care over FTF dermatology care. Findings from this study also reported that 77% of 501 veterans who received teledermatology care showed a higher level of satisfaction with the services and 83% would suggest this type of care to the others.

Whited et al. [164] surveyed 93 teledermatology patients and found that 82% of 101 survey respondents were satisfied with the consultation outcomes and 41.5% of 101 patients expressed their preference for S&F teledermatology. The researchers also reported that 27% of 93 patients were concerned because they needed to wait for the diagnostic results for long periods of time.

A number of studies have been conducted to assess the relationship between LI teledermatology and the patient's satisfaction [144,165]. A total of 126 patients were recruited in a study aimed to compare the effectiveness of LI teledermatology and that of face-to-face consultations in diagnosing skin diseases [165]. The study revealed that most of the patients thought that the system had positive results in time efficiency (92%, 113 out of 122 patients) and financial saving (63%, 75 out of 119 patients). However 18% of 122 patients felt discomfort from being in a videoconferencing session.

23.7.5.2 Referring Clinicians' Satisfaction

Studies have shown that 100% of 5 nurses agreed that the S&F teledermatology was convenient and easy to use [166] and 74% of 19 referring clinicians would like to recommend the system as an alternative way of providing dermatology consultations [167]. A study by Armstrong et al. [168] has revealed that 100% of 10 primary care providers (PCPs) had agreed on the importance of teledermatology care for their patients. The providers also highlighted some benefits of providing teledermatology, such as reduction in consultation costs (30%), shorter turnaround time for consultation (20%) and finally improvement in workflow (30%), communication with dermatologist (20%) and technology (10%). Referring clinicians believed that LI dermatology has provided educational benefits [165]. However, referring clinicians have reported that both the S&F [165,167] and LI [169] teledermatology procedures can be very time consuming.

23.7.5.3   Consultants' Satisfaction

Positive experiences when using teledermatology have been reported by dermatologist consultants [144,164].

One study showed that the dermatologist consultants completely agreed that the S&F teledermatology is reliable and easy to implement for triaging patients with skin diseases [164]. Moreover, another study reported that 70% of the dermatologist consultants were pleased with the quality of consultation that they delivered [170].

Approximately 80% of 113 dermatologist consultants have considered that LI teledermatology was as good as clinic-based care [144]. Another study also revealed that the dermatologist consultants built a good rapport with their patients when providing diagnostic services via LI teledermatology for around 95% of 130 cases [122].

Negative responses to the use of both models are usually associated with the quality of the taken images [118] and the inability to inspect and palpate the skin to evaluate lesions [171].

23.8   Future Developments

Teledermatology made a significant impact on the healthcare sector by providing dermatological care for those living in remote and medically underserved areas [172]. There are certain stages that need to be considered when establishing teledermatology in the world. First, medical staff (including dermatologists, nurses, and physicians) will require continuing training and education on providing either S&F or LI teledermatology services [172–173]. Second, scientists and engineers need to design and develop affordable and robust platforms for teledermatology so that the practice and service effectiveness can be improved [173]. Third, researchers need to conduct large-scale randomized controlled trials for various types of skin diseases that could determine the accuracy and reliability of the teledermatology system [172]. This stage is quite important especially when associated with a high level of morbidity and mortality [172]. Fourth, there is a fundamental need for conducting studies that investigate cost effectiveness of teledermatology and end users' satisfaction [172]. Finally, there are a number of barriers that need to be overcome such as limited existing ICT infrastructure, poor-quality images, delayed reimbursement, and potential medicolegal risks [173].

References

1. F. Sampogna, D. Raskovic, L. Guerra, C. Pedicelli, S. Tabolli, L. Leoni, L. Alessandroni, and D. Abeni, "Identification of categories at risk for high quality of life impairment in patients with vitiligo," *British Journal of Dermatology*, vol. 159, no. 2, pp. 351–359, 2008.
2. F. O. Nestle, D. H. Kaplan, and J. Barker, "Psoriasis," *The New England Journal of Medicine*, vol. 361, no. 5, pp. 496–509, 2009.
3. A. Ahmed, A. Leon, D. C. Butler, and J. Reichenberg, "Quality-of-life effects of common dermatological diseases," *Seminars in Cutaneous Medicine and Surgery*, vol. 32, no. 2, pp. 101–109, 2013.

4. J. M. Gelfand, S. R. Feldman, R. S. Stern, J. Thomas, T. Rolstad, and D. J. Margolis, "Determinants of quality of life in patients with psoriasis: A study from the US population," *Journal of the American Academy of Dermatology*, vol. 51, no. 5, pp. 704–708, 2004.

5. E. J. Horn, K. M. Fox, V. Patel, C. F. Chiou, F. Dann, and M. Lebwohl, "Association of patient-reported psoriasis severity with income and employment," *Journal of the American Academy of Dermatology*, vol. 57, no. 6, pp. 963–971, 2007.

6. A. T. Baibergenova, M. A. Weinstock, and N. H. Shear, "Mortality from acquired bullous diseases of skin in Canadian adults 2000–2007," *International Journal of Dermatology*, vol. 51, no. 11, pp. 1325–1328, 2012.

7. WHO, *Summary: Deaths (Thousands) by Cause and by WHO Region*. World Health Organization, 2013.

8. S. Morris, B. Cox, and N. Bosanquet, "Cost of skin cancer in England," *The European Journal of Health Economics*, vol. 10, no. 3, pp. 267–273, 2009.

9. D. R. Bickers, H. W. Lim, D. Margolis, M. A. Weinstock, C. Goodman, E. Faulkner, C. Gould, E. Gemmen, and T. Dall, "The burden of skin diseases: 2004: A joint project of the American Academy of Dermatology Association and the Society for Investigative Dermatology," *Journal of the American Academy of Dermatology*, vol. 55, no. 3, pp. 490–500, 2006.

10. S. E. Schultes, *Milady's Standard: Nail Technology*. Cengage Learning, 2002.

11. E. M. Sullivan, D. Brown, and D. Vetrosky, *Physician Assistant: A Guide to Clinical Practice: Expert Consult*—Online and Print. Elsevier Health Sciences, 2013.

12. C. Erikson, K. Jones, and C. Tilton, *2012 Physician Specialty Data Book*. Association of American Medical Colleges, 2012.

13. J. S. Resneck, "The influence of controllable lifestyle on medical student specialty choice," *Virtual Mentor*, vol. 8, no. 8, pp. 529–532, 2006.

14. K. E. Edison, D. S. Ward, J. A. Dyer, W. Lane, L. Chance, and L. L. Hicks, "Diagnosis, diagnostic confidence, and management concordance in live-interactive and store-and-forward teledermatology compared to in-person examination," *Telemedicine and e-Health*, vol. 14, no. 9, pp. 889–895, 2008.

15. E. Krupinski, A. Burdick, H. Pak, J. Bocachica, L. Earles, K. Edison, M. Goldyne, T. Hirota, J. Kvedar, K. McKoy et al., "American Telemedicine Association's practice guidelines for teledermatology," *Telemedicine and e-Health*, vol. 14, no. 3, pp. 289–302, 2008.

16. A. S. Pathipati, L. Lee, and A. W. Armstrong, "Health-care delivery methods in teledermatology: Consultative, triage and direct-care models," *Journal of Telemedicine and Telecare*, vol. 17, no. 4, pp. 214–216, 2011.

17. K. Tran, M. Ayad, J. Weinberg, A. Cherng, M. Chowdhury, S. Monir, M. El Hariri, and C. Kovarik, "Mobile teledermatology in the developing world: Implications of a feasibility study on 30 Egyptian patients with common skin diseases," *Journal of the American Academy of Dermatology*, vol. 64, no. 2, pp. 302–309, 2011.

18. N. Shaikh, C. U. Lehmann, P. H. Kaleida, and B. A. Cohen, "Efficacy and feasibility of teledermatology for paediatric medical education," *Journal of Telemedicine and Telecare*, vol. 14, no. 4, pp. 204–207, 2008.

19. C. M. Williams, I. Kedar, L. Smith, H. A. Brandling-Bennett, N. Lugn, and J. C. Kvedar, "Teledermatology education for internal medicine residents," *Journal of the American Academy of Dermatology*, vol. 52, no. 6, pp. 1098–1099, 2005.

20. R. Wootton, *Teledermatology*. RSM Press, 2002.

21. D. Moreno-Ramirez, L. Ferrandiz, A. Ruiz-de Casas, A. Nieto-Garcia, P. Moreno-Alvarez, R. Galdeano, and F. M. Camacho, "Economic evaluation of a store-and-forward teledermatology system for skin cancer patients," *Journal of Telemedicine and Telecare*, vol. 15, no. 1, pp. 40–45, 2009.

22. A. Oakley, P. Kerr, M. Duffill, M. Rademaker, P. Fleischl, and N. Bradford, "Patient cost–benefits of realtime teledermatology—A comparison of data from northern Ireland and New Zealand," *Journal of Telemedicine and Telecare*, vol. 6, no. 2, pp. 97–101, 2000.

23. H. S. Pak, S. K. Datta, C. A. Triplett, J. H. Lindquist, S. C. Grambow, and J. D. Whited, "Cost minimization analysis of a store-and-forward teledermatology consult system," *Telemedicine and e-Health*, vol. 15, no. 2, pp. 160–165, 2009.

24. J. D. Whited, S. Datta, R. P. Hall, M. E. Foy, L. E. Marbrey, S. C. Grambow, T. K. Dudley, D. L. Simel, and E. Z. Oddone, "An economic analysis of a store and forward teledermatology consult system," *Telemedicine Journal and e-Health*, vol. 9, no. 4, pp. 351–360, 2003.

25. J. Bensouilah and P. Buck, *Aromadermatology: Aromatherapy in the Treatment and Care of Common Skin Conditions*. Radcliffe Publishing, 2006.

26. W. De Craecker, N. Roskams, and R. O. de Beeck, *Occupational Skin Diseases and Dermal Exposure in the European Union (EU-25): Policy and Practice Overview*. European Agency for Safety and Health at Work, 2008.

27. A. P. Koesoema, "Skin," 2014.

28. D. H. Chu, *Development and Structure of Skin*. McGraw-Hill, 2008, Ch. 7, pp. 57–73.

29. J. M. B. Myers and G. K. K. Hershey, "Eczema in early life: Genetics, the skin barrier, and lessons learned from birth cohort studies," *The Journal of Pediatrics*, vol. 157, no. 5, pp. 704–714, 2010.

30. T. Bieber, "Atopic dermatitis," *New England Journal of Medicine*, vol. 358, no. 1, pp. 1483–1494, 2008.

31. S. P. Shelov and R. H. Hannemann, *Caring for Your Baby and Young Child: Birth to Age Five*. Bantam Books, 2004.

32. C. van Hees and B. Naafs, *Common Skin Diseases in Africa*. Reinier de Graaf Groep, 2009.

33. CDC, "ID: 13325," 1969. [Online]. Available: http://phil.cdc.gov, accessed on Aug. 18, 2014.

34. CDC and A. E. Kaye, "ID: 3305," 1969. [Online]. Available: http://phil.cdc.gov, accessed on March 18, 2014.

35. K. A. Horii, S. D. Simon, D. Y. Liu, and V. Sharma, "Atopic dermatitis in children in the United States, 1997–2004: Visit trends, patient and provider characteristics, and prescribing patterns," *Pediatrics*, vol. 120, no. 3, pp. e527–e534, 2007.

36. D. Laughter, J. A. Istvan, S. J. Tofte, and J. M. Hanifin, "The prevalence of atopic dermatitis in Oregon school children," *Journal of the American Academy of Dermatology*, vol. 43, no. 4, pp. 649–655, 2000.

37. F. S. Larsen and J. M. Hanifin, "Epidemiology of atopic dermatitis," *Immunology and Allergy Clinics of North America*, vol. 22, no. 1, pp. 1–24, 2002.

38. C. N. Ellis, L. A. Drake, M. M. Prendergast, W. Abramovits, M. Boguniewicz, C. R. Daniel, M. Lebwohl, S. R. Stevens, D. L. Whitaker-Worth, J. W. Cheng et al., "Cost of atopic dermatitis and eczema in the United States," *Journal of the American Academy of Dermatology*, vol. 46, no. 3, pp. 361–370, 2002.

39. C. L. Carroll, R. Balkrishnan, S. R. Feldman, A. B. Fleischer, and J. C. Manuel, "The burden of atopic dermatitis: Impact on the patient, family, and society," *Pediatric Dermatology*, vol. 22, no. 3, pp. 192–199, 2005.

40. A. J. Mancini, K. Kaulback, and S. L. Chamlin, "The socioeconomic impact of atopic dermatitis in the United States: A systematic review," *Pediatric Dermatology*, vol. 25, no. 1, pp. 1–6, 2008.

41. HPA, *Fungal Diseases in the UK—The Current Provision of Support for Diagnosis and Treatment: Assessment and Proposed Network Solution*. Health Protection Agency, 2006.

42. C. Koba, C. Koga, T. Cho, and J. Kusukawa, "Determination of *Candida* species nestled in denture fissures," *Biomedical Reports*, vol. 1, no. 4, pp. 529–533, 2013.

43. A. Glöckner, A. Steinbach, J. J. Vehreschild, and O. A. Cornely, "Treatment of invasive candidiasis with echinocandins," *Mycoses*, vol. 52, no. 6, pp. 476–486, 2009.

44. T. E. Zaoutis, J. Argon, J. Chu, J. A. Berlin, T. J. Walsh, and C. Feudtner, "The epidemiology and attributable outcomes of candidemia in adults and children hospitalized in the United States: A propensity analysis," *Clinical Infectious Diseases*, vol. 41, no. 9, pp. 1232–1239, 2005.

45. J. Morgan, M. I. Meltzer, B. D. Plikaytis, A. N. Sofair, S. Huie-White, S. Wilcox, L. H. Harrison, E. C. Seaberg, R. A. Hajjeh, and S. M. Teutsch, "Excess mortality, hospital stay, and cost due to candidemia: A case-control study using data from population-based candidemia surveillance," *Infection Control and Hospital Epidemiology*, vol. 26, no. 6, pp. 540–547, 2005.

46. M. Pfaller and D. Diekema, "Epidemiology of invasive candidiasis: A persistent public health problem," *Clinical Microbiology Reviews*, vol. 20, no. 1, pp. 133–163, 2007.

47. D. E. Lubbe, "HIV and ENT: Main topic." *CME: Your SA Journal of CPD: Ears, Noses and Throats*, vol. 22, no. 5, pp. 250–254, 2004.

48. CDC and S. Lindsley, "ID: 15679," 1977. [Online]. Available: http://phil.cdc.gov.

49. CDC, J. Molinari, and S. S. Jr., "ID: 6067," 1987. [Online]. Available: http://phil.cdc.gov.

50. D. L. Stulberg, M. A. Penrod, and R. A. Blatny, "Common bacterial skin infections," *American Family Physician*, vol. 66, no. 1, pp. 119–128, 2002.

51. AIHW, *Australian Hospital Statistics 2011–12*. Australian Institute of Health and Welfare, 2013.

52. G. Phoenix, S. Das, and M. Joshi, "Diagnosis and management of cellulitis," *BMJ*, vol. 345, no. 1, p. e4955, 2012.

53. C. Cole and J. Gazewood, "Diagnosis and treatment of impetigo," *American Family Physician*, vol. 75, no. 6, 2007.

54. M. H. Motswaledi, "Impetigo in children: A clinical guide and treatment options," *South African Family Practice*, vol. 53, no. 1, pp. 44–46, 2011.

55. D. Gawkrodger and M. R. Ardern-Jones, *Dermatology: An Illustrated Colour Text*. Elsevier Health Sciences, 2012.

56. J. Hirschmann, "Impetigo: Etiology and therapy." *Current Clinical Topics in Infectious Diseases*, vol. 22, pp. 42–51, 2001.

57. S. Koning, A. Verhagen, L. van Suijlekom-Smit, A. Morris, C. Butler, and J. Van der Wouden, "Interventions for impetigo," *Cochrane Database of Systematic Reviews*, vol. 2, no. 1, pp. 1–94, 2003.

58. L. J. Shallcross, I. Petersen, J. Rosenthal, A. M. Johnson, N. Freemantle, and A. C. Hayward, "Use of primary care data for detecting impetigo trends, United Kingdom, 1995–2010," *Emerging Infectious Diseases*, vol. 19, no. 10, pp. 1646–1648, 2013.

59. CDC, "ID: 5153," 1975. [Online]. Available: http://phil.cdc.gov, accessed on June 15, 2015.

60. L. Lugović-Mihić, F. Barisić, V. Bulat, M. Buljan, M. Situm, L. Bradić, and J. Mihić, "Differential diagnosis of the scalp hair folliculitis," *Acta Clinica Croatica*, vol. 50, no. 3, pp. 395–402, 2011.

61. O. Braun-Falco, G. Plewig, H. Wolff, and W. H. C. Burgdorf, *Dermatology*. Springer-Verlag, Berlin, 2000.

62. J. Heilman, "Hot tub folliculitis," 2013. [Online]. Available: http://commons.wikimedia.org/wiki/File:Folliculitis.JPG, accessed on June 15, 2015.

63. M. J. Kuehnert, D. Kruszon-Moran, H. A. Hill, G. McQuillan, S. K. McAllister, G. Fosheim, L. K. McDougal, J. Chaitram, B. Jensen, S. K. Fridkin et al., "Prevalence of staphylococcus aureus nasal colonization in the United States, 2001–2002," *Journal of Infectious Diseases*, vol. 193, no. 2, pp. 172–179, 2006.

64. AIHW, *A Patient Guide: HPV in Perspective*. American Social Health Association, 1998.

65. A. K. Chaturvedi, "Beyond cervical cancer: Burden of other HPV-related cancers among men and women," *Journal of Adolescent Health*, vol. 46, no. 4, pp. S20–S26, 2010.

66. K. A. Workowski and W. C. Levine, "Sexually transmitted diseases treatment guidelines 2002," *Morbidity and Mortality Weekly Report*, vol. 51, no. 6, pp. 1–80, 2002.

67. A. N. Burchell, R. L. Winer, S. de Sanjosé, and E. L. Franco, "Epidemiology and transmission dynamics of genital HPV infection," *Vaccine*, vol. 24, no. Supplement 3, pp. S52–S61, 2006.

68. C.-J. Yang, S.-X. Liu, J.-B. Liu, Z.-Y. Wang, D.-F. Luo, G.-L. Zhang, and X.-J. Zhang, "Holmium laser treatment of genital warts: An observational study of 1500 cases," *Acta Dermatovenereologica*, vol. 88, no. 2, pp. 136–138, 2008.

69. S. K. Kjær, T. T. Nam, P. Sparen, L. Tryggvadottir, C. Munk, E. Dasbach, K.-L. Liaw, J. Nygård, and M. Nygård, "The burden of genital warts: A study of nearly 70,000 women from the general female population in the 4 Nordic countries," *Journal of Infectious Diseases*, vol. 196, no. 10, pp. 1447–1454, 2007.

70. R. P. Insinga, E. J. Dasbach, and E. R. Myers, "The health and economic burden of genital warts in a set of private health plans in the United States," *Clinical Infectious Diseases*, vol. 36, no. 11, pp. 1397–1403, 2003.

71. S. K. Sukhbir, "Neonatal herpes simplex infection," *ISRN Infectious Diseases*, vol. 2013, no. 1, pp. 1–7, 2012.

72. R. M. Doctor, A. P. Kahn, and C. Adamec, *The Encyclopedia of Phobias, Fears, and Anxieties.* Infobase Publishing, 2009.

73. D. W. Kimberlin, "Neonatal herpes simplex infection," *Clinical Microbiology Reviews*, vol. 17, no. 1, pp. 1–13, 2004.

74. CDC and Dancewiez, "ID: 6478," 1972. [Online]. Available: http://phil.cdc.gov, accessed on June 15, 2015.

75. A. K. Leung, and B. Barakin, "Herpes zoster in childhood." *Open Journal of Pediatrics*, vol. 5, no. 1, pp. 39–44, 2015.

76. K. Owusu-Edusei Jr., H. W. Chesson, T. L. Gift, G. Tao, R. Mahajan, M. C. B. Ocfemia, and C. K. Kent, "The estimated direct medical cost of selected sexually transmitted infections in the United States, 2008," *Sexually Transmitted Diseases*, vol. 40, no. 3, pp. 197–201, 2013.

77. V. J. Fraser, L. Burd, E. Liebson, G. Y. Lipschik, and M. Peterson, *Diseases and Disorders.* Marshall Cavendish, 2007.

78. D. Modi, "Parasites and the skin," *CME*, vol. 27, no. 6, pp. 254–260, 2009.

79. J. S. Davis, S. McGloughlin, S. Y. Tong, S. F. Walton, and B. J. Currie, "A novel clinical grading scale to guide the management of crusted scabies," *PLoS Neglected Tropical Diseases*, vol. 7, no. 9, p. e2387, 2013.

80. CDC, "ID: 4800," 1975. [Online]. Available: http://phil.cdc.gov.

81. O. Chosidow, "Scabies," *New England Journal of Medicine*, vol. 354, no. 16, pp. 1718–1727, 2006.

82. R. Hay, A. Steer, D. Engelman, and S. Walton, "Scabies in the developing world—Its prevalence, complications, and management," *Clinical Microbiology and Infection*, vol. 18, no. 4, pp. 313–323, 2012.

83. M. Muhammad Zayyid, R. Saidatul Saadah, A. Adil, M. Rohela, and I. Jamaiah, "Prevalence of scabies and head lice among children in a welfare home in Pulau Pinang, Malaysia," *Tropical Biomedicine*, vol. 27, no. 3, pp. 442–446, 2010.

84. J. Heukelbach, S. F. Walton, and H. Feldmeier, "Ectoparasitic infestations," *Current Infectious Disease Reports*, vol. 7, no. 5, pp. 373–380, 2005.

85. R. Gretha de Beer, M. A. Miller, L. Tremblay, and J. Monette, "An outbreak of scabies in a long-term care facility: The role of misdiagnosis and the costs associated with control," *Infection Control and Hospital Epidemiology*, vol. 27, no. 5, pp. 517–518, 2006.

86. Z. Grabel, "Vitiligo: A literature review and analysis of variants in TYR gene in multiplex families," PhD dissertation, University of Florida, 2011.

87. G. D. Oger, "Vitiligo," 2009. [Online]. Available: http://commons.wikimedia.org/wiki/File :Vitiligo.JPG, accessed on Aug. 18, 2014.

88. E. M. Shajil, D. Agrawal, K. Vagadia, Y. S. Marfatia, and R. Begum, "Vitiligo: Clinical profiles in Vadodara, Gujarat," *Indian Journal of Dermatology*, vol. 51, no. 2, pp. 100–104, 2006.

89. S. Dogra, D. Parsad, S. Handa, and A. J. Kanwar, "Late onset vitiligo: A study of 182 patients," *International Journal of Dermatology*, vol. 44, no. 3, pp. 193–196, 2005.

90. A. N. Onunu and E. P. Kubeyinje, "Vitiligo in the Nigerian African: A study of 351 patients in Benin City, Nigeria," *International Journal of Dermatology*, vol. 42, no. 10, pp. 800–802, 2003.

91. M. Augustin, A. Gajur, C. Reich, S. Rustenbach, and I. Schaefer, "Benefit evaluation in vitiligo treatment: Development and validation of a patient-defined outcome questionnaire," *Dermatology*, vol. 217, no. 2, pp. 101–106, 2008.

92. M. Dolatshahi, P. Ghazi, V. Feizy, and M. Rezaei Hemami, "Life quality assessment among patients with vitiligo: Comparison of married and single patients in Iran," *Indian Journal of Dermatology, Venereology & Leprology*, vol. 74, no. 6, p. 705, 2008.

93. L. Borimnejad, Z. Parsa Yekta, A. Nikbakht-Nasrabadi, and A. Firooz, "Quality of life with vitiligo: Comparison of male and female Muslim patients in Iran," *Gender Medicine*, vol. 3, no. 2, pp. 124–130, 2006.

94. H. Yahya, "Acne vulgaris in Nigerian adolescents—Prevalence, severity, beliefs, perceptions, and practices," *International Journal of Dermatology*, vol. 48, no. 5, pp. 498–505, 2009.

95. D. Purvis, E. Robinson, and P. Watson, "Acne prevalence in secondary school students and their perceived difficulty in accessing acne treatment," *The New Zealand Medical Journal*, vol. 117, no. 1200, pp. 1–8, 2004.

96. K. Abulnaja, "Changes in the hormone and lipid profile of obese adolescent Saudi females with acne vulgaris," *Brazilian Journal of Medical and Biological Research*, vol. 42, no. 6, pp. 501–505, 2009.

97. Z. El-Akawi, N. Nemr, K. Abdul-Razzak, and M. Al-Aboosi, "Factors believed by Jordanian acne patients to affect their acne condition," *Eastern Mediterranean Health Journal*, vol. 12, no. 6, p. 840, 2006.

98. A. A. Al Robaee, "Prevalence, knowledge, beliefs and psychosocial impact of acne in university students in central Saudi Arabia," *Saudi Medical Journal*, vol. 26, no. 12, pp. 1958–1961, 2005.

99. B. A. Yentzer, J. Hick, E. L. Reese, A. Uhas, S. R. Feldman, and R. Balkrishnan, "Acne vulgaris in the United States: A descriptive epidemiology," *Cutis*, vol. 86, no. 2, pp. 94–99, 2005

100. H. Pak, C. A. Triplett, J. H. Lindquist, S. C. Grambow, and J. D. Whited, "Store-and-forward teledermatology results in similar clinical outcomes to conventional clinic-based care," *Journal of Telemedicine and Telecare*, vol. 13, no. 1, pp. 26–30, 2007.

101. E. Warshaw, N. Greer, Y. Hillman, E. Hagel, R. MacDonald, I. Rutks, and T. Wilt, *Teledermatology for Diagnosis and Management of Skin Conditions: A Systematic Review of the Evidence*. U.S. Department of Veterans Affairs, 2010.

102. H. S. Pak, K. E. Edison, and J. D. Whited, *Teledermatology: A User's Guide*. Cambridge University Press, 2008.

103. A. Pathipati and A. Armstrong, *Teledermatology: Outcomes and Economic Considerations*. InTech, 2011.

104. F. Kaliyadan, "Teledermatology update: Mobile teledermatology," *World Journal of Dermatology*, vol. 2, no. 2, pp. 11–15, 2013.

105. K. Feroze, "Teledermatology in India: Practical implications," *Indian Journal of Medical Sciences*, vol. 62, no. 5, pp. 208–214, 2008.

106. R. Wootton, S. Bloomer, R. Corbett, D. Eedy, N. Hicks, H. Lotery, C. Mathews, J. Paisley, K. Steele, and M. Loane, "Multicentre randomised control trial comparing real time teledermatology with conventional outpatient dermatological care: Societal cost-benefit analysis," *BMJ*, vol. 320, no. 7244, pp. 1252–1256, 2000.

107. F. Kaliyadan and S. Venkitakrishnan, "Teledermatology: Clinical case profiles and practical issues," *Indian Journal of Dermatology, Venereology & Leprology*, vol. 75, no. 1, pp. 32–35, 2009.

108. ITU, *The World in 2015: ICT Facts and Figures*. International Telecommunication Union, 2015. Available at http://www.itu.int/en/ITU-D/Statistics/Pages/facts/default.aspx. Accessed on August 19, 2015.

109. R. S. Istepanian, E. Jovanov, and Y. Zhang, "Guest editorial introduction to the special section on m-health: Beyond seamless mobility and global wireless health-care connectivity," *IEEE Transactions on Information Technology in Biomedicine*, vol. 8, no. 4, pp. 405–414, 2004.

110. C. Perera and R. Chakrabarti, "The utility of mHealth in medical imaging," *Journal of Mobile Technology in Medicine*, vol. 2, no. 3, pp. 4–6, 2013.

111. R. P. Braun, J. L. Vecchietti, L. Thomas, C. Prins, L. E. French, A. J. Gewirtzman, J.-H. Saurat, and D. Salomon, "Telemedical wound care using a new generation of mobile telephones: A feasibility study," *Archives of Dermatology*, vol. 141, no. 2, pp. 254–258, 2005.

112. C. Massone, G. P. Lozzi, E. Wurm, R. Hofmann-Wellenhof, R. Schoellnast, I. Zalaudek, G. Gabler, A. Di Stefani, H. Kerl, and H. P. Soyer, "Cellular phones in clinical teledermatology," *Archives of Dermatology*, vol. 141, no. 10, p. 1319, 2005.

113. C. Massone, R. Hofmann-Wellenhof, V. Ahlgrimm-Siess, G. Gabler, C. Ebner, and H. P. Soyer, "Melanoma screening with cellular phones," *PloS One*, vol. 2, no. 5, p. e483, 2007.

114. C. Massone, A. M. Brunasso, T. M. Campbell, and H. P. Soyer, "Mobile teledermoscopy— Melanoma diagnosis by one click?," *Seminars in Cutaneous Medicine and Surgery*, vol. 28, no. 3, pp. 203–205, 2009.

115. S. Kroemer, J. Frühauf, T. Campbell, C. Massone, G. Schwantzer, H. P. Soyer, and R. Hofmann-Wellenhof, "Mobile teledermatology for skin tumour screening: Diagnostic accuracy of clinical and dermoscopic image tele-evaluation using cellular phones," *British Journal of Dermatology*, vol. 164, no. 5, pp. 973–979, 2011.

116. J. Frühauf, G. Schwantzer, C. M. Ambros-Rudolph, W. Weger, V. Ahlgrimm-Siess, W. Salmhofer, and R. Hofmann-Wellenhof, "Pilot study on the acceptance of mobile teledermatology for the home monitoring of high-need patients with psoriasis," *Australasian Journal of Dermatology*, vol. 53, no. 1, pp. 41–46, 2012.

117. G. Romero, P. Cortina, and E. Vera, "Telemedicine and teledermatology (I): Concepts and applications," *Actas Dermo-Sifiliográficas (English Edition)*, vol. 99, no. 7, pp. 506–522, 2008.

118. S. Chilukuri and A. Bhatia, "Practical digital photography in the dermatology office," *Seminars in Cutaneous Medicine and Surgery*, vol. 27, no. 1, pp. 83–85, 2008.

119. L. Barco, M. Ribera, and J. Casanova, "Guide to buying a camera for dermatological photography," *Actas Dermo-Sifiliográficas (English Edition)*, vol. 103, no. 6, pp. 502–510, 2012.

120. N. Lasierra, A. Alesanco, Y. Gilaberte, R. Magallón, and J. García, "Lessons learned after a three-year store and forward teledermatology experience using internet: Strengths and limitations," *International Journal of Medical Informatics*, vol. 81, no. 1, pp. 332–343, 2012.

121. ATA, *Quick Guides for Store-Forward Teledermatology and Live-Interactive Teledermatology.* American Telemedicine Association, 2012.

122. M. H. Lowitt, I. I. Kessler, C. L. Kauffman, F. J. Hooper, E. Siegel, and J. W. Burnett, "Teledermatology and in-person examinations: A comparison of patient and physician perceptions and diagnostic agreement," *Archives of Dermatology*, vol. 134, no. 4, pp. 471–476, 1998.

123. M. Gray and A. Bowling, "Considerations for a successful teledermatology application," in *Proceedings of the Ninth Annual Health Informatics Conference and Exhibition*, 2010, p. P11.

124. M. Ribera Pibernat, P. F. Peñas, and L. Barco Nebreda, "La teledermatología hoy," *Piel*, vol. 16, no. 5, pp. 225–237, 2001.

125. H. A. Miot, M. P. Paixão, and F. M. Paschoal, "Basics of digital photography in dermatology," *Anais Brasileiros de Dermatologia*, vol. 81, no. 2, pp. 174–180, 2006.

126. T. Williams, C. May, A. Esmail, C. Griffiths, N. Shaw, D. Fitzgerald, E. Stewart, M. Mould, M. Morgan, L. Pickup, and S. Kelly, "Patient satisfaction with teledermatology is related to perceived quality of life," *British Journal of Dermatology*, vol. 145, no. 6, pp. 911–917, 2001.

127. A. E. Burdick and B. Berman, "Teledermatology," *Advances in Dermatology*, vol. 12, no. 1, pp. 19–45, 1997.

128. M. A. Epstein, M. S. Pasieka, W. P. Lord, S. T. Wong, and N. J. Mankovich, "Security for the digital information age of medicine: Issues, applications, and implementation," *Journal of Digital Imaging*, vol. 11, no. 1, pp. 33–44, 1998.

129. H. Boesch and G. Airaghi, "Secure transfer of medical data over the Internet: From regulatory data protection jam to framework-based requirements," *Current Problems in Dermatology*, vol. 32, no. 1, pp. 71–75, 2003.

130. L. Puig, "Confidencialidad y seguridad informática en dermatología," *Piel*, vol. 14, no. 1, pp. 128–132, 1999.

131. D. C. Kibbe, "What you need to know about HIPAA now," *Family Practice Management*, vol. 8, no. 3, p. 43, 2001.

132. D. Moreno Ramirez, L. Ferrandiz Pulido, A. Ruiz de Casas, and M. A. Nieto Garcia, "Teledermatologia: Metodologias de trabajo y aplicaciones," *Monografias de Dermatologia*, vol. 19, no. 1, pp. 348–355, 2006.

133. S. W. Dill and J. J. Digiovanna, "Changing paradigms in dermatology: Information technology," *Clinics in Dermatology*, vol. 21, no. 5, pp. 375–382, 2003.

134. G. n. Burg, U. Hasse, C. Cipolat, R. Kropf, V. Djamei, H. P. Soyer, and S. Chimenti, "Teledermatology: Just cool or a real tool?," *Dermatology*, vol. 210, no. 2, pp. 169–173, 2005.

135. J. D. Whited, "Teledermatology research review," *International Journal of Dermatology*, vol. 45, no. 3, pp. 220–229, 2006.

136. N. Eminović, N. De Keizer, P. Bindels, and A. Hasman, "Maturity of teledermatology evaluation research: A systematic literature review," *British Journal of Dermatology*, vol. 156, no. 3, pp. 412–419, 2007.

137. P. J. Karanicolas, M. Bhandari, H. Kreder, A. Moroni, M. Richardson, S. D. Walter, G. R. Norman, and G. H. Guyatt, "Evaluating agreement: Conducting a reliability study," *The Journal of Bone & Joint Surgery*, vol. 91, no. 3, pp. 99–106, 2009.

138. A. M. Oakley, F. Reeves, J. Bennett, S. H. Holmes, and H. Wickham, "Diagnostic value of written referral and/or images for skin lesions," *Journal of Telemedicine and Telecare*, vol. 12, no. 3, pp. 151–158, 2006.

139. C. Ebner, E. M. Wurm, B. Binder, H. Kittler, G. P. Lozzi, C. Massone, G. Gabler, R. Hofmann-Wellenhof, and H. P. Soyer, "Mobile teledermatology: A feasibility study of 58 subjects using mobile phones," *Journal of Telemedicine and Telecare*, vol. 14, no. 1, pp. 2–7, 2008.

140. G. R. Aguilera, P. Cortina de la Calle, E. V. Iglesias, P. S. Caminero, M. G. Arpa, and J. A. Martin, "Interobserver reliability of store-and-forward teledermatology in a clinical paractice setting," *Actas Dermo-Sifiliograficas*, vol. 105, no. 6, pp. 605–613, 2014.

141. E. Tan, A. Yung, M. Jameson, A. Oakley, and M. Rademaker, "Successful triage of patients referred to a skin lesion clinic using teledermoscopy (image it trial)," *British Journal of Dermatology*, vol. 162, no. 4, pp. 803–811, 2010.

142. C. S. Silva, M. B. Souza, I. A. Duque, L. M. d. Medeiros, N. R. Melo, C. d. A. Araújo, and P. R. Criado, "Teledermatology: Diagnostic correlation in a primary care service," *Anais Brasileiros de Dermatologia*, vol. 84, no. 5, pp. 489–493, 2009.

143. C. Ruiz, C. Gaviria, M. Gaitán, R. Manrique, Á. Zuluaga, and A. Trujillo, "Concordance studies of a web based system in teledermatology," *Colombia Médica*, vol. 40, no. 3, pp. 259–270, 2009.

144. E. Nordal, D. Moseng, B. Kvammen, and M. Løchen, "A comparative study of teleconsultations versus face-to-face consultations," *Journal of Telemedicine and Telecare*, vol. 7, no. 5, pp. 257–265, 2001.

145. J. L. Lesher Jr., L. S. Davis, F. W. Gourdin, D. English, and W. O. Thompson, "Telemedicine evaluation of cutaneous diseases: A blinded comparative study," *Journal of the American Academy of Dermatology*, vol. 38, no. 1, pp. 27–31, 1998.

146. M. Loane, R. Corbett, S. Bloomer, D. Eedy, H. Gore, C. Mathews, K. Steele, and R. Wootton, "Diagnostic accuracy and clinical management by realtime teledermatology: Results from the Northern Ireland arms of the UK multicentre teledermatology trial," *Journal of Telemedicine and Telecare*, vol. 4, no. 2, pp. 95–100, 1998.

147. H. S. Pak, D. Harden, D. Cruess, M. L. Welch, R. Poropatich, and the National Capital Area Teledermatology Consortium, "Teledermatology: An intraobserver diagnostic correlation study, part I," *Cutis*, vol. 71, no. 5, pp. 399–403, 2003.

148. V. A. Heffner, V. B. Lyon, D. C. Brousseau, K. E. Holland, and K. Yen, "Store-and-forward teledermatology versus in-person visits: A comparison in pediatric teledermatology clinic," *Journal of the American Academy of Dermatology*, vol. 60, no. 6, pp. 956–961, 2009.

149. G. Lozzi, H. Soyer, C. Massone, T. Micantonio, B. Kraenke, M. Fargnoli, R. Fink-Puches, B. Binder, A. Di Stefani, R. Hofmann-Wellenhof et al., "The additive value of second opinion teleconsulting in the management of patients with challenging inflammatory, neoplastic skin diseases: A best practice model in dermatology?," *Journal of the European Academy of Dermatology and Venereology*, vol. 21, no. 1, pp. 30–34, 2007.

150. C. M. Barnard and M. E. Goldyne, "Evaluation of an asynchronous teleconsultation system for diagnosis of skin cancer and other skin diseases," *Telemedicine Journal and e-Health*, vol. 6, no. 4, pp. 379–384, 2000.

151. W. A. High, M. S. Houston, S. D. Calobrisi, L. A. Drage, and M. T. McEvoy, "Assessment of the accuracy of low-cost store-and-forward teledermatology consultation," *Journal of the American Academy of Dermatology*, vol. 42, no. 5, pp. 776–783, 2000.

152. E. M. Warshaw, F. A. Lederle, J. P. Grill, A. A. Gravely, A. K. Bangerter, L. A. Fortier, K. A. Bohjanen, K. Chen, P. K. Lee, and H. S. Rabinovitz, "Accuracy of teledermatology for pigmented neoplasms," *Journal of the American Academy of Dermatology*, vol. 61, no. 5, pp. 753–765, 2009.

153. E. M. Warshaw, F. A. Lederle, J. P. Grill, A. A. Gravely, A. K. Bangerter, L. A. Fortier, K. A. Bohjanen, K. Chen, P. K. Lee, H. S. Rabinovitz et al., "Accuracy of teledermatology for nonpigmented neoplasms," *Journal of the American Academy of Dermatology*, vol. 60, no. 4, pp. 579–588, 2009.

154. N. Eminovic, N. F. de Keizer, J. C. Wyatt, G. ter Riet, N. Peek, H. C. van Weert, C. A. Bruijnzeel-Koomen, and P. J. Bindels, "Teledermatologic consultation and reduction in referrals to dermatologists: A cluster randomized controlled trial," *Archives de Dermatology*, vol. 145, no. 1, pp. 558–564, 2009.

155. M. A. Loane, S. E. Bloomer, R. Corbett, D. J. Eedy, C. Evans, and N. Hicks, "A randomized controlled trial assessing the health economics of real-time teledermatology compared with conventional care: An urban versus rural perspective," *Journal of Telemedicine and Telecare*, vol. 7, no. 1, pp. 108–118, 2001.

156. M. T. Hsueh, K. Eastman, L. V. McFarland, G. J. Raugi, and G. E. Reiber, "Teledermatology patient satisfaction in the Pacific Northwest," *Telemedicine and eHealth*, vol. 18, no. 5, pp. 377–381, 2012.

157. E. Kahn, S. Sossong, A. Goh, D. Carpenter, and S. Goldstein, "Evaluation of skin cancer in Northern California Kaiser Permanente's store-and-forward teledermatology referral program," *Telemedicine and eHealth*, vol. 19, no. 10, pp. 780–785, 2013.

158. R. M. Berghout, N. Eminović, N. F. de Keizer, and E. Birnie, "Evaluation of general practitioner's time investment during a store-and-forward teledermatology consultation," *International Journal of Medical Informatics*, vol. 76, no. 3, pp. S384–S391, 2007.

159. M. Loane, A. Oakley, M. Rademaker, N. Bradford, P. Fleischl, P. Kerr, and R. Wootton, "A cost-minimization analysis of the societal costs of realtime teledermatology compared with conventional care: Results from a randomized controlled trial in New Zealand," *Journal of Telemedicine and Telecare*, vol. 7, no. 4, pp. 233–238, 2001.

160. M. Loane, S. Bloomer, R. Corbett, D. Eedy, C. Evans, N. Hicks, P. Jacklin, H. Lotery, C. Mathews, J. Paisley et al., "A randomized controlled trial assessing the health economics of realtime teledermatology compared with conventional care: An urban versus rural perspective," *Journal of Telemedicine and Telecare*, vol. 7, no. 2, pp. 108–118, 2001.

161. A. W. Armstrong, D. J. Dorer, N. E. Lugn, and J. C. Kvedar, "Economic evaluation of interactive teledermatology compared with conventional care," *Telemedicine and e-Health*, vol. 13, no. 2, pp. 91–99, 2007.

162. M. R. Baze, "Application and evaluation of teledermatology in an underserved area of Honduras," PhD dissertation, Virginia Polytechnic Institute and State University, 2011.

163. K. Collins, S. Walters, and I. Bowns, "Patient satisfaction with teledermatology: Quantitative and qualitative results from a randomized controlled trial," *Journal of Telemedicine and Telecare*, vol. 10, no. 1, pp. 29–33, 2004.

164. J. D. Whited, R. P. Hall, M. E. Foy, L. E. Marbrey, S. C. Grambow, T. K. Dudley, S. K. Datta, D. L. Simel, and E. Z. Oddone, "Patient and clinician satisfaction with a store-and-forward teledermatology consult system," *Telemedicine Journal and e-Health*, vol. 10, no. 4, pp. 422–431, 2004.

165. E. Gilmour, S. M. Campbell, M. A. Loane, A. Esmail, C. E. M. Griffiths, M. O. Roland, E. J. Parry, R. O. Corbett, D. Eedy, H. E. Gore, C. Matthews, K. Steel, and R. Wootton, "Comparison of teleconsultations and face-to-face consultations: Preliminary results of a United Kingdom multicentre teledermatology study," *British Journal of Dermatology*, vol. 139, no. 1, pp. 81–87, 1998.

166. J. Lavanya, K. Goh, Y. Leow, M. Chio, K. Prabaharan, E. Kim, Y. Kim, and C. Soh, "Distributed personal health information management system for dermatology at the homes for senior citizens," in *Proceedings of the 28th Annual International Conference of the IEEE Engineering in Medicine and Biology Society*, pp. 6312–6315, 2006.

167. M. A. Weinstock, F. Q. Nguyen, and P. M. Risica, "Patient and referring provider satisfaction with teledermatology," *Journal of the American Academy of Dermatology*, vol. 47, no. 1, pp. 68–72, 2002.

168. A. W. Armstrong, M. W. Kwong, E. P. Chase, L. Ledo, T. S. Nesbitt, and S. L. Shewry, "Teledermatology operational considerations, challenges, and benefits: The referring providers' perspective," *Telemedicine and eHealth*, vol. 18, no. 8, pp. 580–584, 2012.

169. D. Jones, C. Crichton, A. Macdonald, S. Potts, D. Sime, J. Toms, and J. McKinlay, "Teledermatology in the highlands of Scotland," *Journal of Telemedicine and Telecare*, vol. 2, no. 1, pp. 7–9, 1996.

170. H. S. Pak, M. Welch, and R. Poropatich, "Web-based teledermatology consult system: Preliminary results from the first 100 cases," *Studies in Health Technology and Informatics*, vol. 64, pp. 179–184, 1998.

171. D. Eedy and R. Wootton, "Teledermatology: A review," *British Journal of Dermatology*, vol. 144, no. 4, pp. 696–707, 2001.

172. A. M. Lowie, "Teledermatology: A tool for nurse practitioner practice?," *The Journal for Nurse Practitioners*, vol. 8, no. 8, pp. 617–620, 2012.

173. A. W. Armstrong, M. W. Kwong, L. Ledo, T. S. Nesbitt, and S. L. Shewry, "Practice models and challenges in teledermatology: A study of collective experiences from teledermatologists," *PloS One*, vol. 6, no. 12, p. e28687, 2011.

24

Teleaudiology

Robert H. Eikelboom and De Wet Swanepoel

CONTENTS

24.1 Introduction

24.1.1 The Demand for Hearing Health Services

Hearing loss and chronic ear disease are among the most common health disorders in the world. The prevalence of these conditions is highest in the developing world, where populations also face the disadvantage of poor access to health services. Over the past 15 to 20 years, the use of telehealth technologies has been shown to be effective in breaking down some of the barriers to delivery of ear and hearing health services to the underserved.

24.1.2 Prevalence of Hearing Loss and Incidence of Ear Disease

The World Health Organization has regularly reported on the high prevalence of hearing loss. Most recently, permanent disabling hearing loss has been estimated to affect 360 million people worldwide [1]. Broadening the criteria of hearing loss to include those with significant hearing loss, close to 10% of the global population are affected [2]. The prevalence of significant hearing loss can be expected to increase with increasing life expectancy.

Otitis media (OM) is usually categorized as acute or chronic. The incidence of acute OM is estimated to be 709 million per year, 51% of these children [3]. Chronic OM (OM that lasts for at least 6 weeks and accompanied by a perforated tympanic membrane) affects 31 million people each year [3]. Incidence rates are highest in central Africa, the Indian subcontinent, Southeast Asia, and Papua New Guinea. While acute OM usually has no long-term health implications, whereas chronic OM is associated with many long-term consequences, including permanent hearing impairment [4], affecting development, education, and employment [5]. Deaths associated with chronic OM are estimated to number 21,000 per year [3].

24.1.3 Number and Distribution of Ear and Hearing Health Professionals

There appears to be a chronic shortage of health professionals to deal with the large number of cases of hearing loss and ear disorders. The rates of ear specialists and audiologists per capita are highest in the developed high-income countries, e.g., Sweden, United States, Australia, Israel, and Canada, and lowest in developing countries, e.g., most countries in Africa, South and Southeast Asia, and Central America [6–7].

However, there is also evidence that there are also inequalities in delivering health services to disadvantaged populations in developed countries, e.g., Australia [8], United States [9], and Canada [10].

Added to these inequalities is the fact that the current audiological workforce in developed countries is not sufficient to meet the current demand [11], and that training rates of audiologists are insufficient to keep up with the increasing demand for audiological services [11–12].

24.1.4 The Role of Telehealth in Audiology

Telehealth for the delivery of ear and hearing health services has been promoted, explored, and validated since the early 2000s. The two areas of early research were on the diagnosis and management of ear disease using otoscope images [13] and on the remote assessment of hearing [14]. Since then the telehealth assessment of hearing has covered a range of activities ranging from screening of newborns, children, and adults for hearing loss; workplace screening to establish a baseline of hearing thresholds; detailed diagnostic assessment of hearing; and provision of interventions [15].

As well as extending the reach of services to the underserviced, telehealth has the potential to increase the efficiency of audiological service delivery. Coupled with automated technologies, e.g., automated audiometry [15–17] and computer-aided diagnosis [18], telehealth has the potential to increase the efficiency of the delivery of audiology services [12] by task-shifting some routine assessments away from highly trained audiologists.

24.2 Synchronous and Asynchronous Teleaudiology

Telehealth systems are often categorized as delivering services synchronously (live or real time) or asynchronously, also known as store and forward.

Synchronous telehealth provides a direct and live connection between two sites, usually between a medical specialist and a patient at times in the presence of a clinician, or between two medical practitioners. Although live telehealth is often considered as a consultation that utilizes videoconferencing, a telephone call can be considered the earliest and simplest form of live telehealth [19]. It requires no additional infrastructure; is reliable; requires little or no training, being a ubiquitous tool in society; and enables matters to be addressed immediately. Conference calls with more than two parties are also relatively straightforward to arrange.

The addition of live video can add value to a voice-only consultation. Medical practitioners are able to observe clinically important information, e.g., wounds and other external pathology, and can help to build a level of trust and rapport between patients and medical practitioners. A number of other examples illustrate the role that live telehealth can play in delivering an ear and hearing health service:

- Live video consultations allow practitioners to demonstrate techniques or use of equipment, e.g., hearing aid, to the patient.
- With sufficient bandwidth, a live stream from medical imaging equipment, e.g., video otoscope, can be relayed to an ear specialist.
- A live hearing test can be undertaken, with an audiologist remotely controlling an audiometer.

Live telehealth consultations in ear and hearing health can be problematic. Sufficient bandwidth for a good-quality videoconferencing is often not available. Videoconferencing can suffer from poor synchrony between voice and video; even when optimal, videoconferencing is usually challenging for those with a hearing impairment.

Another challenge is that videoconferencing relies on having two or more people and videoconferencing facilities available at the same time and, therefore, may not be time efficient.

Ear and hearing health services are, however, ideally suited to asynchronous telehealth solutions, as ear conditions are rarely life threatening; the overwhelming amount of cases do not require urgent attention.

Store-and-forward telehealth systems collect information, such as clinical history, images, and test results, locally and then send these at a convenient time to a clinician. The advantages of store-and-forward systems are as follows:

- Store-and-forward telehealth has low bandwidth requirements.
- Providing ear and hearing health services in rural and remote areas is often opportunistic; therefore, booking live teleconsultations is not possible.
- All relevant clinical information can be carefully collated and reviewed and is stored for future references.
- Ear specialists can review information and provide advice in groups at convenient times.

Ultimately, the choice of using either or both modes of telehealth depends on the clinical requirements and the technology and human resources available.

24.3 Requirements for Providing Primary through Tertiary Ear and Hearing Health Services

Utilizing telehealth for the provision of ear and hearing health services requires a number of items to be addressed: planning, personnel, protocols, equipment, and training.

24.3.1 Barriers to Success and Planning for Success

Telehealth may be considered a simple technological extension of the use of the telephone for medical consultations. However, a number of factors are at play that may contribute to a failure of telehealth programs:

- Facilitators of technology and IT may be barriers to rolling out or supporting services [20].
- Telehealth requires the use of equipment that is not ordinarily used for other everyday tasks. This unfamiliarity requires training to use equipment and troubleshooting.
- If ear and hearing health services have not been available previously, the community and health service may not be aware of the importance of ear and hearing health and the way that diagnostic, treatment, and rehabilitation services are delivered.
- Ear specialists and audiologists may be reluctant participants, e.g., due to workload and lack of resources [21].

- In most jurisdictions, there is no funding for telehealth consultations.
- Patients may prefer face-to-face consultations, although there is some evidence to suggest that for the most part they are willing to use telehealth for hearing services [22–23].

Therefore, to maximize chances of success, the planning process should carefully consider potential barriers and consider the following: demand for services, consultation with community, consultation with local health services and key clinicians, access to computer and communication resources for implementation and support, local skill levels and training requirements, referral pathway, and funding.

24.3.2 Local Personnel

Local community health workers form the local touch point in telehealth services. In the role as local champions, they have been recognized as key factors in the early success of telehealth implications [24]. They are often the providers of primary healthcare in the absence of general practitioners, audiologists, and medical specialists. Their local knowledge can be invaluable in aspects such as personal knowledge of patients, trust, ability to provide opportunistic care, and local language skills. Their role as facilitators has been demonstrated [25]. While they will not be qualified to make diagnoses or interpretations, with proper training they are qualified to facilitate the information exchange for the specific telehealth encounter of the patient and clinician.

However, the skill and education levels of local health workers may be low. Furthermore, they often work in isolated locations with little peer support. Therefore, significant attention must be paid to continued training and support.

24.3.3 Training and Support

Training and support will have to be designed and implemented to suit the local requirements and the existing skills and knowledge levels of the service providers in the telehealth service. Key items for facilitators will be:

- An understanding of the importance of good ear and hearing health in children and adults (this knowledge is essential in order for clinicians to understand the priority of their work);
- A basic understanding of ear anatomy and the hearing system, the consequences of ear disease and hearing loss, the terminology used in ear and hearing health, and a recognition of normal and abnormal conditions of the ear;
- An understanding of preventive and rehabilitation measures;
- The correct and safe use of equipment, including setup, care, and maintenance; and
- The position of the telehealth encounter in the referral pathway, including the source and destination of referrals.

These may be delivered in the most practical medium available, including the option of videoconferencing. However, training in the use of equipment should ideally be conducted in person, preferably on site. Training should commence under controlled conditions, with participants testing on each other. Closely supervised training in a clinical setting with constant reinforcement and correction where needed should then be undertaken. This practical training should preferably be for at least 1 day. Support should be planned and agreed upon before the service commences. The form and frequency will vary but should at least commence within weeks of starting the service.

Accreditation of the course is desirable as this places value on the training and adds to the career progression of the individuals who are trained.

Clinicians such as audiologists, otolaryngologists, speech therapists, general practitioners, and pediatricians may be involved in a telehealth service. They may be assessing images and other clinical information that has been sent to them electronically and providing advice back to a remote site. Others may be involved in a live telehealth consultation; yet others will be receiving referrals for a face-to-face consultation after someone else provided a telehealth consultation.

Key training items for other clinicians in a telehealth service are the use of telehealth software and hardware and an understanding of their role in a telehealth service.

24.3.4 Equipment

Audiological practice is largely reliant on computer-operated equipment for screening, diagnosis, and intervention (assistive devices), with the potential to be highly compatible with telehealth service provision. Additional equipment may be required for specific telehealth functions. When designing, developing, or purchasing audiological equipment used for clinical assessment for use in telehealth services, the telecommunication and computer technologies to be utilized must be considered.

Telephone networks (plain old telephone service [POTS]) were until recent years the most common telecommunication medium for telehealth. Aside from being suitable for normal telephone and fax communication, these fixed lines were also used for data transmission at restricted speeds. The early days of telehealth's use of the Internet was facilitated by use of modems, which, although limited in bandwidth, were suitable for transmission of clinical information and images using e-mail or other information management systems.

Digital communication lines (e.g., integrated services digital network [ISDN] and asymmetric digital subscriber line [ADSL]) at local, national, and international levels have enabled data transmission speeds to increase. Furthermore, high-speed access to telecommunication networks is now also possible through mobile or cellular networks. These now facilitate videoconferencing, although quality of service is often not assured. With use of dedicated lines, quality of service can be managed, and transmission of live high-quality images of, e.g., surgery and video otoscope, is possible. Current challenges include the cost of installing and using these high-speed networks and negotiating measures that institutions have in place to maintain security, e.g., firewalls.

Methods of managing the data generated in telehealth should be implemented. This is essential in order that good records are kept and data are not lost and to ensure that activity is recorded, that past telehealth interactions can be reviewed, and that patients are followed through the referral pathway. While traditional paper-based record keeping may still be suitable for face-to-face consultations, it has limitations when those involved in the process are not in the same physical location and when activity may take place in isolation from the patient, e.g., in asynchronous telehealth. E-mailing of data and images from one site to another should ideally be avoided, as information is not aggregated and e-mail is not a secure form of data transfer. National and international standards should be consulted regarding the security of data that are stored and transferred to other sites and users.

24.3.5 Protocols and Referral Pathway

It is important that a set of protocols be developed and agreed upon before a telehealth service is put in place listed below.

All participants should read and sign off on the protocols. Participants include (i) medical practitioners providing the primary services, e.g., nurses, health workers, facilitators, and local general practitioners; (ii) medical practitioners providing the specialist services, e.g., audiologists and otolaryngologists; (iii) health service organizations employing the medical practitioners; (iv) funding agencies; and (v) support personnel, e.g., IT support.

The following is a suggested list of items to include in teleaudiology service protocols:

- The purpose and scope of the service: This may define a segment of the population and the geographical area covered.
- A list of institutions, clinicians, and other personnel involved in the service: Names of individuals should be included and updated as required. Accreditation and licensure information should also be included, as well as a statement of who has the responsibility of the patients.
- Equipment, covering make and model, setup, storage, maintenance, and calibration.
- Communications infrastructure, covering suppliers/providers, minimum specifications, and support.
- Training and support for clinicians and other personnel.
- Referral pathway, including information on clinical protocols that will be used.
- Funding of service.
- Interaction with patients.
- Review of service and review.
- Governance of service.

24.4 Teleaudiology Functions

Audiological services are increasingly being subjected to trials and implemented using telehealth service delivery models [15]. These audiological services extend across the audiological functions of screening and diagnosis, to interventions and continuing professional development.

24.4.1 Screening

Screening for ear disease or hearing loss occurs in a number of forms and can easily be performed by telehealth [15].

Screening for hearing loss in newborns is becoming routine in many countries because congenital or early-onset hearing loss is the most common birth defect, affecting between 1 and 3 per 1000 live births [26]. Screening tests include two electrophysiological procedures, automated otoacoustic emissions (OAEs) that measure the integrity of the outer hair cells in the cochlea and automated auditory brain stem response (AABR) testing that measures auditory responses in the EEG generated by the neural activation of the auditory system. The tests are noninvasive and can be conducted within 2 to 3 min per ear. They are normally conducted within a few days of birth by trained nurses or dedicated screeners.

Several reports of teleaudiology screening applications were identified in a systematic review of teleaudiology applications [15]. These included real-time screening of newborns with OAE and AABR [27]. Unsurprisingly, screening results were similar for remote and on-site screening. These applications, however, would seem best mediated through a store-and-forward (asynchronous) telehealth model but these "proof-of-concept" studies support the viability of remote screening. In fact most newborn hearing screening programs are a form of asynchronous telehealth with screening conducted by trained screeners and result uploaded to secure servers for further follow-up.

Hearing screening of schoolchildren has been shown to have the same sensitivity as conventional pure-tone audiometry, but lower specificity [28].

Smartphone technologies have also been used to implement hearing screening of children in school settings [16,29], although to date calibration to international standards has been addressed by only one application [16]. Self-screening of adults has utilized the telephone or Internet, enabling widespread access with minimal infrastructure [30]. This solution is important as it overcomes the need for absolute calibration of test signals. Technological solutions such as these are important as the number of older people increases as a proportion of the population, especially as the large cohort of the Baby Boomer population moves beyond age 55 years [31].

24.4.2 Diagnosis

Utilizing telehealth in the diagnostic process is common but probably not recognized as such in some cases. Asking for a second opinion or interpretation of results from a colleague or specialist by using the telephone or by e-mail are forms of synchronous telehealth and asynchronous telehealth, respectively. Participating in a live consultation where tests are administered and the results of which are collected by a facilitator and reviewed offline by an audiologist are examples of where telehealth enables a diagnosis to be made.

24.4.3 Case History

A key part of the assessment of patients is the taking of a case history. In a live telehealth session this is facilitated through questions from the clinician. Standardized and custom designed forms and surveys can also be utilized prior to consultations, administered on paper to be stored electronically by scanning or data entry, or entered directly by patients on standard or tablet PCs. Standardized questionnaires are available for many aspects of ear and hearing health, for example, hearing loss handicap [32], tinnitus [33], and balance [34].

24.4.4 Otoscopy and Video Otoscopy

Examination of the eardrum and ear canal is essential in all ear and hearing health assessments. Digital still images and videos can be recorded at the remote site and transmitted to another site for assessment by ear specialists [13,35]. Careful selection of a suitable video otoscope is required [36]. Still images can be used to follow up surgery after tympanostomy tube insertion [37], store-and-forward technologies for the assessment of ear disease have been validated [25,38–40], cost–benefits have been demonstrated [41], live video consultations have been incorporated into a telehealth service [40,42], and a reduction in waiting times for specialist services has been shown [43].

The assessment of any person with an ear and hearing complaint should commence with the examination of the external ear and tympanic membrane (eardrum) with an otoscope. The standard otoscope illuminates the ear canal and eardrum and provides the user with a view through a simple lens system.

The examiner first determines if there are any obstructions in the ear canal, such as a buildup of cerumen (earwax); infectious discharge from the middle ear; or a foreign object, e.g., bead, seed, or insect. These obstructions should be carefully removed by a trained doctor or nurse.

The eardrum should be examined for perforations, redness, swelling, signs of OM or middle ear dysfunction, and other abnormalities. All observations should be recorded.

Use and care of an otoscope requires a short training session, especially when used in young children. Recognition of abnormalities requires significantly more training as views through the otoscope can be fleeting, and abnormalities are often difficult to distinguish and differentiate.

Technology advances in the past decade have seen the emergence of the video otoscope as an inexpensive and better method of otoscopy. Conventional otoscopy provides an image that can be viewed only by the clinician and cannot be shared or stored. On the other hand, a video otoscope captures a live image which can be displayed on a video or computer screen. The view from the video otoscope can be shared with the patient and family members and also with students in teaching situations. Short video clips or still images can also be stored. These can be stored for later review and also sent to a consulting ear specialist or audiologist. With suitable a network connection, a live video stream can be sent to another location for a live telehealth consultation.

A hands-on training course in otoscopy is essential to ensure safe practices, as video otoscopes have the potential to cause discomfort and even injury to the ear canal and eardrum. Some of the points to note are the following:

- Some video otoscopes have a fixed focus, while others require manual focusing. Manual focusing while the otoscope is in the ear canal can be cumbersome and even a safety hazard; therefore, prefocusing the otoscope before use is preferred. Ear canals are approximately the same length; therefore, once the correct focus setting has been found, refocusing will not be necessary in most cases.

- Both the patient and the operator should be seated. In the case of video otoscopy, the operator should be positioned so that only a slight head movement is needed to observe the otoscope entering the ear canal and the video screen. Having the patient watch the views of the otoscope is recommended, as this assists in keeping the head still. When changing to the other ear, the operator should move to the other side of the patient and not turn the patient around.

- In the case of infants, otoscopy must be performed as quickly as possible. Infants can be held firmly against the body of the mother. If an infant or a child becomes agitated, otoscopy should be immediately abandoned on that day.

- The otoscope speculum should be wiped with a suitable antiseptic solution. This may have to be repeated a number of times if the tip of the otoscope touches earwax or moisture in the ear canal.

- The operator should hold the otoscope with one hand with finger extended to brace against the side of the head in case of sudden movements. The other hand should retract the pinna.

- The operator should ensure that they are holding the otoscope so that the image is correctly orientated. If this is not done, adjustments that have to be made to negotiate through the ear canal will be hazardous.

- The operator should watch the otoscope tip as it is inserted slowly into the entrance of the ear canal, and then divert their view to the video screen as the otoscope is slowly inserted further into the ear canal.

- A foot switch or another person should be used to control image capture on a computer should this be required.

24.4.4.1 Video Otoscope Selection

Video otoscopes are available for a wide price range. When selecting a video otoscope, the following factors should be taken into account: image quality (color, resolution, focus, and field of view), interoperability with existing image capture software or utility of image capture software supplied, exchangeable speculums, manual or fixed focus, and ergonomics. The use of video otoscope on patients is illustrated in Figure 24.1.

24.4.5 Tympanometry

Middle ear functioning is usually determined by a tympanometer (illustrated in Figure 24.2). The tympanometer provides an indication of the compliance (or movement) of the eardrum, enabling the presence of effusion in the middle ear; Eustachian tube dysfunction; or a perforation of the eardrum. The tympanometer probe is inserted into the entrance of the ear canal, and with slight pressure a seal is formed. A continuous probe tone is presented into the ear and changes in the sound intensity level in the ear canal is measured while air is pumped into and out of the ear canal to generate slight positive and negative pressures. The nature of the sound conduction through the middle ear and the ear canal volume is reported and interpreted. For example, no movement with normal ear canal volume indicates an effusion in the middle ear (also known as "glue ear"), and no movement with a large ear canal volume indicates a perforation of the eardrum.

Depending on the features of tympanometer used, the data generated can be printed out; sent via cable, Bluetooth, or wireless link directly into a computer record; or transposed into paper or electronic patient case notes.

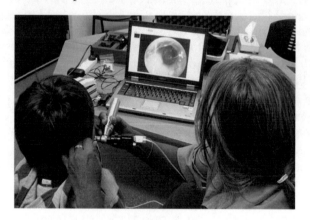

FIGURE 24.1
A video otoscope being used to examine the ear of a child.

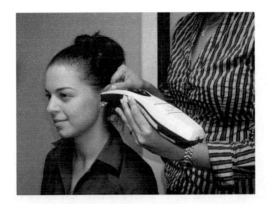

FIGURE 24.2
Image of tympanometer in use.

This procedure is normally performed by audiologists only, but it provides valuable information. It is a procedure that can be carried out by primary care providers in a telehealth setting after appropriate training. Forming an adequate seal with the ear canal presents the most difficulty for new users.

Tympanometry should not be performed when the ear canal is blocked with earwax or when the ear is discharging. In the case of earwax this should be removed by a trained person. In the case of an ear discharge, tympanometry can cause fluid to be sucked into the device, hence damaging it.

A tympanometer should be locally calibrated on a daily basis and calibrated and serviced annually by a calibration agency.

24.4.5.1 Tympanometer Selection

A wide range of tympanometers are available. Portable models should be preferred for telehealth applications.

24.4.6 Audiometry

Assessment of the hearing of a patient is conducted with an audiometer, which determines the nature of suspected hearing loss. Audiometers generate short tones of varying frequencies and intensities, which are delivered to either ear of the patient with a set of earphones. The patient is asked to response to the tones. Two types of tests can be conducted.

A screening audiometer is the simplest test and determines whether the hearing may fall outside of normal limits. A small range of frequencies are tested for each ear, at least 1000 and 4000 Hz for children, but also 500 and 1000 Hz if possible. Normally the loudness of the tones starts at 40 dB and is reduced to 30 dB and then to the softest at 20 or 25 dB if the tones are heard. A screening audiometer is suitable for use in quiet rooms where ambient sound levels are low. It can be easily performed by primary care providers after a short training session.

Diagnostic audiometry is more complex (illustrated in Figure 24.3), and is normally performed only by an audiologist. It is required to establish the severity and shape of a hearing loss and assists in identifying the cause of the hearing loss. A protocol, known as the modified Hughson–Westlake method, is used to determine the threshold of hearing at a range of frequencies. As well as presenting the tones through a set of earphones (to test

FIGURE 24.3
KUDUwave automated audiometer in use.

conduction through the air, viz., air-conduction audiometry), tones can also be presented through a bone-conducting transducer which directly transmits sound to the cochlear (viz., bone-conduction audiometry). Ambient noise levels must be controlled when conducting diagnostic audiometry, normally through the use of an acoustically treated room or booth.

Both screening and diagnostic audiometry have traditionally been conducted manually and in a face-to-face setting. Several innovations in the past decade have enabled audiometry to be included in telehealth practices:

1. Remote control: The first innovation utilized modern audiometers that were able to be controlled via a computer [14]. Via a network connection and "remote-desktop" software, such as pcAnywhere, an audiometer was controlled remotely by an audiologist [44]. Results of studies on children [45] and adults [46] have been reported.

2. Automated audiometer: The clearly defined Hughson–Westlake protocols for audiometry make them adaptable to an automatic test. This has been implemented by a number of manufacturers and validated in various studies [17,47–50].

24.4.6.1 Audiometer Selection

Audiometers suitable for manual screening and diagnostic audiometry are relatively inexpensive compared to tympanometers. Features to consider are connectivity to computers and portability including provision for battery operation. It is possible that screening audiometry is performed outside of an acoustically treated room; therefore, testing with over-the-ear earphones should be considered.

There are currently a few options for automated audiometers. However, only one of these has been independently validated and is available commercially, while others implement only automated air-conduction audiometry.

Audiometers should be calibrated annually by a professional agency.

24.4.7 Speech Audiometry

While pure-tone audiometry is essential to determine the thresholds of hearing, a more real-world measure of hearing ability is speech perception or recognition. This is usually conducted at calibrated sound levels for the speech signal and at times also at a calibrated noise level. Testing for speech perception is challenging to conduct by telehealth, because videoconferencing quality may be so low as to distort audio signals, those from both the assessor and from the patient. Despite these limitations, one reported study showed remarkable accuracy compared to that of face-to-face testing of speech perception [51]. Further research and development is required, pursuing options such as presenting the patient with high-quality prerecordings of words or sentences and forced-choice answers entered via a touch screen.

24.4.8 Auditory Evoked Responses

OAEs and auditory brain stem responses (ABRs) are objective assessments of the hearing system. In the case of OAEs the probe has to be placed correctly in the ear and electrodes have to be placed carefully on the patient for ABRs. Some validation studies have been performed where these are positioned by a facilitator and when the clinician guides the facilitator by videoconferencing [27,52–53]. This facility may overcome the shortage of pediatric audiologists and allows them to reach underserved areas [54], even across state and international borders.

24.4.9 Intraoperative Monitoring

Audiology resources are at times required during surgery, for example, during excision of acoustic schwannomas when the auditory nerve may have to monitored, or during cochlear implant surgery when the integrity of the electrodes can be tested. Intraoperative monitoring may be a substantial drain on time, with often substantial waiting times between preparation and waiting for the pertinent part of the surgery. Connecting monitoring devices to a communication network has been used to allow an audiologist to control and monitor equipment from their office [55].

24.4.10 Balance Assessment

The balance system is anatomical and physiologically close to the hearing system. Assessment of balance usually involves physical manipulation of a patient and, to prevent injuring the patient, significant training of a facilitator should telehealth be undertaken. To date only one case study of vestibular assessment by telehealth has been reported, in this case to observe eye movements by a videoconference link [56].

24.4.11 Intervention

Intervention by a clinician often follows a diagnosis of disease or impairment. This may include counseling, fitting, and maintenance of hearing assistive devices, hearing aids or implants, and tinnitus treatment. Although there is potential for telehealth to improve the delivery of these services to underserved areas [57], there has to date been limited research and implementation in this area.

24.4.11.1 Counseling

Videoconferencing is able to connect individuals with live audio and video feeds and real-time counseling without the need for expensive equipment. E-mail and telephone exchanges can also play an important role in this area. Internet-based tools have been used to monitor patients with hearing aids [58], to educate hearing aid users [59–60], and to treat patients with tinnitus [61–64].

Families require early intervention services for infants and young children with hearing loss to ensure acquisition of developmental milestones. Families can be empowered to meet the needs of their child and can help minimize the need for special education services. These needs can be well catered by videoconferencing [65–67].

24.4.11.2 Hearing Aids and Assistive Devices

Hearing aids are the most common form of rehabilitation. A number of steps involve the audiologist: (i) selection of the most suitable aid; (ii) fitting, which may include taking a physical mold of the outer ear and ear canal; (iii) programming; (iv) counseling and training; and (v) maintenance. The steps of device selection, counseling, and training are suitable for a videoconference consultation. Taking a physical mold is still a challenge as this does pose some significant safety risks for the patient if the person making the mold is not properly trained. 3D imaging technologies have been proposed but are not in use. Hearing aids are now almost exclusively digitally programmed to match the hearing loss of the patient and the situations in which it is used and may be conducted remotely. A limited number of studies have explored some of these steps, demonstrating the potential of telehealth in providing and maintaining hearing aids [68–71].

24.4.11.3 Hearing Implants

Hearing implants are used to rehabilitate the hearing in cases of severe and profound hearing loss. Devices are surgically implanted, usually with an electrode placed in the cochlea or a transducer placed in the middle ear. A recent innovation in teleaudiology is the provision of hearing implant–related services. The usual procedure commenced with a thorough assessment of suitability that includes audiometric assessment and counseling. If suitable surgery takes place, the device is switched on after a few weeks. Electrodes are tested, and each is programmed (or mapped) to produce the best hearing outcome. Regular mapping sessions take place in the first year as the recipient becomes used to the device. Rehabilitation programs are also put in place. Thereafter, annual mapping sessions are recommended. All these functions, other than surgery, have been delivered by telehealth.

Most reported implementations of telehealth services to date have used off-the-shelf products for remote control of programming software and for videoconferencing facilities and a trained facilitator [72–75]. This solution is ideal for situations where there are large regional populations not currently provided with hearing implant services, but where audiologists or other hearing health clinicians are able to provide some local clinical and technical support as assistance to the remotely located implant audiologist.

More recently, solutions have been developed that allow the delivery directly to an individual of services that do not need specialized support personnel but utilize a family member or friend [76]. This caters to people who are unable to easily access a local clinic, for example, those who are in nursing homes or housebound, those who are in very remote locations, or those who would prefer a service they can access directly from home.

One implant manufacturer provides a device that wirelessly communicates with an implant [77]. Although it was made available to some users to control volume and environmental settings, this device can be used as an interface for the implant to be controlled and programmed remotely by an audiologist.

24.4.12 Continued Professional Education

Continuing education and professional development is a key user of telehealth infrastructure. Primary care providers in rural and remote areas often lack access to local peer-level support and opportunities for traditional forms of professional development. Services in rural and remote areas often also suffer from high turnover, therefore increasing the need to provide training and support services [78]. Telehealth opportunities are involved in grand rounds [79], mentoring [80], and training [81].

24.5 Case Studies of Teleaudiology

A number of cases will illustrate the use of telehealth to address the need for an audiology service. The first example is of service delivered in a developing country to an underserved sector of the community, where the primary drivers are a high demand for services and a shortage of clinicians. The second describes a service in a developed country faced with providing diagnostic and surgical otology services to a small scattered population with poor transportation links. The third example is of a teleaudiology service provided by clinics in a developed country to a remote Pacific island faced with the handicaps of distance and lack of local specialist services.

24.5.1 Teleaudiology in Witkoppen, South Africa

A pilot teleaudiology project has been initiated at a primary healthcare clinic, Witkoppen Health and Welfare Centre, serving the severely underserved population of Diepsloot, a densely populated settlement outside of Johannesburg, South Africa. This settlement is made up of government-subsidized housing, brick houses built by landowners, and shacks made from scrap metal, wood, plastic, and cardboard. More than 90% of the estimated 150,000 population are likely to be unemployed, with no access to basic services such as running water, sewage, and the rubbish removal unavailable to many families.

To access ear and hearing health services requires travel to distant hospital-based centers at significant travel costs and time with several visits back and forth likely. In an effort to decentralize basic ear and hearing health services, a teleaudiology clinic was established as part of collaboration between Witkoppen Health and Welfare Centre, a private company (eMoyoDotNet, Johannesburg, South Africa) and the University of Pretoria, South Africa. A trained facilitator manages the teleaudiology clinic. The facilitator has no formal healthcare training but is a community member who has been trained to facilitate patient encounters using a predominantly asynchronous service-delivery model. Services provided include automated testing with a diagnostic air- and bone-conduction audiometer incorporating real-time noise monitoring and double attenuation (insert earphones covered by circumaural earphones) to allow for testing outside a sound booth [48,82]. From the approximately 500 patient visits per day at the clinic, those with ear and hearing health concerns are sent to the teleaudiology clinic, where they received automated hearing testing facilitated by the trained facilitator. Results are automatically uploaded through a cellular network to a secure server from where a remote audiologist makes the necessary interpretation and recommendations to be sent back to the clinic. Further to a hearing assessment, video otoscopy has been used to provide a remote diagnosis on ear disease by an otologist [25,83]. Based on these findings, limited medical treatment can be initiated at the clinic if recommended, and for audiological intervention, patients can be referred directly for hearing aid fittings with their audiogram in hand.

24.5.2 Teleotology in Alaska

The geography of Alaska poses large challenges to health services, with many small communities often inaccessible by road. The Alaska Federal Health Care Access Network (AFHCAN) was initiated in the early 2000s by the Alaskan Native Tribal Health Consortium to deliver store-and-forward telehealth services. A teleotoscopy service has been in the forefront of their services to over 200 communities across Alaska. They developed a hardware cart (afhcan.org/cart.apsx) that includes a video otoscope and tympanometer, as well as devices for dental, cardiac, and respiratory system assessment and for vital-signs monitoring. The cart also contains equipment for power management, network connectivity, scanner, and videoconferencing. The telehealth capabilities are managed by tConsult (afhcan.org/tconsult.aspx), a software suite that captures data and images and manages store-and-forward telehealth consultations with ear specialists.

Participating clinicians and researchers have published studies showing the efficacy of using video otoscope images before [84] and after [85] ventilation tube surgery and the positive impact on surgical wait times [43,86]. Recommendations for surgical treatment and time for surgery were made using store-and-forward clinical information, tympanometry, and video otoscope images and also in a face-to-face consultation. These were compared to actual surgical procedures performed and time for surgery. No significant differences were found [84]. After using teleotoscopy for 6 years, the average waiting times for a face-to-face consultation in cases where this was required reduced by 50% [43].

Problems with image quality control due to the focusing of the Welch Allyn video otoscope led to the development of a focusing tool [87].

24.5.3 Telepractice between Sydney and Western Samoa

The Pacific island of Western Samoa lies some 4300 km northwest from Sydney, Australia. The challenge to support the rehabilitation of a small group of children with cochlear implants on the island was addressed using telepractice by hearing implant specialists at the Sydney Cochlear Implant Centre (SCIC) and the Royal Institute for Deaf and Blind Children (RIDBC) in Sydney, Australia. These children had received implants as a result of charitable donations; however, there was no system in place for postsurgical care, and the cost of travel precluded visits of the recipient and a family member to Australia.

Technology was put in place in 2010 initially to provide ongoing review and programming of the implant recipients' device, as well as training and mentoring of local health professionals. Two computer systems were utilized at both sites, one pair to provide videoconferencing facilities and the other pair to conduct the programming of the implant. This configuration is similar to that reported by a number of sites in Europe [72–74] and the United States [75]. Figure 24.4 illustrates initial trials of telepractice for cochlear implant mapping.

Improvements in network infrastructure have allowed the service to progress from being conducted in the offices of the sole Internet provider in Western Samoa to being conducted at the children's school with wireless connectivity. The use of tablet computers has provided utility to the system.

FIGURE 24.4
Initial trials of telepractice for cochlear implant mapping between Apia, Western Samoa, and Sydney, Australia, March 2010.

Training and mentoring has resulted in skills improvements in local health professional so that in 2014 they are now able to provide independent management of patients. The success of the service has also facilitated more implants in children and provision of services to adults.

To date over 70 people, from 9 months to 70 years of age, have utilized the service.

Recipient satisfaction questionnaires are utilized and show high overall satisfaction, but the elderly adults are least satisfied, consistent with problems with network connectivity.

Associated with this program was the development of online training modules for hearing health professionals [88–89] and a set of guidelines for telepractice [90].

24.6 Conclusions

Ear and hearing health has to date been well served with research and developments in telehealth services. There is a large and growing demand for ear and hearing services that cannot be met by traditional modes of service delivery. However, although some sustainable services are in place, there are still many opportunities for implementation. Many challenges remain, especially those regarding developing a workforce to be involved in delivering telehealth services, building financially sustainable services, and developing effective speech perception tests and methods of delivering interventions.

Acknowledgment

Colleen Psarros, of Sydney Cochlear Implant Centre, New South Wales, is acknowledged for contributing the case study on service to Western Samoa and providing the accompanying photograph.

Partial List of Manufacturers and Suppliers

Video Otoscopes

- MedRx video otoscope, MedRx, http://www.medrx-usa.com/WebSite/pages/medrx-Video-Otoscope.html, accessed on September 6, 2014
- Digital MacroView otoscope, Welch Allyn, http://www.welchallyn.com/en/products/categories/physical-exam/ear-exam.html, accessed on September 6, 2014
- Flexicam Mobile medical camera, Inline, http://www.inline.com.au/otoscope-OX1, accessed on September 6, 2014

Tympanometers

- Otowave tympanometer, MedRx, http://www.medrx-usa.com/WebSite/pages/medrx-Otowave-Tymp.html, accessed on September 6, 2014

Audiometers

- GSI 18, Grason Stadler, http://www.grason-stadler.com, accessed on September 6, 2014
- Maico MA 25 screening audiometer with automated option (air conduction only), Maico, http://www.maico-diagnostic.com/products/audiometers/ma-25/, accessed on June 8, 2015
- KUDUwave, eMoyoDotNet, http://www.emoyo.net, accessed on September 6, 2014
- Avant A2D+, MedRx, http://www.medrx-usa.com/website/pages/medrx-A2D.html, accessed on September 6, 2014
- Interacoustics AD229e, Interacoustics AD629e, Interacoustics, http://www.interacoustics.com/ad629, accessed on June 8, 2015
- AMTAS, Audiology Incorporated, http://audiologyincorporated.com, accessed on September 6, 2014

References

1. WHO. Millions of people in the world have hearing loss that can be treated or prevented. Geneva: World Health Organization 2013. Available from: www.who.int/pbd/deafness/news/Millionslivewithhearingloss.pdf. Accessed on September 6, 2014.
2. WHO. The global burden of disease: 2004 update. World Health Organization. 2008. Available from: http://www.who.int/healthinfo/global_burden_disease/GBD_report_2004update_full.pdf. Accessed on September 6, 2014.
3. Monasta L, Ronfani L, Marchetti F et al. Burden of disease caused by otitis media: Systematic review and global estimates. *PloS One* 2012; **7**: e36226.
4. Berman S. Otitis media in developing countries. *Pediatrics* 1995; **96**: 126–131.
5. Acuin J. Chronic suppurative otitis media—Burden of illness and management options. Geneva: World Health Organization; 2004.
6. Goulios H and Patuzzi RB. Audiology education and practice from an international perspective. *Int J Audiol* 2008; **47**: 647–664.

7. Fagan JJ and Jacobs M. Survey of ENT services in Africa: Need for a comprehensive intervention. *Global Health Action* 2009; **2**.

8. Gunasekera H, Morris PS, Daniels J et al. Otitis media in Aboriginal children: The discordance between burden of illness and access to services in rural/remote and urban Australia. *J Paediatr Child Health* 2009; **45**: 425–430.

9. Zuckerman S, Haley J, Roubideaux Y, and Lillie-Blanton M. Health service access, use, and insurance coverage among American Indians/Alaska Natives and Whites: What role does the Indian Health Service play? *Am J Public Health* 2004; **94**: 53–59.

10. Newbold KB. Aboriginal physician use in Canada: Location, orientation and identity. *Health Economics* 1997; **6**: 197–207.

11. Margolis RH and Morgan DE. Automated pure-tone audiometry: An analysis of capacity, need, and benefit. *Am J Audiol* 2008; **17**: 109–113.

12. Windmill IM and Freeman BA. Demand for audiology services: 30-yr projections and impact on academic programs. *J Am Acad Audiol* 2013; **24**: 407–416.

13. Blakeslee DB, Grist WJ, Stachura ME, and Blakeslee BS. Practice of otolaryngology via telemedicine. *Laryngoscope* 1998; **108**: 1–7.

14. Givens GD, Blanarovich A, Murphy T et al. Internet-based tele-audiometry system for the assessment of hearing: A pilot study. *Telemed J E Health* 2003; **9**: 375–378.

15. Swanepoel D and Hall JW, 3rd. A systematic review of telehealth applications in audiology. *Telemed J E Health* 2010; **16**: 181–200.

16. Swanepoel D, Myburgh H, Howe D et al. Smartphone-based hearing screening with integrated quality control and data management. *Int J Audiol* 2014; **52(12)**: 841–849.

17. Eikelboom RH, Swanepoel D, Motakef S, and Upson GS. Clinical validation of the AMTAS automated audiometer. *Int J Audiol* 2013; **52**: 342–349.

18. Goggin LS, Eikelboom RH, and Atlas MD. Clinical decision support systems and computer-aided diagnosis in otology. *Otolaryngol Head Neck Surg* 2007; **136**: S21–S26.

19. Eikelboom RH. The telegraph and the beginnings of telemedicine in Australia. *Stud Health Tech Informat* 2012; **182**: 67–72.

20. Wade VA and Hamlyn JS. The relationship between telehealth and information technology ranges from that of uneasy bedfellows to creative partnerships. *J Telemed Telecare* 2013; **19**: 401–404.

21. Wade VA, Eliott JA, and Hiller JE. Clinician acceptance is the key factor for sustainable telehealth services. *Qualitative Health Research* 2014; **24**: 682–694.

22. Eikelboom RH and Atlas MD. Attitude to telemedicine, and willingness to use it, in audiology patients. *J Telemed Telecare* 2005; **11**(Suppl 2): S22–S25.

23. Eikelboom RH, Jayakody DMK, Swanepoel D et al. Validation of remote mapping of cochlear implants. *J Telemed Telecare* 2014; **20**: 171–177.

24. Wade V and Eliott J. The role of the champion in telehealth service development: A qualitative analysis. *J Telemed Telecare* 2012; **18**: 490–492.

25. Biagio L, Swanepoel de W, Adeyemo A et al. Asynchronous video-otoscopy with a telehealth facilitator. *Telemed J E Health* 2013; **19**: 252–258.

26. Eiserman WD, Hartel DM, Shisler L et al. Using otoacoustic emissions to screen for hearing loss in early childhood care settings. *Int J Pediatr Otorhinolaryngol* 2008; **72**: 475–482.

27. Krumm M, Huffman T, Dick K, and Klich R. Telemedicine for audiology screening of infants. *J Telemed Telecare* 2008; **14**: 102–104.

28. Lancaster P, Krumm M, Ribera J, and Klich R. Remote hearing screenings via telehealth in a rural elementary school. *Am J Audiol* 2008; **17**: 114–122.

29. Handzel O, Ben-Ari O, Damian D et al. Smartphone-based hearing test as an aid in the initial evaluation of unilateral sudden sensorineural hearing loss. *Audiology & Neuro-otology* 2013; **18**: 201–207.

30. Smits C, Merkus P, and Houtgast T. How we do it: The Dutch functional hearing-screening tests by telephone and Internet. *Clin Otolaryngol* 2006; **31**: 436–440.

31. Swanepoel D, Eikelboom RH, Friedland P et al. Self-reported hearing loss in Baby Boomers from the Busselton Health Study–Audiometric correspondence and predictive value. *J Am Acad Audiol* 2013; **24**: 514–521.
32. Newman CW, Weinstein BE, Jacobson GP, and Hug GA. The Hearing Handicap Inventory for Adults: Psychometric adequacy and audiometric correlates. *Ear Hear* 1990; **11**: 430–433.
33. Wilson PH, Henry J, Bowen M, and Haralambous G. Tinnitus reaction questionnaire: Psychometric properties of a measure of distress associated with tinnitus. *J Speech Hearing Res* 1991; **34**: 197–201.
34. Jacobson GP, Newman CW, Hunter L, and Balzer GK. Balance function test correlates of the Dizziness Handicap Inventory. *J Am Acad Audiol* 1991; **2**: 253–260.
35. Crump WJ and Driscoll B. An application of telemedicine technology for otorhinolaryngology diagnosis. *Laryngoscope* 1996; **106**: 595–598.
36. Mbao MN, Eikelboom RH, Atlas MD, and Gallop MA. Evaluation of video-otoscopes suitable for tele-otology. *Telemed J E Health* 2003; **9**: 325–330.
37. Patricoski C, Kokesh J, Ferguson AS et al. A comparison of in-person examination and video otoscope imaging for tympanostomy tube follow-up. *Telemed J E Health* 2003; **9**: 331–344.
38. Eikelboom RH, Mbao MN, Coates HL et al. Validation of tele-otology to diagnose ear disease in children. *Int J Pediatr Otorhinolaryngol* 2005; **69**: 739–744.
39. Kokesh J, Ferguson AS, Patricoski C, and LeMaster B. Traveling an audiologist to provide oto-laryngology care using store-and-forward telemedicine. *Telemed J E Health* 2009; **15**: 758–763.
40. Smith AC, Dowthwaite S, Agnew J, and Wootton R. Concordance between real-time telemedicine assessments and face-to-face consultations in paediatric otolaryngology. *Med J Aust* 2008; **188**: 457–460.
41. Craemer R, Eikelboom RH, Ellis I et al. Ear telehealth in the Pilbara: Costs and benefits. 2009. Available at: research.esia.org.au.
42. Smith AC, Williams J, Agnew J et al. Realtime telemedicine for paediatric otolaryngology pre-admission screening. *J Telemed Telecare* 2005; **11**(Suppl 2): S86–S89.
43. Hofstetter PJ, Kokesh J, Ferguson AS, and Hood LJ. The impact of telehealth on wait time for ENT specialty care. *Telemed J E Health* 2010; **16**: 551–556.
44. Krumm M, Ribera J, and Klich R. Providing basic hearing tests using remote computing technology. *J Telemed Telecare* 2007; **13**: 406–410.
45. Ciccia AH, Whitford B, Krumm M, and McNeal K. Improving the access of young urban children to speech, language and hearing screening via telehealth. *J Telemed Telecare* 2011; **17**: 240–244.
46. Swanepoel D, Koekemoer D, and Clark J. Intercontinental hearing assessment—A study in tele-audiology. *J Telemed Telecare* 2010; **16**: 248–252.
47. Swanepoel D and Biagio L. Validity of diagnostic computer-based air and forehead bone con-duction audiometry. *J Occup Environ Hyg* 2011; **8**: 210–214.
48. Swanepoel D, Mngemane S, Molemong S et al. Hearing assessment-reliability, accuracy, and efficiency of automated audiometry. *Telemed J E Health* 2010; **16**: 557–563.
49. Margolis RH, Frisina R, and Walton JP. AMTAS®: Automated method for testing auditory sen-sitivity: II. Air conduction audiograms in children and adults. *Int J Audiol* 2011; **50**: 434–439.
50. Margolis RH and Moore BC. AMTAS®: Automated method for testing auditory sensitivity: III. Sensorineural hearing loss and air-bone gaps. *Int J Audiol* 2011; **50**: 440–447.
51. Ribera JE. Interjudge Reliability and validation of telehealth applications of the hearing in noise test. *Seminars in Hearing* 2005; **26**: 13.
52. Swanepoel D, Louw B, and Hugo R. A novel service delivery model for infant hearing screen-ing in developing countries. *Int J Audiol* 2007; **46**: 321–327.
53. Towers AD, Pisa J, Froelich TM, and Krumm M. The reliability of click-evoked and frequency-specific auditory brainstem response testing using telehealth technology. *Seminars in Hearing* 2005; **26**: 26–34.
54. Polovoy C. Audiology telepractice overcomes inaccessibility. *ASHA Leader*. 2008.

55. Shapiro WH, Huang T, Shaw T et al. Remote intraoperative monitoring during cochlear implant surgery is feasible and efficient. *Otol Neurotol* 2008; **29**: 495–498.
56. Virre E, Warner D, Balch D, and Nelson JR. Remote medical consultation for vestibular disorders: Technological solutions and case report. *Telemed J E Health* 1997; **3**: 53–58.
57. Swanepoel D, Clark JL, Koekemoer D et al. Telehealth in audiology: The need and potential to reach underserved communities. *Int J Audiol* 2010; **49**: 195–202.
58. Laplante-Levesque A, Pichora-Fuller MK, and Gagne JP. Providing an Internet-based audiological counselling programme to new hearing aid users: A qualitative study. *Int J Audiol* 2006; **45**: 697–706.
59. Thoren E, Svensson M, Tornqvist A et al. Rehabilitative online education versus Internet discussion group for hearing aid users: A randomized controlled trial. *J Am Acad Audiol* 2011; **22**: 274–285.
60. Manchaiah VK, Stephens D, Andersson G et al. Use of the "patient journey" model in the Internet-based pre-fitting counseling of a person with hearing disability: Study protocol for a randomized controlled trial. *Trials* 2013; **14**: 25.
61. Andersson G, Carlbring P, Kaldo V, and Strom L. Screening of psychiatric disorders via the Internet: A pilot study with tinnitus patients. *Nord J Psychiatr* 2004; **58**: 287–291.
62. Andersson G and Kaldo V. Internet-based cognitive behavioral therapy for tinnitus. *J Clin Psychol* 2004; **60**: 171–178.
63. Kaldo V, Larsen HC, Jakobsson O, and Andersson G. Cognitive behavior therapy via Internet: Patients with tinnitus are helped to manage their problem—simpler and cheaper. *Lakartidningen* 2004; **101**: 556–560.
64. Kaldo V, Levin S, Widarsson J et al. Internet versus group cognitive-behavioral treatment of distress associated with tinnitus: A randomized controlled trial. *Behavior Therapy* 2008; **39**: 348–359.
65. Cason J. Telehealth opportunities in occupational therapy through the Affordable Care Act. *Am J Occup Ther* 2012; **66**: 131–136.
66. McCarthy M, Munoz K, and White KR. Teleintervention for infants and young children who are deaf or hard-of-hearing. *Pediatrics* 2010; **126**(Suppl 1): S52–S58.
67. Constantinescu G. Satisfaction with telemedicine for teaching listening and spoken language to children with hearing loss. *J Telemed Telecare* 2012; **18**: 267–272.
68. Campos PD and Ferrari DV. Teleaudiology: Evaluation of teleconsultation efficacy for hearing aid fitting. *Jornal da Sociedade Brasileira de Fonoaudiologia* 2012; **24**: 301–308.
69. Wesendahl T. Hearing aid fitting: Application of telemedicine in audiology. *Int Tinnitus J* 2003; **9**: 56–58.
70. Ferrari DV and Bernardez-Braga GR. Remote probe microphone measurement to verify hearing aid performance. *J Telemed Telecare* 2009; **15**: 122–124.
71. Pearce W, Ching T, and Dillon H. A pilot investigation into the provision of hearing services using tele-audiology to remote areas. *Aust New Zeal J Audiol* 2009; **31**: 96–100.
72. Ramos A, Rodriguez C, Martinez-Beneyto P et al. Use of telemedicine in the remote programming of cochlear implants. *Acta Otolaryngol* 2009; **129**: 533–540.
73. Wesarg T, Wasowski A, Skarzynski H et al. Remote fitting in Nucleus cochlear implant recipients. *Acta Otolaryngol* 2010; **130**: 1379–1388.
74. Wasowski A, Skarzynski H, Lorens A et al. Remote fitting of cochlear implant system. *Cochlear Implants Int* 2010; **11**: 489–492.
75. McElveen JT, Jr., Blackburn EL, Green JD, Jr. et al. Remote programming of cochlear implants: A telecommunications model. *Otol Neurotol* 2011; **31**: 1035–1040.
76. Eikelboom RH, Jayakody DM, Swanepoel DW et al. Validation of remote mapping of cochlear implants. *J Telemed Telecare* 2014; **20**: 171–177.
77. Cochlear Ltd. Nucleus® 6 Remote Assistant and Remote Hearing Manager. Sydney: Cochlear Ltd. 2014. Available from: http://cochlear.com/wps/wcm/connect/in/home/discover/cochlear -implants/nucleus-6-system/nucleus-6-for-adults/nucleus-6-remote-assistant-and-remote -control Accessed on September 6, 2014.

78. Smith JD, O'Dea K, McDermott R et al. Educating to improve population health outcomes in chronic disease: An innovative workforce initiative across remote, rural and Indigenous communities in northern Australia. *Rural and Remote Health* 2006; **6**: 606. Available from: http://www.rrh.org.au/Articles/subviewnew.asp?ArticleID=606 Accessed on September 6, 2014.

79. Allen M. Evaluation of videoconferenced grand rounds. *J Telemed Telecare* 2002; **8**: 210–216.

80. Glazebrook RM and Harrison SL. Obstacles and solutions to maintenance of advanced procedural skills for rural and remote medical practitioners in Australia. *Rural and Remote Health* 2006; **6**: 502. Available from: http://www.rrh.org.au/articles/subviewnew.asp?articleid=502 Accessed on September 6, 2014.

81. Curran VR. Tele-education. *J Telemed Telecare* 2006; **12**: 57–63.

82. Maclennan-Smith F, Swanepoel D, and Hall JW, 3rd. Validity of diagnostic pure-tone audiometry without a sound-treated environment in older adults. *Int J Audiol* 2013; **52**: 66–73.

83. Biagio L, Swanepoel D, Adeyemo A et al. Asynchronous video-otoscopy by a telehealth facilitator: Telemedicine and e-Health. *Telemed J e-Health* 2013; **19**(4): 252–258.

84. Kokesh J, Ferguson AS, and Patricoski C. Preoperative planning for ear surgery using store-and-forward telemedicine. *Otolaryngol Head Neck Surg* 2010; **143**: 253–257.

85. Kokesh J, Ferguson AS, Patricoski C et al. Digital images for postsurgical follow-up of tympanostomy tubes in remote Alaska. *Otolaryngol Head Neck Surg* 2008; **139**: 87–93.

86. Kokesh J, Ferguson AS, and Patricoski C. The Alaska experience using store-and-forward telemedicine for ENT care in Alaska. *Otolaryngol Clin North Am* 2011; **44**: 1359–1374, ix.

87. Patricoski C, Ferguson AS, and Tooyak A, Jr. A focus tool as an aid to video-otoscopy. *J Telemed Telecare* 2003; **9**: 303–305.

88. HEARnet. Remote Cochlear Implant Mapping. HEARnet (no date). Available from: http://www.hearnet.org.au/health-professionals-area/hearnet-learning/teleaudiology/remote-cochlear-implant-mapping/ Accessed on October 26, 2014.

89. HEARnet. A Home Based Model for Cochlear Implantation: The Role of Telepractice. HEARnet (no date). Available from: http://www.hearnet.org.au/health-professionals-area/hearnet-learning/teleaudiology/a-home-based-model-for-cochlear-implantation-the-role-of-telepractice/ Accessed on October 26, 2014.

90. North J, McCarthy M. RIDBC Teleschool Guiding Principles for Telepractice Sydney; 2012.

25

Teleoncology

Natalie K. Bradford and Helen Irving

CONTENTS

25.1 Introduction

Teleoncology refers to the use of technology to provide treatment, care, or services related to oncology. While teleoncology as a field has been around for as long as telemedicine has, there remains limited evidence to support the practice of teleoncology; as such, compared to other modalities of telemedicine, teleoncology is not common in routine practice.

In this chapter, the current experiences and evidence regarding teleoncology are investigated, with an aim of presenting the potential role of the teleoncology for low and middle income nations.

25.2 Oncology Overview

Oncology is a broad medical term regarding the study of cancer and it encompasses the biology, prevention, screening, and treatment of cancer, including surgery, radiotherapy, chemotherapy, immunotherapy, hormonal therapy, and supportive and palliative care [1]. The term *cancer* is used to describe a broad range of diseases which arise following unregulated cell growth. Cancer cells can originate in any organ of the body; cells divide and multiply in an uncontrolled fashion, form tumors, invade other parts of the body, and spread through the lymphatic or blood system. There are over 200 different types of cancer, which are classified according to the cell of origin, extent of spread, or stage of disease [2].

Cancer is generally diagnosed following investigations which are prescribed as a result of progressive signs or symptoms of disease or through specific cancer-screening programs. Diagnostic tests commonly include blood tests; radiological imaging, including X-rays (Figure 25.1); and CT or MRI scans. If left untreated, most cancers will cause significant morbidity and, ultimately, death. Depending upon the type of cancer, treatment may include surgery, radiotherapy, and chemotherapy. In the last decade, treatment of cancer has also rapidly evolved to include targeted therapies and immunotherapies as a result of increased understanding of the human genome and epigenetics. Treatment for cancer can be complex and highly individualized; thus, an oncologist—a specialist medical practitioner with specific training in treating cancer—usually oversees treatment. There are also subspecialists within the field of oncology for specific disease types or patient groups (e.g., pediatrics). Around the globe, management of a patient's disease usually involves a multidisciplinary team (MDT) including medical and radiation oncologists; surgeons; organ-specific specialists (e.g., gynecologists); and allied health, clinical research, and nursing personnel.

The incidence of cancer around the globe is rising. This is often attributed to the aging population but can also be attributed to the underlying prevalence of cancer becoming identified as a result of the rise of screening programs, surveillance, and greater public awareness of cancer risk. It is predicted that the global incidence of cancer will increase by 50% over the next 20 years and that low- and middle-income countries will bear the brunt of the cancer burden [3]. Despite low- and middle-income countries having constrained health systems which have traditionally rendered cancer treatment as a luxury,

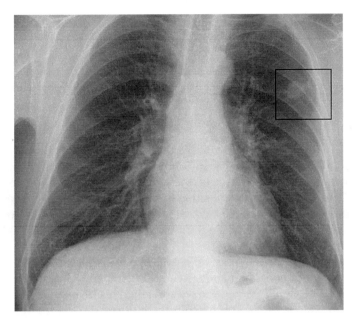

FIGURE 25.1
X-ray image of detection of lung cancer. (From Varni, J.W., and C.A. Limbers, *Pediatric Clinics of North America*, 2009. **56**(4): pp. 843–863.)

only available for the "rich," there are expanding programs which aim to prevent and treat cancer through education and use of off-patent drugs financed through public–private partnerships [3].

25.3 Teleoncology

Telemedicine is the use of telecommunications to advance health. Teleoncology can be described as *"the use of a variety of telecommunications technologies to provide clinical oncology services"* [5]. Teleoncology, therefore, encompasses any application of telemedicine to advance cancer care, including diagnostics (radiology, pathology, and laboratory), treatment (surgery, radiotherapy, chemotherapy, or other drug therapy), and supportive care, including palliative care and follow-up care [6].

Telemedicine in oncology was first described in 1994 with the use of videoconferencing to link oncologists at a large teaching hospital in Kansas with patients and clinicians in rural hospital settings [7]. Most evidence supporting teleoncology has focused on efficacy, costs and benefits, and patient and clinician satisfaction. Over the years, there has been a slow increase in reports of telemedicine being used in oncology to enable healthcare clinicians to collaborate and share information, knowledge, and experiences to improve patient care. Electronic patient records, participating in multidisciplinary team meetings by using technology, and sharing of test results and treatment plans electronically are some examples that can be found in the literature [8]. However, despite being adopted by some health services, there remains a paucity of evidence to support telemedicine in oncology, particularly so in low- and middle-income countries [9], where teleoncology could potentially play a pivotal role in enhancing services for cancer care. Where formal evaluation of teleoncology services does exist, studies are small and methodologically weak [8]; thus, teleoncology has not been adopted as a routine modality of providing care by many cancer services.

25.3.1 Rationale for Teleoncology

Across the globe, there are disparities in cancer care as a result of patient location and health service resources. Most health services and patients receiving treatment desire treatment to be available as close as possible to their home location. However, successful treatment of cancer relies on specialist knowledge and treatment. Teleoncology is seen as a potential mechanism for improving equity of access to specialist care, thus optimizing patient outcomes.

Successful treatment of cancer relies on a multimodal approach which involves many specialists including surgeons, hematologists, and medical and radiation oncologists as well as specialized and highly trained nurses and allied health support. Around the world it is uncommon for this array of specialists, which offers the highest-quality care available, to be located anywhere but in a major tertiary hospital. In low- and middle-income countries, there may not even be access to an oncologist, let alone the multidisciplinary care team. Indeed, in Latin America, cancer is the second most common cause of death in children, despite an 80% cure rate in high-income countries [10]. Globally it is estimated that fewer than 30% of children with cancer have access to contemporary treatment [11]. In high-income countries, there are also disparities in care; many patients with cancer will not live in close proximity to a tertiary facility and, thus, may not receive the best-quality care. Indeed, there is evidence to suggest that in high-income countries, patients living in rural or regional areas have a lower survival rate compared to patients in metropolitan areas [12–13]. Furthermore, in the United States, it is predicted there will be a shortage of 2393 oncologists by the year 2025, compounding the problems further. The American Society of Clinical Oncology states that unless oncologists' productivity can be enhanced, the shortage will strain the ability to provide quality cancer care across the whole of the United States [6].

With increasing pressures on healthcare systems, there is an increasing propensity to manage oncology patients as outpatients rather than as inpatients in a hospital setting. Where a patient would once be admitted for a course of chemotherapy spanning several days, it is now commonplace for patients to be managed in a day therapy oncology department, where they receive therapy and supportive care and are discharged at the end of each day to return the following day to continue the treatment. One of the downfalls of this shift is that there is a reduced opportunity for patients to receive the appropriate education about their disease management and of the expected side effects of treatment.

Thus, there is scope for using telemedicine to improve cancer care, by systematically and effectively using telecommunications between tertiary facilities and remote or resource-poor facilities and to support patients in the ambulatory care setting. By linking facilities with different levels of funding and expertise and providing education and services, it is thereby possible to increase equity of access to care and extend high-quality cancer care to rural and regional locations as well as across borders to resource-poor nations with constrained health systems [6].

25.4 Telecommunications and Models of Cancer Care

Teleoncology can use a variety of communication technologies including synchronous (real-time) interactive videoconferencing, web conferencing protocols, and asynchronous (store-and-forward) communication, e.g., SMS programs, e-mail, or transmission of digital images or other data. Each of these technologies will be briefly described with examples of how they are used to provide teleoncology care.

Videoconferencing systems are one of the most common forms of teleoncology. ISDN systems are often networked throughout a district, state, or nation and are expensive to install and maintain but enable high-quality audiovisual communication between multiple locations to occur. Videoconferencing may be used by clinicians only, e.g., for education or multidisciplinary team meetings, or may include the patient and family, e.g., for oncologist consultation. Videoconferencing can be used to manage all aspects of a patient's cancer care and this is discussed in greater detail in the following sections.

However, before considering using videoconferencing to manage patient care, the technical aspects of managing the systems must be considered; there are multiple examples of systems not well utilized due to poor availability of technical support [14]. Clinicians are unlikely to have the time, knowledge, or inclination to troubleshoot problems that invariably occur; indeed, it would be a poor use of a clinician's time, which could otherwise be spent providing clinical care.

More affordable videoconferencing systems may also be an option, and, particularly in low- and middle-income countries use of the Internet can be an option for videoconferencing, although the Internet is unlikely to be able to consistently provide high-quality audio and video. There are reports of health services using the Internet for video transmission and alternative sources (such as a telephone) for audio transmission of a conversation [15]; this may be an option for increasing the quality of communication by using the Internet.

Web conferencing protocols are a relatively new technology which offers advantages, particularly for informal education and interaction between clinicians. Compared with videoconferencing systems, web conferencing protocols are inexpensive and do not require the same level of technical expertise. However, they do not offer the same quality as videoconferencing; images cannot be manipulated (e.g., zoom in or out) and participants may not all be able to see each other. St. Jude Children's Research Hospital has used web conferencing protocols for many years to support their International Outreach program and their website Cure4Kids (http://www.cure4kids.org) boasts a huge array of educational materials and presentations as well as having the ability to host live synchronous video education sessions for specific diseases, which participants can access from distant locations [16]. The goal is to improve healthcare for children with cancer around the world with access to information, technology, and transfer of organizational skills.

Asynchronous communication uses software to transmit, retrieve, and store data or images. This can be a highly practical and cost-efficient way of accessing specialists or obtaining second opinions between different locations. Examples of sustained programs that utilize this form of telemedicine in oncology include PACS radiology, which revolutionized radiology by facilitating around-the-clock services for many facilities. In Europe, the continent-wide project Trans-European Network for Positron Emission Tomography (TENPET) uses asynchronous communication to support the interpretation of PET scans, which is particularly useful for cancer centers which may not otherwise have access to experts [17].

Broad recommendations that cover considerations for all models of teleoncology services involving patients include

- Patient and family orientation on telemedicine processes prior to consultation;
- Inclusion of a translator service for all patients who require assistance with English;
- Inclusion of a telemedicine coordinator to manage all technical aspects of the consultation; and
- Rapid reporting and documentation of outcomes from consultation to the referring clinician/team.

25.5 Traditional Models of Cancer Care for Regional or Remote Locations

Traditionally, as previously discussed, cancer care has been organized and coordinated by larger, tertiary hospitals with access to the multidisciplinary care team needed to provide comprehensive cancer care and treatment. For patients who live outside the area serviced by these institutions, where cancer treatment and care is limited, there are three common models of care [12]:

1. Medical oncologists will travel by road or air to towns and supervise the treatment, including the administration of chemotherapy to patients. These visits may range in frequency from weekly to monthly, depending upon the services offered to a particular region and their capabilities and resources.
2. Patients travel by road or air to a larger facility where they see the specialist oncologist; receive treatment; and then return to the local facility, where they may also receive some treatment, including chemotherapy. This is particularly common in pediatrics.
3. Patients travel by road or air to larger facilities where they see the specialist oncologist and receive all of their treatment. The patient may not return to their home location until they have fully recovered from treatment. This is also commonly seen in childhood cancer as some smaller facilities may not have pediatric services [18].

25.6 Potential Benefits and Disadvantages of Teleoncology

Teleoncology offers new models of care which can supplement or even replace the traditional models of care, offering support in locations that were not previously possible. The potential benefits and disadvantages to the patients, clinicians, and healthcare service are summarized in Table 25.1 [19].

TABLE 25.1

Benefits and Disadvantages of Teleoncology for Different Stakeholders

	Potential Benefits	Potential Disadvantages
Patient	• Improved access to appropriate care • Improved safety and quality of care • Less travel time • Reduced financial costs associated with travel, accommodation, and meals • Less time lost from work • Improved flexibility with scheduling of appointments • Increased time spent with family • Improved social support • High satisfaction with the care provided via telemedicine • Reassurance	• Depersonalization • Lack of physical examination • Concerns over substitution of other clinicians for specialist oncologist • Stress associated with technology • Preference for seeing the oncologist in person • May not be suitable for patients with hearing, visual, or cognitive impairments
Clinicians—including nurses, allied health personnel, and specialist oncologists	• Decreased need to travel—increased safety • Potential for increased productivity • Optimized time spent on professional duties • Opportunities for peer-to-peer support, peer review, collaboration, and education • Ability to participate in multidisciplinary team meetings	• Scheduling difficulties—may be time consuming • Depersonalization • Legal implications—perceived increased clinical risk • Reimbursement issues • Privacy and confidentiality • Communication barrier—difficulty of expressing empathy
Healthcare system and other providers	• Achieve better health outcomes for equal or lower costs • Improved efficiency of processes and medical services • Reduction of unnecessary referrals • Reduction in unnecessary tests and procedures • Expanded networking opportunities	• High initial cost outlay • Other efficiency-related problems may emerge • Increase in referrals that could be managed locally

25.7 Teleoncology and Cancer Prevention

As more understanding regarding the pathology of disease is generated, preventive medicine is a developing and important field. While most cancer-preventive services are provided by primary care healthcare clinicians, for genetic cancers, or high-risk patient groups, preventive oncology services may be provided by specialists. Where there are genetic risk factors for certain types of cancer, such as breast cancer, individuals may benefit from access to specialist oncology services regarding their options for reducing their risk, which can be provided through telemedicine. This may include increased surveillance or preventive surgery as well as genetic counseling. The Arizona Telemedicine Program offers such a service. This program is an established tele–genetic outreach program, which has demonstrated efficacy and high patient satisfaction; patients have reported that their satisfaction is so high with services offered via telemedicine that there is no need to seek in-person genetic counseling [20]. Several other studies have found that the use of telemedicine is a useful alternative to face-to-face cancer genetic counseling [21].

Environmental risk factors are responsible for many of the preventable cancers and many of these risk factors are controllable. Approximately 33% of cancers are attributed to tobacco, diet, alcohol, air pollution, lack of exercise, or sexually transmitted diseases and, thus, are preventable [22]. As many cancers can be prevented, there is a role for considering the potential of telemedicine to be used for cancer prevention. Health promotion activities generally lend themselves well to technology-based solutions. Examples such as use of SMS for health promotions, web-based learning activities, or "chat sites" have demonstrated efficacy and high levels of patient satisfaction in reducing smoking in patients at risk of heart disease. These types of technologies can bring the support of a healthcare clinician directly to the patient's home to support their behavior change [21]. Indeed a recent Cochrane review identified that interventions which used mobile phone technology were *the* most effective at reducing tobacco smoking [23]. The same concept has been applied to nutritional advice and programs on increasing physical activity. While these initiatives are aimed at reducing the risk of heart disease, they are applicable to promoting a healthy lifestyle, which also reduces cancer risk.

25.8 Teleoncology and Cancer Diagnosis

To obtain an accurate diagnosis, it is essential that tumors are histologically examined and identified by a pathologist. The pathologist may be involved during surgery to ensure that there is an adequate sample to establish a diagnosis [24]. However, many hospitals do not have sufficient access to a pathologist; and when one is not available on site, telepathology enables the electronic transmission of microscopic slide images to an available qualified pathologist for immediate consultation, discussion, and review. Multiple studies have demonstrated the validity of telepathology for both concordance and accuracy compared to conventional practice [25–27]. Many institutions have now integrated telepathology into mainstream practice because of the clear benefits of being able to access a pathologist when required either for a second opinion or for primary diagnosis.

In the United States, UltraClinics uses store-and-forward technologies to speed up the process of diagnosis of breast and prostate cancers. The program was developed as a result of an identified clinical need; breast care was highly fragmented in the state of Arizona and the wait for results of an ambiguous biopsy reportedly could take as long as 1 month, causing extreme anxiety in patients and families [28]. The ultrarapid breast care (URBC) pathway is now streamlined; after a biopsy, the tissue is rapidly processed and the glass slides are digitized and transmitted to a pathologist, who may be in another city or even another state. The service relies on having access to panels of off-site pathologists who are organized into virtual practice groups. Once the pathologist has made a diagnosis, a report is electronically sent to the clinician who performed the biopsy. If the diagnosis is cancer, an appointment is made with an oncologist, which can be undertaken using videoconferencing if there is no oncologist available in the patient's vicinity [28]. The founders of URBC clinic concept believe that this model represents a significant paradigm shift in practice and has become a new standard for cancer care in general.

25.9 Teleoncology and Cancer Treatment

25.9.1 Teleconsultation

In Australia, Sabesan et al. [29] have successfully managed all aspects of a patient's cancer management for patients living in poorly serviced areas, including supervision of medical therapy by using telemedicine. The team, based in a regional part of Queensland, believes that a primary objective of telemedicine should be to improve access to specialist care, thereby reducing the need for unnecessary long-distance travel. To overcome the issue of not being able to examine patients, Sabesan et al. [29] recommend having either a nurse or a medical practitioner present during the consultation at the patient end. Sabesan et al. [29] suggest to physicians who are hesitant about using telemedicine to manage patient's cancer treatments to start by selecting simple cases with the intention of building to more complex management once familiarity and confidence in using telemedicine has developed.

In Canada, another team of specialists has reported their experiences with teleoncology medical consultations. Faced with challenges such as long winters, difficult access to regional areas, high costs of travel and accommodation, and only a few expert oncologists available in metropolitan areas, teleoncology has grown exponentially since 2001 and is the second largest component of the whole telehealth program [19]. The team uses specialized equipment such as handheld fiber-optic scopes to examine eyes, ears, and the throat, as well as high-definition cameras to capture images of wounds and cutaneous tumor deposits. Appointments are scheduled in advance and trained oncology nurses attend at the patient end. Contrary to Sabesan et al.'s [29] experiences, the team reports that teleoncology is more useful for monitoring than for initial consultations or for supervising medical treatments such as chemotherapy [19]. They argue that for certain cancer conditions, such as head and neck or cervical cancer, the visual findings and examinations form a critical part of diagnosis; thus, treatment and teleoncology may not be appropriate during treatment for these diseases. Thus, it is still necessary to have access to traditional models of cancer care. However, the majority of patients receiving follow-up care are reviewed exclusively via telemedicine in this service [19].

There have been discrepancies between reports of satisfaction with teleconsultation in various oncology studies. Earlier studies reported high levels of satisfaction [2,30–32], while studies in more recent times are less enthusiastic [33–35]. This could be attributed to the novelty of the technology dissipating over time, while expectations of safety and quality of healthcare have increased over time [36]. Alternatively, it has been suggested that the high psychosocial burden of a cancer diagnosis and its treatment may be a factor which limits the usefulness of video consultation for both patients and clinicians [37]. However, despite these findings, there is evidence to support the usefulness and effectiveness of teleconsultation. Several studies have identified that teleconsultation and face-to-face consultations were comparable in their ability to make a diagnosis, inform treatment decisions, or decide on a management plan [8,32,38]. Additionally there is evidence that suggests that more patients can be seen in a teleconsultation clinic compared to those in a traditional clinic visit, resulting in time efficiencies and reduction of waiting times to see specialists [39].

25.9.2 Virtual Multidisciplinary Team Meetings

MDT meetings (also known as tumor boards) have long been regarded as the optimal process for reducing unwanted variation in cancer care by ensuring consistent management with best practice, leading to improved outcomes for patients [40–42]. In an MDT meeting, all the relevant clinicians (e.g., oncologist, surgeon, hematologist, nurse, and allied health personnel) are gathered together to discuss a patient's management. MDT meetings are standard practice in oncology and are used to make key decisions regarding patient management, including recruiting patients for research clinical trials; this is important as substantial evidence exists that participating in a research clinical trial improves outcomes for cancer patients [43]. MDT meetings are used to ensure that patients receive the appropriate diagnosis and treatment and that interventions are evidence based. Meetings facilitate the exchange of communication between all clinicians and provide an opportunity for informal education.

MDT meetings can present an inconvenience to clinicians, disrupting clinical activities, and can come at a significant cost. Thus, if meetings were able to be conducted in the virtual space, where not all clinicians were required to be physically present, the virtual MDT may present an effective alternative to the traditional MDT [44]. With the ability to use videoconferencing, the case for having virtual MDT meetings is strong. A virtual MDT involves clinicians who participate in the care of a patient and who interact with other members of the team by using shared clinical data [42].

For clinicians in regional areas, a virtual MDT offers the possibility of presenting and discussing cases with a larger group of specialists, often located at a tertiary facility [45]. Additionally, with the increasing subspecialization of medicine including oncology, world leaders in a particular area may be located and asked to participate in a live consultation. This may contribute to better management for very rare or complex cases [1].

However, virtual MDTs are not common and they face other challenges in addition to the already complex nature of traditional MDTs. Understanding the social nature of meetings, navigating technology, and issues on confidentiality and data protection are all facets that can impede the success of a virtual MDT.

25.9.3 Clinical Trials

Advances in cancer survival over the last 30 years are largely attributed to clinical trials which have sought to determine the most effective treatments in a scientifically robust manner. As a result of participating in rigorously run clinical trials, reductions in errors relating to diagnosis, staging, and appropriate treatment delivery have all reduced. Many clinical trials are run nationally or internationally and telemedicine can be used to confirm diagnosis and eligibility for a clinical trial, as well as to receive expert opinion regarding the staging and treatment plans. Technology is also used to ensure that the integrity of a clinical trial is not compromised by data management systems and to use algorithms in ensuring compliance. These processes have subsequently been reported to build capacity and expertise in cancer care [10].

Telemedicine could also be used to develop the capacity of smaller cancer centers to participate in clinical trials through mentorship and training or to allow participants in regional centers to participate in a clinical trial run by a larger center, expediting accrual to trials. Particularly for patients with rare diseases, this could provide opportunities to participate in a trial and gain access to treatments otherwise not available.

25.9.4 Integrated Decision-Support Systems

Increasingly, clinical pathways, guidelines, or algorithms are used in oncology to standardize clinical practice according to best evidence. These decision-support systems lend themselves easily to technology. In the United Kingdom a team developed a novel clinical decision-support platform for breast cancer known as Multidisciplinary Meeting Assistant and Treatment Selector (MATE). This platform, available through the Internet, demonstrated potential for improving the rate of identification of patients eligible for clinical trials and ensuring that the MDT recommendations were followed in practice [46].

Another project that used integrated decision-support systems was the Technology Exchange for Cancer Health Network (TECH-Net) project in Tennessee, United States. This project linked six independent cancer outreach clinics with specialists based at tertiary hospitals to provide comprehensive and coordinated care [47]. The project consisted of electronic health records which were networked with a dedicated system to provide comprehensive communication between sites as well as the option for videoconferencing. While there were numerous outcomes expected to be reported from this project, such as adherence to medication, quality of life, and satisfaction, like many other telemedicine projects, after initial funding and pilot data results, there was lack of evidence that this project continued to operate beyond the project phase.

Other examples of decision-support systems include electronic chemotherapy-prescribing systems and clinical reminder systems for medication [48–49]. There is also evidence emerging of using technology to prospectively monitor patient's self-reported symptoms with alerts to clinical staff if intervention is required [50]. In Singapore, ambulatory cancer patients have benefitted from access to pharmacists through an SMS service. Patients complete an algorithm by SMS for 5 days following their chemotherapy, which directs them to the appropriate management of nausea and vomiting according to set clinical guidelines. If patients responded with either moderate or severe nausea, the pharmacists are alerted and contact the patient via telephone. Benefits of this service identified by the study authors included reduction in unnecessary hospitalizations, increased compliance with medication management, and increased patient satisfaction with follow-up care [9]. There are plans to incorporate the management of other side effects to enhance this service for ambulatory cancer patients.

25.9.5 Telesurgery

Surgery is traditionally a very conservative field; however, examples of telesurgery have been found in oncology, including remote expert supervision during surgery, preoperative surgery planning [1], and surgical simulation for training purposes.

25.9.6 Teleradiotherapy

While radiotherapy must be delivered where radiation medical devices are available (often major tertiary hospitals), the significant amount of planning and expert supervision which is required during irradiation treatment can be provided at a distance [1]. Patient data and radiological scans can be electronically transmitted to a treating center, where therapy can be planned and advice provided. This may offer advantages to both the patient and health service, with reduced travel requirements and longer time periods in the patients' home setting.

25.10 Supportive Care

25.10.1 Discharge Planning

Videoconferencing has been used successfully to provide follow-up support and ongoing education to patients, as illustrated in Figure 25.2, following treatment for cancer.

FIGURE 25.2
Typical example of videoconferencing for patient care (Supplied by authors).

Examples include using videoconferencing to teach patients how to change a colostomy pouch [8].

25.10.2 Access to Allied Health

Videoconferencing is a convenient option for allied health support for patients living in regional and rural areas. Services such as nutrition support, counseling, and even physiotherapy or speech therapy are all allied health services that can be provided through telemedicine [45]. It is recommended that for any service provision which is provided via telemedicine, a nurse accompanies the patient and is able to ensure that the appropriate follow-up occurs.

25.10.3 Palliative Care

Palliative care refers to the philosophy of care of a patient (and family) when a life-limiting or life-threatening disease is diagnosed. Cancer and its treatment cause significant suffering, particularly pain, nausea, dyspnea, and lethargy, which requires active management. Increasingly palliative care is considered an appropriate model of care to offer to any patient diagnosed with cancer, irrespective of life expectancy, as it responds to the suffering experienced by patients. However, in many parts of the world, resources are not available to offer such care to all patients and palliative care is reserved for those whose disease is terminal. Palliative care focuses on interventions for reducing symptoms which may be physical, emotional, spiritual, and psychosocial rather than interventions aimed at cure with the goal to maximize quality of life.

There has been a significant body of research undertaken in the area of telemedicine and palliative care [51] and the term *telehospice* has been coined to describe the provision of hospice care (synonymous with palliative care in some countries) via telemedicine [52].

In Norway, telemedicine is routinely used to facilitate palliative care consultations. A study undertaken by this group found that after the telemedicine service was introduced, the mean duration of waiting for a consultation after referral reduced from 10.2 to 1.9 days ($N = 108$, $p < 0.001$) [53]. Anecdotally the study authors estimated that transfer to tertiary hospitals was reduced from 13% to 6%. There are clear benefits to patients and families with this decrease in time to see a palliative care team, and with the increased support, these experiences in Norway suggest that patients are able to stay in their preferred location.

25.11 Enhancing Cancer Care in Low- and Middle-Income Nations

There is the potential to use telemedicine to expand established and well-resourced cancer care centers to cancer centers in other nations which do not have the resources or infrastructure to comprehensively manage cancer. Termed *twinning programs*, these initiatives link centers and have demonstrated improved outcomes for cancer survival in nations with limited access to oncologists. Twinning programs utilize teleoncology networks and partnerships to share knowledge. By focusing on a pragmatic selection of goals which are achievable for the local health service, twinning programs are able to maximize the effective use of resources [10]. The model, goals, and objectives of the teleoncology service should be directed by the available resources. Importantly, these programs should be individually designed; a model for twinning teleoncology programs in Vietnam will have very different needs from those for a program in Nigeria, for example. It is recommended that specific cancer diseases are chosen based upon the complexity of treatment required and the available infrastructure. An example of a program which has been successful is in the Solomon Islands. At the time, this nation did not have access to a pathologist or reliable Internet. Investing in expensive videoconferencing was deemed to be a poor use of limited resources; instead, through partnering with cancer centers in Switzerland, Germany, and Australia, a program was developed using existing telepathology services in Switzerland and supported by e-mail communication.

The only requirements for this program were electricity, a computer, Internet access for e-mail exchange, and a camera. The framework enabled clinicians to submit data to a server, where experts from around the world could reply. This process of structured communication is efficient, simple, and sustainable and has proven to be an economical and practical program which has enhanced the Solomon Islands' cancer services without requiring huge investment in infrastructure or resources [27]. The development of local skills has empowered healthcare clinicians to be able to provide care, which has the potential to markedly improve the quality and effectiveness of care provided.

The use of expensive technologies and specialized equipment is not recommended for nations which do not have the resources (human, financial, and technical) to support its use. Indeed, even in high-income nations, there are examples of expensive technologies resulting in mismanagement of funds and not achieving the goal of improving access to cancer care. Thus, the common denominator in technology should be the basis of a twinning teleoncology program and simple technologies such as e-mail and store-and-forward communication may be the most appropriate choice.

There may also be opportunities to collaborate with other facilities and share expensive equipment such as videoconferencing units. Institutions such as banks, universities, or other government sectors may have equipment available, and there may be opportunities to share equipment rather than purchase equipment specifically for one area of a health service.

In low- and middle-income nations, there has also been success with using a hub-and-spoke model where tertiary cancer centers serve as hubs for regional centers, sharing their expertise and advice. In India this has improved the regional center's ability to manage patients, building capacity within the health service, which has resulted in reduced transfers to tertiary centers and an improvement in cancer care in regional centers [54].

The long-held belief that cancer treatment is possible only in countries where specialists are available is starting to shift, as teleoncology enables programs to develop and education and treatment to be facilitated in lower- and middle-income countries [10]. There is the distinct possibility that the disparities in cancer care as attributed to location or national economic status could be reduced using telemedicine—a goal which we have a global responsibility to consider.

25.12 Economics

There have been some reports of the viability of teleoncology clinics for replacing the model of oncologists visiting rural or remote locations. Many of the cost-effectiveness studies reported in the literature have focused on establishing and maintaining equipment compared to the costs of a conventional oncology clinic. In one project, teleoncology was seen as viable when the volume of patients required an oncologist on site for more than 1 day per week. Unless the volume of patients exceeded this amount, teleoncology was not seen as efficient. This same project quantified that to break even financially, the teleoncology service needed to save a minimum of 5 hours of travel time per month for the visiting oncologist [45]. In Ontario in Canada, it was estimated that 25% of all travels for health purposes could have been avoided if their telemedicine system were used. This would result in an average cost avoidance per patient of CAD$433 per patient (one way) for an average distance of 579 km [19].

Norway was the first country to recognize the potential health and economic benefits of telemedicine and commenced reimbursement for telemedicine services in 1996 [55]. While other nations—Scotland, Germany, Australia, and parts of the United States—have followed suit, reimbursement remains one of the main barriers to telemedicine services not reaching their full potential. If health clinicians are not able to receive the same payments for seeing a patient via telemedicine as they can for a face-to-face consultation or service, there is no incentive to use telemedicine.

25.13 Medicolegal Recommendations

Concerns over privacy and confidentiality with telemedicine are often raised and clinicians and patients may be cautious as a result. Such concerns, however, are not limited to telemedicine; healthcare professionals are always bound to the requirement to protect sensitive patient information and maintain privacy. Clinicians and patients should have the opportunity to discuss these concerns and have questions answered.

Another medicolegal concern clinicians may have regard for is the risk of litigation because of decisions made via teleoncology. This fear stems from the concern that not all information required to make clinical decisions may be available through teleoncology. Thus, from the clinician's perspective, there remains a hesitancy to use telemedicine to initiate oncology treatment due to concerns over patient safety. The inability to perform a physical examination remains the most frequently cited barrier. However, in several studies it has been concluded that a physical exam would have made no difference to the outcome or management plan [35,53]; hence, the barrier is more aligned with clinicians' satisfaction and confidence using technology. The technology itself can be challenging; there are limited precedents for working in this way, and innately many clinicians are conservative in their practice. The risks of litigation with services provided through teleoncology are in fact no different from the risk of litigation from services provided face to face; clinicians must always use their clinical judgment to provide appropriate services or advice.

Aspects that can reduce potential medicolegal risks for teleoncology services are the same as those for face-to-face services:

- There should be written policies and clear procedures on managing all aspects of patient care.
- Policies and procedures should be ratified by all facilities involved in patient care.
- Documentation of all services provided should be communicated between clinicians and filed in the patient medical records.

25.14 Training and Education

Healthcare providers in regional and rural settings often report impeded access to educational opportunities for cancer care [18,56]. Videoconferencing is an effective way to deliver education, in-services on specific cases or diseases, and even participation in grand rounds and symposia [57–59]. Indeed, many aspects of teleoncology have the potential to facilitate easy access to existing knowledge and to disseminate knowledge and build capacity within health services.

However, it is not as simple as setting up regular videoconferencing sessions or providing access to integrated decision-support systems. Clinicians at both the local and the distant sites need to be aware of the need for education, to identify the objectives of providing education and training using technology and to engage with the education. Experience has demonstrated that the technology itself is just a medium, for education or training to be successful; there is a reliance on human relationships and good communication skills [32].

For the clinicians delivering patient care using telemedicine, there is also a requirement for education. Clinicians need to feel comfortable with the technology and there should be support available to troubleshoot technical issues. While it is desirable that a clinician is able to manage certain aspects of the technology, e.g., to be able to zoom in a camera or to terminate a videoconference call, other aspects of the technology should not be the clinician's responsibility. Additionally, with videoconferencing, training is required regarding how to engage with a patient and how to manage silences and convey empathy [19]. Indeed there are growing concerns regarding how to effectively manage the psychosocial aspects of cancer care, and more information is needed to understand how the efficacy of interpersonal communication can be maximized and emotional distance minimized using technology.

25.15 Barriers to Progress

25.15.1 Human Factors

Human factors remain the main obstacles to the progress of teleoncology [32]. There have been many failed attempts to introduce teleoncology services by simply purchasing equipment and expecting clinicians to use it; supply of equipment alone will not result in a service. Clinicians must drive a teleoncology service and decide on which services will be offered. If a service is not developed because of a clinical problem or if equipment does not meet the clinical requirements, it will not be used. Similarly, if there are competing clinical priorities, financial disincentives, or political oppositions, equipment will not be used. Failure to use equipment for a teleoncology service has been experienced in all settings; low- and middle-income nations have mirrored experiences in high-income nations, where fully set-up videoconferencing units and networks are underutilized. Reasons for slow adoption in low- and middle-income nations are similar to those reported in high-income nations and are related to human factors: fear of change, reluctance to seek a second opinion, and conflicts over political and professional power [32].

25.15.2 Education and Support

Clinicians need to receive appropriate training to use telemedicine technologies, including how to communicate effectively using technology, and should not be expected to troubleshoot or resolve problems with the technology.

Satisfaction with using technology to provide services can also be another barrier, from both the clinicians' and the patients' ends. One study revealed that 50% of clinicians believed that there were improvements required in their teleoncology service, specifically to improve delays in reporting and documentation; issues related to continuity of care; absence of sufficient process to complete physical exams of patients; and patient comfort with technology [34].

25.15.3 Costs

Costs are another frequently cited barrier. Clinicians need to receive appropriate reimbursement for services provided through telemedicine, and telemedicine equipment can be costly. It is not only the cost of purchasing appropriate equipment that needs to be considered but also the resources required to maintain the equipment and coordinate the delivery of care [20]. In low- and middle-income nations, costs prohibit services and teleoncology is not widely available. However, this is certain to change; technology is becoming more commonplace through the world, and most individuals, even in the poorest parts of the world such as Africa, have access to mobile telephones and the Internet.

25.16 Conclusions

Teleoncology offers a promising option for improving access to high-quality cancer care that is well received by patients in a rural or remote location. It can be economical and offers improvement in health through ongoing monitoring, early identification of problems, and provision of access to a wider range of health professionals. However, teleoncology services need to be carefully introduced and used wisely. It is crucial to match the appropriate clinical question or clinical need with the appropriate telemedicine solution. The human factors must be addressed and the use of telemedicine should be driven by clinical need with the goal of improving cancer care services.

Teleoncology services are able to enhance services in poorly supported locations, facilitate the rapid exchange of medical information, and can play an important role in education and the improvement of the quality of cancer care. Particularly for locations where there is a diverse geographical spread of patients and a limited oncology workforce, teleoncology has the potential to reduce the need for travel and enable services to be delivered in the patients' local area. However, services must be balanced with the overarching goal of maintaining safety and high-quality patient care. For these reasons, there is a need for more evidence to support the routine introduction of teleoncology as a mainstream modality for cancer treatment. Teleoncology is, however, effective for patient education and for monitoring patients in between treatments for side effects, as well as for providing follow-up and supportive care for patients who have completed treatment.

As globalization continues to shape the world in which we live, telemedicine is likely to shape the provision of healthcare. Thus, there is a great need to continue to evaluate services and ensure that teleoncology services meet the needs of clinicians, patients, and families.

Acknowledgment

The authors wish to acknowledge the ongoing support of the Sporting Chance Cancer Foundation.

References

1. Wysocki, W.M., A.L. Komorowski, and M.S. Aapro, The new dimension of oncology: Teleoncology ante portas. *Critical Reviews in Oncology/Hematology*, 2005. **53**(2): pp. 95–100.
2. Allen, A., and J. Hayes, Patient satisfaction with teleoncology: A pilot study. *Telemedicine Journal*, 1995. **1**(1): pp. 41–46.
3. Farmer, P. et al., Expansion of cancer care and control in countries of low and middle income: A call to action. *The Lancet*, 2010. **376**(9747): pp. 1186–1193.
4. Varni, J.W., and C.A. Limbers, The pediatric quality of life inventory: Measuring pediatric health-related quality of life from the perspective of children and their parents. *Pediatric Clinics of North America*, 2009. **56**(4): pp. 843–863.
5. Ricke, J., and H. Bartelink, Telemedicine and its impact on cancer management. *European Journal of Cancer*, 2000. **36**(7): pp. 826–833.
6. Yang, W. et al., Projected supply of and demand for oncologists and radiation oncologists through 2025: An aging, better-insured population will result in shortage. *Journal of Oncology Practice*, 2014. **10**(1): pp. 39–45.
7. Allen, A., and J. Hayes, Patient satisfaction with telemedicine in a rural clinic. *American Journal of Public Health*, 1994. **84**(10): pp. 1693–1693.
8. Kitamura, C., L. Zurawel-Balaura, and R. Wong, How effective is video consultation in clinical oncology? A systematic review. *Current Oncology*, 2010. **17**(3): p. 17.
9. Yap, K.Y.-L. et al., Feasibility and acceptance of a pharmacist-run tele-oncology service for chemotherapy-induced nausea and vomiting in ambulatory cancer patients. *Telemedicine and e-Health*, 2013. **19**(5): pp. 387–395.
10. Hazin, R., and I. Qaddoumi, Teleoncology: Current and future applications for improving cancer care globally. *The Lancet Oncology*, 2010. **11**(2): pp. 204–210.
11. St. Jude Children's Research Hospital. Available from: http://www.stjude.org/international.
12. Sabesan, S., and S. Brennan, Tele Oncology for Cancer Care in Rural Australia, in *Telemedicine Techniques and Applications*, Graschew, G., and Rakowsky, S. (Eds.). InTech, 2011. pp. 289–306.

13. Australian Institute of Health and Welfare and Australasian Society of Cancer Registries, *Cancer in Australia: In Brief, Cancer Series* 2010, AIHW: Canberra.

14. Taylor, P., Evaluating telemedicine systems and services. *Journal of Telemedicine and Telecare*, 2005. **11**(4): pp. 167–177.

15. Bradford, N. et al., Safety for home care: The use of Internet video calls to double-check interventions. *Journal of Telemedicine and Telecare*, 2012. **18**(8): pp. 434–437.

16. Ribeiro, R.C., Improving survival of children with cancer worldwide: The St. Jude International Outreach Program approach. *Studies in Health Technology and Informatics*, 2012. **172**: pp. 9–13.

17. Kontaxakis, G. et al., Integrated telemedicine applications and services for oncological positron emission tomography. *Oncology Reports*, 2006. **15**(4): pp. 1091–1100.

18. Bradford, N. et al., Paediatric palliative care services in Queensland: An exploration of the barriers, gaps and plans for service development. *Neonatal, Paediatric & Child Health Nursing*, 2012. **15**(1): p. 2.

19. Brigden, M. et al., Strengths and weaknesses of teleoncology. *Oncology Exchange*, 2008. **7**: pp. 8–12.

20. Datta, S.K. et al., Telemedicine vs in-person cancer genetic counseling: Measuring satisfaction and conducting economic analysis. *Comparative Effectiveness Research*, 2011. **1**.

21. Lopez, A.M., Telemedicine, Telehealth, and e-Health Technologies in Cancer Prevention, in *Fundamentals of Cancer Prevention*. Springer, 2014. pp. 259–277.

22. World Health Organization. *Cancer Prevention*. February 17, 2014; Available from: http://www.who.int/cancer/prevention/en/.

23. Whittaker, R. et al., Mobile phone-based interventions for smoking cessation. *Cochrane Database of Systematic Reviews*, 2012. **11**.

24. Alsharif, M. et al., Telecytopathology for immediate evaluation of fine-needle aspiration specimens. *Cancer Cytopathology*, 2010. **118**(3): pp. 119–126.

25. Kaplan, K.J. et al., Use of robotic telepathology for frozen-section diagnosis: A retrospective trial of a telepathology system for intraoperative consultation. *Modern Pathology*, 2002. **15**(11): pp. 1197–1204.

26. Cross, S., T. Dennis, and R. Start, Telepathology: Current status and future prospects in diagnostic histopathology. *Histopathology*, 2002. **41**(2): pp. 91–109.

27. Brauchli, K. et al., Diagnostic telepathology: Long-term experience of a single institution. *Virchows Archiv*, 2004. **444**(5): pp. 403–409.

28. Weinstein, R.S. et al., The innovative bundling of teleradiology, telepathology, and teleoncology services. *IBM Systems Journal*, 2007. **46**(1): pp. 69–84.

29. Sabesan, S. et al., Practical aspects of telehealth: Are my patients suited to telehealth? *Internal Medicine Journal*, 2013. **43**(5): pp. 581–584.

30. Kunkler, I.H. et al., A pilot study of tele-oncology in Scotland. *Journal of Telemedicine and Telecare*, 1998. **4**(2): pp. 113–119.

31. Allen, A. et al., A pilot study of the physician acceptance of tele-oncology. *Journal of Telemedicine and Telecare*, 1995. **1**(1): pp. 34–37.

32. Bohnenkamp, S.K. et al. Traditional versus telenursing outpatient management of patients with cancer with new ostomies. *Oncology Nursing Forum*, 2004. **31**(5): pp. 1005–1010.

33. Weinerman, B. et al., Can subspecialty cancer consultations be delivered to communities using modern technology?—A pilot study. *Telemedicine Journal & E-Health*, 2005. **11**(5): pp. 608–615.

34. Brigden, M. et al., A survey of recipient client physician satisfaction with teleoncology services originating from Thunder Bay Regional Health Sciences Centre. *Telemedicine and e-Health*, 2008. **14**(3): pp. 250–254.

35. Taylor, M. et al. The use of telemedicine to care for cancer patients at remote sites. *Journal of Clinical Oncology (Meeting Abstracts)*. 2007. **25**(18S): Abstract 6538.

36. Halligan, M., and A. Zecevic, Safety culture in healthcare: A review of concepts, dimensions, measures and progress. *BMJ Quality & Safety*, 2011. **20**(4): pp. 338–343.

37. Laila, M. et al., Videophones for the delivery of home healthcare in oncology. *Studies in Health Technology and Informatics*, 2008. **136**: p. 39.

38. Stalfors, J. et al., Satisfaction with telemedicine presentation at a multidisciplinary tumour meeting among patients with head and neck cancer. *Journal of Telemedicine and Telecare*, 2003. **9**(3): pp. 150–155.
39. Whitten, P., D.J. Cook, and G. Doolittle, An analysis of provider perceptions for telehospice™. *American Journal of Hospice and Palliative Medicine*, 1998. **15**(5): pp. 267–274.
40. Patkar, V. et al., Cancer multidisciplinary team meetings: Evidence, challenges, and the role of clinical decision support technology. *International Journal of Breast Cancer*, 2011. **2011**.
41. Saini, K. et al., Role of the multidisciplinary team in breast cancer management: Results from a large international survey involving 39 countries. *Annals of Oncology*, 2012. **23**(4): pp. 853–859.
42. Kesson, E.M. et al., Effects of multidisciplinary team working on breast cancer survival: Retrospective, comparative, interventional cohort study of 13 722 women. *BMJ: British Medical Journal*, 2012. **344**.
43. Caldwell, P.H.Y. et al., Clinical trials in children. *The Lancet*, 2004. **364**(9436): pp. 803–811.
44. Munro, A., and S. Swartzman, What is a virtual multidisciplinary team (vMDT)? *British Journal of Cancer*, 2013. **108**(12): pp. 2433–2441.
45. Hede, K., Teleoncology gaining acceptance with physicians, patients. *Journal of the National Cancer Institute*, 2010. **102**(20): pp. 1531–1533.
46. Patkar, V. et al., Using computerised decision support to improve compliance of cancer multidisciplinary meetings with evidence-based guidance. *BMJ Open*, 2012. **2**(3).
47. Waters, T. et al., *Technology Exchange for Cancer Health Network (TECH-Net)*. 2008.
48. Pearson, S.-A. et al., Do computerised clinical decision support systems for prescribing change practice? A systematic review of the literature (1990–2007). *BMC Health Services Research*, 2009. **9**(1): p. 154.
49. Kralj, B. et al., The impact of computerized clinical reminders on physician prescribing behavior: Evidence from community oncology practice. *American Journal of Medical Quality*, 2003. **18**(5): pp. 197–203.
50. McCann, L. et al., Patients' perceptions and experiences of using a mobile phone-based advanced symptom management system (ASyMS©) to monitor and manage chemotherapy related toxicity. *European Journal of Cancer Care*, 2009. **18**(2): pp. 156–164.
51. Hebert, M.A. et al., Analysis of the suitability of "video-visits" for palliative home care: Implications for practice. *Journal of Telemedicine and Telecare*, 2007. **13**(2): pp. 74–78.
52. Whitten, P., G. Doolittle, and M. Mackert, Providers' acceptance of telehospice. *Journal of Palliative Medicine*, 2005. **8**(4): pp. 730–735.
53. Donnem, T. et al., Bridging the distance: A prospective tele-oncology study in Northern Norway. *Supportive Care in Cancer*, 2012. **20**(9): pp. 2097–2103.
54. Sood, S.P. et al., Differences in public and private sector adoption of telemedicine: Indian case study for sectoral adoption. *Studies in Health Technology and Informatics*, 2007. **130**: p. 257.
55. Hartvigsenab, G. et al. Challenges in telemedicine and eHealth: Lessons learned from 20 years with telemedicine in Tromso, in *Medinfo 2007: Proceedings of the 12th World Congress on Health (Medical) Informatics [Brisbane, Australia, August 20–24, 2007]*. 2007. IOS Press.
56. Doorenbos, A.Z. et al., Enhancing access to cancer education for rural healthcare providers via telehealth. *Journal of Cancer Education*, 2011. **26**(4): pp. 682–686.
57. McCrossin, R., Successes and failures with grand rounds via videoconferencing at the Royal Children's Hospital in Brisbane. *Journal of Telemedicine and Telecare*, 2001. **7**(suppl 2): pp. 25–28.
58. Ricci, M.A. et al., The use of telemedicine for delivering continuing medical education in rural communities. *Telemedicine Journal & e-Health*, 2005. **11**(2): pp. 124–129.
59. Haik, B.G., Retinoblastoma Management: Connecting Institutions with Telemedicine, in *Digital Teleretinal Screening*. Springer, 2012. pp. 181–192.

26

Telepathology System Development and Implementation

Ronald S. Weinstein

CONTENTS

26.1 Introduction

Establishment of telepathology services has been, arguably, one of the more difficult technical challenges in telemedicine. Conventional surgical pathology diagnoses are made by pathologists examining glass histopathology and cytopathology slides through conventional light microscopes. Light microscopy carried out in person, hands on, is the counterpart of telepathology where diagnoses are rendered at a distance. Light microscopy-based diagnoses are universally recognized as the gold standard for imaging-based medical diagnostic services. Surgical pathology diagnostic accuracy standards are also very high compared with the diagnostic accuracy of most other medical diagnostic imaging modalities, including computerized tomography and ultrasound. The acceptable level of diagnostic accuracy for surgical pathology diagnoses is generally set at 97%, or higher. This makes the margin for error for surgical pathology diagnoses extremely narrow for telepathology system developers, especially for a process involving complex decisions made by human observers. Nevertheless, pathologists rendering diagnoses with telepathology systems have been expected to equal, or even exceed, this very high standard. Not unexpectedly, this has been difficult to achieve and, quite literally, has been 30 years in the making

[1–2]. Also, almost any application in medicine that involves using robotics to manipulate specimens, including glass slides upon stained tissue sections are mounted can be inherently challenging for that reason alone. Robotic technologies are used in several of the types of telepathology systems introduced by Weinstein's group, but not all [1–12].

The technology for telepathology is compared directly with conventional light microscopy, the primary technology used for conventional pathology diagnostic services, for purposes of medical device approval by the U.S. FDA. Conventional light microscopy is used by surgical pathologists who are specially trained medical doctors who make their living by using light microscopes to render diagnoses on tissue sections prepared from surgically removed patient tissues. The light microscopes have single-axis optical pathways consisting of a series of high-quality, precisely aligned glass lenses. Today's high-quality single-axis optical light microscopes have been used in clinical practice for nearly a century and are relatively inexpensive as compared with radiology multisection scanning equipment (i.e., CT and MRI scanners), yet these yield higher levels of diagnostic accuracy for many diseases. Most of the conventional light microscope's controls, including focus, objective lens selection, and slide movements on a mechanical stage, are manipulated by the pathologist's hands. Such light microscopes have no electrical components other than an illumination lamp. Image quality produced by pathologist's light microscopes has not been substantially improved for many decades. Interestingly, most practicing pathologists today trained on the same types of high-quality light microscopes used for training their senior colleagues, often one or two generations ago. Often, many senior pathologists will have a special fondness for their personal light microscope, somewhat like violinists' loyalty to their violin. They would naturally resist switching to another light microscope, let alone a complex telepathology system. In our experience, pathologists can also be brand loyal, another issue that may influence their opinion when new imaging modalities are discussed with experienced pathologists being asked to evaluate novel telepathology systems.

Most of today's commercially available telepathology systems are built around the same types of single-axis optical lens systems that are used in conventional light microscopes today. Typically, telepathology systems are judged to be of high quality when the images they produce very closely resemble the ordinary light microscope images of the same surgical pathology slides when viewed in person with a conventional light microscope. That comparison can be very challenging for telepathology system developers. Several technologies come into play in order to carry out a telepathology histopathology slide examination, inserted between the glass slide and its digital imaging camera–mounted optical system in the slide processing laboratory, with the distant telepathologist remotely viewing the slide on a video monitor miles away. Digital imaging of light microscope slides, large digital image file assembly, and the transmission and the viewing of images on a video monitor by the remote telepathologist will each add another layer of complexity to the pathology slide imaging process. Each step along the line has the potential to introduce its own imperfections (i.e., artifacts) on what the telepathologist actually visualizes downstream on a video monitor [13]. It is unlikely that any electronic medical imaging system in use today can exactly duplicate the histopathology image as it would be seen by a pathologist using a conventional light microscope in everyday practice, but the evolving systems are getting closer. Despite such pathology image imperfections, a pathologist using any type of telepathology system is expected to achieve the same high level of diagnostic accuracy for both telepathology and traditional light microscopy, in order to have the telepathology system deemed "approvable" by the FDA as a medical imaging device suitable for rendering primary pathology diagnoses. This has proved to be very challenging for telepathology system designers. Digital pathology works only because conventional light microscopes are arguably overengineered and produce images that exceed the quality of images necessary to render most histopathology diagnoses, with the exception of hematopathology diagnoses, which can be challenging to make even with the highest-quality light microscope. (Hematolopathology is the study of blood disorders.)

On the other hand, the potential utility and value of telepathology systems in medical practice is also high, as championed by early innovators in the telepathology field [1–2,5–6,14]. Historically, telepathology became a focus of renewed medical research interest in the mid-1980s, well before the emergence of what is frequently referred to as the "modern era of telemedicine," starting in the mid-1990s. An important driver for the creation of the new field of telepathology in the United States was the intractable problems encountered by pathologists in diagnosing and subclassifying urinary bladder cancers. These diagnostic pathology challenges were impairing important U.S. cancer clinical trials of new therapies being tested for treating urinary bladder cancer. Ironically, many practicing pathologists were quite unaware of the problems they were causing for cancer researchers.

In order to understand this problem in its clinical context, the critical role of the U.S. National Cancer Institute (NCI) in identifying this problem is important to acknowledge. In 1973, distinguished urologists and pathologists informed the U.S. NCI of the fact that surgical pathology diagnoses on patients with urinary bladder cancer were a problem and were compromising urinary bladder cancer clinical trials. This was complicated by the fact that most urinary bladder cancers recur, that is, reappear after the surgical removal of the initial tumor or tumors. At issue were the origins of these bladder cancer recurrences and the possibility that bladder cancers are initially a monoclonal disease, suggesting that recurrences occur because some cancer cells remain in the urinary bladder after therapy and, thus, can reseed new tumors in the urinary bladder wall. Inaccuracies in pathology diagnoses being rendered on histopathology and cytopathology specimens were tarnishing the results of some of these national clinical trials, even rendering some large clinical trials unpublishable because of pathology diagnostic inaccuracies. What was actually of concern was that subclassifications of bladder cancers were not always reproducible by other pathologists, again at the 97% accuracy level of concordance. Although confirming, by light microscopy, that a urology patient has a primary diagnosis of bladder cancer was not difficult for pathologists and met this 97% diagnostic accuracy guideline, subclassifying the cancers by light microscopy proved to be challenging in a surprisingly high number of cases. What is especially noteworthy is that these subclassifications of bladder cancers are critically important in determining the eligibility of bladder cancer patients for treatments using specific experimental therapies. Subclassifications of bladder cancers includes the "staging" and "grading" of the cancers, typically carried out by a pathologist using conventional light microscopy to analyze histopathology slides. Staging of a urinary bladder cancer (i.e., assessing the depth of invasion of the cancer) and grading of a urinary bladder cancer (i.e., assessing the degrees of anaplasia of the cancer cells) are challenging to reproduce among practicing pathologists for a significant minority of cases, even for experts in the uropathology field. Staging and grading errors became a major challenge since such diagnostic errors could result in patients receiving inappropriate therapies, undermining expensive and unique clinical trials and, most importantly, possibly leading to poor clinical outcomes for such bladder cancer patients.

In 1973, in order to address this and other challenges in urinary bladder cancer management, the U.S. NCI created a National Bladder Cancer Group (NBCG) and funded its Central Pathology Laboratory (CPL). The CPL was assigned the task of doing urinary bladder cancer surgical pathology case rereviews. It is noteworthy that bladder cancer was the only type of cancer for which the NCI created a permanent pathology laboratory to rereview the pathology for all patients being entered into national therapy clinical trials. This reflected the magnitude of the problem encountered in subclassifying bladder cancers in patients with recurrent disease, a seemingly intractable problem at that time. This bladder cancer–dedicated CPL had expert pathologists reviewing all surgical pathology specimens (glass histopathology slides) and reports, and all cytopathology slides and reports, for all patients being placed on NCI-funded NBCG clinical trial protocols. The CPL was initially located in Worchester, Massachusetts, and then relocated and housed in the 1980s at Rush–Presbyterian St. Luke's Medical Center in Chicago, Illinois, where Ronald S. Weinstein, MD, was chair of pathology and then named director of the CPL. The CPL pathologists documented, rereviewed the cases, amended the reports, and documented the diagnostic accuracy rates for the other participating pathologists, all of whom were experts on bladder cancer themselves. Diagnostic accuracy rates were variable but higher than expected for some participating institutional pathologists. On rereview, the error rates for staging and grading bladder cancers ranged from a low of 4% to over 30%. It turned out that the challenge of having pathologists make reproducible diagnoses on the surgical pathology evaluations of bladder cancer cases was greater than anticipated.

The director of the CPL had become familiar with television microscopy as a pathology resident at the Massachusetts General Hospital (MGH) in the late 1960s. Robotic telepathology, which was nonexistent at the time, was proposed by the director, Ronald S. Weinstein, MD, as a possible solution to the interobserver variable issue for classifying bladder cancers. The director came up with the initial idea of exploring the possibility of creating a robotic telepathology system. Robotic telepathology would allow for real-time imaging and up-and-down focusing, which were regarded as being especially important for accurate bladder cancer grading and staging. The director then invented, patented, validated, and commercialized robotic telepathology [1–3]. Telepathology systems based on his patents are still in use in everyday practice in the United States [11].

A few years later, around 1987, the CPL pathologists discovered that other pathologists, in Europe, had been exploring the possibility of creating static-image (SI) telepathology services, aimed at rendering pathology second opinions at a distance. Their primary interest was somewhat different. They recognized that subspecialty surgical pathology specialization was already important in a few areas of pathology practice, especially dermatopathology, renal pathology, and neuropathology, but was about to undergo rapid growth in additional subspecialty areas. Maldistribution of subspecialty surgical pathology expertise is a major issue in medicine today [12,14–16].

Because of the importance and urgency of these problems, and the enormous potential size of the worldwide market for telepathology systems and telepathology services, hundreds of millions of dollars have been invested in developing and marketing telepathology systems in recent years by large corporations [17].

Stepping back in time, the history of telepathology actually dates back to the 1950s. Television microscopy (Video microscopy) was first developed and tested by engineers at the David Sarnoff Radio Corporation of America (RCA) Laboratories in Princeton, New Jersey, around 1951 [17]. Television microscopy, also called video microscopy, was widely used in science research laboratories by 1960 [15]. In 1968, television microscopy was implemented at the MGH, in Boston to provide, by television microscopy, clinical light microscopy services (i.e., examination of blood smears and urine samples on glass slides) in a first-of-a-kind walk-in, multispecialty telemedicine clinic that had recently been established at the Logan International Airport 2.7 miles away from the MGH campus. Implementation of television microscopy in the MGH–Logan Airport clinical setting was regarded as a "nonevent" at the MGH (Weinstein, unpublished observation, 1968).

In order to utilize television microscopy, by the MGH–Logan International Airport tele-medicine program, a black-and-white television camera was mounted on a conventional light microscope to obtain a steady stream of real-time television analog images. The glass slide was moved around on the microscope stage by the clinic's nurse–clinic manager located in the Logan International Airport walk-in telemedicine clinic.

26.1.1 Telepathology Systems

Telepathology systems have three basic components: a video camera for capturing digital images of glass histopathology slides or cytopathology slides; a telecommunication system linking the digital imaging system with a distant consultant's computer workstation (i.e., the telepathologist); and the telepathologist's video-enabled workstation [2–3,13].

There are major differences among the various classes of telepathology systems related to differences in their imaging modalities. In radiology, major types of scanners from vari-ous vendors meet certain standards and can be quite similar. A radiologist trained on one brand of scanner can usually diagnose cases coming from different brands of radiology systems, for a class of scanners, such as CT scanners. The situation for telepathology is dif-ferent. There are many different telepathology systems on the market. Some telepathology equipment companies have multiple types of telepathology systems for sale, ranging from entry-level whole-slide imaging (WSI) processors, designed for education programs, to high-end dynamic robotic telepathology (RT)–WSI systems intended for clinical laboratory use. Although single digital images from many of these systems can appear very similar and may even closely resemble what is visualized through a light microscope, the way the video image stream is handled varies among different telepathology systems (Figure 26.1). A major distinction is between static-image (i.e., store-and-forward image) systems and real-time (i.e., dynamic robotic) pathology imaging systems. Static imaging is done in the

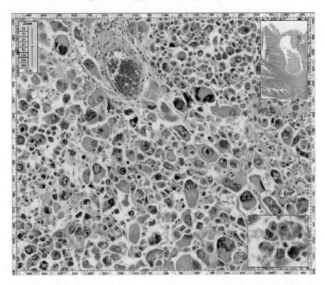

FIGURE 26.1
Whole-slide image of paraffin-embedded hematoxylin and eosin (H&E)-stained brain specimen. Upper right shows a thumbnail image of the glass slide at low magnification. The graphic-imposed square marks the area displayed in the main viewing field. Lower right of the specimen image in the main window shows a small mag-nified image. Tissue section was magnified ×10, using an Aperio digitizer. This is a highly anaplastic lung car-cinoma, metastatic to brain. (Aperio slide scanner, Leica Microsystems, Inc., Buffalo Grove, Illinois, United States.)

asynchronous imaging mode, whereas real-time imaging is done in the synchronous mode. This is a major distinction and has critically important implications for the requirements, for the respective systems, and for glass slide storage and glass slide retrieval mechanisms. Today, most of telepathology system vendors offer so-called hybrid systems that combine various telepathology imaging modalities into single telepathology systems [5,7–8].

26.1.2 Telepathology System Development

The development of telepathology systems is a complicated topic. A helpful approach to gaining an understanding of the range of telepathology products on the market is to examine a set of three specific telepathology classification systems, published at various times for various audiences, and then to use them as a framework for discussion.

Consideration of three specific telepathology system classifications, published between 2001 and the present time, provides practical insights into the development of the telepathology systems used by the laboratory industry today. These three classifications, published in 2001, 2012, and 2014, also reflect the time lines for various specific innovations in telepathology systems (Tables 26.1 through 26.3). It should be noted that some earlier types of telepathology systems, which are no longer commercially available, are still in use in some pathology laboratories scattered around the world and are not considered obsolete by their system users. This needs to be taken into account in assessing the more recent telepathology literature which includes papers by authors using telepathology systems which other investigators might regard as being obsolete systems. There are five so-called pull-through telepathology systems that survived the next wave of innovations in telepathology systems and are still in use (Table 26.3).

It is also instructive to study the evolution of some individual categories of telepathology system over time. For example, the miniature microscope array described as a class 4 telepathology system as listed in the *Human Pathology* telepathology system classification (Table 26.1) could potentially be redeployed as a digital imaging engine in next-generation genomic scanners [8–9,16].

Tables 26.1 and 26.2 show the first two telepathology classifications, the first directed primarily at engineers and the second, a more recent classification, targeted at practicing pathologists. These were published a decade apart.

Table 26.1 shows one of the first comprehensive classifications of telepathology systems. This appeared in the second of three telepathology minisymposiums, published in the journal *Human Pathology* between 1997 and 2009 [7–9]. The classification shown in Table 26.1 was intended primarily for use by engineers. At the time it was published there were over 30 telepathology equipment companies worldwide but fewer than 50 sustainable telepathology services, handling more than a few dozen telepathology cases a year. The full-time engineers outnumbered the full-time telepathologists! This classification represented an attempt to be comprehensive with respect to the types of telepathology systems on the market. The year 2001 was a time when new, entry-level telepathology system vendors were introducing innovative telepathology products at a bewildering rate. This was also evidenced by a sharp upturn in the amount of floor space occupied by telepathology companies in the exhibit hall at the 100th anniversary meeting, in 2002, of the parent organization of the United States and Canadian Academy of Pathology (USCAP), the International Academy of Pathology. This jumped from less than 20% of the exhibitor floor space the year before to nearly 60% of the USCAP exhibitor hall floor space, with several exhibitors (i.e., Aperio and Olympus) occupying large tentlike booths that attracted heavy foot traffic. However, this spike in enthusiasm was short lived and the floor space occupied by

TABLE 26.1

Human Pathology Telepathology System Classification

Dates	Class	Symbol	Category	Enabling Technologies
1968–1989 (first generation)	1A	DNR	Dynamic nonrobotic[a]	Videomicroscopy
	1B	DR	Dynamic robotic	Robotic microscopy
	2A	SFNR	Store-and-forward nonrobotic	Image grabbing boards
	2B	SFR	Store-and-forward robotic	High-definition television (HDTV)
1989–2000 (second generation)	2C	SFSR	Store-and-forward stitch/robotic	Electronic stitch software
	3A	HDSF-NR	Hybrid dynamic/store-and-forward/nonrobotic[b]	
	3B	HDSF-R	Hybrid dynamic/store-and-forward robotic	
	4A	VSA	Virtual-slide/automatic/nonrobotic processor[c]	
	4B	VSI	Virtual-slide/interactive processor	
2000–2001 (third generation)	5A	HVS	Hybrid virtual-slide processor	Combined automatic and interactive
	5B	RVS	Rapid virtual-slide processor	Continuous stage motion and strobe illumination
2001 (fourth generation)	5C	UVS	Ultrarapid virtual-slide processor	Miniaturized array microscope

[a] Dynamic refers to the real-time imaging component of the system.
[b] Hybrid (combined) dynamic and store-and-forward image (*store-and-forward image* and *static image* are synonyms).
[c] A virtual slide is a fully digitized slide stored in computer memory. *Virtual-slide image* and *whole-slide image* (WSI) are synonyms.

TABLE 26.2

Acta Pathologica Microbiologica et Immunologica Scandinavica (APMIS) Telemedicine System Classification

Imaging System	Year
Real-Time Imaging Telepathology	
Television microscopy	1952
Research applications	1955
Clinical applications	1968
Dynamic robotic telepathology	1986
Static-Image Telepathology	
Static-image telepathology	1987
Static robotic image telepathology	1989
Whole-slide image (automated)	1991
Whole-slide image (operator directed)	1994
Multimodality Digital Image Telepathology	
Hybrid dynamic robotic/static image[a]	1989
Dual-image dynamic robotic/static image[b]	2011

[a] Also called static-image enhanced dynamic robotic telepathology.
[b] Also called dynamic robotic telepathology-enhanced whole-slide image (or RT-WSI, which is pronounced "ritzy").

TABLE 26.3

Practitioners' Telepathology System Classification[a]

Imaging System	Year
Real-Time Imaging Telepathology (RT)	
Television microscopy[b]	1952
Dynamic robotic telepathology	1986
Static-Image Telepathology (SI)	
Static-image telepathology	1987
Whole-slide imaging	1991
Multimodality Digital Image Telepathology	
Hybrid dynamic robotic/static imaging[c]	1989
Dynamic robotic telepathology/WSI[d]	2014

[a] Weinstein R.S., unpublished data, 2014.
[b] More recently called virtual microscopy.
[c] Also called static-image enhanced dynamic robotic telepathology.
[d] Also called dynamic robotic telepathology-enhanced whole-slide image (RT-WSI, pronounced "ritzy").

telepathology companies dwindled back to an estimated 30% of exhibit hall floor space 2 years later at the annual USCAP meeting, as the burgeoning telepathology industry encountered increasing regulatory pressures from the FDA. Since then, the stories of companies attempting to get approval for rendering primary diagnoses using their telepathology systems has dragged on with success seeming to be just around the corner.

The 2001 *Human Pathology* telepathology system classification, shown in Table 26.1, lists four generations of telepathology systems, according to the time intervals in which various combinations of telepathology systems were initially created and tested. Table 26.2 simplifies the information shown in Table 26.1 and also deletes the technology research notations, which were in the form of designations of classes of telepathology systems and symbolic designations for individual types of telepathology systems. Once the telepathology system field was consolidated, and preferred systems for practicing pathologists emerged, these designations were omitted. Table 26.3 shows the most recent version of our telepathology system classifications. The principal target audience is practicing pathologists.

26.2 Telepathology System Classifications

Since 1990, dozens of static-image telepathology systems and dynamic robotic telepathology systems have been developed and commercialized [8]. Early simple classifications of telepathology systems began to appear after the introduction of the first hybrid telepathology systems which combined the two principal imaging modalities, static-image and dynamic robotic telepathology systems, into a single hybrid telepathology system [5,7,14–15]. By 1992, the two primary imaging modalities, dynamic robotic imaging (i.e., real-time imaging) and static imaging (i.e., store-and-forward imaging), were being used routinely by the first sustainable telepathology laboratory services that are still clinically active to this day [5,10]. A third category of telepathology imaging system, whole-slide imaging (also called virtual slides), was developed and patented by Bacus and implemented clinically before 2000 [8]. WSI is a direct extension of static imaging. With WSI, a gallery of static-image files, usually representing an entire glass slide in aggregate, are electronically stitched together to form a single giant digital image file. This can be remotely navigated from a distant computer workstation [8–9].

26.2.1 *Human Pathology* Telepathology System Classification (2001)

The 2001 *Human Pathology* telepathology system classification represents a consolidation of information drawn from the engineer, computer science, telecommunications, and pathology literatures. Static-image telepathology (also called store-and-forward telepathology) and dynamic robotic telepathology (also called real-time imaging) represent the two independent fountainheads of innovation for the initial 5 years of telepathology system development, starting around 1984.

"Generations" of systems are included in Table 26.1 along with the approximate time frames in which new innovations first appeared within these generations. Each time frame ends with the introduction of a new *breakthrough innovation* that initiated the introduction of a set of new classes of telepathology systems. For example, videomicroscopy defined the start of the first generation of system. Television microscopy was the first real-time remote pathology viewing system. The development of consumer-grade electronic stitch software initiated the second generation of telepathology systems. The combination of automatic and interventional (i.e., interactive) WSI enabled the third generation WSI systems. These then became the first *rapid* WSI systems. Also within this rapid WSI class is a second technical approach to rapid WSI, namely, continuous stage-motion scanning enabled by the use of strobe illuminations. Fourth-generation telepathology systems are characterized by the achievement of 1-minute WSI scanning. The first ultrarapid slide digital image processing systems used novel miniature array microscopes in their digital imaging engines [16]. Today, many single-axis optical WSI processors can digitally image an entire glass slide in less than 1 minute. Glass slide digital image processing throughput times have increased by one to two orders of magnitude in the last 15 years [12]. Thus, the class 4 system listed in Table 26.1 can be collapsed back into class 3, the rapid virtual-slide processor category.

Classes of telepathology systems were included in the *Human Pathology* telepathology system classification [8] to help track progress in the parallel developments of innovations in each of the two individual static image (SI) and real time (RT) tracks, again within specific time frames.

Symbols were also included in the *Human Pathology* telepathology system classification [8] to help deal with challenging telepathology system nomenclature issues. Much of the innovation and testing of telepathology systems was taking place in Germany, France, and Japan, as well as the United States [14]. Communication among engineers, and chief executive officers (CEOs), in these companies was a formidable challenge since many of the engineers were handicapped by what we identified as "their lack of proficiency in English." Of course, the German, French, and Japanese engineers were equally concerned over our "lack of proficiencies in their languages." We found that the use of symbols to refer to specific types of telepathology systems helped simplify and improve communication when there were a dozen types of telepathology systems in play. These symbols were cast aside when the telepathology system equipment field became more streamlined as it began to mature.

Note that at the time the initial telepathology system classification was published in 2001, it was commonly thought that the telepathology equipment field would continue to develop and expand with the addition of new features and classes of systems well into the foreseeable future. In retrospect, it turns out that the level of diversity of telepathology systems was probably near its peak in 2001, with the exception of the future introduction of commercialized ultrarapid WSI systems and, more recently, the incorporation of robotic telepathology imaging modules into WSI systems. In the next phase of the telepathology system field evolution, as reflected in Tables 26.2 and 26.3, simplification of classifications of telepathology systems took place, in part due to the jettisoning of obsolete features. This can be accounted for by rapid increases in WSI scanning rates, which were accurately predicted in 2001 but nevertheless remained to be achieved, and by the increasing availability of broadband telecommunications networks and cloud computing [8–9,12].

It is important and also noteworthy that at the time that the *Human Pathology* telepathology system classification was published in 2001, the telepathology literature was growing at the rate of 35 to 50 papers per year [8]. The larger field of digital pathology, of which telepathology is a service component, was gaining traction as well and accruing additional publications, many of which had direct relevance to telepathology [9,15]. By mid-2014, the telepathology literature had grown from the first telepathology paper, published in 1986, to over 1000 papers published in the world literature, coming from over 400 laboratories in over 30 countries [1,12]. Telepathology had become a legitimate area for academic study, paralleling the growth of the larger telemedicine literature, which exceeds 15,000 publications.

In a related area, the intellectual property arena, since the early 1990s, there has been a steady increase in the numbers of U.S. telepathology patents and digital pathology patents [17]. The first U.S. patent application for a telepathology system was submitted to the U.S. Patent and Trademark Office in 1987 and is titled "Telepathology Diagnostic Networks" [2]. Corabi International Telemetrics, Inc., was the assignee for the first telepathology patent and took responsibility for advocating for the patent at the U.S. Patent and Trademark Office. Corabi International Telemetrics, Inc., was cofounded by Dr. Weinstein, who borrowed his wife Mary's maiden name, Corabi, to name the company. In the negotiations for approval of the initial patent with the U.S. Patent and Trademark Office, a decision was made by lawyers and the patent examiner to split the original patent into two "divisional" patents, with starting dates in 1993 and 1994. Both patents were granted for 17-year periods and they expired in 2010 and 2011, respectively. These patent expiration dates are relevant since immediately upon expiration of the U.S. patents, major telepathology equipment vendors cannibalized the intellectual property in the patents and quickly incorporated robotic dynamic telepathology modules into their high-end WSI processors [12,18]. Corabi International Telemetrics, Inc. (a company no longer in business), was aware that this was happening since dynamic robotic telepathology–enabled WSI systems were already being marketed in Europe in 2009, in anticipation of the patent expirations the following year in the United States. Expiration of the patents led to the creation and marketing of a new category of telepathology system in the United States, the dynamic robotic telepathology–enabled whole-slide image processor [12]. The dynamic robotic telepathology–enabled whole-slide image processing, also called robotic telepathology–whole-slide imaging, was recently shown to be a clinically useful technology [18]. As of 2014, the U.S. Patent and Trademark Office had issued its 100th telepathology patent and many more in the digital pathology arena [17].

Several other factors added to the complexity of the *Human Pathology* telepathology system classification [8]. This classification was created at a time of frenetic activity among telepathology system inventors and designers. Some of the systems appearing between

2000 and 2004 (i.e., classes 2B and 4B) and some unique features (i.e., class 5C) were introduced to overcome specific barriers: those that were derivative and had unacceptably slow WSI processing times; the limited bandwidth availability, especially in rural and small urban communities; digital image file store issues; and concerns over the so-called z-axis issues, that is the perceived need of pathologists to have the ability to focus "up and down" during histopathology and cytopathology slide viewing. Convenience and efficiency were thought to be important to gain pathologist's acceptance of the technology.

Even more important were pathologists' concerns over the diagnostic accuracy achievable with telepathology systems. By 2000, many studies had shown that surgical pathology services using telepathology were "almost as good" if not "as good" as conventional light microscopy [8]. The almost-as-good designation became a concern and a substantial barrier to implementation, especially in the litigation-prone United States practice environment. The FDA also holds surgical pathology to a higher standard than diagnostic radiology, because surgical pathologists render final diagnoses, unlike the provisional diagnoses rendered by radiologists. The FDA became an impediment to telepathology over the past decade because it appeared to keep increasing the threshold for accepting telepathology for primary surgical pathology diagnoses. Lack of a z-axis through-focus module became a greater issue, and even a reason for rejection of telepathology, once robotic modules became incorporated into high-end WSI systems. Major vendors jumped on the bandwagon and incorporated a dynamic robotic telepathology module into their WSI systems. The downside was the actual need for a robotic module, since this negatively affected the efficiency equation [12]. It is relatively burdensome to use any telemedicine application in a synchronous mode. Synchronous telepathology encounters, if instituted for use on a routine basis, would require the use of more expensive equipment, additional technical support at the distant site, and the use of sophisticated glass slide archiving warehouses.

Another challenge came from dealing with the wide variety of types of telepathology system equipment that were becoming commercially available. This was very confusing to practicing pathologists. It dawned on them that "telepathology" is actually more than one entity. Entering the field would require dedication to the approach and the need to go through a significant learning curve with specialized training. Reimbursement does not favor telepathology by recognizing the incremental costs involved in supporting a telepathology service. Telepathology services, even with the fastest case throughputs were still less efficient than conventional light microscopy–supported surgical pathology practices. The time differential per case, comparing telepathology with conventional light microscopy practice, might be measured in minutes, or even seconds, but the total times expended add up, especially if the surgical case load is 50 to 100 additional cases per workday.

Then what happened between 2000 and 2014 was that WSI scanning throughput rates increased by one to two orders of magnitude. This had been anticipated in the year 2000 [8]. The dramatic increases in WSI scanning throughput rates rendered some early distinctions between various telepathology system products less important, as most of the larger companies could now offer systems with 1 min WSI scan rates [8,16]. The need for an operator-interactive component also disappeared as the new high scan rates for virtual-slide processors eliminated the need for the class 5A/HVS system listed in Table 26.1. The justification for the class 5C/UVS ultrarapid virtual-slide processor, which incorporated groundbreaking array microscopy optics, was also diminished once 1-minute WSI scan times were achievable using conventional single-axis optical light-microscope-type optics. The class 4B system, the virtual-slide/interactive (robotic) processor, that made a grand entrance at a European telepathology meeting in the 1990s was eclipsed by the introduction, in the early 2000s, of rapid and ultrarapid virtual-slide scanners. This interactive robotic component feature had enabled a system operator to select, on the fly, which areas of a glass slide should be digitized next, with the system operator controlling the movements of the digital scanner stage remotely from a distant site. This was visualized by the system operator, who could manage the microscope stage embedded in the WSI processor at a distance in what appeared, to spectators, to be in a very dramatic fashion. At that early stage in the development of WSI systems, digitizing an entire glass slide routinely took many hours, or even days. This operator-controllable slide scanning feature enabled the telepathology system operator/telepathologist to "cut to the chase" and proactively control the digitizing process for serially acquiring a stream of regions of interest (ROIs). Of course, this was a synchronous process, with a stained tissue section beneath the system's objective lens as the system operator viewed the processing of the digital image of the slide in real time. Once rapid scanning became a reality, and the entire virtual-slide processing (i.e., whole-slide imaging) could be accomplished in a relatively brief period of time, less than a few minutes, class 4B systems became outmoded and are now absent from the telepathology system classification scheme.

Finally, with respect to the 2001 *Human Pathology* telepathology system classification, the term *virtual slide* is becoming obsolete. The term *whole-slide imaging* is the preferred nomenclature today. It turned out that the term *virtual*, when used to refer to digitized histopathology slides, confused many people ranging from surgical pathologists to insurance companies and government regulatory agencies. The term *virtual slide* was jettisoned by the year 2012.

Once the term *WSI* caught on, some strong proponents of this term advocated substituting *WSI* for the term *telepathology* [12]. This has not happened except in a few institutions where WSI enthusiasts publish a steady stream of WSI abstracts and papers. Currently, this confuses government regulatory agencies. The fact that the term *telepathology* has actually held its ground is evidenced by the recent publication of the American Telemedicine Association's "Clinical Guidelines for Telepathology," which became available in August 2014. This is an important advance since the term *telepathology* is now officially accepted as an inclusive term that describes both the imaging modality and the clinical practice aspects of the activity, whereas WSI is simply an imaging mode. The next onslaught from the alternative terminology groups will be on the term *telepathology* itself. They would substitute the term *digital pathology* for *telepathology* as well, another questionable proposal at this time. It would be preferable to discuss changes in nomenclature once telepathology systems have been approved for rendering primary diagnoses and reimbursement for telepathology services is firmly in place, for all payers.

26.2.2 *APMIS* Telepathology System Classification (2012)

Table 26.2 shows a more recent, simplified classification of telepathology systems. It shows an updated listing of classes of telepathology systems, along with the approximate year of implementation for each of the individual classes of systems. Class abbreviations and symbols, shown in Table 26.1, have been deleted. Inaugural clinical use of a telepathology system and the publication date can vary by years. This classification was published in a Scandinavian pathology journal, *Acta Pathologica Microbiologica et Immunologica Scandinavica* [12], as part of a minisymposium on telepathology. We chose to publish it there to acknowledge the early contributions of Drs. Thor Eide and Ivar Nordrum, Norwegian pathologists, who were pioneers in the telepathology field, having established the first sustainable telepathology service, in 1989, based on their development of the first hybrid dynamic robotic/static-image system [5]. Their hybrid system became the aspirational system for the telepathology field in the 1990s, and it is still a workhorse telepathology system in telepathology-enabled surgical pathology laboratories, a quarter of a century later [5–6,19]. Nordrum and Eide's classic paper is atop the reference list for our *APMIS* minisymposium paper [12].

26.2.3 American Telemedicine Association Clinical Guidelines (2014)

It is noteworthy that the American Telemedicine Association, the largest telemedicine/telehealth organization in the world and an international leader in producing telemedicine clinical standards and guidelines, issued the clinical guidelines for telepathology in 2014. The ATA guidelines for telepathology adopted the *APMIS* telepathology system [12] listed above, with only minor tweaks. The main difference between the *APMIS* telepathology system classification and the one reproduced in the ATA clinical guidelines for telepathology (2014) is the deletion of the two footnotes. These are restored in the original version of the *APMIS* telepathology system classification, as shown above. Nevertheless, these differences are worth noting since they point to important nomenclature issues.

The ATA clinical guidelines for telepathology can be accessed at the following website: http://www.americantelemed.org/docs/default-source/standards/clinical-guidelines-for -telepathologyFBFDA18D6793.pdf?sfvrsn=4.

26.3 Telepathology System Classification: 2014 Practitioners

For purposes of authoring this third classification, the following rules were established. The designated *primary imaging modality*, listed first in a string of modalities for multimodality WSI systems, is the modality used by the telepathologist to initially survey all of the diagnostic materials on the slide, irrespective of which microscope objective lens is used for the survey. A well-trained pathologist first surveys the entire slide at low magnification, the entire contents of diagnostic materials mounted on a glass slide before progressing from the scanning mode, using a low-power objective lens (or low-magnification electronic digital imaging), to higher magnifications. A critically important role of pathologists is the inclusive sampling of all biological materials on a glass histopathology or cytopathology slide. "Sins of omission" are unacceptable. We are choosing not to identify the primary imaging modality as the one used to render the final diagnosis (often at higher magnification), although it is understood that an argument could be made for taking that approach, as the reasonable alternative. This choice is somewhat subjective. Nevertheless, the initial tissue survey is of paramount importance in surgical pathology practice. Failure to examine the entire specimen at first viewing of a surgical pathology slide, either by conventional light microscopy or with WSI, could lead to a failure to identify the most important ROIs.

Redundancies, unused types of telepathology systems, and superfluous information have been stripped away from Table 26.3. For now, we proposed that this classification become the standard telepathology classification for use by practicing pathologists.

Several other features of this classification in Table 26.3 are also worth highlighting. First, all six of the listed systems are in actual clinical use at this time. Even television microscopy (i.e., videomicroscopy), consisting of a video camera mounted on a light microscope linked to the Internet for distant image transmission, is currently used in some clinical services. Today, most dynamic robotic telepathology systems would have a static-imaging module for static-image archiving of regions of interest or for showing actual diagnostic histopathology fields to colleagues. Static-image telepathology is used to "grab images" for inclusion in surgical pathology reports and to show difficult cases to peers or to save images of interest for clinical conferences. Whole-slide imaging (without a dynamic-robotic component) is classified in Table 26.3 under static-image telepathology. This is in recognition of the fact that whole-slide imaging is the imaging of galleries of static images seamlessly electronically stitched together into giant digital image files. In Table 26.3, two types of multimodality digital image telepathology are listed. Hybrid dynamic robotic/static-image telepathology has been the workhorse telepathology technology for the past 25 years as the system of choice for several model programs. Dynamic robotic telepathology/WSI is a relatively new entrant. The use this module for z-axis exploration on histopathology and cytopathology slides is discussed elsewhere (Weinstein, unpublished data, 2014).

A future addition to the list of classes of telepathology systems will be mobile telepathology (i.e., mTelepathology). This will be listed in the next version of the telepathology system classification. Several groups have shown that smartphones can be used for both static-image telepathology and real-time image acquisition as well as for displaying static images and dynamic real-time images obtained at a distance. Smartphones are rapidly becoming the workstations of choice for many telemedicine applications [19].

26.4 Conclusions

An important observation of telepathology, as compared to other new medical imaging technologies, was that telepathology did not actually represent the creation of a new medical imaging modality, like CT or MRI for radiology. Rather, telepathology was the adaptation of a proven technology, conventional light microscopy, so that surgical pathology could be performed at a distance. Because surgical pathology diagnoses, rendered using traditional light microscopy, had been the gold standard for medical diagnosis for a century and the diagnostic accuracy of surgical pathology was already higher than the diagnostic accuracy of any other diagnostic technology, the challenge for telepathology was to equal the quality of diagnoses readily available by simply placing a glass histopathology slide on the stage of an ordinary light microscope in a conventional surgical pathology laboratory and rendering a diagnosis. The acceptable margin for error, about 3%, was narrow. It proved very challenging to equal the level of accuracy of conventional light microscopy, let alone improve on it, while simultaneously imposing extra layers of technology including video imaging cameras and telecommunications linkages and the added burden of having the surgical pathologist view the slides on a video monitor. Initially, the use of robotics slowed down the viewing processes even more and further compromised the pathologist's user satisfaction with the systems.

In contrast, a new medical imaging modality may be heralded a breakthrough once novel structures and pathologic lesions are better visualized. Innovators with the ultimate goal of duplicating an established gold-standard imaging modality, within the context of a telemedicine service, are likely be frustrated no matter how meritorious the rationale may be for implementing the service. The road to success in telepathology system development is littered with brilliant innovations and impressive solutions to technical barriers that ultimately failed to take hold. Yet dozens of pathologists have stubbornly stuck with their telepathology research and developments programs, ongoing worldwide for decades. The U.S. FDA's anticipated landmark approval of telepathology systems for use in rendering primary surgical pathology diagnosis in the United States now appears to be close at hand. Then broad acceptance of telepathology, with all of its contingencies, and its successful insertion into routine laboratory services may become a reality. In the meantime, a number of laboratories do use telepathology on a daily basis and find it to be of high value [11,20]. The fact does remain that, although hundreds of labs worldwide have published the results of successful internal validation studies for telepathology, relatively few laboratories have actually implemented telepathology services, reflecting, in part, reservations that pathologists have concerning the technology. Nevertheless, with many hundreds of laboratories fully aware of telepathology, with staff members who have been engaged in telepathology research for years, the expansion of telepathology services worldwide could be expedited and telepathology could become a standard of care quite rapidly in the foreseeable future.

Partial List of Manufacturers and Suppliers

Aperio/Leica Optical Microsystems, Inc.
1700 Leider Lane
Buffalo Grove, IL 60089
United States
Office phone: +1-800-248-0123
Fax: +1-847-236-3009
aperio.com

Abbreviation and Nomenclature

FDA Food and Drug Administration
Also, see Table 26.1.

Acknowledgments

I thank Kristine A. Erps and Angelette Holtrust for editing the manuscript.

References

1. Weinstein RS. Prospects for telepathology (Editorial). *Hum Pathol* 17:433–434, 1986.
2. Weinstein RS, Bloom KJ, and Rozek LS. Telepathology and the networking of pathology diagnostic services. *Arch Pathol Lab Med* 111:646–652, 1987.

3. Weinstein RS, Bloom KJ, and Rozek LS. Telepathology: System design and specifications. *SPIE Proceedings Visual Comm Image Processing* 845:404–407, 1987.

4. Weinstein RS, Bloom KJ, and Rozek LS. Static and dynamic imaging in pathology. *IEEE Proceedings Image Management Comm* 1:77–85, 1990.

5. Nordrum I, Engum B, Rinde E et al. Remote frozen section service: A telepathology project to northern Norway. *Hum Pathol* 22:514–518, 1991.

6. Weinstein RS. Telepathology comes of age in Norway (Editorial), *Hum Pathol* 22:511–513, 1991.

7. Weinstein RS, Bhattacharyya AK, Graham AR et al. Telepathology: A ten-year progress report. *Hum Pathol* 28:1–7, 1997.

8. Weinstein RS, Descour MR, Liang C et al. Telepathology overview: From concept to implementation. *Hum Pathol* 32:1283–1299, 2001.

9. Weinstein RS, Graham AR, Richter LC et al. Overview of telepathology, virtual microscopy, and whole slide imaging: Prospects for the future. *Hum Pathol* 40:1057–1069, 2009.

10. Dunn BE, Almagro UA, Choi H et al. Dynamic-robotic telepathology: Department of Veterans Affairs feasibility study. *Hum Pathol* 28:8–12, 1997.

11. Dunn BE, Choi H, Recla DL et al. Robotic surgical telepathology between the Iron Mountain and Milwaukee Department of Veterans Affairs Medical Centers: A 12-year experience. *Hum Pathol* 40:1092–1099, 2009.

12. Weinstein RS, Graham AR, Lian F et al. Reconciliation of diverse telepathology system designs: Historical issues and implications for emerging markets and new applications. *APMIS* 120:256–275, 2012.

13. Krupinski EA. Virtual slide telepathology workstation-of-the-future: Lessons learned from teleradiology. *Semin Diagn Pathol* 26:194–205, 2009.

14. Kayser K, Szymas J, and Weinstein R. *Telepathology: Telecommunication, Electronic Education and Publication in Pathology.* Springer, Berlin, 1999, pp. 1–186.

15. Kaplan KJ, Weinstein RS, and Pantanowitz L. Telepathology, in *Pathology Informatics: Modern Practice & Theory for Clinical Laboratory Computing.* Pantanowitz L, Balis U, and Tuthill M, eds. American Society for Clinical Pathology Press, 2013, pp. 257–272.

16. Weinstein RS, Descour MR, Liang C et al. An array microscope for ultrarapid virtual slide processing and telepathology: Design, fabrication, and validation study. *Hum Pathol* 35:1303–1314, 2004.

17. Cucoranu LC, Vepa S, Parwani A et al. Digital pathology: A systematic evaluation of the patent landscape. *J Pathol Inform* 5:16, 2014.

18. Thrall MJ, Rivera AL, Takei H, and Powell SZ. Validation of a novel robotic telepathology platform for neuropathology intraoperative touch preparations. *J Pathol Inform* 5:21, 2014.

19. Zangbar B, Pandit V, Rhree PI et al. Smartphone surgery: How technology can transform practice. *Telemedicine & eHealth* 20:590–592, 2014.

20. Braunhut BL, Lian F, Webster PD et al. Subspecialty surgical pathologist's performances as triage pathologists in a telepathology-enabled quality assurance (QA) surgical pathology service: A human performance study. *J Pathol Inform* 5:18, 2014.

Further Information

Kayser K, Szymas J, and Weinstein RS. *Telepathology and Telemedicine: Communication, Electronic Education and Publication in e-Health.* VSV Interdisciplinary Medical Publishing, Berlin, 2005, pp. 1–257.

Kayser K, Molnar B, and Weinstein RS. *Digital Pathology Virtual Slide Technology in Tissue-Based Diagnosis, Research and Education.* VSV Interdisciplinary Medical Publishing, Berlin, 2006, pp. 1–193.

27

Acute Care Telemedicine

Nigel R. Armfield and Tim Donovan

CONTENTS

27.1 Introduction

While there is no single agreed definition of *acute care* in the medical literature, it is commonly understood to be care which is provided with some kind of time consideration, i.e., care which if provided in a timely fashion may mean the difference between life and death or life lived with a substantial disability for the patient. Examples of acute care include providing thrombolysis for the treatment of stroke, antibiotics for the treatment of sepsis, or the use of assisted ventilation for a premature newborn infant with respiratory distress.

Hirshon et al. [1] suggest six domains of acute care:

1. Trauma care and acute care surgery.
2. Emergency care.
3. Urgent care.
4. Short-term stabilization.
5. Prehospital care.
6. Critical care.

In this chapter, we propose a definition of *acute care telemedicine* and explore 12 case examples of the role of telemedicine in each of the domains identified by Hirshon et al. [1]. Here, our focus is strongly on the clinical application of telemedicine, rather than the underlying technology used.

27.2 Definitions

27.2.1 Telemedicine

Firstly, it is important to consider what *telemedicine* is. While often described as a technology, in our view, it is more helpful to think of telemedicine as a *process*, albeit one that depends directly on the use of technology. This technology allows clinical processes to be conducted at a distance; hence, it is an enabler, but in itself, the technology is not telemedicine.

Telemedicine can be thought of as the tasks that the clinician carries out (such as observing, consulting, interpreting, and providing opinions), assisted by ICT, in circumstances where there is distance between the patient and provider. Put succinctly, telemedicine is simply "medicine at a distance" [2].

Telemedicine is not new—it has a substantial literature base [3], yet the formal evidence base is still limited and substantial research is still required [4].

27.2.2 Acute Care Telemedicine

Marrying the notion of the temporal nature of acute care with the above reflections on telemedicine, we define *acute care telemedicine* as

> the practice of medicine at a distance, enabled by technology, for patients requiring urgent or emergent care.

The characteristics of acute care telemedicine reflect the circumstances that bring the patients to care—the care is typically unscheduled, resulting from trauma, sudden exacerbation of an existing condition, or unexpected illness. As we will show in the case examples, unlike in many other clinical applications of telemedicine, in acute care the interactions are usually between providers rather than between a provider and patient.

So, revisiting the examples of acute care that we mentioned in Section 27.1, telemedicine could have a role in a specialist center providing remote supervision of thrombolysis for a stroke patient or a consultation to discuss test results and plan the treatment of a patient with sepsis, to observe ventilator settings and chest recession, or to discuss medical evacuation of a premature newborn with respiratory distress.

27.3 Rationale for Telemedicine in Acute Care

In general terms, telemedicine aims to improve access to care. This is achieved by

1. Providing remote access to clinical expertise that is not available locally, and/or
2. Reducing delay in providing appropriate care for the patient.

In most health systems, specialist acute care services are provided in areas of large population and depending on the jurisdiction, these services may be operated as private ventures (i.e., fee for service and/or covered by private health insurance) and/or may be provided as a public service without charge at time of use.

In some health systems, to optimize clinical outcomes in an economical way, such services may be regionalized or centralized in centers of specialist expertise. Whichever way services are organized, they will favor some areas of the population and disadvantage others. Thus, the influence of health service organization on specialist care *availability* and *accessibility* is important when considering the potential aims and role of acute care telemedicine.

27.3.1 Service Availability

The availability of health services describes capacity and it is typically expressed as a supply ratio. Common examples in the acute care context include the ratio of neonatal intensive care unit (NICU) cots to live births; acute care hospital beds per 1000 population; and number of intensivists per 1000 population.

27.3.2 Service Accessibility

In addition to service capacity, it is important to consider accessibility. Accessibility relates to any distance and time impediments to reaching care. These impediments may be related to distance or the unavailability of immediate transport. In some cases, the nearest center that can provide definitive care may be some distance away.

For instance, in a perinatal emergency, while live birth to cot ratios may be favorable (i.e., the service is available), it is not helpful to a mother and infant if the NICU is inaccessible because it is many miles away, transport is difficult, and local health facilities are limited.

In summary, appropriate acute care services may be available but not accessible and vice versa, i.e., accessible but with insufficient capacity.

27.3.3 Health System Responses to Access Impediments

Health systems use a range of responses for overcoming access impediments. Some approaches are suited only to planned care, while others are specifically designed for acute care. Some of the responses may be used in combination. Standard approaches include the following:

1. **Outreach:** Whereby teams periodically travel to provide health services in the community. While this is appropriate for many health problems, it is not useful for unexpected acute care needs.
2. **Nonurgent patient transport:** Patients may be transported by ambulance, by specialist nursing/medical transport, or by commercial airlines. In many cases, patients may be expected to make their own travel arrangements for routine appointments.
3. **Emergency transport:** Whereby patients are stabilized at the scene and subsequently transported to definitive care. In some cases these services may be provided by dedicated transport teams; in other cases they are provided as "retrieval" services where teams travel from a specialist center to the patient and after having stabilized them, evacuate them back to the center. This is an important component of acute care services but is not necessarily rapid. For instance in Australia, large distances can lead to very long round-trip times for patient evacuations, even by air.

4. **Telephone consultation:** In many regionalized systems, large centers may provide advice to smaller centers by using the telephone. While this allows rapid advice to be provided, it depends on verbal descriptions and on the experience of the clinician who is attending the patient. It may be that information is missed or that information is incorrectly communicated.

5. **Telemedicine:** Telemedicine adds visual information to consultations that would otherwise be conducted by telephone. This visual information may be still images, video clips, real-time video consultations, or the availability of remote imaging and laboratory results. In contrast to the telephone, a specialist may observe the patient and their information at first hand, without mediation through a remote clinician.

27.3.4 Telemedicine as an Adjunct

In acute care, telemedicine does not replace in-person care; rather, it operates as an adjunct to existing care and existing processes. Reviewing the health system responses to access impediments described above, telemedicine can be thought of as a jigsaw piece that connects appropriate local care and specialized transport services. In the simplest sense, it can substitute for the use of the telephone and provide additional clinically useful visual information.

27.4 Potential Benefits of Telemedicine

The benefits of acute care telemedicine can be considered from the perspective of the patient and their family, the clinician, and the healthcare system.

27.4.1 Patient and Families

From the patient and family perspective, the benefits of telemedicine are threefold: (i) they are provided with improved timely access to specialist services, potentially leading to important interventions or the saving of time to intervention; (ii) the reassurance that arises from knowing that they, or their family member, have discussed with a clinician at a specialist center and a management plan agreed upon; and (iii) in some cases, the potential that their care may be managed locally, close to home and family, with the support of clinicians located at a remote specialist center.

27.4.2 Clinician

Telemedicine offers advantages to both the providing (i.e., specialist end) and the receiving clinicians. Providing clinicians may receive more accurate and descriptive information relating to the patient, their problems, and their current status. For instance, they may view the patient, work of breathing, tube positions, the monitor, ventilator settings, imaging, and procedures being conducted without relying on verbal descriptions. This more complete picture of the patient leads to improved decision making. Further, video-based telemedicine may enhance the professional relationship and teamwork between referring and specialist centers much more so than use of the telephone allows.

For the receiving clinician, the main advantage of telemedicine is to be able to fully present their patient and, hence, to receive more complete advice from the specialist center. In some cases, this specialist advice may be only to offer reassurance that appropriate care is already being provided, but importantly that reassurance can be confidently given based on complete visually acquired information.

For both the providing and the receiving clinicians, there is an educational benefit of telemedicine—providing clinicians learn more about the cases that referring clinicians are presented with and the care that they are able to provide. For receiving clinicians, they learn from the advice provided by their specialist center colleagues, thus improving the local capabilities.

27.4.3 Health System

From the health system perspective, telemedicine can support the high level of clinical care achieved through service regionalization and the appropriate and economical use of hospital and transport resources. It does this in the following ways:

1. By more reliably identifying which patients should be transferred to specialist care and which ones may be safely managed locally.
2. By avoiding the risks and expense of unnecessary or futile medical evacuation.
3. By avoiding limited specialized transport resources being committed unnecessarily, leaving them available for more urgent cases.
4. By avoiding unnecessary admission to more expensive care (e.g., avoided admission at a tertiary specialist center where costs are higher than those in referring hospitals).
5. By improving clinical coordination and continuity of care between specialist and referring centers.

From the workforce perspective, telemedicine provides a way to support less experienced or less specialized staff at referring hospitals. Allowing patients to be managed locally by supporting clinicians by telemedicine is a way for health systems to invest and build capability in their referring services and clinical staff.

A final note, which is relevant only to private-sector healthcare systems, offering telemedicine-based acute care consultations may be a way to compete for patients from referring centers.

27.4.4 Summary

The role of telemedicine in acute care is the remote delivery of clinical advice and management to patients who have urgent or emergent needs. In many cases, there are distance and time impediments to accessing an appropriate level of care. Telemedicine can help by allowing rapid access to specialist expertise and by providing a more complete picture of the patient for decision making. There are benefits to the patient, their family, clinicians, and the health system accrued through time saving and through the ability to better plan care and manage limited resources.

27.5 Domains of Acute Care and Telemedicine

Having defined acute care telemedicine and its potential benefits, using Hirshon et al.'s [1] six domains of acute care as a framework, we will introduce the practical role of telemedicine and explore its use by using 12 illustrative case examples.

27.5.1 Trauma Care and Acute Care Surgery

Hirshon et al. characterized the trauma care and acute care surgery domain as the "treatment of individuals with acute surgical needs, such as life-threatening injuries, acute appendicitis or strangulated hernias" [1].

The characteristics of telemedicine that are directly relevant to this domain are timeliness and the ability to provide higher-level diagnosis and intervention remotely. In many acute trauma cases, for example, burns injury, the initial period spent in a remote healthcare setting can be crucial as initial resuscitation during the first 1 to 2 hours has been associated with longer-term outcomes. Telemedicine provides rapid communication and visualization of the patient, their vital signs, and the relevant investigations such as X-rays where conventionally these are available by verbal report only. Interventions can be supported by the remote specialist and a number of studies have addressed the accuracy of the visual component of this assessment in specific clinical situations, e.g., radiologic diagnosis of fractures.

Two relevant systematic reviews were published in this area in 2012: Lewis et al. reviewed telemedicine in acute-phase injury management [5] and Wallace et al. reviewed the evidence for telemedicine in burn care [6].

In the primary literature, studies reporting the assessment and treatment of trauma or acute surgical conditions have not been published as frequently as in other categories of telemedicine application to healthcare. From 2004 to 2014, there were 9730 reports of telemedicine use over all medical disciplines, and of these, 51 studies examined trauma care and acute surgical conditions (MEDLINE search, March 13, 2014). These studies were almost exclusively of an observational design or were informal descriptive reports. In their review, Lewis et al. identified one large randomized controlled trial (RCT) by Wong et al. [7], conducted in 2006, which compared telephone, videoconferencing (VC), and teleradiology-only consultations on process-of-care, clinical, and economic outcomes for patients with neurosurgical conditions. Diagnostic accuracy favored VC and teleradiology over the telephone, although the nontelephone consultations took longer to conduct. The investigators found no difference in transfer rate or cost per patient. The use of teleradiology led to a statistically significant reduction in mortality compared to both VC and telephone consultations. There appeared to be no difference in mortality between VC and telephone consultations. In terms of mortality, the results suggest that tertiary access to remote radiology services is beneficial but that VC and telephone are comparable. In this study, a VC consultation took longer and appeared to have no advantage over using the telephone.

In their review of burn care, Wallace et al. [6] found that most studies demonstrated that telemedicine was feasible and acceptable and could potentially reduce costs, but they found no randomized studies and the clinical effectiveness of the approach has not been demonstrated [6].

Among the primary studies reported in the literature, the commonest outcomes assessed were comparisons of diagnosis and interventions with telemedicine either using a historical control period or with telemedicine consultation followed by in-person examination. A small number of studies examined the ability of telemedicine to avoid transfer to a higher-level facility but only Wong et al.'s study [7] used a randomized design. Overall, studies support the view that telemedicine is able to provide rapid communication and patient visualization with sufficient diagnostic accuracy to provide improved patient outcomes in patients with acute trauma or requiring acute surgical management. However, the overall quality of the evidence of clinical effectiveness is poor and further RCTs are required. While formal evidence in this domain is limited, telemedicine is nonetheless being used as illustrated by the following case examples in distant burns and trauma care.

27.5.1.1 Case Example 1: Triage and Referral of Patients with Acute Burn Injuries

Saffle et al. [8] performed a retrospective audit of burns patients who had been transported by air to the Intermountain Burn Center at the University of Utah during 2000 and 2001. The purpose of the review was to assess a priori whether it was likely that telemedicine could assist with remote evaluation and triage of burns patients. In their review they compared referring physician assessments of burn size and decision making regarding endotracheal intubation. They found that telemedicine would likely have significantly changed care. In 21 of 225 cases, they found that telemedicine assessment would likely have changed a transport decision [8].

The team subsequently implemented a telemedicine program [9] for the evaluation of acute burns. In a pre–post design, they compared transport decisions for the 2 years following implementation with the 2 years prior to implementation (the control period). Following the implementation of telemedicine, 44.3% of patients with burns received air transport, compared with 100% of patients in the control period. For the avoided air transports, patients were either transported by family or treated locally. The investigators found that patients transported by air had larger burn sizes and longer lengths of stay, suggesting that the telemedicine process effectively triaged the more serious cases for air transport. They also found that both patients and providers were satisfied with the use of telemedicine.

While this study had an informal design, it does suggest that telemedicine is a feasible method for triaging burns patients, leading to improved transport decisions. In turn, this leads to more appropriate use of limited air transport and specialist center resources.

27.5.1.2 Case Example 2: Rural Acute Trauma Care

Duchesne et al. [10] examined the use of telemedicine for managing rural patients who initially presented at local community hospitals. As with Saffle et al.'s study of burn triage [9], the investigators used a pre–post design for their evaluation.

The study used live videoconferencing to link the University of Mississippi Medical Center with seven rural hospital emergency departments. Over 5 years, the study recruited 814 patients with acute trauma. In the pre period, 351 patients (100%) were transferred to the university specialist medical center. In the post period, 51 of the 463 patients (11%) of patients were transferred, the remainder being discharged home or managed in the local community hospital. Only one patient died in the rural emergency department.

As with the burns case example, telemedicine appeared to be a feasible and safe way to triage patients for transport. The number of patients requiring transport reduced significantly after the implementation of telemedicine, without an increase in negative outcomes at the local community hospitals. From the economic perspective, hospital costs relating to the management of trauma patients reduced from US$7.6 million to US$1.1 million.

Once again, the study design was informal; however, the results do indicate benefit for patients who require review and potential transfer by a specialist center.

27.5.2 Emergency Care

Hirshon et al. characterized the domain of emergency care as "treatment of individuals with acute life-or-limb threatening medical and potentially surgical needs, such as acute myocardial infarctions (AMIs) or acute cerebrovascular accidents, or evaluation of patients with abdominal pain" [1].

In this domain, the most commonly reported application of telemedicine is the remote management of acute cerebrovascular ischemia (stroke), with over 400 articles reported in the literature over the last 10 years. This application, often referred to as telestroke, is perhaps one of the fastest growing areas of telemedicine in contemporary times.

In the literature, the majority of reports have been from developed healthcare systems, and study designs have been predominantly observational and institution based, or regional cohorts of acute presentations of cerebrovascular ischemia, with some being descriptive of a telemedicine cohort only, but a number of others compared outcomes for a historical control (nontelemedicine epoch) to those for an intervention group with the telemedicine application.

There has been a concerted effort to apply evidence-based practices to improve post-stroke outcomes by using telemedicine. A recent review of the use of telemedicine in acute stroke care found a high interrater agreement with a bedside assessment using the National Institutes of Health Stroke Scale (NIHSS), enhancement of correct thrombolysis decision making compared with telephone-only consultation, and that remote telemedicine use in acute stroke management was cost effective [11].

Unlike many other applications of telemedicine, evaluation in the area of stroke has been aided by widely accepted benchmarks, such as the time from presentation to appropriate thrombolysis, and a standardized outcome reporting scale.

A second application of telemedicine in this domain is in the management of AMI. De Waure et al. [12] conducted a systematic review and meta-analysis to critically appraise the evidence for telemedicine in reducing the mortality from AMI. Their review identified five studies, all of which used the transmission of ECGs as the intervention. Only one of the trials was an RCT and that trial did not report its randomization or blinding procedures. The remaining observational studies also varied in methodological quality. All of the studies suggested that remote transmission and review of ECGs was efficacious in reducing mortality. The meta-analysis produced a pooled risk ratio estimate of 0.65 (95% confidence interval [CI] 0.42–0.99). However, observational studies were pooled with a single experimental study, and the upper bound of the confidence interval almost reaches unity (no effect). Thus, while there is an indication of benefit, it falls short of formal evidence. This is acknowledged by the authors, who suggest that further research is needed to assess the clinical effectiveness of the intervention.

As with other applications of telemedicine, despite there being limited formal evidence, it is being used in routine clinical care. The following two case examples illustrate applications in stroke care and in the management of AMI.

27.5.2.1 Case Example 3: Thrombolysis for Acute Cerebrovascular Ischemia

Amorim et al. [13] reported the findings of a telestroke implementation in Pittsburgh, United States. In their model, a consultant neurologist at the University of Pittsburgh Medical Center reviewed patients who presented with suspected stroke at any of 12 community hospitals in western Pennsylvania. The purpose of the review was to assess the suitability of the patient for intravenous tissue plasminogen activator administration (IV tPA), this being the standard therapy for the treatment of acute ischemic stroke. During the consultations, the consultant assessed the patient and scored them using NIHSS. In addition, CT scans and laboratory studies were available for remote review and intracerebral hemorrhage was graded.

The team reviewed the number of patients who received IV tPA at the referring hospitals pre- and postimplementation of the telestroke model. In the pre period, the rate of IV tPA administration was 2.68% (n = 26). Following the implementation of telestroke, the rate was 6.8% (n = 113). Some secondary outcome measures were analyzed, including incorrect treatment decisions with a rate of 0.2% (n = 2) in the pre period and 0.3% (n = 5) in the post period; and symptomatic intracranial hemorrhage (n = 1 in both periods). No differences were found in mortality (either in hospital or short term), or in functional outcomes. This is likely because the study also found no difference in timing to IV tPA administration in either period. Once again, this study had an informal design; however, outcomes were comparable, and not worse, suggesting that a telemedicine review for eligibility for IV tPA was safe. Therefore, telemedicine, along with reducing delays in presentation to treatment, may be a useful part of the jigsaw in acute stroke care. Further formal studies are required to develop an evidence base.

27.5.2.2 Case Example 4: Telemedicine Diagnosis and Treatment of Acute Myocardial Infarction

In new work, Mehta et al. [14] describe the development of a comprehensive telemedicine network known as Lumen Americas Telemedicine Infarct Network (LATIN) in South America, an area that is medically underserved and has high rates of unfavorable health outcomes.

The hub-and-spoke network, which is currently in its pilot phase, operates a five-step telemedicine-enabled protocol to diagnose, triage, and treat patients with suspected ST elevation myocardial infarction (STEMI). These stages are (1) ECG and prehospital management; (2) teleconsultation triage; (3) prehospital/early thrombolysis; (4) pharmacoinvasive pathway; and (5) STEMI intervention pathway–door-to-balloon (D2B) interventions.

In this network, telemedicine consultations were provided by three tertiary facilities in Chile, Brazil, and Columbia. Rapid assessments of ECGs were conducted remotely by cardiologists and results communicated to ambulances and receiving hospitals by using the telemedicine network. As with the telestroke example, remote consultation was also available to provide guidance during triage and thrombolysis/pharmacoinvasive treatment.

The key role of telemedicine is to improve communication, including allowing rapid assessment of an ECG and the communication of the results, prealerting the receiving hospital, and to support the early delivery of treatment.

While the model is ambitious and as yet untested in its final form, if successful it will transform the quality of care of STEMI patients in South America.

27.5.3 Urgent Care

Hirshon et al. characterized the domain of urgent care as "ambulatory care in a facility delivering medical care outside a hospital emergency department, usually on an unscheduled, walk-in basis. Examples include evaluation of an injured ankle or fever in a child" [1].

Telemedicine applications in this domain have largely been reported from developed health systems that have examined the potential to improve health outcomes by providing access to community nonhospital sites, e.g., school or child-care facilities; penal institutions; and telemedicine platforms based on mobile access points, e.g., vehicle platforms to indigenous remote settlements.

With the difficulty in defining robust limits to this broad domain of acute telemedicine, the publication rate is not clear, but it is clear that fewer reports are available in this area. Most reports in this domain have provided observational evidence of system usage only but few have included outcomes of diagnostic comparisons between in-person and telemedicine uses and economic evaluations, particularly examinations of cost effectiveness; and a few have examined specific health system outcomes, e.g., changes in emergency department attendance with telemedicine utilization in a community cohort.

Among the most detailed examinations in this domain have been those from telemedicine platforms that are mobile (vehicle based) that are driven to remote communities and provide consultation for specific high-risk morbidities, for example, vehicle-based otolaryngology telemedicine for remote indigenous communities.

27.5.3.1 Case Example 5: Primary Care Pediatric Telemedicine Consultation for Acute Illness in Child-Care and School Settings

McConnochie et al. reported a novel telemedicine program that provided telemedicine consultations to child-care centers and schools [15]. Their program, known as Health-e-Access, used both real-time VC and store-and-forward exchange of still images and video, including views of the skin, eyes, and tympanic membrane. A digital stethoscope also allowed remote auscultation. The program was designed to provide triage, diagnosis, and treatment for acute cases that otherwise would require visit to a hospital emergency department.

To evaluate the program, the investigators matched 1216 children who attended institutions with telemedicine access (matched for age, gender, and socioeconomic status) with children attending institutions without telemedicine access. Using health insurance data, utilization of telemedicine, hospital emergency department, or primary care was then assessed.

Consultations were provided from 10 primary healthcare centers to 22 child-care and school sites and over 6500 consultations were delivered.

The study found that the in-person rate of attendance at hospital emergency departments was 27.3% less for children who had telemedicine available. The rate of attendance at primary healthcare was 3.3% less for children at institutions with access to telemedicine.

The setting of this study was novel in that it provided acute care directly to child-care centers and schools. The findings showed that many potential emergency visits could be avoided with remote assessments. The effect on primary care attendance was less marked.

27.5.3.2 Case Example 6: Community-Based Ear, Nose, and Throat (ENT) Screening with Telemedicine-Based Review and Community-Based Surgical Outreach for Children at High Risk of Ear Disease

In 2009, Smith et al. [16] designed a mobile school-based program to screen indigenous children at high risk of ear disease. In this program, an indigenous health worker (IHW) visited schools in a van that was specially equipped to conduct ear screenings, including otoscopy, tympanometry, and audiometry. Following each day of screening, the IHW uploaded results by store-and-forward method to a database for remote specialist review at a tertiary children's hospital.

On a monthly basis, a specialist ENT consultant reviewed the results for children who had failed their community-based screening. During this review, the consultant could remotely triage those children who needed intervention at the next surgical outreach visit. By using this simple approach, children who had acute ear problems could be identified quickly and intervention planned without the child needing to leave the community.

Prior to the establishment of the mobile telemedicine-enabled service, screening rates in the community were very low (estimated to be less than 30%), problems were difficult to triage, and surgery was difficult to target. Surgery was available only at the tertiary children's hospital some considerable distance away and there was a high failure to attend rate (51%).

Following the implementation of telemedicine, 80% of children now receive an annual ear health screening, with an attendance rate of 86% for surgical procedures in the community [17]. This case example demonstrates that telemedicine need not be high tech; often, very simple approaches, which integrate telemedicine with usual referral mechanisms, may be the most effective.

27.5.4 Short-Term Stabilization

Hirshon et al. characterized the domain of urgent care as "treatment of individuals with acute needs before delivery of definitive treatment. Examples include administering intravenous fluids to a critically injured patient before transfer to an operating room" [1].

The application of telemedicine in this domain has to date been very limited but has included providing support for resuscitation [18], short-term supervision of life support in trauma cases [19], and examination of the potential of cardiopulmonary resuscitation (CPR) assistance by video [20]. There have been few formal studies in this area outside of simulation.

27.5.4.1 Case Example 7: Ambulance Dispatchers Monitoring Simulated Cardiac Arrest Calls from an Untrained Bystander with either Mobile Phone Video or Standard Call Interaction; Video Input Improved Understanding of Rescuer and Facilitated Assistance to Bystander

In a study of providing CPR assistance with mobile phones, Johnsen and Bolle [20] simulated 10 cardiac arrest scenarios with six emergency services dispatchers. For half of the scenarios, the callers were guided through CPR by using a conventional telephone call. The remaining half of the scenarios were guided using video-based calls. In their analysis, the investigators found that the visual information provided during the video calls was beneficial and supported their guidance. They also reported that their guidance was easier to achieve by video and that the procedure may have been of better quality when guided by video than by an audio-only call. Overall the dispatchers found that video provides the dispatchers with a better understanding of the situation.

With the increasing prevalence of high-quality camera–equipped mobile phones, it seems possible they will have an increasing role in assisting laypersons to provide immediate care. The dispatchers noted, though, that in this study, the protocol for providing remote guidance needed to be modified to suit video, and this is an area that needs further research.

27.5.4.2 Case Example 8: Qualitative Outcome Improved in Remote Canadian
Trauma Patients with Stabilization Using Telesonography

In a 12-month trial, Al-Kadi et al. [21] investigated the use of remote real-time ultrasound (telesonography) to mentor the stabilization of trauma patients. During the trial 23 remote ultrasound examinations were conducted, 20 during acute resuscitations and 3 during live-patient simulations. The trial assessed the use of both the focused assessment with sonography for trauma (FAST) and the extended focused assessment with sonography for trauma (EFAST) which included pneumothorax assessment with 23 patients receiving FAST examinations. EFAST examinations were completed in 17 cases.

By surveying the clinicians involved in the study, the project examined three main outcomes: satisfaction, improvement of collegiality, and improvement of ultrasound skills. Fourteen clinicians completed the survey, of which 93% were satisfied with the use of the technology for performing assessments. The same proportion believed that the project had improved collegiality. 71% of respondents believed that they had improved their ultrasound skills. While the study did not look at outcomes, it provides preliminary information on the usefulness of real-time ultrasound in acute trauma, which may be explored in further studies.

27.5.5 Prehospital Care

The prehospital care domain of acute healthcare as defined by Hirshon et al. encompasses "care provided in the community until the patient arrives at a formal health-care facility capable of giving definitive care. Examples include delivery of care by ambulance personnel or evaluation of acute health problems by local health-care providers" [1].

Interventions using telemedicine in this area of prehospital care are perhaps most difficult as equipment must be installed in ambulances or similar mobile platforms, yet still achieve sufficient diagnostic precision to alter outcomes or allow an earlier directed treatment when compared with routine hospital care.

Telemedicine applications in prehospital care have ranged from simple tertiary review of ECG in patients with ST elevation while they are still in an ambulance, to complex mobile stroke assessment, including remote assessment of mobile CT scans.

Studies in this area of telemedicine have reported a number of outcomes, some of which are clearly clinically important, such as 30-day survival rate, while other reports have used more proxy outcomes such as the time to thrombolysis in myocardial ischemia. Studies examining cost in this area of telemedicine have almost exclusively been cost comparisons between a prehospital telemedicine care and standard hospital treatment. In those cost comparisons reported there is little evidence of cost increases with the telemedicine cointervention.

27.5.5.1 Case Example 9: A Well-Designed Randomized Study of Mobile
Stroke Assessment and Treatment (Including Telemedicine Use)
on Time from Alarm to Therapy Decision in 100 Patients

In Germany in 2012, Walter et al. described an evaluation of mobile telemedicine-enabled stroke assessment [22]. Their mobile stroke unit was a specially equipped ambulance with CT scanner, laboratory, and telemedicine connection to a tertiary center.

The aim of the study was to assess whether commencing diagnosis and treatment early (i.e., at the prehospital stage) had clinical benefits. The rationale for the study was that only a very small proportion of patients (between 2% and 5%) who have a stroke receive throm-bolysis because of delays in reaching hospital. Thus, the mobile stroke unit may have a benefit by reducing time to diagnosis and treatment.

The investigators randomized patients aged between 18 and 80 years of age who had one or more stroke symptoms in the last 2.5 h to receive either mobile stroke care or con-ventional hospital stroke treatment. The primary outcome measure was time of call to therapy decision. Secondary outcome measures included time of call to time of completion of CT scan, time of call to time of completion of laboratory analysis, the number of patients receiving thrombolysis, time of call to delivery of thrombolysis, and neurological outcome.

The study found that care provided by the mobile stroke unit substantially reduced the median delay from time of call to decision to deliver thrombolysis: in the mobile stroke unit group, the median duration was 35 min (interquartile range [IQR] 31–39), whereas in the standard care group, the median delay was 76 min (IQR 63–94), a median difference of 41 min (95% CI 36–48 min). The difference was statistically significant ($p < 0.0001$). Results for the secondary outcome measures also strongly favored the mobile stroke unit over conventional care.

The results of this well-designed study demonstrate that a telemedicine-enabled mobile stroke unit is feasible and has clinical benefits. Further investigation of the scalability and cost-effectiveness of the model would be beneficial.

27.5.5.2 Case Example 10: Cost Analysis of Remote ECG Reading by Telemedicine

While the feasibility, clinical efficacy, and acceptability of telemedicine applications are often examined, detailed examinations of the economics are less common.

Brunetti et al. conducted a cost analysis of prehospital reading of ECGs by an on-call car-diologist [23]. In their study, patients with suspected acute cardiac disease were recruited when they called the emergency medical service by telephone. These patients received an ECG at the scene. The ECG was read remotely by a cardiologist who also provided a tele-medicine consultation.

To examine any potential savings, the costs of the prehospital triage were compared with the costs of conventional hospital-based triage (in-hospital ECG and review by a cardiologist).

For the prehospital care, the cost was €16.79 per event (ECG and teleconsultation). For in-hospital care, the cost ranges from €24.80 to €55.20, indicating that the prehospital care saved €8.10 to €38.40 per event. For the year of 2012, this equates to savings of €891,760 to €4,219,380 for the health system.

Using data from prior studies, the investigators suggested that 69 lives per year could be saved with prehospital diagnosis, with a cost per quality–adjusted life year gain of €1927 (€990–€2508).

This study suggests that prehospital telemedicine assessment by an on-call cardiologist is feasible with economic benefits and that it could save lives at a moderate cost.

27.5.6 Critical Care

According to Hirshon et al.'s definitions of the domains of acute care, critical care is "the specialized care of patients whose conditions are life-threatening and who require com-prehensive care and constant monitoring, usually in intensive care units. Examples are patients with severe respiratory problems requiring endotracheal intubation and patients with seizures caused by cerebral malaria" [1].

Within acute care, the domain of critical care (or intensive care, as it is referred to in many countries) has received the most attention in the literature, with reports of work across neonatal [24–32], pediatric [18,33–38], and adult intensive care [39–42].

The interest is unsurprising; critical care is complex and expensive and services are generally organized in regional centers of expertise, often supported by specialized transport services. In this context, there are a number of potential roles for telemedicine, including:

1. Providing specialist advice to referring centers.
2. Optimizing transport decisions.
3. Providing remote clinical oversight of ICUs, and
4. Providing specialized education and training at a distance.

A systematic review and meta-analysis published in by Young et al. in 2011 [43] suggested that telemedicine can reduce ICU mortality and length of stay. However, there was considerable uncertainty in the results [44]. While firm evidence is still limited, many telemedicine applications have been successfully incorporated into routine care for critically ill patients and this is illustrated in the following two case examples.

27.5.6.1 Case Example 11: Advice and Retrieval Management for Critically Ill Infants

In 2004 in Australia, Armfield et al. embarked on a research program to investigate the role of telemedicine for acute care neonatal consultations. The setting was Queensland's public health system that had three tertiary NICUs, three of which were in the state capital of Brisbane and the third was 1400 km to the north. While Queensland has excellent facilities for the care of high-risk newborns, many infants are born some distance from tertiary care and round-trip times to retrieve infants can be as long as 15 h. Thus, it seemed that telemedicine may have a role to overcome distance and time impediments by providing timely advice, to manage retrieval, and to support the care of infants until the retrieval team arrives.

Telemedicine for newborn infants presented some unique challenges—the patients were very small and typically being cared for in an incubator (Figure 27.1). Space within the nurseries was very limited. Conventional business VC systems were not a good fit to the problem and so the team designed a custom telemedicine system known as Neonatal Examination and Management Online (NEMO) [26]. The system comprised a low-footprint trolley (see Figure 27.2), with a camera mounted to give views of the infant from above, through the wall of the incubator. A second camera allowed audio and video communication between clinicians. At the tertiary hospital site, a PC-based system allowed full remote pan–tilt–zoom control of the cameras, capture of stills, adjustment of brightness, etc. (Figure 27.3). Clinical efficacy and usability of the system was assessed using a room-to-room method comparison study in a tertiary NICU [24,25].

Subsequently, the team assessed the effectiveness of NEMO in routine use by using a multicenter trial. The results showed that telemedicine provided new useful clinical information and that retrieval could be avoided in around 30% of cases. The transport savings were subsequently modeled [27] and clinician acceptance of telemedicine was assessed [28].

NEMO was found to be useful and have the potential to realize significant savings for the health system. However, not all clinicians were willing to incorporate it into their practice, instead preferring to continue to use the telephone to provide advice and to manage retrievals.

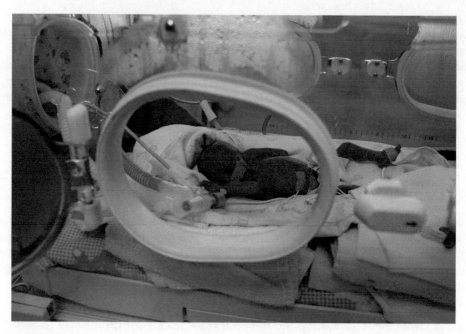

FIGURE 27.1
A newborn incubator.

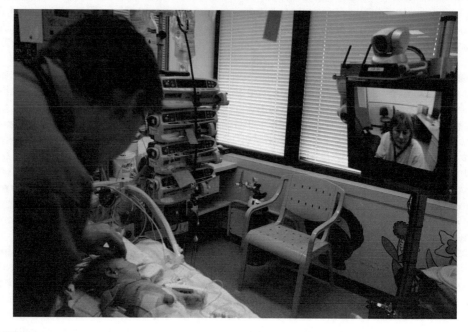

FIGURE 27.2
The camera arrangement for remote consultation.

FIGURE 27.3
Remote control of incubator camera by physicians.

This work, conducted as a partnership between the University of Queensland and Queensland Health, was recognized by two awards for innovation by Queensland's state health department in 2011.

27.5.6.2 Case Example 12: Providing Remote Oversight of Intensive Care Units (e-ICU)

Tele-ICU and *e-ICU* are names given to a model of remote care that has become popular in the United States. In this model, care at a number of ICUs may be managed from a central location by using telemedicine. This may be used to cover staff shortages, to provide out-of-hours cover, or as routine care in the absence of intensivists.

Many tele-ICU programs use an approach originally pioneered in 2000 by Rosenfeld et al. [45], whereby clinicians provide remote oversight of patients in a remote ICU. In their study, four intensivists provided round-the-clock monitoring of a 450-bed surgical ICU from their homes. Equipment to allow real-time consultation and the transmission/reception of vital signs was installed in the ICU and in the consultants' homes.

Using two 16-week baseline periods during the year prior to implementation, the investigators used a pre–post design to assess changes in mortality, length of stay, ICU complication rates, and ICU complication-associated costs. The study found significant reductions in ICU mortality (68% and 48% reduction for the intervention period compared with each baseline) and hospital mortality (33% and 30% compared with each baseline period). Similarly, ICU length of stay was reduced by 34% and 30% compared with baselines. The investigators also observed a reduction in ICU complication rate (a reduction of 44% and 50% compared with each baseline). These reductions in complications led to a reduction in ICU costs of 33% and 36% against each baseline period [45].

The work of the research team continued [46–49] and resulted in a spin-off telemedicine system architecture and business known as VISICU (and the trademarking of the term *eICU*). In 2007, the VISICU business was acquired by Philips Healthcare and the eICU program is now marketed as a commercial product and provides the foundation for many tele-ICU programs in the United States.

27.5.7 Summary

As illustrated by these case examples, telemedicine has a useful role in all of the domains of acute care. Some areas, such as critical care and telestroke, are becoming more developed and have received attention from researchers. Other areas remain embryonic with much scope for research.

27.6 Conclusions

Applications of telemedicine in acute care are still relatively new with scheduled or planned care applications dominating the literature. However, telemedicine shows great promise for patients requiring urgent or emergent care—this care is typically time critical, complex, and expensive. As we have illustrated in our case examples, telemedicine has the potential to save time and to provide care economically by helping to optimize the use of limited transport resources. Several studies, although informal, have shown significant reductions in mortality through the use of acute care telemedicine.

There are some remaining barriers to progress: formal evidence is still limited, yet research in this area is difficult to conduct because RCTs, the gold standard for assessing clinical effectiveness, are expensive and not always practical or ethical to conduct.

Some clinicians remain to be convinced of the merit of telemedicine in acute care. Further research is required to understand clinician perspectives and motivations toward telemedicine. In some jurisdictions there are also problems of licensing and reimbursement that are yet to be dealt with. There are also a lack of formal guidelines or education and training for clinicians in the area of distant acute care.

Nonetheless, while the formal evidence base is still somewhat limited and there are some barriers to growth, acute care telemedicine is being adopted by providers, important examples being the growth of tele-ICU services in the United States and worldwide interest and success of telestroke services.

We believe that with continued service development and evaluation, telemedicine will be shown to have a leading role in providing acute care services, resulting in significant individual, health system, and societal benefits.

Abbreviations and Nomenclature

AMI   acute myocardial infarction
CI   confidence interval

CPR	cardiopulmonary resuscitation
CT	computed tomography
D2B	door-to-balloon
ECG	electrocardiogram
EFAST	extended focused assessment with sonography for trauma
ENT	ear, nose, and throat
FAST	focused assessment with sonography for trauma
ICT	information and communication technology
ICU	intensive care unit
IHW	indigenous health worker
IQR	interquartile range
IV	intravenous
LATIN	Lumen Americas Telemedicine Infarct Network
NEMO	Neonatal Examination and Management Online
NICU	neonatal intensive care unit
NIHSS	National Institutes of Health Stroke Scale
PICU	pediatric intensive care unit
RCT	randomized controlled trial
STEMI	ST elevation myocardial infarction
tPA	tissue plasminogen activator
VC	videoconferencing

References

1. Hirshon, J.M. et al., Health systems and services: The role of acute care. *Bull World Health Org*, 2013. 91(5): pp. 386–388.
2. Telemedicine: Fad or future? *Lancet*, 1995. 345(8942): pp. 73–74.
3. Armfield, N.R. et al., Telemedicine—A bibliometric and content analysis of 17,932 publication records. *Int J Med Inform*, 2014. 83(10): pp. 715–725.
4. Armfield, N.R. et al., Telemedicine—Is the cart being put before the horse? *Med J Aust*, 2014. 200(9): pp. 530–533.
5. Lewis, E.R. et al., Telemedicine in acute-phase injury management: A review of practice and advancements. *Telemed e-Health*, 2012. 18(6): pp. 434–445.
6. Wallace, D. et al., A systematic review of the evidence for telemedicine in burn care: With a UK perspective. *Burns*, 2012. 38(4): pp. 465–480.
7. Wong, H.-T. et al., The comparative impact of video consultation on emergency neurosurgical referrals. *Neurosurgery*, 2006. 59(3): pp. 607–613.
8. Saffle, J.R., L. Edelman, and S.E. Morris, Regional air transport of burn patients: A case for telemedicine? *J Trauma-Injury Infection Crit Care*, 2004. 57(1): pp. 57–64.
9. Saffle, J.R. et al., Telemedicine evaluation of acute burns is accurate and cost-effective. *J Trauma Acute Care Surg*, 2009. 67(2): pp. 358–365.
10. Duchesne, J.C. et al., Impact of telemedicine upon rural trauma care. *J Trauma-Injury Infection Crit Care*, 2008. 64(1): pp. 92–98.
11. Rubin, M.N., and B.M. Demaerschalk, The use of telemedicine in the management of acute stroke. *Neurosurg Focus*, 2014. 36(1): p. E4.
12. de Waure, C. et al., Telemedicine for the reduction of myocardial infarction mortality: A systematic review and a meta-analysis of published studies. *Telemed e-Health*, 2012. 18(5): pp. 323–328.

13. Amorim, E. et al., Impact of telemedicine implementation in thrombolytic use for acute ischemic stroke: The University of Pittsburgh Medical Center telestroke network experience. *J Stroke and Cerebrovasc Dis*, 2013. 22(4): pp. 527–531.
14. Mehta, S. et al., A tale of two cities: STEMI interventions in developed and developing countries and the potential of telemedicine to reduce disparities in care. *J Interv Cardiol*, 2014. 27(2): pp. 155–166.
15. McConnochie, K.M. et al., Acute illness care patterns change with use of telemedicine. *Pediatrics*, 2009. 123(6): pp. e989–e995.
16. Smith, A.C. et al., A mobile telemedicine-enabled ear screening service for Indigenous children in Queensland: Activity and outcomes in the first three years. *J Telemed Telecare*, 2012. 18(8): pp. 485–489.
17. Smith, A.C. et al., Changes in paediatric hospital ENT service utilisation following the implementation of a mobile, indigenous health screening service. *J Telemed Telecare*, 2013. 19(7): pp. 397–400.
18. Kon, A.A., and J.P. Marcin, Using telemedicine to improve communication during paediatric resuscitations. *J Telemed Telecare*, 2005. 11(5): pp. 261–264.
19. Tachakra, S. et al., Supervising trauma life support by telemedicine. *J Telemed Telecare*, 2000. 6(Suppl. 1): pp. S7–S11.
20. Johnsen, E., and S.R. Bolle, To see or not to see—Better dispatcher-assisted CPR with video-calls? A qualitative study based on simulated trials. *Resuscitation*, 2008. 78(3): pp. 320–326.
21. Al-Kadi, A. et al., User's perceptions of remote trauma telesonography. *J Telemed Telecare*, 2009. 15(5): pp. 251–254.
22. Walter, S. et al., Diagnosis and treatment of patients with stroke in a mobile stroke unit versus in hospital: A randomised controlled trial. *Lancet Neurol*, 2012. 11(5): pp. 397–404.
23. Brunetti, N.D. et al., Prehospital telemedicine electrocardiogram triage for a regional public emergency medical service: Is it worth it? A preliminary cost analysis. *Clin Cardiol*, 2014. 37(3): pp. 140–145.
24. Armfield, N. et al., An evaluation of the usability of a system for neonatal teleconsultation. *J Telemed Telecare*, 2007. 13(Suppl. 3): pp. 101–110.
25. Armfield, N.R. et al., Preliminary evaluation of a system for neonatal teleconsultation. *J Telemed Telecare*, 2007. 13(Suppl. 3): pp. 4–9.
26. Armfield, N.R. et al., Mobile Telemedicine: Robots, Fish and other Stories, in *ECEH*. 2007.
27. Armfield, N.R. et al., The costs and potential savings of telemedicine for acute care neonatal consultation: Preliminary findings. *J Telemed Telecare*, 2012. 18(8): pp. 429–433.
28. Armfield, N.R., T. Donovan, and A.C. Smith, Clinicians' perceptions of telemedicine for remote neonatal consultation. *Stud Health Technol Inform*, 2010. 161: pp. 1–9.
29. Fang, J.L. et al., Real-time video communication improves provider performance in a simulated neonatal resuscitation. *Resuscitation*, 2014.
30. Kim, E.W. et al., Telemedicine collaboration improves perinatal regionalization and lowers statewide infant mortality. *J Perinatol*, 2013. 33(9): pp. 725–730.
31. Scheans, P., Telemedicine for neonatal resuscitation. *Neonatal Netw*, 2014. 33(5): pp. 283–287.
32. Wenger, T.L. et al., Telemedicine for genetic and neurologic evaluation in the neonatal intensive care unit. *J Perinatol*, 2014. 34(3): pp. 234–240.
33. Dharmar, M., and J.P. Marcin, A picture is worth a thousand words: Critical care consultations to emergency departments using telemedicine. *Pediatr Crit Care Med*, 2009. 10(5): pp. 606–607.
34. Dharmar, M. et al., Impact of critical care telemedicine consultations on children in rural emergency departments. *Crit Care Med*, 2013. 41(10): pp. 2388–2395.
35. Dharmar, M. et al., The financial impact of a pediatric telemedicine program: A children's hospital's perspective. *Telemed J E Health*, 2013. 19(7): pp. 502–508.
36. Dharmar, M. et al., Telemedicine for children in need of intensive care. *Pediatr Ann*, 2009. 38(10): pp. 562–566.
37. Labarbera, J.M. et al., The impact of telemedicine intensivist support and a pediatric hospitalist program on a community hospital. *Telemed J E Health*, 2013. 19(10): pp. 760–766.

38. Marcin, J.P., Telemedicine in the pediatric intensive care unit. *Pediatr Clin North Am*, 2013. 60(3): pp. 581–592.

39. Boots, R.J., S.J. Singh, and J. Lipman, The tyranny of distance: Telemedicine for the critically ill in rural Australia. *Anaesth Intensive Care*, 2012. 40(5): pp. 871–874.

40. Lilly, C.M. et al., A multicenter study of ICU telemedicine reengineering of adult critical care. *Chest*, 2014. 145(3): pp. 500–507.

41. Sadaka, F. et al., Telemedicine intervention improves ICU outcomes. *Crit Care Res Pract*, 2013. 2013: p. 456389.

42. Willmitch, B. et al., Clinical outcomes after telemedicine intensive care unit implementation. *Crit Care Med*, 2012. 40(2): pp. 450–454.

43. Young, L.B. et al., Impact of telemedicine intensive care unit coverage on patient outcomes: A systematic review and meta-analysis. *Arch Intern Med*, 2011. 171(6): pp. 498–506.

44. Smith, A.C., and N.R. Armfield, A systematic review and meta-analysis of ICU telemedicine reinforces the need for further controlled investigations to assess the impact of telemedicine on patient outcomes. *Evidence Based Nursing*, 2011. 14(4): pp. 102–103.

45. Rosenfeld, B.A. et al., Intensive care unit telemedicine: Alternate paradigm for providing continuous intensivist care. *Crit Care Med*, 2000. 28(12): pp. 3925–3931.

46. Breslow, M., Assessing ICU performance using administrative data. *J Crit Care*, 2001. 16(4): pp. 189–195.

47. Breslow, M.J., Remote ICU care programs: Current status. *J Crit Care*, 2007. 22(1): pp. 66–76.

48. Breslow, M.J. et al., Effect of a multiple-site intensive care unit telemedicine program on clinical and economic outcomes: An alternative paradigm for intensivist staffing. *Crit Care Med*, 2004. 32(1): pp. 31–38.

49. Celi, L.A. et al., The eICU: It's not just telemedicine. *Crit Care Med*, 2001. 29(Suppl. 8): pp. N183–N189.

28

Monitoring for Elderly Care: The Role of Wearable Sensors in Fall Detection and Fall Prediction Research

Kejia Wang, Stephen J. Redmond, and Nigel H. Lovell

CONTENTS

28.1 Introduction

Falls are a prominent cause of injuries and injury related deaths for today's growing population of older adults, posing a major health risk in even the most ordinary settings of daily life. A fall is defined by the World Health Organization as "inadvertently coming to rest on the ground, floor or other lower level, excluding intentional change in position to rest on furniture, wall or other objects" [1]. Approximately one third of community dwelling residents aged over 65 experience at least one fall a year, with the chance of falling increasing with age [1–4]. Although the specific incidence of falls varies between countries in Asia, Europe, Australasia, and the Americas, the trends are the same.

Often leading to injuries such as hip or femur fractures, illnesses from "long lies," or even deaths [5], falls are associated with significant costs. Direct costs were conservatively estimated at A\$648.2 million for Australia for 2007–2008 [6], US\$23.3 billion for the United States, US\$1.6 billion for the United Kingdom [7], and even more when including indirect costs. A study among beneficiaries of Medicare (the United States' largest healthcare insurer) revealed that compared to those who reported no falls, the total aggregate healthcare costs per year were 29% higher for the estimated 3.7 million older adults who reported one fall in 2002, and 79% higher among those reporting recurrent falls [8].

The psychological impact of falls, such as developing a fear of falling, can also have severe effects when an individual limits their physical mobility and living independence due to loss of confidence and increased awareness of frailty [3]. This may, in turn, place additional pressures on the faller's family [5]. Public health services are burdened with the consequences of fall related cases, especially in the face of a growing elderly population and projections of continuing mortality decline over the coming decades [9]. Therefore, there is significant motivation to reduce the impact of injurious falls, through preventive measures and by fall detection.

The field of fall detection and fall prevention research is burgeoning, particularly with the advent of cheaper microelectronics and a growing elderly population [1]. Fall detection methods, designed by medical engineering researchers, are generally based on triggering a sensor based system, whether of the wearable sort or within a system installed in a home or residential dwelling. Clinicians, on the other hand, have devised over decades many different fall risk assessments and surveys [3–5,10–12], to identify elderly at high risk of falling and prescribe intervention strategies where suitable. Fall risk factors range from motor, gait, and balance abnormalities to impaired vision and the use of psychotropic medications [13]. However, the multifactorial nature of fall risk comes with issues of assessment subjectivity—between target populations, between tests, and between clinicians who administer the same assessment. This has led to a recent focus on developing quantitative protocols for assessing fall risk, by the use of body-worn inertial sensors, pressure mats, and motion capture systems.

This chapter will discuss the convergence of wearable sensor technologies with the traditionally clinical orientated research field of fall assessment and prevention in older adults. First, the two existing areas for reducing the impact of falls, namely, fall detection and fall risk assessments for fall prevention, will be broadly introduced. The convergence of these two areas will then be discussed, with an exploration on research trends toward a common aim of fall prediction, in terms of sensor selection, signal analysis techniques, and model training and validation strategies. Current issues regarding the complexity of conducting meta-analyses due to statistical heterogeneity between studies are addressed. A glimpse into the future of uniting telehealth and personal health management systems to streamline fall risk assessments and personalized intervention capabilities concludes the chapter.

28.2 Personal Alarms

The most basic approach to reducing the impact of a fall is with the use of personal emergency alarms, which the wearer can use to call for help. Such systems offer peace of mind and can assist in reducing the occurrence of long lies after a fall, where the faller is unable to get up and several hours or more can pass before help is received. Wearable button-activated alarms that can communicate with emergency services are already commercially available (Figure 28.1, Ref. [14]), allowing seniors to obtain medical attention with a simple button press [15–19]. Their deployment worldwide is high: the Philips Lifeline medical alert service alone having served over 7 million subscribers in the United States to date [20]; Grupo Neat, the leading supplier of care phones, home monitoring, and personal alarm equipment in many European countries, have sold more than 400,000 such devices throughout Europe, South Africa, South America, and Asia; and in Australia, over 62,000 people are assisted every year by Silver Chain alarms [21], 13,000 alarm units are serviced by Tunstall, 36,000 by VitalCall, and over 20,000 by the Department of Veterans' Affairs through various service providers. Despite these numbers, Lawson and Lowe reported that according to Tunstall, the proportion of adults aged over 65 years who own a personal alarm stands at only 1% in Australia, 4% in New Zealand, and up to 8% in the United

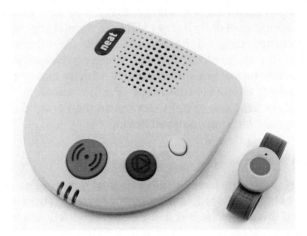

FIGURE 28.1
Left: the Grupo Neat NEO home care phone (From Legrand, "Legrand continues to expand in the assisted living market and signs joint venture agreement with NEAT," *Legrand*, 2014. [Online]. Available at http://www.legrand.com/files/fck/News/files/groupe/Neat/Neat_HD.jpg, accessed: Apr. 22, 2014), which can be programmed to call different locations, such as an alarm central or a relative. *Right*: wearable portable Atom radio trigger for activating alarm when the wearer is in need of support or help; it can also be used to remotely answer incoming calls on NEO (From Grupo Neat, "Home Units," *Grupo Neat*. [Online]. Available at http://www.gruponeat.com, accessed: Jan. 28, 2014).

Kingdom [22]. However, the services are expensive, with an initial set-up cost as well as ongoing monitoring costs [17]; monitoring is limited to within the range of the base unit; and loss of consciousness means the alarm cannot be raised. Such limitations, therefore, call for the development of automatic alert systems and fall prevention strategies, as discussed in the following sections.

28.3 Fall Detection and Activity Monitoring

In situations where a faller is rendered unconscious or otherwise unable to activate their wearable alarm, the ability to automatically detect a fall is advantageous, especially when such capabilities are encapsulated within technologies that are minimally intrusive in an individual's everyday life. Meanwhile, to detect a fall requires a fall to be measured and classified as a "positive" event by the automatic detection algorithm. Both Shany et al. [23] and Schwickert et al. [24] conducted surveys on the literature on wearable sensors for fall detection and activity monitoring. Simple fall detection algorithms commonly employed have been threshold based, where a quantitative parameter calculated in real time from inertial signals triggers an event by exceeding a defined threshold. Other algorithms combine details from two or more phases of detection—for example, the impact phase and the postfall phase—to make a more informed trigger decision. The level of heterogeneity between studies has been extensive; the extent of validation using real world falls from older adults and geriatric patients, as opposed to young healthy subjects simulating falls onto mats, has been minimal [23,24]. This lack of published evidence extends even to commercially available devices [24], revealing the true juvenility of this field.

The recurrently significant rate of false positive alarms in fall detection algorithm performance [23] has caused researchers to suggest that it may be beneficial to quantify and distinguish between activities of daily living (ADLs), or normal human movements [25]. As part of this, researchers have targeted the issues of activity classification and automatic fall detection in a number of ways in recent decades, which can be categorized into two main approaches. Unobtrusive approaches aim to identify a fall by monitoring the individual(s) within a setting, leading to the conceptualization of smart homes. Body-based approaches involve the individual wearing one or more devices on their body. Each have their advantages and disadvantages, which are analyzed below.

28.3.1 Unobtrusive Sensors

Unobtrusive monitoring systems have the advantage of requiring no action on the part of the person or people being monitored, as they are "out of sight, out of mind." Inconsistency in wearer compliance is a major incentive for developing fall detection systems in this format; their accuracy is independent of personal preferences, unlike body-worn sensors whose effectiveness is affected by personal sensor-wearing habits and idiosyncrasies. However, these methods are plagued with their own pitfalls.

Ariani et al. [26] implemented a proof-of-concept wireless sensor network consisting of passive infrared sensors and pressure mats installed around a home, for fall detection at nighttime. Zhang et al. [27] tested this in reality, having a healthy subject undertake a series of predefined movements to simulate typical nighttime activities at home. A 100% specificity and a positive predictive value were reported, but perfect accuracy scores sometimes suggest overlearning or an ungeneralized model. Other groups used vision-based networks for fall detection. Lee and Mihailidis [28] tested a small digital video camera on the ceiling of a bedroom, which determined body postures from pixel count thresholds, and Yu et al. identified human shape change on the plane of the video frame [29]. These works are still in their early stages and validation using larger populations of truly natural movements about the environment is required, as simulated or planned human movements used in testing may not replicate natural movement well.

Although unobtrusive, smart home systems are limited in their scope—confined to their area of installation and within this, areas of sensor coverage. The signals measured have lower resolution, compared to body-worn sensors, in quantifying specific kinematics of the body. Image-analysis-based systems generally involve more memory-intensive computations and require more storage capacity than inertial sensor systems, especially when using high resolution images. Recording or photographing subjects at home also raises issues of privacy.

In contrast, body-worn sensors can give much more insight into the specific physics of the wearer, so may be useful not only in fall detection but also in quantifying fall risk assessments [30–32], with much scope for further research.

28.3.2 Wearable Sensors

Investigations into wearable sensors are currently a major focus in the field. Activity classification has been the first milestone tackled in several fall detection research pipelines. Mathie et al. [33], Karantonis et al. [34], Bianchi et al. [35], and Mathie et al. [36] began with classification of ADL, whereas Bourke et al. [37] focused on posture, impact, and velocity in ADLs and in simulated falls. Accelerometry is a highly popular measurement approach for this, whether alone [34,38,39] or augmented with a gyroscope for angular velocity [9,40], pressure sensor [35], or magnetometer [41]. So diverse are the options in sensor choice, body placement (Figure 28.2), and signal processing techniques that the field is

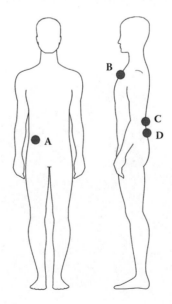

FIGURE 28.2
Common placements for single devices: (A) over (right) iliac crest of pelvis; (B) lanyard on chest, over sternum; (C) over L3 segment of lumbar spine; and (D) over sacrum.

scattered with permutations of activity classifying and fall detecting systems. Resultantly, difficulties in normalization across the field toward a converged research direction have been mentioned as one of the greatest issues facing the area [23], which will be further discussed in Section 28.8. However, while intrusive into daily life, wearable sensing technologies offer a rich body of data and are unrestricted in where they can operate. The ability to raise an alert when the faller is unable to call for help themselves is also a major advantage.

These, sensors have become smaller, lighter, and cheaper over the years; they have been made to be more easily incorporated onto the body in various ways. While some groups have designed sensors to be attached to the body via a strap [34,35,42,43] (Figures 28.3 and 28.4), a clip [37,45], or a lanyard around the neck [19], others have incorporated them into garments such as a vest [46]. Modes of attachment are an important consideration, as the

FIGURE 28.3
Waist-worn triaxial accelerometer device, designed by Lovell and Redmond's group [41].

FIGURE 28.4
Waist-worn triaxial accelerometer device [46], circled, worn by a study participant.

inertial measurements would change with sensor orientation and position. In light of this, a comprehensive examination of the robustness of activity classification and fall detection algorithms with respect to sensor location is warranted. Another issue is that the sensors must be worn during all waking hours by the user to achieve constant monitoring, which may be an optimistic expectation, especially if the device is uncomfortable, hassling, unsightly, or easily forgotten.

In an attempt to reduce falls at their source, clinicians use fall risk assessments in the form of surveys and physical tests, outlined next, to identify individuals with a high fall risk, and prescribe intervention programs or strategies for minimizing falls, where suitable. Here, we refer to fall prevention on the order of months and years; real-time sensor-based predictions of an imminent fall on the order of seconds and accompanying injury-minimization methods (such as via wearable airbags) are not discussed [47].

28.4 Prevention and Fall Risk Assessments

The predictive, preventive approach aims to reduce the chance of a person falling in the first place, according to the idiom "prevention is better than cure." This approach spans a somewhat defined length of time on the order of a few years, first predicting the likelihood of an individual suffering at least one fall in the future, then reassessing after physiotherapeutic interventions to monitor any reductions in fall risk.

Fall prevention strategies such as tailored physiotherapy and exercise sessions are targeted toward older patients with a high risk of falling. This approach requires first a fall risk assessment, to estimate which individuals are more likely to fall in the near future—a complicated task in itself, given the multifactorial nature of falls [47,48]. The complexity of risk assessments used in clinics worldwide ranges greatly across the board. The simplest tests, being survey based, such as the St. Thomas's Risk Assessment Tool in Falling Elderly Inpatients (STRATIFY) [11] and the Morse [49] fall risk assessment tool, require no specialized equipment or test space. More comprehensive procedures involve analysis of one's medical history as well as physical tests of balance, vision, reaction time, and body strength—for example, the physiological profile assessment (PPA) [50], which provides a composite estimate of fall risk based on a range of physiological tests, and the Tinetti test [51–53].

Although these qualitative fall risk assessments mentioned above are widely used and the factors they consider have been identified as correlating to fall risk [50,53], they face a series of limitations that ought to be addressed. In survey-based assessments, it is up to the judgment of the assessor to rate a patient according to criteria, so the results are likely to be subjective. The numerical risk value derived from these surveys is, thus, also subjective in nature. Even in the more thorough assessments, huge variation has been found across equipment, scoring, and cutoff values [51]. Subject behavioral changes influenced by the laboratory nature of the setting, different levels of personal confidence, or white coat syndrome, may also affect the outcomes so that they may not genuinely reflect one's true ability [54]. Furthermore, requiring specialized equipment and a clinician to supervise the entirety of the test is time and labor intensive. Authorized professionals and space are usually scarce, so only one patient is assessed at a time. The predictive ability of the tests is also imperfect—with PPA reporting accuracies of 75% and 79% in distinguishing subjects between multiple fallers and nonmultiple fallers during 1 year after assessment [50]. Finally, researchers must be aware of the limitations of any assessment that make qualitative medical decisions based on quantitative scores—true of all assessments mentioned above—as levels of physical and physiological ability are continuous in reality, but thresholds are defined sometimes subjectively.

A dichotomy exists in the high-fall risk group, where older adults with the lowest and the highest levels of daily activity are at the greatest risk of a fall, but for different reasons. Sedentary, frail adults may present poorer levels of strength, reaction time, balance, and other fall related factors, which increase their fall risk. Meanwhile, highly active adults who are constantly on the move may be more prone to falls, due not to their physical condition but to their highly active lifestyle and more opportunities for trip- or slip-induced falls. This issue highlights the importance of fall risk assessments that identify specific risk factors, to enable more customized intervention, as discussed by Shany et al. [23]. For example, while frail older adults may benefit from strength-training exercises, highly active older adults may find lifestyle changes, such as better footwear, more relevant.

In conclusion, while still useful tools, current fall risk assessments may be improved by incorporating sensor-based technologies. Certain physiological aspects can no doubt be quantified, while opportunities for human error can be reduced, and it would be sensible to take advantage of the range of ever-improving motion analysis techniques available today. There is also the potential to uncover previously undetected aspects of a person's condition, further adding to the utility of such assessments.

28.4.1 Sensor-Based Fall Prediction

The advantages of augmenting miniature wearable sensors with fall risk assessments have been recognized across the field in recent years [23,55,56]. To solve the issue of subjectivity in fall risk assessments, quantitative measurements of physical movement, coupled with purposeful signal analysis, have been suggested. So far, the use of instrumentation to aid in estimating fall risk in elderly or hospitalized patients is being investigated by a number of groups in Australia [30,32,57], across Europe [40,58–62], and in Israel [63,64].

Typically, inertial signal data are collected from subjects while they wear miniature sensors on one or more locations on their body and perform activities ranging from simple standing [65–69] and walking [68,70–75] to specific exercises such as the timed up and go test (TUG) or sit-to-stand-five-times test [31,32,64,70,73,76–79]. Researchers either use commercially available sensors [65,67,79] or design and build their own prototype devices [57,60,76]. Similarly as with fall detection methods, accelerometers have been the most frequently used sensor type in fall risk estimation, either alone [32,57,64,70,73,75–77,79,80] or in combination with gyroscopes [59,65,81]. Experimental protocols that call for measures of the distribution or center of pressure also use a force plate or pressure mat as a reference [65,67,69]. Interestingly, highly differing signal analysis techniques have been chosen across the board, characteristic of the variation in the field.

The time-series data sets obtained in these studies contain a rich body of information, related not only to the activity performed by the subject at the time [25,82] but also to parameters of physical and cognitive function related to falling, including gait speed [62], balance, sway [69], and measures of trunk stability [70,83]. These data are then processed in a variety of ways to extract a set of *features*—such as temporal, spectral, or energy-related parameters. Then, using statistical analysis with regression or machine learning, a model is created, correlating the features with measures of fall likelihood, including the subjects' histories of falls [64,79,80,84–86], fall risk scores derived from clinical assessments [32,57,60,65], and counts of future falls from prospective follow-up studies [64,70,71,74,75,78]. This process is summarized in Figure 28.5. Overall, different combinations of features,

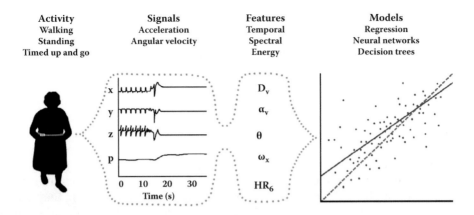

FIGURE 28.5
Left to right: summary of data pipeline, from activity measurement to signal processing, the extraction of features, and the use of features in modeling.

models and validation methods make for a field with an immeasurable number of factors and variables.

One of the common limitations of research here is that the sensor data are often treated as a treasure trove of hidden correlations, processed with only partial consideration of its origin in human movement; and the shift toward taking a more physiologically relevant, biomechanical approach has been gradual. Regardless, providing quantitative patient data will be useful in supplementing current clinical tests (Section 28.4) and to help quantify fall risk by a regulated standard, so it is worth investigating how signal processing and modeling techniques currently employed in the falls field can be used for this application.

28.5 Measurements and Sensors

As previously mentioned in Sections 28.3 and 28.4, several sensors are relevant in the development of wearable devices for movement analysis, activity monitoring, fall detection, and fall prediction. Each provides a different set of physical parameters which may have varying levels of usefulness in monitoring.

28.5.1 Accelerometry

Accelerometers dominate in choice of sensors for falls research. Accelerometers are an inertial sensor that measure linear acceleration along up to three axes over a certain range of frequencies [23,87], and classified according to whether they are suited for AC response (dynamic) or DC response (static) [88]. Given that the bulk of the acceleration power during normal ambulation is concentrated below a relatively low 15 Hz cutoff [89], DC-response accelerometers are more suitable for this application. Responding to both gravitational (frequency = 0 Hz) and movement accelerations, they can be used to measure not only body movements but also orientations with respect to earth [33].

A major advantage of accelerometry is the availability of both movement intensity and frequency information, superior to binary-type switching sensors often incorporated into pedometers [25]. Numerous research groups use accelerometers in prototype fall detection devices, usually in the form of a triaxial accelerometer waist-worn activity-measuring device. Mathie et al. [38] developed such a device for free-living monitoring, consisting of two orthogonally mounted biaxial accelerometers and a waist belt clip for attachment. Similar lightweight triaxial accelerometry devices were designed by Chao et al. [90] for fall detection, by Bourke et al. [37] for assessment of scripted and unscripted ADLs and falls both from the waist (later extended to include a gyroscope [40]) and at three different body locations [39], and by De Bruin et al. [81] for assessment of motor function. In all cases, the accelerometers chosen responded to sufficiently high frequencies of several hundred hertz, enabling down-sampling, filtering, and other refinements. Rich in content, they raise the issue of how to best process their signals to obtain the most valuable information without excessive computation—giving rise to diverse signal processing techniques employed in the field. However, researchers must also be aware of the implications on the signals due to interwearer placement variation. Narayanan et al. [57] noted that the inherent uncertainty in the placement of their device precluded the usability of any signal parameters relying on orientation. To remedy this, some researchers have augmented accelerometry with gyroscopes, discussed below.

28.5.2 Gyroscopy

A gyroscope measures angular velocity by using the Coriolis effect [23]. With a gyroscope, the orientation of the sensor body can be measured independently of the gravitational acceleration vector to deduce postural transitions [9], by single integration. Because this smoothens high-frequency noise, gyroscope signals tend to be cleaner than accelerometry signals [91]. However, error drift can be a significant problem [56], as integrating low-frequency noise in the angular velocity signal creates signal drift in the resultant calculated angle [91]. Gyroscopes also draw significantly more power than accelerometers [40]—raising issues of power-supply method and battery life in the design of wireless body-worn devices.

Najafi et al. [9] and De Bruin et al. [81] designed a chest-mounted sensor with a single-axis gyroscope in the sagittal plane, augmented with two accelerometers in the vertical and frontal directions. Elements of this work were then employed by Godfrey et al. [40], who combined three uniaxial rate gyroscopes for a triaxial gyroscope. After a discrete wavelet transform of the signal to remove the low frequency error drift, the angle of the trunk tilt could be calculated. Godfrey et al. noted, however, that a simpler sensor design with longer activity measurement capabilities could be implemented through simplification of the algorithm, such that gyroscopes were no longer needed. Further investigations into the usefulness of gyroscopes may be warranted, to improve system robustness for activity classification and quantification of angular kinematic parameters [32].

28.5.3 Other Sensors

Other sensor types have also been augmented with a main inertial sensor. Bianchi et al. [35] included a barometric pressure sensor in their waist-worn activity classification device, which helped greatly in reducing the occurrence of false positives in detection of simulated falls [35]—previously a significant problem faced by many in the field [92,93]. This was possible by assuming that a measured increase in barometric pressure signified a drop in the person's altitude associated with falling to a lower level (Figure 28.6).

Magnetometers, sometimes called compasses, measure the orientation of a body segment relative to the earth's magnetic field. This makes use of world coordinates, similarly as accelerometers do, whereas gyroscopes measure only in local coordinates relative to the sensor itself [91]. Unfortunately, they are sensitive to local environmental magnetic fluctuations, resulting in noisy signals [94]. Tilt compensation is also required to account for the orientation of the sensor itself relative to the magnetic field—which is usually achieved by sensor fusion with an accelerometer and gyroscope [91].

Magnetometers have been considered by a few groups for incorporation into wearable sensor devices. It was noted by Pärkkä et al. in activity classification studies that their magnetometer and accelerometer signals were similar in appearance and information content, meaning no further information was added by adding the magnetometer [41]. However, they could be useful where information on the yaw of the sensor/wearer (the horizontal direction it is facing) is desirable [95]. Potential applications include horizontal tracking of the wearer—for example, if the shape of a person's spatial path relative to earth was of particular interest, then a magnetometer could be used to correct for drift in the gyroscope measurements of yaw.

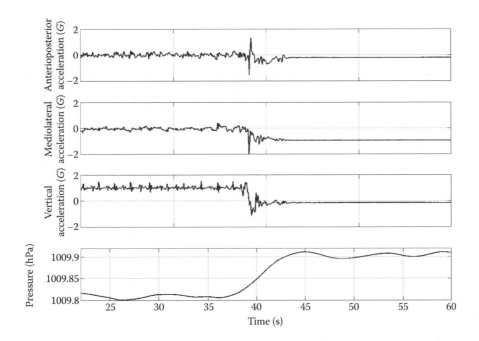

FIGURE 28.6
Signals of walking to a simulated fall from experiments conducted by Bianchi et al. [35], depicting the accelerations in three orthogonal directions ($1G \approx 9.81$ m/s^2) and a low-pass-filtered barometric pressure signal. (From F. Bianchi et al., *IEEE Trans. Neural Syst. Rehabil. Eng.*, vol. 18, no. 6, pp. 619–627, Dec. 2010.)

28.6 Feature Extraction and Analysis

The next phase in the research process of ambulatory monitoring for fall detection and fall risk estimation after signal measurement (Section 28.5) is generally feature extraction, after which extracted features are later used as inputs into decision making models. Feature extraction entails signal processing to obtain pertinent parameters that correlate with the aim of the model; the choice of one affects the choice of the other. Quantitative methods for estimating fall risk are similar, as previously described in Subsection 28.4.1.

The selection of highly predictive features is important in improving the efficiency of the resultant classification algorithms, and minimizing the number of features is desirable for designing well generalized activity classifiers [96]. Commonly extracted features tend to fall under four categories: temporal, spectral, wavelet based, and energy based. In this section, various feature extraction methods are investigated.

28.6.1 Features in Activity Monitoring

Activity classifiers have been developed with signal analysis both in real-time and off-line, postmeasurement. Systems designed to function wirelessly in real time are placed under several hardware and software limitations, such as low power consumption and limited on-board memory. For example, Karantonis et al. [34] chose a hierarchical binary movement classifier [33] that accepted posture angle (relative to the gravitational vector) and simple energy features such as the signal magnitude area (SMA), for a real-time movement classifier. Periods of activity and rest could be distinguished using the SMA, with reasonable accuracy between 74% and 100%. Cyclic activities such as walking could not be detected in real time, due to an inability to buffer more than 3 s worth of data for the required magnitude spectrum [34].

Similarly, Bourke et al. [39] made the assumption that during a fall, trunk and thigh accelerometer signals would reach lower and upper peak values markedly distinct from signals during an ADL. This is reasonably justified under the assumption that an ADL involves no sudden movements. Interestingly, thresholds for the algorithm were chosen based on their experimental data to achieve 100% accuracy, which assumed that the kinematic behaviors of their young subjects were wholly representative of older fallers. In later work, comparisons were made between combinations of algorithms using impact, velocity, and posture parameters, which were derived from a single waist-worn device [37], with the most accurate algorithm utilizing all three parameters. Kangas et al. [97] used similar threshold-based algorithms, comprising these same parameters together with start of fall, detected by monitoring the subthreshold total sum vector of acceleration. While Kangas et al. achieved mixed results, with accuracies ranging from 37% (for falls detected at the wrist by using all four parameters) to 98% (falls detected at the head by using only impact and posture), Bourke et al. reported very high accuracies (near and at 100%) for all algorithms tested, likely due to their scenario-specific choice of thresholds.

Systems where data analysis is performed off-line have less stringent hardware and software restrictions. Najafi et al. [9] noted the importance of being able to identify change in frequency content over time in a signal, an advantage that wavelet transforms have over solely time- or frequency-based analysis. Physically, changes in signal frequency occur in varied walking patterns and sit-to-stand transitions or vice versa. Wavelet transform derived coefficients were, therefore, used together with threshold-based decision algorithms to detect different postural transitions or physical activities. Wavelet decomposition has also been used to reduce noise and error drift, such as in gyroscope measurements [9,40]. Limitations noted by Najafi et al. [9] included the explicit choice of decision rules which boosted algorithm accuracy, the use of a small sample of volunteers not likely representative of an elderly population, and the fact that processing occurred offline.

Conclusively, researchers have been able to classify activities and postures based on features representative of their dynamics. The use of such physically meaningful parameters is promising, but there is much room for advancement, not only in fall detection but also in the investigation of aspects of how a person moves and its relation to their risk of falling. Feature extraction approaches to fall risk estimation and fall prediction models are discussed in the next section.

28.6.2 Features in Fall Risk Estimation and Fall Prediction

While the signal parameters used in fall detection have often been derived based on an understanding of the kinematics of falling, fall risk estimation research has been conducted with a little less emphasis on justifying the features [98] and investigating their meanings. For example, the frequency ratio of the first six harmonics of the TUG in the x axis was selected as one of the regressor parameters by Liu et al. [32], but what does this reveal about the person's physical condition? As discussed by Schultz [99] and Bautmans et al. [62], the field of fall prevention would benefit from understanding the clinical value of features shown to be predictors of falls: intervention programs could be designed and targeted to improve specific fall-related aspects of gait, balance, or muscle ability. In the studies reviewed, many researchers have attached a physiological meaning to some of the signal features used in their models, as discussed below. Naturally, the location of the sensor on the body also determines how a parameter translates into an aspect of biomechanical function: for example, the fundamental frequency of vertical acceleration at the waist would be twice that at a foot, for a given gait cycle.

A frequently derived parameter in sensor-based gait and fall risk studies is the *harmonic ratio* (HR) of a signal:

$$\text{harmonic ratio (HR)} = \frac{\sum \text{Fourier coefficients of even harmonics of signal}}{\sum \text{Fourier coefficients of odd harmonics of signal}}. \qquad (28.1)$$

Doi et al. [70] treated the HR as an indication of trunk stability and smoothness, where the acceleration signals were measured at the C7 and L3 spinous processes (back of neck and lower back). Meanwhile, Yack and Berger [83] described HR as "an index of smoothness of the gait pattern" from the spine of the upper trunk. Bellanca et al. [100] reviewed the use of HRs in gait studies as a global signal variable, stating that the even harmonics of a gait acceleration signal embody the biphasic symmetry of walking in the anteroposterior and vertical directions, while the odd harmonics represent deviation from step symmetry. However, because phase is ignored in the HR, the location of deviations from symmetry is lost, making it difficult to distinguish different pathologies that affect phase differently but have the same HR.

A large set of temporal- and energy-related signal features were extracted from accelerometry signals and used in a linear least-squares model as estimators of fall risk by Narayanan et al. [57]. Frequency-based features were added to Narayanan et al.'s work by Liu et al. [32], including harmonic ratios. Some of the many features were also considered linked to physiological factors, such as high signal periodicity due to a strong, stable walking pattern. Liu et al. [32] anticipated that stable walking generates strongly periodic signals and can be perfectly represented as a sum of harmonics, while a flatter magnitude spectrum would be demonstrated by less periodic or stable gait, potentially indicative of a higher risk of falling.

While Narayanan et al. [57] and Liu et al. [32] focused on temporal and spectral features of acceleration, others have focused on the level of energy or vigor for a given task. Giansanti [60] investigated the energetic biomechanical changes in trunk posture. An interesting estimation of the squared angular was used to relate to rotational kinetic energy by the following equation:

$$\text{rotational kinetic energy} = \frac{1}{2} \times I \times \omega^2, \qquad (28.2)$$

where I is inertia momentum and ω is angular velocity. However, due to the difficulty of knowing I for each subject, ratios of ω^2 were used as model parameters instead.

Hamacher et al. presented a review of kinematic measures for assessing gait stability, noting that some parameters (such as swing and stance time) distinguish better between fallers and nonfallers, while others identify better between old and young adults (such as variability in step width and velocity) [55]. It was also discussed that nonlinear measures of variability and dynamic stability in gait, such as Lyapunov exponents, have become popular in recent years. By treating human gait as a dynamical system, the behavior of a person under small perturbations indicates their gait stability. Consequently, measures of stability elucidate in part the level of dynamic error correction—for example, corrections to maintain support for the center of mass via suitable foot placement. Variability parameters are, therefore, assumed to relate to fall risk, as increased variability would theoretically bring the dynamic state of a person closer to their stability limits.

It is apparent that there is still much scope for investigation into the signal analysis techniques that can be applied to extract features. Such features would not only correlate with fall risk or classify subjects according to their tendency to fall but could, if appropriately chosen, also provide some clinical significance and be of use to physiotherapists and clinicians. Oliver et al. [85] mentioned the need to look for reversible risk factors in patients, for effective fall risk research that would affect intervention strategies. Such an approach should definitely play a part in directing future fall research, as existing work suggests there is great potential in incorporating biomechanical analysis, especially with available technologies such as optical motion capture systems [60,81].

28.7 Fall Risk Model Training and Validation Strategies

The pipeline of inertial data from wearable sensors culminates in model building and validation, to test the performance of those features (Section 28.6) identified as being predictive or enlightening. A purely research focused study was conducted by Robinovitch et al. [101], in which videos of 227 falls from 130 participants were captured over 3 years in two Canadian long-term care facilities, in order to reveal the true circumstances under which falls occurred. However, most research groups have avoided this kind of strategy for obtaining true fall data, given the extensive time, money, and effort input required to set up and maintain such an environment. Instead, the most popular reference fall risk measures used for training and validating fall risk estimators and fall predictor models have included the subjects' histories of falls [64,79,80,84]—which has been treated with justification as a risk factor for falling in itself [85,86]; fall risk scores derived from clinical assessments [32,57,60]; and counts of future falls from prospective follow-up studies [64].

28.7.1 Validation by Clinical Fall Risk Assessment Scores

Existing clinical fall risk assessment scores are an easily obtained point of reference for model validation: either in the form of a numerical score [11,50,51] or as an ordinal categorical label, such as high, moderate, and low risks of falling. Numerous studies have built fall risk estimator models based on clinical scores as a reference [32,57,62,64,76,77,102], with varying degrees of accuracy.

The limitation of using clinical risk scores as the gold standard to which models should aspire is that clinical risk scores are not, in fact, perfect indications of a person's fall likelihood, as noted by several discussions [76,77]. Their advantage is that with a logical progression of low to high fall risk encoded in a numerical scale or some quantifiable parameter, such as the continuous numerical fall risk score from the PPA [50] or the time taken to complete the TUG [84], standard linear-regression modeling techniques can be used [32,57]. However, when evaluating the ability of a clinical assessment to predict falls via prospective studies, the assessments themselves typically report a reasonable but less than perfect accuracy rate, previously mentioned in Section 28.4. Greene et al. [78] found that a number of their sensor derived parameters showing a strong correlation with fall risk (determined by prospective falls) did not correlate with the TUG or the Berg balance scale. This was said to indicate that the same movement patterns that correlated with clinical fall risk are not captured by these conventional clinical measures, which are limited in their scope of assessment. In addition, Marschollek et al. [76] found that model diagnostics such as accuracy, specificity, and sensitivity vary drastically depending on the reference fall risk classifier used. Nevertheless, clinical fall risk scores have been used by many people as their desired model output, as they are easier to obtain and incorporate into scientific studies—for example, many subjects already have existing fall risk records held by hospitals and fall clinics.

As the ultimate goal is to predict real-world falls, the ideal way to validate a model's ability to estimate a person's fall risk is not with clinical fall risk data from screening and evaluation assessments, but with large samples of real prospective fall data [76]—although that comes with its own pitfalls, such as self-reported falls being subject to credibility and recall faults [23]. Ultimately, clinical fall risk scores should be treated as an intermediate indicator of model behavior, rather than the desired end point of a fall risk estimator.

28.7.2 Validation by Fall History

Another common reference for building fall likelihood models is subject fall history—the number of falls a person has had during a defined period of time in the past. Fall count data have been taken from time periods from 6 months [62] to 1 year [64,73,80,83] to 5 years [65,79,84] prior to assessment by sensors. With such data, it is useful to build classifier models, which have taken the form of logistic regression [62,64,76,79,103], classification trees [103], and a range of machine learning algorithms [66,73]. Alternatively, several researchers conducted nonparametric statistical tests to identify pertinent features [64,65,84]. These various models have been developed to categorize subjects into groups of nonfallers (no history of falls) versus fallers (have fallen at least once) [64,65,73,84], or low-risk fallers (have fallen zero or one time) versus high-risk fallers (have fallen more than once) [79,80], where the fall history is considered either over some time frame or over their entire life thus far.

Unfortunately, studies often rely on subject memory to obtain fall history records, which may be prone to errors due to memory recall errors and/or the ability of subjects to under-report falls [104–107]. Others were more thorough in their data search, with Greene et al. [78] matching the fall history reported by subjects with information provided by their relatives, as well as from fall clinic and hospital admission records. Such data checking across multiple sources is certainly advisable, especially given the fallibility of self-reported falls, to reduce the possibility of errors and improve data integrity.

The development of fall risk classifier models is a positive step toward understanding fall risk factors and their embodiment in sensor signals. However, identifying factors that have predictive capabilities would be more useful in aiding fall prevention strategies [69,78,99]—ultimately assisting in cutting the impact and costs of falls at the source prior to any actual fall.

28.7.3 Fall Risk Estimation Using Sensors on Supervised Activities

The gold standard validator of fall risk and fall predictor models is prospective fall data, as it is precisely the future occurrence of falls which determines a person's risk of falling and which we are trying to forecast. Moreover, these falls must be "natural" and occur ideally during everyday life—not induced in any way, to model "normality" and for ethical reasons. As a result, there has been a recent increase in prospective studies, where the study cohort is followed and monitored for falls, for a defined period of time into the future. These studies model subjects' prospective falls by using signal data measured while they were undertaking a directed routine of activities [65,70,71,74,75,78,106].

A common approach is to determine a parameter's classification threshold based on data exploration. For example, Schwesig et al. [71] constructed receiver operating characteristic (ROC) curves for various parameters, to determine their optimal boundaries for distinguishing between fallers and nonfallers. This approach effectively assumes no knowledge of how the physiology of that parameter can be quantified, which would otherwise help make a theory-based decision rather than one dictated by the sample of data from the study.

From a study by Marschollek et al. [106], the predictive abilities of the STRATIFY, TUG, or a combined professional opinion alone were all mediocre—with classification accuracies of 48%, 50%, and 55%, respectively. However, when combined together into an automatically generated logistic regression model, the classification performance improved to 72% accuracy. Whether this was a significant improvement, given that the latter model was built on more data, is worth discussing. A separate logistic regression model comprising sensor-based parameters performed with a similar level of accuracy at 70%, suggesting that a multitude of sensor data may capture the multifactorial nature of fall risk more adequately than a single clinical test. Moreover, an advantage of sensor-based fall risk assessment using a small unobtrusive device is that an expert clinician need not be present during assessment. Unfortunately, a limitation of all sensor-based technologies is the implication of data loss, battery life, and acceptance or compliance issues by the elderly subjects.

Loss to follow-up can be a significant issue in prospective studies that effectively results in data losses. Marschollek et al. [106] experienced a loss of 58% of subjects from their initial cohort of 119 patients, due to death, no answering, being untraceable, withdrawing previous consent, cognitive impairment, and deafness. Consequently, prospective fall data was available from only 46 subjects. Schwesig et al. [71] fared better, with 97% subject retention, while Paterson et al. [74] retained 94% of eligible subjects, and Greene et al. [78] received an 86% response rate to phone interviews with the baseline cohort. Regardless, this highlights the importance of the recruitment phase in such studies, as factors contributing to loss to follow-up are out of the control of researchers and compromise the statistical power of the study.

Prospective fall count is subject to reporting errors, just as fall history reports are. One way to combat this, aside from checking with family and hospital records, would be to continue activity monitoring with a wearable fall detector and augment these data with the fall records provided by subjects. However, neither method is perfect, as fall detection algorithms are still in development and suffer from false positives [37] or optimistic validation [35], discussed in Section 28.3.

Ultimately, sensor-based fall risk assessments utilizing directed routines still require supervision. Even if the sensor based methods appear to improve on qualitative assessments, they are confined to data from only those movements and postures activated by the particular test activity. The trend toward estimating fall risk from data obtained in free-living situations, with no imposition on subject activity, is discussed next.

28.7.4 Fall Risk Estimation Based on Activities of Daily Living

Most of the studies mentioned in Subsections 28.7.1 through 28.7.3 have involved measuring the subjects' physical motion in rigorously defined activities while being supervised in a somewhat artificial laboratory environment. However, it is important to assess an individual in their natural setting, where falls are likely to occur [63,108]—and how an individual performs on a physical test in a laboratory may not accurately reflect their true condition due to confounding behavioral changes or the white coat syndrome [63]. The advantages of breaking free from laboratory and directed routine based fall risk assessments, toward assessment by quantitative analysis of activities of daily living, include a reduced need for a supervising healthcare professional and specialist laboratory equipment, plus the reasons mentioned above. Shany et al. [47] discussed the advantages and disadvantages of unsupervised sensor-based fall risk assessments. For at-home assessments that require a particular physical routine to be performed, major limitations include the difficulty of regulating whether the routine is completed correctly and ensuring the safety of the patient. Correct placement of the sensor device(s) and maintaining the battery charge must also be considered [65]. However, by completing a fall risk assessment using the inertial data collected from a person carrying out their daily living routine, irrespective of time, place, or manner, the former limitations of unsupervised assessments become largely irrelevant.

Recent studies have collected sensor data from subjects during their everyday life to estimate fall risk based on prospective fall data. Weiss et al. [63] analyzed aspects of bouts of walking, measured by a triaxial accelerometer sensor worn for 3 days. When associating with fall history, their binary logistic regression model achieved an accuracy of 67.2%, where those who had fallen at least twice in the past year were labeled "fallers." Conversely, 84.2% accuracy was achieved when classifying subjects based instead on prospective fall count in the next 6 months. While a step in the right direction, this study was limited in length of assessment (only 3 days)—reportedly due to battery life restrictions—and excluded analysis of other potentially valuable data, such as postural or orientation transitions. Similarly, Van Schooten et al. [109] conducted a prospective study where adding features from daily gait acceleration measurements at the waist to multivariate logistic regression models based on clinical risk measures, improved estimations of fall risk. Dadashi et al. [110] presented foot clearance parameters from an inertial sensor on the shoe to enable assessment of risky gait and fall risk. A pilot study by De Bruin et al. [81] also analyzed a range of ADLs in older adults to highlight the importance of long-term ADL monitoring.

One of the difficulties facing the construction of signal processing algorithms to analyze unsupervised sensor data is that it is difficult to regulate ADLs. The high variability in movement and, thus, inertial signals between individuals as well as within an individual under different circumstances presents a challenge for algorithm validation. While it can be argued that this variability is desired in our data sets to ensure model generalizability, it does not make the already complicated task any easier. Observing subjects in their own home, to facilitate signal annotation either in person or with a camera, raises logistical and privacy issues. Therefore, the development of activity classification algorithms, using simulated ADL, may be a feasible alternative reference standard. Activity classification algorithms have been developed with success in the past [33,34,40]; a rigorously validated algorithm for classifying activities from undirected movements should, therefore, perform well if built using reliable pseudo-unsupervised movement data.

It is projected that the field of falls research will be trending toward more attention being invested in developing free-living based fall prediction methods. These have the potential to work in parallel with fall detectors and other personal telehealth devices that are becoming increasingly ubiquitous among both older and younger generations today. We have seen in recent years a multitude of sleek, body-worn fitness devices and functionally equivalent smartphone-based applications that allow individuals to monitor their personal physical activity statistics and basic health indicators blossom on the market [111–113]. Meanwhile, many older adults are becoming equipped with personal alarms and fall detectors, as the desire to maintain independence coupled with their own and their family's concerns regarding safety on a daily basis have opened a market for technologically based self-management and protection tools [15,114] (previously mentioned in Section 28.2). Although the level of engagement between an older adult and their personal device may vary [115] and although the accuracy of currently available commercial devices are unexceptional [111–113], one can expect an increase in usage with an increase in accuracy and user-friendliness as the culture of telehealth becomes the norm [116]. The development of highly accurate and reliable technologies to enable the aforementioned future will, however, require a combined effort between researchers to overcome current bottlenecks in the field. Palumbo et al. [117] insightfully noted the problem of overoptimism in newly developed [98] fall prediction tools. They stated how excellent but unlikely results are frequently reported and how there has been a concerningly low number of externally validated fall prediction studies [117]. Furthermore, widespread statistical heterogeneity between studies was highlighted as affecting the predictive properties of fall risk assessment tools [117,118], which is concerning, given that some of these tools are being used in clinical applications on real people. Section 28.8 addresses these concerns.

28.8 Statistical Heterogeneity and Diversity in the Field

Limitations that are common to the general body of work in this area are the heterogeneity between studies, in the cohort characteristics, sample size, and overall study design. The issue of "more is better" when dealing with population sampling will not be discussed—as that is a general statistical issue. Rather, we focus on the existence of clinical and methodological heterogeneity that make meta-analyses and comparisons between fall studies an extremely complex task.

A review of the literature in fall prediction very quickly reveals the range of criteria that researchers have deemed necessary of their study cohort, as well as diverse experimental methodologies and analysis techniques. A summary of study design variables contributing toward clinical heterogeneity are presented in Table 28.1. It is clear that there is great variation between studies in the various aspects of experimental design. High variation exists also in the results obtained by each of these studies, which are not presented here but summarized by Liu [119].

With multiple sources of heterogeneity in mind, researchers must be wary of how extensively or generally conclusions can be inferred: the results extracted from one study on a particular subset of the entire older population, as defined by certain characteristics, may be skewed in a manner relating to those chosen cohort characteristics, when compared with the actual entire population. The main sources of variation and the issues they raise are summarized in the following sections.

28.8.1 Clinical Heterogeneity: Health Status, Ability, and Origin of Cohort

The heterogeneity within a cohort is often restricted by specifying subject eligibility conditions to control for confounding variables or those of lesser interest. These conditions are often similar, but not standardized, between studies. Where researchers draw the line for which participants to recruit varies between studies and complicates interstudy comparability.

Many studies have recruited healthy older adults with "no history of serious neurological diagnoses that affect gait," such as Parkinson's or cerebellar disease [62,64,70–72,74], and have the ability to complete certain basic physical tasks, such as walking [65,68–72,78], specifically 10 [74], 20 [62], or 25 m [73], or complete the TUG [76], as required in their study methodologies. While it is important that all participants are able to complete the required tasks for the study, those that did not meet the criteria may be predisposed to more risk factors for falling—even if related to existing medical conditions—and may equally benefit from fall prevention measures targeted toward otherwise healthy older adults.

Some studies were inclusive of individuals who required a walking aid [65,68,69], while others excluded them [62,64,65,71–75]. Given that older adults often do not need walking aids but sometimes use them anyway for the added security, as noted by participants in a recent memory and aging study (unpublished work), it is debatable what the most clinically useful approach toward cohort selection is, if there is any. Sample age brackets have spanned middle age [59,76] to near centenarians [69,76], with not all studies stratifying by age. Sample sources have been population pools of independent adults in the community [62,65,75], hospital patients [62,76,78], senior organizations [62], and retirement/care homes. Sample sizes for sensor-based studies also vary between as low as 20 to as high as 264 (Table 28.1), with an even larger range when including studies that do not investigate wearable sensors [55]. All of these variants of clinical heterogeneity illustrate the diversity within the falls research field, which mean that rather than focusing on detailed interstudy comparisons, it may be more useful and accurate to make more general conclusions from meta-analyses regarding risk factors or intervention effects [120].

TABLE 28.1

Summary of Clinical Heterogeneity among Papers That Investigate Sensor-Based Methods for Estimating Fall Risk, Classifying Subjects into Different Risk-Level Classes and Predicting Falls[a]

Authors	Year, Study Type	Cohort Classes	Sample Size	Age of Cohort	Eligibility Criteria	Walking Aid?
Doi et al. [70]	2013, pro	Faller = experienced at least 1 fall during follow-up	73	65+	History of serious neurological diagnoses affecting gait, e.g., Parkinson, were excluded; needed adequate hearing, vision, and speech; could participate in clinical exams	No
Weiss et al. [63]	2013, pro	Faller = 2+ falls in past year, based on self-report	71	65–87	Not previously clinically diagnosed with any gait or balance disorders; cognitively intact; MMSE > 24	Not stated
Weiss et al. [64]	2011, ret	(a) Idiopathic faller (2+ self-reported unexplained falls in previous year); (b) nonfaller	41 (a, 23; b, 18)	50–80	No history of clinically significant stroke, brain surgery, dementia, depression, Parkinson's, traumatic head injury, diabetes mellitus, and other diseases with impact on gait; 2+ falls that could not be explained; else no falls over period	No
Marschollek et al. [76]	2008, ret	High/low risk from TUG, STRATIFY, Barthel index	110 (81; F, 29 M)	45–96; mean: 80	Hospital admission during study and can do TUG	Not stated
Gietzelt et al. [77]	2009, ret	(a) Healthy nonfallers; (b) geriatric nonfallers; (c) geriatric faller group	214. (a, 131; b, 20; c, 90)	Means: a, 64.1; b, 77.2; c, 80.7	History of serious neurological diagnoses affecting gait, e.g., Parkinson's, were excluded; needed adequate hearing, vision, and speech; could participate in clinical exams	No
Greene et al. [65]	2012, ret	(a) Faller (self-reported, in last 5 y); (b) nonfaller	120 (57 M, 63 F)	60+; mean: 73.7; sd: 5.8	Can walk independently with or without aids; community dwelling	N/a
Greene et al. [78]	2012, pro	Recurrent fallers = 2+ falls in follow-up period	226 (62 M, 164 F)	60+	Referred by hospital/clinic/practitioner or self-referred; over 60; able to walk independently with/without walking aid; cognitively intact; can provide informed consent	No
Greene et al. [84]	2010, ret	(a) Faller (self-reported, in past 5 y); (b) nonfaller	264; 77 M (a, 32; b, 45); 119 F < 75 y (a, 72; b, 47); 68 F > 75 y (a, 45; b, 23)	60+	Over 60; able to walk independently with or without help; able to provide informed consent	Not stated

(Continued)

TABLE 28.1 (CONTINUED)

Summary of Clinical Heterogeneity among Papers That Investigate Sensor-Based Methods for Estimating Fall Risk, Classifying Subjects into Different Risk-Level Classes and Predicting Falls[a]

Authors	Year, Study Type	Cohort Classes	Sample Size	Age of Cohort	Eligibility Criteria	Walking Aid?
Narayanan et al. [31]	2008, ret	PPA numerical fall risk	36 (25 F, 11 M)	72–86; mean: 78.8; sd: 3.87	Not mentioned	Not stated
Narayanan et al. [57] Liu et al. [32] Liu et al. [80]	2010, ret 2011, ret 2011, ret	PPA numerical fall risk [32,57]; multiple, nonmultiple fallers [80]	68 (22 M, 46 F)	72–91; mean: 80.1; sd: 4.42	Able to follow instructions for and perform PPA	N/a
Giansanti [60] Giansanti et al. [66]	2006, ret 2008, ret	(a) Low fall risk <65; (b) low fall risk 65+; (c) high fall risk 65+	60 (30 F, 30 M) training set; 200 (100 F, 100 M) validation		Not mentioned	Not stated
Giansanti et al. [59]	2008, ret	Tinetti score levels 1, 2, 3	90 training (30 per group); 100 validation	Overall range 42–84	Not mentioned	Not stated
Schwesig et al. [71]	2013	(a) High risk (3+ falls during follow-up); (b) low risk (<3)	141 followed; most F	62–101; mean: 82.7	60+; no neurological impairment affecting gait/posture, e.g., Parkinson's/cerebellar diseases	No
Doheny et al. [79]	2013	(a) Faller (multiple falls, or 1 fall requiring medical attention during 12 months prior to assessment); (b) nonfaller	39; a, 19 (7 M, 12 F); b, 20 (9 M, 11 F)	Fallers, 66–88 (mean: 74.9; sd: 7); nonfallers, 61–87 (mean: 68.4; sd: 6.2)	Fell regardless of whether injury sustained, and not as result of major intrinsic event or overwhelming hazard	Not stated
Maki [68]	1997	Same as below	75 (14 F, 61 M)	Mean: 82; sd: 6	Same as below	Not stated
Maki et al. [69]	1994	(a) Faller; (b) nonfaller	96 (17 M, 79 F)	62–96; mean: 83; sd: 6	First 100 volunteers (2 self-care residences) who could stand unaided for 90 s, walk 10 m with or without aid, understand verbal instructions, no falls within 1 month prior to testing	Not stated

(Continued)

TABLE 28.1 (CONTINUED)

Summary of Clinical Heterogeneity among Papers That Investigate Sensor-Based Methods for Estimating Fall Risk, Classifying Subjects into Different Risk-Level Classes and Predicting Falls[a]

Authors	Year, Study Type	Cohort Classes	Sample Size	Age of Cohort	Eligibility Criteria	Walking Aid?
Caby et al. [73]	2011	Fell in last 12 months; fall risk via Tinetti and MMT	20 (14 F, 6 M)	70+; mean: 80.85; sd: 5.18	>70; stable medical condition; can walk 25 m; mental abilities to follow instructions	No
Paterson et al. [74]	2011	(a) Multiple faller (2+); (b) nonmultiple faller	97 F	Mean: 68.73; sd: 7.07	Women 55+; can walk 10 m; participate in >30 min exercise on 1+ days/week; exclude fell in previous month, pain during walking, medical condition/medication affecting gait/severe cognitive impairment	No
Laessoe et al. [75]	2007	Not specified, but assumed faller = 1+ falls over follow-up	94 (26% M)	70–80; mean: 73.7; sd: 2.9	Community-dwelling healthy 70–80; invited by senior community center announcements and verbal contacts; excluded if major musculoskeletal disorder, pain limiting daily functions, dependence on gait auxiliaries, ear infection ≤2 weeks prior to test, dependence on special care to stay in community, uncorrected visual/vestibular problems, cognitive impairment	No
Bautmans et al. [62]	2011	(a) Geriatric with higher fall risk (6 months self-reported history, TUG, Tinetti); (b) geriatric control; (c) young control	121; (a, 40; b, 41; c, 40)	65+ old; 18–30 young	Recruited from geriatric department of university hospital, research department database of volunteers, senior organizations; young recruited from staff/students of university; excluded if cognitively deficient, cannot understand/perform test instructions/procedures, walk 20 m, had Parkinson's or cerebrovascular accidents with locomotor disability	No

[a] Study type indicates whether the fall data were collected retrospectively from fall history or from an existing clinical fall risk assessment (ret) or prospectively from a follow-up study (pro). MMSE = mini-mental state examination; MMT = mini-motor test. Sample sizes, gender distributions (F = female, M = male), and ages (y = years, sd = standard deviation) for each cohort class are listed where available.

28.8.2 Methodological Heterogeneity: Subject Classes, Data Collection Techniques, Analysis Methods, and Models

Methodological heterogeneity is statistical heterogeneity due to differences in study designs, which can lead to bias and imprecision in determining effects and correlations or risk factors. Like clinical heterogeneity, methodological heterogeneity also contributes toward the overall statistical heterogeneity or variability in reported results. In the field of fall risk research, this is presented in a number of ways.

Studies that use fall history as the reference measure, discussed previously in Subsection 28.7.2, vary in their definition and distinction between classes of subjects in relation to how often, and to what severity, they have fallen. Subject classes have tended to be distinguished according to low versus high risk fallers, fallers versus non-fallers, or multiple fallers versus nonmultiple fallers, where fall count has been taken within the last n months prior to, or immediately preceding, the experimental assessment. The third column in Table 28.1 details the varying class definitions used in each study. Differences in prospective fall reporting methods also suffer from sources of error to different degrees, whether it be a one-off self-report at the end of follow-up [78], regular fall diaries [63,68,69,75], researcher-initiated phone calls/meetings [68–70,75], and/or hospital/caregiver records [71,78]. Finally, the range of modeling and data analysis techniques as summarized in Section 28.6 and Subsections 28.7.1 through 28.7.4 contribute yet another dimension of diversity to further complicate meta-analysis efforts. Schwickert et al. published a similar summary table of characteristics of fall detection studies, including study size and design, sensor placement and characteristics, algorithm type, and decision logic [24].

Often, differences in the results between studies may indeed be due to methodological diversity. Greene et al. raised a number of valid points regarding difference between their results and those of others [78]. These included the self-reported nature of the follow-up fall data obtained via phone calls; the fact that the cohort comprised predominantly self-referrals, making it "a sample of convenience rather than a representative sample," thus possibly containing sample bias which may explain differences from other studies using cohorts of hospital patients and nursing home residents; and the use of relatively small and/or specific participant populations (such as those with vestibular dysfunction). While controlling for methodological heterogeneity between studies may better facilitate comparisons between studies, one final issue, explained in Subsection 28.8.3, is precisely this limitation of using a relatively small cohort to address such a large and complex issue.

28.8.3 Sample Sizes for Training and Validation

While sample size is an easily criticized factor in a field that is still developing, the nature of the problem lies closer to the freedom that using inertial sensors grants us, in terms of data interpretation. The sheer quantity of data obtainable from even a short temporal recording of a physical activity can be overwhelming. This includes not only raw data but also the potentially infinite number of features obtainable from a signal via standard mathematical and signal processing techniques, in the temporal or the spectral domain, or neither, as addressed in Section 28.6.

Large repositories of inertial data have inspired a tendency in researchers to mine the data for any features that correlate with fall-related targets (clinical risk, history, and prospective count). In most cases, sensible thought has been put into the choice of features, such that a particularly outstanding correlation or significant result can be explained in qualitative, physically meaningful terms. Approaching the data with an open mind may even be necessary, as there is still much to be discovered in the area. However, feature selection rules for model building have not necessarily been strict enough to diminish the chance of finding significance out of luck. Table 28.2 illustrates the variability in number of features considered in each study, with their sample size displayed for comparison. This statistical issue is not exclusive to the field of falls, but it is highly relevant and we should always be mindful of the fact, to maintain a desired level of statistical power in our studies. With each addition of a feature to a model, the data space becomes increasingly sparse—a concept termed "the curse of dimensionality"—and sample sizes in falls studies are rarely high enough to justify more than a few features, from a modeling perspective. Moreover, the sample size is often further unavoidably reduced by setting aside a portion of data for model training and using the remaining portion for validation. Ultimately, this results in an issue summarizable as "too many features, too few samples." Yet, given the complexity of falls, whether we will ever converge on a select few quantitative inertial features that can significantly predict falls is uncertain.

Statistical heterogeneity and low sample sizes are bottlenecks in advancing falls research. Where the literature is not sufficiently homogeneous, in terms of participants

TABLE 28.2

Summary of Varying Starting Number of Features among the Same Papers as in Table 28.1

Authors	Year, Study Type	Sample Size after Subject Exclusions	Number of Features Considered
Doi et al. [70]	2013, pro	73	6
Weiss et al. [63]	2013, pro	71	31
Weiss et al. [64]	2011, ret	41	28
Marschollek et al. [76]	2008, ret	110	9?
Gietzelt et al. [77]	2009, ret	214	4?
Greene et al. [65]	2012, ret	120	44
Greene et al. [78]	2012, pro	226	44
Greene et al. [84]	2010, ret	264	44
Narayanan et al. [31]	2008, ret	36	17
Narayanan et al. [57]	2010, ret	68	54,126
Liu et al. [32]	2011, ret		
Liu et al. [80]	2011, ret		
Giansanti [60]	2006, ret	60 training; 200 validation	2?
Giansanti et al. [66]	2008, ret		
Giansanti et al. [59]	2008, ret	90 training (30 per group); 100 validation	2?
Schwesig et al. [71]	2013, pro	141	12
Doheny et al. [79]	2013, ret	39	70
Maki [68]	1997, pro	75	11
Maki et al. [69]	1994, pro	96	37
Caby et al. [73]	2011, ret	20	67
Paterson et al. [74]	2011, pro	97	8
Laessoe et al. [75]	2007, pro	94	9
Bautmans et al. [62]	2011, ret	121	24

Note: ?, indicates the no. of features considered was not reported.

and outcomes, only general summaries can be made to get result in a particularly detailed meta-analysis, [120]. With the aforementioned issues in mind, initiatives such as the European Fall Repository for the Design of Smart and Self-Adaptive Environments Prolonging Independent Living (FARSEEING) consortium have been set up to enhance collaborations between falls research groups, and to build "an open access meta-database for real-world fall signals and to develop fall-related technologies" [121]. Overall, a combined, international effort from current researchers should facilitate outcomes of impact to a greater extent and enable a more streamlined integration of fall prediction technologies into the community, more so than grinding through smaller-scale, overlapping works in relative isolation.

28.9 Fusing Fall Detection, Daily Monitoring, and Risk Assessments

Despite its limitations, there is clearly great scientific and clinical potential for fall prediction and fall risk assessments based on inertial measurements on ADLs. Meanwhile, wearable sensors and devices are becoming increasingly present in telehealth related applications today, from physical activity monitoring of energy expenditure [95] to automatic fall detection. Devices specifically for the care of older persons, such as emergency alert buttons worn on the wrist or on a lanyard [16–19], play an important role in the spread of individuals equipped with electronic health devices, especially in first-world countries. Other health status–monitoring technologies, such as the TeleMedCare health monitor [122], are also projected to become increasingly used in communities where older individuals desire to be cared for without incessant, direct interaction with doctors' services. Coupled with smartphones and similar personal electronic devices increasingly pervading society, future generations of older persons are expected to be highly equipped with body-worn devices by personal choice.

Consequently, the fusing of fall detection and fall risk estimation functionalities together with existing physiological monitoring technologies may be a natural one. Shany et al. compared the different potential modes of fall risk assessments that utilize sensors for more objective and low-effort evaluation [47]. The advantages of home-based fall risk assessments incorporated into sensor-based devices already used in the community for fall detection include the ability to serve a larger population—potentially every older person in the community. From a public health perspective, this would allow for more effective and timely interventions to reach a wider population with a spectrum of fall risk levels. With long-term monitoring, changes in a person's functional status and fall risk over time can be detected and targeted. The same ADL data used in a real-time fall detection algorithm may also potentially be the basis for fall risk estimation strategies. Achieving this is currently a research goal for a number of groups [63]. It would certainly be ideal to take advantage of the high volume of ADL data available from existing daily monitoring devices.

Mathie et al. [36] and Najafi et al. [9] also began to address the broader functionality of activity classification in individuals, where fall detection is only one of its many uses. Chronic diseases and illnesses such as arthritis and cardiovascular and neurodegenerative diseases often result in a limitation of physical activity and mobility; reliable quantification of daily physical activity and energy expenditure by a wearable sensor-based device could serve as a useful measure of quality of life, progression of disease, and effects

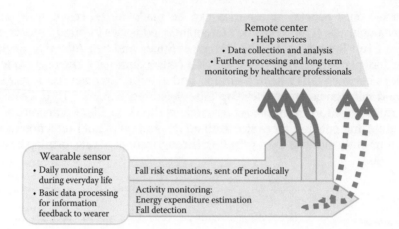

FIGURE 28.7
Schematic diagram of data flow from the wearable sensor (rounded rectangle) to the remote center (trapezium), where data for different purposes can be collated from the same sensors sharing the same on-board algorithm and microprocessing infrastructure (wide arrows). Narrow arrows indicate data transmission to the remote help center, medical center, and/or research facility.

of treatments. Similarly, existing activity classification capabilities achieved in research [9,33,34,56,82] may be expanded from basic postural transitions and activity identifiers to others, such as wandering, for the monitoring of dementia sufferers. The data collected and algorithms implemented could easily be shared across different device functions (Figure 28.7), which would place wearable sensor units centrally in the all-encompassing context of modern telehealth. Therefore, the application of wearable sensor-based devices can expand beyond fall detection and fall risk estimation, providing individuals with a multifunctional well-being monitoring tool to assist in both day-to-day and long-term health management.

28.10 Conclusion

There is undoubtedly significant scope for combining sensor-based activity detection schemes with fall risk assessment and monitoring of the elderly. Inertial measurements may help us gain insight into quantifiable factors that current clinical assessments may not identify. This calls for further investigation into the relationships between physical fall risk factors and how they are detectable, if at all, in inertial sensor data. The development of a wireless wearable sensor device, capable of long-term fall risk and activity monitoring, would enable early detection of any rapid increase in risk, while the predictive power of current clinical fall risk assessments could also be improved.

The focus over the coming years should be directed at conducting prospective trials with activity monitoring over extensive periods, with rigorous follow-up on falls. Investigations ought to be directed toward monitoring free living and ADLs, using a sample of elderly patients large enough for (1) sufficiently high statistical power and (2) building and independent validation of the model. A systematic physiological justification of model parameters is valuable, to prevent model building from being just a purely numerical exercise.

Analyzing inertial signals within a biomechanical modeling framework can aid in identifying fall risk factors in context. Meanwhile, it is also important to maintain a balance between model accuracy and simplicity. Ultimately, this should open up possibilities of developing, at least in part, fall risk assessments that can be completed away from the clinical setting, with less demand for clinical time and resources and less subjectivity.

Abbreviations

ADL	activity of daily living
HR	harmonic ratio
PPA	Physiological Profile Assessment
SMA	signal magnitude area
STRATIFY	St. Thomas's Risk Assessment Tool in Falling Elderly Inpatients
TUG	timed up and go test

References

1. A. Kalache and D. Fu, *WHO global report on falls prevention in older age*, 2008.
2. T. Gill, T. Marin, L. Laslett, C. Kourbelis, and A. Taylor, *An epidemiological analysis of falls among South Australian adults*, Department of Health, Population Research & Outcome Studies Unit, Adelaide, 2009.
3. S. R. Lord, C. Sherrington, and H. B. M. Menz, *Falls in older people: Risk factors and strategies for prevention*, Cambridge University Press, Cambridge, pp. 3–16, 2001.
4. "New South Wales Falls Prevention Baseline Survey 2009 Report," Sydney, 2009.
5. C. Bradley, "Hospitalisations due to falls by older people, Australia 2009–2010," Canberra, 2013.
6. C. Bradley, "Hospitalisations due to falls by older people, Australia 2007–2008," Canberra, 2012.
7. J. C. Davis, M. C. Robertson, M. C. Ashe, T. Liu-Ambrose, K. M. Khan, and C. A. Marra, "International comparison of cost of falls in older adults living in the community: A systematic review," *Osteoporos. Int.*, vol. 21, no. 8, pp. 1295–1306, Aug. 2010.
8. A. Shumway-Cook, M. A. Ciol, J. Hoffman, B. J. Dudgeon, K. Yorkston, and L. Chan, "Falls in the Medicare population: Incidence, associated factors, and impact on health care," *Phys. Ther.*, vol. 89, no. 4, pp. 324–332, Apr. 2009.
9. B. Najafi, K. Aminian, A. Paraschiv-Ionescu, F. Loew, C. J. Büla, and P. Robert, "Ambulatory system for human motion analysis using a kinematic sensor: Monitoring of daily physical activity in the elderly," *IEEE Trans. Biomed. Eng.*, vol. 50, no. 6, pp. 711–723, Jun. 2003.
10. C. Bradley, "Hospitalisations due to falls by older people, Australia 2008–2009," Canberra, 2012.
11. D. Oliver, M. Britton, P. Seed, F. C. Martin, and A. H. Hopper, "Development and evaluation of evidence based risk assessment tool (STRATIFY) to predict which elderly inpatients will fall: Case-control and cohort studies," *BMJ*, vol. 315, no. 7115, pp. 1049–1053, Oct. 1997.
12. M. Tinetti, "Preventing falls in elderly persons," *N. Engl. J. Med.*, vol. 348, no. 1, pp. 42–49, 2003.
13. M. E. Tinetti, M. Speechley, and S. F. Ginter, "Risk factors for falls among elderly persons living in the community," *N. Engl. J. Med.*, vol. 319, no. 26, pp. 1701–1707, Dec. 1988.

14. Legrand, "Legrand continues to expand in the assisted living market and signs joint venture agreement with NEAT," *Legrand*, 2014. [Online]. Available: http://www.legrand.com/files/fck /News/files/groupe/Neat/Neat_HD.jpg. [Accessed: Apr. 22, 2014].

15. Grupo Neat, "Home Units," *Grupo Neat*. [Online]. Available: http://www.gruponeat.com. [Accessed: Jan. 28, 2014].

16. "Medical Alarm Equipment," *St. John*. [Online]. Available: http://www.stjohn.org.nz/Medical -Alarms/Medical-Alarm-Devices/. [Accessed: June 13, 2013].

17. "Seniors," *Life Link*. [Online]. Available: http://www.lifelinkresponse.com.au/seniors/#personal1. [Accessed: June 13, 2013].

18. "Sensors," *Silver Chain*, 2013. [Online]. Available: http://www.silverchain.org.au/wa/health -care/alarm-and-sensors/sensors/. [Accessed: July 13, 2013].

19. "Product Range: Caresse," *Tunstall*. [Online]. Available: http://www.tunstallamerica.com/3_1_3 Caresse.htm. [Accessed: June 13, 2013].

20. "Philips Lifeline," *Philips*, 2014. [Online]. Available: http://www.lifelinesys.com/content/life line-products/auto-alert. [Accessed: Feb. 4, 2014].

21. "Maintain Your Independence with Silver Chain Alarm and Sensors," *Silver Chain*, no. 8. Silver Chain, 2014.

22. H. Lawson and L. Lowe, "Personal Alert Victoria—Evaluation," Melbourne, Victoria, 2006.

23. T. Shany, S. J. Redmond, M. R. Narayanan, and N. H. Lovell, "Sensors-based wearable systems for monitoring of human movement and falls," *IEEE Sens. J.*, vol. 12, no. 3, pp. 658–670, Mar. 2012.

24. L. Schwickert, C. Becker, U. Lindemann, C. Maréchal, A. Bourke, L. Chiari, J. L. Helbostad, W. Zijlstra, K. Aminian, C. Todd, S. Bandinelli, and J. Klenk, "Fall detection with body-worn sensors: A systematic review," *Z. Gerontol. Geriatr.*, vol. 46, no. 8, pp. 706–719, Dec. 2013.

25. M. J. Mathie, A. C. F. Coster, N. H. Lovell, and B. G. Celler, "Accelerometry: Providing an integrated, practical method for long-term, ambulatory monitoring of human movement," *Physiol. Meas.*, vol. 25, no. 2, pp. R1–R20, Apr. 2004.

26. A. Ariani, S. J. Redmond, D. Chang, and N. H. Lovell, "Software simulation of unobtrusive falls detection at night-time using passive infrared and pressure mat sensors," *Conf. Proc. Annu. Int. Conf. IEEE Eng. Med. Biol. Soc.*, vol. 2010, pp. 2115–2118, Jan. 2010.

27. Z. Zhang, U. Kapoor, M. Narayanan, N. H. Lovell, and S. J. Redmond, "Design of an unobtrusive wireless sensor network for nighttime falls detection," *Conf. Proc. Annu. Int. Conf. IEEE Eng. Med. Biol. Soc. IEEE Eng. Med. Biol. Soc. Conf.*, vol. 2011, pp. 5275–5278, Jan. 2011.

28. T. Lee and A. Mihailidis, "An intelligent emergency response system: Preliminary development and testing of automated fall detection," *J. Telemed. Telecare*, vol. 11, no. 4, pp. 194–198, Jan. 2005.

29. X. Yu, X. Wang, and P. Kittipanya-Ngam, "Fall detection and alert for ageing-at-home of elderly," *Lecture Notes Comp. Sci.*, vol. 5597, pp. 209–216, 2009.

30. S. J. Redmond, M. E. Scalzi, M. R. Narayanan, S. R. Lord, S. Cerutti, and N. H. Lovell, "Automatic segmentation of triaxial accelerometry signals for falls risk estimation," *Conf. Proc. Annu. Int. Conf. IEEE Eng. Med. Biol. Soc.*, vol. 2010, pp. 2234–2237, Jan. 2010.

31. M. R. Narayanan, M. E. Scalzi, S. J. Redmond, S. R. Lord, B. G. Celler, and N. H. Lovell, "A wearable triaxial accelerometry system for longitudinal assessment of falls risk," *Conf. Proc. Annu. Int. Conf. IEEE Eng. Med. Biol. Soc.*, vol. 2008, pp. 2840–2843, Jan. 2008.

32. Y. Liu, S. J. Redmond, N. Wang, F. Blumenkron, M. R. Narayanan, and N. H. Lovell, "Spectral analysis of accelerometry signals from a directed-routine for falls-risk estimation," *IEEE Trans. Biomed. Eng.*, vol. 58, no. 8, pp. 2308–2315, 2011.

33. M. J. Mathie, B. G. Celler, N. H. Lovell, and A. C. F. Coster, "Classification of basic daily movements using a triaxial accelerometer," *Med. Biol. Eng. Comput.*, vol. 42, no. 5, pp. 679–687, 2004.

34. D. M. Karantonis, M. R. Narayanan, M. Mathie, N. H. Lovell, and B. G. Celler, "Implementation of a real-time human movement classifier using a triaxial accelerometer for ambulatory monitoring," *IEEE Trans. Inf. Technol. Biomed.*, vol. 10, no. 1, pp. 156–167, Jan. 2006.

35. F. Bianchi, S. J. Redmond, M. R. Narayanan, S. Cerutti, and N. H. Lovell, "Barometric pressure and triaxial accelerometry-based falls event detection," *IEEE Trans. Neural Syst. Rehabil. Eng.*, vol. 18, no. 6, pp. 619–627, Dec. 2010.

36. M. J. Mathie, A. C. F. Coster, N. H. Lovell, B. G. Celler, S. R. Lord, and A. Tiedemann, "A pilot study of long-term monitoring of human movements in the home using accelerometry," *J. Telemed. Telecare*, vol. 10, no. 3, pp. 144–151, Jan. 2004.

37. A. K. Bourke, P. van de Ven, M. Gamble, R. O'Connor, K. Murphy, E. Bogan, E. McQuade, P. Finucane, G. ÓLaighin, and J. Nelson, "Evaluation of waist-mounted tri-axial accelerometer based fall detection algorithms during scripted and continuous unscripted activities," *J. Biomech.*, vol. 43, no. 15, pp. 3051–3057, 2010.

38. M. J. Mathie, a C. F. Coster, N. H. Lovell, and B. G. Celler, "Detection of daily physical activities using a triaxial accelerometer," *Med. Biol. Eng. Comput.*, vol. 41, no. 3, pp. 296–301, May 2003.

39. A. K. Bourke, J. V. O'Brien, and G. M. Lyons, "Evaluation of a threshold-based tri-axial accelerometer fall detection algorithm," *Gait Posture*, vol. 26, no. 2, pp. 194–199, Jul. 2007.

40. A. Godfrey, A. K. Bourke, G. M. Olaighin, P. van de Ven, and J. Nelson, "Activity classification using a single chest mounted tri-axial accelerometer," *Med. Eng. Phys.*, vol. 33, no. 9, pp. 1127–1135, Nov. 2011.

41. J. Pärkkä, M. Ermes, P. Korpipää, J. Mäntyjärvi, J. Peltola, and I. Korhonen, "Activity classification using realistic data from wearable sensors," *IEEE Trans. Inf. Technol. Biomed.*, vol. 10, no. 1, pp. 119–128, 2006.

42. M. Kangas, I. Vikman, J. Wiklander, P. Lindgren, L. Nyberg, and T. Jämsä, "Sensitivity and specificity of fall detection in people aged 40 years and over," *Gait Posture*, vol. 29, no. 4, pp. 571–574, Jun. 2009.

43. M. Kangas, I. Vikman, L. Nyberg, R. Korpelainen, J. Lindblom, and T. Jämsä, "Comparison of real-life accidental falls in older people with experimental falls in middle-aged test subjects," *Gait Posture*, vol. 35, no. 3, pp. 500–505, Mar. 2012.

44. "Estimating risk of falling in the elderly by monitoring daily activities," 2013. [Online]. Available: https://research.unsw.edu.au/projects/estimating-risk-falling-elderly-monitoring-daily-activities. [Accessed: June 28, 2013].

45. C.-C. Yang and Y.-L. Hsu, "Development of a wearable motion detector for telemonitoring and real-time identification of physical activity," *Telemed. e-Health*, vol. 15, no. 1, pp. 62–72, 2009.

46. A. K. Bourke, P. W. J. van de Ven, A. E. Chaya, G. M. OLaighin, and J. Nelson, "Testing of a long-term fall detection system incorporated into a custom vest for the elderly," *Conf. Proc. IEEE Eng. Med. Biol. Soc.*, vol. 2008, pp. 2844–2847, Jan. 2008.

47. T. Shany, S. J. Redmond, M. Marschollek, and N. H. Lovell, "Assessing fall risk using wearable sensors: A practical discussion; A review of the practicalities and challenges associated with the use of wearable sensors for quantification of fall risk in older people," *Z. Gerontol. Geriatr.*, vol. 45, no. 8, pp. 694–706, Dec. 2012.

48. L. A. Lipsitz, P. V. Jonsson, M. M. Kelley, and J. S. Koestner, "Causes and correlates of recurrent falls in ambulatory frail elderly," *J. Gerontol.*, vol. 46, no. 4, pp. M114–M122, July 1991.

49. D. A. Ganz, C. Huang, D. Saliba, V. Shier, D. Berlowitz, C. V. Lukas, K. Pelczarski, K. Schoelles, L. C. Wallace, and P. Neumann, "Preventing Falls in Hospitals: A Toolkit for Improving Quality of Care." RAND Corporation, Boston University School of Public Health and ECRI Institute under Contract No. HHSA290201000017I TO #1, Rockville, Maryland, 2013.

50 S. R. Lord, H. B. Menz, and A. Tiedemann, "A physiological profile approach to falls risk assessment and prevention," *Phys. Ther.*, vol. 83, no. 3, pp. 237–252, Mar. 2003.

51. S. Köpke and G. Meyer, "The Tinetti test: Babylon in geriatric assessment," *Z. Gerontol. Geriatr.*, vol. 39, no. 4, pp. 288–291, Aug. 2006.

52. M. E. Tinetti, T. F. Williams, and R. Mayewski, "Tinetti Balance Assessment Tool," 1986, available at http://www.bhps.org.uk/falls/documents/TinettiBalanceAssessment.pdf. Retrieved on July 25, 2015.

53. M. E. Tinetti, T. F. Williams, and R. Mayewski, "Fall risk index for elderly patients based on number of chronic disabilities," *Am. J. Med.*, vol. 80, no. 3, pp. 429–434, Mar. 1986.

54. T. Shany, "Clinical Evaluation of a Body-Worn Sensor-Based System for Fall Risk Testing," University of New South Wales, 2014.

55. D. Hamacher, N. B. Singh, J. H. Van Dieën, M. O. Heller, and W. R. Taylor, "Kinematic measures for assessing gait stability in elderly individuals: A systematic review," *J. R. Soc. Interface*, vol. 8, no. 65, pp. 1682–1698, Dec. 2011.

56. S. J. Preece, J. Y. Goulermas, L. P. J. Kenney, D. Howard, K. Meijer, and R. Crompton, "Activity identification using body-mounted sensors—A review of classification techniques," *Physiol. Meas.*, vol. 30, no. 4, pp. R1–R33, Apr. 2009.

57. M. R. Narayanan, S. J. Redmond, M. E. Scalzi, S. R. Lord, B. G. Celler, and N. H. Lovell, "Longitudinal falls-risk estimation using triaxial accelerometry," *IEEE Trans. Biomed. Eng.*, vol. 57, no. 3, pp. 534–541, Mar. 2010.

58. D. Giansanti, S. Morelli, G. Maccioni, and G. Costantini, "Toward the design of a wearable system for fall risk detection in telerehabilitation," *Telemed. J. e-health*, vol. 15, no. 3, pp. 296–299, Apr. 2009.

59. D. Giansanti, G. Maccioni, S. Cesinaro, F. Benvenuti, and V. Macellari, "Assessment of fall risk by means of a neural network based on parameters assessed by a wearable device during posturography," *Med. Eng. Phys.*, vol. 30, no. 3, pp. 367–372, Apr. 2008.

60. D. Giansanti, "Investigation of fall risk using a wearable device with accelerometers and rate gyroscopes," *Physiol. Meas.*, vol. 27, no. 11, pp. 1081–1090, Nov. 2006.

61. A. Godfrey, R. Conway, D. Meagher, and G. ÓLaighin, "Direct measurement of human movement by accelerometry," *Med. Eng. Phys.*, vol. 30, no. 10, pp. 1364–1386, 2008.

62. I. Bautmans, B. Jansen, B. Van Keymolen, and T. Mets, "Reliability and clinical correlates of 3D-accelerometry based gait analysis outcomes according to age and fall risk," *Gait Posture*, vol. 33, no. 3, pp. 366–372, Mar. 2011.

63. A. Weiss, M. Brozgol, M. Dorfman, T. Herman, S. Shema, N. Giladi, and J. M. Hausdorff, "Does the evaluation of gait quality during daily life provide insight into fall risk? A novel approach using 3-day accelerometer recordings," *Neurorehabil. Neural Repair*, vol. 20, no. 10, Jun. 2013.

64. A. Weiss, T. Herman, M. Plotnik, M. Brozgol, N. Giladi, and J. M. Hausdorff, "An instrumented timed up and go: The added value of an accelerometer for identifying fall risk in idiopathic fallers," *Physiol. Meas.*, vol. 32, no. 12, pp. 2003–2018, Dec. 2011.

65. B. R. Greene, D. McGrath, L. Walsh, E. P. Doheny, D. McKeown, C. Garattini, C. Cunningham, L. Crosby, B. Caulfield, and R. A. Kenny, "Quantitative falls risk estimation through multi-sensor assessment of standing balance," *Physiol. Meas.*, vol. 33, no. 12, pp. 2049–2063, Dec. 2012.

66. D. Giansanti, V. Macellari, and G. Maccioni, "New neural network classifier of fall risk based on the Mahalanobis distance and kinematic parameters assessed by a wearable device," *Physiol. Meas.*, vol. 29, no. 3, pp. N11–N19, Mar. 2008.

67. P. M. Deshmukh, C. M. Russell, L. E. Lucarino, and S. N. Robinovitch, "Enhancing clinical measures of postural stability with wearable sensors," *Conf. Proc. Annu. Int. Conf. IEEE Eng. Med. Biol. Soc.*, vol. 2012, pp. 4521–4524, Jan. 2012.

68. B. Maki, "Gait changes in older adults: Predictors of falls or indicators of fear," *J. Am. Geriatr. Soc.*, vol. 45, no. 3, pp. 313–320, 1997.

69. B. E. Maki, P. J. Holliday, and A. K. Topper, "A prospective study of postural balance and risk of falling in an ambulatory and independent elderly population," *J. Gerontol.*, vol. 49, no. 2, pp. M72–M84, Mar. 1994.

70. T. Doi, S. Hirata, R. Ono, K. Tsutsumimoto, S. Misu, and H. Ando, "The harmonic ratio of trunk acceleration predicts falling among older people: Results of a 1-year prospective study," *J. Neuroeng. Rehabil.*, vol. 10, no. 1, p. 7, Jan. 2013.

71. R. Schwesig, D. Fischer, A. Lauenroth, S. Becker, and S. Leuchte, "Can falls be predicted with gait analytical and posturographic measurement systems? A prospective follow-up study in a nursing home population," *Clin. Rehabil.*, vol. 27, no. 2, pp. 183–190, Feb. 2013.

72. C. J. Lamoth, F. J. van Deudekom, J. P. van Campen, B. A. Appels, O. J. de Vries, and M. Pijnappels, "Gait stability and variability measures show effects of impaired cognition and dual tasking in frail people," *J. Neuroeng. Rehabil.*, vol. 8, no. 1, p. 2, Jan. 2011.

73. B. Caby, S. Kieffer, M. de Saint Hubert, G. Cremer, and B. Macq, "Feature extraction and selection for objective gait analysis and fall risk assessment by accelerometry," *Biomed. Eng. Online*, vol. 10, no. 1, p. 1, Jan. 2011.

74. K. Paterson, K. Hill, and N. Lythgo, "Stride dynamics, gait variability and prospective falls risk in active community dwelling older women," *Gait Posture*, vol. 33, no. 2, pp. 251–255, Feb. 2011.

75. U. Laessoe, H. C. Hoeck, O. Simonsen, T. Sinkjaer, and M. Voigt, "Fall risk in an active elderly population—Can it be assessed?," *J. Negat. Results Biomed.*, vol. 6, p. 2, Jan. 2007.

76. M. Marschollek, K.-H. Wolf, M. Gietzelt, G. Nemitz, H. Meyer zu Schwabedissen, and R. Haux, "Assessing elderly persons' fall risk using spectral analysis on accelerometric data—A clinical evaluation study," *Conf. Proc.: Ann. Int. Conf. IEEE Eng. Med. Biol. Soc.*, 2008, vol. 2008, pp. 3682–3685.

77. M. Gietzelt, G. Nemitz, K.-H. Wolf, H. Meyer Zu Schwabedissen, R. Haux, and M. Marschollek, "A clinical study to assess fall risk using a single waist accelerometer," *Inform. Health Soc. Care*, vol. 34, no. 4, pp. 181–188, Dec. 2009.

78. B. R. Greene, E. P. Doheny, C. Walsh, C. Cunningham, L. Crosby, and R. A. Kenny, "Evaluation of falls risk in community-dwelling older adults using body-worn sensors," *Gerontology*, vol. 5, no. 5, pp. 472–480, Jan. 2012.

79. E. P. Doheny, C. Walsh, T. Foran, B. R. Greene, C. W. Fan, C. Cunningham, and R. A. Kenny, "Falls classification using tri-axial accelerometers during the five-times-sit-to-stand test," *Gait Posture*, vol. 38, no. 4, pp. 1021–1025, Sept. 2013.

80. Y. Liu, S. J. Redmond, M. R. Narayanan, and N. H. Lovell, "Classification between non-multiple fallers and multiple fallers using a triaxial accelerometry-based system," *Conf. Proc. Annu. Int. Conf. IEEE Eng. Med. Biol. Soc.*, vol. 2011, pp. 1499–1502, Jan. 2011.

81. E. D. De Bruin, B. Najafi, K. Murer, D. Uebelhart, and K. Aminian, "Quantification of everyday motor function in a geriatric population," *J. Rehabil. Res. Dev.*, vol. 44, no. 3, p. 417, 2007.

82. M. Ermes, J. Pärkkä, J. Mäntyjärvi, and I. Korhonen, "Detection of daily activities and sports with wearable sensors in controlled and uncontrolled conditions," *IEEE Trans. Inf. Technol. Biomed.*, vol. 12, no. 1, pp. 20–26, 2008.

83. H. J. Yack and R. C. Berger, "Dynamic stability in the elderly: Identifying a possible measure," *J. Gerontol.*, vol. 48, no. 5, pp. M225–M230, Sep. 1993.

84. B. R. Greene, A. O'Donovan, R. Romero-Ortuno, L. Cogan, C. N. Scanaill, and R. A. Kenny, "Quantitative falls risk assessment using the timed up and go test," *IEEE Trans. Biomed. Eng.*, vol. 57, no. 12, pp. 2918–2926, Dec. 2010.

85. D. Oliver, F. Daly, F. C. Martin, and M. E. T. McMurdo, "Risk factors and risk assessment tools for falls in hospital in-patients: A systematic review," *Age Ageing*, vol. 33, no. 2, pp. 122–130, Mar. 2004.

86. K. L. Perell, A. Nelson, R. L. Goldman, S. L. Luther, N. Prieto-Lewis, and L. Z. Rubenstein, "Fall risk assessment measures: An analytic review," *J. Gerontol. A. Biol. Sci. Med. Sci.*, vol. 56, no. 12, pp. M761–M766, Dec. 2001.

87. "Accelerometers and How They Work," Texas Instruments, Dallas, Texas.

88. A. Chu, "Engineer's Circle: Choosing the Right Type of Accelerometers," Measurement Specialties, Aliso Viejo, California, pp. 1–5, 2012.

89. E. Antonsson and R. Mann, "The frequency content of gait," *J. Biomech.*, vol. 8, no. 1, pp. 39–47, 1985.

90. P.-K. Chao, H.-L. Chan, F.-T. Tang, Y.-C. Chen, and M.-K. Wong, "A comparison of automatic fall detection by the cross-product and magnitude of tri-axial acceleration," *Physiol. Meas.*, vol. 30, no. 10, pp. 1027–1037, Oct. 2009.

91. D. Sachs, "Sensor Fusion on Android Devices: A Revolution in Motion Processing," 2010, Uploaded by GoogleTechTalks, available at https://www.youtube.com/watch?v=C7JQ7Rpwn2k, accessed on July 25, 2015.

92. S. Abbate, M. Avvenuti, P. Corsini, and A. Vecchio, "Monitoring of human movements for all detection and activities recognition in elderly care using wireless sensor network: A survey," in *Wireless Sensor Networks: Application-Centric Design*, edited by Geoff V. Merrett and Yen Kheng Tan, Chapter 9, pp. 1–20, InTech, Rijeka, 2010.

93. J. Klenk, C. Becker, F. Lieken, S. Nicolai, W. Maetzler, W. Alt, W. Zijlstra, J. M. Hausdorff, R. C. van Lummel, L. Chiari, and U. Lindemann, "Comparison of acceleration signals of simulated and real-world backward falls," *Med. Eng. Phys.*, vol. 33, no. 3, pp. 368–373, Apr. 2011.

94. J. F. S. Lin and D. Kulić, "Human pose recovery using wireless inertial measurement units," *Physiol. Meas.*, vol. 33, no. 12, pp. 2099–2115, Dec. 2012.

95. J. Wang, S. J. Redmond, M. Voleno, M. R. Narayanan, N. Wang, S. Cerutti, and N. H. Lovell, "Energy expenditure estimation during normal ambulation using triaxial accelerometry and barometric pressure," *Physiol. Meas.*, vol. 33, no. 11, pp. 1811–1830, Nov. 2012.

96. O. Banos, M. Damas, H. Pomares, A. Prieto, and I. Rojas, "Daily living activity recognition based on statistical feature quality group selection," *Expert Syst. Appl.*, vol. 39, no. 9, pp. 8013–8021, Jul. 2012.

97. M. Kangas, A. Konttila, P. Lindgren, I. Winblad, and T. Jämsä, "Comparison of low-complexity fall detection algorithms for body attached accelerometers," *Gait Posture*, vol. 28, no. 2, pp. 285–291, Aug. 2008.

98. J. Howcroft, J. Kofman, and E. D. Lemaire, "Review of fall risk assessment in geriatric populations using inertial sensors," *J. Neuroeng. Rehabil.*, vol. 10, no. 1, p. 91, Aug. 2013.

99. A. B. Schultz, "Mobility impairment in the elderly: Challenges for biomechanics research," *J. Biomech.*, vol. 25, no. 5, pp. 519–528, May 1992.

100. J. L. Bellanca, K. A. Lowry, J. M. Vanswearingen, J. S. Brach, and M. S. Redfern, "Harmonic ratios: A quantification of step to step symmetry," *J. Biomech.*, vol. 46, no. 4, pp. 828–831, Feb. 2013.

101. S. N. Robinovitch, F. Feldman, Y. Yang, R. Schonnop, P. M. Leung, T. Sarraf, J. Sims-Gould, and M. Loughin, "Video capture of the circumstances of falls in elderly people residing in long-term care: An observational study," *Lancet*, vol. 381, no. 9860, pp. 47–54, Jan. 2013.

102. B. Najafi, K. Aminian, F. Loew, Y. Blanc, and P. A. Robert, "Measurement of stand-sit and sit-stand transitions using a miniature gyroscope and its application in fall risk evaluation in the elderly," *IEEE Trans. Biomed. Eng.*, vol. 49, no. 8, pp. 843–851, Aug. 2002.

103. M. Marschollek, A. Rehwald, K. H. Wolf, M. Gietzelt, G. Nemitz, H. Meyer Zu Schwabedissen, and R. Haux, "Sensor-based fall risk assessment—An expert 'to go,'" *Methods Inf. Med.*, vol. 50, no. 5, pp. 420–426, Jan. 2011.

104. J. Brach, J. Berlin, J. Van Swearingen, A. Newman, and S. Studenski, "Too much or too little step width variability is associated with a fall history in older persons who walk at or near normal gait speed," *J. Neuroeng. Rehabil.*, vol. 2, no. 1, p. 21, 2005.

105. F. Riva, M. J. P. Toebes, M. Pijnappels, R. Stagni, and J. H. van Dieën, "Estimating fall risk with inertial sensors using gait stability measures that do not require step detection," *Gait Posture*, vol. 38, no. 2, pp. 170–174, Jun. 2013.

106. M. Marschollek, A. Rehwald, K.-H. Wolf, M. Gietzelt, G. Nemitz, H. M. zu Schwabedissen, and M. Schulze, "Sensors vs. experts—A performance comparison of sensor-based fall risk assessment vs. conventional assessment in a sample of geriatric patients," *BMC Med. Inform. Decis. Mak.*, vol. 11, no. 1, p. 48, Jan. 2011.

107. D. A. Ganz, T. Higashi, and L. Z. Rubenstein, "Monitoring falls in cohort studies of community-dwelling older people: Effect of the recall interval," *J. Am. Geriatr. Soc.*, vol. 53, no. 12, pp. 2190–2194, Dec. 2005.

108. S. R. Lord, H. B. Menz, and C. Sherrington, "Home environment risk factors for falls in older people and the efficacy of home modifications," *Age Ageing*, vol. 35, Suppl. 2, Table 1, pp. ii55–ii59, Sep. 2006.

109. K. S. Van Schooten, S. M. Rispens, P. J. M. Elders, P. Lips, J. H. Van Dieën, and M. Pijnappels, "Quantity and quality of daily life gait as predictors of falls in older adults," *J. Gerontol* 70:608–615, 2015.

110. F. Dadashi, B. Mariani, S. Rochat, C. J. Büla, B. Santos-Eggimann, and K. Aminian, "Gait and foot clearance parameters obtained using shoe-worn inertial sensors in a large-population sample of older adults," *Sensors (Basel)*, vol. 14, no. 1, pp. 443–457, Jan. 2013.

111. F. Guo, Y. Li, M. S. Kankanhalli, and M. S. Brown, "An evaluation of wearable activity monitoring devices," in *Proceedings of the First ACM International Workshop on Personal Data Meets Distributed Multimedia—PDM '13*, 2013, pp. 31–34.

112. K. L. Dannecker, N. A. Sazonova, E. L. Melanson, E. S. Sazonov, and R. C. Browning, "A comparison of energy expenditure estimation of several physical activity monitors," *Med. Sci. Sports Exerc.*, vol. 45, no. 11, pp. 2105–2112, Dec. 2013.

113. J. Takacs, C. L. Pollock, J. R. Guenther, M. Bahar, C. Napier, and M. A. Hunt, "Validation of the Fitbit One activity monitor device during treadmill walking," *J. Sci. Med. Sport*, Oct. 2013.

114. "About Lifeline Medical Alarm Systems; Philips Lifeline®." [Online]. Available: http://www.lifelinesys.com/content/company-areas. [Accessed: Apr. 12, 2014].

115. R. Steele, A. Lo, C. Secombe, and Y. K. Wong, "Elderly persons' perception and acceptance of using wireless sensor networks to assist healthcare," *Int. J. Med. Inform.*, vol. 78, no. 12, pp. 788–801, Dec. 2009.

116. J. Cohen, "The patient of the future," *Technol. Rev.*, vol. 115, no. 2, pp. 60–63, 2012.

117. P. Palumbo, L. Palmerini, and L. Chiari, "A probabilistic model to investigate the properties of prognostic tools for falls," *Methods Inf. Med.*, ePub ahead of print, 2013.

118. D. Oliver, A. Papaioannou, L. Giangregorio, L. Thabane, K. Reizgys, and G. Foster, "A systematic review and meta-analysis of studies using the STRATIFY tool for prediction of falls in hospital patients: How well does it work?," *Age Ageing*, vol. 37, no. 6, pp. 621–627, Nov. 2008.

119. Y. Liu, "The Development of a Body-Worn Sensor-Based System for Fall Risk Assessment," University of New South Wales, 2013.

120. J. J. Deeks, J. P. T. Higgins, and D. G. Altman, "What is heterogeneity?," in *Cochrane Handbook for Systematic Reviews of Interventions*, 5.1.0 ed., J. P. T. Higgins and S. Green, eds., The Cochrane Collaboration, 2011.

121. S. Mellone, C. Tacconi, L. Schwickert, J. Klenk, C. Becker, and L. Chiari, "Smartphone-based solutions for fall detection and prevention: The FARSEEING approach," *Z. Gerontol. Geriatr.*, vol. 45, no. 8, pp. 722–727, Dec. 2012.

122. "TMC Health Monitor," *Telemedcare*, 2009. [Online]. Available: https://www.telemedcare.com.au/index.php/equipment. [Accessed: Dec. 5, 2013].

Index

Page numbers followed f and t indicate figures and tables, respectively.